Fifth Edition
Environmental Geology

Carla W. Montgomery
Northern Illinois University

**WCB
McGraw-Hill**

Boston, Massachusetts   Burr Ridge, Illinois   Dubuque, Iowa
Madison, Wisconsin   New York, New York   San Francisco, California   St. Louis, Missouri

WCB/McGraw-Hill

A Division of The McGraw·Hill Companies

ENVIRONMENTAL GEOLOGY, Fifth Edition

 This book is printed on recycled, acid-free paper containing 10% postconsumer waste.

1 2 3 4 5 7 8 9 0 QPD/QPD 9 0 9 8 7

ISBN 0–697–34278–6

Editorial director: *Kevin T. Kane*
Publisher: *Edward E. Bartell*
Sponsoring editor: *Lynne M. Meyers*
Developmental editor: *Daryl Bruflodt*
Project manager: *Peggy J. Selle*
Production supervisor: *Mary E. Haas*
Compositor: *Shepherd, Inc.*
Typeface: *Cochin*
Printer: *Quebecor Printing Book Group, Inc.*

Cover image: AP/World Wide Photos
All photographs, unless otherwise indicated, provided by Carla W. Montgomery.

Library of Congress Catalog Card Number: 96–79604

www.mhhe.com

Environmental Geology is affectionately dedicated to the memory of Ed Jaffe, whose confidence in an unknown author made the first edition possible.

-CWM

Contents

Chapter 11
Soil as a Resource

Chapter 12
Mineral Resources

Chapter 13
Energy Resources—Fossil Fuels

Contents

Boxes

Preface

The *environment* is the sum of all the features and conditions surrounding an organism that may influence it. An individual's physical environment encompasses rocks and soil, air and water, such factors as light and temperature, and other organisms present. One's social environment might include a network of family and friends, a particular political system, and a set of social customs that affect one's behavior.

Geology is the study of the earth. Since the earth provides the basic physical environment in which we live, all of geology might in one sense be regarded as environmental geology. However, the term *environmental geology* is usually restricted to refer particularly to geology as it relates directly to human activities, and that is the focus of this book. Environmental geology is geology applied to living. We will examine how geologic processes and hazards influence human activities (and sometimes the reverse), the geologic aspects of pollution and waste-disposal problems, and several other topics.

Why environmental geology? One reason for studying environmental geology might simply be curiosity about the way the earth works, about the *how* and *why* of natural phenomena. Another reason is that we are increasingly faced with environmental problems to be solved and decisions to be made, and in many cases, an understanding of one or more geologic processes is essential to finding an appropriate solution.

Of course, many environmental problems cannot be fully assessed and solved using geologic data alone. The problems vary widely in size and in complexity. In a specific instance, data from other branches of science (such as biology, chemistry, or ecology), as well as economics, politics, social priorities, and so on may have to be taken into account. Because a variety of considerations may influence the choice of a solution, there is frequently disagreement about which solution is "best." Our personal choices will often depend strongly on our beliefs about which considerations are most important.

An introductory text cannot explore all aspects of environmental concerns. Here, the emphasis is on the physical constraints imposed on human activities by the geologic processes that have shaped and are still shaping our natural environment. In a real sense, these are the most basic, inescapable constraints; we cannot, for instance, use a resource that is not there, or build a secure home or a safe dam on land that is fundamentally unstable. Geology, then, is a logical place to start in developing an understanding of many environmental issues. The principal aim of this book is to present the reader with a broad overview of environmental geology. Because geology does not exist in a vacuum, however, the text, from time to time, introduces related considerations from outside geology to clarify other ramifications of the subjects discussed. Likewise, the present does not exist in isolation from the past and future; occasionally, the text looks both at how the earth developed into its present condition and where matters seem to be moving for the future. It is hoped that this knowledge will provide the reader with a useful foundation for discussing and evaluating specific environmental issues, as well as for developing ideas about how the problems should be solved.

About the Book

This text is intended for an introductory-level college course. It does not assume any prior exposure to geology or college-level mathematics or science courses. The metric system is used throughout, except where other units are conventional within a discipline. (For the convenience of students not yet "fluent" in metric units, a conversion table is included in appendix D, and in some cases, metric equivalents in English units are included within the text.)

Each chapter opens with an introduction that sets the stage for the material to follow. In the course of the chapter, important terms and concepts are identified by boldface type, and these terms are collected as "Terms to Remember" at the end of the chapter for quick review. To emphasize the present relevance of the material in the text and to illustrate the variety of current environmental problems, many chapters include actual case histories or specific examples. To these, each reader could no doubt add others from personal experience. Additional supplementary information is included in boxes set off from the main body of the text. Each chapter concludes with review questions and exercises that allow students to test their comprehension of text material and to apply that knowledge to real-world situations, and also suggested readings and pertinent references.

A new feature with this edition is the inclusion of "Net-Notes" at the end of each chapter. These are modest collections of Internet sites to which interested students (or instructors) can turn for additional information and/or images. An effort has been made to concentrate on sites with material at an appropriate level for the book's intended audience, and also on sites likely to be relatively stable in the very fluid world of the Internet (government, agency, educational-institution, or professional-association sites). Space limitations certainly preclude exhaustive listings, and of course new sites are constantly being created. This selection, however, should particularly help the novice user get started exploring. We'll try to update this information on our Web page as appropriate.

The book starts with some background information: a brief outline of earth's development to the present, and a look at one major reason why environmental problems today are so pressing—the large and rapidly growing human population. This is followed by a short discussion of the basic materials of geology—rocks and minerals—and some of their physical properties, which introduces a number of basic terms and concepts that are used in later chapters.

The next several chapters treat individual processes in detail. Some of these are large-scale processes, which may involve motions and forces in the earth hundreds of kilometers below the surface, and may lead to dramatic, often-catastrophic events like earthquakes and volcanic eruptions. Other processes—such as the flow of rivers and glaciers, or the blowing of the wind—occur only near the earth's surface, altering the landscape and occasionally causing their own special problems. In some cases, geologic processes can be modified, deliberately or accidentally; in others, human activities must be adjusted to natural realities.

A subject of increasing current concern is the availability of resources. A series of five chapters deals with water resources, soil, minerals, and energy, the rates at which they are being consumed, probable amounts remaining, and projections of future prospects. In the case of energy resources, we consider both those sources extensively used in the past, and new sources that may or may not successfully replace them in the future.

Increasing population and increasing resource consumption seem to lead to an increasing volume of waste to be disposed of; thoughtless or inappropriate waste disposal commonly leads to increasing pollution. Three chapters examine the interrelated problems of air and water pollution and the strategies available for the disposal of various kinds of wastes.

The final few chapters deal with a more diverse assortment of subjects. Environmental problems spawn laws intended to solve them; the environmental-law chapter looks briefly at a sampling of laws related to geologic matters discussed earlier in the book, as well as at some of the problems with such laws. The land-use planning and engineering geology chapter examines geologic constraints on construction schemes and the broader issue of trying to determine the optimum use(s) for particular parcels of land. Geomedicine, a relatively new field covered in the last chapter, concerns the relationship between health and the geologic setting in which we live.

Relative to the length of time we have been on earth, humans have had a disproportionate impact on this planet. Appendix A explores the concept of geologic time and its measurement, and looks at the rates of geologic and other processes by way of putting human activities in temporal perspective. Appendix B gives an introduction to topographic and geologic maps and satellite imagery. Appendix C provides short reference keys to aid in rock and mineral identification. The Glossary collects definitions of boldface terms and select additional terms for quick reference.

Available with this text is an Instructor's Manual containing over 750 test questions. The test questions found in the Instructor's Manual are also available on Wm. C. Brown's classroom testing software, for use with the Macintosh® and IBM® PC computers. Also available are 100 acetate transparencies of key text illustrations and 100 color slides. These are designed to aid instructors in class presentations and to enhance student learning activities.

Acknowledgments

A great many people have contributed to the development of one or another edition of this book. Portions of the manuscript of the first edition were read by Colin Booth, Lynn A. Brant, Arthur H. Brownlow, Ira A. Furlong, David Huntley, John F. Looney, Jr., Robert A. Matthews, and George H. Shaw, and the entire book was reviewed by Robert A. Marston and Donald J. Thompson. The second edition was enhanced through suggestions from William N. Mode, Laura L. Sanders, Jeffrey J. Gryta, Martin Reiter, Robert D. Hall, Robert B. Furlong, David Gust, and Stephen B. Harper; the third was further refined with the assistance of Susan M. Cashman, Robert B. Furlong, William N. Mode, Frank Hanna, Laura L. Sanders, Paul Nelson, and Michael A. Velbel; and the fourth was strengthened through the input of reviewers Pascal de Caprariis, James Cotter, John Vitek, Paul Schroeder, Steven Lund, Barbara Ruff, Gordon Love, Herbert Adams, Michael McKinney, Thomas E. Hendrix, Clifford Thurber, Ali Tabidian, Dru Germanoski, and Randall Scott Babcock. This latest revision has benefited still further from reviews by Barbara L. Ruff, University of Georgia; Michael Whitsett, University of Iowa; John F. Hildenbrand, Clover Park Technical College; Alvin S. Konigsberg, State University of New York at New Paltz; Vernon P. Scott, Oklahoma State University; Jim Stimson, Carroll College; Kevin Cole, Grand Valley State University; Gilbert N. Hanson, State University of New York at Stony Brook; Doreen Zaback, Vanderbilt University; and Ann E. Holmes, Columbia University. The thoughtful suggestions of all of the foregoing individuals, and many other users who have offered additional advice by personal letter or as feedback to McGraw-Hill sales staff, have substantially improved the text, and their help is most gratefully acknowledged. Any remaining shortcomings are, of course, my own responsibility.

M. Dalechek, C. Edwards, I. Hopkins, and J. McGregor, at the USGS Photo Library in Denver, have provided invaluable assistance with the photo research over the years. The encouragement of a number of my colleagues—particularly Colin Booth, R. C. Flemal, Donald M. Davidson, Jr., R. Kaufmann, and Eugene C. Perry, Jr.—was a special help during development of the first edition. The ongoing support and interest of fellow author, deanly colleague, and ecologist Jerrold H. Zar has, in turn, helped immensely to make the revision cycles survivable. Thanks are also due to the more than two thousand environmental-geology students I have taught, many of whom in the early years suggested that I write a text, and whose classes collectively have provided a testing ground for many aspects of the presentations herein.

My family has been supportive of this undertaking from the inception of the first edition. A very special vote of appreciation goes to my husband Warren—ever-patient sounding board, occasional photographer and field assistant—in whose life this book has so often loomed so large.

Last, but assuredly not least, I would like to express my deep gratitude to the entire McGraw-Hill book team, for their enthusiasm, professionalism, and just plain hard work, without which successful completion of each succeeding edition of this book would have been impossible.

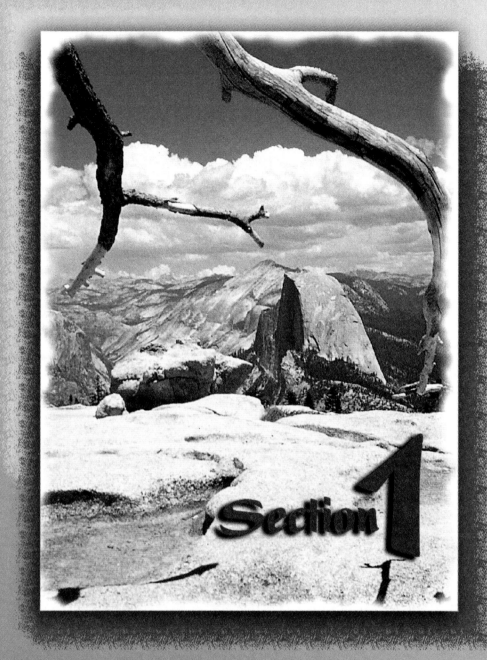

Section 1

The granite domes of Yosemite National Park were formed deep in the crust from the crystallization of a hot melt, then uplifted, and exposed by erosion; now they are being shaped by the actions of water, wind, and gravity.

A sense of historical perspective helps us to appreciate current problems and to anticipate future ones. Many modern environmental problems, such as acid rain and groundwater pollution, have come upon us very recently. Others, such as the hazards posed by earthquakes, volcanoes, and landslides, have always been with us. Recognition of the impact of natural hazards is worldwide: the 1990s have been designated the United Nations Decade for Natural Disaster Reduction, and in recent years, we have seen unprecedented international cooperation on issues such as preservation of the ozone layer.

Chapter 1 briefly summarizes the major events in the earth's development, and allows us to begin to see where human activities fit in. It provides some information about the solar system to help the reader judge the degree to which other planets might provide solutions to such problems as lack of resources and living space. It also introduces the concept of cyclicity in natural processes, and points out that the interrelationships among natural processes may be complex.

The size and growth of earth's human population bear strongly on the ways and extent to which geology and people interact, which is what environmental geology is all about. In fact, many of our problems are as acute as they are simply because of the sheer number of people who now live on the earth. This point will be particularly evident in discussions of resources, pollution, and waste disposal. Again, the global impact of the issues is underscored by the fact that in 1992, more than 170 nations came together in Rio de Janeiro for the United Nations Conference on Environment and Development, to address such issues as global climate change, sustainable development, and environmental protection.

It is difficult to talk for long about geology without discussing rocks and minerals, the stuff of which the earth is made. Chapter 2 introduces these materials and some of their basic properties. Specific physical and chemical properties of rocks and soils are important in considering such diverse topics as resource identification and recovery, waste disposal, assessment of volcanic or landslide hazards, weathering processes and soil formation, and others.

An Overview of Our Planetary Environment

Billions of years ago, out of a swirling mass of gas and dust, evolved a system of varied planets hurtling around a nuclear-powered star—our solar system. One of these planets, and one only, gave rise to complex life forms. Over time, a tremendous diversity of life forms and ecological systems developed, while the planet too evolved and changed, its interior churning, its landmasses shifting, its surface constantly being reshaped. Within the last several million years, the diversity of life on earth has included humans, increasingly competing for space and survival on the planet's surface. With the control over one's surroundings made possible by the combination of intelligence and manual dexterity, humans have found most of the land on the planet inhabitable; they have learned to use not only plant and animal resources, but minerals, fuels, and other geologic materials; in some respects, humans have even learned to modify natural processes that inconvenience or threaten them. As we have learned how to study our planet in systematic ways, we have developed an ever-increasing understanding of the nature of the processes shaping, and the problems posed by, our geological environment. As the human population grows, however, it becomes increasingly difficult for that population to survive on the resources and land remaining, to avoid those hazards that cannot be controlled, and to refrain from making irreversible and undesirable changes in environmental systems. The urgency of perfecting our understanding, not only of natural processes but also of our impact on the planet, is becoming more and more apparent.

Tree roots forcing their way into cracks in rock contribute to the sculpting of Bryce Canyon; now, continued erosion and weathering of the rock in turn threatens this tree's hold, as the earth's surface constantly changes.

The Early Solar System

In recent decades, scientists have been able to construct an ever-clearer picture of the origins of the solar system and, before that, of the universe itself. Most astronomers now accept some sort of "Big Bang" as the origin of today's universe. At that time, enormous quantities of matter were created and flung violently apart across an ever-larger volume of space. The time of the Big Bang can be estimated in several ways. Perhaps the most direct is the back-calculation of the universe's expansion to its apparent beginning. Other methods depend on astrophysical models of creation of the elements or the rate of evolution of different types of stars. Most age estimates overlap in the range of 15 to 20 billion years, although a few researchers suggest an age closer to 10 billion years.

From the debris of the Big Bang formed the stars, as locally high concentrations of mass were collected together by gravity, and some became large and dense enough that energy-releasing atomic reactions were set off deep within them. Stars are not permanent objects. They are constantly losing energy and mass as they burn their nuclear fuel. The mass of material that initially formed the star determines how rapidly the star burns; some stars burned out billions of years ago, while others are probably forming now from the original matter of the universe mixed with the debris of older stars.

Our sun and its system of circling planets, including the earth, are believed to have formed from a rotating cloud of gas and dust, starting nearly 5 billion years ago. Most of the mass of the cloud coalesced to form the sun, which became a star and began to "shine," or release light energy, when its interior became so dense and hot from the crushing effects of its own gravity that nuclear reactions were triggered inside it. Meanwhile, dust condensed from the gases remaining in the flattened cloud disk rotating around the young sun. The dust clumped into planets, the formation of which was essentially complete over $4\frac{1}{2}$ billion years ago.

The Planets

The compositions of the planets formed depended largely on how near they were to the hot sun (figure 1.1). The planets formed nearest to the sun contained mainly high-temperature materials: metallic iron and a few minerals with very high melting temperatures, with little water or gas. Somewhat farther out, where temperatures were lower, the developing planets incorporated much larger amounts of lower-temperature minerals, including some that contain water locked within their crystal structures. (This development later made it possible for the earth to have liquid water at its surface.) Still farther from the sun, temperatures were so low that nearly all of the materials in the original gas cloud condensed—even materials like methane and ammonia, which are gases at normal earth surface temperatures and pressures.

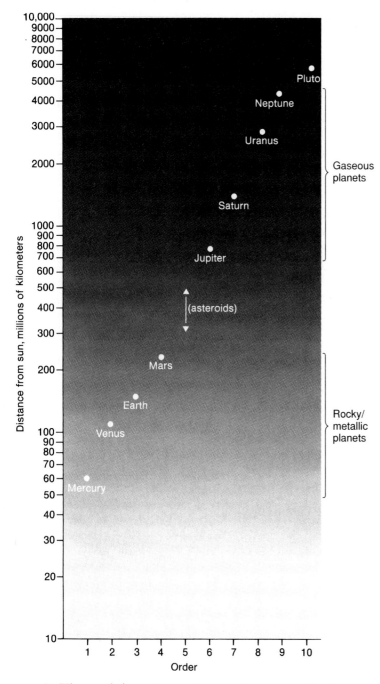

Figure 1.1 As this graph shows, the spacing of the planets' orbits exhibits a geometric regularity. Note that the distance scale is logarithmic—outer planets are MUCH farther out, proportionately, and formed at much colder temperatures.

Table 1.1 Some Basic Data on the Planets			
Planet	Mean distance from Sun (millions of km)	Equatorial diameter, relative to Earth	Density* (g/cu. cm)
Mercury	58	0.38	5.4
Venus	108	0.95	5.2
Earth	150	1.00	5.5
Mars	228	0.53	3.9
Jupiter	778	11.19	1.3
Saturn	1,427	9.41	0.7
Uranus	2,870	4.06	1.2
Neptune	4,479	3.88	1.7
Pluto	5,900	0.23	1.1

Mercury–Mars: Predominatly rocky/metal planets
Jupiter–Pluto: Gaseous planets

Source: Data from Paul Henderson, *Inorganic Geochemistry*, p. 4. Copyright © 1982 by Pergamon Press Ltd., Oxford, England.

*No other planets have been extensively sampled to determine their compositions directly, though we have some data on their surfaces. Their approximate bulk compositions are inferred from the assumed starting composition of the solar nebula and the planets' densities. For example, the higher densities of the inner planets reflect a significant iron content and relatively little gas.

ets could be mined for needed minerals. Both the basic chemistry of these other bodies and the kinds of ore-forming or other resource-forming processes that might occur on them would differ considerably from those on earth, and may not lead to products we would find useful. (This is leaving aside any questions of the economics or technical practicability of such mining activities!) In addition, our principal current energy sources required living organisms to form, and so far, no such life forms have been found on other planets or moons. Even Venus—close to Earth in space, similar in size and density—shows marked differences: Its dense, cloudy atmosphere is thick with carbon dioxide, producing planetary surface temperatures hot enough to melt lead through runaway greenhouse-effect heating (see chapter 9).

Earth, Then and Now

The earth has changed continuously since its formation, undergoing some particularly profound changes in its early history. The early earth was very different from what it is today, lacking the modern oceans and atmosphere and having a much different surface from its present one, probably more closely resembling the barren, cratered surface of the moon. The planet was heated by several processes: the impact of the colliding dust particles and meteorites as they came together to form the earth, compression of the interior by gravity (that materials heat up when compressed can be demonstrated by pumping up a bicycle tire and then feeling the barrel of the pump), and energy release from decay of

The result was a series of planets with a variety of compositions, most quite different from that of earth. This is confirmed by observations and measurements of the planets. For example, the planetary densities listed in table 1.1 are consistent with a higher metal content in planets close to the sun and a larger proportion of ice and gas in the planets farther from the sun (see also figure 1.2). These differences should be kept in mind when it is proposed that other plan-

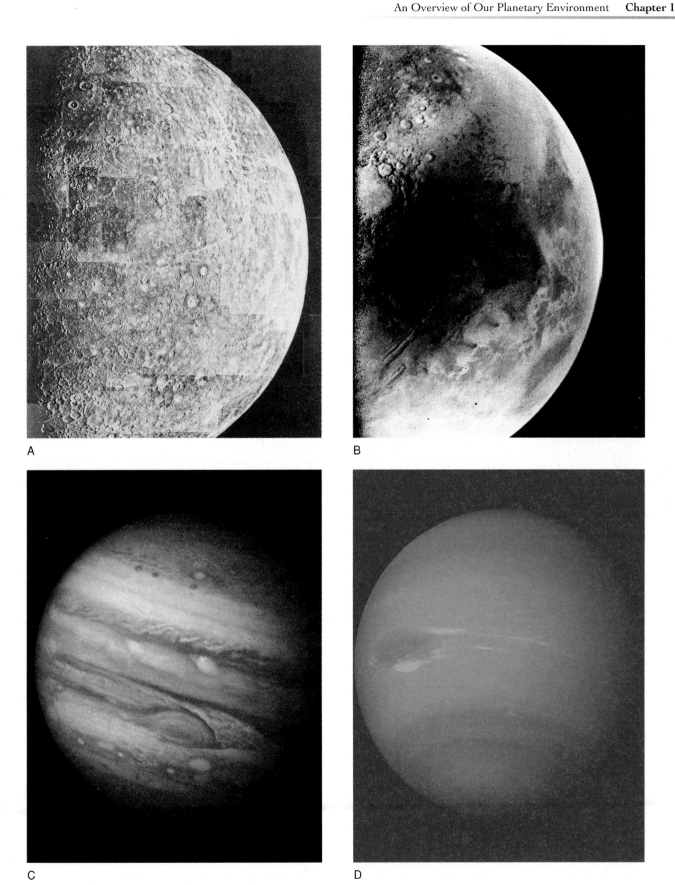

Figure 1.2 The planets of the solar system vary markedly in both composition and physical properties. Mercury (*A*) is rocky, iron-rich, dry, and pockmarked with craters; Mars (*B*) has a ruddy, windswept surface, polar caps of carbon-dioxide ice ("dry ice") and some landforms—like the prominent canyon Valles Marineris, bottom center—that resemble those of earth; Jupiter (*C*) is a huge gas ball, with no solid surface at all, and a system of sixteen moons of ice and rock that circle it to mimic the solar system in miniature; Neptune (*D*) is another, smaller gas ball, with eight orbiting moons.   © NSSDC Goddard Space Flight Center.

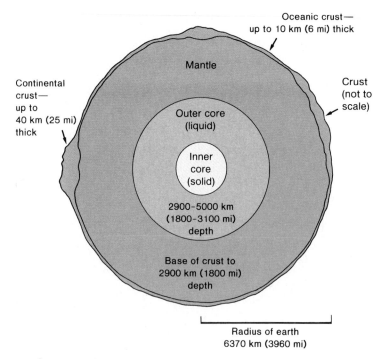

Continental crust— up to 40 km (25 mi) thick

Oceanic crust— up to 10 km (6 mi) thick

Mantle

Crust (not to scale)

Outer core (liquid)

Inner core (solid)

2900–5000 km (1800–3100 mi) depth

Base of crust to 2900 km (1800 mi) depth

Radius of earth 6370 km (3960 mi)

Figure 1.3 A chemically differentiated earth (crust is not drawn to scale, but is greatly exaggerated vertically to be visible at this scale). The core consists mostly of iron; the outer part is molten. The mantle, the largest zone, is made up primarily of ferromagnesian silicates (see chapter 2) and, at great depths, of oxides of iron, magnesium, and silicon. The crust forms a thin skin around the earth. Oceanic crust, which forms the sea floor, has a composition somewhat like that of the mantle, but is richer in silicon. Continental crust is both thicker and less dense. It rises above the oceans and contains more light minerals rich in calcium, sodium, potassium, and aluminum.

the small amounts of several naturally radioactive elements that the earth contains. (See chapters 14 and 15 for discussions of radioactivity.)

These three heat sources combined to raise the earth's internal temperature enough that parts of it, perhaps eventually most of it, melted, although it was probably never molten all at once. Dense materials, like metallic iron, would have tended to sink toward the middle of the earth. As cooling progressed, lighter, low-density minerals crystallized and floated out toward the surface. The eventual result was an earth differentiated into several major compositional zones: the central **core;** the surrounding **mantle;** and a thin **crust** at the surface (see figure 1.3). The process was complete before 4 billion years ago.

Although only the crust and a few bits of uppermost mantle that are carried up into the crust by volcanic activity can be sampled and analyzed directly, we nevertheless have a good deal of information on the composition of the earth's interior. First, scientists can estimate from analyses of stars the starting composition of the cloud from which the solar system formed. Geologists can also infer aspects of the earth's bulk composition from analyses of certain meteorites believed to have formed at the same time as, and under conditions similar to, the earth. Geophysical data demonstrate that the

Table 1.2 Most Common Chemical Elements in the Earth

Whole earth		Crust	
Element	Weight percent	Element	Weight percent
Iron	32.4	Oxygen	46.6
Oxygen	29.9	Silicon	27.7
Silicon	15.5	Aluminum	8.1
Magnesium	14.5	Iron	5.0
Sulfur	2.1	Calcium	3.6
Nickel	2.0	Sodium	2.8
Calcium	1.6	Potassium	2.6
Aluminum	1.3	Magnesium	2.1
(All others, total)	.7	(All others, total)	1.5

(Compositions cited are averages of several independent estimates.)

earth's interior is zoned and also provide information on the densities of the different layers within the earth, which further limits their possible compositions (see chapter 4). These and other kinds of data indicate that the earth's core is made up mostly of iron, with some nickel and a few minor elements, and that the mantle consists mainly of iron, magnesium, silicon, and oxygen combined in varying proportions in several different minerals. The earth's crust is much more varied in composition and very different chemically from the average composition of the earth (see table 1.2). As is evident from that table, many of the metals we have come to prize as resources are relatively uncommon elements in the crust.

The heating and subsequent differentiation of the early earth led to another important result: formation of the atmosphere and oceans. Many minerals that had contained water or gases in their crystals released them during the heating and melting, and as the earth's surface cooled, the water could condense to form the oceans. Without this abundant surface water, which in the solar system is unique to earth, most life as we know it could not exist.

The earth's early atmosphere was quite different from the modern one, aside from the effects of modern pollution. The first atmosphere had little or no free oxygen in it. It probably consisted dominantly of nitrogen and carbon dioxide (the gas most commonly released from volcanoes, aside from water) with minor amounts of such gases as methane, ammonia, and various sulfur gases. Humans could not have survived in this early atmosphere. Oxygen-breathing life of any kind could not exist before the first simple plants—the single-celled blue-green algae—appeared in large numbers to modify the atmosphere. Their remains are found in rocks several billion years old. They manufacture food by photosynthesis, using sunlight for energy, consuming carbon dioxide, and releasing oxygen as a by-product. In time, enough oxygen accumulated that the atmosphere could support oxygen-breathing organisms.

Figure 1.4 The "geologic spiral." Important plant and animal groups appear where they first occurred in significant numbers. All complex organisms—especially humans—have developed relatively recently in the geologic sense.   Source: Modified after Geological Survey Publication, *"Geologic Time."*

Life on Earth

The rock record shows when different plant and animal groups appeared. Some are represented schematically in figure 1.4. The earliest creatures left very few remains because they had no hard skeletons, teeth, shells, or other hard parts that could be preserved in rocks. The first multicelled oxygen-breathing creatures probably developed about 1 billion years ago, after oxygen in the atmosphere was well established. By about 600 million years ago, marine animals with shells had become widespread.

The development of organisms with hard parts—shells, bones, teeth, and so on—greatly increased the number of preserved animal remains in the rock record; consequently, biological developments since that time are far better understood. Dry land was still barren of large plants or animals half a billion years ago. In rocks about 500 million years old is the first evidence of animals with backbones—the fish—and soon thereafter, early land plants developed, before 400 million years ago. Insects appeared approximately 300 million years ago. Later, reptiles and amphibians moved onto the continents. The dinosaurs appeared about 200 million years ago and the first mammals at nearly the same time. Warm-blooded animals took to the air with the development of birds about 150 million years ago, and by 100 million years ago, both birds and mammals were well established.

Such information has current applications. Certain energy sources have been formed from plant or animal remains (see chapter 13). Knowing the times at which particular groups of organisms appeared and flourished is helpful in assessing the probable amounts of these energy sources available and in concentrating the search for these fuels on rocks of appropriate ages.

On a time scale of billions of years, human beings have just arrived. The most primitive human-type remains are no more than 3 to 4 million years old, and modern, rational humans (*Homo sapiens*) developed only about half a million years ago. Half a million years may sound like a long time, and it is if compared to a single human lifetime. In a geologic sense, though, it is a very short time. Nevertheless, we humans have had an enormous impact on the earth, at least at its surface, an impact far out of proportion to the length of time we have occupied it. Later in this chapter, we will see that our impact is likely to continue to increase rapidly as the population does likewise.

Geology, Past and Present

Two centuries ago, geology was mainly a matter of observing and describing natural processes and their products. The subject has become both more quantitative and more interdisciplinary through time. Modern geoscientists draw on the principles of chemistry to interpret the compositions of geologic materials, apply the laws of physics to explain these materials' physical properties and behavior, use the biological sciences to develop an understanding of ancient life forms, and rely on engineering principles to design safe structures in the presence of geologic hazards. The emphasis on the "why," rather than just the "what," has increased.

The Geologic Perspective

Geologic observations now are combined with laboratory experiments, careful measurements, and calculations to develop theories of how natural processes operate. Geology is especially challenging because of the disparity between the scientist's laboratory and nature's. In the research laboratory, conditions of temperature and pressure, as well as the flow of chemicals into or out of the system under study, can be carefully controlled. One then knows just what has gone into creating the product of the experiment. In nature, the geoscientist is often confronted only with the results of the "experiment" and must deduce the starting materials and processes involved.

Another complicating factor is time. The laboratory scientist must work on a time scale of hours, months, years, or, at most, decades. Natural geologic processes may take a million or a billion years to achieve a particular result, by stages too slow even to be detected in a human lifetime (table 1.3). Not only does this make it difficult to arrive at a full understanding of how geologic processes operated in the past. It dictates caution, too, as we try to project, from a few years' data on global changes associated with human activities, the long-range impacts we may be causing.

Also, the laboratory scientist may conduct a series of experiments on the same materials, but the experiments can be stopped and those materials examined after each stage. Over the vast spans of geologic time, a given mass of earth material may have been transformed a half-dozen times or more, under different conditions each time. The history of the rock that ultimately results may be very difficult to decipher from the end product alone.

Table 1.3	Some Representative Illustrations of Geologic Process Rates

Process	Occurs over a time span of about this magnitude
Rising and falling of tides	1 day
"Drift" of a continent by 2–3 centimeters (about 1 inch)	1 year
Accumulation of energy between large earthquakes on a major fault zone	10–100 years
Rebound (rising) by 1 meter of a continent depressed by ice sheets during the Ice Age	100 years
Flow of heat through 1 meter of rock	1000 years
Deposition of 1 centimeter of fine sediment on the deep-sea floor	1000–10,000 years
Life span of a small volcano	100,000 years
Ice sheet advance and retreat during an ice age	100,000 years
Life span of a large volcanic center	1–10 million years
Creation of an ocean basin such as the Atlantic	100 million years
Duration of a major mountain-building episode	100 million years
History of life on earth	Over 1 billion years

Geology and the Scientific Method

The **scientific method** is a means of discovering basic scientific principles. One begins with a set of observations and/or a body of data, based on measurements of natural phenomena or on experiments. One or more **hypotheses** are formulated to explain the observations or data. A hypothesis can take many forms, ranging from a general conceptual framework or model describing the functioning of a natural system, to a very precise mathematical formula relating several kinds of numerical data. What all hypotheses have in common is that they must all be susceptible to testing.

In the classical conception of the scientific method, one uses a hypothesis to make a set of predictions. Then one devises and conducts experiments to test each hypothesis, to determine whether experimental results agree with predictions based on the hypothesis. If they do, the hypothesis gains credibility. If not, if the results are unexpected, the hypothesis must be modified to account for the new data as well as the old. Several cycles of modifying and retesting hypotheses may be required before a hypothesis that is consistent with all the observations and experiments that one can conceive is achieved. A hypothesis that is repeatedly supported by new experiments advances in

Figure 1.5 Collapsed parking structure at California State University at Northridge after 1994 earthquake: After the interior failed, the walls collapsed inward.   Photograph by M. Celebi, courtesy U.S. Geological Survey.

Figure 1.6 Mount Pinatubo in eruption, June, 1991. Photograph by R. Batalon, USAF, courtesy NOAA.

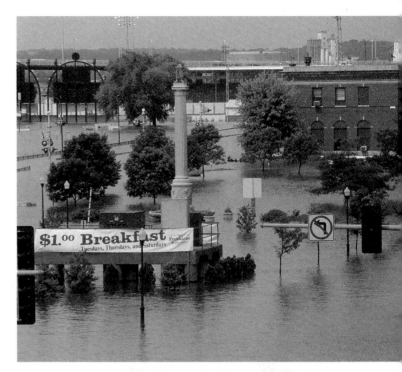

Figure 1.7 A major river like the Mississippi floods when a large part of the area that it drains is waterlogged by too much rain or snowmelt in too short a time. Such floods—like that in summer, 1993, shown here drenching Davenport, Iowa—can be correspondingly long-lasting.   © Doug Sherman/Geofile.

time to the status of a **theory,** a generally accepted explanation for a set of data or observations.

This approach is not strictly applicable to many geologic phenomena because of the difficulty of experimenting with natural systems. For example, one may be able to conduct experiments on a single rock, but not to construct a whole volcano in the laboratory. In such cases, hypotheses are often tested entirely through further observations and modified as necessary until they accommodate all the relevant observations. This broader conception of the scientific method is well illustrated by the development of the theory of plate tectonics, discussed in chapter 3. Even a well-established theory, however, may ultimately be proved incorrect. In the case of geology, complete rejection of an older theory is most often caused by the development of new analytical or observational techniques, which make available wholly new kinds of data that were unknown at the time the original theory was formulated.

The Motivation to Find Answers

In spite of the difficulties inherent in trying to explain geologic phenomena, the search for explanations goes on, spurred not only by the basic quest for knowledge, but also by the practical problems posed by geologic hazards, the need for resources, and concerns about possible global-scale human impacts, such as ozone destruction and global warming.

The hazards may create the most dramatic scenes and headlines, the most abrupt consequences: The 1989 Loma Prieta (California) earthquake caused more than $5 billion in damage; the smaller 1990 Iranian earthquake claimed over 40,000 lives; in 1994, the Northridge (California) earthquake (figure 1.5) cost close to $15 billion, and the 1995 Kobe (Japan) earthquake, similar in magnitude to Loma Prieta, caused over 5,200 deaths and about $100 billion in property damage. The 18 May 1980 eruption of Mount St. Helens took even the scientists monitoring the volcano by surprise, and the 1991 eruption of Mount Pinatubo in the Philippines (figure 1.6) not only devastated local residents but caught the attention of the world through a marked decline in 1992 summer temperatures. Efforts are underway to learn to predict earthquakes and volcanic eruptions reliably enough to save lives, if not property. Likewise, improved understanding of stream dynamics, and more prudent land use, can together reduce the damages from flooding (figure 1.7), which amount in the United States to over $1 billion a year, and

Figure 1.8 Slope failure on a California hillside under-cuts homes.   Photograph by J. T. McGill, USGS Photo Library, Denver, CO.

Figure 1.9 Unreclaimed hillside disturbed by open-pit mining, Elko County, Nevada, is subject to landslides. Photograph courtesy of USGS Photo Library, Denver, CO.

the loss of dozens of lives annually. Landslides (figure 1.8) take a similar toll in property damage, which could be reduced by more attention to slope stability and improved engineering practices. It is not only the more dramatic hazards that are costly: on average, the cost of structural damage from unstable soils each year approximately equals the combined costs of landslides, earthquakes, and flood damages in this country.

Our demand for resources of all kinds continues to grow, and so do the consequences of resource use. In the United States, average per capita water use is 1,565 gallons per day; in many places, groundwater supplies upon which we have come to rely heavily are being measurably depleted. Worldwide, water-resource disputes between nations are increasing. As we mine more extensively for mineral resources, we face the problem of how to minimize associated damage to the mined lands (figure 1.9). The grounding of the *Exxon Valdez* in 1989, dumping 11 million gallons of oil into Prince William Sound, Alaska, was a reminder of the negative consequences of petroleum exploration, just as the 1991 war in Kuwait was a reminder of U.S. dependence on imported oil.

As we consume more resources, we create more waste. In the United States, total waste generation is estimated at close to 300 million tons per year—or more than a ton per person. Careless waste disposal, in turn, leads to pollution. The Environmental Protection Agency continues to identify toxic-waste disposal sites in urgent need of cleanup through Superfund (discussed in chapter 15); by 1991, some 1,236 so-called priority sites had been identified, and only 33 had been cleaned up, at an average cleanup cost per site in the $20–30 million range. As fossil fuels are burned, carbon

dioxide in the atmosphere rises, and modelers of global climate strive to understand what that may do to global temperatures, weather, and agriculture decades in the future.

These are just a few of the kinds of issues that prompt the study of geologic processes from more than academic interest.

Wheels Within Wheels: Earth Cycles and Systems

The earth is a dynamic, constantly changing planet—its crust shifting to build mountains; lava spewing out of its warm interior; ice and water and windblown sand and gravity reshaping its surface, over and over. Some changes proceed in one direction only: for example, the earth has been cooling progressively since its formation, though considerable heat remains in its interior. Many of the processes, however, are cyclic in nature.

Consider, for example, such basic materials as water or rocks. Streams drain into oceans and would soon run dry if not replenished; but water evaporates from oceans, to make the rain and snow that feed the streams to keep them flowing. This describes just a part of the *hydrologic* (water) *cycle*, explored more fully in chapters 6 and 7. Rocks, despite their appearance of permanence in the short term of a human life, participate in the *rock cycle*, introduced in chapter 2 and revisited in chapter 3. The kinds of evolutionary paths rocks may follow through this cycle are many, but consider this illustration: A volcano's lava (figure 1.10) hardens into rock; the rock is weathered into sand and dissolved chemicals; the debris, deposited in an ocean basin, is solidified into a new rock of quite different type; and that new rock is carried into the mantle via plate tectonics, to be melted into a new lava.

Figure 1.10 Molten lava pours from Kilauea, and solidifies to make new rock.   Photograph by Hawaii Volcano Observatory, USGS Photo Library, Denver, CO.

The time frame over which this process occurs is generally much longer than that over which water cycles through atmosphere and oceans, but the principle is similar. The Rocky Mountains as we see them today (figure 1.11) are not the Rocky Mountains as they formed, tens of millions of years ago; they are much eroded from their original height, and in turn contain rocks formed in water-filled basins and deserts from material eroded from more ancient mountains before them.

Chemicals, too, cycle through the environment. The carbon dioxide that we exhale into the atmosphere is taken up by plants through photosynthesis, and when we eat those plants for food energy, we release CO_2 again. The same exhaled CO_2 may also dissolve in the oceans from the atmosphere, to make dissolved carbonate which then contributes to the formation of carbonate shells and carbonate rocks in the ocean basins; those rocks may later be exposed and weathered by rain, releasing CO_2 back into the atmosphere or dissolved carbonate into streams that carry it back to the ocean. The cycling of chemicals and materials in the environment may be complex, as we will see in later chapters.

Furthermore, these processes and cycles are often interrelated, and seemingly local actions can have distant consequences. We dam a river to create a reservoir as a source of irrigation water and hydroelectric power, inadvertently trapping stream-borne sediment at the same time; downstream, patterns of erosion and deposition in the stream channel change, and at the coast, where the stream pours into the ocean, coastal erosion of beaches increases because a part of their sediment supply, from the stream, has been cut off. The volcano that erupts the lava to make the volcanic rock also releases water vapor into the atmosphere, and sulfur-rich gases and dust that influence the amount of sunlight reaching earth's surface to heat it, which in turn can

Figure 1.11 The Beartooth Mountains near Yellowstone National Park, a part of the U.S. Rocky Mountains.

Figure 1.12 World population, currently approaching 6 billion, is projected to reach about 10 billion by the year 2050. Most of that population increase will occur in developing countries.   Sources: Figure from World Resources Institute; data from United Nations Population Division and Population Reference Bureau, 1993.

alter the extent of evaporation and resultant rainfall, which will affect the intensity of landscape erosion and weathering of rocks by water. . . . So although we divide the great variety and complexity of geologic processes and phenomena into more manageable, chapter-sized units for purposes of discussion, it is important to keep such interrelationships in mind. We will note a number of examples throughout the text. And superimposed on, influenced by, and subject to all these natural processes are humans and human activities.

Nature and Rate of Population Growth

Animal populations, as well as primitive human populations, are generally quite limited both in the areas that they can occupy and in the extent to which they can grow. They must live near food and water sources. The climate must be one to which they can adapt. Predators, accidents, and disease take a toll. If the population grows too large, disease and competition for food are particularly likely to cut it back to sustainable levels.

The human population grew relatively slowly for hundreds of thousands of years. Not until the middle of the last century did the world population reach 1 billion. However, by then, a number of factors had combined to accelerate the rate of population increase. The second, third, and fourth billion were reached far more quickly; the world population is now over 5.7 billion, and is expected nearly to double, to about 10 billion, by 2050 (figure 1.12).

Humans are no longer constrained to live only where conditions are ideal. We can build habitable quarters even in extreme climates, using heaters or air conditioners to bring the temperature to a level we can tolerate. In addition, people need not live where food can be grown or harvested or where there is abundant fresh water: The food and water can be transported, instead, to where the people choose to live.

Growth Rates: Causes and Consequences

Population growth occurs when new individuals are added to the population faster than existing individuals are removed from it. On a global scale, the population increases when its birthrate exceeds its death rate. In assessing an individual nation's or region's rate of population change, immigration and emigration must also be taken into account. Improvements in nutrition and health care typically increase life expectancies, decrease mortality rates, and thus increase the rate of population growth. (Worldwide, life expectancy is about 65 years and is expected to increase to

Table 1.4 World and Regional Population Growth and Projections (in millions)

Year	World	North and Central America	South America	Africa	Europe	Asia	Oceania
1950	2,520	220	112	224	549	1,403	12
1990	5,285	424	293	633	722	3,186	26
1995	5,716	454	320	728	727	3,458	28
2025 (projected)	8,294	615	462	1,496	718	4,960	41
Growth rate (%/year) 1990–95	1.6	1.4	1.7	2.8	0.2	1.6	1.5
Doubling time (years)	43	50	41	25	346	43	46

Source: Most data from *World Resources 1996–97* (NY: Oxford University Press, 1996). Doubling times calculated from 1990–95 growth rates.

over 70 years by the year 2025.) Increased use of birth-control or family-planning methods reduces birthrates and, therefore, also reduces the rate of population growth; in fact, a population could begin to decrease if birthrates were severely restricted.

The sharply rising rate of population growth over the past few centuries, as shown in figure 1.12, can be viewed another way. It took until about A.D. 1830 for the world's population to reach 1 billion. The population climbed to 2 billion in the next 100 years, to 3 billion in 30 more years, and so on. It has taken less and less time to add each successive billion people, as ever more people contribute to the population growth and individuals live longer. Now, an additional billion people will be added to the world's population within less than a decade.

There are wide differences in growth rates among regions (table 1.4; figures 1.12 and 1.13.). The reasons for this are many. Religious or social values may cause larger or smaller families to be regarded as desirable in particular regions or cultures. High levels of economic development are commonly associated with reduced rates of population growth; conversely, low levels of development are often associated with rapid population growth. The impact of improved education, which may accompany economic development, can vary: It may lead to better nutrition, prenatal and child care, and thus to increased growth rates, but it may also lead to increased or more effective practice of family-planning methods, thereby reducing growth rates. A few governments of nations with large and rapidly growing populations have considered encouraging or mandating family planning; India and the People's Republic of China have taken active measures.

Even when population growth rate is constant, the number of individuals added per unit of time increases over

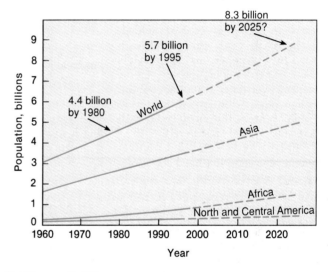

Figure 1.13 World population growth, past and projected; growth from 1960 to 1995, with projections to 2025.

time. This is called **exponential growth,** a concept to which we will return when discussing resources in section IV; see box 1.1.

Growth Rate and Doubling Time

Another way to look at the rapidity of world population growth is to consider the expected **doubling time,** the length of time required for a population to double in size. Doubling time (D) in years may be estimated from growth rate (G), expressed in percent per year, using the simple relationship $D = 70/G$, which is derived from the equation for exponential growth (see "For Further Thought" question 3 at the chapter's end). The higher the growth rate, the shorter the doubling time of the population (see again table 1.4). By region,

Box 1.1 Exponential Growth Examined

The effect of exponential growth is similar to interest compounding. If one invests $100 at 10 percent per year, compounded annually, then, after one year, $10 interest is credited, for a new balance of $110. At the end of the second year, the interest is $11 (10 percent of $110), and the new balance is $121. By the end of the tenth year (assuming no withdrawals), the interest for the year is $23.58, but the interest *rate* has not increased. Similarly, with a population of 1 million growing at 5 percent per year: In the first year, 50,000 persons are added; in the tenth year, the population grows by 77,566 persons. The result is that a graph of population versus time steepens over time, even at a constant growth rate. If the growth rate itself also increases, the curve rises still more sharply, as in figure 1.12.

For many mineral and fuel resources, consumption has been growing very rapidly, even more rapidly than the population. Over the long term, this has certainly been true. Annual consumption of many minerals, for instance, has been growing exponentially at rates of several percent per year. The effects of exponential in-

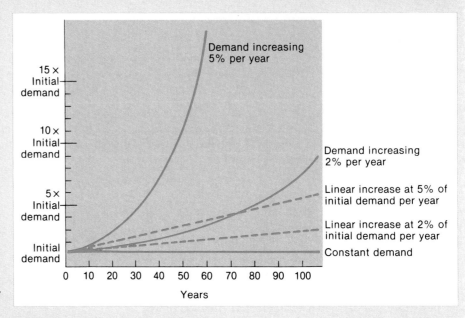

Figure 1 Graphical comparison of the effects of linear and exponential demand growth on consumption of minerals, fuels, water, and other consumable commodities. Exponential population growth produces similar patterns.

creases in resource demand are like the effects of exponential population growth. If demand increases by 2 percent per year, it will double not in 50 years, but in 35. A demand increase of 5 percent per year leads to a doubling

in demand in 14 years, and a tenfold increase in demand in 47 years! (See figure 1.) In other words, a prediction of how soon mineral or fuel supplies will be used up is very sensitive to the assumed rate of change of demand.

---

the most rapidly growing segment of the population today is that of Africa. Its population, estimated at 728 million in 1995, grew at about 2.8 percent per year from 1990–95. The largest segment of the population, that of Asia, is increasing at 1.6 percent per year, and since the nearly $3\frac{1}{2}$ billion people there represent well over half of the world's total population, this leads to a relatively high global average growth rate. The slowest population growth—about 0.2 percent per year—is occurring in Europe, which has less than 10 percent of the world's population. Thus, the fastest growth in general is taking place in the largest segments of the population.

The average worldwide population growth rate is about 1.6 percent per year. This may sound moderate, but it

corresponds to a relatively short doubling time of about forty-one years. At that, the present population growth rate actually represents a decline from nearly 2 percent per year in the mid-1960s. The United Nations projects that growth rates will continue to decrease gradually over the next several decades, probably as a result of scarcer resources coupled with increased population-control efforts. However, a decreasing growth rate is not at all the same as a decreasing population. Depending upon projected fertility rates, U.N. estimates of world population in the year 2050 vary by several billion; 10 billion is an estimate using a medium fertility rate. Figure 1.14 illustrates how those people will be distributed by region, considering differential growth rates from place to place.

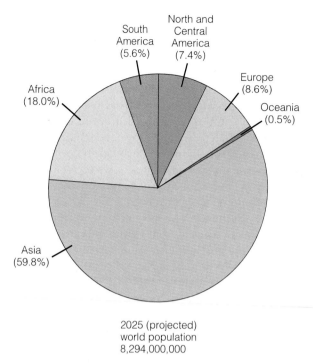

Figure 1.14 Population distribution by region in 1995 with projections to the year 2025. Size of circle reflects relative total population.   Source: Data from *World Resources 1996–97*, World Resources Institute, 1996.

Impacts of the Human Population

The problems posed by a rapidly growing world population have historically been discussed most often in the context of food: that is, how to produce sufficient food and distribute it effectively to prevent the starvation of millions in overcrowded countries or in countries with minimal agricultural development. This is a particularly visible and immediate problem, but it is also only one facet of the population threat.

Farmland and Food Supply

Whether or not the earth can support five billion people, or eight, or eleven, is uncertain. In part, it depends on the quality of life, the level of technological development, and other standards that societies wish to maintain. Yet even when considering the most basic factors, such as food, it is unclear just how many people the earth can sustain. Projections about the adequacy of food production, for example, require far more information than just the number of people to be fed and the amount of available land. The total arable land (land suitable for cultivation) in the world has been estimated at 7.9 billion acres, or about 1.5 acres per person of the present population. The major limitation on this figure is availability of water, either as rainfall or through irrigation. (Chapter 10 explores in more detail the critical problem of availability of water for all purposes.) Further considerations relating to the nature of the soil include the soil's fertility, water-holding capacity, and general suitability for farming. Soil character varies tremendously, as will be seen in chapter 11, and its productivity is similarly variable, as any farmer can attest. Moreover, farmland can deteriorate through loss of nutrients and by wholesale erosion of topsoil, and this degradation must be considered when making production predictions.

There is also the question of what crops can or should be grown. Today, this is often a matter of preference or personal taste, particularly in farmland-rich (and energy- and water-rich) nations. The world's people are not now always being fed in the most resource-efficient ways. To produce one ton of corn requires about 250,000 gallons of water; a ton of wheat, 375,000 gallons; a ton of rice, 1,000,000 gallons; a ton of beef, 7,500,000 gallons. Some new high-yield crop varieties may require irrigation whereas native varieties did not. The total irrigated acreage in the world has more than doubled in three decades. However, water resources are dwindling in many places; in time, the water costs of food production will have to be taken into greater account.

Food production as practiced in the United States is also a very energy-intensive business. The farming is heavily mechanized, and much of the resulting food is extensively processed, stored, and prepared in various ways requiring considerable energy. The products are elaborately packaged and often transported long distances to the consumer. Exporting the same production methods to poor, heavily populated countries short on energy and the capital to buy fuel, as well as food, may be neither possible nor practical. Even if possible, it would substantially increase the world's energy demands.

15

Population and Nonfood Resources

Food is at least a renewable resource. Within a human life span, many crops can be planted and harvested from the same land and many generations of food animals raised. By contrast, the supplies of many of the resources considered in later chapters—minerals, fuels, even land itself—are finite. There is only so much oil to burn, rich ore to exploit, and suitable land on which to live and grow food. When these resources are exhausted, alternatives will have to be found or people will have to do without.

As the population has grown, consumption of mineral and energy resources has grown even faster. As we will see in chapters 12–14, the earth's supply of many such materials is severely limited, especially considering the rates at which these resources are presently being used. Many could be effectively exhausted within decades. Yet most people in the world are consuming very little in the way of minerals or energy. Current consumption is strongly concentrated in a few highly industrialized societies. Per capita consumption of most mineral and energy resources is higher in the United States than in any other nation. For the world population to maintain a standard of living comparable to that of the United States, mineral production would have to increase about fourfold, on the average. There are neither the recognized resources nor the production capability to maintain that level of consumption for long, and the problem will become more acute as the population grows. This strongly implies that standards of living will be lower in the future, one point that will be considered again in section IV.

Some scholars believe that we are already on the verge of exceeding the earth's **carrying capacity,** its ability to sustain its population at a basic, healthy, moderately comfortable standard of living. Whether or not this is the case depends, in part, on what one considers an acceptable standard of living. Estimates of sustainable world population made over the last two decades range from under 7 billion to over 100 billion persons. The wide range is attributable to considerable variations in model assumptions. (For instance, the upper ranges include such assumptions as very high productivity of farmland, comparable to present record levels, with no land reserved for recreation or urbanization.) If the lower estimates of carrying capacity are accurate, this means that, even if we could somehow halt population growth within a few decades, no possible redistribution of resources could guarantee everyone that basic standard of living. Certainly, given global resource availability, even the present world population could not enjoy the kind of high-consumption lifestyle to which the average resident of the U.S. has become accustomed.

It is true that, up to a point, the increased demand for minerals, fuels, and other materials associated with an increase in population tends to raise prices and promote exploration for these materials. The short-term result can be an apparent increase in the resources' availability, as more ex-

A

B

Figure 1.15 Development along the California coast. (*A*) Area prior to significant development. Local geologic hazards include faulting, landslides, and coastal erosion. (*B*) Ten years later, the same area has been almost fully developed with little or no thought to geologic hazards or land-use controls.   Photographs courtesy of USGS Photo Library, Denver, CO.

ploration leads to discoveries of more oil fields, ore bodies, and so on. However, the quantity of each of these resources is finite. The larger and more rapidly growing the population, the more rapidly limited resources are consumed, and the sooner those resources will be exhausted.

Land is clearly a basic resource. Five, eight, or eleven billion people must be *put* somewhere. Already, the global average population density is more than 32 persons per square kilometer of land surface (83 persons per square mile), and that is counting *all* the land surface, including Antarctica, jungles, deserts, and mountain ranges. The ratio of people to readily inhabitable land is plainly much higher. Land is also needed for manufacturing, energy production, transportation, and a variety of other uses. Large numbers of people consuming vast quantities of materials generate vast quantities of wastes. Some of these wastes can be recovered and recycled, but others cannot. It is essential to find places to put the latter, and ways to isolate the harmful materials

from contact with the growing population. This effort claims still more land and, often, resources. All of this is why land-use planning—making the most of every bit of land available—is becoming increasingly important. At present, it is too often true that the ever-growing population settles in areas that are demonstrably unsafe or in which the possible problems are imperfectly known (figures 1.15 and 1.16). We will look further into land-use planning in chapter 19 and in chapters relating to geologic hazards.

Uneven Distribution of People and Resources

Even if global carrying capacity were ample in principle, that of an individual region may not be. None of the resources—livable land, arable land, energy, minerals, or water—is uniformly distributed over the earth. Neither is the population (see figure 1.17). In 1995, the population density in Singapore was 170 times as high as the U.S. average.

Figure 1.16 The landslide hazard to the structures below Cantagalo Rock, Rio de Janeiro, is obvious; they were seriously damaged in 1966–67.   Photograph by F. O. Jones, USGS Photo Library, Denver, CO.

Figure 1.17 Global population density; the darker the shading, the higher the population density.   From Waldo Tobler, Uwe Deichmann, Jon Gottsegen, and Kelly Maloy in Technical Report TR–95–6. Copyright © 1995 National Center for Geographic Information and Analysis, Santa Barbara, CA. Reprinted by permission.

Many of the most densely populated countries are resource-poor. In some cases, a few countries control the major share of one resource. Oil is a well-known example, but there are many others. Thus, economic and political complications enter into the question of resource adequacy. Just because one nation controls enough of some commodity to supply all the world's needs does not necessarily mean that the country will choose to share its resource wealth or to distribute it at modest cost to other nations. Some resources, like land, are simply not transportable and therefore cannot readily be shared. Full discussion of such complexities is beyond the scope of this book, but some will be highlighted in subsequent chapters.

Disruption of Natural Systems

Natural systems tend toward a balance or equilibrium among opposing factors or forces. When one factor changes, compensating changes occur in response. If the disruption of the system is relatively small and temporary, the system may, in time, return to its original condition, and evidence of the disturbance will disappear. For example, a coastal storm may wash away beach vegetation and destroy colonies of marine organisms living in a tidal flat, but when the storm has passed, new organisms will start to move back into the area, and new grasses will take root in the dunes. The violent eruption of a volcano like Mount Pinatubo may spew ash and gases high into the atmosphere, partially blocking sunlight and causing the earth to cool, but within a few years, the ash will have settled back to the ground and normal temperatures will be restored. Dead leaves falling into a lake constitute one kind of natural pollution, but they also provide food for the microorganisms that within weeks or months will break the leaves down and eliminate them.

This is not to say that permanent changes never occur in natural systems. The size of a river channel reflects the maximum amount of water it normally carries. If long-term climatic or other conditions change so that the volume of water regularly reaching the stream increases, the larger quantity of water will, in time, carve out a correspondingly larger channel to accommodate it. The soil carried downhill by a landslide certainly does not begin moving back upslope after the landslide is over; the face of the land is irreversibly changed. Even so, a hillside forest uprooted and destroyed by the slide may, within decades, be replaced by new growth in the soil newly deposited at the bottom of the hill.

Human activities can cause or accelerate permanent changes in natural systems. The impact of humans on the global environment is broadly proportional to the size of the population, as well as to the level of technological development achieved. This can be illustrated especially easily in the context of pollution. The smoke from one campfire pollutes only the air in the immediate vicinity; by the time that smoke is dispersed through the atmosphere, its global impact is negligible. The collective smoke from a century of industrialized society, on the other hand, has caused measurable increases in several atmospheric pollutants worldwide, and these pollutants continue to pour into the air from many sources. It was once assumed that the seemingly vast oceans would be an inexhaustible "sink" for any extra CO_2 that we might generate by burning fossil fuels, but decades of steadily climbing atmospheric CO_2 levels have proven that in this sense, at least, the oceans are not as large as we thought (figure 1.18). Likewise, six people carelessly dumping wastes into the ocean would not appreciably pollute that huge volume of water. The prospect of 6 *billion* people doing the same thing, however, is quite another matter.

Is Extraterrestrial Colonization a Solution?

In recent decades, space travel has become feasible, if not yet commonplace. This has inspired suggestions that space travel could alleviate many of the problems related to

Figure 1.18 We have historically thought of the earth and its atmosphere and oceans as vast, but the large human population can have global-scale impacts.   Photograph © NASA

earth's ever-growing population. Colonization of other planets would indeed provide more land for people to live on, in theory; we could mine other planets' ores to supply our needs; giant solar-energy collectors in space could solve our energy needs; wastes, such as radioactive wastes, could be propelled into space away from the earth; and so on. Unfortunately, the weight of evidence suggests that this simply cannot be done on the scale needed to solve major environmental problems, given existing population pressures, and the time factor.

A major objection to colonization is that none of the other planets or other bodies in the solar system is really suitable for human habitation. It might be possible on a few such bodies to build small, controlled environments within which a few people could live; there is presently talk of establishing a permanent outpost on earth's moon. Examination of the dust-dry, barren, airless lunar surface (figure 1.19) makes clear how constrained life on the moon would be. On most planets or moons, even this limited habitation would be impossible. In addition, all such outposts would almost certainly have to be constructed using resources from earth. As noted earlier in this chapter, the other planets all have different compositions from the earth's, most of them very different. The types of processes that have shaped them since they formed have also differed. It is quite unlikely that suitable ores or fuels could be found on them in any quantity. Mars is the least hostile alternative environmentally, but it is cold and lacks both surface water and a breathable atmosphere. Even if the living conditions elsewhere in the solar system were more hospitable, the cost to transport any significant number of people to another planet, and the quantity of earth's materials needed to support them, would be prohibitive. It looks, in short, as though we will have to solve our environmental problems right here. And time is of the essence: Every hour, now, the world population increases by more than *10,000 people!*

Figure 1.19   The lunar surface is not an environment in which humans could live outside protective structures, with high energy and resource costs.   Photograph © NASA

Overview

Summary

The solar system formed about $4\frac{1}{2}$ billion years ago. The earth is unique among the planets in its chemical composition, abundant surface water, and oxygen-rich atmosphere. The earth passed through a major period of internal differentiation early in its history, which led to the formation of the atmosphere and the oceans. Earth's surface features have continued to change throughout the last 4+ billion years, through a series of processes that are often cyclical in nature, and commonly interrelated. The oldest rocks in which remains of simple organisms are recognized are more than 3 billion years old. The earliest plants were responsible for the development of free oxygen in the atmosphere, which, in turn, made it possible for oxygen-breathing animals to survive. Human-type remains are unknown in rocks over 3 to 4 million years of age. In a geologic sense, therefore, human beings are quite a new addition to the earth's cast of characters, but they have had a very large impact. Geology, in turn, can have an equally large impact on us.

The world population, now over 5.7 billion, can be expected to exceed 6 billion by the year 2000 and to reach about 10 billion by the year 2050. Even our present population cannot entirely be supported at the level customary in the more developed countries, given the limitations of land and resources. Extraterrestrial resources cannot realistically be expected to contribute substantially to a solution of this problem. Future chapters will examine some of the limitations of earth's resources in more detail.

Terms to Remember

carrying capacity 16   hypothesis 8
core 6   mantle 6
crust 6   scientific method 8
doubling time 13   theory 9
exponential growth 13

Exercises

Questions for Review

1. Describe the process by which the solar system is believed to have formed, and explain why it led to planets of different compositions, even though the planets formed simultaneously.

2. How old is the solar system? How recently have human beings come to influence the physical environment?

3. Explain how the newly formed earth differed from the earth we know today.

4. What kinds of information are used to determine the internal composition of the earth?

5. How were the earth's atmosphere and oceans formed?

6. What are the differences among facts, scientific hypotheses, and scientific theories?

7. Many earth materials are transformed through processes that are cyclical in nature. Describe one example.

8. The size of the earth's human population directly affects the severity of many environmental problems. Explain this idea in the context of (a) resources and (b) pollution.

9. If earth's population has already exceeded its carrying capacity, what are the implications for achieving a comfortable standard of living worldwide? Explain.

10. What is the world's present population, to the nearest billion? How do recent population growth rates (over the last few centuries) compare with earlier times? Why?

11. Explain the concept of doubling time. How has population doubling time been changing through history? What is the approximate doubling time of the world's population at present?

12. What regions of the world currently have the fastest rates of population growth? The slowest?

13. Briefly evaluate the feasibility of space colonization as a means of alleviating land and natural-resource shortages.

For Further Thought

1. Investigate the geologic history of some particular part of the country. (A good starting point might be the state geological survey(s) in the region of interest.) Was the area ever an ocean basin, a desert, subject to volcanic activity? How long ago was it first inhabited by humans?

2. The urgency of population problems can be emphasized by calculating such "population absurdities" as the time at which there will be one person per square meter or per square foot of land, and the time at which the weight of people will exceed the weight of the earth. Try calculating these "population absurdities" by using the world population projections from figure 1.12 or table 1.4 and the following data concerning the earth:

Mass $5,976 \times 10^{21}$ kg*
Land surface area (approx.)   149,000,000 sq. km
Average weight of human body (approx.)   75 kg

3. Derive the relation between doubling time and growth rate, starting from the exponential-growth relation

$$N = N_0\, e^{(G/100) \cdot t}$$

where N = the growing quantity (number of people, tons of iron ore consumed annually, dollars in a bank account, or whatever); N_0 is the initial quantity at the start of the time period of interest; G is growth rate expressed in percent per year, and "percent" means "parts per hundred"; and t is the time in years over which growth occurs. Keep in mind that doubling time, D, is the length of time required for N to double.

4. Select a single country or small set of related countries; examine recent and projected population growth rates in detail, including the factors contributing to the growth rates and trends in those rates. Compare with similar information for the U.S. or Canada.

Suggested Readings/References

Badash, L. 1989. The age-of-the-earth debate. *Scientific American* 261 (August): 90–96.

Blaxter, K. 1986. *People, food, and resources.* Cambridge, England: Cambridge University Press.

Chapman, C. K. 1982. *Planets of rock and ice.* New York: Scribner.

Deardorff, K. E., and Montgomery, P. 1996; National Population Trends [online], U.S. Census Bureau http://www.census.gov/ftp/pub/poplation/ pop-profile/nattrend.html

Eicher, D. L., McAlester, A. L., and Rottman, M. L. 1984. *The history of the earth's crust.* Englewood Cliffs, N.J.: Prentice-Hall.

Greeley, R. 1985. *Planetary landscapes.* London: Allen and Unwin.

Hays, W. W. Natural hazards research and response. *U.S. Geological Survey Yearbook 1990.*

Kaufmann, W. J. III. 1985. *Universe.* New York: W. H. Freeman.

Keyfitz, N. 1989. The growing human population. *Scientific American* 261 (September): 118–26.

Lewis, J. S., and Prinn, R. G. 1984. *Planets and their atmospheres: Origin and evolution.* New York: Academic Press.

Mason, B., and Moore, C. B. 1982. *Principles of geochemistry.* 4th ed. New York: John Wiley and Sons.

Report on the FAO/UNFPA expert consultation on land resources for populations of the future. 1980. Rome: Food and Agriculture Organization of the United Nations.

Simons, E. L. 1989. Human origins. *Science* 245 (September): 1343–50.

---

*This number is so large that it has been expressed in scientific notation, in terms of powers of 10. It is equal to 5,976 with twenty-one zeroes after it. For comparison, the land surface area could also have been written as 149×10^6 sq. km, or $149,000 \times 10^3$ sq. km, and so on.

Steinhart, J. S., and Steinhart, C. E. 1982. Energy use in the U.S. food system. In *Perspectives on energy*, 3d ed., edited by L. C. Ruedisili and M. W. Firebaugh. New York: Oxford University Press.

United Nations Department of International Economic and Social Affairs. 1981. *World population prospects as assessed in 1980.* New York: United Nations.

World Resources Institute. 1996. World Resources 1996–97. New York: Oxford University Press. Also accessible online http://www.wri.org/wri/wr-96-97/index.html

Environment Canada's State of the Environment Fact Sheet 95-1, on the environmental implications of fast food:
   http://199.212.18.12/folio.pgi/osoeeng/query=/doc/{344,0,0,0}?

As a starting point for information on the planets:

National Aeronautics and Space Administration Online Educational Resources
   http://quest.arc.nasa.gov/OER/EDRC22.html

National Space Sciences Data Center (NSSDC) home page
   http://nssdc.gsfc.nasa.gov/nssdc/nssdc_home.html
   Photos of planets and moons from many missions (Apollo, Mariner, Voyager, Galileo, and more)
   http://nssdc.gsfc.nasa.gov/photo_gallery/

Additional population information:

United Nations Population Information Network (POPIN)
   http://www.undp.org/popin/popin.html
Frequently Requested Population Tables (U.S. Census), includes demographic information
   http://www.census.gov/stat_abstract/pop.html

General resource/population/environmental information:
World Resources Institute home page
   http://www.wri.org
Teachers' resources of the WRI Environmental Education Project
   http://www.wri.org/wri/enved/index.html

"Ask-A-Geologist" e-mail service: U.S. Geological Survey staff and others answer geologic questions
   http://walrus.wr.usgs.gov/docs/ask-a-ge.html

Chapter 2

Considering the limited number of chemical elements in nature, the variety of substances found on earth and the diversity of their physical properties are astonishing. The same is true even of just the rocks and minerals. Most of these are made up of an even smaller subset of elements, yet they are very diverse in color, texture, and other properties. The differences in the physical properties of rocks, minerals, and soils determine their suitability for different purposes—extraction of water or of metals, construction, manufacturing, waste disposal, agriculture, and other uses. For this reason it is helpful to understand something of the nature of these materials. Also, each rock contains clues to the kinds of processes that formed it and to the geologic setting where it is likely to be found. For example, we will see that the nature of a volcano's rocks may indicate what hazards it presents to us; our search for new ores or fuels is often guided by an understanding of the specialized geologic environments in which they occur.

Cubes of fluorite covered with smaller calcite crystals; from Cave-In-Rock, Illinois. © Doug Sherman/Geofile

Atomic Structure

All natural and most synthetic substances on earth are made from the ninety naturally occurring chemical elements (box 2.1). An **atom** is the smallest particle into which an element can be divided while still retaining the chemical characteristics of that element (see figure 2.1). The **nucleus,** at the center of the atom, contains one or more particles with a positive electrical charge (**protons**) and usually some particles of similar mass that have no charge (**neutrons**). Circling the nucleus are the negatively charged **electrons.**

The number of protons in the nucleus determines what chemical element that atom is. Every atom of hydrogen contains one proton in its nucleus; every oxygen atom contains eight protons; every carbon atom, six; every iron atom, twenty-six; and so on. The characteristic number of protons is the **atomic number** of the element.

Elements and Isotopes

With the exception of the simplest hydrogen atoms, all nuclei contain neutrons, and the number of neutrons is similar to or somewhat greater than the number of protons. The number of neutrons in atoms of one element is not always the same. The sum of the number of protons and the number of neutrons in a nucleus is the atom's **atomic mass number.** Atoms of a given element with different atomic mass numbers—in other words, atoms with the same number of protons but different numbers of neutrons—are distinct **isotopes** of that element. Some elements have only a single isotope, while others may have ten or more. (The reasons for

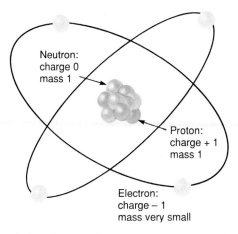

Figure 2.1 Schematic drawing of atomic structure (greatly enlarged and simplified). The nucleus is actually only about 1/1000th of the overall size of the atom. Also, the electrons are not balls with spheroidal orbital paths, nor do electrons always behave quite like particles.

25

Box 2.1 The Periodic Table

Some idea of the probable chemical behavior of elements can be gained from a knowledge of the **periodic table** (figure 1). The Russian scientist Dmitri Mendeleyev first observed that certain groups of elements showed similar chemical properties, which seemed to be related in a regular way to their atomic numbers. At the time (1869) that Mendeleyev published the first periodic table, in which elements were arranged so as to reflect these similarities of behavior, not all the elements had even been discovered, so there were some gaps. The addition of elements identified later confirmed the basic concept. In fact, some of the missing elements were found more easily because their properties could to some extent be anticipated from their expected positions in the periodic table.

We now can relate the periodicity of chemical behavior to the electronic structures of the elements. For example, those elements in the first column, known as the alkali metals, have one electron in the outermost shell of the neutral atom. Thus, they all tend to form cations of +1 charge by losing that odd electron. Outermost electron shells become increasingly full from left to right across a row. The next-to-last column, the halogens, are those elements lacking only one electron in the outermost shell, and they thus tend to gain one electron to form anions of charge –1. In the right-hand column are the inert gases, whose neutral atoms contain all fully filled electron shells.

The elements that occur naturally on earth have now been known for decades. Additional new elements must be created, not simply discovered, as these very heavy elements, with atomic numbers above 92 (uranium), are too unstable to have lasted from the early days of the solar system to the present. Some, like plutonium, are by-products of nuclear-reactor operation; others are created by nuclear physicists who, in the process, learn more about atomic structure and stability.

12 ←	Atomic number
Mg ←	Chemical symbol
24.31 ←	Atomic weight

* Elements heavier than uranium synthesized experimentally

1a																	0
1 H 1.008	IIa											IIIa	IVa	Va	VIa	VIIa	2 He 4.00
3 Li 6.94	4 Be 9.01											5 B 10.81	6 C 12.01	7 N 14.00	8 O 15.99	9 F 18.99	10 Ne 20.18
11 Na 22.99	12 Mg 24.31	IIIb	IVb	Vb	VIb	VIIb	VIII		Ib	IIb		13 Al 26.98	14 Si 28.09	15 P 30.97	16 S 32.06	17 Cl 35.45	18 Ar 39.95
19 K 39.10	20 Ca 40.08	21 Sc 44.96	22 Ti 47.90	23 V 50.94	24 Cr 51.99	25 Mn 54.94	26 Fe 55.85	27 Co 58.93	28 Ni 58.71	29 Cu 63.54	30 Zn 65.37	31 Ga 69.72	32 Ge 72.59	33 As 74.92	34 Se 78.96	35 Br 79.91	36 Kr 83.80
37 Rb 85.47	38 Sr 87.62	39 Y 88.91	40 Zr 91.22	41 Nb 92.91	42 Mo 95.94	43 Tc (99)	44 Ru 101.97	45 Rh 102.91	46 Pd 106.4	47 Ag 107.87	48 Cd 112.40	49 In 114.82	50 Sn 118.69	51 Sb 121.75	52 Te 127.60	53 I 126.90	54 Xe 131.30
55 Cs 132.91	56 Ba 137.34	57–71 see below	72 Hf 178.49	73 Ta 180.95	74 W 183.85	75 Re 186.2	76 Os 190.2	77 Ir 192.2	78 Pt 195.09	79 Au 196.97	80 Hg 200.59	81 Tl 204.37	82 Pb 207.19	83 Bi 208.98	84 Po (210)	85 At (210)	86 Rn (222)
87 Fr (223)	88 Ra (226)	89–103 see below	104* Rf (261)	105* Ha (260)	106* 263												

57 La 138.91	58 Ce 140.12	59 Pr 140.91	60 Nd 144.24	61 Pm (147)	62 Sm 150.35	63 Eu 151.96	64 Gd 157.25	65 Tb 158.92	66 Dy 162.50	67 Ho 164.93	68 Er 167.26	69 Tm 168.93	70 Yb 173.04	71 Lu 174.97
89 Ac (227)	90 Th 232.04	91 Pa (231)	92 U 238.03	93* Np (237)	94* Pu (242)	95* Am (243)	96* Cm (247)	97* Bk (247)	98* Cf (251)	99* Es (254)	100* Fm (253)	101* Md (256)	102* No (254)	103* Lw (257)

Figure 1 The periodic table.

these phenomena involve principles of nuclear physics and the nature of the processes by which the elements are produced in the interiors of stars, and we will not go into them here!)

For most applications, we are concerned only with the elements involved, not with specific isotopes. When a particular isotope is to be designated, this is done by naming the element (which, by definition, specifies the atomic number, or number of protons) and the atomic mass number (protons plus neutrons). Carbon, for example, has three natural isotopes. By far the most abundant is carbon-12, the isotope with six neutrons in the nucleus in addition to the six protons common to all carbon atoms. The two rarer isotopes are carbon-13 (six protons plus seven neutrons) and carbon-14 (six protons plus eight neutrons). Chemically, all behave alike. The human body cannot, for instance, distinguish between sugar containing carbon-12 and sugar containing carbon-13.

Other differences between isotopes may, however, make a particular isotope useful for some special purpose. Carbon-14 is naturally radioactive, which means that, over a period of time, carbon-14 nuclei decay at a known rate; this fact allows carbon-14 to be used to date materials containing carbon. (See appendix A.) Differences in the properties of two uranium isotopes are important in understanding nuclear power options (chapter 14): only one of the two common uranium isotopes is suitable for use as reactor fuel, and must be extracted and concentrated from natural uranium as it occurs in uranium ore. The fact that radioactive elements will inexorably decay—releasing energy—at their own fixed, constant rates is part of what makes radioactive-waste disposal such a challenging problem (chapter 15).

Ions

In an electrically neutral atom, the number of protons and the number of electrons are the same; the negative charge of one electron just equals the positive charge of one proton. Most atoms, however, can gain or lose some electrons. When this happens, the atom has a positive or negative electrical charge and is called an **ion.** If it loses electrons, it becomes positively charged, since the number of protons then exceeds the number of electrons. If the atom gains electrons, the ion has a negative electrical charge. Positively and negatively charged ions are called, respectively, **cations** and **anions.** Both solids and liquids are, overall, electrically neutral, with the total positive and negative charges of cations and anions balanced. Moreover, free ions do not exist in solids; cations and anions are bonded together. In a solution, however, individual ions may exist and move independently. Many minerals break down into ions as they dissolve in water. Individual ions may then be taken up by plants as nutrients or react with other materials. The concentration of hydrogen ions determines a solution's acidity.

Compounds

The electrical attraction between oppositely charged ions can cause them to become chemically bonded together. This is called **ionic bonding.** Bonds between atoms may also form if the atoms share electrons. This is **covalent bonding.** Whatever the nature of the bonding, when atoms or ions of different elements combine in this way, they form **compounds.** A compound is a chemical combination of two or more chemical elements, in particular proportions, that has a distinct set of physical properties, usually very different from those of any of the individual elements in it. Consider, as a simple example, common table salt (sodium chloride). Sodium is a silver metal, and chlorine is a greenish gas that is poisonous in large doses. When equal numbers of sodium and chlorine atoms combine to form table salt, or sodium chloride, the resulting compound forms colorless crystals that do not resemble either of the component elements.

Minerals—General

Minerals Defined

A **mineral** is a naturally occurring, inorganic, solid element or compound with a definite chemical composition and a regular internal crystal structure. *Naturally occurring*, as distinguished from synthetic, means that minerals do not include the thousands of chemical substances invented by humans. *Inorganic* means not produced solely by living organisms or by biological processes. That minerals must be *solid* means that the ice of a glacier is a mineral, but liquid water is not. Chemically, minerals may consist either of one element—like diamonds, which are pure carbon—or they may be compounds of two or more elements. Some mineral compositions are very complex, consisting of ten elements or more. Minerals have a definite chemical composition or a compositional range within which they fall. The presence of certain elements in certain proportions is one of the identifying characteristics of each mineral (see box 2.2). Finally, minerals are crystalline, at least on the microscopic scale. **Crystalline** materials are solids in which the atoms are arranged in regular, repeating patterns (figure 2.2). These patterns may not be apparent to the naked eye, but most solid compounds are crystalline, and their crystal structures can be recognized and studied using X rays and other techniques. Examples of noncrystalline solids include glass (discussed later in the chapter) and plastic.

Identifying Characteristics of Minerals

The two fundamental characteristics of a mineral that together distinguish it from all other minerals are its chemical composition and its crystal structure. No two minerals are identical in both respects, though they may be the same in one. For example, diamond and graphite (the "lead" in a lead pencil) are

Box 2.2 Chemical Symbols: A Scientific Shorthand

Each chemical element is denoted by a one- or two-letter symbol. Many of these symbols make sense in terms of the English name for the element—O for oxygen, He for helium, Si for silicon, and so on. Other symbols reflect the fact that, in earlier centuries, scientists were generally versed in Latin or Greek: The symbols Fe for iron and Pb for lead, for example, are derived from *ferrum* and *plumbum*, respectively, the Latin names for these elements.

The chemical symbols for the elements can express the compositions of substances very precisely. Subscripts after a symbol indicate the number of atoms/ions of one element present in proportion to the other elements in the formula. For example, the formula $Fe_3Al_2Si_3O_{12}$ represents a compound in which, for every twelve oxygen atoms, there are three iron atoms, two aluminum atoms, and three silicon atoms. (This happens to be a variety of the mineral garnet.) The chemical formula is much briefer than describing the composition in words. It is also more exact than the mineral name "garnet," for there are several compo-sitions of garnets with the same basic kind of formula and crystal structure: Other examples include a calcium-aluminum garnet with the formula $Ca_3Al_2Si_3O_{12}$ and a calcium-chromium garnet, $Ca_3Cr_2Si_3O_{12}$. Moreover, chemical formulas are understood by all scientists, while mineral names are known primarily to geologists. Sample formulas of some common minerals appear in table 1.

Formulas can become very complex, especially when different elements can substitute for each other in the same site in the crystal structure. Iron and magnesium often do this in silicates. Biotite, a common, dark-colored mica, may be rich in iron and have a formula of $KFe_3AlSi_4O_{10}(OH)_2$, or it may be rich in magnesium and have a formula of $KMg_3AlSi_4O_{10}(OH)_2$, or, more commonly, it may contain some iron and some magnesium, which together total three atoms per formula. The generalized formula is then $K(Fe,Mg)_3AlSi_4O_{10}(OH)_2$.

Table 1   Sample Formulas of Some Common Minerals

Mineral	Chemical composition	Chemical formula
quartz	silicon dioxide	SiO_2
microcline (a potassium feldspar)	potassium aluminum silicate	$KAlSi_3O_8$
calcite	calcium carbonate	$CaCO_3$
hematite	ferric iron oxide	Fe_2O_3
pyrite	iron disulfide	FeS_2

chemically the same—both are made up of pure carbon. Their physical properties, however, are vastly different because of the differences in their internal crystalline structures. In a diamond, each carbon atom is firmly bonded to every adjacent carbon atom in every direction. In graphite, the carbon atoms are bonded strongly in two dimensions into sheets, but the sheets are only weakly held together in the third dimension. Diamond is clear, transparent, colorless, and very hard, and a jeweler can cut it into beautiful precious gemstones. Graphite is black, opaque, and soft, and its sheets of carbon atoms tend to slide apart as the weak bonds between them are broken.

A mineral's composition and crystal structure can usually be determined only by using sophisticated laboratory equipment. When a mineral has formed large crystals with well-developed shapes, a trained mineralogist may be able to infer some characteristics of its internal atomic arrangement, but most mineral samples do not show large symmetric crys-tal forms by which they can be recognized with the naked eye. Moreover, many minerals share similar external forms (figure 2.3). No one can look at a mineral and know its chemical composition without first recognizing what mineral it is. Thus, when scientific instruments are not at hand, mineral identification must be based on a variety of other physical properties that in some way reflect the mineral's composition and structure. These other properties are often what make the mineral commercially valuable. However, they are rarely unique to one mineral and often are deceptive. A few examples of such properties follow.

Other Physical Properties of Minerals

Perhaps surprisingly, color is often not a reliable guide to mineral identification. While some minerals always appear the same color, many vary from specimen to specimen.

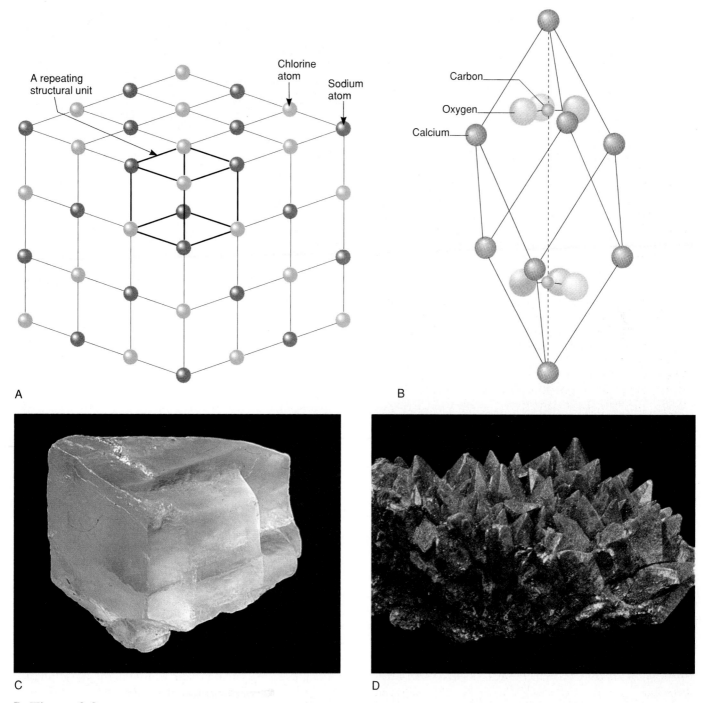

Figure 2.2 Examples of crystal structures. (*A* and *B*) In real crystals, atoms are not round balls connected by sticks, of course, but these diagrams—halite (*A*) and calcite (*B*)—illustrate the basic geometry of some crystals. (*C* and *D*) The photographs—halite (*C*) and calcite (*D*)—show how the crystal form of the mineral can reflect its internal structure. (See also the calcite in the chapter-opener photo.) The similarity is not always so obvious.   Photographs (*C*) and (*D*)   © The McGraw-Hill Companies Inc./Bob Coyle, photographer.

Variation in color is usually due to the presence of small amounts of chemical impurities in the mineral that have nothing to do with the mineral's basic characteristic composition, and such variation is especially common when the pure mineral is light-colored or colorless. The very common mineral quartz, for instance, is colorless in its pure form. However, quartz also occurs in other colors, among them

rose pink, golden yellow, smoky brown, purple (amethyst), and milky white. Clearly, quartz cannot always be recognized by its color, or lack of it.

Another example is the mineral corundum (figure 2.4), a simple compound of aluminum and oxygen. In pure form, it too is colorless, and quite hard, which makes it a good abrasive. It is often used for the grit on sandpaper. Yet a little

A

B

C

Figure 2.3 Many minerals share the same crystal form; here, (*A*) galena (PbS), (*B*) fluorite (CaF_2), and halite (NaCl) all form cubes.   Photographs © The McGraw-Hill Companies Inc./Doug Sherman, photographer.

Figure 2.4 The many colors of these corundum gemstones—including ruby and sapphire—illustrate why color is a poor guide in mineral identification.   Photograph by Dane A. Penland, © 1992 Smithsonian Institution.

color from trace impurities not only disguises corundum, it can transform this utilitarian material into highly prized gems: Blue-tinted corundum is also known as sapphire, and red corundum is ruby. Even when the color shown by a mineral sample is the true color of the pure mineral, it is probably not unique. There are more than two thousand minerals, so there are usually many of any one particular color.

Hardness, the ability to resist scratching, is another easily measured physical property that can help to identify a mineral, although it usually does not uniquely identify the mineral. Classically, hardness is measured on the Mohs hardness scale (table 2.1), in which ten common minerals are arranged in order of hardness. Unknown minerals are assigned a hardness on the basis of which minerals they can scratch and which minerals scratch them. A mineral that scratches gypsum and is scratched by calcite is assigned a hardness of $2\frac{1}{2}$ (the hardness of an average fingernail). Because a diamond is the hardest natural substance known on earth, and corundum the second-hardest mineral, these minerals might be identifiable from their hardnesses. Among the thousands of "softer" (more readily scratched) minerals, however, there are many of any particular hardness, just as there are many of any particular color.

A number of other physical properties may individually be common to many minerals. Only by considering a whole set of such nonunique properties as color, hardness, density, cleavage (the way a mineral breaks apart when struck), and others can a mineral be identified without complex instruments.

Table 2.1 The Mohs Hardness Scale

Mineral	Assigned hardness
talc	1
gypsum	2
calcite	3
fluorite	4
apatite	5
orthoclase	6
quartz	7
topaz	8
corundum	9
diamond	10

For comparison, the approximate hardnesses of some common objects, measured on the same scale, are: fingernail, $2\frac{1}{2}$; copper penny, $3\frac{1}{2}$; glass, 5 to 6; pocketknife blade, 5 to 6.

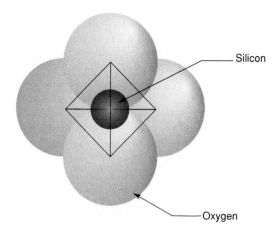

Figure 2.5 The basic silica tetrahedron, building block of all silicate minerals.

Unique or not, the physical properties arising from minerals' compositions and crystal structures are often what give minerals value from a human perspective—the slickness of talc (main ingredient of talcum powder), the malleability and conductivity of copper, the durability of diamond, and the rich colors of tourmaline gemstones are all examples. Some minerals have several useful properties: table salt (halite), a necessary nutrient, also imparts a taste we find pleasant, dissolves readily to flavor liquids but is soft enough not to damage our teeth if we munch on crystals of it sprinkled on solid food, and will serve as a food preservative in high concentrations, among other helpful qualities.

Types of Minerals

As was indicated earlier, minerals can be grouped or subdivided on the basis of their two fundamental characteristics—composition and crystal structure. In this section, we will briefly review some of the basic mineral groups. A comprehensive survey of minerals is well beyond the scope of this book, and the interested reader should refer to standard mineralogy texts for more information. A summary of physical properties of selected minerals is found in appendix C.

Silicates

In chapter 1, we noted that the two most common elements in the earth's crust are silicon and oxygen. It comes as no surprise, therefore, that by far the largest group of minerals is the **silicate** group, all of which are compounds containing silicon and oxygen, and most of which contain other elements as well. Because this group of minerals is so large, it is subdivided on the basis of crystal structure, by the ways in which the silicon and oxygen atoms are linked together. The

basic building block of all silicates is a tetrahedral arrangement of four oxygen atoms (anions) around the much smaller silicon cation (figure 2.5). In different silicate minerals, these *silica tetrahedra* may be linked into chains, sheets, or three-dimensional frameworks by the sharing of oxygen atoms. Some of the physical properties of silicates and other minerals are closely related to their crystal structures (see figure 2.6). In general, however, we need not go into the structural classes of the silicates in detail. It is more useful to mention briefly a few of the more common, geologically important silicate minerals.

While not the most common, *quartz* is probably the best-known silicate. Compositionally, it is the simplest, containing only silicon and oxygen. Quartz is found in a large variety of rocks and soils. Commercially, the most common use of pure quartz is in the manufacture of glass, which also consists mostly of silicon and oxygen. Quartz-rich sand and gravel are used in very large quantities in construction.

The most abundant group of minerals in the crust is a set of chemically similar minerals known collectively as the *feldspars*. They are composed of silicon, oxygen, aluminum, and either sodium, potassium, or calcium, or some combination of these three. Again, logically enough, these common minerals are made from elements abundant in the crust. They are used extensively in the manufacture of ceramics.

Iron and magnesium are also among the more common elements in the crust and are therefore found in many silicate minerals. **Ferromagnesian** is the general term used to describe those silicates—usually dark-colored (black, brown, or green)—that contain iron and/or magnesium. These silicates may or may not contain other elements also. In general, ferromagnesian minerals are relatively susceptible to weathering. Rocks containing a high proportion of ferromagnesian minerals, then, tend to weather easily, which is an important consideration in construction. Individual ferromagnesium minerals may be important in particular contexts. *Olivine,* a simple ferromagnesian mineral, is a major

A

B

Figure 2.6 Another relationship between structures and physical properties is cleavage. Because of their internal crystalline structures, many minerals break apart preferentially in certain directions. (*A*) Micas are silicates in which atoms are tightly bonded into two-dimensional sheets, which are, in turn, less strongly bonded together. Micas show a tendency to break between sheets of atoms. (*B*) Halite has a cubic structure like galena and breaks into cubic or rectangular pieces.   Photographs © The McGraw-Hill Companies Inc./Bob Coyle, photographer.

constituent of earth's mantle; gem-quality olivines from mantle-derived volcanic rocks are the semiprecious gem *peridot*. *Asbestos* is not a single mineral, but a general term for various minerals that occur in a form that can be separated into thin, strong fibers; however, the several different asbestos minerals are all ferromagnesian silicates.

Like the feldspars, the *clays* are another group of several silicate minerals with similar physical properties, compositions, and crystal structures. Clays are sheet silicates, built on an atomic scale of stacked-up sheets of linked silicon and oxygen atoms. Because the bonds between sheets are relatively weak, the sheets tend to slide past each other, a characteristic that contributes to the slippery texture of many clays and related minerals. Clays are somewhat unusual among the silicates in that their structures can absorb or lose water, depending on how wet conditions are. Absorbed water may increase the slippery tendencies of the clays. Also, some clays expand as they soak up water and shrink as they dry out. A soil rich in these "expansive clays" is a very unstable base for a building, as we will see in later chapters. On the other hand, clays also have important uses, especially in making ceramics and in building materials. Other clays are useful as lubricants in the muds used to cool the drill bits in oil-drilling rigs.

Nonsilicates

Just as the silicates, by definition, all contain silicon plus oxygen as part of their chemical compositions, each nonsilicate mineral group is defined by some chemical constituent or characteristic that all members of the group have in common. Most often, the common component is the same negatively charged ion or group of atoms. Discussion of some of the nonsilicate mineral groups with examples of common or familiar members of each follows. See also table 2.2.

The **carbonates** all contain carbon and oxygen combined in the proportions of one atom of carbon to three atoms of oxygen (written CO_3). The carbonate minerals all dissolve relatively easily, particularly in acids, and the oceans contain a great deal of dissolved carbonate. Geologically, the most important, most abundant carbonate mineral is calcite, which is calcium carbonate. Precipitation of calcium carbonate from seawater is a major process by which marine rocks are formed (see the discussion under "Sediments and Sedimentary Rocks" later in this chapter). Another common carbonate mineral is dolomite, which contains both calcium and magnesium in approximately equal proportions. Carbonates may contain many other elements— iron, manganese, or lead, for example. The limestone and marble we use extensively for building and sculpture consist mainly of carbonates, generally calcite; calcite is also an important ingredient in cement.

The **sulfates** all contain sulfur and oxygen in the ratio of 1:4 (SO_4). A calcium sulfate—gypsum—is the most important, for it is both relatively abundant and commercially useful, particularly as the major constituent in plaster of paris. Sulfates of many other elements, including barium, lead, and strontium, are also found.

When sulfur is present without oxygen, the resultant minerals are called **sulfides.** A common and well-known sulfide mineral is the iron sulfide *pyrite*. Pyrite (figure 2.7) has also been called "fool's gold" because its metallic golden color often deceived early gold miners and prospectors into thinking they had struck it rich. Pyrite is not a commercial source of iron because there are richer ores of this metal. Nonetheless, sulfides comprise many economically important metallic ore minerals. An example that may be familiar is the lead sulfide mineral *galena*, which often forms in silver-colored cubes. The rich lead ore deposits near Galena,

Table 2.2 *Some Nonsilicate Mineral Groups**

Group	Compositional characteristic	Examples
carbonates	metal(s) plus carbonate (1 carbon + 3 oxygen atoms, CO_3)	calcite, calcium carbonate, $CaCO_3$ dolomite, calcium-magnesium carbonate, $CaMg(CO_3)_2$
sulfates	metal(s) plus sulfate (1 sulfur + 4 oxygen atoms, SO_4)	gypsum, calcium sulfate, with water, $CaSO_4 \cdot 2H_2O$ barite, barium sulfate, $BaSO_4$
sulfides	metal(s) plus sulfur, without oxygen	pyrite, iron disulfide, FeS_2 galena, lead sulfide, PbS cinnabar, mercury sulfide, HgS
oxides	metal(s) plus oxygen	magnetite, iron oxide, Fe_3O_4 hematite, ferric iron oxide, Fe_2O_3 corundum, aluminum oxide, Al_2O_3 spinel, magnesium-aluminum oxide, $MgAl_2O_4$
hydroxides	metal(s) plus hydroxyl (1 oxygen + 1 hydrogen atom, OH)	gibbsite, aluminum hydroxide, $Al(OH)_3$ (found in aluminum ore) brucite, magnesium hydroxide, $Mg(OH)_2$ (one ore of magnesium)
halides	metal(s) plus halogen element (fluorine, chlorine, bromine, or iodine)	halite, sodium chloride, $NaCl$ fluorite, calcium fluoride, CaF_2
native elements	mineral consists of a single chemical element	gold (Au), silver (Ag), copper (Cu), sulfur (S), graphite (carbon, C)

*Other groups exist, and some complex minerals contain components of several groups (carbonate and hydroxyl groups, for example).

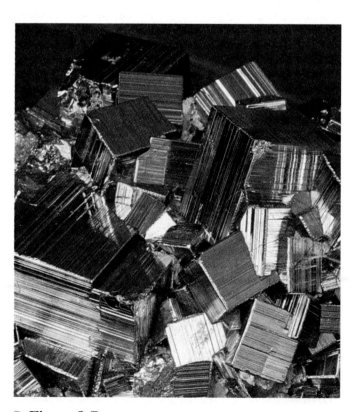

Figure 2.7 Pyrite ("fool's gold"), an iron sulfide (FeS_2), a very common sulfide mineral.   Photograph © The McGraw-Hill Companies, Inc./Doug Sherman, photographer.

Illinois, gave the town its name. Sulfides of copper, zinc, and numerous other metals may also form valuable ore deposits (see chapter 12). Sulfides may also be problematic: when pyrite associated with coal is exposed by strip-mining and weathered, the result is sulfuric acid in runoff water from the mine.

Minerals containing just one or more metals combined with oxygen, and lacking the other elements necessary to classify them as silicates, sulfates, carbonates, and so forth, are the *oxides*. Iron combines with oxygen in different proportions to form more than one oxide mineral. One of these, magnetite, is, as its name suggests, magnetic, which is relatively unusual among minerals. Magnetic rocks rich in magnetite were known as "lodestone" in ancient times and were used as navigational aids like today's more compact compasses. Another iron oxide, hematite, may sometimes be silvery black but often has a red color and gives a reddish tint to many soils. Iron oxides on Mars's surface are believed to be responsible for that planet's orange hue. Many other oxide minerals also exist, including corundum, the aluminum oxide mineral mentioned earlier.

Native elements, as shown in table 2.2, are even simpler chemically than the other nonsilicates. Native elements are minerals that consist of a single chemical element, and the minerals' names are usually the same as the corresponding elements. Not all elements can be found, even rarely, as native elements. However, some of our most highly prized materials, such as gold, silver, and platinum, often occur as native elements. Diamond and graphite are both examples of native carbon, but two mineral names are needed to distinguish these two very different forms of the same element. Sulfur may occur as a native element, either with or without associated sulfide minerals. Some of the richest copper ores contain native copper. Other metals that may occur as native elements include tin, iron, and antimony.

Rocks

A **rock** is a solid, cohesive aggregate of one or more minerals, or mineral materials (for example, volcanic glass, discussed later). This means that a rock consists of many individual mineral grains (crystals)—not necessarily all of the same mineral—or crystals plus glass, which are firmly held together in a solid mass. Because the many mineral grains of beach sand fall apart when handled, sand is not a rock, although, in time, sand grains may become cemented together to form a rock. The properties of rocks are important in determining their suitability for particular applications, such as for construction materials or for the base of a building foundation. Each rock also contains within it a record of at least a part of its history, in the nature of its minerals and in the way the mineral grains fit together. The three broad categories of rocks—*igneous, sedimentary,* and *metamorphic*—are distinguished by the processes of their formation. However, they are also linked, over time, by the *rock cycle.*

The Rock Cycle

In chapter 1, we noted that the earth is a constantly changing body. Mountains come and go; seas advance and retreat over the faces of continents; surface processes and processes occurring deep in the crust or mantle are constantly altering the planet. One aspect of this continual change is that rocks, too, are always subject to change. We do not have a single sample of rock that has remained unchanged since the earth formed. Therefore, when we describe a rock as being of a particular type, or give it a specific rock name, it is important to realize that we are really describing the form it has most recently taken, the results of the most recent processes acting on it.

For example, as will be described in the next column, an *igneous rock* is one crystallized from molten material, formed at high temperatures. But virtually any rock can be melted, and in the process, its previous characteristics may be obliterated. So when the melt crystallizes, we have an igneous rock—but what was it before? A *sedimentary rock* is formed from sediment, debris of preexisting rocks deposited at low temperatures at the earth's surface. But the sediment, in turn, is derived by weathering—physical and chemical breakdown of those rocks—and a given deposit of sediment may include bits of many different rocks. (Look closely at beach sand or river gravel.) The essence of the concept of the **rock cycle,** shown in very simplified form in figure 2.8, is that rocks, far from being the permanent objects we may imagine them to be, are continually being changed by geological processes. Over the vast span of geologic time, billions of years, a given

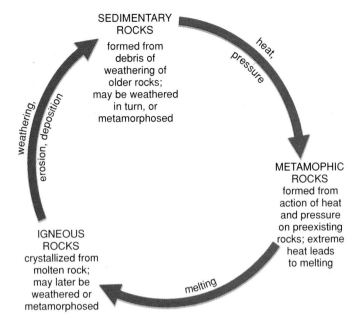

Figure 2.8 A simplified view of principal processes in the rock cycle. While this sketch is not meant to suggest that rocks progress only from sedimentary to metamorphic to igneous, back to sedimentary, and so on, it captures the idea that rocks are involved in processes leading to continual changes in earth materials.

bit of material may have been subject to many, many changes, and may have been part of many different rocks. The following sections examine in more detail the various types of rock-forming processes involved, and some of the rocks they form.

Igneous Rocks

At high enough temperatures, rocks and minerals can melt. **Magma** is the name given to naturally occurring hot, molten rock material. Silicates are the most common minerals, so magmas are usually rich in silicate. They also contain some dissolved water and gases and generally have some solid crystals suspended in the melt. An **igneous** rock is a rock formed by the solidification and crystallization of a cooling magma. (*Igneous* is derived from the Latin term *ignis,* meaning "fire.")

Since high temperatures are required to melt silicates, magmas form at some depth below the surface. The molten material may or may not reach the surface before it cools enough to crystallize and solidify. The depth at which a magma crystallizes will affect how rapidly it cools and the sizes of the mineral grains in the resultant rock.

If a magma remains well below the surface during cooling, it cools relatively slowly, insulated by overlying rock and soil. It may take hundreds of thousands of years or more to crystallize completely. Under these conditions, the crystals have ample time to form and to grow very large, and the rock eventually formed has mineral grains large enough to be seen individually with the naked eye. A rock formed in this way is a **plutonic** igneous rock. (The name is derived from Pluto, the Greek god of the lower world.) *Granite* is probably the most widely known example of a plutonic rock (figure 2.9A). Compositionally, a typical granite consists principally of quartz and feldspars, and it usually contains some ferromagnesian minerals or other silicates. The proportions and compositions of these constituent minerals may vary, but all granites show the coarse, interlocking crystals characteristic of a plutonic rock. Much of the mass of the continents consists of granite or of rock of granitic composition.

A magma that flows out on the earth's surface while still wholly or partly molten is called **lava.** Lava is a common product of volcanic eruptions, and the term **volcanic** is given to an igneous rock formed at or close to the earth's surface. Magmas that crystallize very near the surface cool more rapidly. There is less time during crystallization for large crystals to form from the magma, so volcanic rocks are typically fine-grained, with most crystals too small to be distinguished with the naked eye. In extreme cases, where cooling occurs very fast, even tiny crystals may not form before the magma solidifies. The resulting clear, noncrystalline solid is a natural **glass** (figure 2.9B). The most common volcanic rock is *basalt,* a dark rock rich in ferromagnesian minerals and feldspar (figure 2.9C). The ocean floor consists largely of basalt. Occasionally, a melt begins to crystallize slowly at depth, growing some large crystals, and then is subjected to rapid cooling (following a volcanic eruption, for instance). This results in coarse crystals in a fine-grained groundmass, a *porphyry* (figure 2.9D).

Though there are fundamental similarities in the origins of magmas and volcanic rocks, there are practical differences too. As will be explored in chapter 5, differences in the chemical compositions of magmas lead to differences in the behavior of the volcanoes from which they erupt, explaining why the Hawaiian volcano Kilauea erupts so quietly that tourists can almost walk up and touch the lava flows in safety, while Mount St. Helens and the Philippine volcano Pinatubo are prone to violent, sudden, and devastating explosions.

Regardless of the details of their compositions or cooling histories, all igneous rocks have some textural characteristics in common. If they are crystalline, their crystals, large or small, are tightly interlocking or intergrown (unless they are formed from loose material such as volcanic ash). If glass is present, crystals tend to be embedded in or closely surrounded by the glass. The individual crystals tend to be angular in shape, not rounded. There is usually little pore space, little empty volume that could be occupied by such fluids as water. Structurally, most igneous rocks are relatively strong unless they have been fractured, broken, or weathered.

Sediments and Sedimentary Rocks

At the lower end of the temperature spectrum are the **sedimentary** rocks. **Sediments** are loose, unconsolidated accumulations of mineral or rock particles that have been transported by wind, water, or ice, or shifted under the influence of gravity, and redeposited. Beach sand is a kind of sediment; so is the mud on a river bottom. Soil is a mixture of mineral sediment and organic matter. Most sediments originate, directly or indirectly, through the weathering of pre-existing rocks—

either by physical breakup into finer and finer fragments, or by solution, followed by precipitation of crystals out of solution. The physical properties of sediments and soils bear on a broad range of environmental problems, from the stability of slopes and building foundations, to selection of optimal waste-disposal sites, to how readily water drains away after a rainstorm and how likely that rain is to produce a flood.

When sediments are compacted or cemented together into a solid cohesive mass, they become sedimentary rocks. The set of processes by which sediments are transformed into rock is collectively described as **lithification** (from the Greek word *lithos*, meaning "stone"). The resulting rock is generally more compact and denser, as well as more cohesive, than the original sediment. Sedimentary rocks are formed at or near the earth's surface, at temperatures close to ordinary surface temperatures. They are subdivided into two groups—clastic and chemical.

Clastic sedimentary rocks (from the Greek word *klastos*, meaning "broken") are formed from the products of the mechanical breakup of other rocks. Natural processes continually attack rocks exposed at the surface. Rain and waves pound them, windblown dust scrapes them, frost and tree roots crack them, water dissolves them—these and other processes are all part of the weathering of rocks, which is described more fully in chapter 11. In consequence, rocks are broken up into smaller and smaller pieces and ultimately, perhaps, into individual mineral grains. The resultant rock and mineral fragments may be transported by wind, water, or ice, and accumulate as sediments in streams, lakes, oceans, deserts, or soils. Later geologic processes can cause these sediments to become lithified. Burial under the weight of more sediments may pack the loose particles so tightly that they hold firmly together in a cohesive mass. Except with very fine-grained sediments, however, compaction alone is rarely enough to transform the sediment into rock. Water seeping slowly through rocks underground also carries dissolved minerals, which may precipitate out of solution to stick the sediment particles together with a natural mineral cement.

Clastic sedimentary rocks are most often named on the basis of the average size of the particles that form the rock. *Sandstone,* for instance, is a rock composed of sand-sized sediment particles, 1/16 to 2 millimeters (0.002 to 0.08 inches) in diameter. *Shale* is made up of finer-grained sediments, and the individual grains cannot be seen in the rock with the naked eye. *Conglomerate* is a relatively coarse-grained rock, with fragments above 2 mm (0.08 inches) in diameter, and sometimes much larger. Regardless of grain size, clastic sedimentary rocks tend to have, relatively, considerable pore space between grains. This is a logical consequence of the way in which these rocks form, by the piling up of preexisting rock and mineral grains. Also, as sediment particles are transported by water or other agents, they may become more rounded and thus not pack together very tightly or interlock as do the mineral grains in an igneous rock. Many clastic sedimentary rocks are therefore not particularly strong structurally, unless they have been extensively cemented.

A

B

C

D

Figure 2.9 Igneous rocks, crystallized from melts. (*A*) Granite, a plutonic rock. (*B*) Obsidian (volcanic glass). (*C*) Basalt, a volcanic rock. (*D*) Porphyry, an igneous rock with coarse crystals in a finer matrix. (*B*), (*C*), and (*D*) Photographs © The McGraw-Hill Companies, Inc./ Bob Coyle, photographer.

Chemical sedimentary rocks form not from mechanical breakup and transport of fragments, but from crystals formed by precipitation or growth from solution. A common example is *limestone*, composed mostly of calcite (calcium carbonate). The chemical sediment that makes limestone may be deposited from fresh or salt water; under favorable chemical conditions, thick limestone beds, perhaps hundreds of meters thick, may form. Another example of a chemical sedimentary rock is *rock salt*, made up of the mineral halite, which is the mineral name for ordinary table salt (sodium chloride). A salt deposit may form when a body of salt water is isolated from an ocean and dries up.

Some chemical sediments have a large biological contribution. For example, many organisms living in water have shells or skeletons made of calcium carbonate or of silica (chemically equivalent to quartz). The materials of these shells or skeletons are drawn from the water in which the organisms grow. In areas where great numbers of such creatures live and die, the "hard parts"—the shells or skeletons—may pile up on the bottom, eventually to be buried and lithified.

Gravity plays a role in the formation of all sedimentary rocks. Mechanically broken-up bits of materials accumulate when the wind or water is moving too weakly to overcome gravity and move the sediments; repeated cycles of transport and deposition can pile up, layer by layer, a great thickness of sediment. Minerals crystallized from solution, or the shells of dead organisms, tend to settle out of the water under the force of gravity, and again, in time, layer on layer of sediment can build up. Layering, then, is a very common feature of sedimentary rocks and is frequently one way in which their sedimentary origins can be identified. Figure 2.10 shows several kinds of sedimentary rocks; note the layering in figure 2.10E.

Sedimentary rocks can yield information about the settings in which the sediments were deposited. For example, the energetic surf of a beach washes away fine muds and leaves coarser sands and gravels; sandstone may, in turn, reflect the presence of an ancient beach. Distribution of glacier-deposited sediments of various ages contributes to understanding not only of global climate change but of continental drift (see next chapter).

Metamorphic Rocks

The name *metamorphic* comes from the Greek for "changed form." A **metamorphic** rock is one that has formed from another, preexisting rock that was subjected to heat and/or pressure. The temperatures required to form metamorphic rocks are not as high as magmatic temperatures. Significant changes can occur in a rock at temperatures well below melting. Heat and pressure commonly cause the minerals in the rock to recrystallize. The original minerals may form larger crystals that perhaps interlock more tightly than before. Also, some minerals may break down completely, while new minerals form under the new temperature and pressure conditions. Pressure may cause the rock to become deformed—compressed, stretched, folded, or compacted. All of this occurs while the rock is still solid.

The sources of the elevated pressures and temperatures of metamorphism are many. An important source of pressure is simply burial under many kilometers of overlying rock. The weight of the overlying rock can put great pressure on the rocks below. One source of elevated temperatures is the fact that temperatures increase with depth in the earth. In general, in most places, crustal temperatures increase at the rate of about 30 degrees centigrade per kilometer of depth (close to 60° F per mile)—which is one reason deep mines have to be air-conditioned! Deep in the crust, rocks are subjected to enough heat and pressure to show the deformation and recrystallization characteristic of metamorphism. Another heat source is a cooling magma. When hot magma formed at depth rises to shallower depths in the crust, it heats the adjacent, cooler rocks, and they may be metamorphosed; this is **contact metamorphism.** Metamorphism can also result from the stresses and heating to which rocks are subject during mountain-building or plate-tectonic movement (see chapter 3). Such metamorphism on a large scale, not localized around a magma body, is **regional metamorphism.**

Any kind of preexisting rock can be metamorphosed. Some names of metamorphic rocks reflect their compositions and suggest what the earlier rock may have been. *Marble* is metamorphosed limestone in which the individual calcite grains have recrystallized and become tightly interlocking. The remaining sedimentary layering that the limestone once showed may be folded and deformed in the process, if not completely obliterated by the recrystallization. *Quartzite* is a quartz-rich metamorphic rock, often formed from a very quartz-rich sandstone. The quartz crystals are typically much more tightly interlocked in the quartzite, and the quartzite is a more compact, denser, and stronger rock than the original sandstone. Other metamorphic rock names describe the characteristic texture of the rock, regardless of its composition. In a rock subjected to directed stress, minerals that form elongated or platy crystals may be lined up parallel to each other. The resultant texture is described as **foliation,** from the Latin for "leaf" (as in the parallel leaves, or pages, of a book). *Slate* is a metamorphosed shale that has developed foliation under stress. The resulting rock tends to break along the foliation planes, parallel to the alignment of those minerals. The same characteristic is observed in *schist,* a coarser-grained, mica-rich metamorphic rock in which the mica flakes are similarly oriented. In other metamorphic rocks, different minerals may be concentrated in irregular bands, often producing a striped appearance. Such a rock is called a *gneiss* (pronounced "nice"). Because such terms as *schist* and *gneiss* are purely textural, the rock name can be modified by adding key features of the rock composition: "biotite-garnet schist," "granitic gneiss," and so on. Several examples of metamorphic rocks are illustrated in figure 2.11.

The Rock Cycle Revisited

It should be evident from the descriptions of the major rock types and how they form that rocks of any type can be transformed into rocks of another type or into another distinct rock of the same general type through the appropriate geologic processes. A sandstone may be weathered until it breaks up; its fragments may then be transported, redeposited, and lithified to form another sedimentary rock. It might instead be deeply buried, heated, and compressed, which could tranform it into the metamorphic rock quartzite; or it could be heated until some or all of it melted, and from that melt an igneous rock could be formed. Likewise, a schist could be broken up into small bits, forming a sediment that might eventually become sedimentary rock; more intense metamorphism

Figure 2.10 Sedimentary rocks, formed at low temperatures. (*A*) Limestone. (*B*) Shale. (*C*) Sandstone (enlarged to show rounded grains). (*D*) Conglomerate, a coarser-grained rock similar to sandstone (note quarter for scale.) (*E*) Bryce Canyon, Utah, was carved through layers of sedimentary rock. Note characteristic layering.   (*A*) Photograph by I. J. Witkind, USGS Photo Library, Denver, CO. (*B*) © The McGraw-Hill Companies, Inc./Bob Coyle, photographer. (*C*) © The McGraw-Hill Companies, Inc./Doug Sherman, photographer.

A

B

C

D

Figure 2.11 Metamorphic rocks have undergone mineralogical, chemical, and/or structural change. (*A*) Marble, metamorphosed limestone. (*B*) Quartzite, metamorphosed sandstone. (*C*) Schist, showing characteristic coarse mica flakes aligned in parallel. (*D*) Gneiss.   (*B*) © The McGraw-Hill Companies, Inc./Bob Coyle, photographer.

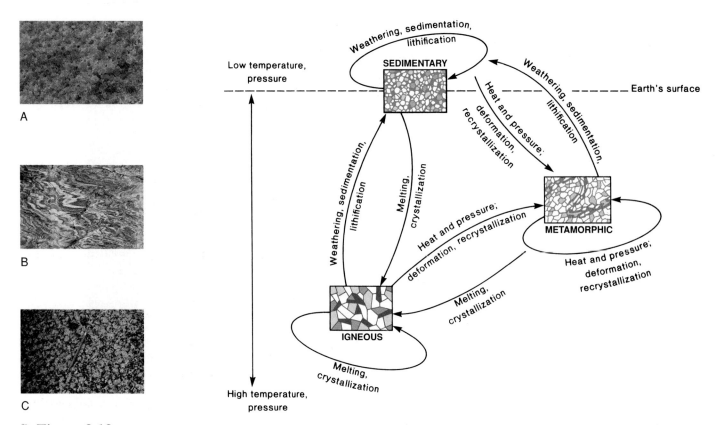

A

B

C

Figure 2.12   The rock cycle—a schematic view. Basically, a variety of geologic processes can transform any rock into a new rock of the same or a different class. The geologic environment is not static; it is constantly changing.   (A) and (B) © The McGraw-Hill Companies, Inc./Doug Sherman, photographer.

could transform it into a gneiss; or extremely high temperatures could melt it to produce a magma from which a granite could crystallize. Crustal rocks can be carried into the mantle and melted; fresh magma cools and crystallizes to form new rock; erosion and weathering processes constantly chip away at the surface. A more comprehensive view of the links among different rock types is shown in generalized form in figure 2.12. In the next chapter, we will look at the rock cycle again, but in a somewhat different way, in the context of plate tectonics. Most interactions of people with the rock cycle involve the sedimentary and volcanic components of the cycle.

Overview

Summary

The smallest possible unit of a chemical element is an atom. Isotopes are atoms of the same element that differ in atomic mass number; chemically, they are indistinguishable. Atoms may become electrically charged ions through the gain or loss of electrons. When two or more elements combine chemically in fixed proportions, they form a compound.

Minerals are naturally occurring inorganic solids, each of which is characterized by a particular composition and internal crystalline structure. By far the most abundant minerals in the earth's crust and mantle are the silicates. They can be subdivided into groups on the basis of their crystal structures, by the ways in which the silicon and oxygen atoms are arranged. The nonsilicate minerals are generally grouped on the basis of common chemical characteristics.

Rocks are cohesive solids formed from rock or mineral grains or glass. The way in which rocks form determines how they are classified into one of three major groups: igneous rocks, formed from magma; sedimentary rocks, formed from low-temperature accumulations of particles or by precipitation from solution; and metamorphic rocks, formed from preexisting rocks through application of heat and pressure. Through time, geologic processes acting on older rocks change them into new and different ones, so that, in a sense, all kinds of rocks are interrelated. This concept is the essence of the rock cycle.

Terms to Remember

anion 27	lava 35
atom 25	lithification 35
atomic mass number 25	magma 34
atomic number 25	metamorphic 37
carbonate 32	mineral 27
cation 27	native element 33
chemical sedimentary	neutron 25
rock 36	nucleus 25
clastic sedimentary	oxide 33
rock 35	periodic table 26
compound 27	plutonic 34
contact metamorphism 37	proton 25
covalent bonding 27	regional
crystalline 27	metamorphism 37
electron 25	rock 34
ferromagnesian 31	rock cycle 34
foliation 37	sediment 35
glass 35	sedimentary 35
igneous 34	silicate 31
ion 27	sulfate 32
ionic bonding 27	sulfide 32
isotope 25	volcanic 35

Exercises

Questions for Review

1. Briefly define the following terms: *ion, isotope, compound, mineral,* and *rock.*
2. What two properties uniquely define a particular mineral?
3. Give the distinctive chemical characteristics of each of the following mineral groups: silicates, carbonates, sulfides, oxides, and native elements.
4. What is an igneous rock? How do volcanic and plutonic rocks differ in texture?
5. What are the two principal classes of sedimentary rocks?
6. Name several possible sources of the heat or pressure that can cause metamorphism.
7. What is the rock cycle?

For Further Thought

1. It is generally true that the more ancient rocks are less widely found and that their histories are more difficult to interpret. Consider why this should be so in the context of the rock cycle.

2. What kinds of rocks underlie your region of the country? (Your local geological survey could assist in providing the information.) What does this tell about the region's history? Are there currently any significant geologic hazards in the area? Any identified resources?

Suggested Readings/References

Berry, L. G., Mason, B., and Dietrich, R. V. 1983. *Mineralogy.* 2d ed. San Francisco: W. H. Freeman.

Dana, J. D. 1985. *Manual of mineralogy.* 20th ed. Revised by C. S. Hurlbut, Jr. and C. Klein. New York: John Wiley & Sons.

Dietrich, R. V., and Skinner, B. J. 1979. *Rocks and rock minerals.* New York: John Wiley & Sons.

Ehlers, E. G., and Blatt, H. 1982. *Petrology: Igneous, sedimentary, and metamorphic.* San Francisco: W. H. Freeman.

Ernst, W. G. 1969. *Earth materials.* Englewood Cliffs, N.J.: Prentice-Hall.

Mears, B., Jr. 1978. Chapters 3 and 4 in *Essentials of geology.* New York: D. Van Nostrand.

O'Donoghue, M. 1976. *VNR color dictionary of minerals and gemstones.* New York: Van Nostrand Reinhold.

Sinkankas, J. 1964. *Mineralogy for amateurs.* Princeton, N.J.: D. Van Nostrand.

Zoltai, T., and Stout, J. H. 1984. *Mineralogy: Concepts and principles.* Minneapolis, Minn.: Burgess.

A listing of state, U.S., provincial, and various international geological surveys and associations
> http://www.lib.berkeley.edu/EART/surveys.html

Information from/about the U.S. Geological Survey
> http://geology.usgs.gov
> The National Geologic Map Database
> http://geology.usgs.gov/ngmdb/

The Mineralogical Society of America's Teaching Resource Site, with links to other sites dealing with minerals
> http://geology.smith.edu/msa/Teaching.html

Information about the Smithsonion Institution Gem and Mineral Collection
> http://galaxy.einet.net/images/gems/gems-icons/html

Rockhounds Information Page, a very diverse site with links to educational resources, geologic professional societies, commercial suppliers, and more
> http://www.perspective.com/infodyn/rockhounds/rockhounds.html

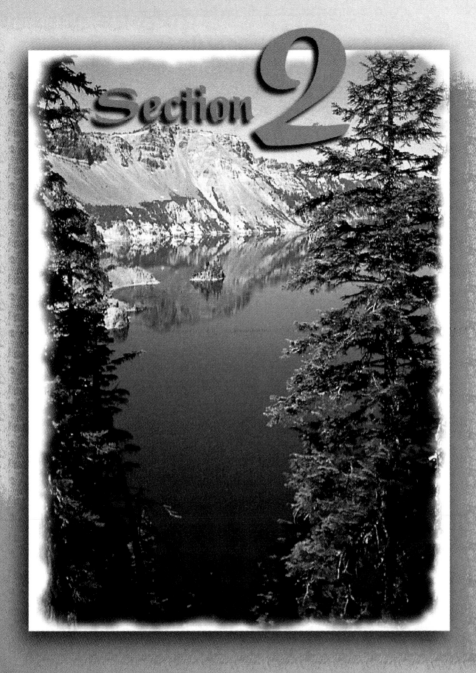

Section 2

The peacefulness of Crater Lake today belies the crater's explosive volcanic origin.

The processes that shape the earth can be broadly divided into "internal processes" and "surface processes." In this section, we will explore the causes and effects of the former. While internal processes produce effects at the earth's surface—volcanic eruptions are one example—they are "internal" in the sense that they appear to be mainly caused or driven, directly or indirectly, by the heat of the earth's interior. The effects of internal processes frequently extend to considerable depths; for instance, the seismic waves from earthquakes may pass through the whole earth. Some of the earth's internal heat is left over from this planet's accretion and differentiation, and more heat is continually produced by the decay of radioactive elements in the earth.

Plate tectonics, described in detail in chapter 3, provides the conceptual framework for understanding many aspects of earthquakes and volcanoes, which are discussed in chapters 4 and 5. We will see that, far from being haphazard occurrences, most of these phenomena occur in predictable zones on the earth. They are the logical consequences of the shifting of the earth's crust and the resultant creation and destruction of the crust and underlying uppermost mantle rocks. Knowledge of plate tectonics and its implications is an important aid to understanding how many kinds of mineral deposits are formed, which, in turn, is helpful in attempting to find new supplies of limited mineral resources.

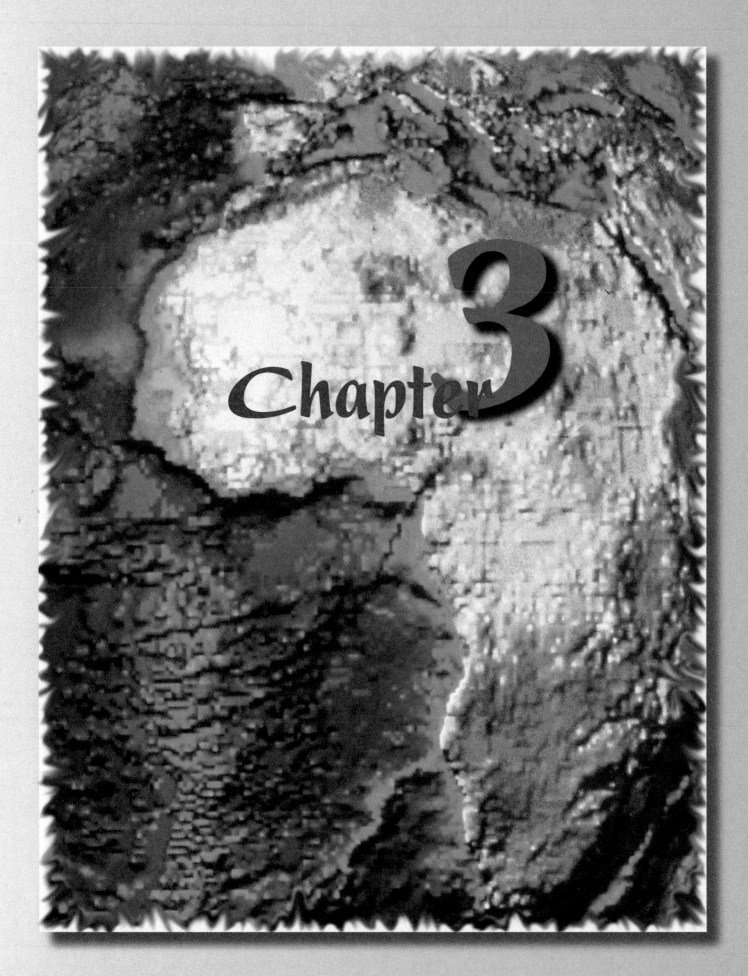

Chapter **3**

Plate Tectonics

Several centuries ago, observers looking at global maps noticed the similarity in outline of the eastern coast of South America and the western coast of Africa (figure 3.1). In 1855, Antonio Snider went so far as to publish a sketch showing how the two continents could fit together, jigsaw-puzzle fashion. Such reconstruction gave rise to the bold suggestion that perhaps these continents had once been part of the same landmass, which had later broken up.

This concept of **continental drift** had an especially vocal champion in Alfred Wegener, who began to publish his ideas in 1912 and continued to do so for more than two decades. Several other prominent scientists found the idea plausible. However, most people, scientists and non-scientists alike, had difficulty visualizing how something as massive as a continent could possibly "drift" around on a solid earth, or why it should do so. There were obvious physical objections to solid continents plowing through solid ocean basins, and no evidence of the expected resultant damage in crushed and shattered rock on continents or sea floor. The majority of reputable scientists either scoffed at the idea, or, at best, politely ignored it.

As it turns out, the supporting evidence was simply undiscovered or unrecognized at the time. Beginning in the 1960s, data of many different kinds began to accumulate that indicated that the continents have indeed moved. Continental drift turned out to be just one aspect of a broader theory known as *plate tectonics*, which has evolved over the last two decades. **Tectonics** is the study of large-scale movement and deformation of the earth's outer layers. **Plate tectonics** relates such deformation to the existence and movement of rigid "plates" over a weak or partly molten layer in the earth's upper mantle.

Spreading ridges and rifts are one type of plate boundary; they occur not only in the ocean, but also on the continents, as in East Africa.

Plate Tectonics—General Principles

One of the many obstacles to accepting the hypothesis of continental drift was imagining solid continents moving over solid earth. However, as noted in chapter 1, the earth is not completely solid from the surface to the center of the core. In fact, a plastic, locally partly molten zone lies relatively close to the surface, so that, in a sense, a thin solid skin of rock floats on a semisolid layer below. The existence of plates, and the occurrence of earthquakes in them, are related to the way rocks respond to stress.

Stress and Strain in Geological Materials

An object is under **stress** when force is being applied to it. The stress may be **compressive,** tending to squeeze or compress the object, or it may be **tensile,** tending to pull the object apart. A **shearing stress** is one that tends to cause different parts of the object to move in different directions across a plane or to slide past one another, as when a deck of cards is spread out on a tabletop by a sideways sweep of the hand.

Strain is deformation resulting from stress. It may be either temporary or permanent, depending on the amount and type of stress and on the strength of the material to resist it. If **elastic deformation** occurs, the amount of deformation is proportional to the stress applied, and the material returns to its original size and shape when the stress is removed. A gently stretched rubber band shows elastic behavior. Rocks, too, may behave elastically, although much greater stress is needed to produce detectable strain. Once the **elastic limit** of a material is reached, the material may go through a phase of **plastic deformation** with increased stress. During this stage, relatively small added stresses yield

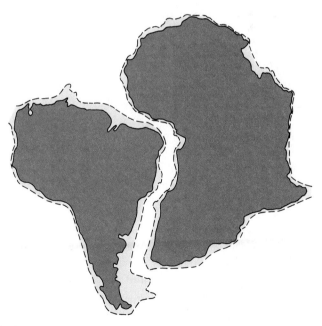

Figure 3.1 The jigsaw-puzzle fit of South America and Africa suggests that they were once joined together and subsequently separated by continental drift. (Blue shaded area is the continental shelf.)

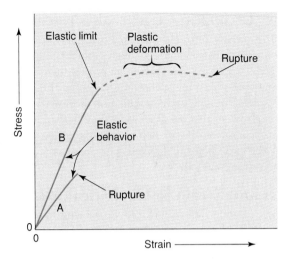

Figure 3.2 A stress-strain diagram. Elastic materials deform in proportion to the stress applied (straight line segments). A very brittle material (curve A) may rupture before the elastic limit is reached. Other materials subjected to stresses above their elastic limits deform plastically until rupture (curve B).

large corresponding strains, and the changes are permanent: the material does not return to its original size and shape after removal of the stress. A glassblower, an artist shaping clay, a carpenter fitting caulk into cracks around a window, and a blacksmith shaping a bar of hot iron into a horseshoe are all making use of the plastic behavior of materials.

If stress is increased further, solids eventually break, or **rupture.** In **brittle** materials, rupture may occur before there is any plastic deformation. Brittle behavior is characteristic of most rocks at near-surface conditions, with low temperature and low confining pressures, and is illustrated by A in figure 3.2. At greater depths, where temperatures are higher and rocks are confined and, in a sense, supported by surrounding rocks at high pressure, rocks may behave plastically (dashed section of curve B in figure 3.2). The effect of temperature can be seen in the behavior of cold and hot glass. A rod of cold glass is brittle and snaps under stress before appreciable plastic deformation occurs, while a warmed glass rod may be bent and twisted without breaking. Plastic deformation of rocks is often reflected in the production of folds, such as that illustrated in figure 3.3A. Brittle failure results in production of fractures and faults (figure 3.3B) as rocks rupture rather than flow.

The physical behavior of a rock, then, is affected by external factors, such as temperature or confining pressure, as well as by the intrinsic characteristics of the rock itself. Also, rocks respond differently to different types of stress. Most are far stronger under compression than tension; a given rock may rupture under a tensile stress only one-tenth as high as the compressive stress required to break it at the same pressure and temperature conditions. Consequently, the term *strength* has no single simple meaning when applied to rocks unless all of these variable are specified. Even when they are, rocks in their natural settings do not always exhibit

the same behavior as carefully selected samples tested in the laboratory. The natural samples may be weakened by fracturing or weathering, for example. Several rock types of quite different properties may be present at the same site. Considerations such as these complicate the work of the geological engineer.

Time is also a factor in the physical behavior of a rock. Materials may respond differently to given stresses, depending on the period of time over which the stress is applied. (This can be seen on a human timescale, in the phenomenon of fatigue: a machine component or structural material may fail under stresses well below its rupture strength if the stress has been applied repeatedly over a period of time.) Rocks can behave similarly, showing elastic or brittle behavior if stressed suddenly, as by passing seismic waves from an earthquake, but plastic behavior in response to prolonged stress, as from the weight of overlying rocks.

Lithosphere and Asthenosphere

The earth's crust and uppermost mantle are solid, somewhat brittle and elastic. This outer solid layer is called the **lithosphere,** from the Greek word *lithos,* meaning "rock." The lithosphere varies in thickness from place to place on the earth. It is thinnest underneath the oceans, where it extends to a depth of about 50 kilometers (about 30 miles). The lithosphere under the continents is thicker, extending in places to over 100 kilometers (60 miles).

The layer below the lithosphere is the **asthenosphere,** which derives its name from the Greek word *asthenes,* meaning "without strength." The asthenosphere extends to an average depth of about 300 kilometers (close to 200 miles) in the mantle. Its lack of strength or rigidity results from a combination of high temperatures and moderate confining pressures that allows the rock to flow plastically under stress. Below the asthenosphere, as pressures increase faster than temperatures with depth, the mantle again becomes more rigid and elastic.

The asthenosphere was discovered by studying the behavior of seismic waves from earthquakes (see chapter 4). Its presence makes the concept of continental drift more plausible. The continents need not drag across solid rock; instead, they can be pictured as sliding over a softened, deformable layer underneath the lithospheric plates. The relationships among crust, mantle, lithosphere, and asthenosphere are illustrated in figure 3.4.

Recognition of the existence of the plastic asthenosphere made plate motions more plausible, but it did not prove that they had occurred. Much additional information had to be accumulated before the rates and directions of plate movements could be documented and before most scientists would accept the concept of plate tectonics.

Locating Plate Boundaries

The distribution of earthquakes and volcanic eruptions indicates that these phenomena are far from uniformly distributed over the earth (figure 3.5). They are, for the most part, concentrated in belts or linear chains. This suggests that

A

B

Figure 3.3 (*A*) Folding such as this occurs while rocks are deeper in the crust, at elevated temperatures and confining pressure. (*B*) This rock suffered brittle failure under stress, which produced a small fault. (Uneven thicknesses of layers at bottom suggest previous plastic deformation.)   (*A*) Photograph by M. R. Mudge, USGS Photo Library, Denver, CO. (*B*) Photograph by W. B. Hamilton, USGS Photo Library, Denver, CO.

47

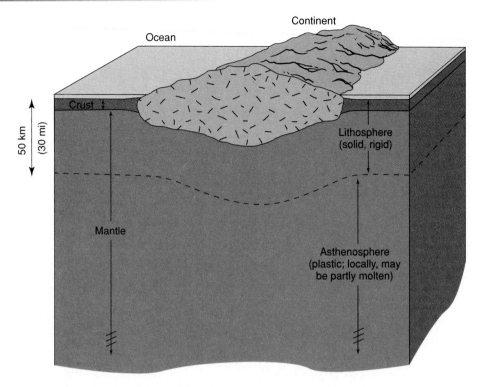

Figure 3.4 The outer zones of the earth (not to scale). The terms *crust* and *mantle* have compositional implications (see chapter 1); *lithosphere* and *asthenosphere* describe physical properties (the former, solid and rigid; the latter, plastic and, locally, partly molten). The lithosphere includes the crust and uppermost mantle. The asthenosphere lies entirely within the upper mantle. Below it, the rest of the mantle is more rigidly solid again.

Figure 3.5 Locations of modern volcanoes and earthquakes around the world.   Source: Map plotted by the Environmental Data and Information Service of NOAA; earthquakes from U.S. Coast and Geodetic Survey.

the rigid shell of lithosphere is cracked in places, broken up into pieces, or plates. The volcanoes and earthquakes are concentrated at the boundaries of these lithospheric plates, where plates jostle or scrape against each other. (The effect is somewhat like ice floes on an arctic sea: most of the grinding and crushing of ice and the spurting-up of water from below occur at the edges of the blocks of ice, while their solid central portions are relatively undisturbed.) About half a dozen very large lithospheric plates and many smaller ones have now been identified (figure 3.6).

Direction of plate motion

Subduction zone
(sawteeth point toward
overriding plate)

Spreading ridge

Transform fault

Figure 3.6 Principal world lithospheric plates, inferred from information such as that shown in figure 3.5 and other data. Arrows denote approximate relative directions of plate movement.   Source: After W. Hamilton, U.S. Geological Survey.

Plate Movements—Accumulating Evidence

The Topography of the Sea Floor

If South America and Africa really have moved apart, one might expect to see some evidence of this between them, some feature or features on the sea floor to indicate the continents' passage. A topographic map of the floor of the Atlantic Ocean shows an obvious ridge running north-south about halfway between those continents (figure 3.7). This mid-ocean ridge might be the seam from which the two continents moved apart. Similar ridges are found on the floors of other oceans, as seen in the same figure. In other places, frequently along the margins of continents, there are trenches several kilometers deep (see figures 3.7 and 3.19). It remained for scientists studying the ages and magnetic properties of seafloor rocks, as described below, to demonstrate the significance of the ocean ridges to plate tectonics. The trenches, too, can be explained by plate tectonics.

Magnetism in Rocks—General

Most iron-bearing minerals are at least weakly magnetic at surface temperatures. Each magnetic mineral has a **Curie temperature,** below which it remains magnetic, but above which it loses its magnetic properties. The Curie temperature varies from mineral to mineral, but it is always below

the mineral's melting temperature. A hot magma is therefore not magnetic, but as it cools and solidifies, and ferromagnesian silicates and other iron-bearing minerals crystallize from it, those magnetic minerals tend to line up in the same direction. Like tiny compass needles, they align themselves parallel to the lines of force of the earth's magnetic field, which run north-south, and they point to the magnetic north pole. They retain their internal magnetic orientation unless they are heated again. This is the basis for the study of **paleomagnetism,** "fossil magnetism" in rocks.

Magnetic north, however, has not always coincided with its present position. In the early 1900s, scientists investigating the direction of magnetization of a sequence of volcanic rocks in France discovered some flows that appeared to be magnetized in the opposite direction from the rest: their magnetic minerals pointed south instead of north. Confirmation of this discovery in many places around the world led to the suggestion in the late 1920s that the earth's magnetic field had "flipped," or reversed polarity; that is, that the north and south poles had switched places. During the time those surprising rocks had crystallized, a compass needle would have pointed to the magnetic south pole, not north.

Today, the phenomenon of magnetic reversals is well documented. Rocks crystallizing at times when the earth's field was in the same orientation as it is at present are said to be normally magnetized; rocks crystallizing when the field was oriented the opposite way are described as reversely

Figure 3.7 Shaded relief map of the world, including seafloor topography.   Source: After poster MGG-5 from the Marine Geology and Geophysics Division of the NOAA National Geophysical Data Center.

RED Highest
Elevation

magnetized. Over the history of the earth, the magnetic field has reversed many times, at variable intervals: sometimes, the polarity remained constant for close to a million years before reversing, while other reversals are separated by only a few tens of thousands of years. Through the combined use of magnetic measurements and age determinations on the magnetized rocks, geologists have been able to reconstruct the reversal history of the earth's magnetic field in some detail.

The explanation for magnetic reversals must be related to the origin of the magnetic field. The outer core is a metallic fluid, consisting mainly of iron. Motions in an electrically conducting fluid can generate a magnetic field, and this is believed to be the origin of the earth's field. (The simple presence of iron in the core is not enough to account for the magnetic field, as core temperatures are far above the Curie temperature of iron.) Perturbations or changes in the fluid motions, then, could account for reversals of the field. The details of the reversal process remain to be deduced.

Paleomagnetism and Seafloor Spreading

The ocean floor is made up largely of basalt, a volcanic rock rich in ferromagnesian minerals. During the 1950s, the first large-scale surveys of the magnetic properties of the sea floor produced an entirely unexpected result. The floor of the ocean was found to consist of alternating "stripes" or bands of normally and reversely magnetized rocks, symmetrically arranged around the ocean ridges. At first, this seemed so incredible that it was assumed that the instruments or measurements were faulty. However, other studies consistently obtained the same results. For several years, geoscientists strove to find a convincing explanation for these startling observations.

Then, in 1963, an elegant explanation was proposed by the team of F. J. Vine and D. H. Matthews, and independently by L. W. Morley. The magnetic stripes could be explained as a result of **seafloor spreading,** the moving apart of lithospheric plates at the ocean ridges. If the oceanic lithosphere splits and plates move apart, a rift in the lithosphere begins to open. But the result is not a 50-kilometer-deep crack. As rifting begins, the release of pressure from above allows extensive melting to occur, producing magma that escapes from the asthenosphere. The magma rises, cools, and solidifies to form new basaltic rock, which becomes magnetized in the prevailing direction of the earth's magnetic field. If the plates continue to move apart, the new rock will also split and part, making way for more magma to form still younger rock, and so on.

If, during the course of seafloor spreading, the polarity of the earth's magnetic field reverses, the rocks formed after the reversal are polarized oppositely from those formed before it. The ocean floor is a continuous sequence of basalts formed over tens or hundreds of millions of years, during which time there have been dozens of polarity reversals. The basalts of the sea floor have acted as a sort of magnetic tape recorder throughout that time, preserving a

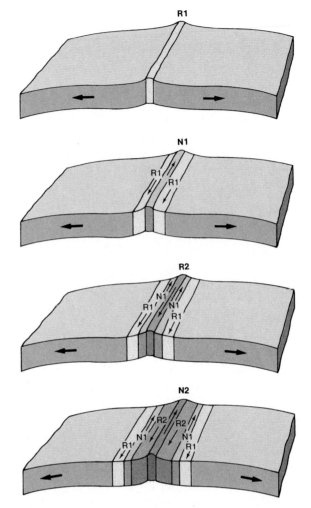

Figure 3.8 Formation of magnetic stripes on the sea floor. As each new piece of sea floor forms at the ridge, it becomes magnetized in a direction dependent on the orientation of the earth's magnetic field at that time. The magnetism is "frozen in," or preserved, in the rock thereafter. Past reversals of the magnetic field are reflected in alternating bands of normally (N) and reversely (R) magnetized rocks. Numbers reflect the order in which bands were formed.

record of polarity reversals in the alternating bands of normally and reversely magnetized rocks. The process is illustrated schematically in figure 3.8.

Age of the Ocean Floor

The ages of seafloor basalts themselves lend further support to this model of seafloor spreading. Specially designed research ships can sample sediment from the deep-sea floor and drill into the basalt beneath. An extensive program of seafloor drilling was carried out by the JOIDES consortium (Joint Oceanographic Institutions for Deep Earth Sampling).

The time at which an igneous rock, such as basalt, crystallized from its magma can be determined by methods

Figure 3.9 Age distribution of the sea floor. Note relative spreading rates of Mid-Atlantic Ridge (slower) and East Pacific Rise (faster), shown by wider color bands in the Pacific. Seafloor relief is superimposed in grey on age data.   Source: Marine Geology and Geophysics Division of the NOAA National Geophysical Data Center.

described in appendix A. When this is done for many samples of seafloor basalt, a pattern emerges. The rocks of the sea floor are youngest close to the ocean ridges and become progressively older the farther away they are from the ridges on either side (see figure 3.9). Like the magnetic stripes, the age pattern is symmetric across each ridge. It thus appears that, as seafloor spreading progresses, previously formed rocks are continually spread apart and moved farther from the ridge, while fresh magma rises from the asthenosphere to form new lithosphere at the ridge. The oldest rocks recovered from the sea floor, well away from the ridges, are about 200 million years old.

Polar-Wander Curves

Evidence for plate movements does not come only from the sea floor. For reasons outlined later in the chapter, much older rocks are preserved on the continents than in the ocean—some continental samples are close to 4 billion years old—so longer periods of earth history can be investigated through continental rocks. Studies of magnetic orientations of continental rocks can span many hundreds of millions of years and yield quite complex data.

The lines of force of earth's magnetic field not only run north-south, they vary in dip with latitude: vertical at the magnetic poles, horizontal at the equator, at varying intermediate dips in between. When the orientation of magnetic minerals in a rock is determined in three dimensions, one can thus determine not only the direction of magnetic north, but (magnetic) latitude as well, or how far removed that region was from magnetic north at the time the rock's magnetism assumed its orientation.

Magnetized rocks of different ages on a single continent may point to very different apparent magnetic pole positions. The magnetic north and south poles may not simply be reversed, but may be rotated or tilted from the present magnetic north and south. When the directions of magnetization and latitudes of many rocks of various ages from one continent are determined and plotted on a map, it appears that the magnetic poles have meandered far over the surface of the earth—if the position of the continent is assumed to have been fixed on the earth throughout time. The resulting curve, showing the apparent movement of the magnetic pole relative to the continent as a function of time, is the **polar-wander curve** for that continent (figure 3.10).

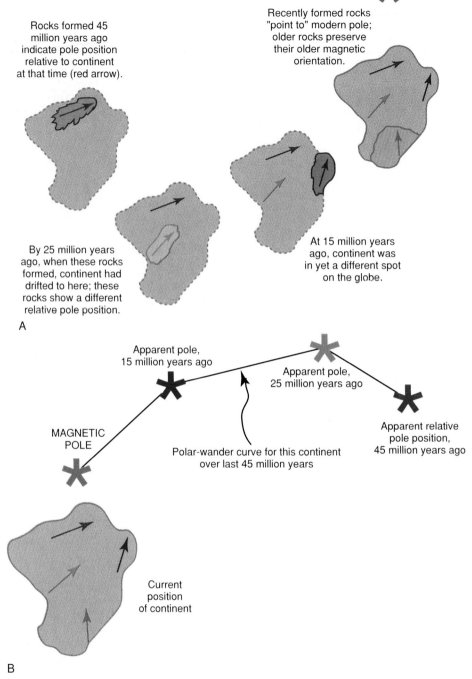

Figure 3.10   Polar-wander curves actually reflect wandering continents. (*A*) As rocks crystallize, their magnetic minerals align with the contemporary magnetic field. But continental drift changes relative position of continent and magnetic pole over time. (*B*) Assuming a stationary continent, the shifting relative pole positions suggest "polar wander."

Rocks formed 45 million years ago indicate pole position relative to continent at that time (red arrow).

MAGNETIC POLE

Recently formed rocks "point to" modern pole; older rocks preserve their older magnetic orientation.

By 25 million years ago, when these rocks formed, continent had drifted to here; these rocks show a different relative pole position.

At 15 million years ago, continent was in yet a different spot on the globe.

A

Apparent pole, 15 million years ago

Apparent pole, 25 million years ago

Apparent relative pole position, 45 million years ago

MAGNETIC POLE

Polar-wander curve for this continent over last 45 million years

Current position of continent

B

The discovery of polar-wander curves was initially puzzling because there are good geophysical reasons to believe that the earth's magnetic poles should remain close to the geographic (rotational) poles. (In particular, the fluid motions in the outer core that cause the magnetic field could be expected to be strongly influenced by the earth's rotation.) Additionally, the polar-wander curves for different continents do not match. Rocks of exactly the same age from two different continents may seem to point to two very different magnetic poles—as seen, for example, in figure 3.11. This confusion can be eliminated, however, if it is assumed that the magnetic poles have always remained close to the geographic poles but that the continents have moved and rotated. The polar-wander curves then provide a way

to map the directions in which the continents have moved through time, relative to the stationary poles and relative to each other.

Other Evidence

Continental rocks also form under a wider range of conditions than seafloor rocks and thus yield more varied kinds of information. Sedimentary rocks, for example, may preserve evidence of the ancient climate of the time and place in which the sediments were deposited. Such evidence shows that the climate in many places has varied widely through time. We find evidence of glaciation in places now located in the tropics, in parts of Australia, southern Africa, and South America (figure 3.12). There are desert sand deposits in the rocks of

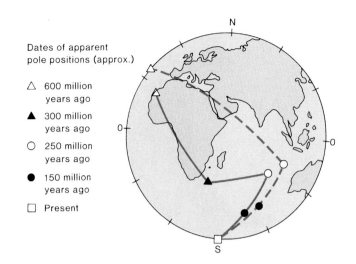

Dates of apparent
pole positions (approx.)

△ 600 million
years ago

▲ 300 million
years ago

○ 250 million
years ago

● 150 million
years ago

□ Present

Figure 3.11 Examples of polar-wander curves. The apparent position of the magnetic pole relative to the continent is plotted for rocks of different ages, and these data points are connected to form the curve. The present positions of the continents are shown for reference. Polar-wander curves for Africa (solid line) and Arabia (dashed line) suggest that these landmasses moved quite independently up until about 250 million years ago, when the polar-wander curves converge. After M. W. McElhinny, *Paleomagnetism and Plate Tectonics.* Copyright © 1973 by Cambridge University Press, New York, NY. Reprinted by permission.

Figure 3.12 Glacial deposits across the southern continents suggest past juxtaposition. Arrows indicate direction of ice flow.
Source: After Arthur Holmes, *Principles of Physical Geology,* 2d ed., Ronald Press, New York, NY, 1965.

A

B

Figure 3.13 Curious distribution of fossil remains of the plant *Glossopteris* (A) and dinosaur *Mesosaurus* (B) could be explained most easily by continental drift (C).

regions that now have moist, temperate climates and the remains of jungle plants in now-cool places. There are coal deposits in Antarctica, even though coal deposits form from the remains of a lush growth of land plants. These observations cannot be explained as the result of global climatic changes, for the continents do not all show the same warming or cooling trends at the same time. However, climate is to a great extent a function of latitude, which strongly influences surface temperatures: Conditions are generally warmer near the equator and colder near the poles. Dramatic shifts in an individual continent's climate might result from changes in its latitude in the course of continental drift.

Sedimentary rocks also preserve fossil remains of ancient life. Some plants and animals, now extinct, seem to have lived only in a few very restricted areas, which now are widely separated geographically on different continents (figure 3.13). Examples include the fossil plant *Glossopteris,* remains of which are found in limited areas of India, southern Africa, and even Antarctica. The fossils of a small dinosaur, *Mesosaurus,* are similarly dispersed across two continents. It is difficult to imagine one distinctive variety of animal or plant developing simultaneously in two or more small areas thousands of kilometers apart, or somehow migrating over vast expanses of ocean. (Besides, *Glossopteris* was a land plant, *Mesosaurus* a freshwater animal.) The theory of continental drift provides a way out of the quandary.

C

The organism may have lived in a single, geographically restricted area, and the rocks in which its remains are now found may subsequently have been separated and moved in several different directions by drifting continents.

As mentioned earlier, the apparent similarity of the coastlines of Africa and South America triggered early speculation about the possibility of continental drift. Actually, the pieces of this jigsaw puzzle fit even better if we look not at the coastlines but at the outer edges of the continental shelves, just where the depth of the ocean begins to increase rapidly toward typical ocean-basin depths (see figures 3.1 and 3.7). Computers have been used to help determine the best possible physical fit of the pieces. The results are imperfect, perhaps because not every bit of continent has been perfectly preserved in the tens or hundreds of millions of years since continental breakup, and volcanic activity has locally added rock as well. Continental reconstructions can be refined using details of continental geology—rock types, rock ages, fossils, ore deposits, mountain ranges, and so on. If two now-separate continents were once part of the same landmass, then the geologic features presently found at the margin of one should have counterparts at the corresponding edge of the other. In short, the geology should match when the puzzle pieces are reassembled.

Efforts to reconstruct the ancient locations and arrangements of the continents have been relatively successful, at least for the not-too-distant geologic past. It has been shown, for example, that a little more than 200 million years ago there was a single, great supercontinent that has been named *Pangaea*, from the Greek for "all lands." The present seafloor spreading ridges are the lithospheric scars of the subsequent breakup of Pangaea (figure 3.14). However, they are not the only kind of boundary found between plates.

Types of Plate Boundaries

Different things happen at the boundaries between lithospheric plates, depending on the relative motions of the plates and on whether continental or oceanic lithosphere is at the edge of each plate.

Divergent Plate Boundaries

At a **divergent plate boundary,** such as a mid-ocean spreading ridge, lithospheric plates move apart, the release of pressure facilitates some melting in the asthenosphere, magma wells up from the asthenosphere, and new lithosphere is created (figure 3.15A). A great deal of volcanic activity thus occurs at spreading ridges. In addition, the pulling-apart of the plates of lithosphere results in earthquakes along these ridge plate boundaries (see also chapter 4).

Continents can be rifted apart, too (figure 3.15B). This is less common, perhaps because continental lithosphere is so much thicker than oceanic lithosphere. In the early stages of continental rifting, volcanoes may erupt along the rift, or great flows of basaltic lava may pour out through the fissures in the continent. If the rifting continues, a new ocean basin will eventually form between the pieces of the continent. Geologists believe that this is what is happening now in east Africa: The easternmost part of the continent is being rifted apart from the rest and an ocean will one day separate the pieces. Lake Tanganyika, formed by this rifting, shows its origins in the complementary shapes of its east and west shores (figure 3.16). The rifting can be seen at continental scale in figure 3.7.

A

B

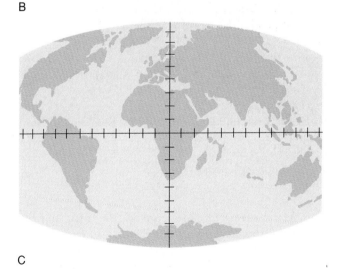

C

Figure 3.14 Reconstructed plate movements during the last 200 million years. *Laurasia* and *Gondwanaland* are names given to the northern and southern portions of Pangaea, respectively. Although these changes occur very slowly in terms of a human lifetime, they illustrate the magnitude of some of the natural forces to which we must adjust. (*A*) 200 million years ago. (*B*) 100 million years ago. (*C*) Today.   Source: After R. S. Dietz and J. C. Holden, "Reconstruction of Pangaea," *Journal of Geophysical Research,* 75:4939–4956, 1970, copyright by the American Geophysical Union.

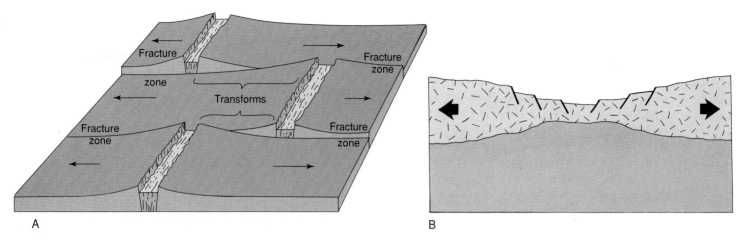

A   B

Figure 3.15 Divergent plate boundaries. (*A*) A seafloor spreading ridge. (*B*) Early stages of continental rifting.

Figure 3.16 Satellite image of Lake Tanganyika, in the East African rift zone. (In this false-color image, green vegetation appears red.)   © NASA.

As seawater circulates through the fresh, hot lithosphere at a spreading ridge, it is heated and reacts with the rock, becoming metal-rich. As it gushes back out of the sea floor, cooling and reacting with cold seawater, it may precipitate potentially valuable mineral deposits. Economically unprofitable to mine now, they may nevertheless prove useful resources for the future; see chapter 12.

Transform Boundaries

The actual structure of a spreading ridge is more complex than a single, straight crack. A close look at a mid-ocean spreading ridge reveals that it is not a continuous rift thousands of kilometers long (see figure 3.7). Rather, ridges consist of many short segments slightly offset from one another. The offset area is a special kind of fault, or break in the lithosphere, known as a **transform fault** (see figure 3.15A). The opposite sides of a transform fault belong to two different plates, and these are moving in opposite directions. As the plates scrape past each other, earthquakes occur along the transform fault.

The famous San Andreas Fault in California is an example of a transform fault that slices a continent sitting along a spreading ridge. The East Pacific Rise, a seafloor spreading ridge off the northwestern coast of North America, disappears under the edge of the continent, to reappear farther south in the Gulf of California (see figures 3.6 and 3.7). The San Andreas is the transform fault between these segments of spreading ridge. Most of North America is part of the North American Plate. The thin strip of California on the west side of the San Andreas Fault, however, is moving northwest with the Pacific Plate. The stress resulting from the shearing displacement across the fault leads to ongoing earthquake activity, described in chapter 4.

Convergent Plate Boundaries

At a **convergent plate boundary,** as the name indicates, plates are moving toward each other. Just what happens depends on what sort of lithosphere is at the leading edge of each plate (figure 3.17). Continental lithosphere is relatively low in density and is therefore buoyant with respect to the dense, iron-rich mantle, so it tends to "float" on the asthenosphere. Oceanic lithosphere is more similar in density to the underlying asthenosphere, so it is more easily forced down into the asthenosphere as plates move together.

In a continent-continent collision, the two landmasses come together, crumple, and deform. One may partially override the other, but the buoyancy of continental lithosphere ensures that neither sinks deep into the mantle, and a very large thickness of continent may result. Earthquakes are frequent during active collision as a consequence of the large stresses involved in the process. The extreme height of the Himalaya Mountains is attributed to just this sort of continent-continent collision. India was not always a part of

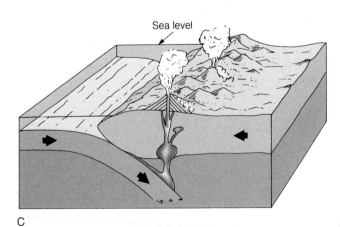

Figure 3.17 Convergent plate boundaries. (*A*) Continent-continent collision. (*B*) Ocean-ocean convergence. (*C*) Subduction zone at ocean-continent convergence.

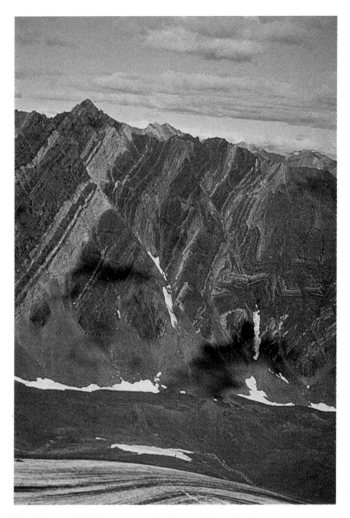

Figure 3.18 Folding and faulting resulting from continental-lithosphere convergence; Cook Inlet, Alaska. Photograph by N. J. Silberling, USGS Photo Library, Denver, CO.

the Asian continent. Paleomagnetic evidence indicates that it drifted northward over tens of millions of years (recall figure 3.14) until it "ran into" Asia, and the Himalayas were built up in this collision. Earlier, the ancestral Appalachian Mountains were built in the same way, as Africa and North America converged prior to the breakup of Pangaea. In fact, many major mountain ranges worldwide represent sites of sustained plate convergence in the past. Figure 3.18 begins to reflect the deformation possible as a result of continental convergence.

More commonly, oceanic lithosphere is at the leading edge of one or both of the converging plates. One plate of oceanic lithosphere may be pushed under the other plate and descend into the asthenosphere. This type of plate boundary, where one plate is carried down below (subducted beneath) another, is called a **subduction zone** (see figure 3.17 B,C). The nature of subduction zones can be demonstrated in many ways, one of which involves earthquake depths, as described in the next chapter.

The subduction zones of the world balance the seafloor equation. If new oceanic lithosphere is constantly being created at spreading ridges, an equal amount must be destroyed somewhere, or the earth would simply keep getting bigger. This excess sea floor is consumed in subduction zones. The subducted plate is heated by the hot asthenosphere, and, in time, becomes hot enough to melt. Meanwhile, at the spreading ridges, other melts rise, cool, and crystallize to make new sea floor. So, in a sense, the oceanic lithosphere is constantly being recycled, which explains why few very ancient seafloor rocks are known. Only rarely is a bit of sea floor caught up in a continent during convergence and preserved. Most often, the sea floor in a zone of convergence is subducted and destroyed. The buoyant continents are not so easily reworked in this way; hence, very old rocks may be preserved on them. Because all continents are part of moving plates, sooner or later they all are inevitably transported to a convergent boundary, as leading oceanic lithosphere is consumed.

Subduction zones are, geologically, very active places. Sediments eroded from the continents may accumulate in the trench formed by the down-going plate, and some of these sediments may be carried down into the asthenosphere to be melted along with the sinking lithosphere. Volcanoes form where the melted material rises up through the overlying plate to the surface. At an ocean-ocean convergence, the result is commonly a line of volcanic islands, an **island arc** (see figure 3.17B). The great stresses involved in convergence and subduction give rise to numerous earthquakes. The bulk of mountain building, and its associated volcanic and earthquake activity, is related to subduction zones. Parts of the world near or above modern subduction zones, and therefore prone to both volcanic and earthquake activity, include the Andes region of South America, western Central America, parts of the northwest United States and Canada, the Aleutian Islands, China, Japan, and much of the rim of the Pacific Ocean basin (see figure 3.6).

How Far, How Fast, How Long, How Come?

Past Motions, Present Velocities

Rates and directions of plate movement can be determined in a variety of ways. As previously discussed, polar-wander curves from continental rocks can be used to determine how the continents have drifted. Seafloor spreading is another way of determining plate movement. The direction of seafloor spreading is usually obvious: away from the ridge. Rates of seafloor spreading can be found very simply by dating rocks at different distances from the spreading ridge and dividing the distance moved by the rock's age (the time it has taken to move that distance from the ridge at which it formed). For example, if a 10-million-year-old piece of sea floor is collected at a distance of 100 kilometers from the ridge, this represents an average rate of movement over that time of 100 kilometers/10 million years. If we convert that to units that are easier to visualize, it works out to about one centimeter (a little less than half an inch) per year.

Another way to monitor rates and directions of plate movement is by using mantle **hot spots.** These are isolated areas of volcanic activity usually not associated with plate boundaries (see also chapter 5). Geologists believe that these volcanoes reflect unusual mantle beneath the hot spots. If we assume that mantle hot spots remain fixed in position while the lithospheric plates move over them, the result should be a string of volcanoes of differing ages with the youngest closest to the hot spot. (Some geoscientists question whether hot spots do, in fact, remain fixed over tens of millions of years; but in many cases, the data are consistent with essentially stationary hot spots.)

A good example can be seen in the north Pacific Ocean (see figure 3.19). A topographic map shows a V-shaped chain of volcanic islands and submerged volcanoes. When rocks from these volcanoes are dated, they show a progression of ages, from about 75 million years at the northwestern end of the chain, to about 40 million years at the bend, through progressively younger islands to the still-active volcanoes of the island of Hawaii, located at the eastern end of the Hawaiian Island group. From the distances and age differences between pairs of points, we can again determine the rate of plate motion. For instance, Midway Island and Hawaii are about 2,700 kilometers (1,700 miles) apart. The volcanoes of Midway were active about 25 million years ago. Over the last 25 million years, then, the Pacific plate has moved over the mantle hot spot at an average rate of 2,700 kilometers/25 million years, or about 11 centimeters (4.3 inches) per year. The orientation of the volcanic chain shows the direction of plate movement—west-northwest. The kink in the chain at about 40 million years ago indicates that the direction of movement of the Pacific plate changed at that time.

Hot spots occur under continents as well as beneath ocean basins. They are, however, somewhat easier to detect in oceanic regions, perhaps because the associated magmas can more readily work their way up through the thinner oceanic lithosphere to build visible volcanoes.

Plate motions can be quantified using other methods, too—for example, considering widths of strips of sea floor produced over different periods of time (recall figure 3.9), or using points of known age and position along a continent's polar-wander curve. Looking at many such determinations from all over the world, geologists find that average rates of plate motion are 2 to 3 centimeters (about 1 inch) per year. In a few places, movement at rates up to about 10 centimeters/year is observed, and, elsewhere, rates may be slower, but a few centimeters per year is typical. This seemingly trivial amount of motion does add up through geologic time. Movement of two centimeters per year for 100 million years means a shift of 2,000 kilometers, or about 1,250 miles!

Figure 3.19 The Hawaiian Islands and other volcanoes in a chain formed over a hot spot. Numbers indicate dates of volcanic eruptions, in millions of years. Movement of the Pacific plate has carried the older volcanoes far from the hot spot, now under the active volcanic island of Hawaii.   *World Ocean Floor* map by Bruce C. Heezen and Marie Tharp, 1977. Copyright © 1977 by Marie Tharp. Reproduced by permission of Marie Tharp, 1 Washington Ave., South Nyack, NY 10960.

New lithosphere forms
at spreading center.

Continental
lithosphere

Oceanic lithosphere

Material spreads, cools,
and sinks.

Warm asthenosphere
rises under spreading
ridge; magma escapes
and intrudes crust.

Cooled material is warmed again
deeper in the asthenosphere.

Dense material skins back deeper
into asthenosphere at subduction zone.

Figure 3.20 A possible driving force behind plate tectonics is slow convection in the partly molten asthenosphere. Alternatively, plate motions may drive asthenospheric circulation.

Why Do Plates Move?

A driving force for plate tectonics has not been definitely identified. For many years, the most widely accepted explanation was that the plates were moved by large **convection cells** slowly churning in the plastic asthenosphere (figure 3.20). According to this scheme, hot material rises at the spreading ridges; some magma escapes to form new lithosphere, but the rest of the rising asthenospheric material spreads out sideways beneath the lithosphere, slowly cooling in the process. As it flows outward, it drags the overlying lithosphere outward with it, thus continuing to open the ridges. When it cools, the flowing material becomes dense enough to sink back deeper into the asthenosphere, for example under subduction zones.

The existence of convection cells in the asthenosphere has not been proven definitively. There is some question, too, whether flowing asthenosphere could exercise enough drag on the lithosphere above to propel it laterally and force it into convergence. An alternative explanation, for which considerable evidence has recently accumulated, is that the weight of the dense, down-going slab of lithosphere in the subduction zone pulls the rest of the trailing plate along with it, opening up the spreading ridges so magma can ooze upward. Mantle convection might then result from lithospheric drag, not the reverse. Another variant of this scheme pictures dense slabs of lithosphere "sliding off" topographic highs (visible in figures 3.7 and 3.9) formed at spreading ridges by rising warm asthenospheric mantle, and perhaps dragging some asthenosphere along laterally. Many geoscientists now believe that a combination of these mechanisms is responsible for plate motion; and there may be contributions from other mechanisms not yet considered. That lithospheric motions and flow in the asthenosphere are coupled seems likely, but which one may drive the other is still unclear.

History of Plate Tectonics

How long plate-tectonic processes have been active is also not entirely clear. The magnetic stripes characteristic of seafloor spreading are apparent over even the oldest, 200-million-year-old ocean floor. From continental rocks, apparent polar-wander curves going back more than a billion years can be reconstructed, although the relative scarcity of undisturbed ancient rocks makes such efforts more difficult for the earth's earliest history. It seems clear that the continents have been shifting in position over the earth's surface for at least 2 billion years, though not necessarily at the same rates or with exactly the same results as at present. Plate-tectonic processes have long played a major role in shaping the earth and are likely to continue doing so for the foreseeable future.

Plate Tectonics and the Rock Cycle

In the previous chapter, we noted that all rocks may be considered related by the concept of the rock cycle. We can also look at the rock cycle in a plate-tectonic context, as illustrated in figure 3.21. New igneous rocks form from magmas rising out of the asthenosphere at spreading ridges or in

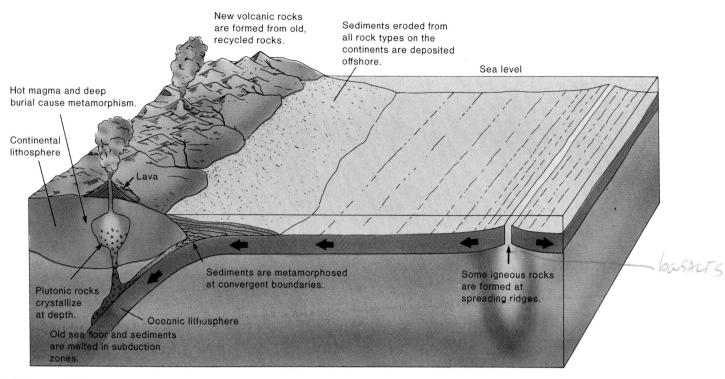

Figure 3.21 The rock cycle as interpreted in plate-tectonic terms.

Box 3.1 Plate Tectonics and Radioactive Waste: New Solutions or Wishful Thinking?

When plate-tectonic theory was being developed and the existence and nature of subduction zones discovered, some began to see subduction zones as the ultimate in waste disposal. In particular, subduction zones were suggested as possible disposal sites for canisters of radioactive wastes.

The appeal of the idea is the image of these potentially hazardous materials being carried deep into the earth, there to vanish, effectively, forever (at least from a human perspective). The principal drawbacks are twofold. First, given the extremely slow rates of plate motion, complete subduction of wastes would take thousands of years at least. Unless the waste canisters were somehow emplaced well into the subducted plate, they would sit exposed to interaction with seawater in the meantime. Seawater, especially seawater warmed by decaying radioactive wastes, is highly corrosive and might breach the canisters, allowing leakage of the wastes. Second, a plate undergoing subduction does not slip quietly under the plate above like a spatula under a pancake. Subduction is accompanied by earthquakes and deformation of the plates and the sediments carried on them. Some of the overlying sediments are not, in fact, subducted but are scraped off onto the overriding plate. It is possible that some waste canisters could be ruptured or caught up in the scraped-off sedimentary pile.

Enthusiasm for this scheme has waned. However, now that the significance of plate tectonics to the formation and destruction of the sea floor has been recognized, there *is* interest in using the stable interiors of large oceanic plates as possible waste-disposal sites. An international committee is presently studying this idea; see chapter 15.

subduction zones. The heat radiated by the cooling magmas can cause metamorphism, with recrystallization at an elevated temperature changing the texture and/or the mineralogy of the surrounding rocks. Some of these surrounding rocks may themselves melt to form new igneous rocks. The forces of plate collision at convergent margins also contribute to metamorphism by increasing the pressures acting on the rocks. As described in chapter 11, weathering and erosion on the continents wear down preexisting rocks of all kinds into sediment. Much of this sediment is eventually transported to the edges of the continents, where it is deposited in deep basins. Through burial under more layers of sediment, it may become solidified into sedimentary rock. Sedimentary rocks, in turn, may be metamorphosed or even melted by the stresses and the igneous activity at the plate margins. Some of these sedimentary or metamorphic materials may also be carried down with subducted oceanic lithosphere, to be melted and eventually recycled as igneous rock. Plate-tectonic activity thus plays a large role in the formation of new rocks from old that proceeds continually on the earth.

Overview

Summary

Rocks subjected to stress may behave either elastically or plastically. At low temperatures and confining pressures, they are more rigid, elastic, and often brittle. At higher temperatures and pressures, or in response to stress applied gradually over long periods, they tend more toward plastic behavior.

The outermost solid layer of the earth is the 50- to 100-kilometer-thick lithosphere, which is broken up into a series of rigid plates. The lithosphere is underlain by a plastic layer of the mantle, the asthenosphere, over which the plates can move. This plate motion gives rise to earthquakes and volcanic activity at the plate boundaries. At seafloor spreading ridges, which are divergent plate boundaries, new sea floor is created from magma rising from the asthenosphere. The sea floor moves in conveyor-belt fashion, ultimately to be destroyed in subduction zones, a type of convergent plate boundary, where the sea floor is carried down into the asthenosphere and eventually remelted.

Evidence for seafloor spreading includes the age distribution of seafloor rocks and the magnetic stripes on the ocean floor. Continental drift can be demonstrated by such means as polar-wander curves, fossil distribution, and evidence of ancient climates revealed in the rock record. Past supercontinents can be reconstructed by fitting together modern continental margins and matching similar geologic features from continent to continent.

Present rates of plate movement average a few centimeters a year. One possible driving force for plate tectonics is slow convection in the asthenosphere; another possibility is gravity pulling cold, dense lithosphere down into the asthenosphere, dragging the asthenosphere along. Plate-tectonic processes appear to have been more or less active for much of the earth's history. They play an integral part in the rock cycle. Later chapters examine in greater detail how plate tectonics can be related to earthquakes, volcanic activity, and formation of mineral deposits.

Terms to Remember

asthenosphere 46	lithosphere 46
brittle 46	paleomagnetism 49
compressive stress 45	plastic deformation 45
continental drift 45	plate tectonics 45
convection cell 62	polar-wander curve 53
convergent plate	rupture 46
boundary 58	seafloor spreading 52
Curie temperature 49	shearing stress 45
divergent plate	strain 45
boundary 57	stress 45
elastic deformation 45	subduction zone 59
elastic limit 45	tectonics 45
hot spot 60	tensile stress 45
island arc 60	transform fault 58

Exercises

Questions for Review

1. What causes strain in rocks? How do elastic and plastic materials differ in their behavior?
2. What is plate tectonics, and how are continental drift and seafloor spreading related to it?
3. Define the terms *lithosphere* and *asthenosphere*. Where are the lithosphere and asthenosphere found?
4. Describe two kinds of paleomagnetic evidence supporting the theory of plate tectonics.
5. Cite at least three kinds of evidence, other than paleomagnetic evidence, for plate tectonics.
6. Explain how a subduction zone forms and what occurs at such a plate boundary.

7. What are hot spots, and how do they help to determine rates and directions of plate movements?

8. Describe how convection in the asthenosphere may drive the motion of lithospheric plates. Alternatively, how might plate motions churn the plastic asthenosphere?

9. Describe the rock cycle in terms of plate tectonics.

For Further Thought

1. The moon's structure is very different from that of earth. For one thing, the moon has an approximately 1,000-kilometer-thick lithosphere. Would you expect plate-tectonic activity similar to that on earth to occur on the moon? Why or why not?

2. The Atlantic Ocean is approximately 5,000 km wide. If each plate has moved away from the Mid-Atlantic Ridge at an average speed of 1.5 cm per year, how long did the Atlantic Ocean basin take to form? What would the spreading rate be in miles per hour?

3. New methods of measuring relative plate motions in progress at present have increased our understanding of plate tectonics. Visit the following site on the World Wide Web, examine the available information on global- and regional-scale plate motions, and compare with figure 3.9. Are the data consistent? Comment.

   http://cddis.gsfc.nasa.gov/926/slrtecto.html

Suggested Readings/References

Bird, J. M., ed. 1980. *Plate tectonics.* 2d ed. Washington, D.C.: American Geophysical Union. (Selected papers from publications of the American Geophysical Union.)

Bonatti, E. 1987. The rifting of continents. *Scientific American* 256 (March): 96–103.

Dietz, R. S., and Holden, J. C. 1970. The breakup of Pangaea. *Scientific American* 223 (April): 30–41.

Hoffman, K. A. 1988. Ancient magnetic reversals: Clues to the geodynamo. *Scientific American* 258 (May): 76–83.

Hurley, P. M. 1968. The confirmation of continental drift. *Scientific American* 218 (April): 52–64.

Kious, W. J., and Tilling, R. I. 1996. *This dynamic earth: The story of plate tectonics.* U.S. Geological Survey. This is an excellent summary, with diagrams. [online] http://geology.usgs.gov/publications/text/dynamic.html

Marvin, U. B. 1973. Continental drift: *The evolution of a concept.* Washington, D.C.: Smithsonian Institution Press.

McElhinny, M. W. 1973. *Paleomagnetism and plate tectonics.* New York: Cambridge University Press.

Menard, H. W. 1986. *The oceans of truth: A personal history of global tectonics.* Princeton, N.J.: Princeton University Press.

Molnar, P., and Tapponier, P. 1977. The collision between India and Eurasia. *Scientific American* 236 (April): 30–41.

Mutter, J. C. 1986. Seismic images of plate boundaries. *Scientific American* 254 (February): 66–75.

Nance, R. D., Worsley, T. R., and Moody, J. B. 1988. The supercontinent cycle. *Scientific American* 259 (July): 72–79.

Raymo, C. 1982. *The crust of our earth: An armchair traveler's guide to the new geology.* Englewood Cliffs, N.J.: Prentice-Hall.

Small, C., and Sandwell, D. 1996. Sight unseen. *Natural History* 105 (3): 28–33.

Vink, G. E., Morgan, W. J., and Vogt, P. R. 1985. The earth's hot spots. *Scientific American* 252 (April): 50–57.

Wegener, A. 1924. *The origin of continents and oceans.* London: Methuen.

Woods Hole Oceanographic Institution. "Mid-Ocean Ridges." *Oceanus* 34 (Winter) 1991–92. (A collection of short articles on different aspects of spreading-ridge geology and ecology.)

General U.S. Geological Survey (USGS) information on plate tectonics

   http://geology.usgs.gov/publications/text/dynamic.html

Information on world plate boundaries, magnetic lineations, and fracture zones

   http://www.aist.go.jp/GSJ/dMG/free/plates/intro.html

U.S. National Geophysical Data Center, data on and images of global relief maps, many of which reveal plate-tectonic features

   http://www.ngdc.noaa.gov/mgg/global/global.html

National Earthquake Information Center information relates earthquakes to plate tectonics; for example

   http://wwwneic.cr.usgs.gov/neis/general/handouts/rift_man.html

An Internet search for "plate tectonics" may reveal some sites in individual geoscience departments with relevant material. (None are provided here because the sites' long-term stability is uncertain.)

Chapter 4

Earthquakes

"The Alaska earthquake on March 27, 1964, commonly referred to as the "Good Friday Earthquake," struck at 5:36 P.M. A few members and staff of the Anchorage Petroleum Club were preparing for an early dinner. I was one of those members. The Petroleum Club was located on the top of the Anchorage-Westward Hotel, a fifteen-story building which, at the time, was the tallest building in Anchorage. The outside walls of the dining room were made up entirely of windows to allow a full view of the mountains to the east.

With some preliminary tremors, the earthquake gave warning to those in the dining room to move away from the windows just in case it became worse—and it did. There was barely time to move to the center of the dining room when the first strong shock hit the building, throwing almost everyone to the floor. For the next three minutes, due to the extreme swaying, it was impossible to do anything but remain on the floor. One could get an idea of the earthquake's strength only by the motion of an outside temporary scaffold elevator. The swaying motion was so great that from the floor, it seemed that the building was about to go over. The noise generated by shock waves hitting the building sounded like the pounding of a giant sledge hammer.

The dining room tables had been set for dinner, and after the earthquake the floor was covered with broken glasses, dishes, and silverware, but no debris from the building itself. Not even one of the dining room windows was broken during the earthquake, and fortunately, no one present was seriously injured. After the earthquake, looking down at the Fourth Avenue area of downtown Anchorage, it was difficult to visualize what that area had looked like before. For several blocks east of the hotel the north side of Fourth Avenue had dropped down due to slumping and sliding during the earthquake.

As soon as possible, the small group went down the fire escape stairway together to ground level. The city seemed so quiet. In addition to the visible destruction, Anchorage was without water, electric power, and natural gas. Fortunately, there were no fires caused by broken gas lines. At least it was still daylight and overall the weather was not bad, the temperature just below freezing.

Those who could retrieve their belongings from the hotel proceeded to the YMCA, where they served as volunteers, setting up dormitories for transients, and serving hot coffee and what food there was on hand. It was the beginning of a long night. The magnitude-8.4 main shock was followed, over the next 24 hours, by ten aftershocks that were over magnitude 6.0 on the Richter scale, and by many smaller ones." (Contributed by Dr. Kenneth G. Smith, Adjunct Professor, Dallas Baptist University)

Strong horizontal displacement during the 1995 Kobe, Japan earthquake toppled this elevated highway, killing 12 drivers.
© Reuter/Sankei Shimbun

Terrifying as the 1964 Alaskan earthquake was, it caused only 115 deaths and $540 million in property damage. With the 1989 Loma Prieta earthquake, despite the more than $5 billion in property damage (figure 4.1A), only sixty-three lives were lost. In the 1994 Northridge earthquake, even fewer lives were lost, but damage estimates are around $30 billion, and the death toll would have been higher had that earthquake not occurred in the early morning. The 1995 Kobe, Japan earthquake caused over 5,200 deaths, and damages estimated at $95–140 billion (figure 4.1B). History records other earthquake disasters claiming over a million lives (table 4.1). The casualties and the property damage result from a variety of causes. This chapter reviews the causes and effects of earthquakes and explores how their devastation can be minimized.

A

B

Figure 4.1 (*A*) Damage from the 1989 Loma Prieta earthquake in California: Collapse of I-880, Oakland, California. (*B*) Overturned section of Hanshin Expressway, National Highway 3, eastern Kobe. This expressway was built in the 1960s to then-existing seismic design standards. Elevated freeways are common space-saving devices in densely populated Japan. (*A*) Photograph by D. Keefer, from U.S. Geological Survey Open-File Report 89–687. (*B*) Photograph by Christopher Rojahn, Applied Technology Council.

Table 4.1 Selected Major Historic Earthquakes

Year	Location	Magnitude*	Deaths	Damages†
70 B.C.	China	(IX)	6,000	unknown
342 A.D.	Turkey		40,000	unknown
365	Crete: Knossos		50,000	unknown
1201	Egypt, Syria		1,100,000	over $25 million
1290	China	6.75	100,000	unknown
1456	Italy	(XI)	40,000+	over $25 million
1556	China	(XI)	830,000	over $25 million
1667	USSR: Shemakh	6.9	80,000	over $25 million
1693	Italy, Sicily		100,000	$5–25 million
1703	Japan	8.2	200,000	$5–25 million
1737	India: Calcutta		300,000	$1–5 million
1755	Portugal: Lisbon	(XI)	62,000	over $25 million
1780	Iran: Tabriz		100,000	unknown
1850	China	(X)	20,600	over $25 million
1854	Japan	8.4	3,200	over $25 million
1857	Italy: Campania	6.5	12,000	unknown
1868	Peru, Chile	8.5	52,000	$300 million
1883	Java		100,000	over $25 million
1906	United States: San Francisco	8.3	700	over $25 million
1906	Ecuador	8.9	1,000	unknown
1908	Italy: Calabria	7.5	58,000	over $25 million
1920	China: Kansu	8.5	200,000	over $25 million
1923	Japan: Tokyo	8.3	143,000	$2.8 billion
1933	Japan	8.9	3,000+	over $25 million
1939	Chile	8.3	28,000	$100 million
1939	Turkey	7.9	30,000	over $25 million
1960	Chile	8.5	5,700	$675 million
1964	Alaska	8.4	115	$540 million
1967	Venezuela	7.5	1,100	over $140 million
1970	N. Peru	7.8	66,800	$250 million
1971	United States: San Fernando	6.4	60	$500 million
1972	Nicaragua	6.2	5,000	$800 million
1976	Guatemala	7.5	23,000	$1.1 billion
1976	NE Italy	6.5	1,000	$8 billion
1976	China: Tangshan	8.0	250,000	over $25 million
1978	Iran	7.4	20,000	unknown
1983	Japan: Honshu	7.8	104	$416 million
1983	Turkey	6.9	2,700	over $25 million
1985	Chile	7.8	177	$1.8 billion
1985	SE Mexico	8.1	5,600+	$5–25 million
1988	USSR: Afghanistan	6.8	25,000	unknown
1989	United States: Loma Prieta, California	7.1	63	$5.6–7.1 billion
1990	Iran	6.4	40,000+	over $7 million
1992	United States: Landers, California	7.4	1	over $100 million
1994	United States: Northridge, California	6.8	57	$13–20 billion
1995	Japan: Kobe	7.2	5,200+	$95–140 billion

Primary source (except 1988 and 1989 entries): *Catalog of Significant Earthquakes 2000 B.C.–1979* and (supplement to 1985), World Data Center A, Report SE–27.
Note: This table is not intended to be a complete or systematic compilation but to give examples of major damaging earthquakes throughout history.
*Where actual magnitude is not available, maximum Mercalli intensity is sometimes given in parentheses (see table 4.3).
†Estimated in 1979 dollars, for pre-1980 entries. For these same entries, amounts "over $25 million" may be far over, but are simply not accurately known.

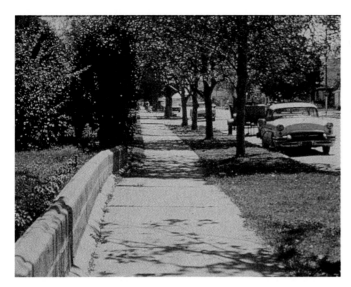

Figure 4.2 Wall and sidewalk offset by creep along the Hayward Fault, Hollister, California. Photograph courtesy of USGS Photo Library, Denver, CO.

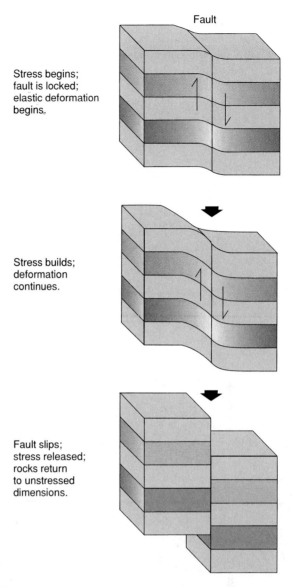

Fault

Stress begins; fault is locked; elastic deformation begins.

Stress builds; deformation continues.

Fault slips; stress released; rocks return to unstressed dimensions.

Figure 4.3 The phenomenon of elastic rebound: After the fault slips, the rocks spring back elastically to an undeformed condition.

Earthquakes—Basic Theory

Basic Terms

Major earthquakes dramatically demonstrate that the earth is a dynamic, changing system. Earthquakes, in general, represent a release of built-up stress in the lithosphere. They occur along **faults,** planar breaks in rock along which there is displacement of one side relative to the other. Sometimes, the stress produces new faults or breaks; sometimes, it causes slipping along old, existing faults. When movement along faults occurs gradually and relatively smoothly, it is called **creep** (figure 4.2). Creep—sometimes termed *aseismic slip,* meaning fault displacement without significant earthquake activity—can be inconvenient but rarely causes serious damage.

When friction between rocks on either side of a fault prevents the rocks from slipping easily, or when the rock under stress is not already fractured, some elastic deformation will occur before failure. When the stress at last exceeds the rupture strength of the rock (or the friction along a pre-existing fault), a sudden movement occurs to release the stress. This is an **earthquake,** or *seismic slip.* With the sudden displacement and associated stress release, the rocks snap back elastically to their previous dimensions; this behavior is called **elastic rebound** (figure 4.3; recall the discussion of elastic behavior from the previous chapter).

Faults come in all sizes, from microscopically small to thousands of kilometers long. Likewise, earthquakes come in all sizes, from tremors so small that even sensitive instruments can barely detect them, to massive shocks that can level cities. (Indeed, the "aseismic" movement of creep is

actually characterized by many microearthquakes, so small that they are typically not felt at all.) The amount of damage associated with an earthquake is partly a function of the amount of accumulated energy released as the earthquake occurs.

The point on a fault at which the first movement or break occurs during an earthquake is called the earthquake's **focus,** or hypocenter (figure 4.4). In the case of a large earthquake, for example, a section of fault many kilometers long may slip, but there is always a point at which the first movement occurred, and this point is the focus. The point on the earth's surface directly above the focus is called the **epicenter.** When news accounts tell where an earthquake occurred, they report the location of the epicenter.

Earthquake Locations

As shown in figure 4.5, the locations of major earthquake epicenters are concentrated in linear belts. These belts correspond to plate boundaries; recall figure 3.6. Not all earthquakes occur at plate boundaries, but most do, because these areas are where plates jostle, collide with, or slide past each

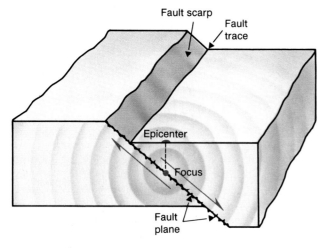

Figure 4.4 Simplified diagram of a fault, showing the focus, or hypocenter (point of first break along the fault), and the epicenter (point on surface directly above the focus). Seismic waves dissipate the energy released by the earthquake as they travel away from the fault zone. Fault scarp is cliff formed along the fault plane at ground surface; fault trace is the line along which the fault plane intersects the ground surface.

other, where relative plate movements may build up very large stresses, where major faults or breaks may already exist on which further movement may occur. (On the other hand, intraplate earthquakes certainly occur, and may be quite severe, for reasons explored later in the chapter.)

A map showing only deep-focus earthquakes (earthquakes with focal depths over 100 kilometers) would look somewhat different: the spreading ridges would have disappeared, while subduction zones would still show by their frequent deeper earthquakes. The explanation is that earthquakes occur in the lithosphere, where rocks are more rigid and brittle and are capable of rupturing or slipping suddenly. In the plastic asthenosphere, material flows, rather than ruptures, under stress. Therefore, the deep-focus earthquakes are concentrated in subduction zones, where brittle lithosphere is pushed deep into the mantle.

Seismic Waves and Earthquake Severity

Seismic Waves

When an earthquake occurs, it releases the stored-up energy in **seismic waves** that travel away from the focus. There are several types of seismic waves. *Body waves* (P waves and S waves) travel through the interior of the earth (box 4.1). **P waves** are compressional waves. As P waves travel through matter, the matter is alternately

Figure 4.5 World seismicity, 1979–1988. Size of symbol corresponds to magnitude of earthquake; color indicates depth of focus. This poster displays selected hypocenters for earthquakes that occurred from January 1979 through December 1988. Symbols are plotted at epicentral locations. The symbol colors represent three depth-of-focus classes as indicated in the table above. Symbol sizes are scaled into three magnitude ranges. Within each magnitude category, the earthquake data base has been thinned by plotting no more than one epicenter in a single cell of a 0.25 degree grid. There are over 23,000 seperate epicenter locations shown on this poster, representing 97,069 earthquakes.   Source: U.S. Geological Survey, National Earthquake Information Center.

Box 4.1 Earthquakes and Seismic Waves as Clues to the Earth's Interior

The behavior of seismic body waves in the earth has allowed geologists to deduce information about the earth's internal structure. For example, the existence of the asthenosphere was discovered in part from the fact that seismic-wave velocities drop when they pass through it. Generally, rocks become denser and seismic waves travel faster with increasing depth in the earth. At a depth of 50 to 100 kilometers (which geologists now recognize as the base of the lithosphere), S-wave velocities decrease. Shear waves cannot travel through a liquid; they are slowed down and propagate less efficiently in the asthenosphere because of its plastic behavior and tendency to deform under stress, rather than store and release energy elastically.

S waves cannot travel through the earth's liquid outer core at all. When a major earthquake occurs, P waves from the quake are detected all over the earth, but S waves do not reach the part of the world on the opposite side from the earthquake. Figure 1 shows representative P-wave and S-wave paths in the earth; the arrows indicate the directions in which the seismic waves are traveling. A "seismic shadow" is being cast by the liquid outer core. Both the core's size and its liquid state can be determined from the existence and size of the shadow zone. Even P waves have a (more limited) shadow zone, due to deflection by the outer core.

Figure 1 The seismic shadow zone phenomenon.

Nearer the surface, seismic data have helped to confirm plate tectonics, not only because earthquakes are concentrated in belts that we now recognize as plate boundaries. Detailed examination of focal depths in subduction zones reveals a pattern of earthquake foci dipping away from the trench into the mantle, in a feature known as a *Benioff zone* (figure 2). The earthquakes occurring in the Benioff zone reveal the presence of the slab of cold, brittle lithosphere being subducted.

—Continued

compressed and expanded. P waves travel through the earth, then, much as sound waves travel through air. A compressional sort of wave can be illustrated with a Slinky® toy by stretching the coil to a length of ten feet or so along a smooth surface, holding one end in place, pushing in the other end suddenly, and then holding that end still also. A pulse of compressed coil travels away from the end moved. **S waves** are shear waves, involving a side-to-side motion of molecules. Shear-type waves can also be demonstrated with a Slinky® by stretching the coil out along the ground as before, but this time twitching one end of the coil sideways (perpendicular to its length). As the wave moves along the length of the Slinky®, the loops of coil move sideways relative to each other, not closer together and farther apart as with the compressional wave. See also figure 4.6.

Seismic surface waves are somewhat similar to surface waves on water, discussed in chapter 7. That is, they cause

Box 4.1—Continued

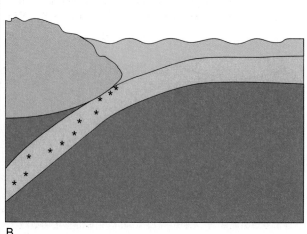

Figure 2 The Benioff zone of progressively deeper earthquakes dipping away from a trench along the subducted slab. (*A*) The data. (*B*) The interpretation in plate-tectonic terms.

Figure 3 Cutaway view of seismic-velocity variations in the earth's interior. Blue (colder?) regions have above-average velocities, red (warmer?) Have below-average. Velocity variations in the upper mantle are for S waves; total variation may be up to 10%. In deeper regions, variations are for P-wave velocities, which vary laterally by less than 2%. Reprinted with permission from A. M. Dziewonski, "Global Images of the Earth's Interior," *Science* 236:37, April 13, 1987. Copyright © 1987 American Association for the Advancement of Science.

Most recently, a new technique—*seismic tomography*—uses high-precision data on seismic-wave velocities in and propagation through the earth to reveal temperature variations and other features of the earth's interior (figure 3). One unexpected discovery has been that the core/mantle boundary appears not to be smooth, but to have considerable relief on it. Why this should be so is not yet understood. Seismic tomography is helping to clarify patterns of mantle convection also.

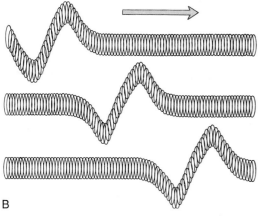

Figure 4.6 (Schematic) Seismic body waves as illustrated by a Slinky® toy. In both cases, the time sequence is from top of figure to bottom; arrows indicate the direction in which the wave is traveling. (*A*) P wave (compressional). (*B*) S wave (shear).

rocks and soil to be displaced in such a way that the ground surface ripples or undulates. Some cause vertical ground motions, like ripples on a pond; others cause horizontal shearing motions. Most structural damage from earthquakes is caused by the surface waves.

Locating the Epicenter

Earthquake epicenters can be located using seismic body waves. Both types of body waves cause ground motions that are detectable using a **seismograph.** P waves travel faster through rocks than do S waves. Therefore, at points some distance from the scene of an earthquake, the first P waves arrive somewhat before the first S waves; indeed, the two types of body waves are actually known as *primary* and *secondary* waves, reflecting these arrival-time differences.

The difference in arrival times of the first P and S waves is a function of distance to the earthquake's epicenter. The effect can be illustrated by considering a pedestrian and a bicyclist traveling the same route, starting at the same time. If the cyclist can travel faster, he or she will arrive at the destination first. The longer the route to be traveled, the greater the difference in time between the arrival of the cyclist and the later arrival of the pedestrian. Likewise, the farther the receiving seismograph is from the earthquake

epicenter, the greater the time lag between the first arrivals of P waves and S waves. This is illustrated graphically in figure 4.7A.

Once several recording stations have determined their distances from the epicenter in this way, the epicenter can be located on a map (figure 4.7C). If the epicenter is 200 kilometers from station *X,* it is located somewhere on a circle with a 200-kilometer radius around point *X.* If it is also found to be 150 kilometers from station *Y,* the epicenter must fall at either of the two points that are both 200 kilometers from *X* and 150 kilometers from *Y.* If the epicenter is also determined to be 100 kilometers from point *Z,* its position is uniquely identified.

In practice, complicating factors, such as inhomogeneities in the crust, make epicenter location somewhat more difficult. Results from more than three stations are usually required, and computers assist in reconciling all the data. The general principle, however, is as described.

Magnitude and Intensity

All of the seismic waves represent energy release and transmission; they cause the ground shaking that people associate with earthquakes. The amount of ground motion is related to the **magnitude** of the earthquake. Earthquake magnitude is most commonly reported in this country using the **Richter magnitude scale,** named after geophysicist Charles F. Richter, who developed it.

A magnitude number is assigned to an earthquake on the basis of the amount of ground displacement or shaking that it produces near the epicenter. The amount of ground motion is measured by a seismograph. The reading is adjusted for the distance of the instrument from the earthquake epicenter (because ground motion naturally decreases with increasing distance from the site of the earthquake) so that different measuring stations in different places will arrive at approximately the same magnitude value, a reflection of the ground displacement as it would have been measured close to the epicenter. The Richter scale is logarithmic, which means that an earthquake of magnitude 4 causes ten times as much ground movement as one of magnitude 3, one hundred times as much as one of magnitude 2, and so on (see figure 4.8). The amount of energy released rises even faster with increased magnitude, by a factor of about 30 for each unit of magnitude: An earthquake of magnitude 4 releases approximately thirty times as much energy as one of magnitude 3, and nine hundred times as much as one of magnitude 2. There is no upper limit to the Richter scale. The largest recorded earthquakes have had magnitudes of about 8.9. Earthquakes of this size have occurred in Japan and Ecuador. Although we only hear of the very severe, damaging earthquakes, there are, in fact, hundreds of thousands of earthquakes of all sizes each year. Table 4.2 summarizes the frequency and effects of earthquakes in different Richter magnitude ranges.

Measurement of earthquake size in terms of Richter magnitude is something of an oversimplification, for the associated rock deformation and ground displacement may be

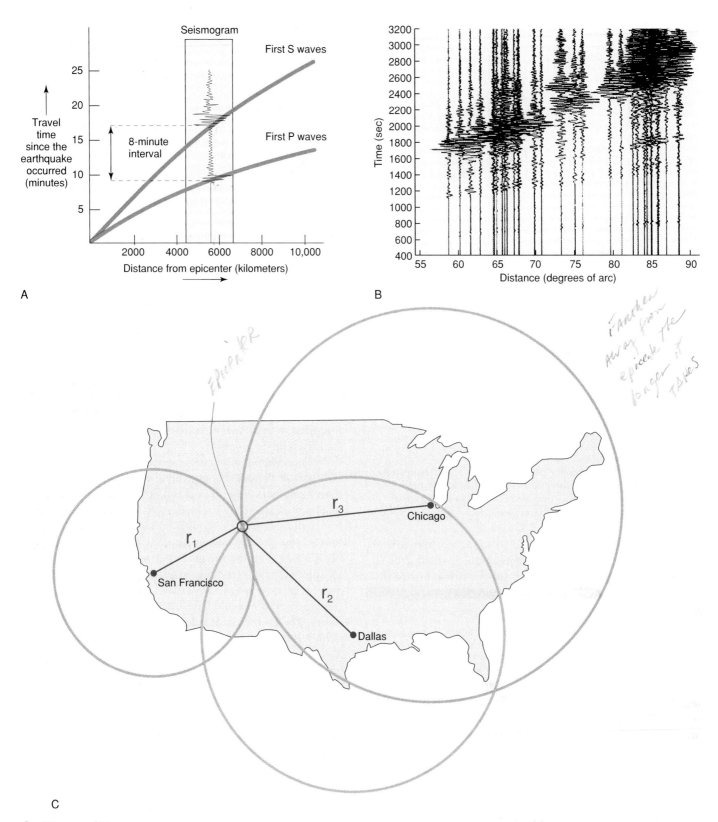

A

B

C

Figure 4.7 Use of seismic waves in locating earthquakes. (*A*) Difference in times of first arrivals of P waves and S waves is a function of the distance from the earthquake focus. This seismogram has a difference in arrival times of 8 minutes; the earthquake occurred about 5,500 km away. (*B*) The same effect is shown in these data from a magnitude-8 earthquake in the Kuril Islands on 3 December 1995. Data are from stations in the U.S. National Seismic Network. (*C*) Triangulation using data from several seismograph stations allows location of an earthquake's epicenter: If the earthquake epicenter is 1,350 km from San Francisco, it is on a circle of that radius around the city. If it is also found to be 1,900 km from Dallas, it must fall at one of only two points. If it is further determined to be 2,400 km from Chicago, its location is uniquely determined: northeastern Utah, near Salt Lake City. (B) Source: From U.S. Geological Survey National Earthquake Information Center.

examined in various ways. One can, for example, look at body-wave magnitude, surface-wave magnitude, and so on, by considering different aspects of the ground motion. However, Richter magnitude is most widely reported outside the scientific community.

An alternative way of describing the size of an earthquake is by the earthquake's **intensity.** Intensity is a measure of the earthquake's effects on humans and on surface features. It is not a unique characteristic of an earthquake, nor is it defined on a precise quantitative basis. The surface effects produced by an earthquake of a given magnitude vary considerably as a result of such factors as local geologic conditions, quality of construction, and distance from the epicenter. A single earthquake, then, can produce effects of many different intensities in different places (figure 4.9), although it will have only one magnitude assigned to it. Intensity is also a somewhat subjective measure in that it is based on direct observation by individuals. Different observers in the same spot may assign different intensity values to a single earthquake. Nevertheless, intensity is a more direct indication of the impact of a particular seismic event on humans in a given place than is magnitude. Several dozen intensity scales are in use worldwide. The most widely applied intensity scale in the United States is the Modified Mercalli Scale, a modern version of which is summarized in table 4.3.

Earthquake-Related Hazards and Their Reduction

Earthquakes can have a variety of harmful effects, some obvious, some more subtle. As noted earlier, earthquakes of the same magnitude occurring in two different places can cause very different amounts of damage, depending on such variables as the nature of the local geology, whether the area affected is near the coast, and whether

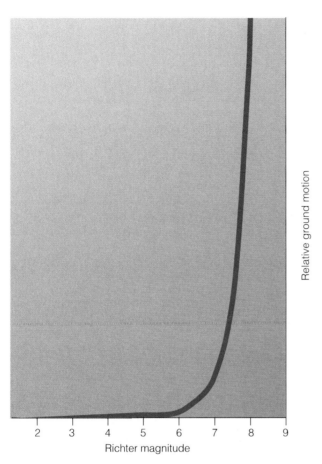

Figure 4.8 Ground motion increases by a factor of 10 for each unit increase in Richter magnitude; so, an earthquake of magnitude 7 causes ten times the displacement (shaking) of one of magnitude 6.

Table 4.2   Frequency of Earthquakes of Various Magnitudes

Description	Magnitude	Number per year	Approximate energy released (ergs)
great earthquake	over 8	1 to 2	over 5.8×10^{23}
major earthquake	7–7.9	18	$2–42 \times 10^{22}$
destructive earthquake	6–6.9	120	$8–150 \times 10^{20}$
damaging earthquake	5–5.9	800	$3–55 \times 10^{19}$
minor earthquake	4–4.9	6,200	$1–20 \times 10^{18}$
smallest usually felt	3–3.9	49,000	$4–72 \times 10^{16}$
detected but not felt	2–2.9	300,000	$1–26 \times 10^{15}$

Source: Frequency data from B. Gutenberg and C. F. Richter, *Seismicity of the Earth and Associated Phenomena.* Copyright © 1954, Princeton University Press, Princeton, N. J.

Note: For every unit increase in Richter magnitude, ground displacement increases by a factor of ten, while energy release increases by a factor of thirty. Therefore, most of the energy released by earthquakes each year is released not by the hundreds of thousands of small tremors, but by the handful of earthquakes of magnitude 7 or larger.

Figure 4.9 Zones of different Mercalli intensity for the Charleston, South Carolina, earthquake of 1886.   Source: Data from U.S. Geological Survey.

Table 4.3	Modified Mercalli Intensity Scale (abridged)	

Intensity	Description
I	Not felt.
II	Felt by persons at rest on upper floors.
III	Felt indoors—hanging objects swing. Vibration like passing of light trucks.
IV	Vibration like passing of heavy trucks. Standing automobiles rock. Windows, dishes, and doors rattle; wooden walls or frame may creak.
V	Felt outdoors. Sleepers wakened. Liquids disturbed, some spilled; small objects may be moved or upset; doors swing; shutters and pictures move.
VI	Felt by all; many frightened. People walk unsteadily; windows and dishes broken; objects knocked off shelves, pictures off walls. Furniture moved or overturned; weak plaster cracked. Small bells ring. Trees and bushes shaken.
VII	Difficult to stand. Furniture broken. Damage to weak materials, such as adobe; some cracking of ordinary masonry. Fall of plaster, loose bricks, and tile. Waves on ponds; water muddy; small slides along sand or gravel banks. Large bells ring.
VIII	Steering of automobiles affected. Damage to and partial collapse of ordinary masonry. Fall of chimneys, towers. Frame houses moved on foundations if not bolted down. Changes in flow of springs and wells.
IX	General panic. Frame structures shifted off foundations if not bolted down; frames cracked. Serious damage even to partially reinforced masonry. Underground pipes broken; reservoirs damaged. Conspicuous cracks in ground.
X	Most masonry and frame structures destroyed with their foundations. Serious damage to dams and dikes; large landslides. Rails bent slightly.
XI	Rails bent greatly. Underground pipelines out of service.
XII	Damage nearly total. Large rock masses shifted; objects thrown into the air.

the terrain is steep or flat. Some of the hazards can be minimized; some can be avoided only by staying clear of the dangerous area.

Ground Motion

Ground shaking and *movement along the fault* are obvious hazards. The offset between rocks on opposite sides of the fault can break power lines, pipelines, buildings, roads, bridges, and other structures that actually cross the fault. (In the 1906 San Francisco earthquake, maximum relative horizontal displacement across the San Andreas Fault was more than 6 meters.) Fault displacement aside, the shaking produced as accumulated energy is released through seismic waves causes damage to and sometimes complete failure of buildings, with the surface waves—especially shear surface waves—responsible for most of this damage. Shifts of even a few tens of centimeters can be devastating, especially to structures made of weak materials such as adobe (figure 4.10), and as shaking continues, damage may become progressively worse. Such effects, of course, are most severe on or very close to the fault, so the simplest strategy would be not to build near fault zones. However, many cities have already developed near major faults. Sometimes, cities are rebuilt many times in such places. The ancient city of Constantinople—now Istanbul—has been leveled by earthquakes repeatedly throughout history. Yet there it sits.

Short of moving whole towns, what else can be done in these cases? Power lines and pipelines can be built with extra slack where they cross a fault zone, or they can be designed with other features to allow some "give" as the fault slips and stretches them. Such considerations had to be taken into account when the Trans-Alaska Pipeline was built, for it crosses several known, major faults along its route; examples of special engineering features in the pipeline are seen in chapter 19.

Designing "earthquake-resistant" buildings is a greater challenge and is a relatively new idea that has developed

Figure 4.10 The nature of building materials used greatly influences how well they survive earthquakes. Hollow clay bricks are commonly used for building walls in Iran; in the 1990 earthquake, many such walls collapsed. Photograph by M. Mehrain, Dames and Moore, courtesy of National Geophysical Data Center.

mainly in the last few decades. Engineers have studied how well different types of building have withstood real earthquakes. Scientists can conduct laboratory experiments on scale models of skyscrapers and other buildings, subjecting them to small-scale shaking designed to simulate the kinds of ground movement to be expected during an earthquake. On the basis of their findings, special building codes for earthquake-prone regions can be developed.

This approach, however, has many complications. There are a limited number of records of just how the ground does move in a severe earthquake. To obtain the best such records, sensitive instruments must be in place near the fault zone beforehand, and those instruments must survive the earthquake. (However, sometimes alternative methods will provide useful information: after the 1995 Kobe earthquake, geoscientists studied videotapes from security cameras, using the recorded motions of structures and furnishings to deduce the ground motions during the earthquake.) Even with good records from an actual earthquake, there are no guarantees that the laboratory experiments accurately

simulate real earthquake conditions. Such uncertainties raise major concerns about the safety of dams and nuclear power plants near active faults. In an attempt to circumvent the limitations of scale modeling, the United States and Japan set up in 1979 a cooperative program to test earthquake-resistant building designs. The tests included experiments on a full-sized, seven-story, reinforced-concrete structure, in which ground shaking was simulated by using hydraulic jacks. Results were not entirely as anticipated on the basis of earlier modeling studies, which indicates that full-scale experiments and observations may be critical to designing optimum building codes in earthquake-prone areas.

A further complication is that the same building codes cannot be applied everywhere. Not all earthquakes produce the same patterns of ground motion. The 1994 Northridge earthquake underscored the dependence of reliable design on accurately anticipating ground motion. The epicenter was close to that of the 1971 San Fernando earthquake that destroyed many freeways (figure 4.11). They were rebuilt to be earthquake-resistant, assuming the predominantly horizontal displacement and ground motion that is characteristic of the San Andreas. The Northridge quake, however, had a strong vertical-displacement component as well—and many of the rebuilt freeways collapsed again. It is also important to consider not only how structures are built, but what they are built *on*. Buildings built on solid rock (bedrock) seem to suffer far less damage than those built on deep soil. In the 1906 San Francisco earthquake, buildings erected on filled land reclaimed from San Francisco Bay suffered up to four times more damage from ground shaking than those built on bedrock; the same general pattern was repeated in the 1989 Loma Prieta earthquake (see box 4.2). The extent of damage in Mexico City resulting from the 1985 Mexican earthquake (figure 4.12) was partly a consequence of the fact that the city is underlain by weak layers of volcanic ash and clay. Most smaller and older buildings lacked the deep foundations necessary to reach more stable sand layers at depth. Many of these buildings collapsed completely. Acapulco suffered far less damage, although it was much closer to the earthquake's epicenter, because it stands firmly on bedrock. These observations are directly linked to the fact that ground shaking is amplified in unconsolidated materials (figure 4.13).

The characteristics of the earthquakes in a particular region also must be taken into account. For example, severe earthquakes are generally followed by many **aftershocks,** earthquakes that are weaker than the principal tremor. The main shock usually causes the most damage, but when aftershocks are many and are nearly as strong as the main shock, they may also cause serious destruction.

The duration of an earthquake also affects how well a building survives it. In reinforced concrete, ground shaking leads to formation of hairline cracks, which then widen and develop further as long as the shaking continues. A concrete building that can withstand a one-minute main shock might collapse in an earthquake in which the main shock lasts three minutes. Many of the California building codes, used

A

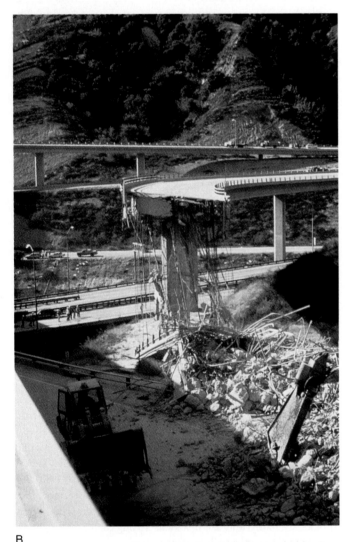

B

Figure 4.11 Freeway damage from the 1971 San Fernando earthquake (*A*) was repeated in the 1994 Northridge quake (*B*) because the design of the rebuilt overpasses did not account for the atypical ground motion experienced in 1994. This interchange of I-5 and C-14 was under construction in 1971; it was not subsequently retrofitted for enhanced earthquake resistance.   (*A*) Photograph by E. V. Leyendecker, (*B*) Photograph by M. Celebi, both courtesy of U.S. Geological Survey.

as models around the world, are designed for a 25-second main shock, but earthquake main shocks can last ten times that long.

A final complication is that even the best building codes are typically applied only to new construction. When a major city is located near a fault zone, thousands of vulnerable older buildings may already have been built in high-risk areas. The costs to redesign, rebuild, or even modify all of these buildings would be staggering. Most legislative bodies are reluctant to require such efforts; indeed, many do nothing even about municipal buildings built in fault zones.

Ground Failure

Landslides can be a serious secondary earthquake hazard in hilly areas. As will be seen in chapter 8, earthquakes are one of the major events that trigger slides on unstable slopes. The best solution is not to build in such areas. Even if a whole region is hilly, detailed engineering studies of rock and soil properties and slope stability may make it possible to avoid the most dangerous sites. Visible evidence of past landslides is another indication of especially dangerous areas.

Ground shaking may cause a further problem in areas where the ground is very wet—in filled land near the coast or in places with a high water table. This problem is **liquefaction.** When wet soil is shaken by an earthquake, the soil particles may be jarred apart, allowing water to seep in between them, greatly reducing the friction between soil particles that gives the soil strength, and causing the ground to become somewhat like quicksand. When this happens, buildings can just topple over or partially sink into the liquefied soil; the soil has no strength to support them. The effects of liquefaction were dramatically illustrated in Niigata, Japan, in 1964. One multistory apartment building tipped over to settle at an angle of 30 degrees to the ground while the structure remained intact! (See figure 4.14.) Liquefaction was likewise a major cause of damage from the Loma Prieta earthquake and the Kobe earthquake. In some areas prone to liquefaction, improved underground drainage systems may be installed to try to keep the soil drier, but little else can be done about this hazard, beyond avoiding the areas at risk. Not all areas with wet soils are subject to liquefaction; the nature of the soil or fill plays a large role in the extent of the danger.

Tsunamis and Coastal Flooding

Coastal areas, especially around the Pacific Ocean basin where so many large earthquakes occur, may also be vulnerable to **tsunamis.** These are seismic sea waves, sometimes improperly called "tidal waves," although they have nothing to do with tides. When an undersea or near-shore earthquake occurs, sudden movement of the sea floor may set up waves traveling away from that spot, like ripples in a pond caused by a dropped pebble. Contrary to modern movie fiction, tsunamis are not seen as huge breakers in the open ocean that topple ocean liners in one sweep. In the open sea, the tsunamis are only unusually broad swells on the water surface. As tsunamis approach land, however, they develop into large breaking

Figure 4.12 Building failure in Mexico City from the 1985 earthquake: pancake-style collapse of fifteen-story, reinforced-concrete structure.   Photograph by M. Celebi, USGS Photo Library, Denver, CO.

Figure 4.13 Ground shaking from the Loma Prieta earthquake, only moderate in bedrock (trace A), was amplified in younger river sediments (B) and especially in more recent, poorly consolidated mud (C).   Source: R. A. Page, et al., *Goals, Opportunities, and Priorities for the USGS Earthquake Hazards Reduction Program*, U.S. Geological Survey Circular 1079, p. 7.

Figure 4.14 Effects of soil liquefaction during an earthquake in Niigata, Japan, 1964. The buildings, which were designed to be earthquake-resistant, tipped over intact. Photograph courtesy of National Geophysical Data Center.

waves, as the undulating waters "touch bottom" near shore. These breakers can easily be over 15 meters high in the case of larger earthquakes. Several such breakers may crash over the coast in succession; between waves, the water may be pulled swiftly seaward, emptying a harbor or bay, and perhaps pulling unwary onlookers along. Tsunamis can travel very quickly—speeds of 1,000 kilometers/hour (600 miles/hour) are not uncommon—and tsunamis set off on one side of the Pacific may still cause noticeable effects on the other side of the ocean. A tsunami set off by a 1960 earthquake in Chile was still vigorous enough to cause 7-meter-high breakers when it reached Hawaii some fifteen hours later, and twenty-five hours after the earthquake, the tsunami was detected in Japan.

Given the speeds at which tsunamis travel, little can be done to warn those near the earthquake epicenter, but people living some distance away can be warned in time to evacuate, saving lives, if not property; see figure 4.15. In 1948, two years after a devastating tsunami hit Hawaii, the U.S. Coast and Geodetic Survey established a Tsunami Early Warning System, based in Hawaii. Whenever a major earthquake occurs in the Pacific region, sea-level data are collected from a series of monitoring stations around the Pacific. If a tsunami is detected, data on its source, speed, and estimated time of arrival can be relayed to areas in danger,

Sea-wave travel times to Honolulu, Hawaii

Figure 4.15 Tsunami travel times (in hours) to Hawaii from various points in the Pacific Basin. Source: After U.S. Geological Survey.

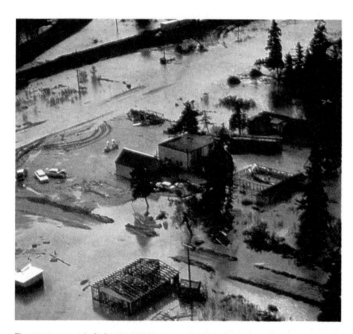

Figure 4.16 Flooding in Portage, Alaska, due to tectonic subsidence during 1964 earthquake.   Photograph by G. Plafker, USGS Photo Library, Denver, CO.

Figure 4.17 Major seismic gaps (shaded) in the Western Hemisphere. These presumably represent locked sections of major fault zones and are, therefore, areas where severe earthquakes are likely to occur in the future. (The Lesser Antilles gap in the Caribbean may be an exception, "lubricated" by thick sediments in the subduction zone.) Seismic gaps are identified by inspecting maps of earthquake epicenters and comparing with plate-boundary locations.

and people can be evacuated as necessary. The system does not, admittedly, always work perfectly. In the 1964 Alaskan earthquake, disrupted communications prevented warnings from reaching some areas quickly enough. Also, residents' responses to the warnings were variable: Some ignored the warnings, and some even went closer to shore to watch the waves, often with tragic consequences. Two tsunami warnings were issued in early 1986 following earthquakes in the Aleutians. They were largely disregarded. Fortunately, in that instance, the feared tsunamis did not materialize.

Even in the absence of tsunamis, there is the possibility of coastal flooding from sudden subsidence as plates shift during an earthquake. Areas that were formerly dry land may be permanently submerged and become uninhabitable (figure 4.16). Conversely, uplift of sea floor may make docks and other coastal structures useless.

Fire

A secondary hazard of earthquakes in cities is fire, which may be more devastating than ground movement. In the 1906 San Francisco earthquake, 70 percent of the damage was due to fire, not simple building failure. As it was, the flames were confined to a 10-square-kilometer area only by dynamiting rows of buildings around the burning section. Fires occur because fuel lines and tanks and power lines are broken, touching off flames and fueling them. At the same time, water lines also are broken, leaving no way to fight the fires effectively. Putting in numerous valves in all water and fuel pipeline systems helps to combat these problems because breaks in pipes can then be isolated before too much pressure or liquid is lost.

Earthquake Prediction and Forecasting

Millions of people already live near major fault zones. For that reason, prediction of major earthquakes could result in many saved lives. Some progress has been made in this direction, but as the complexity of seismic activity is more fully recognized, initial optimism has been tempered somewhat.

Seismic Gaps

Maps of the locations of earthquake epicenters along major faults show that there are stretches with little or no seismic activity, while small earthquakes continue along other sections of the same fault zone. Such quiescent, or dormant, sections of otherwise-active fault zones are called **seismic gaps.** They apparently represent "locked" sections of faults along which friction is preventing slip. The locations of principal seismic gaps in the Americas are shown in figure 4.17. These areas may be sites of future serious earthquakes. On either side of a locked section, stresses are being released by earthquakes. In the seismically quiet locked sections, friction is apparently sufficient to prevent the fault from slipping, so the stresses are simply building up. The fear, of course, is that the accumulated energy will become so great that, when that locked section of fault finally does slip again, a very large earthquake will result.

Recognition of these seismic gaps makes it possible to identify areas in which large earthquakes may be expected in the future. The 1989 Loma Prieta, California, earthquake occurred in what had been a seismic gap along the San Andreas Fault. A major seismic gap along the subduction zone bordering the western side of Central America is likely to be the site of the next major earthquake to cause serious damage to Mexico City.

Earthquake Precursors

In its early stages, earthquake prediction was based particularly on the study of earthquake **precursor phenomena,** things that happen or rock properties that change prior to an earthquake. Many different possibly useful precursor phenomena have been studied. For example, the ground surface may be uplifted and tilted prior to an earthquake. P-wave velocities in rocks near the fault (measured by using artificially caused shocks, such as those from small explosions) may drop, then rise before an earthquake; the same pattern has been seen in the ratio of P-wave to S-wave velocity. Electrical resistivity (the resistance of rocks to electric current flowing through them) has been observed to increase, then decrease before an earthquake. Changes have been observed in the levels of water in wells and/or in the content of radon (a radioactive gas that occurs naturally in rocks as a result of decay of trace amounts of uranium within them). The Chinese have documented a number of examples of anomalous behavior of animals prior to an earthquake.

The hope, with the study of precursor phenomena, is that one can identify patterns of precursory changes that can be used with confidence to issue earthquake predictions that are precise enough as to time to allow precautionary evacuations or other preparations. Unfortunately, the precursors have not proven to be very reliable. Not only does the length of time over which precursory changes are seen before an earthquake vary, but the pattern of precursors—which parameters show changes, and of what sorts—also varies; and, most problematic, some earthquakes seem quite unheralded by recognizable precursors at all. Loma Prieta was one of these; so was Northridge, and so was Kobe. After each of these earthquakes, geoscientists went back to scour their records for precursors they might have overlooked. In the case of Loma Prieta, four years after the earthquake, only a few small phenomena that might have been precursors had been identified in retrospect—anomalous behavior of a geyser a hundred miles from the epicenter, increased magnetic "noise" at some California monitoring sites weeks or months before the quake—but nothing was found for which a clear physical link with the earthquake could be made so as to inspire confidence that it was a useable predictor of future earthquakes.

In the near term, it seems that the more feasible approach is earthquake *forecasting,* identifying levels of earthquake probability in fault zones within relatively broad time windows, as described in the next column. This at least allows for long-term preparations, such as structural improvements. Ultimately, short-term, precise predictions that can save more lives are still likely to require recognition and understanding of precursory changes.

Figure 4.18 Periodicity of earthquakes assists in prediction/forecasting efforts. (*A*) Long-term records of major earthquakes on the San Andreas show both broad periodicity and some tendency toward clustering of 2–3 earthquakes in each active period. Prehistoric dates are based on carbon-14 dating of faulted peat deposits. (*B*) The Parkfield prediction was based on an average interval of 22 years between earthquakes of magnitude 5.5 or greater.   Source: (*A*) Data from U.S. Geological Survey Circular 1079.

The Earthquake Cycle and Forecasting

Studies of the dates of large historic and prehistoric earthquakes along major fault zones have suggested that they may be broadly periodic, occurring at more-or-less regular intervals (figure 4.18). This is interpreted in terms of an **earthquake cycle** that would occur along a (non-creeping) fault: a period of stress buildup, sudden fault rupture in a major earthquake, followed by a brief interval of aftershocks reflecting minor lithospheric adjustments; then another extended period of stress buildup (figure 4.19). The periodicity can be understood in terms of two considerations: First, assuming that the stress buildup is primarily associated with the slow, ponderous, inexorable movements of lithospheric plates, which at least over decades or centuries will move at roughly constant rates, one might reasonably expect an approximately constant rate of buildup of stress (or accumulated strain). Second, the rocks along a given

Figure 4.19 The "earthquake cycle" concept.   Source: After U.S. Geological Survey Circular 1079, fig. 6, p. 15.

fault zone will have particular physical properties, which allow them to accumulate a certain amount of energy before failure, or fault rupture, and that amount would be approximately constant from earthquake to earthquake. Those two factors together suggest that periodicity is reasonable. Once the pattern for a particular fault zone is established, one may use that pattern together with measurements of strain accumulation in rocks along fault zones to project the time-window during which the next major earthquake is to be expected along that fault zone, and to estimate the likelihood of the earthquake's occurrence in any particular time period. Also, if a given fault zone or segment can only store a certain amount of accumulated energy before failure, the maximum size of earthquake to be expected can be estimated.

This is, of course, an oversimplification of a very complex reality. For example, though we speak of "The San Andreas Fault," it is not a single simple, continuous slice through the lithosphere, nor are the rocks it divides either identical or homogeneous on either side. Both stress buildup and displacement are distributed across a number of faults and small blocks of lithosphere (figure 4.20). Earthquake forecasting is correspondingly more difficult. Nevertheless, the U.S. Geological Survey's Working Group on California Earthquake Probabilities published, in 1988, a set of forecasts for different segments of the San Andreas (figure 4.21). Note that the time-

window in question was the 30-year period 1988–2018. Such projections, longer-term analogues of the meteorologist's forecasting of probabilities of precipitation in coming days, are revised as small (and large) earthquakes occur and more data on slip in creeping sections are collected.

Current Status of Earthquake Prediction

Since 1976, the director of the U.S. Geological Survey has had the authority to issue warnings of impending earthquakes and other potentially hazardous geologic events (volcanic eruptions, landslides, and so forth). An Earthquake Prediction Panel reviews scientific evidence that might indicate an earthquake threat and makes recommendations to the director regarding the issuance of appropriate public statements. These statements could range in detail and immediacy from a general notice to residents in a fault zone of the existence and nature of earthquake hazards there, to a specific warning of the anticipated time, location, and severity of an imminent earthquake.

In early 1985, the panel made its first endorsement of an earthquake prediction. They agreed that an earthquake of Richter magnitude about 6 could be expected on the San Andreas Fault near Parkfield, California, in January 1988 ± 5 years. The prediction was based, in part, on cyclic patterns of seismicity in the area: On average, major earthquakes occur there every twenty-two years, and the last was in 1966 (recall figure 4.18B). Local residents expressed neither surprise nor concern at the prediction. Both the degree of confidence and the estimated probability were very high. Yet the critical time period came and went—and no Parkfield earthquake. One suggestion as to why not is that the 1983 Coalinga earthquake, a magnitude-6.5 event 30 kilometers northeast of Parkfield, may have released enough stress to delay the anticipated Parkfield event, perhaps for several years. Research is continuing.

In the People's Republic of China, some 10,000 scientists and technicians and 100,000 part-time amateur observers work on earthquake prediction. Their motivation is great: From 1966 to 1976, eleven earthquakes of magnitude 6.8 or greater occurred in China. In February 1975, after months of smaller earthquakes, radon anomalies, and increases in ground tilt followed by a rapid increase in both tilt and microearthquake frequency, the scientists predicted an imminent earthquake near Haicheng in northeastern China. The government ordered several million people out of their homes into the open. Nine and one-half hours later, a major earthquake struck, and many lives were saved because people were not crushed by collapsing buildings. Over the next two years, earthquake scientists successfully predicted four large earthquakes in the Hebei district. They also concluded that a major earthquake could be expected near Tangshan, about 150 kilometers southeast of Beijing. In the latter case, however, they could only say that the event was likely to occur sometime during the following two months. When the earthquake—magnitude over 8.0 with aftershocks up to magnitude 7.9—did occur, there was no immediate warning, no sudden change in precursor phenomena, and more than 250,000 people died.

Only four nations—Japan, the former Soviet Union, the People's Republic of China, and the United States—

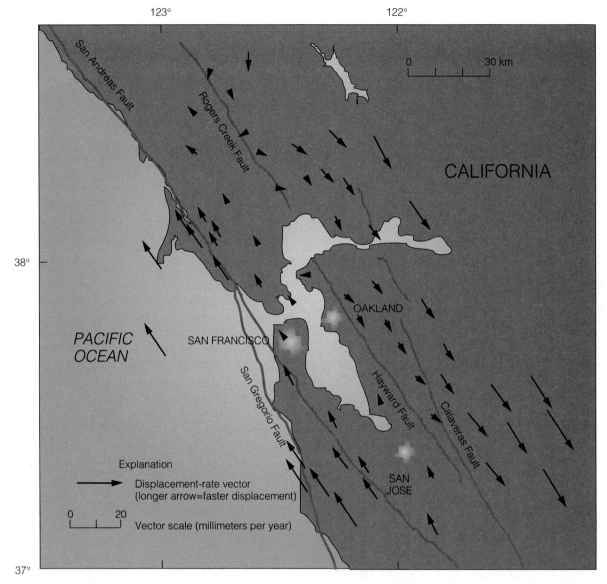

Figure 4.20 Displacement between the Pacific and North American plates is complexly distributed across a number of faults in the San Andreas system.   Source: After U.S. Geological Survey Circular 1079, fig. 12, p. 24.

have had government-sponsored earthquake prediction programs. Such programs typically involve intensified monitoring of active fault zones to expand the observational data base, coupled with laboratory experiments designed to increase understanding of precursor phenomena and the behavior of rocks under stress. Even with these active research programs, scientists can generally only forecast that a major earthquake will occur within a certain period of time—sometimes months or years—in the future. Sometimes they are right, sometimes wrong. Often, they are hampered by lack of personnel or equipment in critical areas. They cannot monitor every area at once. Also, as noted earlier, earthquake precursors are inconsistent and not yet completely understood. There may or may not be any distinctive warning signals shortly before a given earthquake.

In the United States, earthquake predictions have not been regarded as reliable or precise enough to justify such actions as large-scale evacuations. It may someday be possible to predict the timing and size of major earthquakes accurately enough that populated areas can be evacuated in an orderly way when serious earthquakes are imminent, thereby saving many lives. The property damage will still be inevitable as long as people persist in living and building in earthquake-prone areas. In any case, consistently reliable earthquake predictions are probably still more than a decade in the future.

Yet another twist to the problem of earthquake prediction and hazard assessment was provided by the Landers, California, earthquake in June of 1992. It occurred along a recently recognized fault zone east of the San Andreas. Aftershocks also occurred, as would be expected. What was unexpected was the significantly increased seismicity

PROBABILITIES OF LARGE EARTHQUAKES
ALONG SEGMENTS OF THE SAN ANDREAS FAULT
1988–2018

Figure 4.21 Earthquake forecast map issued in 1988 by the Working Group on California Earthquake Probabilities. Height of bar indicates estimated probability of a magnitude 6.5–7 earthquake in the period 1988–2018; color of bar indicates confidence in forecast. Source: After U.S. Geological Survey Circular 1079, fig. 15, p. 29.

observed hours to days after the Landers quake, at distances up to 1,250 kilometers from the site. Aftershocks are commonly interpreted as adjustments to crustal stress shifts resulting from the earthquake, either due to the passing seismic waves or due to the physical displacement of the fault during the earthquake. But the former would be expected to trigger aftershocks almost immediately after the main shock, and the latter mechanism is effective only rather close to the fault rupture of the main shock. In light of the observation that many of the distant aftershocks occurring days after the Landers quake were in areas of geothermal or recent volcanic activity, it has been suggested that these delayed aftershocks were triggered by changes in fluid (water or magma) pressure and distribution that developed following the passage of the seismic waves from Landers. A relationship between fluid pressure and faulting is already well documented, as described below; what is new is the recognition that fluid-triggered aftershocks can happen so far from the main-shock epicenter, which has implications for regional aftershock warnings. Research into the triggering mechanism will surely continue.

Earthquake Control?

Since earthquakes are ultimately caused by forces strong enough to move continents, human efforts to stop earthquakes from occurring would seem to be futile. However,

there has been speculation about the possibility of moderating of some of earthquakes' most severe effects.

Unlocking Locked Faults

If locked faults, or seismic gaps, represent areas in which energy is accumulating, perhaps enough to cause a major earthquake, then releasing that energy before too much has built up might prevent a major catastrophe. At one time, it was suggested that carefully placed nuclear explosions could be used to "unstick" locked faults. This idea is not being considered very seriously anymore, partly because of concern about radiation release, and partly because the sudden, poorly controlled jolt of a nuclear explosion in a locked section of fault with a lot of built-up stress could itself cause the feared large earthquake. What is needed is a method of loosening locked faults gently, in a more controlled way.

Fluid Injection

In the mid-1960s, the city of Denver began to experience small earthquakes. They were not particularly damaging, but they were puzzling. In time, geologist David Evans suggested a connection with an Army liquid-waste-disposal well at the nearby Rocky Mountain Arsenal. The Army denied any possible link, but a comparison of the timing of the earthquakes with the quantities of liquid pumped into the well at different times showed a very strong correlation

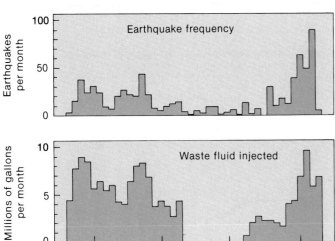

Figure 4.22 Correlation between waste disposal at the Rocky Mountain Arsenal (bottom diagram) and the frequency of earthquakes in the Denver area.   Adapted from David M. Evans, "Man-Made Earthquakes in Denver," *Geotimes*, 10 (9):11–18, May/June 1966. Used by permission.

(figure 4.22). The earthquake foci were also concentrated near the arsenal. It seemed likely that the liquid was "lubricating" old faults, allowing them to slip. In other words, the increased pore pressure resulting from the pumping-in of fluid decreased the shear strength, or resistance to shearing stress, along the fault zone.

Experiments conducted later at an abandoned oil field near Rangely, Colorado, suggested similar possibilities. When the fluid pressure of liquids pumped into the ground exceeded a certain level, old faults apparently were reactivated; existing stresses in the rocks were greater than the fluid-reduced shear strength, and earthquakes occurred as the faulted rocks slipped. Other observations and experiments around the world have supported the concept that fluids in fault zones may facilitate movement along a fault.

Such observations at one time prompted speculation that *fluid injection* might be used along locked sections of major faults to allow the release of built-up stress. Unfortunately, geologists are far from sure of the results to be expected from injecting fluid (probably water) along large, locked faults. There is certainly no guarantee that only small earthquakes would be produced. Indeed, in an area where a fault had been locked for a long time, injecting fluid along that fault could lead to the release of all the stress at once, in a major, damaging earthquake, just as might happen if a nuclear explosion jarred the fault loose. If a great deal of energy had accumulated, it is not even clear that all that energy *could* be released entirely through small earthquakes (recall table 4.2). The possible casualties and damage are a tremendous concern, as are the legal and political consequences of a serious, human-induced earthquake. The technique might be more safely used in major fault zones that have not been seismically quiet for long, where less stress has built up, to prevent that stress from continuing to build. The first at-

tempts are also likely to be made in areas of low population density to minimize the potential danger. Certainly, much more research would be needed before this method could be applied safely and reliably on a large scale.

Earthquake Awareness, Public Response

Concerns Related to Predictions

Assuming that routine earthquake prediction becomes reality, some planners have begun to look beyond the scientific questions to possible social or legal complications of such predictions. For example, individuals have speculated that, if a major earthquake were predicted several years ahead for a particular area, property values would plummet because no one would want to live there. A counterargument to that view, perhaps, is San Francisco: Despite the widespread acceptance of the idea that another large earthquake will strike, the city continues to thrive.

The logistics of evacuating a large urban area on short notice if a near-term earthquake prediction is made is another concern. Many urban areas have hopelessly snarled traffic every rush hour; what happens if everyone in the city wants to leave at once? Some critics say that a rapid evacuation might involve more casualties than the eventual earthquake. And what if a city is evacuated, people are inconvenienced or hurt, perhaps property is damaged by vandals and/or looters, and then the predicted earthquake never comes? Will the issuers of the warning be sued? Will future warnings be ignored? As earthquake prediction accuracy improves, someone will have to wrestle with difficult questions such as these. In principle, prediction seems desirable, but, in practice, it may be fraught with complications. In 1996, the chair of Japan's earthquake prediction panel quit in protest over the fact the panel was not allowed to issue earthquake "warnings"—only definite predictions, which are exceptionally difficult at present.

Public Response to Earthquake Hazards

A step toward making the geologic hazard warning system work would be increasing public awareness of earthquakes as a hazard. Several other nations are far ahead of the United States in this regard. In the People's Republic of China, vigorous public-education programs (and several recent major earthquakes) have made earthquakes a well-recognized hazard; strong government support for and the involvement of large numbers of citizens in earthquake prediction efforts have resulted in high visibility and widespread community support for those efforts. "Earthquake drills," which stress orderly response to an earthquake warning, are held in Japan on the anniversary of the 1923 Tokyo earthquake, in which more than 100,000 people died.

By contrast, in the United States, surveys have repeatedly shown that even people living in such high-risk areas as along the San Andreas Fault are often unaware of the earthquake hazard. Many of those who *are* aware believe that the risks are not very great and indicate that they would not take any special action even if a specific earthquake warning were issued. Past mixed responses to tsunami warnings underscore doubts about public reaction to earthquake predictions.

Many public officials have a corresponding lack of understanding or concern. Earthquake prediction aside, changes in land use, construction practices, and siting could substantially reduce the risk of property damage from earthquakes. Earthquake-prone areas also need comprehensive disaster-response plans. Within the United States, California is a leader in taking necessary actions in these areas. However, even there, most laws aimed at earthquake-hazard mitigation have been passed in spurts of activity following sizeable earthquakes, notably the Long Beach earthquake of 1933 and the San Fernando earthquake of 1971 (see chapter 18). Between crises, efforts drop off markedly. In many other parts of the United States, these efforts are nonexistent. Even understanding of basic safety precautions (table 4.4) is limited.

An important first step to improving public response is education to improve public understanding of earthquake phenomena and hazards. The National Earthquake Hazards Reduction Program includes such an educational component. The need for it was reinforced in 1990 by public reaction to an unofficial earthquake "prediction." In that year, a biologist-turned-climatological-consultant predicted a major earthquake for the New Madrid Fault zone (see below). The "prediction" was based on the fact that on 3 December 1990, the alignment of sun, moon, and earth would result in maximum tidal (gravitational) stress on the solid earth—which, presumably, would tend to trigger earthquakes on long-quiet faults. Geologists had previously found no correlation between tidal stresses and major earthquakes; the "prediction" was not endorsed by the scientific community. Unfortunately, geologists could not declare absolutely that an earthquake would not occur, only that the likelihood was extremely low. Meanwhile, in the New Madrid area, there was widespread panic: stores were stripped of emergency supplies; people scheduled vacations for the week in question (public-service personnel were forbidden to do so); schools were closed for that week. In the end, as some reporters described it, the biggest news was the traffic jam as all the curious left town after the earthquake didn't happen. The geologic community was very frustrated at indiscriminate media coverage that certainly contributed to public concern.

Even in the area of structural design for earthquake resistance, more educational efforts may be needed. Before the Northridge earthquake, many people living and working in buildings designed for earthquake resistance assumed that those structures would be undamaged in an earthquake; they were shocked to find that wasn't the case. But there are different levels of structural reliability for which

Table 4.4 Some Earthquake Safety Rules

During the shaking
1. Do not panic.
2. If you are indoors, stay there. Seek protection under a table or desk, or in a doorway. Stay away from glass. Do not use matches, candles, or any open flame; douse all fires. In a high-rise building, get under a desk; do not rush for exits. (Stairways may be jammed with people, and power for elevators may fail.)
3. If you are outside, move away from buildings and power lines, and stay in the open. Do not run through or near buildings.
4. If you are in a moving car, bring it to a stop as quickly as possible, but stay in it. The car's springs will absorb some of the shaking; the car will offer you protection.

After the shaking
1. Check, but do not turn on, utilities. If you smell gas, open windows, shut off the main valve, and leave the building. Report the leak to the utility, and do not reenter the building until it has been checked out. If water mains are damaged, shut off the main valve. If electrical wiring is shorting, close the switch at the main meter box. Avoid downed power lines.
2. Clean up spilled medicines or potentially harmful materials.
3. Turn on the radio or television, if possible, for emergency bulletins. Stay off the telephone except to report an emergency.
4. Obtain emergency water from water heaters, toilet tanks, melted ice cubes. Do not eat or drink anything from open containers near shattered glass. Check your freezer and, if the power is off, plan to use foods soon that will spoil quickly.
5. Stay out of severely damaged buildings that could collapse in aftershocks.
6. Do not go sightseeing; you will only hamper the efforts of emergency personnel and repair crews.

Source: National Oceanic and Atmospheric Administration.

one might design. For most residential and commercial buildings, "substantial life safety" may be the goal: the building may be extensively damaged, but casualties will be few. Designing for "damage control" means that in addition, the structure should be repairable after the earthquake. The highest (and most expensive) standard is design for "continued operation," meaning little or no structural damage at all; this would be important for hospitals, fire stations, and other buildings housing essential services, as well as for facilities that might present hazards if the buildings failed (nuclear power plants, manufacturing facilities using or making toxic chemicals). It is important that both designers and their clients understand the implications of choices about the desired level of structural integrity in a major earthquake.

A   B

C   D

Figure 4.23 Examples of property damage from the 1964 earthquake in Anchorage, Alaska. (*A*) Fourth Avenue landslide: Note the difference in elevation between the shops and street on the right and left sides. (*B*) Wreckage of Government Hill School, Anchorage. (*C*) Landslide in Turnagain Heights area, Anchorage. Close inspection reveals a number of houses amid the jumble of down-dropped blocks in the foreground. (*D*) Boats washed into the heart of Kodiak by a tsunami.   Photographs *A, B,* and *C* courtesy of USGS Photo Library, Denver, CO.   Photograph *D* courtesy of National Geophysical Data Center.

Further Thoughts on Modern and Future U.S. and Canadian Earthquakes

Areas of Widely Recognized Risk

Southern Alaska sits above a subduction zone. On 27 March 1964, it was the site of the largest earthquake of this century in North America (magnitude estimated at approximately 8.5). The main shock, which lasted three to four minutes, was felt more than 1,200 kilometers from the epicenter. About twelve thousand aftershocks, many over magnitude 6, occurred during the next two months, and more continued intermittently through the following year. Structures over an area greater than 100,000 square kilometers were damaged; ice on rivers and lakes cracked over an area of 250,000 square

kilometers. Many areas experienced permanent uplift or depression. The uplifts—up to 12 meters on land and over 15 meters in places on the sea floor—left some harbors high and dry and destroyed habitats of marine organisms. Downwarping of 2 meters or more flooded other coastal communities. Tsunamis destroyed four villages and seriously damaged many coastal cities; they were responsible for about 90 percent of the 115 deaths. The tsunamis traveled as far as Antarctica, and were responsible for sixteen deaths in Oregon and California. Landslides—both submarine and aboveground—were widespread and accounted for most of the remaining casualties. Levels of water in wells rose or fell abruptly as far away as South Africa. Damage just to roads, airports, and the Alaska Railroad from ground shaking, cracks, landslides, and flooding was estimated at about $100 million. Total damage was estimated at $300 million (see figure 4.23). The death toll

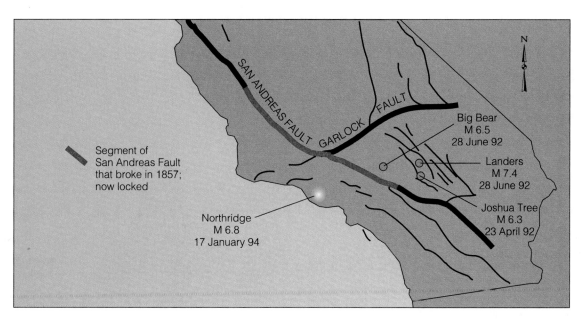

Figure 4.24 The 1994 Northridge earthquake is one of a series clustered around the southern locked segment of the San Andreas Fault.

might have been far higher had the earthquake not occurred in the early evening, on the Good Friday holiday, before peak tourist or fishing season. The area can certainly expect more earthquakes—including severe ones. The coastal regions there will also continue to be vulnerable to tsunamis.

The National Academy of Sciences panel that studied the effects of the 1964 Alaskan earthquake ultimately made a number of recommendations. Among them were the following: (1) structures should be strengthened to withstand earthquakes more effectively; (2) more data on seismic activity and its effects in the area should be gathered; (3) the tsunami warning system should be improved so that alerts can be distributed faster and more effectively; (4) earthquake prediction efforts should be intensified; (5) hazard maps showing the kinds of risks existing in different places should be prepared; (6) the public should be better informed of those risks. Those recommendations need not be considered unique to Alaska. They make basic good sense for any area with a history of or potential for significant earthquakes. Many of the same recommendations were made by a task force following the 1971 San Fernando earthquake. Increased federal funding during the late 1970s for U.S. Geological Survey and National Science Foundation programs has led to some progress, particularly in the areas of hazard identification and earthquake forecasting. However, recommendations such as those for structural improvements can take years to carry out, and, where they lack the force of law, may not be undertaken at all.

When the possibility of another large earthquake in San Francisco is raised, most geologists debate not "whether" but "when?" At present, that section of the San Andreas Fault has been locked for some time. The last major earthquake there was in 1906. At that time, movement occurred along at least 300 kilometers, and perhaps 450 kilometers, of the fault.

If another earthquake of the size of the 1906 event were to strike the San Francisco area during a busy workday, some estimates put the expected number of casualties at fifty thousand or more, including the number of people at risk of drowning in case of dam failures in the area, with tens of billions of dollars in property damage from ground shaking alone. Scientists working on earthquake prediction hope that they will at least be able to give fairly precise warning of the next major Californian tremor by the time it occurs. At present, the precursor phenomena, insofar as they are understood, do not indicate that a great earthquake will occur near San Francisco within the next few years. The more time seismologists have to improve their methods, the more accurate the eventual warning will be. However, the Loma Prieta earthquake (box 4.2) was none too reassuring in this regard.

Events of the past few years have suggested that there may, in fact, be a greater near-term risk along the southern San Andreas, near Los Angeles, than previously thought. In the spring and summer of 1992, a set of three significant earthquakes, including the Landers quake, occurred east of the southern San Andreas (figure 4.24). The 1994 Northridge earthquake then occurred on a previously unrecognized fault. That quake was unexpected in several respects—lack of precursors, occurrence on a "new" fault, and the type of fault displacement. It seems to have been a thrust fault, along which rocks on one side of the fault are thrust up and over those below, so there is vertical displacement and net shortening of the crust, not just horizontal slip. The southern San Andreas in this region, which last broke in 1857, has remained locked through all of this, so the accumulated strain has not been released by this cluster of earthquakes. Some believe that their net effect has been to increase the stress and the likelihood of failure along this section of the San Andreas. Moreover, in the late 1800s, there was a significant increase in numbers of moderate earthquakes around the northern

Box 4.2 Lessons from Loma Prieta

The Loma Prieta earthquake, named for the highest peak in the southern Santa Cruz mountains, near its epicenter, was a good example of an earthquake that was forecast but not predicted. It occurred along a section of the San Andreas Fault that had been locked since the 1906 San Francisco earthquake. Largely on the basis of the amount of fault displacement associated with major earthquakes on the San Andreas here, and the apparent accumulated potential slip since the fault has been locked, the U.S. Geological Survey's Working Group on California Earthquake Probabilities had, in 1988, concluded that this segment had a 30 percent probability of rupturing in the thirty-year period from 1988 to 2018. There had been increased seismicity over a period of sixteen months before the earthquake, which was duly noted. On the other hand, there were no foreshocks or recognized precursory changes immediately before the event, so no short-term prediction was made and no precautionary public response was possible.

Damage resulted from several causes. Ground shaking was one, and as might have been expected, the resultant damage was most severe for structures built on artificial fill (figure 1) and on San Francisco Bay

Figure 1   Some of the most severe damage in San Francisco from the 1989 Loma Prieta earthquake was in the Marina district, built on filled land.   Photograph by M. Bennett, from U.S. Geological Survey Open-File Report 89–687.

mud. The vulnerability of structures built on fill had been recognized since the 1906 earthquake, yet these same areas had nevertheless been rebuilt thereafter; in clearing the 1989 debris in the Marina district in San Francisco, crews unearthed rubble from the 1906 quake. The failure of parts

of I-880 (shown in figure 4.1) was attributed, in part at least, to its founding in Bay mud.

Liquefaction contributed to the structural damage. It was manifested locally by the formation of "sand boils" (figure 2) where liquefied sediment bubbled to the surface. The shaking

San Andreas, followed by the 1906 San Francisco earthquake. The 1992–94 activity may represent a similar pattern of increased activity building up to a major earthquake along the southern San Andreas. Time will tell.

Other Potential Problem Areas

Perhaps an even more dangerous situation is that in which people are not conditioned to regard their area as hazardous or earthquake-prone. The central United States is an example (see figure 4.25). Though most Americans think "California" when they hear "earthquake," the strongest and probably the most devastating series of earthquakes ever in the United

States occurred in the vicinity of New Madrid, Missouri, during 1811–12. The three strongest shocks are each estimated to have had Richter magnitudes over 8, they were spaced over almost two months, from 16 December 1811 to 7 February 1812. Between the December 16th earthquake and the following March 15, some 1,874 shocks in all were noted by one resident engineer even without the aid of sensitive modern seismographs. Some towns were leveled, others drowned by flooding rivers. The most vigorous tremors were felt from Quebec to New Orleans and along the east coast from New England to Georgia. Lakes were uplifted, "blown up," and emptied, while elsewhere the ground sank to make new lakes. Boats were flung

Figure 2 Sand boils in field near Watsonville, California. Photograph by J. Tinsley, from U.S. Geological Survey Open-File Report 89–687.

Figure 3 Landslide blocking Highway 17, Santa Cruz Mountains.   Photograph by T. Holzer, from U.S. Geological Survey Open-File Report 89–687.

jolted loose landslides (figure 3) over an area of about 14,000 square kilometers.

In some respects, the Loma Prieta earthquake was fortuitously timed. The freeway failures would undoubtedly have claimed far more lives had the roads been clogged with their usual rush-hour traffic. But on 17 October 1989, many baseball fans in the area had already gone home to watch the World Series game between San Francisco and Oakland that was to begin shortly. This earthquake was also far from the most violent possible along the fault; the duration of the main shock, for instance, was barely a third of the duration of the 1906 tremor.

The Loma Prieta earthquake served as a reminder of the need for earthquake awareness in planning, policy, and design in the Bay area. That need has not been lessened by the event. It did not release the accumulated strain on the peninsular segment of the fault (recall figure 4.21). Indeed, by releasing the adjacent segment to the south, the Loma Prieta quake may have increased the near-term likelihood of failure on the peninsular segment. The combined probability of a quake of magnitude 7 or greater either on the peninsular segment of the San Andreas or on the nearby Hayward Fault (quiescent since 1868) had been estimated at 50 percent over the next thirty years; that probability estimate has been increased to 67 percent since Loma Prieta.

out of rivers; an estimated 150,000 acres of timberland were destroyed. Aftershocks continued for more than *ten years.* Total damage and casualties have never been accurately estimated.

The potential damage from even one earthquake as severe as the worst of the New Madrid shocks is enormous: 12 million people now live in the area. Yet very few of those 12 million are probably fully aware of the nature and extent of the risk, and recent events may have caused them to take such risks less seriously. The big earthquakes there were a long time ago, beyond the memory of anyone now living. Minor earthquakes along the New Madrid Fault zone in the summer of 1987 only briefly brought the danger to public attention. More attention was focused on the area by the (unfulfilled) amateur earthquake prediction made in late 1990, mentioned earlier, a "prediction" based on no recognized precursors and emphatically not endorsed by official panels or geologic associations. The risk of further serious earthquakes is real. Beneath the midcontinent is a failed rift, a place where the continent began to rift apart, then stopped. Long and deep faults in the lithosphere there remain a zone of weakness in the continent and a probable site of more major tremors. It is just a matter of time. Unfortunately, the press gave far more coverage to the 1990 "earthquake prediction" and to public reaction to it than to the scientific basis for prediction efforts or to the underlying geology and the nature of the hazard. As a result, and because there then was no major New Madrid earthquake in 1990, there is a real danger that any future official scientific earthquake predictions for this area will not be taken seriously enough. The

effects may be far-reaching, too, when the next major earthquake comes. Seismic waves propagate much more efficiently in the solid, compact rocks of the continental interior than in, for example, the complexly faulted, jumbled rocks around the San Andreas Fault (figure 4.26). Therefore, when the next major earthquake occurs in the New Madrid Fault zone, the consequences will be widespread.

There are still other hazards. The 1886 Charleston, South Carolina, earthquake (figure 4.27) was a clear indication that severe earthquakes can occur in the east also. Much of the southeastern United States, however, is blanketed in sediment, which somewhat complicates assessment of past activity and future hazards. Meanwhile, in the Pacific Northwest, there is potential seismicity associated with subduction. Recent measurements among the mountains of Olympic National Park reveal crustal shortening and uplift that may indicate locking along the subduction zone to the west. An earthquake there could be as large as the 1964 Alaskan earthquake—also subduction-related—but with far more potential for damage and lost lives among the more than 10 million people in the Seattle/Portland area.

Canada is, on the whole, much less seismically active than the United States. Still, each year about 300 earthquakes (only four of magnitude greater than 4) occur in eastern Canada, where an apparent zone of crustal weakness exists in the continental interior (figure 4.28A), and numerous other earthquakes—again, mostly small ones—occur in western Canada, primarily associated with the subduction

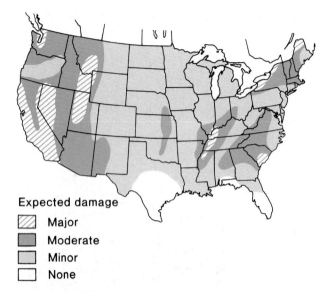

Expected damage

- ▨ Major
- ▦ Moderate
- ▧ Minor
- ▢ None

Figure 4.25 U.S. seismic-risk map. High-risk areas are those in which severely damaging earthquakes have occurred historically, not only those in which earthquakes are frequent. Recent research in fact suggests other areas of increased risk, such as the Pacific Northwest where subduction is occurring. Source: National Oceanic and Atmospheric Administration.

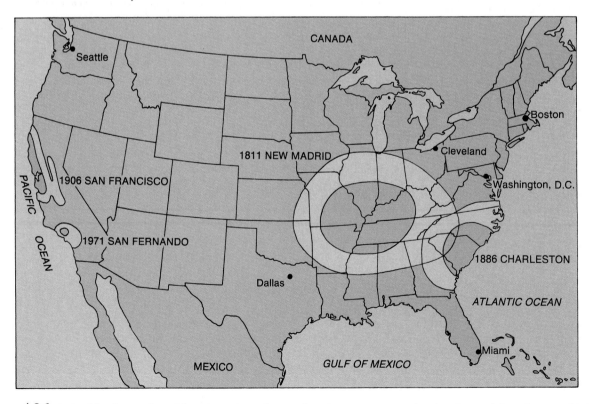

Figure 4.26 Part of the danger that midcontinent, intraplate earthquakes represent is related to the much broader areas likely to be affected, given the much more efficient transmission of seismic energy in the compact rocks that underlie these areas.   Source: After U.S. Geological Survey Professional Paper 1240-B.

zone there (figure 4.28B). Fortunately, severe Canadian earthquakes have been rare: a magnitude-7.2 event, complete with tsunami, off the Grand Banks of Newfoundland in 1929; an earthquake swarm with largest event magnitude 5.7 in central New Brunswick in 1982; a magnitude-7.3 quake

under Vancouver Island in 1946; and a total of seven quakes of magnitude 6 or greater in the area of southwestern Canada in the last century. Still, the potential for significant earthquakes clearly continues to exist, while public awareness, particularly in eastern Canada, is likely to be low.

Figure 4.27 Damage from 1886 Charleston, South Carolina, earthquake.   Photograph by J. K. Hillers, USGS Photo Library, Denver, CO.

A

B

Figure 4.28 Recent Canadian earthquake activity. (*A*) Several hundred earthquakes per year occur in eastern Canada; the seismograph network of the Geological Survey of Canada detects all of magnitude 3 or greater, 2.5 or above in populated areas. (*B*) Earthquakes of magnitude greater than 3.0 in western Canada since 1980.   Source: From the Geological Survey of Canada.

Overview

Summary

Earthquakes result from sudden slippage along fault zones in response to stress. Most earthquakes occur at plate boundaries and are related to plate-tectonic processes. Earthquake hazards include ground rupture and shaking, liquefaction, landslides, tsunamis, coastal flooding, and fires. While earthquakes cannot be stopped, their negative effects can be limited by: (1) designing structures in active fault zones to be more resistant to earthquake damage; (2) identifying and, wherever possible, avoiding development in areas at particular risk from earthquake-related hazards; (3) increasing public awareness of and preparedness for earthquakes in threatened areas; (4) learning enough about patterns of seismicity over time along fault zones, and about earthquake precursor phenomena, to make accurate and timely predictions of earthquakes and thereby save lives. Until precise predictions on a short timescale become feasible, longer-term earthquake forecasts may serve to alert the public to general levels of risk and permit structural improvements and planning for earthquake response.

Terms to Remember

aftershocks 77	magnitude 73
creep 69	precursor phenomena 82
earthquake 69	P waves 70
earthquake cycle 82	Richter magnitude scale 73
elastic rebound 69	seismic gap 81
epicenter 69	seismic waves 70
fault 69	seismograph 73
focus 69	S waves 71
intensity 75	tsunami 78
liquefaction 78	

Exercises

Questions for Review

1. Explain the concept of fault creep and its relationship to the occurrence of damaging earthquakes.
2. What is an earthquake's focus? Its epicenter? Why are deep-focus earthquakes concentrated in subduction zones?
3. Name the two kinds of seismic body waves, and explain how they differ.
4. On what is the assignment of an earthquake's magnitude based? Is magnitude the same as intensity? Explain.
5. List at least three kinds of earthquake-related hazards, and describe what, if anything, can be done to minimize the danger that each poses.
6. What is a seismic gap, and why is it a cause for concern?
7. Describe the concept of earthquake cycles, and their usefulness in forecasting earthquakes.
8. Explain the distinction between earthquake prediction and earthquake forecasting, and describe the kinds of precautions that can be taken on the basis of each.
9. Evaluate fluid injection as a possible means of minimizing the risks of large earthquakes.
10. What mechanism, if any, exists in the United States for warning people of earthquake hazards?
11. Areas identified as high risk on the seismic-risk map of the United States may not have had significant earthquake activity for a century or more. Why, then, are they mapped as high-risk regions?

For Further Thought

1. Research several earthquake predictions made within the last decade in the United States or elsewhere. How specific were the predictions? How accurate? What was the public response, if any?
2. Investigate the history of any modern earthquake activity in your own area or in any other region of interest. What geologic reasons are given for this activity? How probable is significant future activity, and how severe might it be? (The U.S. Geological Survey or state geological surveys might be good sources of such information.)
3. Consider the issue of earthquake prediction from a public-policy standpoint. How precise would you wish a prediction to be before making a public announcement? Ordering an evacuation? What kinds of logistic problems might you anticipate?
4. Check out current/recent seismicity (a) for the U.S. and Canada, or (b) for the world, over a period of several weeks. (See NetNotes for data source locations.) Tabulate by magnitude and compare with table 4.2. Are the recent data consistent? Comment.

Suggested Readings/References

Austin, M. 1981. No more wooden towers for San Francisco, 1906. In *Language of the earth*, edited by R. H. T. Rhodes and R. O. Stone. New York: Pergamon Press.

Bolt, B. A. 1987. *Earthquakes*. New York: W. H. Freeman.

Commission on the Alaskan Earthquake, National Research Council. 1970. *Human ecology* and *Summary and recommendations*. Vols. 1 and 2 of *The great Alaska earthquake of 1964*. Washington, D.C.: National Academy of Sciences.

Earthquakes. 1993. From *The Citizens' Guide to Geologic Hazards*, Amer. Inst. of Professional Geologists, pp. 38–53.

Gere, J. M., and Shah, H. C. 1984. *Terra non firma.* New York: W. H. Freeman.

Hanks, T. C. 1985. *National earthquake hazard reduction program — scientific status.* U.S. Geological Survey Bulletin 1659.

Hays, W. W., ed. 1981. *Facing geologic and hydrologic hazards.* U.S. Geological Survey Professional Paper 1240–B. (See especially section Z, "Hazards from Earthquakes," pp. 6–38.)

Hildenbrand, T. G., et al. 1995. Crustal geophysics gives insight into New Madrid seismic zone. *EOS* 76 (14 February): 65–69.

Johnston, A. C., and Kantner, L. R. 1990. Earthquakes in stable continental crust. *Scientific American* 262 (March): 68–75.

Johnston, M. J. S. 1993. *The Loma Prieta, California earthquake of October 17, 1989—preseismic observations.* U.S. Geological Survey Professional Paper 1550–C.

Kerr, R. A. 1993. Parkfield quakes skip a beat. *Science* 259 (19 February): 1120–22.

Mogi, K. 1985. *Earthquake prediction.* New York: Academic Press.

Page, R. A., Boore, D. M., Bucknam, R. C., and Thatcher, W. R. 1992. *Goals, Opportunities, and Priorities for the USGS Earthquake Hazards Reduction Program.* U.S. Geological Survey Circular 1079.

Penick, J. L., Jr. 1981. *The New Madrid earthquakes.* 2d ed. Columbia, Mo.: University of Missouri Press.

Petroski, H. 1994. Broken bridges. *American Scientist* 82 (July-August): 318–21.

Reid, T. R. 1995. Kobe wakes to a nightmare. *National Geographic* (July): 116–136.

Rial, J. A., Saltzman, N. G., and Ling, H. 1992. Earthquake-induced resonance in sedimentary basins. *American Scientist* 80 (Nov.-Dec.): 566–78.

Somerville, P. 1995. Kobe earthquake: An urban disaster. *EOS* 76 (7 February): 49–51.

Spence, W., Herrmann, R. B., Johnston, A. C., and Reagor, G. 1993. *Responses to Iben Browning's prediction of a 1990 New Madrid, Missouri, earthquake.* U.S. Geological Survey Circular 1083.

Takada, S., Hassani, N., and Abdd-Aziz, M. 1995. Quick report on the 1995 South Hyogo great earthquake. [Online] http://mech.mech.kobe-u.ac.jp/contents/HANSHIN.html

Tank, R., ed. 1983. *Environmental geology.* 3d ed. New York: Oxford University Press. (This collection of readings includes six chapters relating to earthquakes.)

U.S. Geological Survey Staff. 1990. The Loma Prieta, California earthquake: An anticipated event. *Science* 247 (January): 286–93.

Wesson, R. L., and Wallace, R. E. 1985. Predicting the next earthquake in California. *Scientific American* 252 (February): 35–43.

An excellent, comprehensive data source is the National Earthquake Information Center (NEIC) of the U.S. Geological Survey. The NEIC home page includes links to many other sources of data, maps, and information on earthquakes.

> http://wwwneic.cr.usgs.gov/

Data and maps of recent seismic activity (including a listing of all earthquakes over the previous two weeks or so) for the U.S. and Canada

> http://www.geophys.washington.edu/cnss.cat.html

Information on the Canadian National Earthquake Hazards Program—including a link to information on use of seismology in monitoring compliance with the nuclear test ban treaty

> http://www.seismo.emr.ca/
> http://www.seismo.nrcan.gc.ca/welcome.html

Seismological Society of America homepage

> http://www.bdt.com:80/ssa/
> and their resources for teachers
> http://www.geo.purdue.edu/seismology_resources.html

An extensive listing of maps and information on recent earthquakes, assembled from a variety of sources,

> http://www.comet.chv.va.us/earthquake/htm/quakes/htm

"Earthquakes" seem to be a popular topic on the Internet. A search on that term will turn up a number of sources of information, academic and otherwise!

Chapter 5

Volcanoes

On the morning of 18 May 1980, thirty-year-old David Johnston, a volcanologist with the U.S. Geological Survey, was watching the instruments monitoring Mount St. Helens in Washington State. He was one of many scientists keeping a wary eye on the volcano, expecting some kind of eruption, perhaps a violent one, but uncertain of its probable size. His observation post was more than 9 kilometers from the mountain's peak. Suddenly, he radioed to the control center, "Vancouver, Vancouver, this is it!"

It was indeed. Seconds later, the north side of the mountain blew out in a massive blast that would ultimately cost nearly $1 billion in damages and twenty-five lives, with another thirty-seven persons missing and presumed dead. The mountain's elevation was reduced by more than 400 meters. David Johnston was among the casualties. His death illustrates that even the experts still have much to learn about the ways of volcanoes. That fact has been reinforced by a dozen more volcanologists' deaths since 1990—in Japan, Indonesia, Colombia, and Ecuador—as scientists studying active volcanoes were unexpectedly overcome by sudden eruptive activity.

Until about fifteen years ago, many residents of the U.S. barely thought about volcanoes as a hazard. This was, perhaps, a little provincial, for other parts of the world certainly have suffered harm from volcanic phenomena in recent times. Historically, in the U.S. the Hawaiian volcanoes have erupted periodically, although the resulting damage has usually been minimal. Also, Mount Katmai in Alaska exploded in 1912, but it was far enough from heavily populated areas that, again, little damage resulted. It was not until a major explosive eruption occurred in the contiguous forty-eight states—when Mount St. Helens went up with a roar in a cloud of ash—that volcanoes suddenly and dramatically demanded Americans' attention. Many Americans were startled to discover that they lived in areas threatened by volcanoes.

Mount St. Helens is not the only source of volcanic danger in the United States. A dozen or more peaks in the western United States could potentially become active, and other areas also may present some threat. For example, the spectacular geysers, hot springs, and other thermal features of Yellowstone National Park reflect its fiery volcanic past and, some believe, foretell an equally lively future. Another, more immediate concern is the Long Valley–Mammoth Lakes area of California, where defor-

Ropy lava from recent eruptions builds up the island of Hawaii.

Figure 5.1 Lava flowing over a nearly new road in Royal Gardens subdivision, Kilauea, Hawaii, 1983.
Photograph by J. D. Griggs, USGS Photo Library, Denver, CO.

mation of the ground surface and increased seismic activity may be evidence of rising magma and impending eruption. And since 1983, Kilauea has been intermittently active, swallowing roads and homes as it builds Hawaii ever larger (figure 5.1).

Like earthquakes, volcanoes are associated with a variety of local hazards and, in some cases, global impacts. The dangers from any particular volcano, which are a function of what may be called its eruptive style, depend on the kind of magma it erupts and on its geologic and geographic settings. This chapter examines different kinds of volcanic phenomena and ways to minimize the dangers.

Magma Sources and Types

Temperatures at the earth's surface are too low to melt rock, but temperature, like pressure, increases with depth. Some of this internal heat is left over from the earth's formation; more is constantly generated by decay of naturally occurring radioactive elements in the earth. Magmas originate at depths where the temperature is high enough and the pressure is low enough that the rock can melt, wholly or partially. The majority of magmas originate in the upper mantle, at depths between 50 and 250 kilometers (30 and 150 miles), and are typically generated in one of three plate-tectonic settings: (1) at divergent plate boundaries, both ocean ridges and continental rift zones; (2) over subduction zones; and (3) at "hot spots," those isolated areas of volcanic activity that are not associated with current plate boundaries (figure 5.2).

The tectonic setting in which magma is produced (which determines the available source material) and the extent of melting control the composition of the resulting melt. The major compositional variables among magmas are the proportions of silica (SiO_2), iron, and magnesium. In general, the more silica-rich magmas are poorer in iron and

Figure 5.2 Relationships of volcanic activity to plate tectonics.

magnesium, and vice versa. A magma's composition, in turn, influences its physical properties, which determine the way it errupts—quietly or violently—and therefore the particular types of hazards it represents. A magma (or rock) relatively rich in iron and magnesium (and ferromagnesian silicates) is sometimes described as *mafic;* relatively silica-rich rocks tend to be rich in feldspar as well, and such rocks and magmas are therefore sometimes called *felsic.* Figure 5.3 relates the various common volcanic rock types and their plutonic equivalents.

Melting of asthenosphere at a spreading ridge or rift zone may be facilitated by the reduction in pressure from the overlying plates as they split and move apart. The mantle is extremely rich in ferromagnesians ("ultramafic"); the resultant melt is mafic too. This accounts for the **basalt** that forms new sea floor at spreading ridges. At a continental rift zone, the asthenosphere-derived melt is again mafic, but the rising hot basaltic magma may warm the more granitic continental crust enough to produce some silica-rich melt also— **rhyolite,** the volcanic equivalent of granite, or **andesite,** intermediate in composition between the mafic basalt and felsic rhyolite.

Magmatic activity in a subduction zone can be complex. Here, melting may occur not only because lithosphere is being plunged into the deeper, hotter asthenosphere, but also because water is released from the subducted sediments and slab of lithosphere; this water lowers the melting tem-

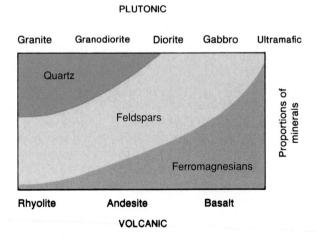

Figure 5.3 Common volcanic rock types and their plutonic equivalents: the rock names reflect varying proportions of silica, iron, and magnesium, and thus of common silicate minerals.

perature of the asthenosphere above the downgoing slab, causing melting, just as salt sprinkled on an icy sidewalk promotes melting of the ice. The composition of magma that reaches the surface depends on the nature and proportions of materials melted. Melting of the basaltic crust and mafic upper mantle of the subducted slab, or of the asthenosphere

Figure 5.4 Volcanoes of the world.   Source: After R. Decker and B. Decker, *Volcanoes*, 1981 W. H. Freeman and Company, New York, NY.

above it, would tend to produce basaltic or perhaps andesitic magma. Sediments derived from continental crust would be relatively silica-rich, so addition of melted sediments to the magma would tend to make it more felsic. Interaction with and incorporation of additional melt from an overriding continent as the magma rises toward the surface adds more silica, and ultimately, an andesitic-to-rhyolitic melt may result.

Finally, the magma typical of a hot-spot volcano in an ocean basin is similar to that of a seafloor spreading ridge: basaltic melt derived from partial melting of a very mafic/ultramafic mantle which will remain basaltic even if it should interact with the basaltic sea floor on its way to the surface.

The silica-poor mafic magmas that produce the basalt of the sea floor and of many volcanoes are characteristically "thin," or low in viscosity, so they flow very easily. (Note that in this context, "silica-poor" still means 45–50% SiO_2; all of the principal minerals in basalt, as in most volcanic rocks, are silicates.) The more felsic (silica-rich) andesitic and rhyolitic magmas (the latter up to 75% SiO_2) are more viscous, thicker and stiffer, and flow very sluggishly. Magmas also contain dissolved water and gases. As magma rises, pressure on it is reduced, and these gases begin to escape. From the more fluid, silica-poor magmas, gas escapes relatively easily. The viscous, silica-rich magmas tend to trap the gases, which may lead, eventually, to an explosive eruption.

Kinds and Locations of Volcanic Activity

Individual Volcanoes—Locations

Most people think of volcanoes as eruptions from the central vent of some sort of mountainlike object. Figure 5.4 is a map of such volcanoes presently or recently active in the world, and it shows a particularly close association between volcanoes and subduction zones. Proximity to an active subduction zone, then, is an indication of possible volcanic, as well as earthquake, hazards. The so-called Ring of Fire, the collection of volcanoes rimming the Pacific Ocean, is really a ring of subduction zones. A few volcanoes are associated with continental rift zones—for instance, Kilimanjaro and other volcanoes of the East African Rift system.

As with earthquakes, a few volcanic anomalies are not associated with plate boundaries. Hawaii is one: an isolated hot spot in the mantle apparently accounts for the Hawaiian Islands, as discussed in chapter 3. Other hot spots may lie under the Galápagos Islands, Iceland, and Yellowstone National Park, among other examples. What, in turn, accounts for the hot spots is not clear. Some have suggested that hot spots represent regions within the mantle that have slightly higher concentrations of heat-producing radioactive elements and, therefore, more melting and more magma. Others have suggested that the cause of hot spots lies in

Figure 5.5 Schematic diagram of a fissure eruption. (At a spreading ridge, the magma has generally solidified before it can spread very far sideways at the surface.)

A

B

Figure 5.6 Flood basalts. (*A*) Areal extent of Columbia River flood basalts. (*B*) Multiple lava flows, one atop another, can be seen in this photograph taken near Coulee City, Washington.   (*B*) © B. F. Molnia/TERRA PHOTOGRAPHICS/BPS.

similar concentrations of radioactive elements in the outer core, and that plumes of warm material rise from the core/mantle boundary to create them. Whatever the cause of hot spots, they are long-lived features; recall from chapter 3 that the path of the Pacific plate over the now-Hawaiian hot spot can be traced for some 70 million years.

Seafloor Spreading Ridges, Fissure Eruptions

Given the definition of volcanic rock—igneous rock formed at or near the earth's surface—most volcanic rock originates at the seafloor spreading ridges, where magma fills cracks in the lithosphere and crystallizes close to the surface. Spreading ridges spread at the rate of only a few centimeters per year, but there are some 50,000 kilometers (about 30,000 miles) of these ridges presently active in the world. All in all, that adds up to an immense volume of volcanic rock. However, most of this activity is out of sight under the oceans, where it is largely unnoticed and presents no dangers to people.

The outpouring of magma at spreading ridges is an example of **fissure eruption,** the eruption of magma out of a crack in the lithosphere, rather than from a single pipe or vent (see figure 5.5). (The structure of any fissure that transects the whole lithosphere, however, is more complex than shown in that figure, with multiple cracks and magma pathways.) There are also examples of fissure eruptions on the continents. One example in the United States is the Columbia Plateau, an area of over 150,000 square kilometers (60,000 square miles) in Washington, Oregon, and Idaho, covered by layer upon layer of basalt, piled up nearly a mile deep in places (figure 5.6). This area may represent the ancient beginning of a continental rift that ultimately stopped spreading apart. It is no longer active but serves as a reminder of how large a volume of magma can come welling up from the asthenosphere when zones of weakness in the lithosphere provide suitable openings. Even larger examples of such "flood basalts" on continents, covering more than a million square kilometers, are found in India and in Brazil.

The volcanoes of figure 5.4, eruptions from pipes or vents rather than from fissures, can be categorized by the kinds of structures they build. The structure, in turn, reflects the kind of volcanic material erupted and, often, the tectonic setting. It also provides some indication of what to expect in the event of future eruptions.

Shield Volcanoes

As noted, the mafic basaltic lavas are comparatively fluid, so they flow very freely and far when erupted. The kind of volcano they build, consequently, is very flat and low in relation to its diameter, and large in areal extent. This low, shieldlike shape has led to the term **shield volcano** for such a structure (figure 5.7). Though the individual lava flows may be thin—a few meters or less in thickness—the buildup of hundreds or thousands of flows through time can produce quite large objects. The Hawaiian Islands are all shield volcanoes. Mauna Loa, the largest peak on the still-active island of Hawaii, rises 3.9 kilometers (about 2.5 miles) above sea level. If measured properly from its true base, the sea floor, it is much more impressive: about 10 kilometers (6 miles)

Volcanoes   Chapter 5

B

C

Figure 5.7 Shield volcanoes and their characteristics. (*A*) Schematic diagram of a shield volcano in cross section. (*B*) Very thin lava flows, like these on Kilauea in Hawaii, are characteristic of shield volcanoes. (*C*) Fluidity of Hawaiian lavas is evident even after they have solidified.   Photograph (*B*) courtesy of USGS Photo Library, Denver, CO.

high and 100 kilometers in diameter at its base. With such broad, gently sloping shapes, the islands do not necessarily even look much like volcanoes from sea level, but seen from above, their character is clear (figure 5.8).

Volcanic Domes

Rhyolitic and andesitic lavas tend to be more viscous and flow less readily. They ooze out at the surface like thick toothpaste from a tube, piling up close to the volcanic vent, rather than spreading freely. The resulting structure is a more compact and steep-sided **volcanic dome.** Mount St. Helens is characterized by this kind of stiff, viscous lava, and a volcanic dome has formed in the crater left by its 1980 explosion (figure 5.9), although the volcano as a whole is a *composite volcano* (see p. 103). Novarupta, near Mount Katmai in Alaska, is another example of a volcanic dome. Such thick, slowly flowing lavas also seem to solidify and stop up the vent from which they erupt before much material has emerged. Volcanic domes, then, tend to be relatively small in areal extent compared to shield volcanoes, individually perhaps only a few tens or hundreds of meters high, although they can contribute to building quite high peaks.

Cinder Cones

When magma wells up toward the surface, the pressure on it is reduced, and dissolved gases try to bubble out of it and escape. The effect is much like popping the cap off a soda bottle: The soda contains carbon dioxide gas under pressure, and when the pressure is released by removing the cap, the gas comes bubbling out.

Sometimes, the built-up gas pressure in a rising magma is released more suddenly and forcefully by an explosion that flings bits of magma and rock out of the volcano. The magma may freeze into solid pieces before falling to earth. The bits of violently erupted volcanic material are described collectively as **pyroclastics,** from the Greek words for fire (*pyros*) and broken (*klastos*) (figure 5.10). The most energetic pyroclastic eruptions are more typical of volcanoes with thicker, more viscous silicic lavas because the thicker lavas tend to trap more gases. Gas usually escapes more readily and quietly from the thinner basaltic lavas, though even basaltic volcanoes may sometimes put out quantities of ash and small fragments.

The fragments of pyroclastic material can vary considerably in size from very fine, flourlike dust, through coarser,

101

A

B

Figure 5.8 Mauna Loa, an example of a shield volcano.
(*A*) View from low altitudes. Note the gently sloping shape.
The summit caldera has been enlarged by collapse. The peak of
Mauna Kea rises at rear of photograph. (*B*) Bird's-eye view of
Hawaii, taken by Landsat satellite, shows its volcanic character
more clearly. The large peak with abundant relatively fresh,
dark lava flows surrounding it is Mauna Loa; the smaller one,
above it, is Mauna Kea.   Photographs courtesy of USGS Photo
Library, Denver, CO.

A

B

Figure 5.9 Volcanic dome formation. (*A*) Schematic of
volcanic dome formation. (*B*) Dome being built in summit crater
of Mount St. Helens.   Photograph (*B*) courtesy of USGS Photo
Library, Denver, CO.

Figure 5.10 Rock (volcanic breccia) formed from welded
hot pyroclastic material from explosive volcanic eruption.

A

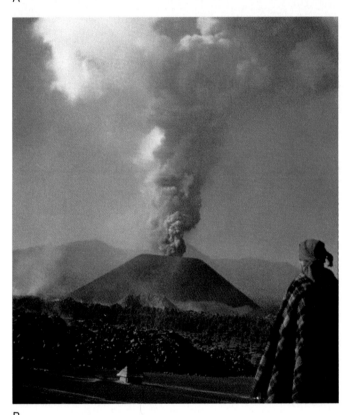

B

Figure 5.11 Paricutín (Mexico), a classic cinder cone. (*A*) Night view shows formation by accumulation of pyroclastics flung out of the vent. (*B*) Shape of the structure revealed by day is typical symmetric form of cinder cones. (*B*) Photograph by K. Segerstrom, both courtesy of USGS Photo Library, Denver, CO.

gritty volcanic ash and cinders ranging up to golf-ball-size pieces, to large volcanic blocks that may be the size of a house. Block-sized blobs of liquid lava may also be thrown from a volcano; these are volcanic "bombs." When the pyroclastics fall close to the vent from which they are thrown, they may pile up into a very symmetric cone-shaped heap known as a **cinder cone** (figure 5.11).

A

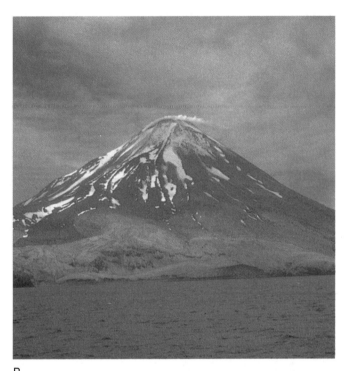

B

Figure 5.12 (*A*) Schematic cross section of a stratovolcano (composite volcano), formed of alternating layers of lava and pyroclastics. (*B*) A composite volcano in the Aleutian islands. Photograph (*B*) by R. E. Wilcox, USGS Photo Library, Denver, CO.

Composite Volcanoes

Many volcanoes erupt somewhat different materials at different times. They may emit some pyroclastics, then some lava, then more pyroclastics, and so on. Volcanoes built up in this layer-cake fashion are called **stratovolcanoes,** or, alternatively, **composite volcanoes,** because they are built up of more than one kind of material (figure 5.12). Most of the potentially dangerous volcanoes of the western United States, in the Cascade Range, are of this type; Mount St. Helens is just one example. They tend to have fairly stiff, gas-charged andesitic lavas that sometimes flow and sometimes trap enough gases to erupt explosively with a rain of pyroclastic material. Also, the very large volume of andesitic magma that can be formed over a subduction zone can make such volcanoes a particular hazard. We will look at these volcanoes in more detail later.

Figure 5.13 Formation of "lava trees" near Kilauea illustrates the effect of quenching lava. As hot lava hits cooler trees—and moisture in trees vaporizes, absorbing more heat—lava is quenched and hardened. Main mass of fluid lava flows on, leaving the lava trees. Photograph by J. D. Griggs, USGS Photo Library, Denver, CO.

Hazards Related to Volcanoes

Lava

Until the eruption of Mount St. Helens, most people in the United States, if they had thought about volcanoes at all, would have regarded lava as the principal hazard. Actually, lava generally is not life-threatening. Most lava flows advance at speeds of only a few kilometers an hour or less, so one can evade the advancing lava readily even on foot. The lava will, of course, destroy or bury any property over which it flows. Lava temperatures are typically over 850° C (over 1550° F) and basaltic lavas are typically over 1,100° C (2,000° F). Combustible materials—houses and forests, for example— burn at such temperatures. Other property may simply be engulfed in lava, which then solidifies into solid rock.

Lavas, like all liquids, flow downhill, so one way to protect property is simply not to build close to the slopes of the volcano. Throughout history, however, people *have* built on or near volcanoes, for many reasons. They may simply not expect the volcano to erupt again (a common mistake even today—consider, for example, Mount St. Helens). Also, soil formed from the weathering of volcanic rock may form only slowly, but it is often very fertile. The Romans cultivated the slopes of Vesuvius and other volcanoes for that reason. Sometimes, too, a volcano is the only land available, as in the Hawaiian Islands or Iceland.

Iceland sits astride the Mid-Atlantic Ridge and is wholly volcanic in origin. The fact that Iceland rises above the water surface may indicate that it is on a hot spot as well as a ridge. Certainly it is a particularly active volcanic area. Off the southwestern coast of Iceland are several smaller volcanic islands, which are also part of that country. One of these tiny islands, Heimaey, accounts for the processing of about 20 percent of Iceland's major export, fish.

Heimaey's economic importance stems from an excellent harbor. In 1973, the island was split by rifting, followed by several months of eruptions. The residents were evacuated within hours of the start of the event. In the ensuing months, homes, businesses, and farms were set afire by falling hot pyroclastics or buried under pyroclastic material or lava. Many buildings were saved only by repeated shoveling of heavy loads of hot pyroclastic fragments from their roofs. When the harbor itself and, therefore, the island's livelihood were threatened by encroaching lava, officials took the unusual step of deciding to fight back.

The key to the harbor's defense was the fact that the colder lava becomes, the thicker, more viscous, and slower-flowing it gets. When cooled enough, it may solidify altogether (figure 5.13). When a large mass of partly solidified lava has accumulated at the advancing edge of a flow, it begins to act as a natural dam to stop or slow the progress of the flow. How could one cool a lava flow more quickly? The answer surrounded the island—water. Using pumps on boats and barges in the harbor, the people of Heimaey directed streams of water at the lava's edge. They also ran metal pipes (prevented from melting, in part, by the cool water running through them) across the solid crust that had started to form on the flow, and positioned them to bring water to places that could not be reached directly from the harbor. And it worked: part of the harbor was filled in, but much of it was saved, along with the fishing industry on Heimaey (see figure 5.14). In fact, the remaining harbor, partially closed off and thus sheltered by the newly formed rock, was in some respects improved by the increased protection.

This bold scheme succeeded, in part, because the needed cooling water was abundantly available and, in part, because the lava was moving slowly enough already that there was time to undertake flow-quenching operations before the harbor was filled. Also, the economic importance of the harbor justified the expense of the effort. Similar efforts in other areas have not always been equally successful.

Where it is not practical to arrest the lava flow altogether, it may be possible to divert it from an area in which a great deal of damage may be done to an area where less valuable property is at risk. Sometimes, the motion of a lava flow is slowed or halted temporarily during an eruption because the volcano's output has lessened or because the flow has encountered a natural or artificial barrier. The magma contained within the solid crust of the flow remains molten for days, weeks, or sometimes months. If a hole is then punched in this crust by explosives, the remaining fluid magma inside can flow out and away. Careful placement of the explosives can divert the flow in a chosen direction.

A procedure of this kind was tried in Italy, where, in early 1983, Mount Etna began another in an intermittent series of eruptions. By punching a hole in a natural dam of old volcanic rock, officials hoped to divert the latest lavas to a broad, shallow, uninhabited area. This would have served the dual purpose of directing the lava away from populated areas and providing it with a wide area in which to spread out, which would ideally cause the lava to cool more rapidly, and thus to slow down. Unfortunately, the effort was only briefly successful. Part of the flow was diverted, but within four days, the lava had abandoned the planned alternate channel and resumed its original flow path. Later, new flows threatened further destruction of inhabited areas.

Lava flows may be hazardous to property, but in one sense, they are at least predictable: Like other fluids, they flow downhill. Their possible flow paths can be anticipated, and once they have flowed into a relatively flat area, they tend to stop. Other kinds of volcanic hazards can be trickier to deal with, and they affect much broader areas.

A

B

Figure 5.14 Impact of lava flows on Heimaey, Iceland. (*A*) Map showing extent of lava filling the harbor of Heimaey after 1973 eruption. (*B*) Lava-flow control efforts on Heimaey. Source: (*A*) Data from R. Decker and B. Decker, *Volcanoes*, Copyright © 1981 by W. H. Freeman and Company. Photograph (*B*) courtesy of USGS Photo Library, Denver, CO.

Pyroclastics

Pyroclastics—fragments of hot rock and spattering lava—are often more dangerous than lava flows. Pyroclastics may erupt more suddenly, explosively, and spread faster and farther. The largest blocks and volcanic bombs present an obvious danger because of their size and weight. For the same reasons, however, they usually fall quite close to the volcanic vent, so they affect a relatively small area.

The sheer volume of the finer ash and dust particles can make them as severe a problem, and they can be carried

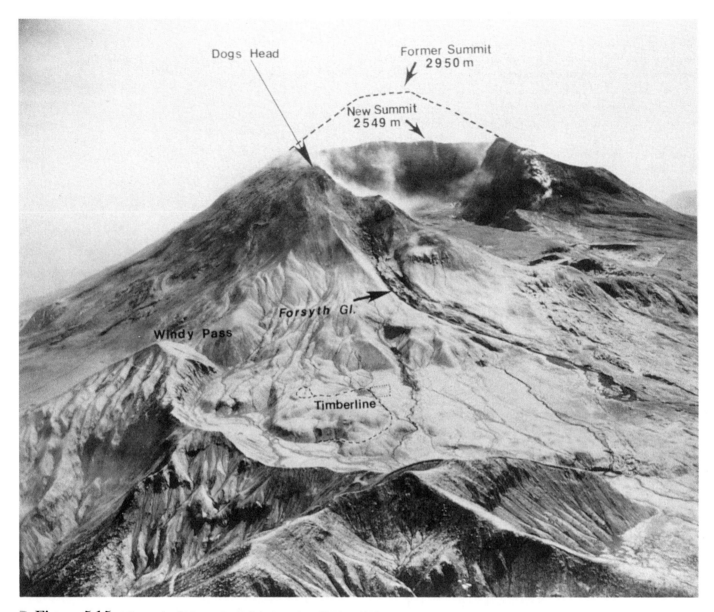

Figure 5.15 Aftermath of Mount St. Helens eruption, 18 May 1980.   Photograph by M. M. Brugman, USGS Photo Library, Denver, CO.

over a much larger area. Also, ashfalls are not confined to valleys and low places. Instead, like snow, they can blanket the countryside. The 18 May 1980 eruption of Mount St. Helens (see aftermath, figure 5.15) was by no means the largest such eruption ever recorded, but the ash from it blackened the midday skies more than 150 kilometers away, and measurable ashfall was detected halfway across the United States. Even in areas where only a few millimeters of ash fell, transportation ground to a halt as drivers skidded on slippery roads and engines choked on airborne dust. Homes, farmland, timberland, cars, and other property were buried under the hot ash. Volcanic ash is also a health hazard, both uncomfortable and dangerous to breathe. The cleanup effort required to clear the debris strewn about by Mount St. Helens was enormous. It has been estimated that about 1 cubic kilometer of material—some fresh pyroclastics, some older bits of the volcano—were blown out in the major 1980 eruption of Mount St. Helens; 600,000 tons of ash landed on the city of Yakima, Washington, more than 100 kilometers away.

Past explosive eruptions of other violent volcanoes have been equally devastating, or more so, in terms of pyroclastics (figure 5.16). When the city of Pompeii was destroyed by Mount Vesuvius in A.D. 79, it was buried not by lava, but by ash. (This is why extensive excavation of the ruins has been possible.) Contemporary accounts suggest that most residents had ample time to escape as the ash fell, though many chose to stay, and died. They and the town were ultimately buried under tons of ash. Pompeii's very existence was forgotten until some ruins were discovered during the late 1600s. In this century, in 1912, Mount Katmai in Alaska erupted violently and emitted an estimated 13 cubic kilometers (about 3 cubic miles) of pyroclastic material, more than ten times what Mount St. Helens spewed out.

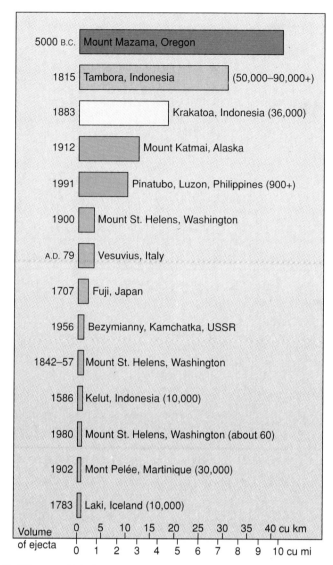

Year	Volcano	Volume of ejecta
5000 B.C.	Mount Mazama, Oregon	
1815	Tambora, Indonesia	(50,000–90,000+)
1883	Krakatoa, Indonesia (36,000)	
1912	Mount Katmai, Alaska	
1991	Pinatubo, Luzon, Philippines (900+)	
1900	Mount St. Helens, Washington	
A.D. 79	Vesuvius, Italy	
1707	Fuji, Japan	
1956	Bezymianny, Kamchatka, USSR	
1842–57	Mount St. Helens, Washington	
1586	Kelut, Indonesia (10,000)	
1980	Mount St. Helens, Washington (about 60)	
1902	Mont Pelée, Martinique (30,000)	
1783	Laki, Iceland (10,000)	

Volume of ejecta: 0 5 10 15 20 25 30 35 40 cu km / 0 1 2 3 4 5 6 7 8 9 10 cu mi

Figure 5.16 Volume of pyroclastics ejected during major explosive eruptions. Numbers of casualties, where available, are given in parentheses. Source: Data from U.S. Geological Survey.

Figure 5.17 The combination of large volumes of ash and heavy typhoon rains at Mount Pinatubo in 1991 proved too much weight for many buildings to bear. Photograph by R. P. Hoblitt, courtesy U.S. Geological Survey.

Figure 5.18 Mudflow and flood damage from eruption of Mount St. Helens, 1980: Damaged homes along the north fork of the Toutle River. Photograph by C. D. Miller, USGS Photo Library, Denver, CO.

Fortunately, there were no densely inhabited areas nearby. Even that event is dwarfed by the 1815 explosion of Tambora in Indonesia, which ejected an estimated 30 cubic kilometers of debris, and by the prehistoric explosion of Mount Mazama, in Oregon, which created the basin for the modern Crater Lake. The 1991 eruption of Mount Pinatubo demonstrated the special problems created by compounding of natural events: heavy ashfall was followed by soaking monsoon rains, and structures that had initally stood up to the weight of the ash collapsed under the greater weight of ash saturated by rain (figure 5.17).

Lahars

The combination of volcanic ash and water can be especially problematic, for the two can combine to create a volcanic mudflow called a **lahar.** Such mudflows, like lava flows, flow downhill. They may tend to follow stream channels, choking them with mud and causing floods of stream waters. Flooding produced in this way was a major source of damage near Mount St. Helens (figure 5.18). In A.D. 79, Herculaneum, closer to Vesuvius than was Pompeii, was partially invaded by volcanic mudflows, which preserved the bodies of some of the volcano's victims in mud casts. With a snow-capped volcano, the heat of the falling ash melts snow and ice on the mountain, producing a lahar. The 1985 eruption of Nevado del Ruiz, in Colombia, was a recent example of the devastation possible as a result of these flows. The swift and sudden mudflows that swept down its steep slopes after its snowy cap was partially melted by hot ash were the principal cause of over 20,000 deaths in towns below the volcano. At Mount Pinatubo in the Philippines, in 1991, tropical rains soaking

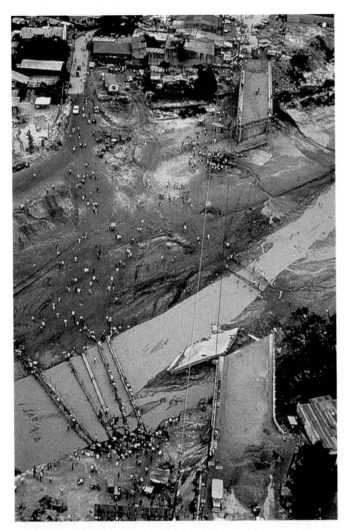

Figure 5.19   Aerial view of Abacan River channel in Angeles City near Clark Air Force Base, Philippines. A mudflow has taken out the main bridge; makeshift bridges allow pedestrians to cross the debris.   Photograph by T. J. Casadevall, U.S. Geological Survey.

Figure 5.20   Pyroclastic flow from Mount St. Helens. Photograph by P. W. Lipman, USGS Photo Library, Denver, CO.

ash deposits on the mountain triggered similar deadly mudflows (figure 5.19), while dry ash might have held on the volcano's slopes.

Even after the immediate danger is past, when the lahars have flowed, long-term impacts remain. They often leave behind stream channels partially filled with mud, their water-carrying capacity permanently reduced, which aggravates ongoing flood hazards in the area. To dredge this mud to restore channel capacity can be very costly, and areas far from the volcano can be thus affected. This was a significant problem in the case of Mount St. Helens.

Pyroclastic Flows—Nuées Ardentes

Another special kind of pyroclastic outburst is a deadly, denser-than-air mixture of hot gases and fine ash that forms a hot **pyroclastic flow,** quite different in character from a rain of ash and cinders. A pyroclastic flow is sometimes known as a *nuée ardente* (from the French for "glowing cloud"). A nuée ardente is very hot—temperatures can be over 1,000° C in the interior—and it can rush down the slopes of the volcano at more than 100 kilometers per hour (60 miles per hour), charring everything in its path. Such pyroclastic flows accompanied the major eruption of Mount St. Helens in 1980 (figure 5.20).

Perhaps the most famous such event in recent history occurred during the 1902 eruption of Mont Pelée on the Caribbean island of Martinique. The volcano had begun erupting weeks before, emitting both ash and lava, but was believed by many to pose no imminent threat to surrounding towns. Then, on the morning of May 8, with no immediate advance warning, a nuée ardente emerged from that volcano and swept through the nearby town of St. Pierre and its harbor. In a period of about three minutes, an estimated twenty-five to forty thousand people died or were fatally injured, burned or suffocated. The single reported survivor in the town was a convicted murderer who had been imprisoned in the town dungeon, underground, where he was shielded from the intense heat. He spent four terrifying days buried alive without food or water before rescue workers dug him out. Figure 5.21 gives some idea of the devastation.

Just as some volcanoes have a history of explosive eruptions, so, too, many volcanoes have a history of eruption of pyroclastic flows. Again, the composition of the lava is

Figure 5.21   St. Pierre, Martinique, West Indies, was destroyed by a nuée ardente from Mont Pelée, 1902. Photograph by Underwood and Underwood, courtesy Library of Congress.

linked to the likelihood of such an eruption. While the emergence of a nuée ardente may be sudden and unheralded by special warning signs, it is not generally the first activity shown by a volcano during an eruptive stage. Steam had issued from Mont Pelée for several weeks before the day St. Pierre was destroyed, and lava had been flowing out for over a week. This suggests one possible strategy for avoiding the dangers of these hot pyroclastic/gas flows: When a volcano known or believed to be capable of such eruptions shows signs of activity, leave.

Sometimes, however, human curiosity overcomes both fear and common sense, even when danger is recognized. A few days before the destruction of St. Pierre, the wife of the U.S. consul there wrote to her sister:

> *This morning the whole population of the city is on the alert, and every eye is directed toward Mont Pelée, an extinct volcano. Everybody is afraid that the volcano has taken into its heart to burst forth and destroy the whole island. Fifty years ago, Mont Pelée burst forth with terrific force and destroyed everything within a radius of several miles. For several days, the mountain has been bursting forth in flame and immense quantities of lava are flowing down its sides.* All the inhabitants are going up to see it. . . . *(Garesche 1902; emphasis added)*

Evacuations can themselves be disruptive. If, in retrospect, the danger turns out to have been a false alarm, people may be less willing to heed a later call to clear the area. There are two volcanoes named La Soufrière in the Caribbean, both similar in character to Mont Pelée, both po-

tentially deadly. In 1976, La Soufrière on Guadeloupe began its most recent eruption. Some seventy thousand people were evacuated and remained displaced for months, but only a few small explosions occurred. Government officials and volcanologists were blamed for disruption of people's lives, needless as it turned out to be. When La Soufrière on Saint Vincent began to erupt in 1979, officials were much less anxious to start issuing evacuation orders, despite recognition that here also was the potential for many casualties: A 1902 eruption of this volcano had killed sixteen hundred people. Luckily, no great harm resulted this time from failure to evacuate.

Fortunately, officials took the 1980 threat of Mount St. Helens seriously after its first few signs of reawakening. The immediate area near the volcano was cleared of all but a few essential scientists and law enforcement personnel. Access to hundreds of square kilometers of terrain around the volcano was severely restricted for both residents and workers (mainly loggers). Others with no particular business in the area were banned altogether. Many people grumbled at being forced from their homes. A few flatly refused to go. Numerous tourists and reporters were frustrated in their efforts to get a close look at the action. When the major explosive eruption came, there were casualties, but far fewer than there might have been. On a normal spring weekend, two thousand or more people would have been on the mountain, with more living or camping nearby. Considering the influx of the curious once eruptions began and the numbers of people turned away at roadblocks leading into the restricted zone, it is possible that, with free access to the area, tens of thousands could have died. Many lives were likewise saved by timely evacuations near Mount Pinatubo in 1991, made possible by careful monitoring and supported by intensive educational efforts by the government.

Toxic Gases

In addition to lava and pyroclastics, volcanoes emit a variety of gases. Many of these, such as water vapor and carbon dioxide, are nontoxic, although, as discussed below, even those gases can be dangerous at high concentrations. Others, including carbon monoxide, various sulfur gases, and hydrochloric acid, are actively poisonous. Many people have been killed by volcanic gases even before they realized the danger. During the A.D. 79 eruption of Vesuvius, fumes overcame and killed some unwary observers. Again, the best defense against toxic volcanic gases is common sense: Get well away from the erupting volcano as quickly as possible.

A bizarre disaster, still not fully explained but possibly attributable to volcanic gases, occurred in Cameroon, Africa, in 1986. Lake Nyos is one of a series of lakes that lie along a rift zone that is the site of intermittent volcanic activity. On 21 August 1986, a massive cloud of carbon dioxide (CO_2) gas from the lake flowed out over nearby towns and claimed seventeen hundred lives. The victims were not poisoned; they were suffocated (figure 5.22). Carbon dioxide is about 50 percent denser than air, so such a cloud would tend to

A   B

Figure 5.22 Catastrophe at Lake Nyos. (*A*) The lake itself. The dense cloud of carbon dioxide flowed downhill (toward rear of photograph) from the lake. (*B*) Animal carcasses attest to the deadliness of the cloud.   Photographs by M. L. Tuttle, USGS Photo Library, Denver, CO.

settle near the ground and to disperse only slowly. Anyone engulfed in that cloud would suffocate from lack of oxygen, and without warning, for CO_2 is both odorless and colorless. The source of the gas and the cause of its sudden release are both subjects of ongoing investigation. Because CO_2 is a common volcanic gas, it may have escaped from near-surface magma and made its way into the lake, where it accumulated in lake-bottom waters. Seasonal overturn of the lake waters as air temperatures changed may have led to its release. Whether this particular incident can be classified correctly as a disaster of volcanic association is thus not yet certain, but it may have been.

Steam Explosions

Some volcanoes are deadly not so much because of any characteristic inherent in the particular volcano but because of where they are located. In the case of a volcanic island, large quantities of seawater may seep down into the rock, come close to the hot magma below, turn to steam, and blow up the volcano like an overheated steam boiler. This is called a **phreatic eruption** or explosion.

The classic example is Krakatoa in Indonesia, which exploded in this fashion in 1883. The force of its explosion was estimated to be comparable to that of 100 million tons of dynamite, and the sound was heard 3,000 kilometers away in Australia. Some of the dust shot 80 kilometers into the air, causing red sunsets for years afterward, and ash was detected over an area of 750,000 square kilometers. Furthermore, the shock of the explosion generated a tsunami over 40 meters high! Krakatoa itself was an uninhabited island, yet its 1883 eruption killed an estimated thirty-six thousand people, mostly in low-lying coastal regions inundated by tsunamis.

In earlier times, when civilization was concentrated in fewer parts of the world, a single such eruption could be even more destructive. During the fourteenth century B.C., the volcano Santorini on the island of Thera exploded and caused a tsunami that wiped out many Mediterranean coastal cities. Some scholars have suggested that this event contributed to the fall of the Minoan civilization, as well as to the legend of the undersea city of Atlantis.

Secondary Effects: Climate and Atmospheric Chemistry

A single volcanic eruption can have a global impact on climate, although the effect may be only brief. Intense explosive eruptions put large quantities of volcanic dust high into the atmosphere. The dust can take several months or years to settle, and in the interim, it partially blocks out incoming sunlight, thus causing measurable cooling. After Krakatoa's 1883 eruption, worldwide temperatures dropped nearly half a degree centigrade, and the cooling effects persisted for almost ten years. The larger 1815 eruption of Tambora in Indonesia caused still more dramatic cooling: 1816 became known around the world as the "year without a summer." Such experiences support fears of a "nuclear winter" in the event of a nuclear war because modern nuclear weapons are powerful enough to cast many tons of fine dust into the air, as well as to trigger widespread forest fires that would add ash to the dust.

The meterological impacts of volcanoes are not confined to the effects of volcanic dust. The 1982 eruption of the Mexican volcano El Chichón did not produce a particularly large quantity of dust, but it did shoot volumes of unusually sulfur-rich gases into the atmosphere. These gases produced clouds of sulfuric acid droplets that spread around the earth. Not only do the acid droplets block some sunlight, like dust, but in time they also fall back to earth as acid rain, the consequences of which are explored in later chapters.

The 1991 eruptions of Mount Pinatubo in the Philippines (figure 5.23) had long-lasting, global effects on climate that were not only measurable by scientists but noticeable to a great many other people around the world. In addition to quantities of visible dust, Pinatubo pumped an estimated 20 million tons of sulfur dioxide (SO_2) into the atmosphere. While ash falls out of the atmosphere relatively quickly (like

Figure 5.23 Ash and gas from Mount Pinatubo was shot into the stratosphere, and had an impact on climate and atmospheric chemistry worldwide. Eruption of 12 June 1991. Photograph by K. Jackson, U.S. Air Force.

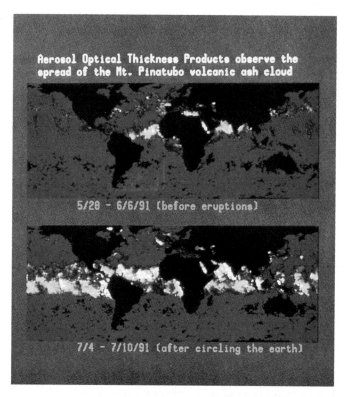

Figure 5.24 Satellites tracked the path of the airborne sulfuric-acid mist formed by SO_2 from Mount Pinatubo; winds slowly spread it into a belt encircling the earth (bright yellow in figure).   Image by G. J. Orme, Department of the Army, Ft. Bragg, NC.

particulate air pollutants described in chapter 17), the mist of sulfuric-acid droplets formed from stratospheric SO_2 stays airborne for several years. This acid-mist cloud slowly encircled the earth, most concentrated close to the equator near Pinatubo's latitude (figure 5.24). At its densest, the mist scattered up to 30 percent of incident sunlight and decreased by 2 to 4 percent the amount of solar radiation reaching earth's surface to warm it. Average global temperatures decreased by about 0.5° C (1° F), but the decline was not uniform; the greatest cooling was observed in the intermediate latitudes of the Northern Hemisphere (where the U.S. is located). Mount Pinatubo thus received the credit—or the blame—for the unusually cool summer of 1992 in the United States. The gradual natural removal of Pinatubo-generated ash and later SO_2 from the atmosphere resulted in a rebound of temperatures in the lower atmosphere by 1994 (figure 5.25).

Figure 5.25 Effect of 1991 eruption of Pinatubo on near-surface air temperatures. Removal of ash and dust from the air was relatively rapid; sulfate aerosols persisted longer. Source: Data from National Geophysical Data Center.

Another more subtle impact of this SO_2 on atmospheric chemistry may present a different danger. The acid seems to aggravate ozone depletion (see discussion of ozone and the "ozone hole" in chapter 17), which means increased risk of skin cancer to those exposed to strong sunlight.

Issues in Predicting Volcanic Eruptions

Classification of Volcanoes by Activity

In terms of their activity, volcanoes are divided into three categories: active; dormant, or "sleeping"; and extinct, or "dead." Unfortunately, the rules for assigning a particular volcano to one category or another are not precise. A volcano is generally considered **active** if it has erupted within recent history. When the volcano has not erupted recently but is fresh-looking and not too eroded or worn down, it is regarded as **dormant:** inactive for the present but with the potential to become active again. Historically, a volcano that has no recent eruptive history and appears very much eroded has been considered **extinct,** or very unlikely to erupt again.

As volcanologists learn more about the frequency with which volcanoes erupt, however, it has become clear that these guidelines are too simplistic. Volcanoes differ widely in their normal patterns of activity. Statistically, a "typical" volcano erupts once every 220 years, but 20 percent of all volcanoes erupt less than once every 1,000 years, and 2 percent erupt less than once in 10,000 years. Long quiescence, then, is no guarantee of extinction. Just as knowledge of past eruptive style allows anticipation of the kinds of hazards possible in future eruptions, so knowledge of the eruptive history of any given volcano is critical to knowing how likely future activity is, and on what scales of time and space.

The first step in predicting volcanic eruptions is monitoring, keeping an instrumental eye on the volcano. However, there are not nearly enough personnel or instruments available to monitor every volcano all the time. There are an estimated three to five hundred active volcanoes in the world (the inexact number arises from not knowing whether some are truly active, or only dormant). Monitoring those alone would be a large task. Commonly, intensive monitoring is undertaken only after a volcano shows some signs of near-term activity (see the discussion of volcanic precursors below). Dormant volcanoes might become active at any time, so, in principle, they also should be monitored; they can sometimes become very active very quickly. Theoretically, extinct volcanoes can be safely ignored, but that assumes that extinct volcanoes can be distinguished from the long-dormant. Vesuvius had been regarded as extinct until it destroyed Pompeii and Herculaneum in A.D. 79. Mount Pinatubo had been classed as dormant prior to 1991—it had not erupted in more than 400 years—then, within a matter of two months, it progressed from a few steam emissions to a major explosive eruption. The detailed eruptive history of many volcanoes is imperfectly known over the longer term.

Volcanic Precursors

What do scientists look for when monitoring a volcano? One common advance warning of volcanic activity is seismic activity (earthquakes). The rising of a volume of magma and gas up through the lithosphere beneath a volcano puts stress on the rocks of the lithosphere, and the process may produce months of small (and occasionally large) earthquakes. In August 1959, earthquakes were detected 55 kilometers below Kilauea, Hawaii, a depth that corresponds to the base of the lithosphere there. Over the subsequent few months, the earthquakes became shallower and more numerous as the magma pushed upward. In late November, by which time up to one thousand small earthquakes were recorded every day, a crack opened on the flank of the volcano and lava poured out. The eruptions of Mount St. Helens have likewise been preceded by increased frequency and intensity of earthquake activity. Sometimes, the shock of a larger earthquake may itself unleash the eruption. The major explosion of Mount St. Helens in 1980 is believed to have been set off indirectly in this way: An earthquake shook loose a landslide from the bulging north slope of the volcano, which lessened the weight confining the trapped gases inside and allowed them to blast forth.

Bulging, tilt, or uplift of the volcano's surface is also a warning sign. It often indicates the presence of a rising magma mass, the buildup of gas pressure, or both. Figure 5.26, which is a record of tilt and of eruptions on Kilauea, shows that eruptions are preceded by an inflating of the volcano as magma rises up from the asthenosphere to fill the shallow magma chamber. Unfortunately, it is not possible to specify exactly when the swollen volcano will crack to release its contents. That varies from eruption to eruption with the pressures and stresses involved and the strength of the overlying rocks. This limitation is not unique to Kilauea. Uplift, tilt, and seismic activity may indicate that an eruption is approaching, but geologists do not yet have the ability to predict its exact timing.

Other possible predictors of volcanic eruptions are being evaluated. Many volcanologists speculate that changes in the mix of gases coming out of a volcano may give clues to impending eruptions; SO_2 content of the escaping gas shows promise as a precursor, perhaps because more SO_2 can escape as magma nears the surface. Surveys of ground surface temperatures may reveal especially warm areas where magma is particularly close to the surface and about to break through. As with earthquakes, there have been reports that volcanic eruptions have been "foreseen" by animals, which have behaved strangely for some hours or days before the event. Perhaps animals are sensitive to some changes in the earth that scientists have not thought to measure.

Certainly, much more work is needed before the exact timing and nature of major volcanic eruptions can be anticipated consistently. Recognition of impending eruptions of Kilauea and Mount St. Helens was obviously successful in saving many lives through evacuations and restrictions on access to the danger zones. Mount Pinatubo displayed increasing evidence of forthcoming activity for two months before its major eruption of 15 June 1991—local earthquakes, increasing in number, and becoming concentrated below a bulge on the volcano's summit a week before that eruption; preliminary ash emissions; a sudden drop in gas output that suggested a plugging of the volcanic "plumbing" that would tend to increase internal gas pressure and explosive potential. Those warnings, combined with aggressive and comprehensive public warning

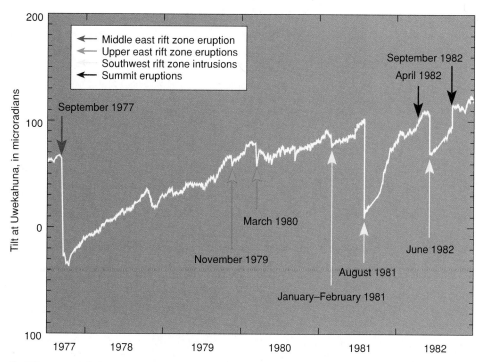

Figure 5.26 Tilt record of Kilauea, Hawaii. The increasing tilt of the volcano's slopes reflects bulging due to rising magma below. Note the deflation after eruptions. Source: Data from "The Pu'u O'o Eruptions of Kilauea Volcano," *Volcanoes in Hawaii*, U.S. Geological Survey.

procedures and, eventually, evacuation of over 80,000 people greatly reduced the loss of life. Indeed, many of the approximately 500 who died were not injured in the explosive eruption, but rather crushed later as the weight of rain-saturated volcanic ash caused roofs to collapse. In 1994, geoscientists monitoring Rabaul volcano in Papua New Guinea took a magnitude-5 earthquake followed by increased seismic activity generally as the signal to evacuate a town of 30,000 people. The evacuation, rehearsed before, was orderly—and hours later, a five-day explosive eruption began. The town was covered by a meter or more of wet volcanic ash, and a quarter of the buildings were destroyed, but no one died.

These successes clearly demonstrate the potential for reduction in fatalities when precursory changes occur and local citizens are educated to the dangers. On the other hand, the exact moment of the Mount St. Helens explosion was not known until seconds beforehand, and it is not currently possible to tell whether or just when a pyroclastic flow might emerge from a volcano like La Soufrière. Nor can volcanologists readily predict the volume of lava or pyroclastics to expect from an eruption or the length of the eruptive phase, although they can anticipate the likelihood of an explosive eruption or other eruption of a particular style.

Response to Eruption Predictions

When data indicate an impending eruption that might threaten populated areas, the safest course is evacuation until the activity subsides. However, a given volcano may remain more or less dangerous for a long time. An active phase, consisting of a series of intermittent eruptions, may continue for months or years. In these instances, either property must be abandoned for prolonged periods, or inhabitants must be prepared to move out not once but many times. Given the uncertain nature of eruption prediction at present, some precautionary evacuations will continue to be shown, in retrospect, to have been unnecessary, or unnecessarily early.

Accurate prediction and assessment of the hazard is particularly difficult with volcanoes reawakening after a long dormancy because historical records for comparison with current data are sketchy or nonexistent. Beside the Bay of Naples is an old volcanic area known as the Phlegraean Fields. In the early 1500s, the town of Tripergole rose up some 7 meters as a chamber of gas and magma bulged below, and the region was shaken by earthquakes. Then, suddenly, a vent opened, and in three days, the town was buried under a new 140-meter-high cinder cone. For the last four centuries, the only thermal activity has been steaming fumaroles and hot springs. But in 1970–72 and again in 1982–84, the nearby town of Pozzuoli has been uplifted nearly 3 meters. During the last episode of bulging, it was shaken by over four thousand earthquakes in a single year, leaving the town so badly damaged that many of its eighty thousand residents had to sleep in "tent cities" outside the town for months. Business could be conducted in town by day, but the town had to be evacuated at night. The threat and the disruption in the people's lives persisted for years. These episodes are believed to be related to the presence of a shallow magma chamber. Volcanologists are still unsure of what sort of volcanic activity can be expected here, how large in scale, or how soon.

And while some volcanoes, like Mount St. Helens or Mount Pinatubo, may produce weeks or months of warning signals as they emerge from dormancy to full-blown eruption, allowing time for public education and reasoned response, not all volcanoes' awakenings may be so leisurely.

Present and Future Volcanic Hazards in the United States

Hawaii

Several of the Hawaiian volcanoes are active or dormant, and eruptions are frequent in some places. It is, however, not practical to evacuate the islands indefinitely. A first, common-sense step toward reducing losses from eruptions might be to prevent any new development in recently active areas. Incredibly, however, even this is not being done (see box 5.1, p. 116).

Cascade Range

Mount St. Helens is only one of a set of volcanoes that threaten the western United States and southwestern Canada. There is subduction beneath the Pacific Northwest, which is the underlying cause of continuing volcanism there (figure 5.27). Several more of the Cascade Range volcanoes have shown signs of reawakening. Lassen Peak was last active between 1914 and 1921, not so very long ago geologically. Its products are very similar to those of Mount St. Helens; violent eruptions are possible. Mount Baker (last eruption, 1870), Mount Hood (1865), and Mount Shasta have shown seismic activity, and steam is escaping from these and from Mount Rainier, which last erupted in 1882. In fact, at least nine of the Cascade peaks are presently showing thermal activity of some kind (steam emission or hot springs). The eruption of Mount St. Helens has in one sense been useful: It has focused attention on this threat. Scientists are watching many of the Cascade Range volcanoes very closely now; there is particular concern about possible mudflow dangers from Mount Rainier (figure 5.28). With skill, and perhaps some luck, they may be able to recognize when others of these volcanoes may be close to eruption in time to avert disaster.

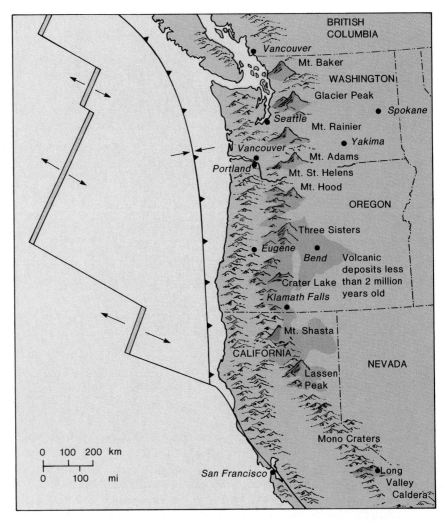

Figure 5.27 The Cascade Range volcanoes and their spatial relationship to the subduction zone and to major cities (plate-boundary symbols as in figure 3.6). Shaded area is covered by young volcanic deposits less than 2 million years old.

The Aleutians

South-central Alaska and the Aleutian island chain sit above a subduction zone. This makes the region vulnerable not only to earthquakes, as noted in the last chapter, but also to volcanic eruptions (figure 5.29A). The volcanoes are typically andesitic with composite cones, and pyroclastic eruptions are common. In 1986, Augustine (figure 5.29B) began a period of eruption; in late 1989, Redoubt volcano followed suit, following twenty-three years of quiescence, with the eruptive phase heralded by a period of increased seismicity and steam emission. Neither volcano lies close to a densely populated area. Nevertheless, they have presented special hazards. Jet aircraft flying through clouds of ash from the erupting Augustine experienced some engine trouble. It was subsequently discovered that the volcano's andesitic ash melts at temperatures close to the operating temperatures of jet engines. Thereafter, pilots were advised to fly clear of these ash clouds, and a warning system was put in place by the National Oceanic and Atmospheric Administration. Despite this, in 1989 a KLM 747 flew through an ash cloud from Redoubt at 37,000 feet,

lost power in all four engines, and plunged over 13,000 feet before pilots were able to restart the engines. This near-disaster perhaps ensured that future ash-cloud warnings will be taken more seriously.

Other Volcanic Areas—Long Valley and Yellowstone Calderas

Finally, there are areas that do not now constitute an obvious hazard, but that are causing some geologists concern because of increases in seismicity or thermal activity. One of these is the Mammoth Lakes area of California. In 1980, the area suddenly began experiencing earthquakes. Within one forty-eight-hour period in May 1980, four earthquakes of magnitude 6 rattled the region, interspersed with hundreds of lesser shocks. Mammoth Lakes lies within the Long Valley Caldera, a 13-kilometer-long oval depression formed during violent pyroclastic eruptions seven hundred thousand years ago (figure 5.30). Smaller eruptions occurred around the area as recently as fifty thousand years ago.

A **caldera** is an enlarged volcanic summit crater, which may be formed either by an explosion enlarging an existing

Figure 5.28 Over the last 5,000 – 6,000 years, huge mudflows have poured down stream valleys and into low-lying areas, tens of miles from Mount Rainier's summit. Even Tacoma or Seattle might be threatened by renewed activity on a similar scale. Source: Data from T. W. Sisson, USGS Open-File Report 95–642.

A

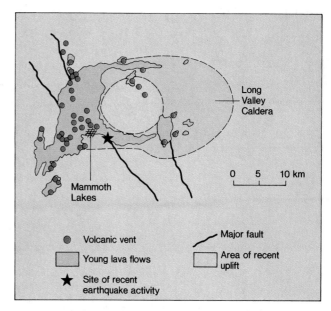

B

Figure 5.29 The Aleutians are a region of active volcanism—fortunately, a rather sparsely populated one. (*A*) Landsat satellite photograph of southwestern Alaska showing major volcanic peaks. (*B*) Augustine volcano in eruption, spewing clouds of andesitic ash. Photograph (*B*) by M. E. Yount. Both photographs courtesy of USGS Photo Library, Denver, CO.

crater (Crater Lake, Mount St. Helens) or by collapse of a volcano after a magma chamber within has emptied. With a caldera formed by explosive eruption, like Long Valley, the concern about renewed activity is fear of further explosions, perhaps massive ones. Geophysical studies have shown that a partly molten mass of magma close to 4 kilometers across currently lies below the Long Valley caldera. Furthermore, since 1975, the center of the caldera has bulged upward more than 25 centimeters (10 inches); at least a portion of the magma appears to be rising in the crust. In 1982, seismologists realized that the patterns of earthquake activity in the caldera were disturbingly similar to those associated with the eruptions of Mount St. Helens. In May 1982, the director of the U.S. Geological Survey issued a formal notice of potential volcanic hazard for the Mammoth Lakes/Long Valley area. A swarm of earthquakes of magnitudes up to 5.6 occurred there in January of 1983, and the seismicity continues. So far, there has been no eruption, but scientists continue to monitor developments very closely, seeking to understand what is happening below the surface so that they can anticipate what volcanic activity may develop and how soon. One of the unknown factors is the magma type in question, and therefore the probable eruptive style: Both basalts and rhyolites are found among the young volcanics in the area.

Long Valley is not the only such feature being watched somewhat warily. As noted earlier, another area of uncertain future is Yellowstone National Park. Yellowstone, at present, is notable for its geothermal features—geysers, hot springs,

Figure 5.30 Map of Mammoth Lakes area, showing Long Valley Caldera, site of recent earthquakes, and area of recent uplift beneath which magma is rising. (See location in figure 5.27.) Source: After R. A. Bailey, U.S. Geological Survey, 1983.

Box 5.1 Life on a Volcano's Flanks

The island of Hawaii is really a complex structure built of five shield volcanoes (figure 1).

The so-called East Rift of Kilauea in Hawaii has been particularly active in recent years, with major eruptions in the 1970s. Less than ten years later, a new subdivision—Royal Gardens—was begun below the East Rift. In 1982–83, lava flows from a renewed round of eruptions reached down the slopes of the volcano and quietly obliterated new roads and several houses. The eruptions have continued intermittently over more than a decade since then. The photographs in figure 2 illustrate some of the results.

When asked why anyone would have built in such a spot, where the risks were so obvious, one native shrugged and replied, "Well, the land was cheap." It will probably remain so, too—a continuing temptation to unwise development in the absence of tighter zoning restrictions.

Of course, when one lives on an active or dormant volcano, no spot may be altogether safe. Still, relative levels of risk can be identified on the basis of past eruptive and seismic activity, and topography (figure 3). While the risks to life are generally low everywhere because of the quiet character of Hawaiian eruptions, property damage could be greatly reduced by greater attention to the level of probable vulnerability to lava flows of areas under consideration for further development.

Figure 1   Map of Hawaii showing five volcanoes and major associated rift zones.   Source: After C. Heliker, *Volcanic and Seismic Hazards of Hawaii,* U.S. Geological Survey, 1991.

Figure 2 (*A*) Lavas invade Royal Gardens subdivision, Hawaii, 1983. Flows from March through July covered 330 lots and destroyed sixteen homes. (*B*) Million-dollar Wahalua visitors' center in Hawaii Volcanoes National Park, aflame as it is engulfed by lava in 1989. (*C*) Ruins of the visitors' center after the lavas cooled.   Photographs (*A*) and (*B*) by J. D. Griggs, courtesy of USGS Photo Library, Denver, CO.

—Continued

Box 5.1—Continued

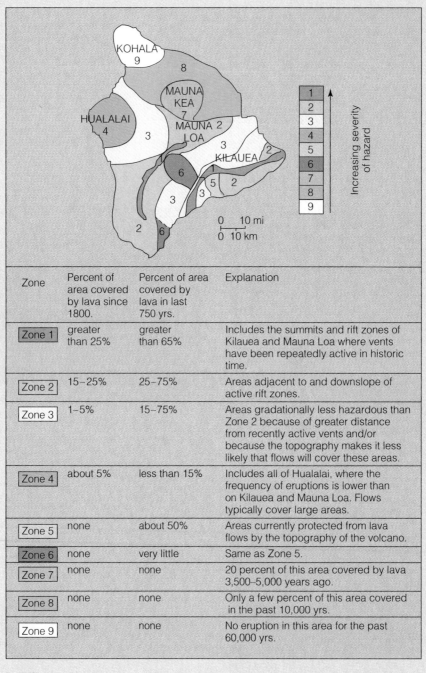

Zone	Percent of area covered by lava since 1800.	Percent of area covered by lava in last 750 yrs.	Explanation
Zone 1	greater than 25%	greater than 65%	Includes the summits and rift zones of Kilauea and Mauna Loa where vents have been repeatedly active in historic time.
Zone 2	15–25%	25–75%	Areas adjacent to and downslope of active rift zones.
Zone 3	1–5%	15–75%	Areas gradationally less hazardous than Zone 2 because of greater distance from recently active vents and/or because the topography makes it less likely that flows will cover these areas.
Zone 4	about 5%	less than 15%	Includes all of Hualalai, where the frequency of eruptions is lower than on Kilauea and Mauna Loa. Flows typically cover large areas.
Zone 5	none	about 50%	Areas currently protected from lava flows by the topography of the volcano.
Zone 6	none	very little	Same as Zone 5.
Zone 7	none	none	20 percent of this area covered by lava 3,500–5,000 years ago.
Zone 8	none	none	Only a few percent of this area covered in the past 10,000 yrs.
Zone 9	none	none	No eruption in this area for the past 60,000 yrs.

Figure 3 The Island of Hawaii is divided into zones of relative hazard from lava flows; 1 = highest degree of risk.   Source: After C. Heliker, *Volcanic and Seismic Hazards of Hawaii*, U.S. Geological Survey, 1991.

and fumaroles (figure 5.31A). All of these features reflect the presence of hot rocks at shallow depths in the crust below the park. Until recently, it was believed that the heat was just left over from the last cycle of volcanic activity in the area, which began 600,000 years ago. Recent studies, however, suggest that in fact, considerable fluid magma may remain beneath the park, which is also still a seismically active area. Yellowstone, then, is an intracontinental hot-spot volcano. One reason for concern is the scale of past eruptions: In the last cycle, 1,000 cubic kilometers (over 200 cubic miles) of pyroclastics were ejected! Indeed, the size of the caldera itself (figure 5.31B) suggests that the potential scale of future eruptions is large. Whether activity on that scale is likely in the foreseeable future, no one knows. The research goes on.

A

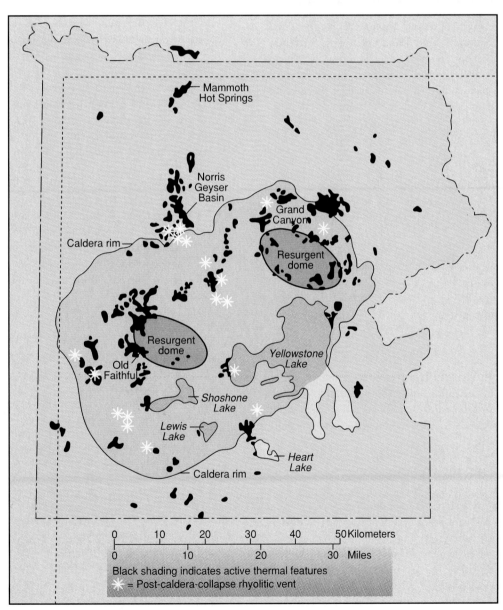

B

Figure 5.31 Continuing thermal activity at Yellowstone National Park (*A*), the size of the caldera (*B*), and scale of past eruptions cause concern about the future of the region. (*B*) Used with permission from *Yellowstone Magmatic Evolution*, Copyright © 1984 by the National Academy of Sciences. Courtesy of the National Academy Press, Washington, D.C.

Overview

Summary

Most volcanic activity is concentrated near plate boundaries. Volcanoes differ in eruptive style and in the kinds of dangers they present. Those along spreading ridges and at hot spots tend to be more placid, usually erupting a fluid, basaltic lava. Subduction-zone volcanoes produce a much more viscous, silica-rich, gas-charged magma, so, in addition to lava, they may emit large quantities of pyroclastics and other deadly products like pyroclastic flows and toxic gases. At present, volcanologists can detect the early signs that a volcano may erupt in the near future, but they cannot predict the exact timing or type of eruption. Individual volcanoes, however, show characteristic eruptive styles (as a function of magma type) and patterns of activity. Therefore, knowledge of a volcano's eruptive history allows anticipation of the general nature of eruptions and of the likelihood of renewed activity in the near future.

Terms to Remember

active volcano 112
andesite 98
basalt 98
caldera 114
cinder cone 103
composite volcano 103
dormant volcano 112
extinct volcano 112
fissure eruption 100

lahar 107
phreatic eruption 110
pyroclastic flow 108
pyroclastics 101
rhyolite 98
shield volcano 100
stratovolcanoes 103
volcanic dome 101

Exercises

Questions for Review

1. Define a fissure eruption, and give an example.
2. The Hawaiian Islands are all shield volcanoes. What are shield volcanoes, and why are they not especially hazardous to life?
3. The eruptive style of Mount St. Helens is quite different from that of Kilauea in Hawaii. Why?
4. What are pyroclastics? Identify a kind of volcanic structure that pyroclastics may build.
5. Describe two strategies for protecting an inhabited area from an advancing lava flow.
6. What is a nuée ardente, and why is a volcano known for producing these hot pyroclastic flows a special threat during periods of activity?
7. Explain the nature of a phreatic eruption, and give an example.
8. How may volcanic eruptions influence global climate?
9. Discuss the distinctions among active, dormant, and extinct volcanoes, and comment on the limitations of this classification scheme.
10. Describe two precursor phenomena that may precede volcanic eruptions.
11. What is the underlying cause of present and potential future volcanic activity in the Cascade Range of the western United States?

For Further Thought

1. Before the explosion of Mount St. Helens in May 1980, scientists made predictions about the probable extent of and kinds of damage to be expected from a violent eruption. Investigate those predictions, and compare them with the actual effects of the blast.
2. How near to you is the closest active volcano? The closest dormant one? Are there any ancient lava flows or pyroclastic deposits in your area, and if so, how old are the youngest of these?
3. As this is written, Kilauea continues an active phase that has lasted more than a decade. Check out the current activity report of the Hawaii Volcanoes Observatory (see NetNotes). Or, check the latest information of the Cascades Volcano Observatory for current status of Mount St. Helens, Mount Rainier, or another Cascade volcano. Or, go exploring other sites for information on other active volcanoes worldwide. Identify the tectonic setting of the volcano(es) and relate to any current activity.

Suggested Readings/References

Alaska Volcano Observatory Staff. 1990. The 1989–1990 eruption of Redoubt Volcano. *EOS, Trans. American Geophysical Union* 71 (13 February): 265–75.

Axelrod, D. I. 1981. *Role of volcanism in climate and evolution.* Geological Society of America Special Paper 185.

Bailey, R. A. 1983. Mammoth Lakes earthquakes and ground uplift: Precursors to possible volcanic activity? U.S. Geological Survey, *Earthquake Information Bulletin* 15 (3): 88–102.

Blong, R. J. 1984. *Volcanic hazards: A source book on the effects of eruptions.* New York: Academic Press.

Bullard, F. M. 1984. *Volcanoes of the earth.* 2d ed. Austin, Tex.: University of Texas Press.

Delaney, P. T., et al. 1990. Deep magma body beneath the summit and rift zones of Kilauea Volcano, Hawaii. *Science* 247 (10 March): 1311–16.

Foxworthy, B. L., and Hill, M. 1982. *Volcanic eruptions of Mount St. Helens: The first 100 days.* U.S. Geological Survey Professional Paper 1249.

Francis, P., and Self, S. 1987. Collapsing volcanoes. *Scientific American* 256 (June): 90–97.

Garesche, W. A. 1902. *Complete story of the Martinique and St. Vincent horrors.* L. G. Stahl.

Gore, R. 1984. The dead do tell tales of Vesuvius. *National Geographic* 165:557–613.

———. 1984. A prayer for Pozzuoli. *National Geographic* 165:614–25.

Grove, N. 1992. Volcanoes: Crucibles of creation. *National Geographic* 173 (December): 5–41.

Heliker, C. 1991. *Volcanic and seismic hazards on the island of Hawaii.* U.S. Geological Survey.

Krakauer, J. 1996. Mount Rainier: Washington's tranquil threat. *Smithsonian* (July): 32–41.

McAlister, S. A., and Ayme, L. A. 1902. *Martinique flood of fire and burning rain.* Philadelphia, Pa.: P. W. Ziegler.

Pinatubo Volcano Observatory Team. 1991. Lessons from a major eruption: Mt. Pinatubo, Philippines. *EOS, Trans. American Geophysical Union* 72 (3 December): 545–55.

Rampino, M. R., and Self, S. 1984. The atmospheric effects of El Chichón. *Scientific American* 250 (January): 48–57.

Saarinen, T. F., and Sell, J. L. 1985. *Warning and response to Mount St. Helens' eruption.* Albany: State University of New York Press.

Sisson, T. W. 1995. Volcano hazards fact sheet: History and hazards of Mount Rainier, Washington. USGS open-file report 95–642 [online] http://vulcan.wr.usgs.gov/Volcanoes/Rainier/Hazard/OFR95-642/OFR95-642.html

Tilling, R. I. (ed.). 1988. *How volcanoes work.* A collection of readings from the *Journal of Geophysical Research.* Washington, D.C.: American Geophysical Union.

Tyson, P. 1996. Under the volcano. *Technology Review* 41 (January): 40–47.

Volcanoes and the earth's interior. 1983. San Francisco: W. H. Freeman. (A selection of readings from *Scientific American,* 1975–82.)

Williams, R. S., Jr., and Moore, J. G. 1973. Iceland chills a lava flow. *Geotimes* 18 (August): 14–17.

The Cascades Volcano Observatory of the U.S. Geological Survey maintains information and links to other sources on all U.S. volcanoes, not just the Cascade volcanoes; its home page is at

http://vulcan.wr.usgs.gov/home.html

The Hawaii Volcanoes Observatory is not only an excellent source of data on (naturally) the Hawaiian volcanoes; they also maintain "Volcano Watch," a weekly newsletter on volcanic activity worldwide, with links to other data sources at

http://www.soest.hawaii.edu/hvo/

"Volcano World" is a very popular site, maintained with support from NASA to promote public access to and use of space science data, with links to a tremendous variety of information about terrestrial volcanoes *and* those on other planets, including lists of current/recent volcanic activity. See

http://volcano.und.nodak.edu/vw.html

Michigan Technological University maintains a particularly informative "Volcanoes Page" including abundant references at

http://www.geo.mtu.edu/volcanoes

More detailed information on relationships between volcanic eruptions and global climate is available from NASA at

http://spso2.gsfc.nasa.gov/NASA_FACTS/volcanoes/volcano.html

For information on volcano observatories worldwide, try

http://www.dartmouth.edu/pages/rox/volcanoes/EvG.html

Good links to assorted volcano-related sites are also found at

http://skye.gsfc.nasa.gov/wwvolcano.html

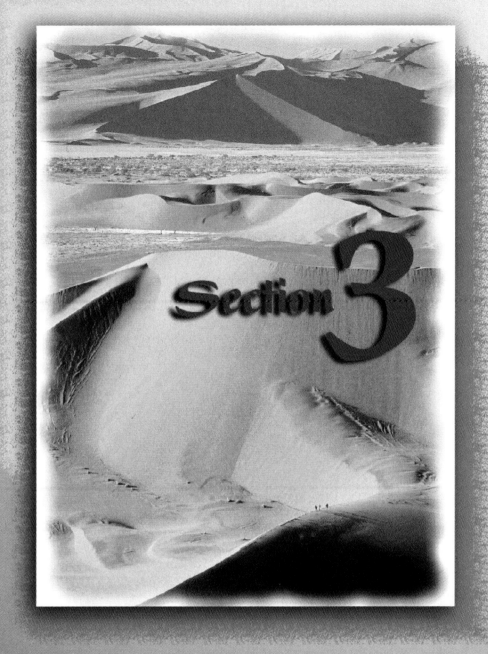

Section 3

Huge sand dunes at Mamimbia, Africa dwarf the walkers in the foreground. © Jess Stock/Tony Stone Images

Many areas are threatened by the hazards associated with the internal geologic process discussed in section II. Every place on earth, however, is affected by various kinds of *surface processes*, those geologic processes with causes and effects found at or near the earth's surface. These processes involve principally the actions of water, ice, wind, and gravity; local climate plays a major role in determining the relative importance of those properties.

The earth's surface is, geologically, a very active place. It is here that one sees the interactions of the internal heat of the earth (which builds mountains and shifts the land), the external heat from the sun (which drives the wind and provides the energy to drive the hydrologic cycle), and the inexorable force of gravity (which constantly tries to pull everything down to the same level).

Chapters 6–9 explore several kinds of surface processes: stream-related processes, including flooding; coastal processes of erosion and mass transport; mass movements, the downhill march of material; and the effects of ice and wind in sculpturing the land. Chapter 9 also examines some relationships among climate, ice, and deserts.

Chapter 6

Streams and Flooding

Water is the single most important agent sculpturing the earth's surface. Mountains may be raised by the action of plate tectonics and volcanism, but they are shaped primarily by water. Streams carve valleys, level plains, and move tremendous amounts of sediment from place to place.

Floods are probably the most widely experienced catastrophic geologic hazards. Statistics compiled by an *ad hoc* working group of the United Nations suggest that in the decade of the 1990s, approximately 100,000 floods will occur worldwide. On the average, in the United States alone, floods annually take over eighty-five lives and cause well over $1 billion in property damage. The 1993 flooding in the Mississippi River basin, while responsible for only 48 deaths, caused an estimated $15–20 billion in damages.

Most floods do not make national headlines, but they may be no less devastating to those affected by them. Some floods are the result of unusual events, such as the collapse of a dam, but the vast majority are a perfectly normal, and to some extent predictable, part of the natural functioning of streams. Before discussing flood hazards, therefore, we will examine how water moves through the hydrologic cycle and also look at the basic characteristics and behavior of streams.

Bond Falls near Paulding, Ontonogan County, in Michigan's Upper Peninsula. © Doug Sherman/Geofile

The Hydrologic Cycle

The **hydrosphere** includes all the water at and near the surface of the earth. Most of it is believed to have been outgassed from the earth's interior early in its history, when the earth's temperature was higher. Now, except for occasional minor additions from volcanoes bringing up "new" water from the mantle, and small amounts of water returned to the mantle with subducted lithosphere, the quantity of water in the hydrosphere remains essentially constant.

All of the water in the hydrosphere is caught up in the **hydrologic cycle,** illustrated in figure 6.1. The largest single reservoir in the hydrologic cycle, by far, consists of the world's oceans, which contain over 97 percent of the water in the hydrosphere; lakes and streams together contain only 0.016 percent of the water. The main processes of the hydrologic cycle involve evaporation into and precipitation out of the atmosphere. Precipitation onto land can reevaporate (directly from the ground surface or indirectly through plants by evapotranspiration), infiltrate into the ground, or run off over the ground surface. Surface runoff may occur in streams or by unchanneled overland flow. Water that sinks into the ground may also flow (see chapter 10) and commonly returns, in time, to the oceans. The oceans are the principal source of evaporated water because of their vast areas of exposed water surface.

The total amount of water moving through the hydrologic cycle is large, more than 100 million billion gallons per year. A portion of this water is temporarily diverted for human use, but it ultimately makes its way back into the natural global water cycle by a variety of routes, including release of municipal sewage, evaporation from irrigated fields, or discharge of industrial wastewater into streams. Water in the hydrosphere may spend extended periods of time—even tens of thousands of years—in storage in one or another of the water reservoirs, but from the longer perspective of geologic history, it is still regarded as moving continually through the hydrologic cycle. In the process, it may participate in other geologic cycles and processes: for example, streams eroding rock and moving sediment, and subsurface waters dissolving rock and transporting dissolved chemicals, are also contributing to the rock cycle.

Streams and Their Features

Streams—General

A **stream** is any body of flowing water confined within a channel, regardless of size. It flows downhill through local topographic lows, carrying away water over the earth's surface. The region from which a stream draws water is its **drainage basin** (figure 6.2). The size of a stream at any point is related in part to the size (area) of the drainage basin upstream from that point, which determines how much water from falling snow or rain can flow into the stream. Its size is also influenced by several other factors, including climate (amount of precipitation and evaporation), vegetation or lack of it, and the underlying geology, all of which are explained in more detail later in the chapter. That is, a stream carves its channel (in the absence of human intervention), and the channel carved is broadly proportional to the volume of water that must typically be accommodated. Long-term, sustained changes in precipitation, land use, or other factors that change the volume of water customarily flowing in the stream will be reflected in corresponding changes in channel geometry.

The size of a stream may be described by its **discharge,** the volume of water flowing past a given point (or, more precisely, through a given cross-section) in a specified length of time. Discharge is the product of channel cross section (area) times stream velocity. Historically, the conventional unit used to express discharge is cubic feet per second; the analogous metric unit is cubic meters per second. Discharge may vary from less than one cubic foot per second on a small creek to millions of cubic feet per second in a major river.

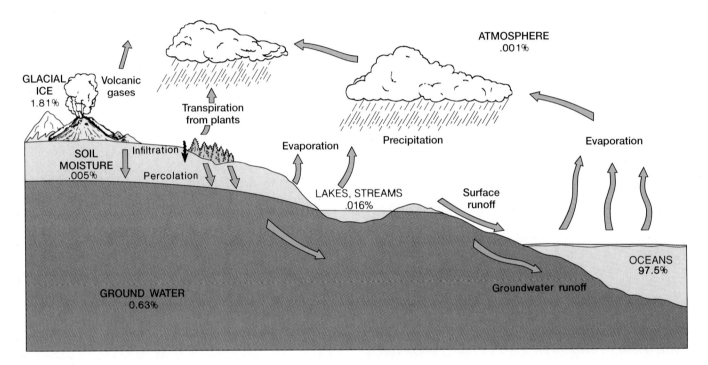

Figure 6.1   Principal processes and reservoirs of the hydrologic cycle.

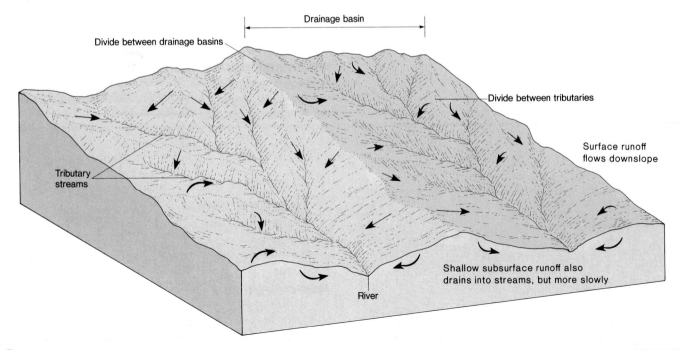

Figure 6.2   Streams and their drainage basins.

Sediment Transport

Water is a powerful agent for transporting material. Streams can move material in several ways. Heavier debris may be rolled or pushed along the bottom of the streambed. This material is then described as the *bed load* of the stream. The *suspended load* consists of material that is light or fine enough to be moved along suspended in the stream, supported by the flowing water. Suspended sediment clouds a stream and gives the water a muddy appearance. Material of intermediate size may be carried in short hops along the streambed by a process called *saltation* (figure 6.3). Finally, some substances may be completely dissolved in the water, to make up the stream's *dissolved load.*

The total quantity of material that a stream transports by all these methods is called, simply, its **load.** Stream **capacity** is a measure of the total load of material a stream can move. Capacity is closely related to discharge: the faster the water flows, and the more water is present, the more material can be moved. How much of a load is actually transported also depends on the availability of sediments or soluble material: a stream flowing over solid bedrock will not be able to dislodge much material, while a similar stream flowing through sand or soil may move considerable material.

Figure 6.3 Schematic diagram of saltation, one process of sediment transport. Particles move in short jumps, jarred loose by water or other particles and briefly carried in the stream flow.

Velocity, Gradient, and Base Level

Stream velocity is related partly to discharge and partly to the steepness (pitch or angle) of the slope down which the stream flows. The steepness of the stream channel is called its **gradient.** All else being equal, the higher the gradient, the steeper the channel (by definition), and the faster the stream flows. Gradient and velocity commonly vary along the length of a stream (figure 6.4), especially if the stream is a large one. Nearer its source, where the first perceptible trickle of water signals the stream's existence, the gradient is usually steeper, and it tends to decrease downstream. Velocity may or may not decrease correspondingly; the effect of decreasing gradient may be counteracted by other factors, including increased water volume as additional tributaries enter the stream, and changes in the channel's width and depth (increasing cross section) overall, discharge does tend to increase downstream.

By the time the stream reaches its end, or mouth, which is usually where it flows into another body of water, the gradient is typically quite low. Near its mouth, the stream is approaching its **base level,** which is the lowest elevation to which the stream can erode downward. For most streams, base level is the water (surface) level of the body of water into which they flow. For streams flowing into the ocean, for instance, base level is sea level. The closer a stream is to its base level, the lower the stream's gradient. It may flow more slowly in consequence, although the slowing effect of decreased gradient may be compensated by an increase in discharge due to widening of the channel and a corresponding increase in channel cross section. The downward pull of gravity causes a stream to cut down toward its base level. Counteracting this erosion is the influx of fresh sediment into the stream from the drainage basin. Over time, natural streams tend toward a balance, or equilibrium, between erosion and deposition of sediments. The **longitudinal**

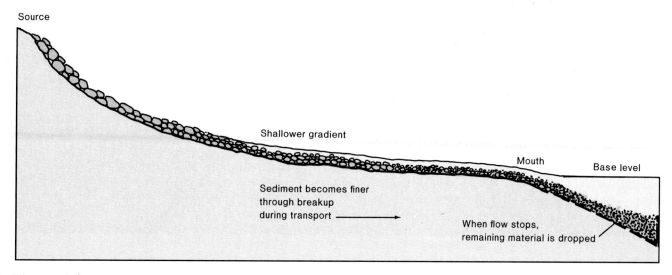

Figure 6.4 Typical longitudinal profile of a stream in a temperate climate. Note that the gradient decreases from source to mouth. Particle sizes may decrease with breakup downstream. Total discharge, capacity, and velocity may all increase downstream, despite the decrease in gradient.

profile of such a stream (a sketch of the stream's elevation from source to mouth) assumes a characteristic concave-upward shape (see figure 6.4).

Velocity and Sediment Sorting

Variations in a stream's velocity along its length are reflected in the sediments deposited at different points. The more rapidly a stream flows, the larger and denser are the particles moved. The sediments found motionless in a streambed at any point are those too big or heavy for that stream to move at that point. Where the stream flows most quickly, it carries gravel and even boulders along with the finer sediments. As the stream slows down, it starts dropping the heaviest, largest particles—the boulders and gravel—and continues to move the lighter, finer materials along. If stream velocity continues to decrease, successively smaller particles are dropped: the sand-sized particles next, then the clay-sized ones. In a very slowly flowing stream, only the finest sediments and dissolved materials are still being carried. If a stream flows into a body of standing water, like a lake or ocean, the stream's flow velocity drops to zero, and all the remaining suspended sediment is dropped.

The relationship between the velocity of water flow and the size of particles moved accounts for one characteristic of stream-deposited sediments: They are commonly **well sorted** by size or density, with materials deposited at a given point tending to be similar in size or weight. If a stream is still carrying a substantial load as it reaches its mouth, and it then flows into still waters, a large fan-shaped pile of sediment, a **delta,** may be built up (figure 6.5A). A similarly shaped

A

B

Figure 6.5 (*A*) The Mississippi River delta (satellite photo). The river deposits a million tons of sediment each day. Here, the suspended sediment shows as a milky, clouded area in the Gulf of Mexico. The sediment deposition is gradually building Louisiana outward into the gulf. (*B*) Overlapping alluvial fans formed as mountain streams flow out onto dry plain near Albuquerque, New Mexico. Photograph (*A*) © NASA.

Moves faster —outside

Moves slower deposit on inside

Figure 6.6 Development of meanders. Channel erosion is greatest at the outside and on the downstream side of curves, at the cut bank; deposition of point bars occurs in the sheltered areas on the inside of curves. Over time, meanders migrate both laterally and downstream.

Figure 6.7 Braided streams, with many channels dividing and rejoining, in an Alaskan glacial-outwash plain.

feature, an **alluvial fan,** is formed when a tributary stream flows into a more slowly flowing, larger stream, or a stream flows from mountains into a dry plain or desert (figure 6.5B).

An additional factor controlling the particle size of stream sediments is physical breakup and/or dissolution of the sediments during transport. That is, the farther the sediments travel, the longer they are subjected to collision and solution, and the finer they tend to become. Stream-transported sediments may thus tend, overall, to become finer downstream, whether or not the stream's velocity changes along its length.

Channel and Floodplain Evolution

Streams do not ordinarily flow in straight lines for very long. Small irregularities in the channel cause local fluctuations in velocity, which result in a little erosion where the water

flows strongly against the side of the channel and some deposition of sediment where it slows down a bit. Bends, or **meanders,** thus begin to form in the stream. Once a meander forms, it tends to enlarge and also to shift downstream (figure 6.6). It is eroded on the outside and downstream side of the meander, the **cut bank,** where the water flows somewhat faster (and the channel, too, tends to be a little deeper there as a result); **point bars,** consisting of sediment deposited on the insides of meanders, build out the banks in those parts of the channel. The rates of lateral movement of meanders can range up to tens or even hundreds of meters per year, although rates below 10 meters/year (35 feet/year) are more common on smaller streams.

Obstacles or irregularities in the channel may slow flow enough to cause localized sediment deposition there. Over a period of time, if the sediment load of the stream is large, these channel islands can build up until they reach the surface, effectively dividing the channel in a process called *braiding.* If the sediment load is very large in relation to water volume, the **braided stream** may develop a complex pattern of many channels that divide and rejoin, shifting across a broad expanse of sediment (figure 6.7).

Over a period of time, the combined effects of erosion on the cut banks and deposition of point bars on the inside banks of meanders, and downstream meander migration, together produce a broad, fairly flat expanse of land covered with sediment around the stream channel proper. This is the stream's **floodplain,** which is the area into which the stream spills over during floods. Sediment deposition during floods can, in fact, be an important additional factor contributing to the formation of this flat area surrounding the channel. The floodplain, then, is a normal product of stream evolution over time (see figures 6.8 and 6.9).

Meanders do not broaden or enlarge indefinitely. Very large meanders represent major detours for the flowing stream. At times of higher discharge, especially during floods, the stream may make a shortcut, or cut off a meander, abandoning the old, twisted channel for a more direct downstream route. The cutoff meanders are called **oxbows.** These abandoned channels may be left dry, or they may be filled with standing water, making *oxbow lakes.*

Flooding

The size of an unmodified stream channel is directly related to the quantity of water that it usually carries. That is, the volume of the channel is approximately sufficient to accommodate the average maximum discharge reached each year. Much of the year, the surface of the water is well below the level of the stream banks. In times of higher discharge, the stream may overflow its banks, or **flood.** This may happen, to a limited extent, as frequently as every two to three years with streams in humid regions. More severe floods occur correspondingly less often.

The vast majority of stream floods are linked to precipitation (rain or snow). When rain falls or snow melts, some of the water **infiltrates,** or sinks into the ground. It may then *percolate* through soil and rock at greater depths. Some evaporates directly into the atmosphere. The rest of the water becomes surface runoff, flowing downhill over the surface under the influence of gravity into a puddle, lake, ocean, or stream.

Factors Governing Flood Severity

Many factors together determine whether a flood will occur. The quantity of water involved and the rate at which it enters the stream system are the major factors. When the water input exceeds the capacity of the stream to carry that water away downstream within its channel, the water overflows the banks. Worldwide, the most intense rainfall events occur in Southeast Asia, where storms have drenched the region with up to 200 centimeters (80 inches) of rain in less than three days. (To put such numbers in perspective, that amount of rain is more than double the average *annual* rainfall for the United States!) In the United States, several regions are especially prone to heavy rainfall events: the southern states, vulnerable to storms from the Gulf of Mexico; the western coastal states, subject to prolonged storms from the Pacific Ocean; and the midcontinent states, where hot, moist air from the Gulf of Mexico can collide with cold air sweeping down from Canada. Streams that drain the Rocky Mountains are likely to flood during snowmelt, especially when rapid spring thawing follows a winter of unusually heavy snow. Even areas that usually have only moderate precipitation can occasionally experience intense storms. For

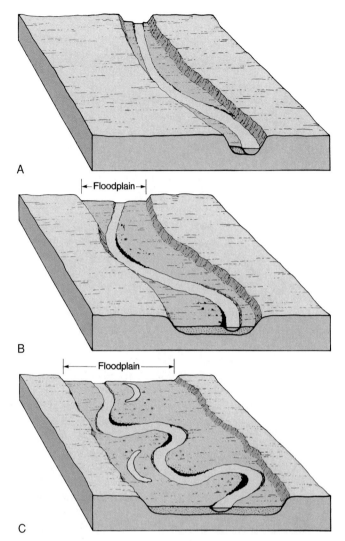

Figure 6.8 Meandering and sediment deposition during floods contribute to development of a floodplain. (*A*) Initially, the stream channel is relatively straight. (*B*) Small bends enlarge and migrate over time. (*C*) Broad, flat floodplain is develped around stream channel.

example, in August 1987, the Chicago area had a record-setting storm that dumped over 9 inches of rain in twenty-four hours, choking storm sewers and paralyzing transportation. (That amount of rain is about one-fourth of the region's average *annual* precipitation.) A period of prolonged heavy rains in December 1996 and January 1997 was primarily responsible for severe flooding in the Pacific Northwest, California, and Nevada.

The rate of surface runoff is influenced by the extent of infiltration, which, in turn, is controlled by the soil type and how much soil is exposed. Soils, like rocks, vary in porosity and permeability. A very porous and permeable soil allows a great deal of water to sink in relatively fast. If the soil is less permeable or is covered by artificial structures, the proportion of water that runs off over the surface increases. Once even permeable soil is saturated with water,

Figure 6.9 Example of a floodplain carved by a meandering stream: Sweetwater River, Wyoming.   Photograph by W. R. Hansen, USGS Photo Library, Denver, CO.

however, any additional moisture is necessarily forced to become part of the surface runoff.

Topography also influences the extent or rate of surface runoff: The steeper the terrain, the more readily water runs off over the surface and the less it tends to sink into the soil. Water that infiltrates the soil, like surface runoff, tends to flow down-gradient, and may, in time, also reach the stream. However, the subsurface runoff water, flowing through soil or rock, generally moves much more slowly than the surface runoff. The more gradually the water reaches the stream, the better the chances that the stream discharge will be adequate to carry the water away without flooding. Therefore, the relative amounts of surface and subsurface (groundwater) runoff, which are strongly influenced by the near-surface geology of the drainage basin, are fundamental factors affecting the severity of stream flooding.

Vegetation may reduce flood hazards in several ways. The plants may simply provide a physical barrier to surface runoff, decreasing its velocity and thus slowing the rate at which water reaches a stream. Also, plant roots working into the soil loosen it, which tends to maintain or increase the

soil's permeability, and hence infiltration, thus reducing the proportion of surface runoff. Plants also absorb water, using some of it to grow and releasing some slowly by evapotranspiration from foliage. All of these factors reduce the volume of water introduced directly into a stream system.

All other factors being equal, local flood hazard may vary seasonally or as a result of meteorologic fluctuations. Just as soil already saturated from previous storms cannot readily absorb more water, so the solidly frozen ground of cold regions prevents infiltration; a midwinter rainstorm in such a region may produce flooding with a quantity of rain that could be readily absorbed by the soil in summer. The extent and vigor of growth of vegetation varies seasonally also, as does atmospheric humidity and thus evapotranspiration.

Flood Characteristics

During a flood, the water level of a stream is higher than usual, and its velocity and discharge also increase as the greater mass of water is pulled downstream by gravity. The higher volume (mass) and velocity together produce the increased force that gives floodwaters their destructive power

131

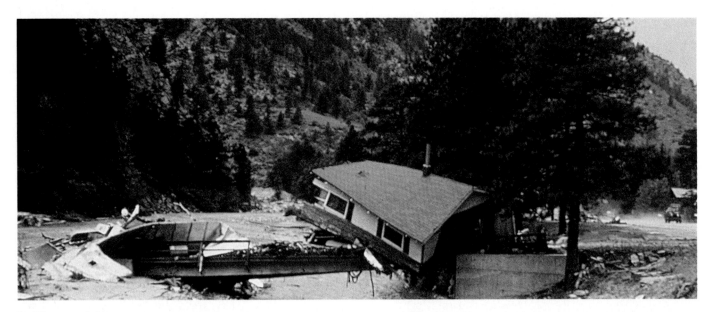

Figure 6.10 Damage from Big Thompson Canyon (Colorado) flooding. The greater force associated with deeper, swifter floodwaters causes severe damage. Rise of stream stage can be particularly pronounced, and rapid, in a narrow canyon such as this, where lateral spread of water is restricted; the flow, too, may be stronger and faster than in flat land, where floodwaters can spread broadly. Photograph courtesy of USGS Photo Library, Denver, CO.

(figure 6.10). The elevation of the water surface at any point is termed the **stage** of the stream. A stream is at *flood stage* when stream stage exceeds bank height. The magnitude of a flood can be described by either the maximum discharge or maximum stage reached. The stream is said to **crest** when the maximum stage is reached. This may occur within minutes of the influx of water, as with flooding just below a failed dam. However, in places far downstream from the water input, or where surface runoff has been slowed, the stream may not crest for several days after the flood episode begins. In other words, just because the rain has stopped does not mean that the worst is over, as was made abundantly evident in the 1993 Midwestern flooding.

Flooding may afflict only a few kilometers along a small stream, or a region the size of the Mississippi River drainage basin, depending on how widely distributed the excess water is. Floods that affect only small, localized areas (or streams draining small basins) are sometimes called **upstream floods.** These are most often caused by sudden, locally intense rainstorms and by events like dam failure. Even if the total amount of water involved is moderate, the rapidity with which it enters the stream can cause it temporarily to exceed the stream channel capacity. The resultant flood is typically brief, though it can also be severe. Because of the localized nature of the excess water during an upstream flood, there is unfilled channel capacity farther downstream to accommodate the extra water. Thus, the water can be rapidly carried down from the upstream area, quickly reducing the stream's stage. The *flash floods* that may follow cloudbursts in desert regions are typical upstream floods; in this case, infiltration into parched ground rather than removal by surface runoff is likely to cause the water level to subside rapidly after the flood crests.

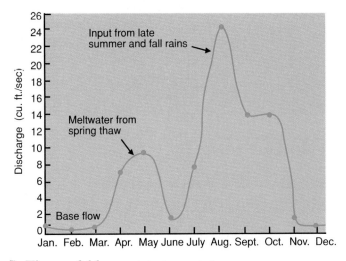

Figure 6.11 Sample hydrograph for one point on Horse Creek near Sugar City, Colorado, showing the stream discharge over a period of one year. Source: U.S. Geological Survey Open-File Report 79–681.

Floods that affect large stream systems and large drainage basins are called **downstream floods.** These floods more often result from prolonged heavy rains over a broad area or from extensive regional snowmelt. Downstream floods usually last longer than upstream floods because the whole stream system is choked with excess water. The spring 1990 flooding in Texas, Oklahoma, and throughout the south-central United States, or the summer 1993 flooding in the Mississippi River basin, are excellent examples of downstream flooding, brought on by weeks of above-average rainfall.

A

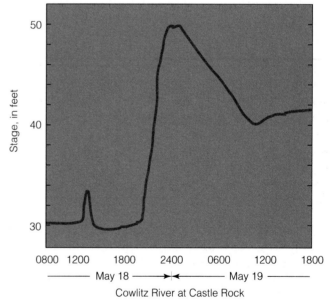

B

Figure 6.12 (*A*) Flood hydrographs for two different points along Calaveras Creek near Elmendorf, Texas. The flood has been caused by heavy rainfall in the drainage basin. In the upstream part of the basin, flooding quickly follows the rain. The larger stream lower in the drainage basin responds more sluggishly to the input. (*B*) Mudflow into the Cowlitz River from the 18 May 1980 eruption of Mount St. Helens caused a sharp rise in stream stage and flooding; infilling of the channel by sediment resulted in a long-term increase in stream stage for given discharge.   Sources: (*A*) Water Resources Division, U.S. Geological Survey. (*B*) After U.S. Geological Survey.

Stream Hydrographs

Fluctuations in stream stage or discharge over time can be plotted on a **hydrograph** (figure 6.11). Hydrographs spanning long periods of time are very useful in constructing a picture of the "normal" behavior of a stream and of that stream's response to flood-causing events. Logically enough, a flood shows up as a peak on the hydrograph. The height and width of that peak and its position in time relative to the water-input event(s) depend, in part, on where the measurements are taken relative to where the excess water is entering the system. Upstream, where the drainage basin is smaller and the water need not travel so far to reach the stream, the peak is likely to be sharper—with a higher crest

and more rapid rise and fall of the water level—and to occur sooner after the influx of water. By the time that water pulse has moved downstream to a lower point in the drainage basin, it will have dispersed somewhat, so that the arrival of the water spans a longer time. The peak will be spread out, so that the hydrograph shows both a later and a broader, gentler peak (see figure 6.12A). A short event like a severe cloudburst will tend to produce a sharper peak than a more prolonged event like several days of steady rain or snowmelt, even if the same amount of water is involved. Sometimes, the flood-causing event also causes permanent changes in the drainage system (figure 6.12B).

Figure 6.13 Flood-frequency curve for Eagle River at Red Cliff, Colorado. Records span forty-six years, so predictions concerning the size of even infrequent floods are likely to be fairly accurate. Source: U.S. Geological Survey Open-File Report 79–1060.

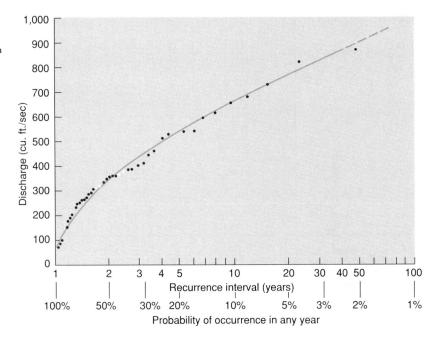

Flood-Frequency Curves

Another way of looking at flooding is in terms of the frequency of flood events of differing severity. Long-term records make it possible to construct a curve showing discharge as a function of recurrence interval for a particular stream or section of one. The sample **flood-frequency curve** shown in figure 6.13 indicates that for this stream, on average, an event producing a discharge of 675 cubic feet/second occurs once every ten years, an event with discharge of 350 cubic feet/second occurs once every two years, and so on. A flood event can then be described by its **recurrence interval:** how frequently a flood of that severity occurs, *on average*, for that stream.

Alternatively, one can refer to the *probability* that a flood of given size will occur in any one year; this probability is the inverse of the recurrence interval. For the stream of figure 6.13, a flood with a discharge of 675 cubic feet/second is called a "ten-year flood," meaning that a flood of that size occurs about once every ten years, or has a 10% ($\frac{1}{10}$) probability of occurrence in any year; a discharge of 900 cubic feet/second is called a "forty-year flood" and has a 2.5% ($\frac{1}{40}$) probability of occurrence in any year, and so on.

Flood-frequency (or flood-probability) curves can be extremely useful in assessing regional flood hazards. If the severity of the one-hundred- or two-hundred-year flood can be estimated in terms of discharge, then, even if such a rare and serious event has not occurred within memory, scientists and planners can, with the aid of topographic maps, project how much of the region would be flooded in such events. They can speak, for example, of the "one-hundred-year floodplain" (the area that would be water-covered in a one-hundred-year flood). Such information is useful in preparing flood-hazard maps, an exercise that is also aided by use of aerial or satellite photography (see box 6.2, figure 3). Such maps, in turn, are helpful in siting new construction projects

to minimize the risk of flood damage or in making property owners in threatened areas aware of dangers.

Unfortunately, the best-constrained part of the curve is, by definition, that for the lower-discharge, more frequent, higher-probability, less-serious floods. Often, considerable guesswork is involved in projecting the shape of the curve at the high-discharge end. Much of the United States has been settled for a century or less. Reliable records of stream stages, discharges, and the extent of past floods typically extend back only a few decades. Many areas, then, may never have recorded a fifty- or one-hundred-year flood. Moreover, when the odd severe flood does occur, how does one know whether it was a sixty-year flood, a one-hundred-year flood, or whatever? The high-discharge events are rare, and even when they happen, their recurrence interval can only be estimated. Considerable uncertainty can exist, then, about just how bad a particular stream's one-hundred- or two-hundred-year flood might be. This problem is explored further in box 6.1.

It is important to remember that the recurrence intervals or probabilities assigned to floods are *averages*. The recurrence-interval terminology may be misleading in this regard. Over many centuries, a fifty-year flood should occur an average of once every fifty years; or, in other words, there is a 2 percent chance that that discharge will be exceeded in any one year. However, that does not mean that two fifty-year floods could not occur in successive years, or even in the same year. The probability of two fifty-year flood events in one year is very low (2% × 2%, or .04%), but it is possible. The Chicago area, in 1986–87, experienced two one-hundred-year floods within a ten-month period. Therefore, it is foolish to assume that, just because a severe flood has recently occurred in an area, it is in any sense "safe" for awhile. Statistically, another such severe flood probably will not happen for some time, but there is no guarantee of that. While it may seem somewhat misleading, the recurrence-interval

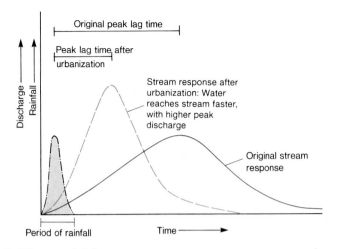

Figure 6.14 Hydrograph reflecting modification of stream response to precipitation following urbanization: Peak discharge increases; lag time to peak decreases.

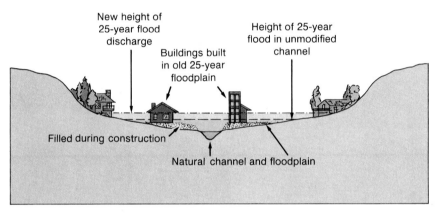

Figure 6.15 How floodplain development increases flood stage for given discharge.

terminology is still in common use by many agencies involved with flood-hazard mapping and flood-control efforts, such as the U.S. Army Corps of Engineers.

Another complication is that streams in heavily populated areas are affected by human activities, as we shall see further in the next section. The way a stream responded to 10 centimeters of rain from a thunderstorm a hundred years ago may be quite different from the way it responds today. The flood-frequency curves are therefore changing with time. Except for measures specifically designed for flood control, most human activities have tended to aggravate flood hazards and to decrease the recurrence intervals of high-discharge events. In other words, what may have been a one-hundred-year flood two centuries ago might be a fifty-year flood, or a more frequent one, today.

The challenge of determining flood recurrence intervals from historical data as described in box 6.1 is not unlike the forecasting of earthquakes by applying the earthquake-cycle model to historic and prehistoric seismicity along a fault zone. The results inevitably have associated uncertainties, and the more complex the particular geologic setting, and the sparser the data, the larger the uncertainties will be.

Consequences of Development in Floodplains

Reasons for Floodplain Occupation

Why would anyone live in a floodplain? One reason might be ignorance of the extent of the flood hazard. Someone living in a stream's one-hundred- or two-hundred-year floodplain may be unaware that the stream could rise that high, if historical flooding has all been less severe. In mountainous areas, floodplains may be the only flat or nearly flat land on which to build, and construction is generally far easier and cheaper on nearly level land than on steep slopes. Around a major river like the Mississippi, the one-hundred- or two-hundred-year floodplain may include a major portion of the land for miles around, as became painfully evident in 1993, and it may be impractical to leave that much real estate entirely vacant. Farmers have settled in floodplains since ancient times because flooding streams deposit fine sediment over the lands flooded, replenishing nutrients in the soil and thus making the soil especially fertile. Where rivers are used for transportation, cities may have been built deliberately as close to the water as possible. And, of course, many streams are very scenic features to live near. Obviously, the more people settle and build in floodplains, the more damage flooding will do. What people often fail to realize is that floodplain development can actually increase the likelihood or severity of flooding.

Effects of Development on Flood Hazards

As mentioned earlier, two factors affecting flood severity are the proportion and rate of surface runoff. The materials extensively used to cover the ground when cities are built, such as asphalt and concrete, are relatively impermeable and greatly reduce infiltration. Therefore, when considerable area is covered by these materials, surface runoff tends to be much more concentrated and rapid than before, increasing the risk of flooding. This is illustrated by figure 6.14. If *peak lag time* is defined as the time lag between a precipitation event and the peak flood discharge (or stage), then, typically, that lag time decreases with increased urbanization, where the latter involves covering land with impermeable materials and/or installing storm sewers. The peak discharge (stage) also increases.

Buildings in a floodplain also can increase flood heights (see figure 6.15). The buildings occupy volume that water formerly could fill, and a given discharge then corresponds to a higher stage (water level). Floods that occur are more serious. Filling in floodplain land for construction similarly decreases the volume available to stream water and further aggravates the situation.

Box 6.1 How Big Is the One-Hundred-Year Flood?

The difficulty of knowing the true recurrence intervals of floods of various magnitudes can be further illustrated by an example.

A common way to estimate the recurrence interval of a flood of given size is as follows: Suppose the records of maximum discharge (or maximum stage) reached by a particular stream each year have been kept for N years. Each of these yearly maxima can be given a rank M, ranging from 1 to n, 1 being the largest, n the smallest. Then the recurrence interval R of a given annual maximum is defined as:

$$R = \frac{(N+1)}{M}$$

This approach assumes that the N years of record are representative of "typical" stream behavior over such a period; the larger N, the more likely this is to be the case.

For example, table 1 shows the maximum one-day mean discharges of the Big Thompson River, as measured near Estes Park, Colorado, for twenty-five consecutive years, 1951–75. If these values are ranked 1 to 25, the 1971 maximum of 1,030 cubic feet/second is the seventh largest and, therefore, has an estimated recurrence interval of

$$\frac{(25+1)}{7} = 3.71 \text{ years,}$$

or a 27 percent probability of occurrence in any year.

Suppose, however, that only ten years of records are available, for 1966–75. The 1971 maximum discharge happens to be the largest in that period of record. On the basis of the shorter record, its estimated recurrence interval is $(10 + 1)/1 = 11$ years,

Table 1   Calculated Recurrence Intervals for Discharges of Big Thompson River at Estes Park, Colorado

Year	Maximum mean one-day discharge (cu. ft. / sec.)	For twenty-five-year record M (rank)	For twenty-five-year record R (years)	For ten-year record M (rank)	For ten-year record R (years)
1951	1,220	4	6.50	3	3.67
1952	1,310	3	8.67	2	5.50
1953	1,150	5	5.20	4	2.75
1954	346	25	1.04	10	1.10
1955	470	23	1.13	9	1.22
1956	830	13	2.00	6	1.83
1957	1,440	2	13.0	1	11.0
1958	1,040	6	4.33	5	2.20
1959	816	14	1.86	7	1.57
1960	769	17	1.53	8	1.38
1961	836	12	2.17		
1962	709	19	1.37		
1963	692	21	1.23		
1964	481	22	1.18		
1965	1,520	1	26.0		
1966	368	24	1.08	10	1.10
1967	698	20	1.30	9	1.22
1968	764	18	1.44	8	1.38
1969	878	10	2.60	4	2.75
1970	950	9	2.89	3	3.67
1971	1,030	7	3.71	1	11.0
1972	857	11	2.36	5	2.20
1973	1,020	8	3.25	2	5.50
1974	796	15	1.73	6	1.83
1975	793	16	1.62	7	1.57

Source: Data from U.S. Geological Survey Open-File Report 79–681.

Measures taken to drain water from low areas can likewise aggravate flooding along a stream. In cities, storm sewers are installed to keep water from flooding streets during heavy rains, and, often, the storm water is channeled straight into a nearby stream. This works fine if the total flow is moderate enough, but by decreasing the time normally taken by the water to reach the stream channel, such measures increase the probability of flooding. The same is true of the use of tile drainage systems in farmland (figure 6.16). Water that previously stood in low spots in the

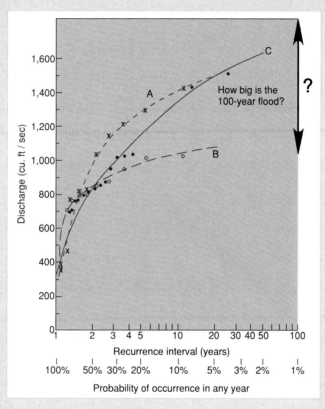

Figure 1   Flood-frequency curves for a given stream can look very different, depending on the length of record available and the particular period studied.

corresponding to a 9 percent probability of occurrence in any year.

Alternatively, if we look at only the first ten years of record, 1951–60, the recurrence interval for the 1958 maximum discharge of 1,040 cubic feet/second (a maximum discharge of nearly the same size as that in 1971) can be estimated at 2.2 years, meaning that it would have a 45 percent probability of occurrence in any one year.

Which estimate is right? Perhaps none of them, but their differences illustrate the need for long-term records to smooth out short-term anomalies in stream-flow patterns.

The point is further illustrated in figure 1. It is rare to have one hundred years or more of records for a given stream, so the magnitude of fifty-year, one-hundred-year, or larger floods is commonly estimated from a flood-frequency curve. Curves *A* and *B* in figure 1 are based, re-

spectively, on the first and last ten years of data from the Big Thompson River (last two columns of table 1). These two data sets give estimates for the size of the one-hundred-year flood that differ by more than 50 percent. These results, in turn, both differ somewhat from an estimate based on the full twenty-five years of data (curve *C*). Figure 1 graphically illustrates how important long-term records can be in the projection of recurrence intervals of larger flood events.

field and only slowly reached the stream after infiltration instead flows swiftly and directly into the stream, increasing the flood hazard for those along the banks.

Both farming and urbanization also disturb the land by removing natural vegetation and, thus, leaving the soil

exposed. The consequences are twofold. As noted earlier, vegetation can decrease flood hazards somewhat by providing a physical barrier to surface runoff, by soaking up some of the water, and through plants' root action, which keeps the soil looser and more permeable. Vegetation

Figure 6.16 How farmland drainage increases the flood hazard in a floodplain. (*A*) Before the installation of tile drainage, some surface runoff was trapped in fields to infiltrate to the stream channel slowly. (*B*) The drainage system transfers water rapidly to the stream and increases the likelihood of flooding.

also can be critical to preventing soil erosion. When vegetation is removed and erosion increased, much more soil can be washed into streams. There, it can fill in, or "silt up," the channel, decreasing the channel's volume and thus reducing the stream's capacity to carry water away quickly.

Strategies for Reducing Flood Hazards

Restrictive Zoning and "Floodproofing"

Short of avoiding floodplains altogether, various approaches can reduce the risk of flood damage. Many of these approaches—restrictive zoning, special engineering practices—are similar to strategies applicable to reducing damage from seismic and other geologic hazards. And commonly, existing structures present particular challenges.

A first step is to identify as accurately as possible the area at risk. Careful mapping coupled with accurate stream discharge data should allow identification of those areas threatened by floods of different recurrence intervals. Land that could be inundated often—by twenty-five-year floods, perhaps—might best be restricted to land uses not involving much building. The land could be used, for example, for livestock grazing pasture or for parks or other recreational purposes.

Unfortunately, there is often economic pressure to develop and build on floodplain land, especially in urban areas. Local governments may feel compelled to allow

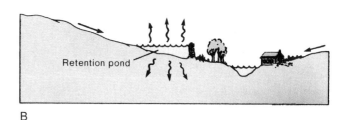

Figure 6.17 Use of retention pond to moderate flood hazard. (*A*) Before the retention pond, fast surface runoff caused flooding along the stream. (*B*) The retention pond traps some surface runoff, prevents the water from reaching the stream quickly, and allows for slow infiltration or evaporation instead.

building at least in the outer fringes of the floodplain, perhaps the parts threatened only by one-hundred- or two-hundred-year floods. In these areas, new construction can at least be designed with the flood hazard in mind. Buildings can be raised on stilts so that the lowest floor is above the expected two-hundred-year flood stage, for example. While it might be better not to build there at all, such a design at least minimizes interference by the structure with flood flow and also the danger of costly flood damage to the building. A major limitation of any floodplain zoning plan, whether it prescribes design features of buildings or bans buildings entirely, is that it almost always applies only to new construction. In many places, scores of older structures are already at risk. Programs to move threatened structures off floodplains are few, and also costly when many buildings are involved. Strategies to modify the stream or the runoff patterns may help to protect existing structures without moving them.

Retention Ponds, Diversion Channels

If open land is available, flood hazards along a stream may be greatly reduced by the use of **retention ponds** (figure 6.17). These ponds are large basins that trap some of the surface runoff, keeping it from flowing immediately into the stream. They may be elaborate artificial structures; old, abandoned quarries; or, in the simplest cases, fields dammed by dikes of piled-up soil. The latter option also allows the land to be used for other purposes, such as farming, except on those rare occasions of heavy runoff when it is needed as a retention pond. Retention ponds are frequently a relatively inexpensive option, provided that ample undeveloped land is available, and they have the added advantage of not altering the character of the stream.

A similar strategy is the use of *diversion channels* that come into play as stream stage rises, and redirect some of the water flow into areas adjacent to the stream where flooding will cause minimal damage. The diversion of water might be into farmland or recreational land, and away from built-up areas, to reduce loss of life and property damage.

Channelization

Channelization is a general term for various modifications of the stream channel itself that are usually intended to increase the velocity of water flow, the volume of the channel, or both. These modifications, in turn, increase the discharge of the stream, and hence the rate at which surplus water is carried away.

The channel can be widened or deepened, especially where soil erosion and subsequent sediment deposition in the stream have partially filled in the channel. Care must be taken, however, that channelization does not alter the stream dynamics too greatly elsewhere. Dredging of Missouri's Blackwater River in 1910 enlarged the channel and somewhat increased the gradient, thereby increasing stream velocities. As a result, discharges upstream from the modifications increased. This, in turn, led to more stream bank erosion and channel enlargement upstream and to increased flooding below.

Alternatively, a stream channel might be rerouted—for example, by deliberately cutting off meanders to provide a more direct path for the water flow. Such measures do tend to decrease the flood hazard upstream from where they are carried out. In the 1930s, the Mississippi River below Cairo, Illinois, was shortened by a total of 185 kilometers (115 miles) in a series of channel-straightening projects, with perceptible reduction in flooding as a result.

A meandering stream, however, often tends to keep meandering or to revert to old meanders. Channelization is not a one-time effort. Constant maintenance is required to limit erosion in the straightened channel sections and to keep the river in the cutoffs. The erosion problem along channelized or modified streams has in some urban areas led to the construction of wholly artificial stream channels made of concrete or other resistant materials. While this may be effective, it is also costly and frequently unsightly as well.

The Mississippi also offers a grand-scale example of the difficulty of controlling a stream's choice of channel: the Atchafalaya. This small branch diverges from the Mississippi several hundred miles above its mouth. Until a logjam blocking it was removed in the last century, it diverted little of the river's flow. Once that blockage was cleared, the water began to flow increasingly down the Atchafalaya, rather than the main branch of the Mississippi. The increased flow in turn scoured a larger channel, allowing more of the water to flow via the Atchafalaya. The water available for navigation on the lower Mississippi, and the sediment it carried to build the Mississippi delta, diminished. In the late 1950s, the U.S. Army Corps of Engineers

Figure 6.18 Levees protect floodplain land from flooding by raising the height of the stream bank.

built the first of a series of structures designed to limit the diversion of water into the Atchafalaya. The 1973 flooding (box 6.2) taxed the engineering, and more substantial structures were built; but whether they represent a permanent solution is unclear.

Finally, a drawback common to many channelization efforts (and one not always anticipated beforehand) is that by causing more water to flow downstream faster, channelization often increases the likelihood of flooding downstream from the alterations. Meander cutoff, for instance, increases the stream's gradient by shortening the channel length over which a given vertical drop occurs.

Levees

The problem that local modifications may increase flood hazards elsewhere is sometimes also a complication with the building of **levees,** raised banks along a stream channel (figure 6.18). Some streams form low, natural levees along the channel through sediment deposition during flood events. These levees may purposely be enlarged (or created where none exist naturally). Because levees raise the height of the stream banks close to the channel, the water can rise higher without flooding the surrounding country. This ancient technique was practiced thousands of years ago on the Nile by the Egyptian pharaohs. It may be carried out alone or in conjunction with channelization efforts, as on the Mississippi.

Confining the water to the channel, rather than allowing it to flow out into the floodplain, however, effectively shunts the water downstream faster during high-discharge events, increasing flood risks downstream. It artificially raises the stage of the stream for a given discharge, which can increase the risks upstream too, as the increase in stage is propagated above the confined section. Another problem is that levees may make people feel so safe about living in the floodplain that, in the absence of restrictive zoning, development will be far more extensive than if the stream were allowed to flood naturally from time to time. If the levees have not, in fact, been built high enough and an unanticipated, severe flood overtops them, or if they simply fail and are breached during a high-discharge event, far more lives and property may be lost as a result. Problems with levees were abundantly demonstrated in the 1993 Mississippi River flooding, when an estimated 80 percent of

Box 6.2 Life on the Mississippi

The Mississippi River is the highest-discharge stream in the United States and the third largest in the world; its drainage basin covers 40 percent of the area of the contiguous forty-eight states (figure 1). It has a relatively long history of settlement, of floods, and of efforts to control those floods. The first levees were built in 1717, after New Orleans was flooded. Further construction of levees continued intermittently over the next century, but the efforts were always local, with no overall regional plan. Following a series of floods in the mid-1800s, there was great public outcry for greater flood-control measures. In 1879, the Mississippi River Commission was formed to oversee and coordinate flood-control efforts over the whole lower Mississippi area. The commission urged the building of more levees. In 1882, when over $10 million had already been spent on levees, severe floods caused 284 breaks in the system and prompted renewed efforts. By the turn of the century, planners were confident that the levees were high enough and strong enough to withstand any flood. Flooding in 1912–13, accompanied by twenty more failures of the levees, proved them wrong. By 1926, there were over 2,900 kilometers (1,800 miles) of levees along the Mississippi, standing an average of 6 meters high; nearly half a billion cubic meters of earth had been used to construct them. Funds totalling more than twenty times the original cost estimates had been spent, and the commission was saying, more cautiously, that the end of the flood-control fight was near.

In 1927, the worst flooding recorded to that date occurred along the Mississippi. More than 75,000 people were forced from their homes; the levees were breached in 225 places; 183 people died; nearly 50,000 square kilometers of land were flooded; and there was an estimated $500 million worth of damage. This was regarded as a crisis of national importance, and thereafter, the federal government took over primary management of the flood-control measures. In the upper parts of the drainage basin, the government undertook more varied efforts—for example, five major flood-control dams were built along the Missouri River in the 1940s in response to flooding there—but the immense volume of the lower Mississippi cannot easily be contained in a few artificial reservoirs. The building up and reinforcing of levees continued to play a major role in flood-control planning.

The fall and winter of 1972–73 were mild and wet over much of the Mississippi's drainage basin. By early spring, the levels in many tributary streams were already high. Prolonged rain in March and April contributed to record-setting flooding in the spring of 1973. The river remained above flood stage for as long as ninety-seven consecutive days in some places. More than 12 million acres (about 50,000 square kilometers) of land were flooded, 50,000 people were evacuated, and over $400 million in damage was done, not counting secondary effects, such as loss of wildlife. During the three-month flood, an estimated 240 million tons of sediment were washed down the Mississippi, much of it fertile soil irrecoverably eroded from farmland.

The devastation of the 1973 floods along the Mississippi occurred in spite

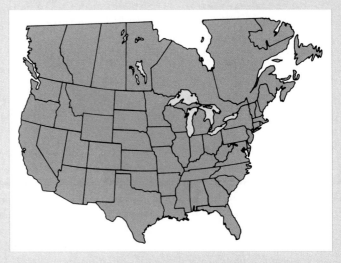

Figure 1 The Mississippi River's drainage basin (darker shading).   Source: U.S. Geological Survey Circular 1001.

of more than 250 years of ever-more-extensive flood-control efforts and still worse flooding was possible (figure 2). A sustained period of heavy rains and thunderstorms in the summer of 1993 led to flooding that caused 50 deaths, over $10 billion in property and crop damage and, locally, record-breaking flood stages and/or levee breaches at various places in the Mississippi basin (figures 3 and 4). The destruction of Vallmeyer, Illinois, was so complete that the town was rebuilt, not on its original site in the floodplain, but on a bluff above it, safe from future flooding at last. Other parts of the Mississippi River drainage basin suffered severe flooding in 1995 and 1996.

Inevitably, the question of what level of flood control is adequate or desirable arises. The greater the flood against which we protect ourselves, the more costly the efforts. We could build still higher levees, bigger dams, and larger reservoirs and spillways, and be safe even from a projected five-hundred-year flood or worse (barring levee failure), but the costs would be astronomical and could exceed the property damage expected from such events. By definition, the probability of a five-hundred-year flood or worse is very low. Precautions against high-frequency flood events seem to make sense. But at what point, for what recurrence interval, does it make more economic or practical sense simply to accept the small but real risk of the rare, disastrous, high-discharge flood events?

Figure 2   The summer 1993 flooding in the upper Mississippi basin was not only unusual in its timing (compare with 1965 and 1973 patterns) but, in many places, broke flood-stage and/or discharge records set in 1973.   Source: Hydrograph data courtesy U.S. Geological Survey.

—Continued

Box 6.2—Continued

Figure 3 Mississippi River Basin flooding, 19 July 1993. The Missouri River (entering from top of image) and Kansas River (left) have spread broadly across their floodplains. Within the flooded areas, the rivers' normal channels are outlined by vegetation (red on image). Kansas City, Missouri, municipal airport is just east of the junction of the rivers. Such images are useful in assessing flood damage. Photograph © NASA.

A

B

Figure 4 Consequences of Mississippi River flooding, 1993. (*A*) Flooding along the riverfront, Davenport, Iowa, summer 1993; note submerged baseball stadium in foreground. (*B*) The flooding turned rural areas into lakes dotted with trees and scattered houses; in places, it was hard to pick the river channel out of the vast expanse of water. Photograph (*A*) © Doug Sherman/Geofile. Photograph (*B*) courtesy of C. S. Melcher, U.S. Geological Survey.

A

B

Figure 6.19 The negative sides of levees as flood-control devices. (*A*) If they are breached, as here in the 1993 Mississippi River floods, those living behind the levees may suddenly and unexpectedly find themselves flooded. (*B*) Breached or overtopped levees dam water behind them. The stream stage may drop rapidly after the flood crests, but the water levels behind the levees remain high. This photograph was taken from the top of a levee, looking downstream; the Kishwaukee River, DeKalb, Illinois, July 1983. Both the levee and sandbagging failed to hold back rising floodwaters. By the time this picture was taken, the water level in the river (right) had dropped several meters. Water trapped behind the levee (left) however, still remained nearly as high as the top of the levee, prolonging the flooding and increasing the damage there.   (*A*) Photograph courtesy C. S. Melcher, U.S. Geological Survey.

privately-built levees failed, as in figure 6.19A (though most federally built levees held). Also, if levees are overtopped, water is then trapped outside the stream channel, where it may stand for some time after the flood subsides until infiltration and percolation return it to the stream (figure 6.19B).

Levees alter sedimentation patterns, too. During flooding, sediment is deposited in the floodplain outside the channel. If the stream and its load are confined by levees, increased sedimentation may occur in the channel. This will raise stream stage for a given discharge—the channel bottom is raised, so stage rises also—and either the levees must be continually raised to compensate, or the channel must be dredged, an additional expense.

Flood-Control Dams and Reservoirs

Yet another approach to moderating stream flow to prevent or minimize flooding is through the construction of flood-control dams at one or more points along the stream. Excess water is held behind a dam in the reservoir formed upstream and may then be released at a controlled rate that does not overwhelm the capacity of the channel beyond. Additional benefits of constructing flood-control dams and their associated reservoirs (artificial lakes) may include availability of the water for irrigation, generation of hydroelectric power at the dam sites, and the development of recreational facilities for swimming, boating, and fishing at the reservoir.

The practice also has its drawbacks, however. Navigation on the river—both by people and by aquatic animals—may be restricted by the presence of the dams. Also, the creation of a reservoir necessarily floods much of the stream valley behind the dam and may destroy wildlife habitats or displace people and their works (for example, a railroad might have been built along a stream valley in a hilly area).

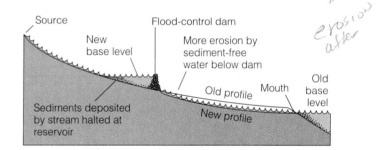

Figure 6.20 Effects of dam construction on a stream: change in base level, sediment deposition in reservoir, and reshaping of the stream channel.

If the stream normally carries a high sediment load, further complications arise. The reservoir represents a new base level for the stream above the dam (figure 6.20). When the stream flows into that reservoir, its velocity drops to zero, and it dumps its load of sediment. Silting-up of the reservoir, in turn, decreases its volume, so it becomes less effective as a flood-control device. Some reservoirs have filled completely in a matter of decades, becoming useless. Others have been kept clear only by repeated dredging, which can be expensive and presents the problem of where to dump the dredged sediment. At the same time, the water released below the dam is free of sediment and thus may cause more active erosion of the channel there. Or, if a large fraction of water in the stream system is impounded by the dam, the greatly reduced water volume downstream may change the nature of vegetation and of habitats there. Recognition of some negative long-term impacts of the Glen Canyon Dam system led, in 1996, to an experiment in controlled flooding of the Grand Canyon (box 6.3).

Box 6.3 Flooding—On Purpose?

The presence of the Glen Canyon Dam upstream from the Grand Canyon on the Colorado River (figure 1) has caused profound changes in dynamics of the river's flow through the Grand Canyon. The dam was built largely for irrigation and generation of hydroelectric power; therefore, a large volume of water was deliberately impounded in Lake Powell, the artificial reservoir behind the dam. The reduced water flow downstream from the dam not only reduced flooding and

sediment transport through the Grand Canyon; it reduced annual fluctuations in water temperature, thereby allowing non-native animal (fish) species to compete with native species, and vegetation to encroach on the river channel, changing the nature of habitats along the river.

A controlled flood, much larger than typical modern streamflow but well below the largest pre-dam floods (figure 2), was planned for early March 1996. Among the many

objectives were the redistribution of sediment, restoration of higher-elevation sand bars and camping beaches, provision of water to higher-elevation vegetation, disruption of life cycles of non-native fishes, and habitat restoration of native fishes.

Stream gages recorded the deliberate flooding (figure 3; note the time-lag of the flood pulse reaching successive gages along the stream), and later studies of channel cross sections (figure 4), and aerial photographs,

Figure 1   An unintended consequence of the Glen Canyon Dam (upper right) is considerable reduction of water flow through the Grand Canyon.   Source: U.S. Geological Survey.

suggested that many of the objectives had been accomplished. Sediment was redistributed, vegetation scoured out of the channel, beaches restored. Assuming that post-flood studies indicate that the objectives were achieved with minimal negative effects, additional periodic deliberate floods will be carried out to maintain the desired effects.

For a more complete description of the rationale for the flooding and the post-flood investigations planned, see

http://h2o.usgs.gov/public/wid/FS_089-96/FS_089-96.html

Real-time streamflow data for the gaging stations in the Grand Canyon shown in figure 1 can be found at

http://wwwdaztcn.wr.usgs.gov/gc_rtsw.html

Post-flood effects on beaches are described at

http://wwwdaztcn.wr.usgs.gov/beach.html

Ongoing investigations of the impacts of the flooding are being conducted, not only by the U.S. Geological Survey, which provided the data for figures in this box; various results will be reported on the Glen Canyon Environmental Studies (GCES) page:

http://phantom.uc.usbr.gov

The purpose of the flooding is further explored at

http://phantom.uc.usbr.gov/pbf.html The GCES site includes links to before-and-after views of selected sites along the river, showing beach changes.

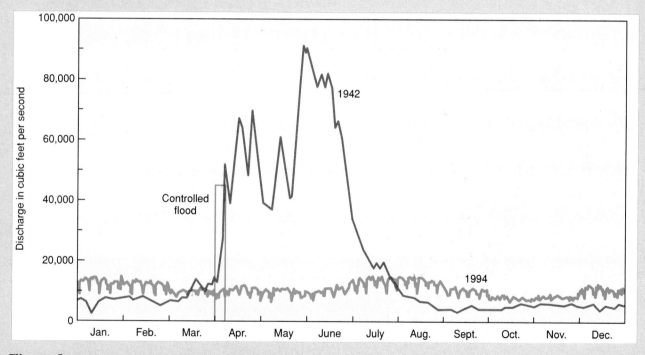

Figure 2   The first of a planned series of controlled floods involved discharges well above modern levels, but below historic flooding of 1942.

—Continued

Box 6.3—Continued

Figure 3 The controlled flood on hydrographs; station locations shown in figure 1. Source: U.S. Geological Survey.

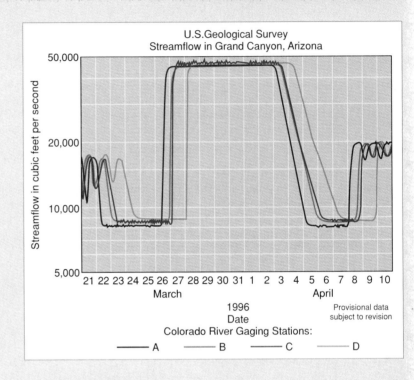

Figure 4 Sample cross section after flooding shows some objectives achieved: Higher water levels allowed point-bar/beach buildup at higher elevations, while higher discharge and faster stream flow during flooding scoured the channel deeper. Source: U.S. Geological Survey.

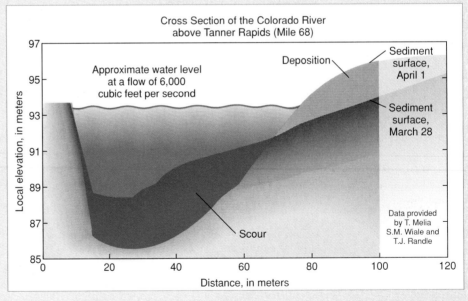

Some large reservoirs, such as Lake Mead behind Hoover Dam, have even been found to cause earthquakes. The reservoir water represents an added load on the rocks, increasing the stresses on them, while infiltration of water into the ground under the reservoir increases pore pressure sufficiently to reactivate old faults. Since Hoover Dam was built, at least ten thousand small earthquakes have occurred in the vicinity. Filling of the reservoir behind the Konya Dam in India began in 1963, and earthquakes were detected within months thereafter. In December 1967, an earthquake of magnitude 6.4 resulted in 177 deaths and considerable damage there. At least ninety other cases of reservoir-induced seismicity are known. Earthquakes caused in this way are usually of low magnitude, but their foci, naturally, are close to the dam. This raises concerns about the possibility of catastrophic dam failure caused, in effect, by the dam itself.

If levees increase residents' sense of security, the presence of a flood-control dam may do so to an even greater extent. People may feel that the flood hazard has been eliminated, and thus neglect even basic floodplain-zoning and other practical precautions. But dams may fail; floods may result from intense rainfall or runoff *below* the dam; and a dam/reservoir complex may not necessarily be managed so as to maximize its flood-control capabilities. For obvious reasons, dams and reservoirs often serve multiple uses (e.g., flood control plus hydroelectric power generation plus water supply). But for maximally effective flood control, one wants maximum water-storage *potential*, while for other uses, it is generally desirable to maximize actual water storage—leaving little extra volume for flood control. The several uses of the dam/reservoir system may thus tend to compete with, rather than to complement, each other.

Overview

Summary

Streams are active agents of sediment transport. They are also powerful forces in sculpturing the landscape, constantly engaged in erosion and deposition, their channels naturally shifting over the earth's surface. Flooding is the normal response of a stream to an unusually high input of water in a short time. Regions at risk from flooding, and the degree of risk, can be identified if accurate maps and records about past floods and their severity are available. However, records may not extend over a long enough period to permit precise predictions about the rare, severe floods. Moreover, human activities may have changed regional runoff patterns or stream characteristics over time, making historical records less useful in forecasting future problems. Strategies designed to minimize flood damage include restricting or prohibiting development in floodplains, controlling the kinds of floodplain development, channelization, and the use of retention ponds, levees, and flood-control dams. Unfortunately, many flood-control procedures have drawbacks, one of which may be increased flood hazards elsewhere along the stream.

Terms to Remember

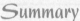

alluvial fan 129	channelization 139
base level 127	crest 132
braided stream 129	cut bank 129
capacity 127	delta 128

discharge 126	levees 139
downstream flood 132	load 127
drainage basin 125	longitudinal profile 127
flood 130	meanders 129
flood-frequency	oxbows 130
curve 134	point bars 129
floodplain 129	recurrence interval 134
gradient 127	retention pond 138
hydrograph 133	stage 132
hydrologic cycle 125	stream 125
hydrosphere 125	upstream flood 132
infiltrates 130	well sorted 128

Exercises

Questions for Review

1. Define stream load, and explain what factors control it.
2. Why do stream sediments tend to be well sorted? (Relate sediment transport to variations in water velocity.)
3. Explain how enlargement and migration of meanders contribute to floodplain development.
4. Discuss the relationship between flooding and (a) precipitation, (b) soil characteristics, (c) vegetation, and (d) season.
5. How do upstream and downstream floods differ? Give an example of an event that might cause each type of flood.

6. What is a flood-frequency curve? What is a recurrence interval? In general, what is a major limitation of these measures?

7. Describe two ways in which urbanization may increase local flood hazards. Sketch the change in stream response as it might appear on a hydrograph.

8. Channelization and levee construction may reduce local flood hazards, but they may worsen the flood hazards elsewhere along a stream. Explain.

9. Outline two potential problems with flood-control dams.

10. List several appropriate land uses for floodplains that minimize risks to lives and property.

For Further Thought

1. Investigate the availability of flood-hazard maps for a nearby stream. On what sorts of data are the maps based? How complete are the data, and how up-to-date are the maps?

2. For a river that has undergone channel modification, investigate the history of flooding before and after that modification, and compare the costs of channelization to the costs of any flood damages. (The Army Corps of Engineers often handles such projects and may be able to supply the information.)

3. Consider the implications of various kinds of flood-plain restrictions and channel modifications that might be considered by the government of a city through which a river flows. List criteria on which decisions about appropriate actions might be based.

4. Check on the latest research results on the Grand Canyon flooding described in box 6.3, and report/comment on how those results relate to the stated flood objectives.

5. Select a major flood event, and search the Internet for (a) data on that flooding, and (b) relevant precipitation data for the same region. Relate the two, commenting on probable factors contributing to the flooding.

Suggested Readings/References

Bolt, B. A., Horn, W. L., MacDonald, G. A., and Scott, R. F. 1975. *Geological hazards.* New York: Springer-Verlag. (Chapter 7 surveys causes of floods, with numerous local examples, and discusses flood-hazard mitigation efforts.)

Chin, E. H., Skelton, J., and Guy, H. P. 1975. *The 1975 Mississippi River basin flood.* U.S. Geological Survey Professional Paper 937.

Floods. 1993. From *The Citizens' Guide to Geologic Hazards,* Amer. Inst. of Professional Geologists, pp. 98–112.

Frank, A. D. 1968. *The development of a federal program of flood control on the Mississippi River.* New York: Columbia University Press.

Illinois Department of Transportation, Division of Water Resources. 1982. *Protect your home from flood damage.* Springfield, Ill.: Illinois Department of Transportation.

Leopold, L. B., and Langbein, W. B. 1966. River meanders. *Scientific American* 214 (June): 60–70.

Mairson, A. 1994. The great flood of '93. *National Geographic* 185 (January): 44–87.

Melcher, N. B., and Parrett, C. 1993. 1993 Upper Mississippi Basin floods. *Geotimes* (December): 15–17.

Morisawa, M. 1985. *Rivers.* New York: Longman.

Petts, G. E., and Foster, G. P. 1985. *Rivers and landscape.* Baltimore, Md.: Edward Arnold.

Richards, K. (ed.). 1987. *River channels: Environment and process.* New York: Blackwell.

Ritter, D. F. 1986. *Process geomorphology.* 2d ed. Dubuque, Ia.: Wm. C. Brown Publishers.

Schultz, E. F., Koelzer, V. A., and Mahmood, K. 1973. *Floods and droughts.* Proceedings of the Second International Symposium in Hydrogeology. Fort Collins, Colo.: Water Resources Publishers.

Tank, R. W. 1983. *Environmental geology, text and readings.* New York: Oxford University Press. (The section headed "Floods," pp. 218–77, includes several readings on responses to flood hazards.)

Watson, B. 1996. A town makes history by rising to new heights. *Smithsonian* (June): 110–120.

Whipple, W., Grigg, N. S., Grizzard, T., Randall, C. W., Shubinski, R. P., and Tucker, L. S. 1983. *Stormwater management in urbanizing areas.* Englewood Cliffs, N.J.:Prentice-Hall.

The Federal Emergency Management Agency (FEMA) offers advice on appropriate actions to take before, during, and after a flood, at

http://www.fema.gov/fema/floodf.html

The U.S. Geological Survey provides a variety of hydrologic and water-resource information. For real-time hydrologic (stream-gaging) data for many states, and links to historical streamflow data and to weather information, see

http://h2o.er.usgs.gov/public/realtime.html

A USGS page specifically related to flood hazards is at http://www.usgs.gov/themes/flood.html

The National Water Conditions site offers streamflow data (current) over the U.S. and southern Canada; it is a joint effort of USGS and Environment Canada's Climate Information Branch, at

http://h2o.usgs.gov/nwc

A series of satellite images pre- and post-flood, for a section of the Missouri River, that demonstrate effects of the 1993 flooding and historical data, can be found at

http://edcwww2.cr.usgs.gov/moberly.html

Links to a number of information sources on the 1993 flooding are assembled at

http://minerva.sws.uiuc.edu/flood.html

Additional images and information on the Scientific Assessment and Strategy Team (SAST), formed to assess the 1993 Mississippi River basin flooding and to develop a long-term floodplain management plan, are found at

http://edcwww.cr.usgs.gov/doc/edchome/sast/sastnews.html

The Dartmouth Flood Observatory at Dartmouth College has created a site through which information on and high-resolution images of floods since 1 January 1994 can be accessed at

http://www.dartmouth.edu/artsci/geog/floods/index.html

Chapter

7

Coastal Zones and Processes

Coastal areas vary greatly in character and in the kinds and intensities of geologic processes that occur along them. They may be dynamic and rapidly changing under the interaction of land and water, or they may be comparatively stable. What happens in coastal zones is of direct concern to a large fraction of the people in the United States: 30 of the 50 states abut a major body of water (Atlantic or Pacific Ocean, Gulf of Mexico, one of the Great Lakes), and those 30 states are home to approximately 85% of the nation's population. To the extent that states become involved in coastal problems, they then affect all those people—in fact, about half of that 85% actually live in the coastal regions, not just in coastal states, and that population is growing rapidly (figure 7.1). Coastal areas provide temporary or permanent homes for 45% of U.S. threatened or endangered species, 30% of waterfowl, and 50% of migratory song birds. In this chapter, we will briefly review some of the processes that occur at shorelines, look at several different types of shorelines, and consider the impacts that various human activities have on them.

Groins and breakwaters trap sediments along the shoreline of Lake Michigan; Lake Forest, Illinois © Doug Sherman/Geofile

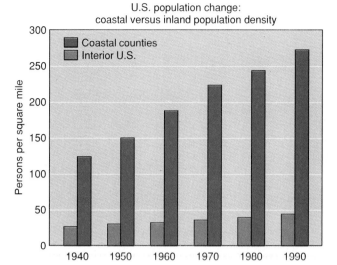

Figure 7.1 The density of population in coastal counties in the U.S. is far greater than in interior counties—and growing rapidly. Source: National Oceanic and Atmospheric Administration, 1990.

Nature of the Coastline

The nature of a coastal region is determined by a variety of factors, including tectonic setting, the materials present at the shore, and the energy with which water strikes the coast.

One factor that influences the geometry of a coastline is plate tectonics. Continental margins can be described as *active* or *passive* in a tectonic sense, depending upon whether there is or is not an active plate boundary at the continent's edge. The western margin of North America is an **active margin,** one at which there is an active plate boundary, with various sections of it involved in subduction or transform faulting. An active margin is often characterized by cliffs above the waterline (figure 7.2), a narrow continental shelf, and a relatively steep drop to oceanic depths offshore. The eastern margin of North America is a **passive margin,** far removed from the active plate boundary at the Mid-Atlantic Ridge. It is characterized by a broad continental shelf and extensive development of broad beaches and sandy offshore islands.

A **beach** is a gently sloping surface washed over by the waves and covered by sediment (figure 7.3). The sand or sediment of a beach may have been produced locally by wave erosion, transported overland by wind from behind the beach, or delivered to the coast and deposited there by streams or coastal currents.

Waves and currents are the principal forces behind natural shoreline modification. Waves are induced by the flow of wind across the water surface, which sets up small undulations in that surface. The shape and apparent motion of waves reflect the changing geometry of the water surface; the actual motion of water molecules is quite different. While a wave may seem to travel long distances across the water, the water molecules are actually rising and falling locally in

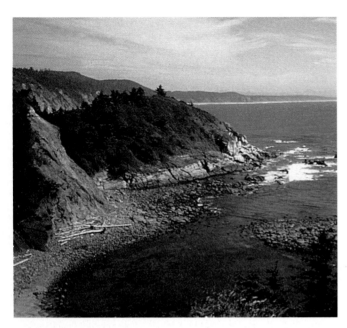

Figure 7.2 The coast of Oregon is marked by cliffs, common along active margins.

circular orbits that grow smaller with depth (figure 7.4A). The depth at which water motion is negligible is the **wave base.** As waves approach the shore and "feel bottom," or in other words when water shallows to wave base, the orbits are disrupted, and, eventually, the waves develop into breakers (figure 7.4B,C).

The most effective erosion of solid rock along the shore is through wave action—either the direct pounding by breakers, or the grinding effect of sand, pebbles, and cobbles propelled by waves, which is called **milling,** or *abrasion.* Logically, the erosive effects are concentrated at the waterline, where breaker action is most vigorous (figure 7.5). If the water is salty, it may cause even more rapid erosion through solution and chemical weathering of the rock.

Water levels vary relative to the elevation of coastal land, over the short term, as a result of tides and storms. *Tides* are the periodic regional rise and fall of water levels as a consequence of the effects of the gravitational pull of the sun and moon on the watery envelope of oceans surrounding the earth. The earth's rotation tends to cause a bulge in this water mass,

Figure 7.3 A beach. (*A*) Schematic beach profile. The beach may be backed by a cliff rather than by sand dunes. (*B*) The various components of the profile are clearly developed on this Hawaiian beach.

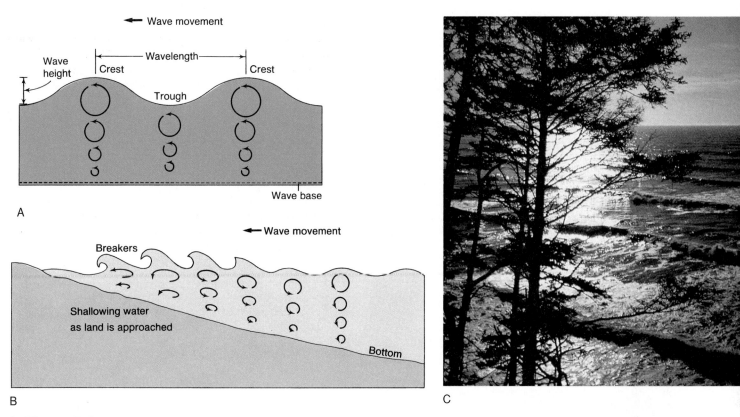

Wave movement →

Wavelength

Wave height

Crest Crest

Trough

Wave base

A

← Wave movement

Breakers

Shallowing water as land is approached

Bottom

B

C

Figure 7.4 (*A*) Waves are undulations in the water surface, beneath which water moves in circular orbits. (*B*) Breakers develop as waves approach the shore and orbits are disrupted. (*C*) Evenly spaced natural breakers developing on a gently sloping shore off the Oregon coast.

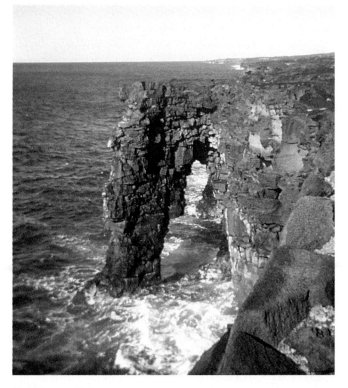

Figure 7.5 Sea arch formed by wave action on lava-flow coastline, Hawaii. Note that erosion is greatest at the waterline, undercutting the rock above.

greatest near the equator where the velocity of the surface is most rapid, and decreasing near the poles. Superimposed on this effect of the earth's rotation are the effects of the gravitational pull of the sun and especially the moon. The closer an object, the stronger its gravitational attraction. The moon therefore pulls most strongly on matter on the side of the earth facing it, least strongly on the opposite side. The combined effects of gravity and rotation cause two bulges in the water envelope: one facing the moon, where the moon's gravitational pull on the water is greatest, and one on the opposite side of the earth, where the gravitational pull is weakest and rotational forces dominate. As the rotating earth spins through these two bulges of water each day, overall water level at a given point on the surface rises and falls twice daily. This is the phenomenon recognized as tides. Tidal extremes are greatest when sun, moon, and earth are all aligned, and the sun and moon are thus pulling together. The resultant tides are *spring tides* (figure 7.6A). (They have nothing to do with the spring season of the year.) When the sun and moon are pulling at right angles to each other, the difference between high and low tides is minimized. These are *neap tides* (figure 7.6B). The magnitude of water-level fluctuations in any one spot is also controlled, in part, by the local underwater topography. Storms can create further variations in water height, and in intensity of erosion, as discussed later in the chapter.

Over longer periods, water levels may shift systematically up or down as a result of tectonic processes or from

A

B

Figure 7.6 Tides. (*A*) Spring tides: sun, moon, and earth are all aligned (near times of full and new moons). (*B*) Neap tides: when moon and sun are at right angles, tidal extremes are reduced.

Figure 7.7 Tectonic uplift of over 3 meters on Latouche Island, Prince William Sound, caused by the 1964 Alaskan earthquake.   Photograph by G. Plafker, USGS Photo Library, Denver, CO.

changes related to glaciation or human activities such as extraction of ground water or oil. These relative changes in land and water height may produce distinctive coastal features, and may also aggravate coastal problems.

Emergent and Submergent Coastlines

Causes of Long-Term Sea-Level Change

As plates move, crumple, and shift, continental margins may be uplifted or dropped down (figure 7.7; recall also figure 4.16). Such movements may shift the land by several meters in a matter of seconds to minutes, and then cease, abruptly changing the geometry of the land/water interface and the patterns of erosion and deposition.

The plastic quality of the asthenosphere means that the lithosphere sits somewhat buoyantly on it, and can sink or rise in response to loading or unloading. For example, in regions overlain and weighted down by massive ice sheets dur-

ing the last ice age, the lithosphere was downwarped by the ice load. Lithosphere is so rigid, and the drag of the viscous mantle so great, that tens of thousands of years later, the lithosphere is still slowly springing back to its pre-ice elevation. Where the thick ice extended to the sea, a consequence is that the coastline is slowly rising relative to the sea. This is well documented, for instance, in Scandinavia, where the rebound still proceeds at rates of up to 2 centimeters per year (close to 1 inch per year). Conversely, in basins being filled by sediment, like the Gulf of Mexico, the sediment loading can cause slow sinking of the lithosphere.

Ice caps, as noted in chapter 9, also represent an immense reserve of water. As this ice melts, sea levels rise worldwide. This is perhaps the most widely experienced ongoing factor in changing coastal water levels; its consequences will be explored further later in the chapter. Global warming could aggravate the sea-level rise in another way, too: warmed water expands, and this effect, multiplied over the vast volume of the oceans, could cause sea-level rise as great as the melting of the ice itself. There is also some concern that, if a large chunk of the Antarctic ice sheet were suddenly to break off and plunge into the ocean, like a giant ice cube dropped into a full glass, a much more sudden sea-level rise would result.

Locally, pumping large volumes of liquid from underground has caused measurable—and problematic—subsidence of the land. In Venice, Italy, groundwater extraction that began in the 1930s had caused subsidence of up to 15 centimeters (about 6 feet) by 1970, when the pumping was stopped—by which time a number of structures had been flooded. Oil extraction in Long Beach, California, caused similar subsidence problems earlier in this century.

Signs of Changing Sea Levels

A distinctive coastal feature that develops where the land is rising and/or the water level is falling is a set of **wave-cut platforms** (figure 7.8). Given sufficient time, wave action

A

B

Figure 7.8 Wave-cut platforms. (*A*) Wave-cut platforms form when land is elevated or sea level falls. (*B*) Wave-cut platforms at Mikhail Point, Alaska.   Photograph (*B*) by J. P. Schafer, USGS Photo Library, Denver, CO.

tends to erode the land down to the level of the water surface. If the land rises in a series of tectonic shifts and stays at each new elevation for some time before the next movement, each rise results in the erosion of a portion of the coastal land down to the new water level. The eventual product is a series of steplike terraces. These develop most visibly on rocky coasts, rather than coasts consisting of soft, unconsolidated material. The surface of each such step—each platform—represents an old water-level marker on the continent's edge.

When, on the other hand, sea level rises or the land drops, one result is that streams that once flowed out to sea now have the sea rising partway up the stream valley from the mouth. A portion of the floodplain may be filled by encroaching seawater, forming a **drowned valley** (figure 7.9). A variation on the same theme is produced by the advance and retreat of coastal glaciers during major ice ages. Glaciers flowing off land and into water do not just keep eroding deeper underwater until they melt. Ice floats; freshwater glacier ice floats especially readily on denser salt water. The carving of glacial valleys by ice, therefore, does not extend long distances offshore. During the last ice age, when sea level worldwide was as much as 100 meters lower than it now is, alpine glaciers at the edges of continental landmasses cut valleys into what are now submerged offshore areas. With the retreat of the glaciers and the concurrent rise in sea level, these old glacial valleys were emptied of ice and partially filled with water. That is the origin of the steep-walled fjords so common in Scandinavian countries.

The particular processes and problems of each coastal region will depend to some extent on the particular characteristics of the region, but some general observations can be made.

Coastal Erosion, Sediment Deposition and Transport

Sand Transport and Beach Erosion

More subtle than the erosion of coastal cliffs is the shifting of sand or its removal from a low-lying beach by **longshore currents** (figure 7.10). When waves approach a beach at an angle, as the water washes up onto and down off the beach, it also moves laterally along the shoreline. Likewise, any sand caught up by and moved along with the flowing water is not carried straight up the beach but is transported at an angle to the shoreline. The net result is **littoral drift,** sand movement along the beach in the same general direction as the motion of the longshore current. Currents tend to move consistently in certain preferred directions on any given beach, which means that, over time, they continually transport sand from one end of the beach to the other. On many beaches where this occurs, the continued existence of the beach is assured by a fresh supply of sediment produced locally by wave erosion or delivered by streams.

Beachfront property owners, concerned that their beaches (and perhaps their houses and businesses also) will wash away, erect structures to try to "stabilize" the beach, which generally end up further altering the beach's geometry. One commonly used method is the construction of one or more groins or jetties (figure 7.11)—long, narrow obstacles set more or less perpendicular to the shoreline. By disrupting the usual flow and velocity patterns of the currents, these structures change the shoreline. Currents slowed by such a barrier tend to drop their load of sand up-current

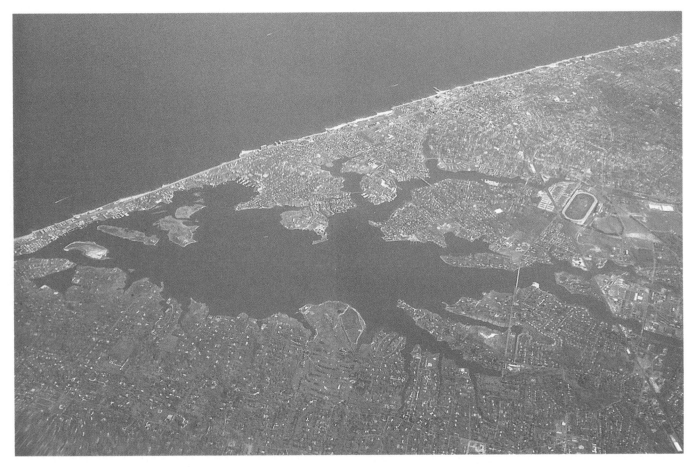

Figure 7.9 A drowned valley along the New England coast.   © David Castillon.

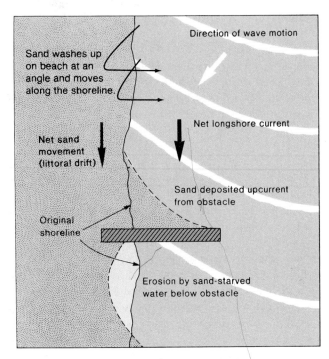

Figure 7.10 Longshore currents and their effect on sand movement. Dashed line indicates shoreline after modification.

from it. Below (down-current from) the barrier, the water picks up more sediment to replace the lost load, and the beach is eroded. The common result is that a formerly stable, straight shoreline develops an unnatural scalloped shape. The beach is built out in the area up-current of the groins and eroded landward below. Beachfront properties in the eroded zone may be more severely threatened after construction of the "stabilization" structures than they were before.

Interference with sediment-laden waters can cause redistribution of sand along beachfronts. A marina built out into such waters may cause some deposition of sand around it and, perhaps, in the protected harbor. Farther along the beach, the now-unburdened waters can more readily take on a new sediment load of beach sand. Breakwaters, too, though constructed primarily to moderate wave action, may cause sediment redistribution (figure 7.12).

Even modifications far from the coast can affect the beach. One notable example is the practice, increasingly common over the last century or so, of damming large rivers for flood control, power generation, or other purposes. As indicated in chapter 6, one consequence of the construction of artificial reservoirs is that the sediment load carried by the stream is trapped behind the dam. Farther downstream, the

A   B

Figure 7.11 Construction of groins and jetties obstructs longshore currents and causes deposition of sand. (*A*) Groins on Rockaway Beach, New York, have produced a scalloped shoreline. (*B*) Cold Spring Inlet, New Jersey. Photograph (*A*) © B. F. Molnia/TERRA PHOTOGRAPHICS/BPS. Photograph (*B*) courtesy of U.S. Army Corps of Engineers.

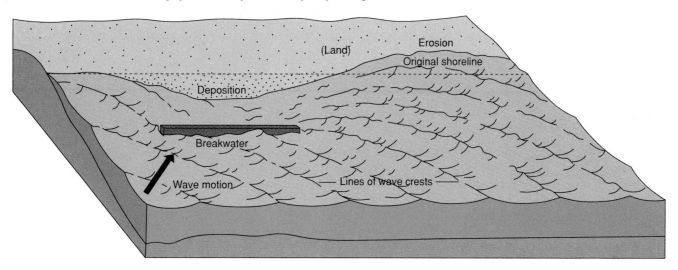

Figure 7.12 Sediment is deposited behind a breakwater; erosion occurs down-current through action of water that has lost its original sediment load. Note that the breakwater may also cause wave refraction.

cutoff of sediment supply to coastal beaches near the mouth of the stream can lead to erosion of the sand-starved beaches. It may be a difficult problem to solve, too, because no readily available alternate supply of sediment might exist near the problem beach areas. Flood-control dams, especially those constructed on the Missouri River, have reduced the sediment load delivered to the Gulf of Mexico by the Mississippi River by more than half over the last thirty-five years, initiating rapid coastal erosion in parts of the Mississippi delta.

Where beach erosion is rapid and development (especially tourism) is widespread, efforts have sometimes been made to import replacement sand to maintain sizeable beaches. The initial cost of such an effort can be very high: for example, in the late 1970s, creation of a new beach

10 miles long and 300 feet wide in Miami Beach cost $65 million. Moreover, if no steps are, or can be, taken to reduce the erosion rates, the sand will have to be replenished over and over, at ever-rising cost, unless the beach is simply to be abandoned. Typically, the reason for beach "nourishment" or "replenishment" efforts is, after all, that natural erosion has removed much of the sand; it is then perhaps not surprising that continued erosion may make short work of the restored beach. Still, it was very likely a shock to those who paid $5.2 million for a 1982 beach-replenishment project in Ocean City, New Jersey, that the new beach lasted only $2\frac{1}{2}$ months. (See also box 7.1.)

In many cases, too, it has not been possible—or has not been thought necessary—to duplicate the mineralogy or

157

Box 7.1 Trouble on the Texas Coast

The Texas coast, like much of the Atlantic margin and Gulf Coast of the United States, is rimmed with barrier islands and barrier peninsulas. The area is particularly vulnerable to alteration by natural forces because it is frequently subject to severe storms, which are accompanied by elevated tides (storm surges, commonly several meters or more above normal high tides), strong winds, and high waves. Over the last century, an average of one tropical storm a year has made landfall somewhere along the Texas coast, and a hurricane has struck more than once every five years. A particularly fierce storm hit Galveston, Texas, in 1900, washing away two-thirds of the city's buildings and causing six thousand deaths. Modern meteorological monitoring and improved communications have greatly reduced the number of hurricane deaths in recent decades, but the shifting shoreline poses a continuing challenge to structures.

A period of quiet weather in the 1950s was accompanied by a rush of development along the Texas coast; another period of accelerated building occurred in the 1970s. The number of structures at risk continues to grow. Meanwhile, the landward retreat of the beaches also continues, at an average rate of 2 to 7 meters (6 to 23 feet) per year. In some especially unstable areas, shoreline changes of more than 20 meters per year have been recorded.

As in other dynamic environments, stabilization efforts have had mixed results. In 1902, in response to the devastation of the 1900 hurricane, the Galveston seawall (figure 1) was

Figure 1   The Galveston seawall. Note absence of beach.   Photograph by W. T. Lee, USGS Photo Library, Denver, CO.

built. The nearly 16-kilometer-long, 6-meter-high structure, built at a cost of $15.5 million (including the cost of raising houses by as much as 4 meters behind the seawall), has offered valuable protection to structures on land, especially during subsequent storms. However, it has also demonstrated some of the permanent changes that can result from seawall construction.

As noted in the chapter, a portion of wave energy is reflected back from smooth-faced seawalls, so the sand in front of them is more actively eroded than it might be in the absence of the wall. Longshore currents are commonly strengthened along a seawall, and the seawall cuts off the supply of sand from any dunes at the back of the beach to the area in front of the

seawall. The result, typically, is gradual loss of the sandy beach in front of the seawall within fifty years of its construction. The beach in front of the Galveston seawall, once up to several hundred meters wide, is virtually gone now. Moreover, the unprotected area at either end of the seawall is being eroded very rapidly. Changes caused by the seawall have made it likely that the seawall will eventually have to be replaced by another, larger, more expensive structure.

Beach replenishment has not often been practiced on the Texas coast, partly for lack of funds and partly for lack of suitable replacement sand. Such an effort was undertaken at Corpus Christi, however, using sand from a nearby river. It has succeeded in the sense that a sandy beach has been preserved there. On the other hand, the river sand is much coarser-grained than the original beach sand. It has stabilized at a steeper slope angle, both above and below the waterline, than characterized the original beach, and the beach has consequently become less suitable for use by small children. Also, of course, the replenishment efforts must continue if the new beach is not to be eroded away in its turn. The vulnerability of the area is illustrated by the breaching of sandy barrier islands by sand washout during storms (figure 2).

Beach-replenishment efforts in barrier-island regions elsewhere on the east coast of the United States have been rather less successful. If "beach life span" is defined as the length of time required for 50% of the new sand to be lost, that life span has been estimated at less than two years in 40% of cases, 2 to 5 years in nearly 50% of cases, and more than five years in only 12% of cases. Such observations have led one investigator to conclude that "(1) the parameters used to design beaches don't work, (2) predictions of beach durability are *always* wrong, and (3) nobody in the coastal engineering community evaluates past projects, so no progress has been made in understanding beach replenishment. In addition the public is unaware of the uncertainties of beach replenishment and, consequently, the taxpayers take it on the chin" (Pilkey 1989 p. 308). Whether or not this is overly pessimistic, experience certainly supports the general truth of two principles: first, that shoreline engineering permanently alters the shoreline, and second, that once shoreline engineering is begun, it must be continued unless structures or beaches are ultimately to be abandoned.

Figure 2 Storm overwash from Hurricane Beulah (1967) reopened old inlets and seriously eroded these barrier islands near Corpus Christi, Texas.   Photograph courtesy of USGS Photo Library, Denver, CO.

Lines of
wave crests

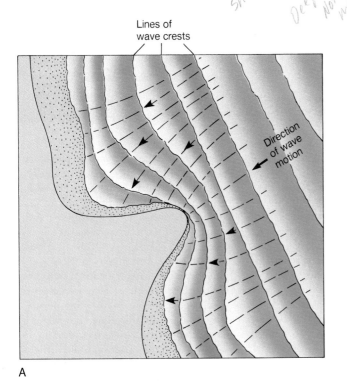

Direction
of wave
motion

A

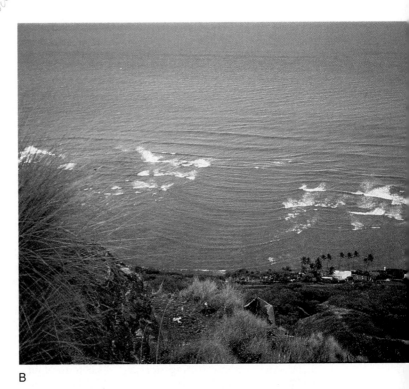

B

Figure 7.13 Wave refraction. (*A*) The energy of waves approaching a jutting point of land is focused on it by refraction. (*B*) Wave refraction caused by nearshore coral reefs, Oahu, Hawaii.

grain size of the sand originally lost. The result has sometimes been further environmental deterioration. When coarse sands are replaced by finer ones, softer and muddier than the original sand, the finer material more readily stays suspended in the water, clouding it. This is not only unsightly, but can also be deadly to organisms. Off Waikiki Beach in Hawaii and off Miami Beach, delicate coral reef communities have been damaged or killed by this extra water turbidity (cloudiness).

Cliff Erosion

Both waves and currents can vigorously erode and transport loose sediments, such as beach sand, much more readily than solid rock. Erosion of sandy cliffs may be especially rapid. Removal of material at or below the waterline undercuts the cliff, leading, in turn, to slumping and sliding of sand sediments and the swift landward retreat of the shoreline.

Not all places along a shoreline are equally vulnerable. Jutting points of land, or headlands, are more actively under attack than recessed bays because wave energy is concentrated on these headlands by **wave refraction,** deflection of the waves around irregularities in the coastline (figure 7.13). Wave refraction occurs because waves "touch bottom" first as they approach the projecting points, which slows the waves down; wave motion continues more rapidly elsewhere. The net result is deflection of the waves around and toward the headlands, where the wave energy is thus focused. (The effect is not unlike the refraction and focusing of light by a lens.) The long-term tendency is toward a

Figure 7.14 As waves approach this section of the Oregon coast, projecting rocks are subject to highest-energy wave action; sand accumulates in the recessed bays.

rounding-out of angular coastline features, as headlands are eroded and sediment is deposited in the lower-energy environments of the bays (figure 7.14).

Various measures, often as unsuccessful as they are expensive, may be tried to combat cliff erosion. A fairly common practice is to place some kind of barrier at the base of

A

B

Figure 7.15 Some shore-protection structures. (*A*) Riprap near El Granada, California. The structure was built in 1973; it may have to be abandoned within decades as erosion of the sandy cliff continues unabated on either side of the riprap. (*B*) Seawall protects estate at Newport, Rhode Island.   (*A*) Photograph courtesy of USGS Photo Library, Denver, CO.

the cliff to break the force of wave impact (figure 7.15 and box 7.1). The protection may take the form of a solid wall (*seawall*) of concrete or other material, or a pile of large boulders or other blocky debris (*riprap*). If the obstruction is placed only along a short length of cliff directly below a threatened structure (as in figure 7.15A), the water is likely to wash in beneath, around, and behind it, rendering it largely ineffective. Erosion continues on either side of the barrier; loss of the protected structure will be delayed but probably not prevented. Wave energy may also bounce off, or be reflected from, a short length of smooth seawall to attack a nearby unprotected cliff with greater force.

Another approach to direct shoreline protection is to erect breakwaters farther away from the shore, parallel to it, to reduce the energy of the pounding waves. This may slow the erosion but, again, is unlikely to stop it, especially if the cliffs are sandy or otherwise made of weak or unconsolidated materials.

Especially Difficult Coastal Environments

Many coastal environments are unstable, but some, including barrier islands and estuaries, are particularly vulnerable either to natural forces or to human interference.

Barrier Islands

Barrier islands are long, low, narrow islands paralleling a coastline somewhat offshore from it (figure 7.16). Exactly how or why they form is not known. Some theories suggest that they have formed through the action of longshore currents on delta sands deposited at the mouths of streams. Their formation may also require changes in sea level. However they have formed, they provide important protection for the water and shore inland from them because they con-

Figure 7.16 Barrier islands: North Carolina's outer banks. Photograph by R. Dolan, USGS Photo Library, Denver, CO.

stitute the first line of defense against the fury of high surf from storms at sea. The barrier islands themselves are extremely vulnerable, partly as a result of their low relief. Water several meters deep may wash right over these low-lying islands during unusually high storm tides, such as occur during hurricanes (figure 7.17).

Because they are usually subject to higher-energy waters on their seaward sides than on their landward sides, most barrier islands are retreating landward with time. Typical rates of retreat on the Atlantic coast of the United States are 2 meters (6 feet) per year, but rates in excess of 20 meters per

Figure 7.17 Consequences of storm surges on North Carolina's outer banks. Photograph by R. Dolan, USGS Photo Library, Denver, CO.

A

year have been noted. Clearly, such settings represent particularly unstable locations in which to build, yet the aesthetic appeal of long beaches has led to extensive development on privately owned stretches of barrier islands (see, for example, figure 7.18). About 1.4 million acres of barrier islands exist in the United States, and approximately 20 percent of this total area has been developed. Thousands of structures, including homes, businesses, roads, and bridges, are at risk.

On barrier islands, shoreline-stabilization efforts—building groins and breakwaters and replenishing sand—tend to be especially expensive and, frequently, futile. At best, the benefits are temporary, as noted in box 7.1. Construction of artificial stabilization structures may easily cost tens of millions of dollars and, at the same time, destroy the natural character of the shoreline and even the beach, which was the whole attraction for developers in the first place. At least as costly an alternative is to keep moving buildings or rebuilding roads ever farther landward as the beach before them erodes. Expense aside, this is clearly a "solution" for the short term only, and in many cases, there is no place left to which to retreat anyway. Considering the inexorable rise of sea level (see next page) and that more barrier-island land is being submerged more frequently, the problems will only get worse.

Estuaries

An **estuary** is a body of water along a coastline, open to the sea, in which the tide rises and falls and in which fresh and salt water meet and mix to create brackish water. San Francisco Bay, Chesapeake Bay, Puget Sound, and Long Island Sound are examples. Some estuaries form at the lower ends of stream valleys, especially in drowned valleys. Others may be tidal flats in which the water is more salty than not.

B

Figure 7.18 Many barrier-island areas are heavily developed but highly vulnerable. (*A*) South Dade County, Florida. (*B*) Homes on North Carolina's outer banks threatened after storm surge; these houses later had to be moved. Photograph (*A*) by P. Zobel, (*B*) by R. Dolan, both courtesy of USGS Photo Library, Denver, CO.

Over time, the complex communities of organisms in estuaries have adjusted to the salinity of that particular water. The salinity itself reflects the balance between freshwater input, usually river flow, and salt water. Any modifications that alter this balance change the salinity and can have a catastrophic impact on the organisms. Also, water circulation in estuaries is often very limited. This makes them especially vulnerable to pollution; because they are not freely flushed out by vigorous water flow, pollutants can accumulate. Unfortunately, many vital wetlands areas are estuaries under pressure from environmental changes. One factor is that many of the world's large coastal cities—San Francisco and New York, for example—are located beside estuaries that may receive polluted effluent water. Another is that demand for the fresh water that would otherwise flow into estuaries is diminishing that flow. The salinity of San Francisco Bay has been rising markedly as fresh water from rivers flowing into the bay is consumed en route.

Land Reclamation from Estuaries

The problem with estuaries is not so much that natural changes in estuaries affect us but that we may be changing the estuaries for the worse. For example, where land is at a premium, estuaries are frequently called on to supply more. They may be isolated from the sea by dikes and pumped dry or, where the land is lower, wholly or partially filled in. Naturally, this further restricts water flow and also generally changes the water's chemistry. In addition, development may be accompanied by pollution. All of this greatly stresses the organisms of the estuary. Depending on the local geology, additional problems may arise. It was already noted in chapter 4 that buildings erected on filled land rather than on bedrock suffer much greater damage from ground shaking during earthquakes. This was abundantly demonstrated in San Francisco in 1906 and again in 1989. Yet the pressure to create dry land for development where no land existed before is often great.

One of the most ambitious projects involving land reclamation from the sea in an estuary is that of the Zuider Zee in the Netherlands (figure 7.19). The estuary itself did not form until a sandbar was breached in a storm during the thirteenth century. Two centuries later, with a growing population wanting more farmland, the reclamation was begun. Initially, the North Sea's access to the estuary was dammed up, and the flow of fresh water into the Zuider Zee from rivers gradually changed the water in it from brackish to relatively fresh. (This was necessary if the reclaimed land was to be used for farming; most crops grow poorly, if at all, in salty ground.) Portions were filled in to create dry land, while other areas were isolated by dikes and then pumped dry. More than half a million acres of new land have been created here.

Certainly, this has been a costly and prolonged effort. The marine organisms that had lived in the Zuider Zee estuary were necessarily destroyed when it was converted to a freshwater system. Continual vigilance over and mainte-

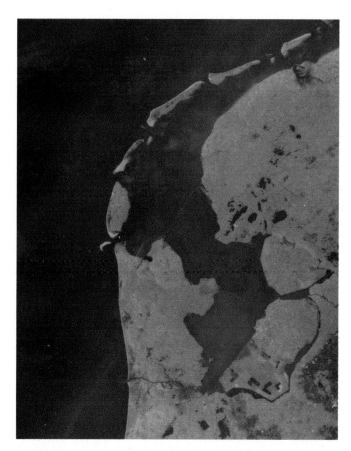

Figure 7.19 The Zuider Zee, a partially filled estuary in the Netherlands (Landsat satellite photo). Patches of reclaimed land (polders) are at lower right. Most recent efforts are aimed at reclaiming the west polder, visible as lighter patch of shallow water at bottom center of photo.   Photograph © NASA.

nance of the dikes is required because if the farmlands were flooded with salt water they would be unsuitable for growing crops for years.

In the Netherlands, where land is at a premium, the benefits of filling in the Zuider Zee may outweigh the costs, environmental and economic. Destruction of *all* estuaries would, however, result in the elimination of many life forms uniquely adapted to their brackish-water environment. Even organisms that do not live their whole lives in estuaries may use the estuary's quiet waters as breeding grounds, and so these populations also would suffer from habitat destruction. Where other land exists for development, it may be argued that the estuaries are too delicate and valuable a system to tamper with.

Further Thoughts on Coastal Dynamics

Present and Future Sea-Level Trends

Much of the coastal erosion presently plaguing the United States is the result of gradual but sustained sea-level rise, probably from the melting of the remaining polar ice. The rise

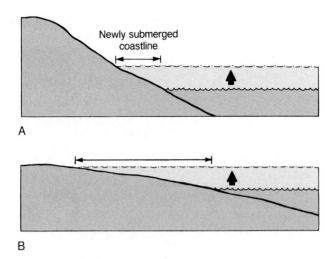

Figure 7.20 Effects of small rises in sea level. (*A*) On steeply sloping shoreline. (*B*) On gently sloping shoreline: A little increase in depth may flood a large area. Dashed lines indicate new, higher sea level.

is currently estimated at about $\frac{1}{3}$ meter (1 foot) per century. While this does not sound particularly threatening, two additional factors should be considered. First, many coastal areas slope very gently or are nearly flat, so that a small rise in sea level results in a far larger inland retreat of the shoreline (figure 7.20). Rates of shoreline retreat from rising sea level have, in fact, been measured at several meters per year in some low-lying coastal areas. A sea-level rise of $\frac{1}{3}$ meter (about 1 foot) would erode beaches by 15 to 30 meters in the northeast, 65 to 130 meters in California, and up to 300 meters in Florida—and the average commercial beach is only about 30 meters wide now. Second, the documented rise of atmospheric carbon-dioxide levels, discussed in chapter 9, suggests that increased greenhouse-effect heating will melt the remaining ice caps more rapidly, accelerating the sea-level rise; some estimates put the anticipated rise in sea level at 1 meter by the year 2100. Aside from the havoc this would create on beaches, it would flood 30 to 70 percent of U.S. coastal wetlands.

Consistently rising sea levels are a major reason why shoreline-stabilization efforts repeatedly fail. It is not just that the high-energy coastal environment presents difficult engineering problems. The problems themselves are intensifying as rising water levels bring waves farther and farther inland, pressing ever closer to and more forcefully against more and more structures developed along the coast.

Storms and Coastal Erosion

Superimposed on long-term sea-level change are other water-level fluctuations of shorter duration. Unconsolidated materials, such as beach sand, are especially rapidly eroded during storms. The low air pressure associated with a storm causes a bulge in the water surface. This, coupled with strong onshore winds, can result in unusually high tides during the storm, a storm **surge.** The temporarily elevated water level associated with the surge, together with unusually energetic wave action, combine to attack the coast with exceptional force. The

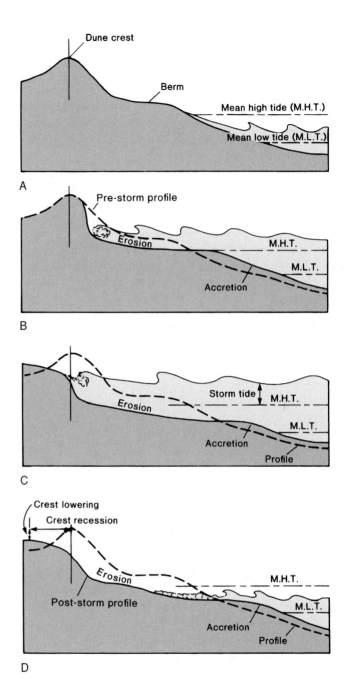

Figure 7.21 Alteration of shoreline profile due to accelerated erosion by unusually high storm tides. (*A*) Normal wave action. (*B*) Initial attack of storm waves. (*C*) Storm wave attack of foredune. (*D*) After storm wave attack, normal wave action.   Source: Shore Protection Guidelines, U.S. Army Corps of Engineers.

gently sloping expanse of beach (*berm*) along the outer shore above the usual high-tide line may suffer complete overwash, with storm waves reaching and beginning to erode dunes or cliffs beyond (figure 7.21). The post-storm beach profile typically shows landward recession of dune crests, as well as removal of considerable material from the zone between dunes and water (figure 7.21D; recall also figure 7.18B).

Hurricanes Hugo in 1989 and Andrew (figure 7.22A) in 1992 illustrated the impact of storm surges especially

A

B

Figure 7.22 Hurricanes and cyclones raise sea levels in storm surges, whip up high waves with high winds. (*A*) Hurricane Andrew, 25 August 1992; this storm caused an estimated $25 billion in damage, making it the most costly natural disaster in U.S. history. Sample of damage at Lakes by the Bay, Florida shown in (*B*). At its peak, Andrew, generally a category 4 hurricane, reached the category 4/category 5 boundary.   Images courtesy of NOAA.

Table 7.1 Saffir-Simpson Hurricane Scale

Category	Wind speed (in miles per hour)	Storm surge (feet above normal tides)	Evacuation
No. 1	74–94	4.5	No.
No. 2	96–110	6–8	Some shoreline residences and low-lying areas, evacuation required.
No. 3	111–130	9–12	Low-lying residences within several blocks of shoreline, evacuation possibly required.
No. 4	131–155	13–18	Massive evacuation of all residences on low ground within 2 miles of shore.
No. 5	155	18	Massive evacuation of residential areas on low ground within 5–10 miles of shore possibly required.

Source: National Weather Service, NOAA.

dramatically. To appreciate the magnitude of the effects, see table 7.1, and note that (considering the gentle slopes of many coastal regions) evacuations miles inland may be necessary in response to an anticipated surge of "only" 10 to 20 feet. At least hurricanes are now tracked well enough that there is warning to make evacuation possible. Still, the damage to immobile property can be impressive (figure 7.22B).

Costs of Construction— and Reconstruction—in High-Energy Environments

U.S. federal disaster relief funds, provided in barrier-island and other coastal areas, can easily run into hundreds of millions of dollars after each major local storm. Increasingly, the

1971   1914   1866

Figure 7.23 Building about to be lost to cliff erosion. This cliff at Moss Beach in San Mateo County, California, has retreated more than 50 meters in a century. Past cliff positions are known from old maps and photographs; arrows indicate the position of the cliff at corresponding dates.   Photograph by K. R. Lajoie, USGS Photo Library, Denver, CO.

question is asked, does it make sense to continue subsidizing and maintaining development in such very risky areas? More and more often, the answer being given is "No!" From 1978 to 1982, $43 million in federal flood insurance (see chapter 18) was paid in damage claims to barrier-island residents, which far exceeded the premiums they had paid. (The houses in figure 7.18B were moved in 1980 with federal flood-insurance program funds.) So, in 1982, Congress decided to remove federal development subsidies from about 650 miles of beachfront property and to eliminate in the following year the availability of federal flood insurance for these areas. It simply did not make good economic sense to encourage unwise development through such subsidies and protection. Still, more than a million flood-insurance policies representing over $1 billion in coverage were still outstanding half a dozen years later in vulnerable coastal areas.

Most barrier-island areas are, in fact, owned by federal, state, or local governments. Even where they are undeveloped, much money has been spent maintaining access roads and various other structures, often including beach-protection structures. It is becoming increasingly obvious that the most sensible thing to do with many such structures is to abandon the costly and ultimately doomed efforts to protect or maintain them, and simply to leave these areas in their dynamic, natural, rapidly changing state, most often as undeveloped recreation areas. Certainly there are many safer, more stable, and more practical areas in which to build.

Recognition of Coastal Hazards

Conceding the attraction of beaches and coastlines and the likelihood that people will want to continue living along them, can at least the most unstable or threatened areas be identified so that the problems can be minimized? That is often possible, but it depends both on observations of present conditions and on some knowledge of the area's history.

The best setting for building near a beach or on an island, for instance, is at a relatively high elevation (5 meters or more above normal high tide, to be above the reach of most storm tides), and in a spot with many high dunes between the proposed building site and the water, to supply added protection. Thick vegetation, if present, will help to stabilize the beach sand. Also, information about what has happened in major storms in the past is very useful. Was the site flooded? Did the overwash cover the whole island? Were protective dunes destroyed? A key factor in determining a "safe" elevation, too, is overall water level. On a lake, not only the short-term range of storm tide heights but also the long-term range in lake levels must be considered. (See box 7.2.) On a seacoast, it would be important to know if that particular stretch of coastline was emergent or submergent over the long term. Around the Pacific Ocean basin, possible danger from tsunamis should not be overlooked.

On either beach or cliff sites, one very important factor is the rate of coastline erosion. Information might be obtained from people who have lived in the area for some time. Better and more reliable guides are old aerial photographs, if available from the U.S. or state geological survey, county planning office, or other sources, or detailed maps made some years earlier that will show how the coastline looked in past times. Comparison with the present configuration and knowledge of when the photos were taken or maps made (information that is usually available) allows estimation of the rate of erosion (figure 7.23). It also should be kept in mind that the shoreline retreat may be more rapid in the future than it has been in the past as a

Box 7.2 Great Lakes Shoreline Erosion on the Rise?

The Great Lakes are freshwater lakes, but they are large enough to share many of the coastal dynamics of the seacoast, including significant shoreline erosion.

The water levels in the Great Lakes fluctuate for a variety of reasons. An obvious factor is variation in the input from precipitation and associated runoff to the lakes. Winds and storm surges temporarily raise local lake levels. Evaporation, water consumption in areas adjacent to the lakes, and outflow reduce water levels. Outflow from one lake to the next is through dams, locks, and power-generation facilities. Lake levels can, therefore, be regulated artificially, and the International Joint Commission makes policy decisions concerning lake-level adjustments.

Periods of high water peaked in 1929, the early 1950s, 1968–69 (on Lakes Superior and Erie), and 1973–74. A 1971 study showed more than one-third of the Great Lakes' coastline to be eroding significantly, despite the presence in some places of shoreline protection structures. The low-lying beaches may expand during periods of low water and shrink as water levels rise, but the march of the cliffs is in one direction only: inland (figure 1). In those places most severely under attack, cliff edge retreat has occurred at rates averaging more than 2 meters (6 feet) per year for over a decade. The costs and impacts of high water levels, especially when areas of extensive development are threatened, are enormous (figure 2). The problems have been recognized, and solutions sought, for many years. But shoreline protection is, at best, expensive and, particularly when the water keeps rising, of limited long-term value.

The mid-1980s were another time of abnormally high water in the Great Lakes. Since 1987, lake levels have subsided somewhat. It remains to be seen if the waters will soon begin to rise again. Still, even if the water continues to subside, the cliffs will hardly restore themselves!

A

B

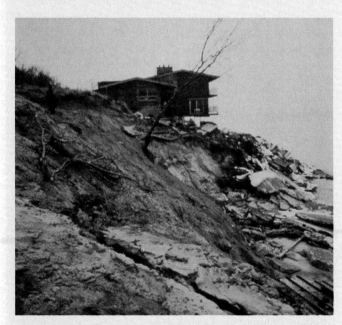

Figure 1   House on a cliff imperiled by erosion, Lake Michigan shoreline.   Photograph by Paul Sequeira, courtesy of EPA/National Archives.

Figure 2   Damages from periods of high water on the Great Lakes. These expenses reflect 1973 price levels. (*A*) Cost of shoreland damage during 1972–74, 1975, and 1976, according to state. The total cost of protective measures plus damages was $401 million. (*B*) Cost of shoreland damage during 1972–74, 1975, and 1976, by lake area. Total damages were $231 million; total cost of protective measures was $170 million.

consequence of a faster-rising sea level, as discussed earlier. On cliff sites, too, there is landslide potential to consider, and in a seismically active area, the dangers are greatly magnified.

It is advisable to find out what shoreline modifications are in place or are planned, not only close to the site of interest but elsewhere along the coast. These can have impacts tens of kilometers away from where they are actually built. Sometimes, aerial or even satellite photographs make it possible to examine the patterns of sediment distribution and movement along the coast, which should help in the assessment of the likely impact of any shoreline modifications. The fact that such structures exist or are being contemplated is itself a warning sign! A history of repairs to or rebuilding of structures not only suggests a very dynamic coastline, but also the possibility that protection efforts might have to be abandoned in the future for economic reasons.

Overview

Summary

Many coastal areas are rapidly changing. Accelerated by rising sea levels worldwide, erosion is causing shorelines to retreat landward in most areas, often at rates of more than a meter a year. Sandy cliffs and barrier islands are especially vulnerable to erosion. Efforts to stabilize beaches are generally expensive, often ineffective over the long (or even short) term, and frequently change the character of the shoreline. They may also cause unforeseen negative consequences to the coastal zone and its organisms. Demand for development has led not only to construction on unstable coastal lands, but also to the reclamation of estuaries to create more land, to the detriment of plant and animal populations.

Terms to Remember

active margin 151	milling 152
barrier islands 161	surge 164
beach 151	passive margin 151
drowned valley 155	wave base 152
estuary 162	wave-cut platform 154
littoral drift 155	wave refraction 160
longshore current 155	

Exercises

Questions for Review

1. High storm tides may cause landward recession of dunes. Explain this phenomenon, using a sketch if you wish.
2. Evaluate the use of riprap and seawalls as cliff protection structures.
3. Explain longshore currents and how they cause littoral drift.
4. Sketch a shoreline on which a jetty has been placed to restrict littoral drift; indicate where sand erosion and deposition will subsequently occur and how this will reshape the shoreline.
5. Discuss the pros and cons of sand replenishment as a strategy for stabilizing an eroding beach.
6. Name three ways in which the relative elevation of land and sea may be altered. What is the present trend in global sea level?
7. Briefly explain the formation of (a) wave-cut platforms and (b) drowned valleys.
8. What are barrier islands, and why have they proven to be particularly unstable environments for construction?
9. What is an estuary? Why do estuaries constitute such distinctive coastal environments?
10. Name at least two ways in which the dynamics of a coastline over a period of years can be investigated.

For Further Thought

1. Choose any major coastal city, find out how far above sea level it lies, and determine what proportion of it would be inundated by a relative rise in sea level of (a) 1 meter and (b) 5 meters. Consider what defensive actions might be taken to protect threatened structures in each case. (It may be instructive to investigate what has happened to Venice, Italy, under similar circumstances.)
2. Consider the vulnerability of barrier-island real estate, and think about what kinds of protections and controls you might advocate: How much should a prospective purchaser of the property be told about beach stability? Should local property owners pay for shoreline-protection structures, or should state or federal taxes be used? Should property owners be prevented from developing the land, for their own protection? What would you consider the most important factors in your decisions? What are the main unknowns?
3. If you live in a near-coastal area, calculate how soon water would reach your home at a rate of sea-level rise of about 1 meter per century.
4. After a major coastal storm (hurricane, typhoon), search the Internet for information on the size and the

physical and economic consequences. A place to start would be the National Oceanic and Atmospheric Administration at

http://www.ncdc.noaa.gov/

Look also for any special sites on the Internet that may have been created specifically for exchange of information about the event.

Suggested Readings/References

American Geological Institute. 1981. Old solutions fail to solve beach problems. *Geotimes* (December): 18–22.

Bird, E. C. F. 1985. *Coastline changes: A global review.* New York: John Wiley & Sons.

Carter, C. H., et al. 1987. *Living with the Lake Erie shore.* Durham, N.C.: Duke University Press.

Carter, W. 1988. *Coastal environments.* New York: Academic Press.

Dolan, R., and Goodell, H. G. 1986. Sinking cities. *American Scientist* 74 (January-February): 38–47.

Environmental Protection Agency. 1989. "Can our coasts survive more growth?" *EPA Journal* 15 (September/October). This whole issue is devoted to coastal problems.

Gray, A. J. 1977. Reclaimed land. Chap. 13 in *The coastline,* edited by R. S. K. Barnes. New York: John Wiley & Sons.

Great Lakes Basin Commission. 1979. The ups and downs of lake level regulation. *Great Lakes Communicator* 9 (March).

Heikoff, J. M. 1980. *Marine and shoreland resources management.* Ann Arbor, Mich.: Ann Arbor Science Publishers.

Horton, T. 1993. Hanging in the balance: Chesapeake Bay. *National Geographic* 183 (June): 3–35.

International Joint Commission. 1995. Information about the Commission's role and activities. [online]: http://www.great-lakes.net:2200/partners/IJC/html/whatis.html

MacLeish, W. H., ed. 1981. The coast. *Oceanus* 23 (4).

Morton, R. A., Pilkey, O. R. Jr., Pilkey, O. R. Sr., and Neal, W. J. 1983. *Living with the Texas shore.* Durham, N.C.: Duke University Press.

Peterson, D., Cayan, D., DiLeo, J., Noble, M., and Dettinger, M. 1995. The role of climate in estuarine variability. *American Scientist* 83 (January-February): 58–67.

Pilkey, O. H. 1989. The engineering of sand. *Journal of Geological Education* 37 (November): 308–11.

Rahn, P. H. 1986. Engineering geology of coastal regions. Ch. 10 in *Engineering geology.* New York: Elsevier.

Schuck-Kolben, R. E., and Kaufman, L. 1990. Hurricane Hugo, September 1989. *U.S. Geological Survey Yearbook 1990,* 6–8.

Thompson, D. A., and Yocom, T. G. 1993. Uncertain ground. *Technology Review* (August-September): 20–29.

U.S. Army Corps of Engineers. 1971. *Shore protection guidelines.* Washington, D.C.: U.S. Government Printing Office.

Walden, D. 1990. Raising Galveston. *American Heritage of Invention and Technology* (Winter): 8–18.

Williams, S. J., and Stewart, R. 1996. Understanding coastal wetlands. *Geotimes* (July): 18–22.

The National Oceanic and Atmospheric Administration (NOAA), alone or in collaboration with other agencies, conducts a great deal of oceanographic and coastal research. For example, see the following:

Marine geology and geophysics
http://www.ngdc.noaa.gov/mgg/index.HTML
Great Lakes Environmental Research Labs
http://www.glerl.noaa.gov/data/data.html
NOAA/National Geophysical Data Center support to coastal programs
http://www.ngdc.noaa.gov/coast.html

USGS Western Region Marine and Coastal Surveys includes links to case studies and illustrations of a variety of coastal and nearshore areas at
http://walrus.wr.usgs.gov/docs/erosion.html

For information on USGS Programs in the Great Lakes— including coastal issues—see
http://water.usgs.gov/public/wid/html/GL.html

The Coastal Ocean Processes (CoOP) home page, supported by NSF, NOAA, and the office of Naval Research, is at
http://www.coop.hpel.und.edu

The U.S. Fish and Wildlife Service Coastal Ecosystems Program deals more with coastal biological environments than coastal erosion; see
http://www.fws.gov/~cep/cepcode.html

Federal Emergency Management Agency Fact Sheet on Hurricanes:
http://femapubl.fema.gov/fema/hurricaf.html

For online images of hurricanes and other major storms:
http://www.ncdc.noaa.gov/pub/data/images/olimages.html

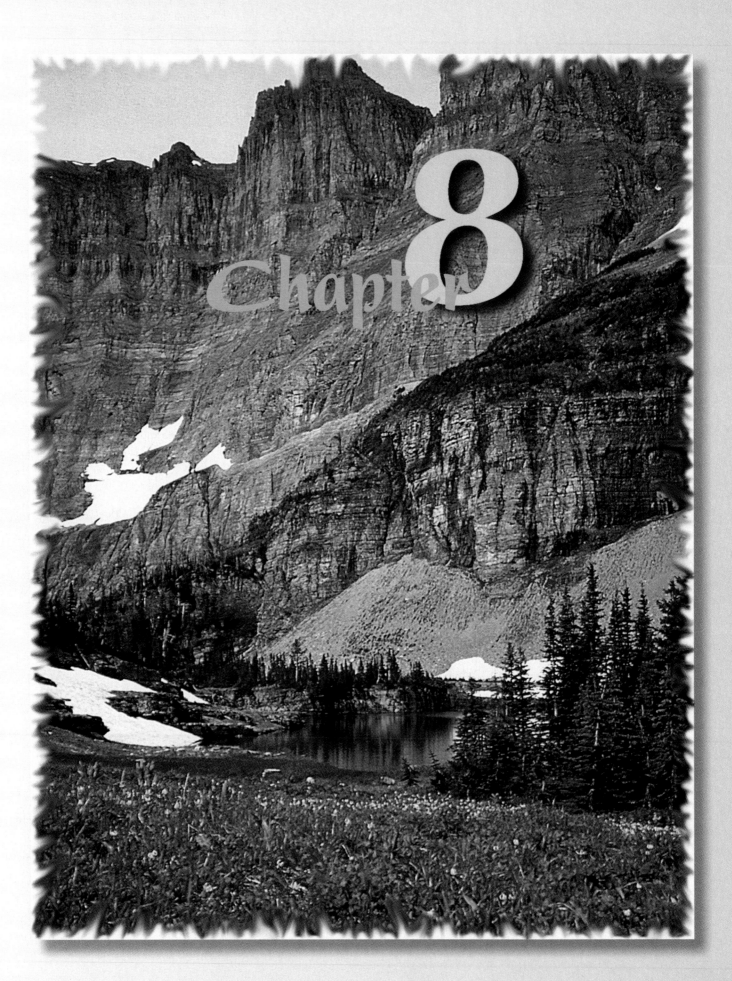

Chapter **8**

Mass Movements

While the internal heat of the earth drives mountain-building processes, just as inevitably, the force of gravity acts to tear the mountains down. Gravity is the great leveler. It tugs constantly downward on every mass of material everywhere on earth, causing a variety of phenomena collectively called **mass wasting,** or **mass movements,** whereby geological materials are moved downward, commonly downslope, from one place to another. The movement can be slow, subtle, almost undetectable on a day-to-day basis but cumulatively large over days or years. Or the movement can be sudden, swift, and devastating, as in a rockslide or avalanche.

Landslide is a general term for the results of rapid mass movements. Overall, landslide damage is greater than one might imagine from the usually brief coverage of landslides in the news media. In the United States alone, landslides and related mass movements cause over $1.5 billion in property damage every year, though losses of human life are relatively small. Many landslides occur quite independently of human activities. In some areas, active steps have been taken to control downslope movement or to limit its damage. On the other hand, certain human activities aggravate local landslide dangers, usually as a result of failure to take those hazards into account.

Large areas of this country—and not necessarily only mountainous regions—are potentially at risk from landslides (figure 8.1). It is important to realize, too, that given the area represented in that figure, it is necessarily a generalization, representing *regional* landslide hazard levels associated with the characteristic regional topography, soil type, etc. One can live in a high-risk spot within an area generally classified as low-risk on this map!

Piles of talus from rockfalls accumulate at the base of steep cliffs in the Rocky Mountains.

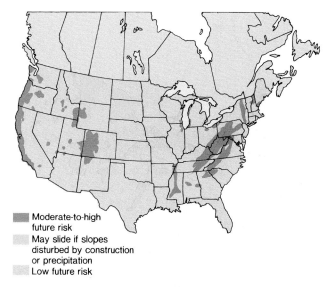

Moderate-to-high future risk

May slide if slopes disturbed by construction or precipitation

Low future risk

Figure 8.1 Landslide potential of regions in the United States. Source: After U.S. Geological Survey Professional Paper 950, 1978.

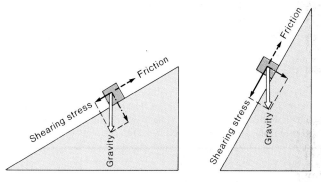

Figure 8.2 Effects of slope geometry on slide potential. The mass of the block and thus the total downward pull of gravity are the same in both cases, but the steeper the slope, the greater the shearing stress component. An increase in pore pressure can decrease frictional resistance to shearing stress.

Factors Influencing Slope Stability

Basically, mass movements occur whenever the downward pull of gravity overcomes the forces—usually frictional—resisting it. The downslope pull tending to cause mass movements, called the **shearing stress,** is related to the mass of material and the slope angle. Counteracting the shearing stress is *friction*, or, in a coherent solid, **shear strength.** When shearing stress exceeds frictional resistance or the shear strength of the material, sliding occurs. Therefore, factors that increase shearing stress, decrease friction, or decrease shear strength tend to increase the likelihood of sliding, and vice versa. Application of these simple principles is key to reducing landslide hazards.

Effects of Slope

All else being equal, the steeper the slope, the greater the shearing stress (figure 8.2), and therefore the greater the likelihood of slope failure. For dry, unconsolidated material,

A

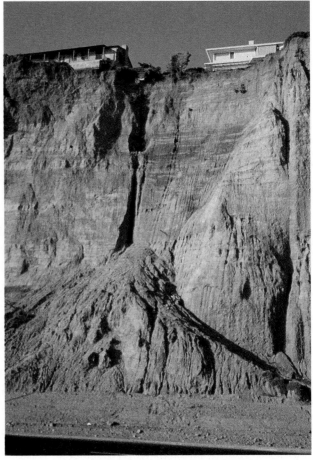

B

Figure 8.3 (*A*) Angle of repose indicates an unconsolidated material's resistance to sliding. Coarser and rougher particles can maintain a steeper stable slope angle. (*B*) This California cliff holds its near-vertical slope only temporarily; eventually the sediment collapses to assume its angle of repose, illustrated by the sloping surface of the sediment pile at the base of the cliff. Photograph by J. T. McGill, USGS Photo Library, Denver, CO.

the **angle of repose** is the maximum slope angle at which the material is stable (figure 8.3). This angle varies with the material. Smooth, rounded particles tend to support only very low-angle slopes (imagine trying to make a heap of marbles or ball bearings), while rough, sticky, or irregular particles can be piled more steeply without becoming unstable. Other properties being equal, coarse fragments can usually maintain a steeper slope angle than fine ones. The tendency of a given material to assume a constant characteristic slope can be seen in such diverse geologic forms as cinder cones (figure 8.4A) and sand dunes (figure 8.4B).

Solid rock can be perfectly stable even at a vertical slope but may lose its strength if it is broken up by weathering or fracturing. Also, in layered sedimentary rocks, there may be weakness along bedding planes, where different rock units are imperfectly held together; some units may themselves be weak or even slippery (clay-rich layers, for example). Such planes of weakness are potential slide or failure planes.

Slopes may be steepened to unstable angles by natural erosion by water or ice. Erosion also can undercut rock or soil, removing the support beneath a mass of material and thus leaving it susceptible to falling or sliding. This is a common contributing factor to landslides in coastal areas and along stream valleys (figure 8.5).

Over long periods of time, slow tectonic deformation also can alter the angles of slopes and bedding planes, making them steeper or shallower. This is most often a significant factor in young, active mountain ranges, such as the Alps or the coast ranges of California. Steepening of slopes by tectonic movements has been suggested as the cause of a large rockslide in Switzerland in 1806 that buried the township of Goldan under a block of rock nearly 2 kilometers long, 300 meters across, and 60 to 100 meters thick. In such a case, past evidence of landslides may be limited in the immediate area because the dangerously steep slopes have not existed long in a geologic sense.

Effects of Fluid

Aside from its role in erosion, water can greatly increase the likelihood of mass movements in other ways. For example, it can seep along bedding planes in layered rock, reducing friction and making sliding more likely. An increase in pore-water pressure in saturated rocks decreases the rocks' resistance to shearing stress, tending to cause faulting and/or sliding. As was described in chapter 4, elevated pore-fluid pressure can trigger earthquakes by facilitating slip in faulted rocks under tectonic stress. It is also effective in promoting sliding in rocks under stress due to gravity.

A

B

Figure 8.4 (*A*) The angle of repose of volcanic cinders is shown not only by the smoothly sloping sides of each individual cinder cone in the summit crater of Haleakala volcano in Hawaii, but also by the consistent slope and shape from cone to cone. (*B*) The uniform slope on the face of this sand dune is maintained by slumping to the sand's angle of repose. Oregon Sand Dunes National Recreation Area.

Figure 8.5 Failure of slope under new road, Floresta Creek Valley, Brazil. Photograph by F. O. Jones, USGS Photo Library, Denver, CO.

While addition of limited moisture to dry soils may increase adhesion, the tendency of the particles to stick together (it takes damp sand to make a sand castle), saturation of unconsolidated materials reduces the friction between particles that otherwise provides cohesion and strength, and the reduced friction can destabilize a slope. The very mass of water in saturated soil may add enough extra weight, enough additional shearing stress, to set off a landslide on a slope that was stable when dry.

Also, the expansion and contraction of water freezing and thawing in cracks in rocks or in soil can act as a wedge to drive chunks of material apart. *Frost heaving*, the expansion of wet soil as it freezes and the ice expands, loosens and displaces the soil, which may then be more susceptible to sliding when it next thaws.

Some soils rich in clays absorb water readily; one type of clay, montmorillonite, may take up twenty times its weight in water and form a very weak gel. Such material fails easily under stress. Other clays expand when wet, contract when dry, and can destabilize a slope in the process (figure 8.6; see also chapter 19).

Sudden, rapid landslides commonly involve a triggering mechanism. Heavy rainfall or rapid melting of abundant snow can be a very effective trigger, quickly adding weight, decreasing friction, and increasing pore pressure. Tropical regions subject to intense cloudbursts may also be subject to frequent slides in thick tropical soils—which may tend to be clay-rich—and in severely weathered rock (figure 8.7).

Effects of Vegetation

Plant roots, especially those of trees and shrubs, can provide a strong interlocking network to hold unconsolidated materials

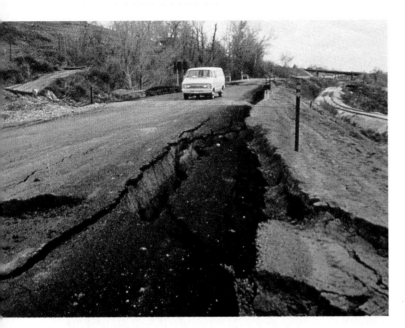

Figure 8.6 Instability in expansive clay seriously damaged this road near Boulder, Colorado. *Photograph courtesy of USGS Photo Library, Denver, CO.*

Figure 8.7 Landslide (mudflow) following heavy rains in Brazil. *Photograph by F. O. Jones, USGS Photo Library, Denver, CO.*

Figure 8.8 Examination of this Alaskan roadcut shows surface soil held in place by grass roots, erosion and slope failure below.

Earthquakes

Landslides are a common consequence of earthquakes in hilly terrain, as noted in chapter 4. Seismic waves passing through rock stress and fracture it. The added stress may be as much as half that already present due to gravity. Ground shaking also jars apart soil particles and rock masses, reducing the friction that holds them in place. In the 1964 Alaskan earthquake, the Turnagain Heights section of Anchorage was heavily damaged by landslides (recall figure 4.23C). California contains not only many fault zones but also many sea cliffs and hillsides prone to landslides during earthquakes.

One of the most lethal earthquake-induced landslides occurred in Peru in 1970. An earlier landslide had already occurred below the steep, snowy slopes of Nevados Huascarán, the highest peak in the Peruvian Andes, in 1962, without the help of an earthquake. It had killed approximately thirty-five hundred people. In 1970, a magnitude-7.7 earthquake centered 130 kilometers to the west shook loose a much larger debris avalanche that buried most of the towns of Yungay and Ranrachirca and more than eighteen thousand people (figure 8.9). Once sliding had begun, escape was impossible: Some of the debris was estimated to have moved at 1,000 kilometers per hour (about 600 miles per hour).

together and prevent flow (figure 8.8). In addition, vegetation takes up moisture from the upper layers of soil and can thus reduce the overall moisture content of the mass, increasing its shear strength. Moisture loss through the vegetation by transpiration also helps to dry out sodden soil more quickly. Commonly, then, vegetation tends to increase slope stability. However, the plants also add weight to the slope. If the added weight is large and the root network of limited extent, the vegetation may have a destabilizing effect instead.

A

Quick Clays

A geologic factor that contributed to the 1964 Anchorage landslides and that continues to add to the landslide hazards in Alaska, California, parts of northern Europe, and elsewhere is a material known as "quick" or "sensitive" clay. **True quick clays** (figure 8.10) are most common in northern polar latitudes. The grinding and pulverizing action of massive glaciers can produce a **rock flour** of clay-sized particles, less than 0.02 millimeter (0.0008 inch) in diameter. When this extremely fine material is deposited in a marine environment, and the sediment is later uplifted above sea level by tectonic movements, it contains salty pore water. The sodium chloride in the pore water acts as a glue, holding the clay particles together. Fresh water subsequently infiltrating the clay washes out the salts, leaving a delicate, honeycomblike structure of particles. Seismic-wave vibrations break the structure apart, reducing the strength of the quick clay by as much as twenty to thirty times, creating a finer-grained equivalent of quicksand that is highly prone to sliding. Nor is a large earthquake necessary to trigger failure; vibrations from passing vehicles can also do it. So-called **sensitive clays** are somewhat similar in behavior to quick clays but may be formed from different materials. Weathering of volcanic ash, for example, can produce a sensitive-clay sediment. Such deposits are not uncommon in the western United States.

B

Figure 8.9 Nevados Huascarán debris avalanche, Peru, 1970. (*A*) Aerial view. (*B*) Closeup of toe of slide. Note that view is rotated about 90° from photograph in (*A*).   Photographs courtesy of USGS Photo Library, Denver, CO.

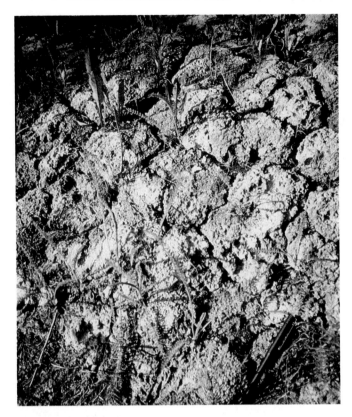

Figure 8.10 Example of quick clay. Turnagain Heights, Anchorage, Alaska.

Impact of Human Activities

Given the factors that influence slope stability, it is possible to see many ways in which human activities increase the risk of mass movements. One way is to clear away stabilizing vegetation. In some instances, where clear-cutting logging operations have exposed sloping soil, for example, landslides of soil and mud have occurred far more frequently or been more severe than before.

Many types of construction lead to oversteepening of slopes. Highway roadcuts, quarrying or open-pit mining operations, and construction of stepped home-building sites on hillsides are among the activities that can cause problems (figure 8.11). Where dipping layers of rock are present, removal of material at the bottom ends of the layers may leave large masses of rock unsupported, held in place only by friction between layers. Slopes cut in unconsolidated materials at angles higher than the angle of repose of those materials are by nature unstable, especially if there is no attempt to plant stabilizing vegetation (figure 8.12). In addition, putting a house above a naturally unstable or artificially steepened slope adds weight to the slope, thereby increasing the shear stress acting on the slope. Other activities connected with the presence of housing developments on hillsides can increase the risk of landslides in more subtle ways. Watering the lawn, using a septic tank for sewage disposal, and even putting in an in-ground swimming pool from which water can seep slowly out are all activities that increase the moisture content of the soil and render the slope more susceptible to slides. On the other hand, the planting of suitable vegetation can reduce the risk of slides.

Recent building-code restrictions have limited development in a few unstable hilly areas, and some measures have been taken to correct past practices that contributed to landslides. (Chapter 18 includes an illustration of the effectiveness of legislation to restrict building practices in Los Angeles County.) However, landslides do not necessarily cease just because ill-advised practices that caused or accelerated them have been stopped. Once activated or reactivated, slides may continue to slip for decades. The area of Portuguese Bend in the Palos Verdes Peninsula in Los Angeles County is a case in point.

In the 1950s, housing developments were begun in Portuguese Bend. In 1956, a 1-kilometer-square area began to slip, though the slope in the vicinity was less than 7 degrees, and within months, there had been 20 meters of movement. What

Figure 8.11 Effects of construction and human habitation on slope stability.

Figure 8.12 Failure of a steep, unvegetated slope along a roadcut threatens the cabin above.

A

B

Figure 8.13 Damage from the Portuguese Bend landslide. (*A*) Sliding has separated driveway from garage. (*B*) Logs prop up house to prevent collapse.   Photographs courtesy of Martin Reiter.

activated this particular slide is unclear. Most of the homes used cesspools for sewage disposal, which added fluid to the ground, potentially increasing fluid pressure. On the other hand, the county highway department, in building a road across what became the top of this slide, added a lot of fill, and thus weight. In any event, the resultant damage was extensive: cracks developed within the sliding mass; houses on and near the slide were damaged or destroyed; a road built across the base of the slide had to be rebuilt repeatedly. Over $10 million in property damage resulted from the slippage (figure 8.13). Worse, the movement has continued for decades since, slow but unstoppable, at a rate of about 3 meters per year. Some portions of the slide have moved 70 meters.

Human activities can increase the hazard of landslides in still other ways. Irrigation and use of septic tanks increase the flushing of water through soils and sediments. In areas underlain by quick or sensitive clays, these practices may hasten the washing-out of salty pore waters and the destabilization of the clays, which may fail after the soils are drained. Artificial reservoirs may cause not only earthquakes, but landslides, too. As the reservoirs fill, pore pressures in rocks along the sides of the reservoir increase, and the strength of the rocks to resist shearing stress can be correspondingly decreased. The case of the Vaiont reservoir disaster (box 8.1) is a classic in this respect. Chapter 19 describes other examples of ways in which engineering activities may influence land stability.

Types of Mass Wasting

When downslope movement is quite slow, even particle by particle, the motion is described as **creep.** Soil creep, which is often triggered by frost heaving, occurs more commonly than rock creep. Though gradual, creep may nevertheless

Box 8.1 The Vaiont Reservoir Disaster: A Landslide of Human Making

The Vaiont River flows through an old glacial valley in the Italian Alps. The valley is underlain by a thick sequence of sedimentary rocks, principally limestones with some clay-rich layers, that were folded and fractured during the building of the Alps. The Alps are a relatively young mountain range, and the rocks appear still to be under some tectonic stress. The sedimentary units of the Vaiont Valley are in the form of a syncline, or trough-shaped fold, so the beds on either side of the valley dip down toward the valley (figure 1). The rocks themselves are relatively weak and particularly prone to sliding along the clay-rich layers. Evidence of old rockslides can be seen in the valley and was noted in drill cores taken in the early 1960s shortly after construction of the dam was completed. Extensive solution of the carbonates by ground water further weakens the rocks.

The Vaiont Dam, built for power generation, is the highest "thin-arch" dam in the world. It is made of concrete and stands over 265 meters (875 feet) high at its highest point. Modern engineering methods were used to stabilize the rocks in which it is based. The capacity of the reservoir behind the dam was 150 million cubic meters—initially.

The obvious history of landslides in the area originally led some to object to the dam site. Later, the very building of the dam and subsequent filling of the reservoir aggravated an already-precarious situation. As the water level in the reservoir rose following completion of the dam, pore pressures of ground water in the rocks of the reservoir walls rose also. This tended to buoy up the rocks and to swell the clays of the clay-rich layers, further decreasing their strength, making sliding easier.

In 1960, a block of 700,000 cubic meters of rock slid from the slopes of Monte Toc on the south wall into the reservoir. Creep was noted over a still greater area. A set of monitoring stations was established on the slopes of Monte Toc to track any further movement. Measurable creep continued. In 1960–61, the rate of creep occasionally reached 25 to 30 centimeters (10 to 12 inches) per week, and the total volume of rock affected by the creep was estimated at about 200 million cubic meters. In the year or two following, creep rates declined, to an average of 1 centimeter per week, which hardly caused great concern.

Late summer and early fall of 1963 were times of heavy rainfall in the Vaiont Valley. The saturated rocks represented more mass pushing downward on zones of weakness. The reservoir level rose by over 20 meters, and pore pressures in the rocks increased. By mid-September 1963, measured creep rates also had increased again, to about 1 centimeter per day. What was still not realized was that the rocks were slipping not in a lot of small blocks, but as a single, coherent mass.

Animals sensed the danger before people did. On October 1, animals that had been grazing on those slopes of Monte Toc moved off the hillside, and reportedly refused to return.

The rains continued, and so did the creep. Creep rates grew to 20 to 30 centimeters per day. Finally, on

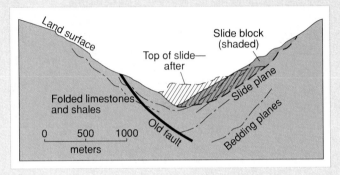

Figure 1 Geologic cross section of the Vaiont valley. From G. A. Kiersch, "The Vaiont Reservoir Disaster," *Mineral Information Service* 18(7):129–138, 1965. State of California, Division of Mines and Geology.

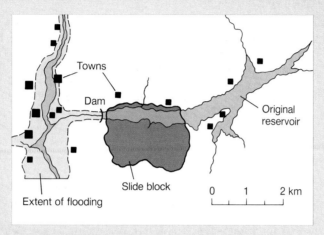

Figure 2 The landslide at the Vaiont reservoir and the resultant flood damage.   From G. A. Kiersch, "The Vaiont Reservoir Disaster," *Mineral Information Service* 18(7):129–138, 1965. State of California Division of Mines and Geology.

October 8, the engineers realized that all the creep-monitoring stations were moving together. They also discovered that a far larger area of hillside was moving than they had thought. They tried to reduce the water level in the reservoir by opening the gates of two outlet tunnels. But the water level continued to rise: The silently creeping mass had begun to encroach on the reservoir, decreasing its capacity significantly. On the night of October 9, in the midst of another downpour, the disaster struck.

A resident of the town of Casso, on the slopes above the reservoir, later reported that, at first, there was just a sound of rolling rocks, ever louder. Then a gust of wind hit his house, smashing windows and raising the roof so that the rain poured in. The wind suddenly died; the roof collapsed.

A chunk of hillside, over 240 million cubic meters in volume, had slid into the reservoir (figure 2). The shock of the slide was detected by seismom-eters in Rome, in Brussels, and elsewhere around Europe. That immense movement, occurring in less than a minute, set off the shock wave of wind that rattled Casso and drew the water 240 meters *upslope* out of the reservoir after it. Displaced water crashed over the dam in a wall 100 meters (over 325 feet) above the dam crest and rushed down the valley below. The water wave was still over 70 meters high some 1.5 kilometers downstream at the mouth of the Vaiont Valley, where it flowed into the Piave River. The energy of the rush-ing water was such that some of it flowed *upstream* in the Piave for more than 2 kilometers. Within about five minutes, nearly three thousand people were drowned, and whole towns wiped out.

It is a great tribute to the designer and builder of the Vaiont Dam that the dam itself held through all of this, re-sisting forces far beyond the design specifications; it still stands today. It is also very much to the discredit of those who chose the site that the dam and reservoir were ever put in such a spot. Ample evidence of slope instability dated back long before the conception of the dam. With more thorough study of the rocks' properties and structure and the possible effects of changing water levels, the Vaiont tragedy could have been avoided.

Types of Mass Movement and Typical Rates of Movement

	Material moves as coherent unit	Chaotic, incoherent movement
Free-falling motion	Rockfall Speed: Rapid (meters/second)	Soil fall (rare) Speed: Rapid (meters/second)

In contact with surface below	Slides:
	  rockslides
	  slump (soil)
	
	Speed: Highly variable (meters/year to meters/second)
		Flows:
		  snow avalanche
		  debris avalanche
		  nuée ardente
		  earthflow
		  mudflow
		  lahar
		
		Speed: Rapid (meters/day to meters/second)

Figure 8.14 Comparison of nature of various types of mass movements.

leave telltale signs of its occurrence so that areas of particular risk can be avoided. For example, building foundations may gradually weaken and fail as soil shifts; roads and railway lines may be disrupted. Often, soil creep causes serious property damage, though lives are rarely threatened.

Mass movements may be subdivided on the basis of the type of material moved and the rate of movement. The material moved can range from unconsolidated, fairly fine material (for example, soil or snow) to large, solid masses of rock. *Landslide* is a nonspecific term for rapid mass movements in rock or soil. There are several types of landslides, and a particular event may involve more than one type of motion. The rate of motion is commonly related to the proportion of moisture: the wetter the material, the faster the

movement. A few examples may clarify the different types of mass movements, which are summarized in figure 8.14.

Falls

A **fall** is a free-falling action in which the moving material is not always in contact with the ground below. Falls are most often **rockfalls** (figure 8.15). They frequently occur on very steep slopes when rocks high on the slope, weakened and broken up by weathering, lose support as materials under them erode away. Falls are common along rocky coastlines where cliffs are undercut by wave action, and also at roadcuts through solid rock. The coarse rubble that accumulates at the foot of a slope prone to rockfalls is **talus**.

Figure 8.15 Accumulation of talus from multiple rockfalls, Devil's Postpile National Monument.

Figure 8.16 "Fallen City" rockslide, complex rockslide in the Bighorn Mountains. Sliding occurred along bedding planes in limestone.

Slumps and Slides

In a **slide,** a fairly cohesive unit of rock or soil slips downward along a clearly defined surface or plane. Rockslides most often involve movement along a bedding plane between successive layers of sedimentary rocks, or slippage where differences in original composition of the sedimentary layers results in a weakened layer or a surface with little cohesion (figure 8.16). Rockslides may be complex, with multiple rock blocks slipping along the same plane.

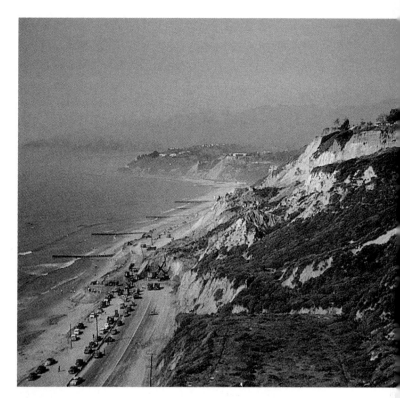

Figure 8.17 This 1956 slump in the Pacific Palisades area, California, covered part of Highway 101; note undisturbed land surface at top of the slump block. Photograph by J. T. McGill, USGS Photo Library, Denver, CO.

Slumps can occur in rock or soil. In a soil slump, a rotational movement of the soil mass typically accompanies the downslope movement. A *scarp* may be formed at the top of the slide. Usually, the surface at the top of the slide is relatively undisturbed. The lower part of the slump block may end in a flow (figure 8.17). A rockslide that does not move very far may also be a slump.

Flows and Avalanches

In a **flow,** the material moved is not coherent but moves in a more chaotic, disorganized fashion, with mixing of particles within the flowing mass, as a fluid flows. Flows of unconsolidated material are extremely common. They need not involve only soils. Snow avalanches are one kind of flow (figure 8.18); nuées ardentes, or pyroclastic flows (see chapter 5), are another, associated only with volcanic activity. Where soil is the flowing material, these phenomena may be described as *earthflows* (fairly dry soil; see figure 8.19) or *mudflows* (if saturated with water), which include the volcanic lahars. A flow involving a wide variety of materials—soil, rocks, trees, and so on—together in a single flow is a **debris avalanche** (figure 8.20). Regardless of the nature of the materials moved, all flows have in common the chaotic, incoherent movement of the particles or objects in the flow.

Figure 8.18 Beginning of a snow avalanche, set off by the slight disturbance of a skier passing over an unstable patch of snow. Photograph by Ludwig, courtesy of U.S. Forest Service/USGS Photo Library, Denver, CO.

Figure 8.19 Recent earthflows in grassland, Gilroy, California. Photograph ©The McGraw-Hill Companies, Inc./Doug Sherman, photographer.

Figure 8.20 Puget Peak debris avalanche following 1964 earthquake, Alaska. Photograph by G. Plafker, USGS Photo Library, Denver, CO.

Scales and Rates of Movements

Mass movements can occur on a variety of scales and at a variety of rates. They may involve a few cubic meters of material or more than a billion cubic meters. Total displacement may be over a distance of only a few centimeters or even millimeters, or the falling, sliding, or flowing material may travel several kilometers. In the most rapid mass movements, which include most rockfalls, avalanches, and mudflows, materials can travel at speeds of hundreds of kilometers per hour, up to several tens of meters per second. There is little time for people to react once these events start, and such rapid events are associated with the greatest proportion of mass-movement casualties. Slides generally move at more moderate rates, in the range of a few meters per week to meters per day; slower slides, slumps, and earthflows may move as slowly as a millimeter per year or even less. These movements are much more likely to cause property damage than casualties.

Consequences of Mass Movements

Just as landslides can be a result of earthquakes, other unfortunate events, notably floods, can be produced by landslides. A stream in the process of cutting a valley may create unstable slopes. Subsequent landslides into the valley can dam up the stream flowing through it, creating a natural reservoir. The filling of the reservoir makes the area behind the earth dam uninhabitable, though it usually happens slowly enough that lives are not lost.

Such flooding was one effect of the largest landslide of recent history. It occurred in 1911, when an earthquake of magnitude 7.0 struck a remote area in the Pamir Mountains of the Soviet Union. The resulting landslide, involving some 2.5 billion cubic meters of material, overwhelmed the village of Usoy, killing fifty-four people. It also dammed the river Murgab. The lake that formed behind the slide eventually grew to be nearly 300 meters (900 feet) deep and 54 kilometers (over 30 miles) long and, in time, inundated the site of another town, Sarez.

Similar problems, though without the loss of life, developed as a consequence of heavy rains and snows in Utah in 1982–83. There were landslides in unprecedented numbers. In one case, an old 3-million-cubic-meter slide near Thistle, Utah, was reactivated. It blocked Spanish Fork Canyon, creating a large lake and cutting off one transcontinental railway line and a major highway route into central Utah (figure 8.21). Shipments of coal and other supplies were disrupted. Debris flows north of Salt Lake City also forced hundreds of residents to evacuate and destroyed some homes, and water levels in reservoirs in the Wasatch Plateau were lowered to reduce dangers of landslides into the reservoirs or breaching of the dams.

A further danger is that the unplanned dam formed by a landslide will later fail. In 1925, an enormous rockslide a kilometer wide and over 3 kilometers (2 miles) long in the valley of the Gros Ventre River in Wyoming blocked that valley, and the resulting lake stretched to 9 kilometers long. After spring rains and snowmelt, the natural dam failed. Floodwaters swept down the valley below. Six people died, but the toll might have been far worse in a more populous area.

Possible Preventive Measures

Avoiding the most landslide-prone areas altogether would greatly limit damages, of course, but as is true with fault zones, floodplains, and other hazardous settings, developments may already exist in areas at risk, and economic pressure for more development can be strong. Population density, too, coupled with lack of "safe" locations, may push development into unsafe areas, as was shown in figures 1.15, 1.16, and 8.7.

Sometimes, local officials just resign themselves to the existence of a mass-movement problem and perhaps take some steps to limit the resulting damage (figure 8.22). In places where the structures to be protected are few or small, and the slide zone is narrow, it may be economically feasible to bridge structures and simply let the slides flow over them. For example, this might be done to protect a railway line or road running along a valley from avalanches—either snow or debris—from a particularly steep slope. This solution would be far too expensive on a large scale, however, and no use at all if the base on which the structure was built were sliding also.

There are some other steps that can be taken to reduce actual landslide hazards.

Figure 8.21 An old, 3-million-cubic-meter slide near Thistle, Utah, was reactivated in the wet spring of 1983. It blocked Spanish Fork Canyon and cut off highway and rail routes. Residents were driven from their homes as the lake behind the slide grew.   Photograph courtesy of USGS Photo Library, Denver, CO.

Slope Reduction

If a slope is too steep to be stable under the load it carries, any of the following steps will reduce slide potential: (1) reduce the slope angle, (2) place additional supporting material at the foot of the slope to prevent a slide or flow at the base of the slope, or (3) reduce the load (weight, shearing stress) on

A

B

C

Figure 8.22   Measures to reduce the consequences of slope instability. (*A*) Chain-link fencing draped over roadcut to protect road from rockfall. (*B*) Shelter built over railroad track or road in snow avalanche area diverts snow flow. (*C*) Example of avalanche-protection structure in the Canadian Rockies, in Canada's Glacier National Park.

A

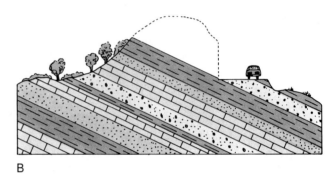

B

Figure 8.23   Slope stabilization by slope reduction and removal of unstable material along roadcut. (*A*) Before: Roadcut leaves steep, unsupported slope. If shale layers (grey) are relatively impermeable, fluid may accumulate along bedding planes, promoting sliding. (*B*) After: Material removed to reduce slope angle and load.

the slope by removing some of the rock or soil (or artificial structures) high on the slope (figure 8.23). These measures may be used in combination. Depending on just how unstable the slope, they may need to be executed cautiously. If earth-moving equipment is being used to remove soil at the top of a slope, for example, the added weight of the equipment and vibrations from it could possibly trigger a landslide.

Retention Structures

To stabilize exposed near-surface soil, ground covers or other vegetation may be planted (preferably fast-growing materials with sturdy root systems). But sometimes plants are insufficient, and the other preventive measures already described impractical, in a particular situation. Then, retaining walls or other stabilization structures can be built against the slope to try to hold it in place (figures 8.24, 8.25). Given the distribution of stresses acting on retaining walls, the greatest successes of this kind have generally been low, thick walls placed at the toe of a fairly coherent slide to stop its movement. High, thin walls have been less successful (see, for example, box 8.2).

Box 8.2 Landslide Control in the Rockies: An Example

Sometimes, whether the terrain is suitable or not, a decision is made to build in a certain location, and the result is often a nonstop battle against nature. Such a spot is the east side of Beartooth Pass in the Rocky Mountains, where the road descends rapidly to the town of Red Lodge, Montana. Roads in this area are few, and it is easy to see why; a sample of the topography is shown in figure 8.28A. The slopes are frighteningly steep; the guardrails are dented from falling boulders. Spring is an especially dangerous time because when the soil is sodden from melted snow, it slides especially readily, often taking sections of road with it. The road is critical to local transportation and has, therefore, been regularly rebuilt after each slide.

Two decades or so ago, an attempt was made to stabilize several large expanses of the most slide-prone slopes by applying concrete to the slope surface (figure 1A–D). Realizing the danger of having water accumulate behind these concrete sheets, building up weight and pore pressure, the engineers added drainage holes (B). A return visit three summers later strongly suggested that this solution was not an unqualified success. The slope had continued to move, and the thin concrete sheets had buckled and cracked in many places (C,D). Furthermore, some very new asphalt below one such barrier (C) showed that the spring outwash had claimed the soil *underneath* the roadway there as well.

Figure 1 Slope stabilization attempts, east side of Beartooth Pass, Montana. (*A*) Retaining cement plastered on hillside. (*B*) Closeup showing drainage holes. (*C*) Broken retaining structure with freshly repaired road below. (*D*) Closeup of failed stabilization structure.

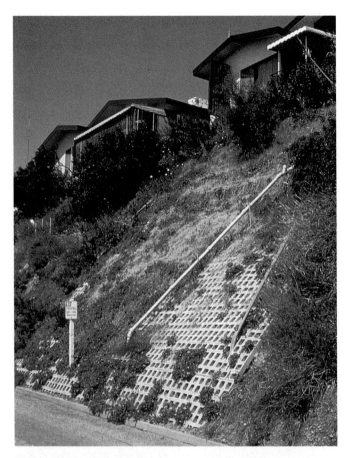

Figure 8.24 Retaining wall of concrete blocks on slope cut in old slide debris, Pacific Palisades area, California. Photograph by J. T. McGill, USGS Photo Library, Denver, CO.

Fluid Removal

Since water can play such a major role in mass movements, the other principal strategy for reducing landslide hazards is to decrease the water content or pore pressure of the rock or soil. This might be done by covering the surface completely with an impermeable material and diverting surface runoff above the slope. Alternatively, subsurface drainage might be undertaken. Systems of underground boreholes can be drilled to increase drainage, and pipelines installed to carry the water out of the slide area (figure 8.26). All such moisture-reducing techniques naturally have the greatest impact where rocks or soils are relatively permeable. Where the rock or soil is fine-grained and drains only slowly, hot air may be blown through boreholes to help dry out the ground. Such moisture reduction reduces pore pressure and increases frictional resistance to sliding.

Other Slope-Stabilization Measures

Other slope-stabilization techniques that have been tried include the driving of vertical piles into the foot of a shallow slide to hold the sliding block in place. The procedure works only where the slide is comparatively solid (loose soils may simply flow between piles), with thin slides (so that piles can be driven deep into stable material below the sliding mass), and on low-angle slopes (otherwise, the shearing stresses may simply snap the piles). So far, this technique has not been very effective.

A

B

Figure 8.25 Slope stabilization: (A) Side view of retaining structure. (B) Closeup shows that somewhat naturalistic appearance is achieved by building structure of rock rubble encased in chain-link fencing.

Figure 8.26 Improved drainage enhances slope stability by reducing both load and pore pressure. (*A*) Before: Water trapped in soil causes movement, pushing down retaining wall. (*B*) After: Water drains through pipe, allowing wall to keep slope stable. (*C*) Drainage of roadcut in California.

The use of rock bolts to stabilize rocky slopes and, occasionally, rockslides has had greater success (figure 8.27). Rock bolts have long been used in tunneling and mining to stabilize rock walls. It is sometimes also possible to anchor a rockslide with giant steel bolts driven into stable rocks below the slip plane. Again, this works best on thin slide blocks of very coherent rocks on low-angle slopes.

Procedures occasionally used on unconsolidated materials include hardening unstable soil by drying and baking it with heat (this procedure works well with clay-rich soils) or by treating with portland cement. By far the most common strategies, however, are modification of slope geometry and load, dewatering, or a combination of these techniques. The more ambitious engineering efforts are correspondingly expensive and usually reserved for large construction projects, not individual homesites.

Determination of the absolute limits of slope stability is an imprecise science. This was illustrated in Japan in 1969, when a test hillside being deliberately saturated with water unexpectedly slid prematurely, killing several of the researchers. The same point is made less dramatically every time a slope-stabilization effort fails.

Recognizing the Hazards

Mass movements occur as a consequence of the basic characteristics of a region's climate, topography, and geology, independent of human activities. Where mass movements are naturally common, they tend to recur repeatedly in the same places. Recognition of past mass movements in a region indicates a need for caution and serious consideration of the hazard in any future development. Such recognition can also save lives.

Recognizing past rockfalls can be quite simple, especially in vegetated areas. Large chunks of rock are inhospitable to most vegetation, so rockfalls tend to remain barren of trees and plants. The few trees that do take hold may only contribute to further breakup by root action. Lack of vegetation may also mark the paths of past debris avalanches or other soil flows or slides (figure 8.28). These scars on the landscape point plainly to slope instability.

Landslides are not the only kinds of mass movements that recur in the same places. Snow avalanches disappear when the snow melts, of course, but historical records of past avalanches can pinpoint particularly risky areas. Records of the character of past volcanic activity and an examination of a volcano's typical products can similarly be used to recognize a particular volcano's tendency to produce pyroclastic flows.

Very large slumps and slides may be less obvious, especially when viewed from the ground. The coherent nature of rock and soil movement in many slides means that vegetation growing atop the slide may not be greatly disturbed by the movement. Aerial photography or high-quality topographic maps can be helpful here. In a regional overview, the mass movement often shows up very clearly, revealed by a scarp at the head of a slump or an area of hummocky, disrupted topography relative to surrounding, more stable areas.

A · B

Figure 8.27 (*A*) Installation of rock bolts to stabilize a slope. The bolts are steel cables anchored in cement. Tightening of the nuts at the surface pulls unstable layers together and anchors them to the bedrock below. (*B*) Rock bolts used together with fencing to stabilize a slope in the Bighorn Mountains, Wyoming.   Photograph © Doug Sherman/Geofile.

A · B

Figure 8.28 (*A*) Areas prone to landslides may be recognized by the failure of vegetation to establish itself on unstable slopes. Rock Creek Valley, Montana. (*B*) Multiple avalanche scars on an Alaskan hillside. The newer scar is unvegetated; the older scar is overgrown with grass but lacks large trees.

A

B

Figure 8.29   Tilted tree trunks are a consequence of creep. (*A*) These curved tree trunks have developed on an imperfectly stabilized slope above a roadcut. Lassen Park, California. (*B*) Other signs of creep: tilted monuments, fences, utility poles.

A

B

Figure 8.30   Evidence of slope instability in the Pacific Palisades of California includes broken walls (*A*), and, in extreme cases, slumped yards (*B*).   Photographs by J. T. McGill, USGS Photo Library, Denver, CO.

With creep or gradual soil flow, individual movements are short-distance and the whole process is slow, so vegetation may continue to grow in spite of the slippage. More detailed observation, however, can reveal the movement. For example, trees are biochemically "programmed" to grow vertically upward. If the soil in which trees are growing begins to creep downslope, tree trunks may be tilted, and this indicates the soil movement. Further tree growth will continue to be vertical. If slow creep is prolonged over a considerable period of time, curved tree trunks may result (figure 8.29A). (Trees growing in slanted fractures in rock, however, may

develop curved trunks through similar growth tendencies in the absence of creep; this is a better indicator of movement in unconsolidated material.) Inanimate objects can reflect soil creep, too. Slanted utility poles and fences and the tilting-over of once-vertical gravestones or other monuments also indicate that the soil is moving (figure 8.29B). The ground surface itself may show cracks parallel to (across) the slope.

A prospective home buyer can look for additional signs that might indicate unstable land underneath (figure 8.30). Ground slippage may have caused cracks in driveways, garage floors, freestanding brick or concrete walls, or buildings;

cracks in walls or ceilings are especially suspicious in newer buildings that would not yet normally show the settling cracks common in old structures. Doors and windows that jam or do not close properly may reflect a warped frame due to differential movement in the soil and foundation. Sliding may have caused leaky swimming or decorative pools, or broken utility lines or pipes. If movement has already been sufficient to cause such obvious structural damage, it is probable that the slope cannot be stabilized adequately, except perhaps at very great expense. Possible sliding should be investigated particularly on a site with more than 15 percent slope (15 meters of rise over 100 meters of horizontal distance), on a site with much steeper slopes above or below it, or in any area where landslides are a recognized problem. (Given appropriate conditions, slides may occur on quite shallow slopes.)

Landslide Warnings?

In early 1982, severe rain-triggered landslides in the San Francisco Bay area killed twenty-five people and caused over $66 million in damages. In response to this event, the U.S. Geological Survey began to develop a landslide warning system. The basis of the warning system was the development of quantitative relationships among rainfall intensity (quantity of water per unit time), storm duration, and a variety of slope and soil characteristics relating to slope stability—slope angle, pore fluid pressure, shear strength, and so on. These relationships were formulated using statistical analyses of observational data on past landslides. For a given slope, it was possible to approximate threshold values of storm intensity and duration above which landsliding would become likely, given the recent precipitation history at the time (how saturated the ground was prior to the particular storm of concern).

The system, though incomplete, was tested in February of 1986. Warnings were broadcast as special weather advisories on local radio and television stations. These warnings were rather specific as to time, more general as to area at risk (necessarily so, given the complexity of the geology and terrain and lack of highly detailed data on individual small areas). Some local government agencies recommended evacuations, and many were able to plan emergency responses to the landslides before they occurred. Many landslides did occur; total estimated landslide damage was $10 million, with one death. Of ten landslides for which the times of occurrence are known precisely, eight took place when forecast.

The models need refinement. The ultimate goal would be *predictions* that are precise both as to time and as to location. To achieve this, too, more extensive data on local geology and topography are needed. Certainly the public is not yet accustomed to landslide warnings, so response is uneven. But the successes of the 1986 efforts do suggest that landslide prediction has considerable potential to reduce casualties and enhance the efficiency of agencies' responses to such events.

Overview

Summary

Mass movements occur when the shearing stress acting on rocks or soil exceeds the shear strength of the material to resist it. Gravity provides the main component of shearing stress. The basic susceptibility of a slope to failure is determined by a combination of factors, including geology (nature and strength of materials), geometry (slope steepness), and moisture content. Sudden failure usually also involves a triggering mechanism, such as vibration from an earthquake, addition of moisture, steepening of slopes (naturally or through human activities), or removal of stabilizing vegetation. Landslide hazards may be reduced by such procedures as modifying slope geometry, reducing the weight acting on the slope, planting vegetation, or dewatering the rocks or soil. Damage and loss of life can also be limited by restricting or imposing controls on development in areas at risk, and where adequate data are available, through landslide warning systems.

Terms to Remember

angle of repose 172	rockfall 180
creep (rock or soil) 177	rock flour 175
debris avalanche 181	sensitive clay 175
fall 180	shearing stress 171
flow 181	shear strength 171
landslide 171	slide 181
mass movement 171	slump 181
mass wasting 171	talus 180
quick clay 175	

Exercises

Questions for Review

1. Explain in general terms why landslides occur. What two factors particularly influence slope stability?
2. Earthquakes are one landslide-triggering mechanism; water can be another. Evaluate the role of water in mass movements.

3. What are quick clays and sensitive clays?

4. Describe at least three ways in which development on hillsides may aggravate landslide hazards.

5. Give two ways to recognize soil creep and one way to identify sites of past landslides.

6. Briefly explain the distinctions among falls, slides, and flows.

7. Common slope-stabilization measures include physical modifications to the slope itself or the addition of stabilizing features. Choose any three such strategies and briefly explain how they work.

For Further Thought

1. Experiment with the stability of a variety of natural unconsolidated materials (soil, sand, gravel, and so forth). Pour the dry materials into separate piles; compare angles of repose. Add water, or put an object (load) on top, and observe the results.

2. Find aerial or other photographs of your area of the country (a local library might be able to provide them). Or, search the Internet for sources—you might start with the USGS photo library page at http://www.usgs.gov/fact-sheets/photographic-library/photographic-library.html
Examine them for evidence of past mass wasting.

3. Consider how you might respond to a known local landslide hazard or a warning. How definite a warning would you look for before evacuating? Would you take any precautions prior to issuance of an official warning if conditions seemed to be increasing the risk?

Suggested Readings/References

Aune, R. A. 1983. Quick clays and California's clays: No quick solutions. In *Environmental geology, text and readings*, 3d ed., edited by R. W. Tank. New York: Oxford University Press.

Bolt, B. A., Horn, W. L., MacDonald, G. A., and Scott, R. F. 1975. Hazards from landslides. Chap. 4 in *Geological hazards*. New York: Springer-Verlag.

Crozier, M. J. 1986. *Landslides: Causes, consequences, and environment*. Dover, N.H.: Croom Helm.

Fleming, R. W., and Taylor, F. A. 1980. *Estimating the cost of landslide damage in the United States*. U.S. Geological Survey Circular 832.

Highland, L. M., and Brown, W. L., III. 1996. Landslides—The natural hazard sleepers. *Geotimes* (January): 16–19.

Keefer, D. K. et al. 1987. Real-time landslide warning during heavy rainfall. *Science* 238:921–25.

Kiersch, G. A. 1965. The Vaiont reservoir disaster. *Mineral Information Service* 18 (7): 129–38.

Leveson, D. 1980. *Geology and the urban environment*. New York: Oxford University Press.

Radbrich-Hall, D. H., Colton, R. B., Davies, W. E., Skipp, B. A., Lucchitta, I., and Varnes, D. J. 1976. Preliminary landslide overview map of the conterminous United States. U.S. Geological Survey Miscellaneous Field Studies Map MF-771.

Schultz, A. P., and Jibson, R. W., eds. 1989. *Landslide processes of the eastern United States and Puerto Rico*. Geological Society of America Special Paper 236.

Schuster, R. L., and Krizek, R. J., eds. 1978. *Landslides, analysis and control*. Washington, D.C.: National Research Council, Transportation Research Board Special Report 176.

Schuster, R. L., Varnes, D. J., and Fleming R. W. 1981. Landslides. In *Facing geologic and hydrologic hazards*, edited by W. W. Hay. U.S. Geological Survey Professional Paper 1240–B.

Utgard, R. O., McKenzie, G. D., and Foley, D. 1978. *Geology in the urban environment*. Minneapolis: Burgess.

Varnes, D. J. 1984. *Landslide hazard zonation: A review of principles and practice*. Paris: United Nations Educational, Scientific, and Cultural Organization (UNESCO).

Voight, B., ed. 1978. *Rockslides and avalanches*. New York: Elsevier Science Publishing.

At the home page for the National Disaster Reference Database, there are links to various reports—and images—of landslides; see

http://ltpwww.gsfc.nasa.gov/ndrd-cgi/ndrd

The U.S. Geological Survey site specific to landslide hazards is at

http://www.usgs.gov/themes/landslide.html

The National Landslide Information Center is found at
http://geohazards.cr.usgs.gov/html_files/nlicsun.html

The Cyberspace Snow and Avalanche Center has education/research page with links to technical information on avalanches, avalanche reports, avalanche safety information, and other avalanche-related sites, at
http://www.csac.org/Education/

A somewhat different site, with a modest collection of mass-movement images (captioned in French) can be found at
http://wasss.entpe.fr/biblima/imaglis.htm

Many earthquake-related sites will include earthquake-triggered-landslide information. Also, for a history of slides and debris flows at Mount Rainier, see USGS Open-File Report 93–111 online at:
http://vulcan.wr.usgs.gov/Volcanoes/Rainier/OFR93-111.html

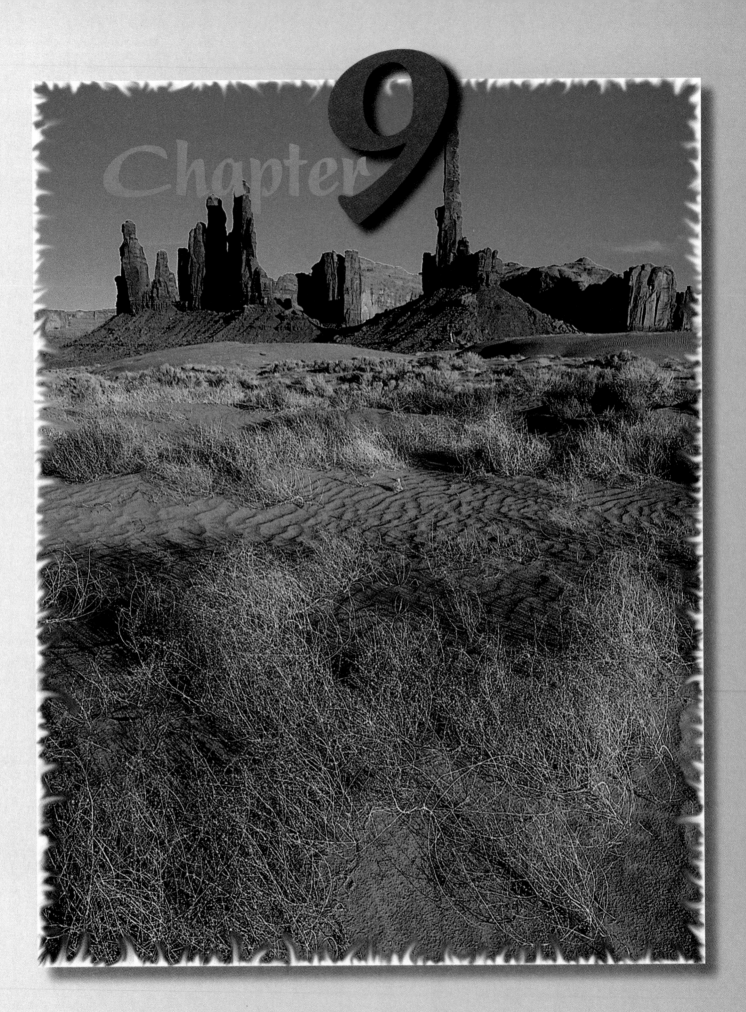

Chapter **9**

Geology and Climate: Glaciers, Deserts, and Global Climate Trends

T his chapter rounds out the discussion of surface processes by considering the actions of ice and wind as geologic agents and aspects of the relationships of both to global climate. As it happens, these agents are most active in some of the more extreme local climates on earth: glaciers in cold regions where moisture, or at least ice, is present year-round, and wind in deserts, sparsely vegetated, typically dry, often very hot.

Glaciers are more than just another agent by which our landscape is shaped. They represent a very large reserve of (temporarily solid) water. Past glacial activity has also left behind sediments that provide gravel for construction, store subsurface water for future consumption, and contribute to farmland fertility.

Except during severe storms, wind is rarely a serious hazard. As an agent of change of the earth's surface, wind is considerably less efficient than water or ice. Nevertheless, it can locally play a significant role, particularly with regard to erosion. The effects of wind are most often visible in deserts, where little vegetation or development is present to mask or counter its effects.

Spires of the DeChelly Sandstone stand as backdrop to dunes in Monument Valley, Arizona. © Doug Sherman/Geofile

Over a human lifetime, the extent of glaciation, the expanse of deserts, and the overall global climate vary little. Over the longer term, significant changes can occur. We have evidence in the geologic record for a number of episodes during which vast ice masses covered major portions of the continents. Today, human activities are demonstrably changing the chemistry of our atmosphere, which may lead to permanent changes in climate, through global warming. One likely consequence is melting of glaciers worldwide, with resultant rise in net sea level, inundating low-lying coastal regions. There is also concern that on a warmer globe, arid lands will be more extensive. Desertification caused by human activities is already a serious concern of those who wonder how the earth's still-growing population will be fed; drought aggravated by global warming may compound the problem. Many causes of global warming and climate change are known. Modeling their exact effects is more difficult.

Glaciers and Glacial Features

A **glacier** is a mass of ice that moves over land under its own weight, through the action of gravity. A quantity of snow sufficient to form a glacier does not fall in a single winter, so for glaciers to develop, the climate must be cold enough that some snow and ice persist year-round. This, in turn, requires an appropriate combination of altitude and latitude. Glaciers tend to be associated with the extreme cold of polar regions. However, temperatures also generally decrease at high altitudes (figure 9.1), so glaciers can exist in mountainous areas even in tropical or subtropical regions. Three mountains in Mexico have glaciers, but these glaciers are all at altitudes above 5,000 meters (15,000 feet). Similarly, there are glaciers on Mount Kenya and Mount Kilimanjaro in east Africa, but only at altitudes above 4,400 meters.

Figure 9.1 The snow line on a mountain range reflects decreasing temperature at higher elevations. Olympic Range, Washington.

Box 9.1 Glaciers as a Water Source

In the context of greenhouse-effect heating (see the discussion later in the chapter), the large quantity of water that glaciers represent appears to be a threat. On the other hand, glaciers can be viewed as an important freshwater reserve. Approximately 75 percent of the fresh water on earth is stored as ice in glaciers. The world's supply of glacial ice is the equivalent of sixty years of precipitation over the whole earth.

In regions where glaciers are large or numerous, glacial meltwater may be the principal source of summer streamflow. By far the most extensive glaciers in the United States are those of Alaska (figure 1), which cover about 3 percent of the state's area. Summer streamflow from these glaciers is estimated at nearly 50 trillion gallons of water! Even in less-glaciated states like Montana, Wyoming, California, and Washington, streamflow from summer meltwater amounts to tens or hundreds of billions of gallons.

Anything that modifies glacial melting patterns can profoundly affect regional water supplies. Dusting the glacial ice surface with something dark, like coal dust, increases the heating of the surface, and hence the rate of melting and water flow. Conversely, cloud seeding over glaciated areas could be used to increase precipitation and the amount of water stored in the glaciers. Increased meltwater flow can be useful not only to increase water supplies but also to achieve higher levels of hydroelectric power production. Techniques for modifying glacial meltwater flow are not now being used in the United States, in part because the majority of U.S. glaciers are in national parks, primitive areas, or other protected lands. However, some of these techniques have been practiced in parts of the former Soviet Union and China. Global warming could be expected to increase the release of meltwater and/or decrease water storage in such glaciers. In the short term, this would produce more abundant surface water, but at the expense of the ice reserve for the future.

Decreased glacier volume may have more subtle water-supply effects. Early in the melt season in alpine regions, some of the meltwater is stored in pore space in glacial snow and ice. As glacial melting proceeds through spring and summer, this stored water adds to glacial meltwater volume, as the glacial snow/ice containing it melts, enhancing water supplies in many regions. If glacier size decreases, so does this water storage capacity. Recent studies of the North Cascade Glacier Climate Project have documented two resultant problems: increased likelihood of spring flooding, and decreased meltwater flow later in the season. In the Andes and Himalayas, where agriculture depends heavily on glacial meltwater runoff, glacial retreat may signal future reductions in crop productivity.

Glacier Formation

For glaciers to form, there must be sufficient moisture in the air to provide the necessary precipitation. (Glaciers are, after all, a large water reservoir; see box 9.1.) In addition, the amount of winter snowfall must exceed summer melting so that the snow accumulates year by year. Most glaciers start in the mountains as snow patches that survive the summer. Slopes that face the poles (north-facing in the Northern Hemisphere, south-facing in the Southern Hemisphere), protected from the strongest sunlight, favor this survival. So do gentle slopes, on which snow can pile up thickly instead of plummeting down periodically in avalanches.

As the snow accumulates, it is gradually transformed into ice. The weight of overlying snow packs it down, drives out much of the air, and causes it to recrystallize into coarser, denser, interlocking ice crystals. The conversion of snow into solid glacial ice may take only a few seasons or several thousand years, depending on such factors as climate and the rate of snow accumulation at the top of the pile, which hastens compaction. Eventually, the mass of ice becomes large enough that, if there is any slope to the terrain, it begins to flow downhill—or, on more level ground, the ice begins to flow outward from the thickest accumulations. It is then a true glacier. Some glaciers may also slide along a meltwater layer at the base of the ice. The total movement may be a nearly imperceptible 20 meters per year or, for brief periods at least, a glacier may "surge" at up to 10 kilometers per year (6 miles per year).

Types of Glaciers

Glaciers are divided into two types on the basis of size and occurrence. The most numerous today are the **alpine glaciers,** also known as mountain or valley glaciers. As these names suggest, they are glaciers that occupy valleys in mountainous terrain, most often at relatively high altitudes (figure 9.2). Most of the estimated 70,000 to 200,000 glaciers in the world today are alpine glaciers.

▌ **Figure 1** Alaska is the most extensively glaciated state in the United States. College Fjord, Alaska. Photograph by W. Montgomery.

The larger and rarer **continental glaciers** are also known as *ice caps* (generally less than 50,000 square kilometers in area) or *ice sheets* (larger) (figure 9.3). They can cover large portions of continents and reach thicknesses of a kilometer or more. Though they are far fewer in number than the alpine glaciers, the continental glaciers comprise far more ice. At present, the two principal continental glaciers are the Greenland and the Antarctic ice sheets. The Antarctic ice sheet is so large that it could easily cover the forty-eight contiguous United States. The geologic record indicates that, at several times in the past, even more extensive ice sheets existed on earth.

Movement and Change of Glaciers

Glaciers do not behave as rigid, unchanging lumps of ice. Their flow is plastic, like that of asphalt on a warm day (or of the convecting upper mantle), and different parts of glaciers move at different rates. At the base of a glacier, or where it scrapes against valley walls, the ice moves more slowly. In the higher, more central portions, it flows more freely and

rapidly. Also, material is continually cycled through a glacier. Where it is cold enough for snow to fall, fresh material accumulates, adding to the weight of ice and snow that pushes the glacier downhill. Parts of a glacier may locally flow upslope for short distances. However, the net flow of the glacier must be downslope, under the influence of gravity. The downslope movement may be either down the slope of the underlying land surface, or in the downslope direction of the *glacier's* surface, from areas of thicker ice to thinner, regardless of the underlying topography (also known as flow "down gradient").

At some point, the advancing edge of the glacier terminates, either because it flows out over water and breaks up, creating icebergs by a process known as **calving,** or, more commonly, because it has flowed to a place that is warm enough that ice loss by melting or evaporation is at least as rapid as the rate at which new ice flows in (figure 9.4). The set of processes by which ice is lost from a glacier is collectively called **ablation.** The **equilibrium line** is the line on the glacial surface where there is no net gain or loss of material.

Figure 9.2 Alpine glacier in Denali National Park, Alaska. For other examples, see box 9.1, figure 1.

Figure 9.4 Schematic of longitudinal cross section of glacier. In the zone of accumulation, the addition of new material exceeds loss by melting or evaporation; the reverse is true in the zone of ablation, where there is a net loss of ice.

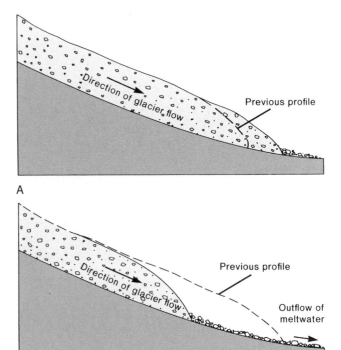

Figure 9.5 Glacial advance and retreat. (*A*) Advance of the glacier moves both ice and sediment downslope. (*B*) Ablation, especially melting, causes apparent retreat of the glacier, accompanied by deposition of sediment from the lost ice volume.

Over the course of a year, the size of a glacier varies somewhat (figure 9.5). In winter, the rate of accumulation increases, and melting and evaporation decrease. The glacier becomes thicker and longer; it is then said to be advancing. In summer, snowfall is reduced or halted and melting accelerates; ablation exceeds accumulation. The glacier is then described as "retreating," although, of course, it does not flow backward, uphill. The leading edge simply melts back faster than the glacier moves forward. Over many years, if the climate remains stable, the glacier achieves a sort of dynamic equilibrium, in which the winter advance just balances the summer retreat, and the average extent of the glacier remains constant from year to year.

Figure 9.3 A continental glacier, or ice sheet—aerial view of a portion of the Antarctic ice sheet. Mount Overlord, in foreground, rises to 3,400 meters (10,000 feet) above sea level, but most of it is covered by ice. Photograph by W. B. Hamilton, USGS Photo Library, Denver, CO.

Table 9.1 Response of Cascade (U.S.), Swiss, and Italian Glaciers to Rising Global Temperature			
	Advancing	**Stationary**	**Retreating**
North Cascades			
1967	7	8	7
1974	9	0	2
1985	5	10	32
1990	1	5	41
1995	0	0	47
Switzerland			
1967	31	14	55
1975	70	3	28
1986	42	9	13
1989	19	3	83
1993	6	0	73
Italy			
1981	25	10	10
1985	25	6	14
1988	26	13	92
1990	9	9	123

Source: Data from M. Pelto, personal communication (1995).

Figure 9.6 Rock Creek Valley, Montana, a classic U-shaped glacial valley.

More significant are the sustained increases or decreases in glacier size that result from consistent cooling or warming trends. An unusually cold or snowy period spanning several years would be reflected in a net advance of the glacier over that period, while a warm period would produce a net retreat. Such phenomena have been observed in recent history. From the mid-1800s to the early 1900s, a worldwide retreat of alpine glaciers far up into their valleys was observed. Beginning about 1940, the trend seemed to reverse, and glaciers advanced. Recent studies suggest that, in the immediate future, a global warming trend can be expected, as we shall see in a later section. It is important to bear in mind in all discussions of "global" climate changes, however, that considerable local climate variations can exist, as a result of local topography, geography, precipitation, winds and water currents, and many other factors. So, for example, though a majority of alpine glaciers worldwide are currently retreating, and over several decades that trend has strengthened, some glaciers are still stable or advancing (table 9.1).

Glacial Erosion and Deposition

A glacier's great mass and strength make it especially effective in erosion. Only because streams far outnumber glaciers is liquid water, overall, a more significant agent in shaping the earth's surface then is ice. Glaciers do, however, leave distinctive features by which their past extent and movements can be tracked, long after the ice is gone.

Like a stream, an alpine glacier carves its own valley. The solid ice carves a somewhat different shape, however: A glacial valley is typically U-shaped (figure 9.6). On a grand scale, thick continental glaciers may gouge out a series of large troughs in the rock over which they move (figure 9.7). On a very small scale, bits of rock that have become frozen into the ice at the base of a glacier act like sandpaper on solid rock below, making fine, parallel scratches, called **striations,** on the underlying surface (figure 9.8). Such signs not only indicate the past presence of glaciers but also the direction in which they flowed. Erosion by the scraping of ice or included sediment on the surface underneath is **abrasion.** Water can also seep into cracks in rocks at the base of the glacier and refreeze, attaching rock fragments to the glacier. As the ice moves on, small bits of the rock are torn away in a process known as *plucking.*

Glaciers also transport and deposit sediment very efficiently. Debris that has been eroded from the sides of an alpine glacier's valley, or that has fallen onto the glacier from peaks above, tends to be carried along on or in the ice, regardless of the size, shape, or density of the sediment. It is not likely to be deposited until the ice carrying it melts or evaporates, and then all the debris, from large boulders to fine dust, is dropped together. Sediment deposited directly from the melting ice is called **till.** Till is characteristically poorly sorted material (figure 9.9), as contrasted with stream-deposited sediment. Some of the till may be transported and redeposited by the meltwater, whereupon the sediment is called **outwash.** Till and outwash together are two varieties of glacial **drift.**

Figure 9.7 Satellite photograph of the Finger Lakes region, New York State. These long, narrow lakes fill gouges in the rock that were carved by a thick advancing ice sheet, widening north-south stream valleys. Photograph © NASA.

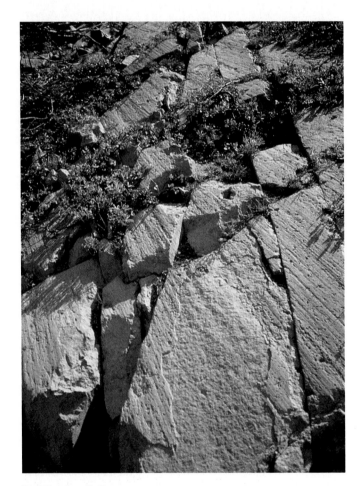

Figure 9.8 Striations on a rock surface show direction of glacial flow. Notice that the grooves extend continuously across the fractures breaking up the outcrop, but are absent where the scratched surface rock has weathered away. Lassen Park, California.

Figure 9.9 Till deposited by an Alaskan glacier shows characteristically poor sorting.

A landform made of till is called a **moraine.** There are many different types of moraine. One of the most distinctive is a curving ridge of till piled up at the nose of a glacier, some of it pushed along ahead of the glacier as it advances, much of it accumulated at the glacier's end by repeated annual advances and retreats. This is an *end moraine.* When the glacier begins a long-term retreat, a *terminal moraine* is left to mark the farthest advance of the ice (figure 9.10). Moraines are useful in mapping the extent of past glaciation (figure 9.11).

Glaciers are responsible for much of the topography and near-surface geology of northern North America. The Great Lakes occupy basins carved and deepened by the most recent ice sheets and were once filled with glacial meltwater as the ice sheets retreated. Much of the basic drainage pattern of the Mississippi River was established during the surface runoff of the enormous volumes of meltwater; the sediments carried along with that same meltwater are partially responsible for the fertility of midwestern farmland (see "Loess" later in the chapter). Infiltration of a portion of the meltwater resulted in groundwater supplies tapped today for drinking water. The importance of the legacy of past glaciers should not be overlooked just because their extent at present is more limited.

Wind and Its Geologic Impacts

Air moves from place to place over the earth's surface mainly in response to differences in pressure, which are often related to differences in surface temperature. Solar radiation falls most directly on the earth near the equator, and, consequently, solar heating of the atmosphere and surface is more intense there. The sun's rays are more dispersed near the poles. On a non-rotating earth of uniform surface, that surface would be heated more near the equator and less near the poles; correspondingly, the air over the equatorial regions would be warmer than the air over the poles.

A

B

Figure 9.10 (*A*) Formation of a terminal moraine occurs through a combination of "bulldozing" by an advancing glacier and repeated cycles of till deposition by melting ice. Advancing glacier deposits till in the zone of ablation and pushes material ahead of ice front; then the glacier retreats, leaving a ridge of till behind. (*B*) This glacier, in Denali National Park, Alaska, shows moraine (foreground) left by the retreat of the glacier. The terminal moraine is overgrown with vegetation. Note the outwash also.

Figure 9.11 Moraines of northeastern Illinois (darker green shading) reflect the advances of many separate tongues of ice over hundreds of thousands of years.   Source: After H. B. Willman and J. C. Frye, *Glacial Map of Illinois*, Illinois State Geological Survey, 1970.

Because warm, less-dense (lower-pressure) air rises, warm near-surface equatorial air would rise, while cooler, denser polar air would move into the low-pressure region. The rising warm air would spread out, cool, and sink. Large circulating air cells would develop, cycling air from equator to poles and back. These would be atmospheric convection cells, analogous to the mantle convection cells associated with plate motions.

The actual picture is considerably more complicated. The earth rotates on its axis, which adds a net east-west component to air flow as viewed from the surface. Land and water are heated differentially by sunlight, with surface temperatures on the continents generally fluctuating much more than temperatures of adjacent oceans. Thus, the pattern of distribution of land and water influences the distribution of high- and low-pressure regions and further modifies air flow. In addition, the earth's surface is not flat; terrain irregularities introduce further complexities in air circulation. Friction between moving air masses and land surfaces can also alter wind direction and speed; tall, dense vegetation may reduce near-surface wind speeds by 30 to 40 percent over vegetated areas. A generalized view of actual global air circulation patterns is shown in figure 9.12. Different regions of the earth are characterized by different prevailing wind directions. Most of the United States is in a zone of westerlies, in which winds generally blow from west/southwest to east/northeast. Local weather conditions and the details of local geography produce regional deviations from this pattern on a day-to-day basis.

Air and water have much in common as agents shaping the land. Both can erode and deposit material; both move material more effectively the faster they flow; both can move particles by rolling them along, by saltation, or in suspension (see chapter 6). Because water is far denser than air, it is much more efficient at eroding rocks and moving sediments, and has the added abilities to dissolve geologic materials and to attack them physically by freezing and thawing. On average worldwide, wind erosion moves only a small percentage of the amount of material moved by stream erosion. Historically, the cumulative significance of wind erosion is approximately comparable to that of glaciers. Nevertheless, like glaciers, winds can be very important in individual locations particularly subject to their effects.

A

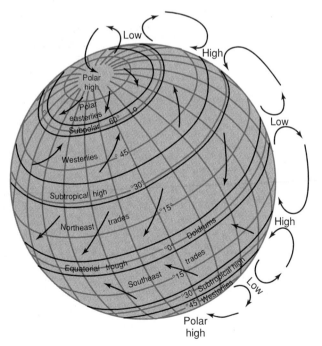

Figure 9.12   Principal present atmospheric circulation patterns.   From Arthur Getis, Judith Getis, and Jerome Fellman, *Introduction to Geography*, 3d ed. Copyright © 1991 McGraw-Hill Higher Education, Dubuque, Iowa. All Rights Reserved. Reprinted by permission.

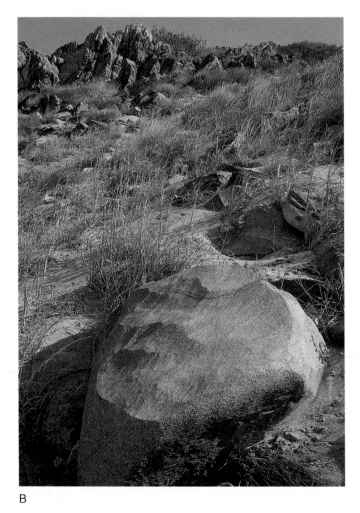

B

Figure 9.13   Wind abrasion on low-lying rocks results in the planing of rock surfaces. (*A*) If the wind is predominantly from one direction, rocks are planed or flattened on the upwind side. With a persistent shift in wind direction, additional facets are cut in the rock. (*B*) Example of ventifact: rock polished and faceted by windblown sand at San Gorgonio Pass near Palm Springs, California.   Photograph © Doug Sherman/Geofile.

Wind Erosion

Like water, wind erosion acts more effectively on sediment than on solid rock, and wind-related processes are especially significant where the sediment is exposed, not covered by structures or vegetation, in such areas as deserts, beaches, and unplanted (or incorrectly planted) farmland. In dry areas like deserts, wind may be the major or even the sole agent of sediment transport.

Wind erosion consists of either abrasion or deflation. *Wind abrasion* is the wearing-away of a solid object by the impact of particles carried by wind. It is a sort of natural sandblasting, analogous to milling by sand-laden waves. Where winds blow consistently from certain directions, exposed boulders may be planed off in the direction(s) from which

A

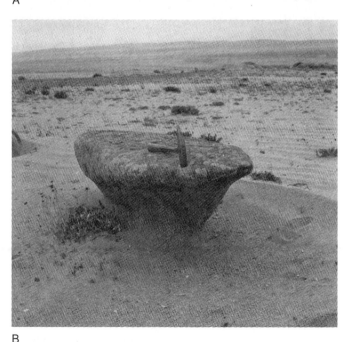

B

Figure 9.14 (*A*) This rock in the Garden of the Gods, Colorado, was sculpted, in part, by wind abrasion. (*B*) Undercutting of granite boulder by near-surface abrasion. Photograph (*B*) by K. Segerstrom, USGS Photo Library, Denver, CO.

they have been abraded, becoming **ventifacts** ("wind-made" rocks; figures 9.13, 9.14). If wind velocity is too low to lift the largest transported particles very high above the ground, tall rocks may show undercutting close to ground level (figure 9.14B). Abrasion can also cause serious property damage. Desert travelers caught in windstorms have been left with cars stripped of paint and windshields so pitted and frosted that they could no longer be seen through. Abrasion likewise scrapes paint from buildings and can erode construction materials such as wood or soft stone.

Deflation is the wholesale removal of loose sediment, usually fine-grained sediment, by the wind. In barren areas, a combination of deflation and erosion by surface runoff may proceed down to some level at which larger rocks are exposed, and these larger rocks protect underlying fine material from further erosion. The resulting surface is called *desert pavement* (figure 9.15). A desert pavement surface, once established, can be quite stable. However, if the protective coarse-rock layer is disturbed—for example, by construction or resource-exploration activities—newly exposed fine sediment may be subject to rapid wind erosion.

The importance of vegetation in retarding erosion was demonstrated especially dramatically in the United States during the early twentieth century. After the Civil War, there was a major westward migration of farmers to the Great Plains. They found a great deal of flat or gently rolling land, much of it covered by prairie grasses and wildflowers rather than by thick forests, so it was easy to adapt for farming. Over much of the area, native vegetation was removed and the land plowed and planted to crops. Elsewhere, grazing livestock cropped the prairie. While adequate rainfall continued, all was well. Then, in the 1930s, several years of drought killed the crops, leaving the soil bare and vulnerable to erosion by the unusually strong winds that followed. This

Figure 9.15 Desert pavement developed on gravel. Death Valley National Monument, California.   Photograph by C. B. Hunt, USGS Photo Library, Denver, CO.

A

B

A

B

Figure 9.17 Dune migration and its consequences. (*A*) Schematic of dune migration. (*B*) Marching sand dunes encroaching on full-grown trees. Oregon Dunes National Recreation Area.

Figure 9.16 The formation of dunes. (*A*) Sand dunes, Sherman County, Oregon. (*B*) Snow dunes on barren ground in winter. Note the ground bared by deflation between dunes. (*A*) Photograph by E. D. McKee, USGS Photo Library, Denver, CO.

We look further at the Dust Bowl and the broader problem of minimizing soil erosion on cropland in chapter 11.

Wind Deposition

Where sediment is transported and deposited by wind, the principal depositional feature is a **dune,** a low mound or ridge, usually made of sand. Dunes start to form when sediment-laden winds slow down. The lower the velocity, the less the wind can carry. The coarser and heavier particles are dropped first. The deposition, in turn, creates an obstacle that constitutes more of a windbreak, causing more deposition. Once started, a dune can grow very large. A typical dune is 3 to 100 meters high, but dunes as high as 200 meters are found. What ultimately limits a dune's size is not known, but it is probably some aspect of the nature of the local winds.

was the "Dust Bowl" period. Hundreds of millions of tons of soil were picked up by the strong winds, transported, and then dumped as the winds lulled, burying homes and farms. The native prairie vegetation had been suitably adapted to the climate, while many of the crops were not. Once the crops died, there was nothing left to hold down the soil and protect it from the west winds sweeping across the plains.

Figure 9.18 Crossbeds in a sand dune. Oregon Dunes National Recreation Area.

When the word *dune* is mentioned, most people's reaction is *sand dune*. However, the particles can be any size, ranging from sand down to fine dust, and they can even be snow or ice crystals, although sand dunes are by far the most common. The particles in a given set of dunes tend to be similar in size. Eolian, or wind-deposited, sediments are well sorted like many stream deposits, and for the same reason: The velocity of flow controls the size and weight of particles moved. The coarser the particles, the stronger was the wind forming the dunes. The orientation of the dunes reflects the prevailing wind direction (if winds show a preferred orientation), with the more shallowly sloping side facing upwind (figure 9.16).

Dune Migration

Dunes move if the wind blows predominantly from a single direction. As noted previously, a dune assumes a characteristic profile in cross section, gently sloping on the windward side, steeper on the downwind side. With continued wind action, particles are rolled or moved by saltation up the shallower slope. They roll down the steeper face, or **slip face,** which tends to assume a slope at the angle of repose of sand (or whatever size of particle is involved; recall figure 8.4B). The net effect of these individual particle shifts is that the dune moves slowly downwind (figure 9.17A). As layer upon layer of sediment slides down the slip face, slanted *crossbeds* develop in the dune (figure 9.18).

Migrating dunes, especially large ones, can be a real menace as they march across roads and even through forests and over buildings (figure 9.17B). During the Dust Bowl era, farmland and buildings were buried under shifting, windblown soil. The costs to clear and maintain roads along sandy beaches and through deserts can be high because dunes can move several meters or more in a year. The usual approach to dune stabilization is to plant vegetation. However, since many dunes exist, in part, because the terrain is too dry to support vegetation, such efforts are likely to be futile. Aside from any water limitations, young plants may be difficult to establish in shifting dune sands because their tiny roots may not be able to secure a hold.

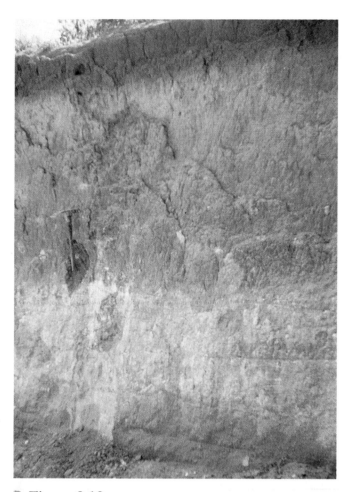

Figure 9.19 Example of loess.   Photograph by L. B. Buck, USGS Photo Library, Denver, CO.

Loess

Rarely is the wind strong enough to move sand-sized or larger particles very far or very rapidly. Fine dust, on the other hand, is more easily suspended in the wind and can be carried many kilometers before it is dropped. A deposit of windblown silt is known as **loess** (figure 9.19). The fine mineral fragments in loess are in the range of 0.01 to 0.06 millimeter (0.0004 to 0.0024 inch) in diameter.

The principal loess deposits in the United States are in the central part of the country, and their spatial distribution provides a clue to their source (figure 9.20). They are concentrated around the Mississippi River drainage basin, particularly on the east sides of major rivers of that basin. Those same rivers drained away much of the meltwater from retreating ice sheets in the last ice age. Glaciers grinding across the continent, then, were apparently the original producers of this finest-sized sediment. The sediment was subsequently washed down the river valleys, and the lightest material was further blown eastward by the prevailing west winds.

As noted in chapter 8, the great mass of ice sheets enables them to produce quantities of finely pulverized *rock flour.* Because dry glacial erosion does not involve as much chemical weathering as stream erosion, many soluble minerals are preserved in glacial rock flour. These minerals provide some plant

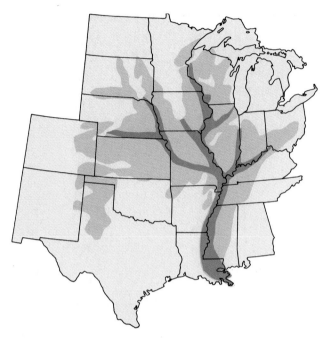

Figure 9.20 Loess distribution in the central United States (shaded) and locations of principal stream valleys supplied by glacial meltwater (purple). Loess to the southwest may have been derived from western deserts.   Source: After J. Thorp and H. T. U. Smith, *"Pleistocene Eolian Deposits of the United States, Alaska, and Parts of Canada,"* Geological Society of America Map, 1952.

nutrients to the farmland soils now developed on the loess. Because newly deposited loess is also quite porous and open in structure, it has good moisture-holding capacity. These two characteristics together contribute to making the farmlands developed on midwestern loess particularly productive.

Not all loess deposits are of glacial origin. Loess deposits form wherever there is an abundant supply of very fine sediment. Loess derived from the Gobi Desert covers large areas of China, for example, and additional loess deposits are found close to other major deserts.

Loess does have drawbacks with respect to applications other than farming. While its light, open structure is reasonably strong when dry and not heavily loaded, it may not make suitable foundation material. Loess is subject to hydrocompaction when wetted, during which it can settle, crack, and become denser and more consolidated, to the detriment of structures built on top of it. The very weight of a large structure can also cause settling and collapse.

Deserts and Desertification

Many of the features of wind erosion and deposition are most readily observed in deserts. A **desert** is a region with so little vegetation that no significant population can be supported on that land. It need not be hot or even, technically, dry. Ice sheets are a kind of desert. In more temperate climates, deserts are characterized by very little precipitation, but they may be consistently hot, cold, or variable in temperature, depending on the season or time of day. The distribution of the arid regions of the world (exclusive of polar deserts) is shown in figure 9.21.

Causes of Natural Deserts

A variety of factors contribute to the formation of a desert, so deserts come in several different types, depending on their origins.

One factor is moderately high surface temperatures. Most vegetation, under such conditions, requires abundant rainfall and/or slow evaporation of what precipitation does fall. The availability of precipitation is governed, in part, by the global air-circulation patterns shown in figure 9.12. Warm air holds more moisture than cold. Similarly, when the pressure on a mass of air is increased, the air can hold more moisture. Air spreading outward from the equator at high altitudes is chilled and at low pressure, since air pressure and temperature decrease with increasing altitude. Thus, the air holds little moisture. When that air circulates downward, at about 30 degrees north and south latitudes, it is warmed as it approaches the surface, and also subjected to increasing pressure from the deepening column of air above it. It can then hold considerably more water, so when it reaches the earth's surface, it causes rapid evaporation. Note in figure 9.21 that many of the world's major deserts fall in belts close to these zones of sinking air at 30 degrees north and south of the equator. These are the tropical-latitude deserts.

Topography also plays a role in controlling the distribution of precipitation. A high mountain range along the path of principal air currents between the ocean and a desert area may be the cause of the latter's dryness. As moisture-laden air from over the ocean moves inland across the mountains, it is forced to higher altitudes, where the temperatures are colder and the air thinner (lower pressure). Under these conditions, much of the moisture originally in the air mass is forced out as precipitation, and the air is much drier when it moves farther inland and down out of the mountains. In effect, the mountains cast a **rain shadow** on the land beyond. Rain shadows cast by the Sierra Nevada of California and, to a lesser extent, by the southern Rockies contribute to the dryness and desert regions of the western United States.

Because the oceans are the major source of the moisture in the air, distance from the ocean (in the direction of air movement) can by itself be a factor contributing to the formation of a desert. The longer an air mass is in transit over dry land, the greater chance it has of losing some of its moisture through precipitation. This contributes to development of deserts in continental interiors. On the other hand, even coastal areas can have deserts under special circumstances. If the land is hot and the adjacent ocean cooled by cold currents, the moist air coming off the ocean will be cool and carry less moisture than air over a warmer ocean. As that cooler air warms over the land and becomes capable of holding still more moisture, it causes rapid evaporation from the land rather than precipitation. This phenomenon is observed along portions of the western coasts of Africa and South America. Polar deserts can also be attributed to the differences in moisture-holding capacity between warm and cold air: Air traveling from warmer latitudes to colder near-polar ones will tend to lose moisture by precipitation, so less

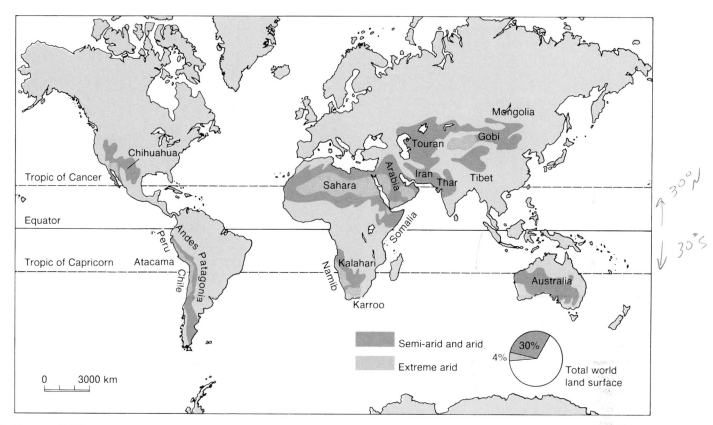

Figure 9.21   Distribution of the world's arid lands.   Source: Data from A. Goudie and J. Wilkinson, *The Warm Desert Environment*. Copyright © 1977 by Cambridge University Press, New York.

remains to fall as snow near the poles, and the limited evaporation from cold high-latitude oceans contributes little additional moisture to enhance local precipitation. Thick polar ice caps, then, reflect effective preservation of what snow does fall, rather than heavy precipitation.

Desertification

Climatic zones shift over time. In addition, topography changes, global temperatures change, and plate motions move landmasses to different latitudes. Amid these changes, new deserts develop in areas that previously had more extensive vegetative cover. The term **desertification,** however, is generally restricted to apply only to the relatively rapid development of deserts caused by the impact of human activities.

The exact definition of the lands at risk is difficult. *Arid* lands are commonly defined as those with annual rainfall of less than 250 millimeters (about 10 inches); *semiarid* lands, 250 to 500 millimeters (10–20 inches); and extremely arid lands, or *deserts*, those areas that may have at least 12 consecutive months without rainfall. The extent to which vegetation will thrive in low-precipitation areas also depends on such additional factors as temperature and local evaporation rates. Many of the arid lands border true desert regions. Desertification does not involve the advance or expansion of desert regions as a result of forces originating within the desert. Rather, desertification is a patchy conversion of dry-but-habitable land to uninhabitable desert as a consequence of land-use practices (perhaps accelerated by such natural factors as drought).

Causes of Desertification

Vegetation in dry lands is, by nature, limited (figure 9.22). At the same time, it is a precious resource, which may in various cases provide food for people or for livestock, wood for shelter or energy, and protection for the soil from erosion. Desertification typically involves severe disturbance of that vegetation. The environment is not a resilient one to begin with, and its deterioration, once begun, may be irreversible and even self-accelerating.

On land used for farming, native vegetation is routinely cleared to make way for crops. While the crops thrive, all may be well. If the crops fail, or if the land is left unplanted for a time, several consequences follow. One, as in the Dust Bowl, is erosion. A second, linked to the first, is loss of soil fertility. The topmost soil layer, richest in organic matter, is most nutrient-rich and also is the first lost to erosion. A third result may be loss of soil structural quality. Under the baking sun typical of many dry lands, and with no plant roots to break it up, the soil may crust over, becoming less permeable. This increases surface runoff, correspondingly decreasing infiltration by what limited precipitation does fall. That, in turn, decreases reserves of soil moisture and ground water on which future crops may depend. All of these changes together make it that much harder for future crops to succeed, and the problems intensify.

Similar results follow from the raising of numerous livestock on the dry lands. In drier periods, vegetation may be reduced or stunted. Yet it is precisely during those periods

Figure 9.22 The inability of deserts to sustain appreciable life is obvious from scarce vegetation. Monument Valley, Arizona.

that livestock, needing the vegetation not only for food but also for the moisture it contains, put the greatest grazing pressure on the land. The soil may again be stripped bare, with the resultant deterioration and reduced future growth of vegetation as previously described for cropland.

Natural drought cycles thus play a role in desertification. However, in the absence of intensive human land use, the degradation of the land during drought is typically less severe, and the natural systems in the arid lands can recover when the drought ends. On a human time scale, desertification—permanent conversion of marginal dry lands to desert—is generally observed only where human activities are also significant.

Desertification in the United States

Though desertification makes news most often in the context of Third World countries, the process is ongoing in many other places. In the United States, a large portion of the country is potentially vulnerable: Much of the western half of the country can be classified as "semiarid" on the basis of its low precipitation. An estimated 1-million-plus square miles of land—more than a third of this low-rainfall area—has undergone severe desertification, characterized by loss of desirable native vegetation, seriously increased erosion, and reduced crop yields. The problems are caused or aggravated by intensive use of surface water and overgrazing and the population of these areas—especially the "Sun Belt"—is growing.

Why, then, so little apparent concern? The main reason is that so far, the impact on humans has been modest. Other parts of the country can supply needed

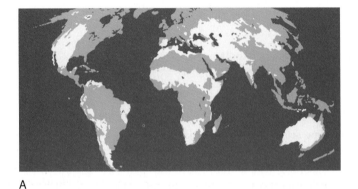

Figure 9.23 The world's "drylands" comprise a variety of climates; precipitation, and therefore the ability to support both vegetation and human and animal populations, is limited. As much as 40% of the world's land surface, in over 100 different countries, fits this broad definition (*A*); degradation—loss of soil, water resources, or biodiversity—has reduced productivity over about 70% of that total (*B*), and the rest is vulnerable.   Base map from UNEP Dryland Degradation Map, 1993. Reprinted by permission.

Figure 9.24 Aerial photograph of the Nebraska Sand Hills area, formerly a desert, shows a well-preserved dune about 75 meters high, now covered by vegetation under moister modern conditions. Photograph by T. S. Ahlbrandt, USGS Photo Library, Denver, CO.

food. The erosion and associated sediment-redistribution problems are not yet highly visible to most people. Groundwater supplies are being tapped both for water supplies and for irrigation, creating the illusion of adequate water; but as will be seen in the next chapter, groundwater users may be living on borrowed time, or at least borrowed water that is destined to run out. As that happens, it will become painfully obvious that too much pressure from human use was brought to bear on dry lands.

Global Impact of Desertification

Desertification is cause for concern because it effectively reduces the amount of arable (cultivatable) land on which the world depends for food. An estimated 600 million people worldwide now live on the arid lands; by recent U.N. definition, close to 40% of the world's land surface is "drylands" (figure 9.23A), and that *excludes* the true deserts. All of those lands, in some measure, are potentially vulnerable to desertification. More than 10% of those 600 million people live in areas identified as actively undergoing desertification now; overall, up to 70% of the drylands are undergoing significant degradation (figure 9.23B). Some projections suggest that, by the end of this century, one-third of the world's once-arable land will be rendered useless for the culture of food crops as a consequence of desertification and attendant soil deterioration. The recent famine in Ethiopia may have been precipitated by a drought, but it has been prolonged by desertification brought on by overuse of land incapable of supporting concentrated human or animal populations.

Global Climate, Past and Future

Climate is the result of the interplay of a number of factors. The main source of energy input to the earth is sunlight, which warms the land surface, which in turn radiates heat into the atmosphere. Globally, how much heating occurs is related to the sun's energy output, and how much of the sun-

light falling on earth actually reaches the surface. Incoming sunlight may be blocked by cloud cover or, as we have previously seen, by dust and sulfuric acid droplets from volcanic eruptions. Heat (infrared rays) radiating outward from earth's surface may or may not be trapped by certain atmospheric gases, as described below.

Locally, at any given time in earth's history a wide variety of local climates can obviously exist simultaneously — icy glaciers, hot deserts, steamy rain forests, temperate regions — and a given region may also be subject to wide seasonal variations in temperature and rainfall. One of the challenges in determining present global climate trends, then, is deciding just how to measure global climate at any given time. Commonly, several different kinds of data are used to characterize present climate, including air temperatures, over land and sea, at various altitudes; snow and ice cover on land, and extent of sea ice; ocean-water temperatures; distribution of atmospheric moisture and precipitation. An understanding of past climatic fluctuations would be helpful in developing models of possible future climate change, but the evidence available is generally less direct and comprehensive.

Evidence of Climates Past

Aspects of local climate may often be deduced from the geologic record, especially the sedimentary-rock record. For example, the now-vegetated dunes of the Nebraska Sand Hills (figure 9.24) are remnants of an arid "sand sea" in the region 18,000 years ago, when conditions must have been much drier than they are now. We have noted that glacial deposits may be recognized in now-tropical regions. As will be seen in chapter 13, coal deposits indicate warm, wet conditions, conducive to lush plant growth, perhaps in a swampy setting. Some idea of global climate at a given time may be gained from compiling such data from a variety of localities (taking into account changes in latitude related to continental drift); but such records are far from complete.

A particular challenge in reconstructing past climate is that we do not have direct measurements of ancient temperatures. These must be determined indirectly using various methods. For example, from marine sediments comes evidence of water-temperature variations. The proportion of calcium carbonate ($CaCO_3$) in Pacific Ocean sediments in a general way reflects water temperature, because the solubility of calcium carbonate is strongly temperature-related.

A particularly powerful tool involves variations in the proportions of oxygen isotopes in geological materials. By far the most abundant oxygen isotope in nature is oxygen-16 (^{16}O); the heaviest is oxygen-18 (^{18}O). Because they differ in mass, so do molecules containing them, and the effect is especially pronounced when oxygen makes up most of the mass, as in H_2O. Certain natural processes, including evaporation and precipitation, produce fractionation between ^{16}O and ^{18}O; see figure 9.25. As water evaporates, the ligher $H_2^{16}O$ evaporates preferentially, and the water vapor will then be isotopically "lighter" (richer in ^{16}O, poorer in ^{18}O) than the residual water. Conversely, as rain or snow condenses and falls, the

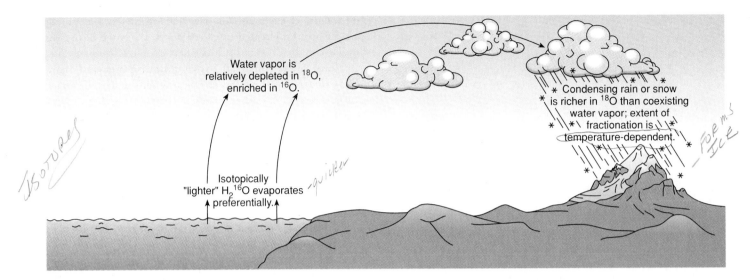

Figure 9.25   Mass diferences between ^{18}O and ^{16}O lead to different $^{18}O/^{16}O$ ratios in coexisting water and water vapor; the size of the difference is a function of temperature.

Figure 9.26   Data from the Vostok ice core indicate past temperature fluctuations, which can be compared with variations in CO_2 (carbon dioxide) and CH_4 (methane) in the atmosphere.   Adapted from "Global Climatic Change," by Richard A. Houghton and George M. Woodwell. Copyright © 1989 by Scientific America, Inc. All Rights Reserved.

precipitation will be relatively enriched in the heavier $H_2^{18}O$. As water vapor evaporated from equatorial oceans drifts toward the poles, depositing $H_2^{18}O$-enriched precipitation, the remaining water vapor becomes progressively lighter isotopically, and so does subsequent precipitation; snow falling near the poles has a much lower $^{18}O/^{16}O$ ratio than tropical rain, or seawater. The fractionation between ^{18}O and ^{16}O in coexisting water vapor and rain or snow is temperature-dependent, more pronounced at lower temperatures. Therefore, at a given latitude, variations in $^{18}O/^{16}O$ ratio of precipitation reflect variations in temperature. (Similarly, variations in $^{18}O/^{16}O$ ratio in the oxygen-rich carbonate [$CaCO_3$] or silica [SiO_2] skeletons of marine microorganisims give evidence of variations in the temperature of the near-surface seawater from which these skeletons precipitated.)

The longevity of the massive continental glaciers—sometimes hundreds of thousands of years—makes them useful in preserving evidence of both air temperature and atmospheric composition in the past. Tiny bubbles of contemporary air may be trapped as the snow is converted to dense glacial ice. The oxygen-isotope composition of the ice reflects seawater O-isotope composition and the air temperature at the time the parent snow fell. If those temperatures can be correlated with other evidence, such as the carbon-dioxide content of the air bubbles or the presence of volcanic ash layers in the ice from prehistoric explosive eruptions, much information can be gained about possible causes of past climatic variability (figure 9.26), and the climate-prediction models may be refined accordingly. The ice sheets themselves are remnants of a climatic shift known as an ice age.

Ice Ages and Their Possible Causes

The term *Ice Age* is used by most people to refer to a time in the relatively recent geologic past when extensive continental glaciers covered millions more square kilometers of area than they now do. There were many cycles of advance and retreat of the ice sheets during this epoch, known to geologists as the Pleistocene, which spanned the time from about two million to ten thousand years ago. At its greatest extent, Pleistocene glaciation in North America covered essentially all of Canada and much of the northern United States (figure 9.27). The ice was well over a kilometer thick over most of that area. Climatic effects extended well beyond the limits of the ice, with cooler temperatures and more precipitation common in surrounding lands, and worldwide sea levels lowered as vast volumes of water were locked up as ice in ice sheets (figure 9.28).

Pleistocene glaciation can be studied in some detail because it was so recent, and the evidence left behind is thus fairly well preserved. It is not, however, the only large-scale continental glaciation in earth's past. There have been at least half a dozen ice ages over the earth's history, going back a billion years or more. As previously noted, the pre-

drift reassembly of the pieces of the continental jigsaw puzzle is aided by matching up features produced by past ice sheets on what are now separate continents.

An underlying question is what kinds of factors might be involved in creating conditions conducive to the formation of such immense ice masses, bearing in mind that ice a kilometer thick represents far more than a few winters' snow. The proposed causes of ice ages fall into two groups: those that involve events external to the earth and those for which the changes arise entirely on earth.

One possible external cause, for example, would be a significant change in the sun's energy output. Present cycles of sunspot activity cause variations in sunlight intensity, which should logically result in temperature fluctuations worldwide. However, the variations in solar-energy output would have to be about ten times as large as they are in the modern sunspot cycle to account even for short-term temperature fluctuations observed on earth. To cause an ice age of major proportions and of thousands of years' duration, any cooling trend would have to last much longer than the eleven years of the present sunspot cycle. Another problem with linking solar activity fluctuations with past ice ages is simply lack of evidence. Although means exist for estimating temperatures on earth in the past, scientists have yet to conceive of a way to determine the pattern of solar activity in ancient times. Thus there is no way to test the theory, to prove or disprove it.

Another proposed cause of ice ages is the observed variation in the tilt of the earth's axis in space. This tilt varies over time relative to the earth's orbital plane. Changes in tilt do not change the total amount of sunlight reaching earth, but they do affect its distribution. If, for a time, the poles were tilted farther away from incident sunlight, the polar regions might become sufficiently colder that a large ice sheet could begin to develop. Some evidence supports this theory. When glaciers are extensive and large volumes of water are locked up in ice, sea level is lowered. The reverse is true as ice sheets melt. It is possible to use ages of Caribbean coral reefs to date high stands of sea level over the last few hundred thousand years, and these periods of high sea level correlate well with periods during which maximum solar radiation fell on the mid-Northern Hemisphere. Conversely, then, periods of reduced sun exposure are apparently correlated with ice advance (lower sea level). However, this precise a record does not go back very far in time. It may also be that changes in tilt are related to changes in precipitation patterns over the oceans, and thus to changes in seawater salinity and mixing patterns, which in turn would alter the distribution of organisms in the oceans and the oceans' effectiveness as a "sink" for atmospheric carbon. (To understand what this has to do with heating, see the later section on the greenhouse effect.)

Since the advent of plate-tectonic theory, a novel suggestion for a cause of ice ages has invoked continental drift.

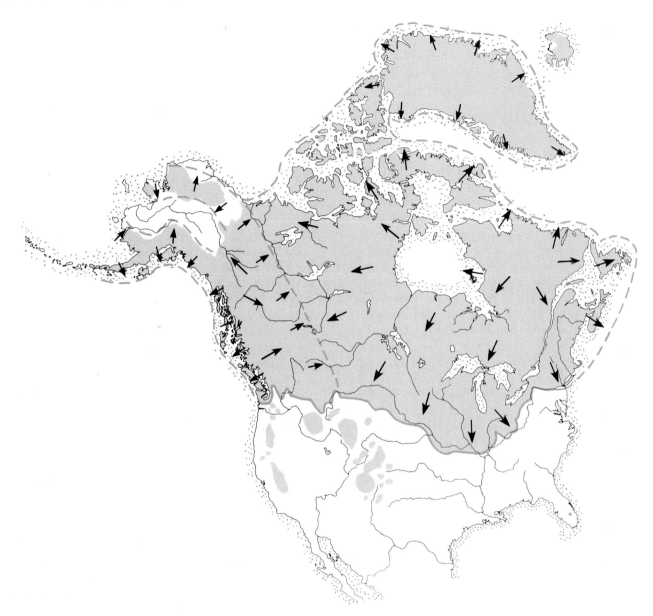

Figure 9.27 Extent of glaciation in North America during the Pleistocene epoch. Arrows show the direction of ice movement.
Source: After C. S. Denny, *National Atlas of the United States,* U.S. Geological Survey.

Prior to the breakup of Pangaea, all the continental landmasses were close together. This concentration of continents meant much freer circulation of ocean currents elsewhere on the globe and more circulation of warm equatorial waters to the poles. The breakup of Pangaea disrupted the oceanic circulation patterns. The poles could have become substantially colder as their waters became more isolated. This, in turn, could have made sustained development of a thick ice sheet possible. A limitation of this mechanism is that it only accounts for the Pleistocene glaciation, while considerable evidence exists for extensive glaciation over much of the Southern Hemisphere—India, Australia, southern Africa, and South America—some 200 million years ago, just *prior* to the breakup of Pangaea. Beyond that, not enough is known about continental positions prior to the

formation of Pangaea to determine whether continental drift and resulting changes in oceanic circulation patterns could be used to explain earlier ice ages.

Another possible earth-based cause of ice ages might involve blocking of incoming solar radiation by something in the atmosphere. The resultant cooling might be adequate to induce the start of an ice age. We have observed that volcanic dust and sulfuric-acid droplets in the atmosphere can cause measurable cooling, but with the modern examples at least, the effects were significant only for a year or two. In none of these cases was the cooling serious enough or prolonged enough to cause an ice age. Even some of the larger prehistoric eruptions, such as those that occurred in the vicinity of Yellowstone National Park or the explosion that produced the crater in Mount Mazama that is now Crater Lake, would

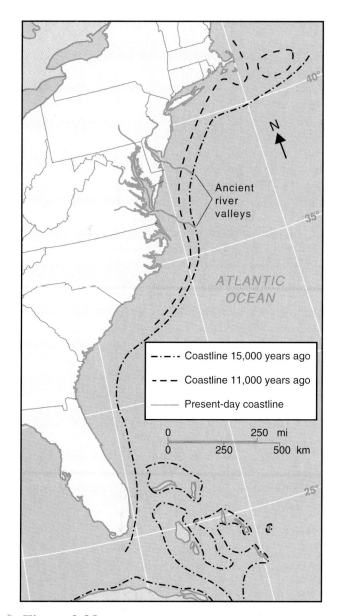

Figure 9.28 Shoreline of eastern North America extended much farther seaward at peak of last major ice advance of the Ice Age. By about 6,000 years ago, melting had raised sea level approximately to its present level.   From Michael Bradshaw and Ruth Weaver, *Physical Geography*. Copyright © 1993 Mosby-Year Book, Inc. Reprinted with permission of McGraw-Hill Higher Education, Dubuque, Iowa. All Rights Reserved.

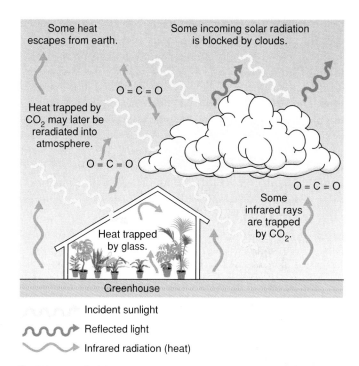

Figure 9.29 The greenhouse effect (schematic). Both glass and air are transparent to visible light. Like greenhouse glass, carbon dioxide molecules in the atmosphere trap infrared rays radiating from the sun-warmed surface of the earth.

have been inadequate individually to produce the required environmental change, and again, the dust and acid droplets from any one of these events would have settled out of the atmosphere within a few years. However, there have been periods of more intensive or frequent volcanic activity in the geologic past. The cumulative effects of multiple eruptions during such an episode might have included a sustained cooling trend and, ultimately, ice-sheet formation.

The foregoing is a brief and incomplete sampling of the many processes and phenomena that may have caused climatic upheavals in the past. The possibilities are many, and, as yet, there is insufficient evidence to demonstrate any single theory correct. Of more immediate concern is the realization that present human activities may also have begun to alter the climate irrevocably in the other direction, toward global warming.

The Greenhouse Effect, Global Warming, Climate Change

On a sunny day, it is much warmer inside a greenhouse than outside it. Light enters through the glass and is absorbed by the ground, plants, and pots inside. They, in turn, radiate heat: infrared radiation, not visible light. Infrared rays, with longer wavelengths than visible light, cannot readily escape through the glass panes; the rays are trapped, and the air inside the greenhouse warms up. The same effect can be observed in a closed car on a bright day.

In the atmosphere, carbon dioxide molecules act similarly to the greenhouse's glass. Light reaches the earth's surface, warming it, and the earth radiates infrared rays back. But the longer-wavelength infrared rays are trapped by the carbon dioxide molecules, and a portion of the radiated heat is thus trapped in the atmosphere. Hence the term **"greenhouse effect."** (See figure 9.29.) As a result of the greenhouse effect, the atmosphere stays warmer than it would if that heat radiated freely back out into space.

As we will see again in chapter 13, the evolution of a technological society has meant rapidly increasing energy consumption. Historically, we have relied most heavily on carbon-rich fuels—wood, coal, oil, and natural gas—to supply that energy. These probably will continue to be important energy sources for several decades at least. One combustion by-product that all of these fuels have in common is carbon dioxide gas (CO_2).

Excess water in the atmosphere readily falls out as rain or snow. Some of the excess carbon dioxide is removed by geologic processes (see chapter 17), but in the past century or so, since the start of the so-called Industrial Age in the mid-nineteenth century, the amount of carbon dioxide in the air has increased by an estimated 25 percent—and its concentration continues to climb (figure 9.30). If the heat trapped by carbon dioxide is proportional to the concentration of carbon dioxide in the air, the increased carbon dioxide, it is feared, will gradually cause increased greenhouse-effect heating of the earth's atmosphere.

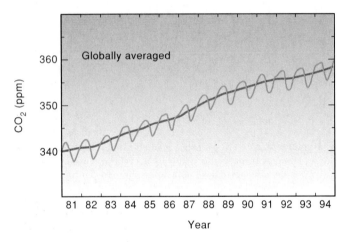

Figure 9.30   Rise in atmospheric CO_2 over past several decades is clear. (Zigzag pattern probably reflects seasonal variations in uptake by plants.) Pre-industrial levels in ice cores were about 280 ppm.   Source: Data from NOAA Carbon Cycle Group.

One potential problem arising from increased greenhouse-effect heating is agricultural. In many parts of the world, agriculture is already only marginally possible because of hot climate and little rain. A temperature rise of only a few degrees could easily make living and farming in these areas impossible. Also, the warmer it gets, the faster the remaining ice sheets will melt. Significant decreases in the extent of ice caps could further alter global weather patterns, compounding the agricultural problems. Recent projections suggest that summer soil-moisture levels in the Northern Hemisphere could drop by up to 40 percent with a doubling in atmospheric CO_2. This is a sobering prospect for farmers in areas where rain is barely adequate now.

Another potential problem is that meltwater from the ice caps would, in turn, be added to the oceans, and the net effect would be a rise in sea level. If *all* the ice melted, sea level could rise by close to 75 meters (about 250 feet). About 20 percent of the world's land area would be submerged. Many millions, perhaps billions, of people now living in coastal or low-lying areas would be displaced, since a large fraction of major population centers grew up along coastlines and rivers. The Statue of Liberty would be up to her neck in water. The consequences to the continental United States and southern Canada are illustrated in figure 9.31. Also, raising of the base levels of streams draining into the ocean would alter the stream channels and could cause significant flooding along major rivers.

Such large-scale melting of ice sheets would take time, perhaps several thousand years. On a shorter time scale, however, the problem could still be significant. Continued intensive fossil-fuel use could easily double the level of

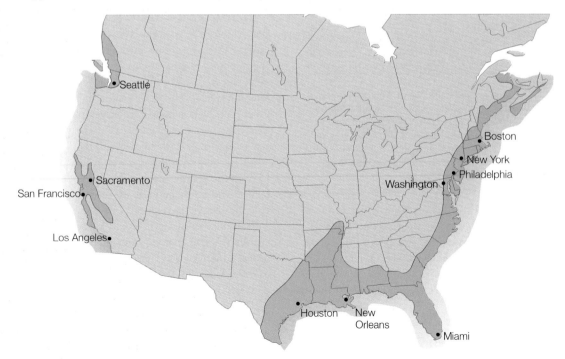

Figure 9.31   The flooding (blue shaded areas) that would result from melting of all of the world's ice sheets. The resultant 75-meter rise in sea level would flood many major cities in the United States and elsewhere.

carbon dioxide in the atmosphere before the middle of the next century. This would produce a projected temperature rise of 4° to 8° C (7°–14° F), which would be sufficient to melt at least the West Antarctic ice sheet completely in a few hundred years. The resulting 3- to 6-meter rise in sea level, though it sounds small, would nevertheless be enough to flood many coastal cities and ports, along with most of the world's beaches. This would be both inconvenient and extremely expensive. (The only consolation is that the displaced inhabitants would have decades over which to adjust to the changes.) Studies of the Ross Ice Shelf in Antarctica have recently raised another specter: sudden breakoff of huge masses of ice, which would plunge into the sea as monstrous icebergs, raising global sea level abruptly and catastrophically by about 6 meters. At present, far too little is known to assess the likelihood of this accurately.

One might reasonably ask how much heating results from particular increased CO_2 levels—in other words, how acute will greenhouse-effect heating become, how soon? Unfortunately, there is no simple answer to such questions. Too

Figure 9.32 Global temperature rise emerges from background noise.   Source: Data from Department of Commerce NOAA/National Environmental Satellite, Data and Information Service/National Climatic Data Center.

Figure 9.33 Global methane concentrations have risen even faster than CO_2 concentrations, but total CH_4 is still below 2 ppm.   Source: Data from Department of Commerce NOAA/National Environmental Satellite, Data and Information Service/National Climatic Data Center.

many variables are involved in global climate modeling. For example, if air temperatures rise, evaporation of water from the oceans would tend to increase. More water in the air means more clouds. More clouds, in turn, would reflect more sunlight back into space before it could reach and heat the surface. (Some speculate that over centuries, the world might then cool off, perhaps even enough to initiate another ice advance or ice age far in the future.) In fact, NASA has identified cloud cover as the single largest source of uncertainty in projections of greenhouse-effect heating. Interactions between atmosphere and oceans (box 9.2) contribute further uncertainty.

Nor is it easy even to define "normal" global temperature, for there is abundant evidence of historic temperature fluctuations over periods of years to centuries similar in magnitude to recent warming trends. Perhaps the best-known of the modern fluctuations is the period known as the "Little Ice Age," which lasted from about A.D. 1450–1850. Worldwide temperatures were not actually vastly colder than at present; in eastern Europe, for example, average temperatures were just 1° to 1.5° C (2°–3° F) lower. However, the practical impact of this seemingly small drop in temperature was profound. Alpine glaciers in the Alps and elsewhere advanced dramatically, as paintings from the period attest. The high mountains of Ethiopia were blanketed in snow. At times, one could walk between Staten Island and Manhattan. The river Thames froze more than two dozen times during the Little Ice Age; it has not completely frozen again since the winter of 1813–14. The early American colonists apparently had to endure winters far more bitter than those routinely experienced today. Reduced evaporation from the colder oceans led to devastating droughts in many parts of the world. Then, beginning about 1850, sustained warming began. The warming trend lasted nearly a century until global temperatures started to decline again in about 1940, perhaps due to increased particulate air pollution; as that pollution was reduced, beginning in the 1960s, temperatures again rose. Data for the last 10,000 years suggest that variations of up to 6° C (11° F) in average annual temperatures have occurred, independently of significant human impact on atmospheric chemistry. Only in the mid-1990s did climatologists feel they had clear evidence of greenhouse-effect heating (beyond the probable range of background temperature variations); see figure 9.32.

Neither present nor future greenhouse-effect heating is likely attributable entirely to CO_2. Water vapor, methane (CH_4), nitrous oxide (N_2O), and the synthetic chlorofluorocarbons are among other recognized "greenhouse gases." Indeed, some, such as methane, are even *more* effective at greenhouse-effect heating of the atmosphere than is CO_2. Methane is produced naturally during some kinds of decay of organic matter and during the digestive processes of many animals. Human activites that increase atmospheric methane include extraction and use of fossil fuels, raising of domestic livestock (cattle, dairy cows, goats, camels, pigs, and others), and the growing of rice in rice paddies, where bacterial decay produces methane in the standing water. Methane concentrations are also rising sharply (figure 9.33), but

Box 9.2 Winds and Currents, Climate and Commerce: El Niño

Discussions of global climate would be incomplete without considering the oceans, but not only because they are the ultimate source of rain and glacial ice. Ocean and atmosphere systems are interrelated in many ways. The oceans are one sink for atmospheric CO_2; it dissolves in seawater to make carbonate from which sediments and organisms' shells and skeletons are made. Ocean currents of warmer or colder water can moderate local climates: for example, the climate of the British Isles is moderated by the warm Gulf Stream, which carries especially warm surface waters across the Atlantic toward northern Europe. And winds can modify ocean circulation also, sometimes with dramatic effects.

Most of the vigorous circulation of the oceans is confined to the near-surface waters. Only the shallowest waters, within 100 to 200 meters of the surface, are well mixed by waves, currents, and winds, and warmed and lighted by the sun. The average temperature of this layer is about 15° C (60° F).

Below the surface layer, temperatures decrease rapidly to about 5° C (40° F) at 500 to 1,000 meters below the surface. Below this is the so-called *deep layer* of cold, slow-moving, rather isolated water. The temperature of this bottommost water is close to freezing and may even be slightly below freezing (the water is prevented from freezing solid by its dissolved salt content and high pressure). This cold, deep layer originates largely in the polar regions and flows very slowly toward the equator.

When winds blow parallel or nearly parallel to a coastline, the resultant currents may cause the warm surface waters to be blown offshore. This, in turn, creates a region of low pressure and may result in *upwelling* of deep waters to replace the displaced surface waters (figure 1). The deeper waters are relatively enriched in dis-

Figure 1 Warm surface water normally overlies colder seawater at depth. When offshore winds blow warm waters away from shore and create local low pressure, upwelling of colder deep waters may occur.

solved nutrients, in part because few organisms live in the cold, dark depths to consume those nutrients. When the nutrient-laden waters rise into the warm, sunlit zone near the surface, they can support abundant plant life and, in turn, animal life that feeds on the plants. Many rich fishing grounds are located in zones of coastal upwelling. The west coasts of North and South America and of Africa are subject to especially frequent upwelling events.

From time to time, however, for reasons not precisely known, the upwelling is suppressed for a period of weeks or longer. Abatement of coastal winds, for one, reduces the pressure gradient driving the upwelling. The reduction in upwelling of the fertile cold waters has a catastrophic effect on the Peruvian anchoveta industry. Such an event is called *El Niño* ("the [Christ] Child") by the fishermen be-

cause it commonly occurs in winter, near the Christmas season.

At one time, El Niño events were believed to be rather sporadic and isolated, local events. Now it is recognized that they are cyclic in nature, occurring every four to seven years as part of the El Niño–Southern Oscillation (referring to the southern oceans). The particularly intense El Niño of 1982–83 illustrates the scale of the phenomenon (figure 2). Extensive shifts in seawater surface temperatures in turn cause large-scale changes in evaporation, and thus precipitation, and may in turn alter subsequent wind-circulation patterns. Climatic changes associated with El Niño events generally include changes in frequency, intensity, and paths of Pacific storms, short-term droughts and floods in various regions of the world, and changes in the timing and intensity of the monsoon season in India. The 1982–83 El Niño is blamed for the loss of thousands of lives and for over $13 billion in damage around the world. The fact that these consequences occur at times of unusually warm surface waters in the eastern Pacific contributes to concern over possible consequences of global warming. In the short term, climatologists hope to use improved understanding of El Niño phenomena as a predictive tool in anticipating upcoming climate-related problems and planning ahead to deal with them.

A

B

C

Figure 2 Maps of sea surface temperature illustrate El Niño (blue = cool, 0 – 12° C; green = moderate, 13 – 24° C; reds and yellows, warm, 25 – 30° C). Diagram A (January 1984) shows normal conditions, with warm water (#1) in the western Pacific, and a tongue of cooler water (#2) along the equator farther east. Diagram B, at height of El Niño (January 1983), shows that cooler-water belt has vanished. Diagram C shows temperature differences between January 1983 and "normal" January 1984: band of abnormally warm water is evident (#3).   Maps courtesy of NASA/Earth Observing System.

currently, CH_4 is still far less abundant than CO_2. If the amounts of minor but potent greenhouse gases continue to rise, they will complicate climate predictions tied primarily to CO_2 projections.

While the magnitude and rate of anticipated heating are debated, and leaving other greenhouse gases aside, there is consensus that increased CO_2 means increased trapping of heat in the atmosphere. In order to moderate that heating, either the rate of input of CO_2 into the atmosphere could be reduced, or the natural "sinks" for atmospheric CO_2 enhanced. A major sink for CO_2 is use by plants in manufacturing food by photosynthesis, and this is one reason for concern about the impacts of large-scale destruction of rain forests. In order to reduce CO_2 emissions, we could reduce dependence on fossil fuels; but this is likely to be a gradual process. Nor will its effects on global warming necessarily be wholly good: Reduction in burning of coal can reduce emissions of both CO_2 *and* sulfur gases (see chapters 13 and 17). However, the sulfur gases, which make sulfuric acid in the atmosphere, may have a *cooling* effect, as we have learned from volcanoes. Such frustrating uncertainties about the precise impacts of various changes in atmospheric chemistry spur continued research to refine the models.

It is important to bear in mind, too, that "global warming" is not the only climate change anticipated, that "global" climate changes do not affect all parts of the world equally, and that not all changes will be negative everywhere. Associated changes in wind-flow patterns and amounts and distribution of precipitation will cause differential impacts in different areas, not all of which will be equally resilient. A region of temperate climate and ample rainfall may not be seriously harmed by a temperature change of a few degrees, or several inches more or less rainfall. A modest temperature rise in colder areas may increase the length of the growing season, and some dry areas may enjoy increased productivity with increased rainfall.

Places where temperature or moisture conditions already make living marginal are more vulnerable. There is reason to believe that global warming may produce more extremes—catastrophic flooding from torrential storms, devastating droughts, "killer heat waves" like that costing hundreds of lives in Chicago in 1995. Because of the many variables involved and the immense complexity of the calculations, global-climate computer models must also limit the resolution at which effects are calculated. Typical models operate on areal units, or cells, that are hundreds of kilometers on a side, meaning that small local climatic anomalies may be missed. United Nations studies further suggest that less-developed nations may be much less able to cope with the stresses of climate change, which may have implications for global political stability as well.

Overview

Summary

Glaciers past and present have sculptured the landscape not only in mountainous regions but over wide areas of the continents. They leave behind U-shaped valleys, striated rocks, piles of poorly sorted sediment (till) in a variety of landforms (moraines), and outwash. Meltwater from glaciers supplies surface and ground water. Most present glaciers are alpine glaciers. The two major ice sheets remaining are in Greenland and Antarctica.

Wind moves material much as flowing water does, but less forcefully. As an agent of erosion, wind is less effective than water but may have significant effects in dry, exposed areas, such as beaches, deserts, or farmland. Wind action creates well-sorted sediment deposits, as dunes or in blankets of fine loess. The latter can improve the quality of soil for agriculture. However, loess typically makes a poor base for construction. Features created by wind are most obvious in deserts, which are sparsely vegetated and support little life. The extent of unproductive desert and arid lands may be increasing through desertification brought on by intensive human use of these fragile lands.

Modern burning of fossil fuels has been increasing the amount of carbon dioxide in the air. The resultant greenhouse-effect heating may begin to melt remaining ice sheets, causing a rise in global sea level and, ultimately, flooding of many coastal areas. Other probable consequences include changes in global weather patterns, including amount and distribution of precipitation, which may have serious implications for agriculture. However, in the short term, the amount of greenhouse-effect heating is difficult to separate from background fluctuations in global temperatures. Longer-term climatic projections are complicated by the number and variety of factors to be considered (including other greenhouse gases; effects of atmospheric dust and sulfuric acid, of both volcanic and human origins; changes in cloud cover as temperatures change; and interactions between oceans and atmosphere).

Terms to Remember

ablation 195	alpine glacier 194
abrasion 197	calving 195

continental glacier 195   loess 203
deflation 201   moraine 198
desert 204   outwash 197
desertification 205   rain shadow 204
drift 197   slip face 203
dune 202   striations 197
equilibrium line 195   till 197
glacier 193   ventifact 201
greenhouse effect 211

Exercises

Questions for Review

1. Discuss ways in which glaciers might be manipulated for use as a source of water.
2. Briefly describe the formation and annual cycle of an alpine glacier.
3. What is a moraine? How can moraines be used to reconstruct past glacial extent and movements?
4. What is an ice age? Choose any two proposed causes of past ice ages and evaluate the plausibility of each. (Is the effect on global climate likely to have been large enough? long enough? Is there any geologic evidence to support the proposal?)
5. How is sunlight falling on the earth's surface a factor in wind circulation?
6. By what two principal processes does wind erosion occur?
7. Briefly describe the process by which dunes form and migrate.
8. What is loess? Must the sediment invariably be of glacial derivation, as much U.S. loess appears to be?
9. Assess the significance of loess to (a) farming and (b) construction.
10. What is desertification? Describe two ways in which human activities contribute to the process.
11. Explain the greenhouse effect and its relationship to modern industrialized society.
12. Why is it difficult to project the impact of increased atmospheric CO_2 on global temperature? Explain.
13. What are the principal concerns related to continued greenhouse-effect heating? On what time scale might they be significant?
14. Climate projections are complicated by the number of factors and processes involved. Name and explain any three factors that modify global air temperature. Describe any two secondary effects of this temperature increase that could influence local or global climates.
15. What is an El Niño event? What is its apparent significance to climate and to the productivity of coastal fishing grounds? Explain.

For Further Thought

1. During the Pleistocene glaciation, a large fraction of the earth's land surface may have been covered by ice. Assume that 10 percent of the 149 million square kilo-

meters was covered by ice averaging 1 kilometer thick. How much would sea level have been depressed over the 361 million square kilometers of oceans? You may find it interesting to examine a bathymetric (depth) chart of the oceans to see how much new land this would have exposed.
2. After the next windstorm, observe the patterns of dust distribution (or, if in a snowy climate and season, snow distribution), and try to relate them to the distribution of obstacles that may have altered wind velocity.
3. The World Meteorological Organization declared 1995 "the warmest year yet" (since 1861). Investigate the year's weather events: What was unusual? What was the status of El Niño? What happened to greenhouse-gas concentrations? Was heating universal; if not, what regions were unusually cool? Was there any general pattern in global precipitation, or were different regions affected in different ways? A place to start is http://www.wmo.ch/web/wcp/wcphtml/statemnt/statmall.html
You might then want to select one region and check recent/current weather and climate conditions; see NetNotes for source sites.

Suggested Readings/References

Ausubel, J. H. 1991. A second look at the impacts of climate change. *American Scientist* 79 (May-June): 210–21.
Bentley, C. R. 1980. If West Sheet melts rapidly—what then? *Geotimes* (August): 20–21.
Berner, E. K., and Berner, R. A. 1996. *Global environment: Water, air, and geochemical cycles.* Upper Saddle River, N.J.: Prentice-Hall.
Berner, R. A., and Lasaga, A. C. 1989. Modeling the geochemical carbon cycle. *Scientific American* 260 (March): 74–81.
Broecker, W. S., and Denton, G. H. 1990. What drives global cycles? *Scientific American* 262 (January): 48–56.
Brookfield, M. E., and Ahbrandt, T. S., eds. 1983. *Eolian sediments and processes.* New York: Elsevier Science Publishing.
Cooke, R. U., and Warren, A. 1973. *Geomorphology in deserts.* Berkeley: University of California Press.
Environmental Protection Agency. 1983. *Projecting future sea level rise.* 2d ed. Washington, D.C.: U.S. Government Printing Office.
Freeley, R., and Iverson, J. D. 1985. *Wind as a geological process.* Cambridge, England: Cambridge University Press.
Houghton, R. A., and Woodwell, G. M. 1989. Global climatic change. *Scientific American* 260 (April):36-44.
Hyde, P. 1987. *Drylands: The deserts of North America.* San Diego: Harcourt Brace Jovanovich.
Imbrie, J., and Imbrie, K. P. 1979. *Ice ages.* Hillside, N.J.: Enslow.
Issar, A. S. 1995. Climatic change and the history of the Middle East. *American Scientist* (July-August): 350–55.
Karl, T. R., and Baker, C. B. 1995. Global warming update. [Online] http://www.ncdc.noaa.gov/gblwrmupd/global.html
Meier, M., and Post, A. 1980. *Glaciers: A water resource.* U.S. Geological Survey.

Pelto, M. S. 1993. Changes in water supply in alpine regions due to glacier retreat. Amer. Inst. of Physics.

Perry, T. S. 1993. Modeling the world's climate. *IEEE Spectrum* 30 (July): 33–42.

Secretariat of the U.N. Conference on Desertification, Nairobi. 1977. *Desertification: Its causes and consequences.* New York: Pergamon Press.

Sharp, R. P. 1988. *Living ice: Understanding glaciers and glaciation.* New York: Cambridge University Press.

Sheridan, D. 1981. *Desertification of the United States.* Washington, D.C.: Council on Environmental Quality.

Turekian, K. K. 1996. *Global environmental change.* Upper Saddle River, N. J.: Prentice-Hall.

U.S. Environmental Protection Agency. 1990. The greenhouse effect: What can we do about it? *EPA Journal* 16 (March/April).

Woods Hole Oceanographic Institution. 1989. "The Oceans and Global Warming." *Oceanus* 32 (Summer).

Zorpetle, G. 1993. Sensing climate change. *IEEE Spectrum* 30 (July): 20–27.

The United Nations Environment Program has assembled close to 100 "Climate Change Fact Sheets," indexed at and accessible via

http://www.unep.ch/iucc/fs-index.html

A source of worldwide climate information, with numerous links to other sites, is the Global Change Master Directory at

http://gcmd.gsfc.nasa.gov/

One of the functions of the Earth Observing System (EOS) of satellites is to monitor climate-related phenomena. EOS imagery and related graphics on such subjects as the greenhouse effect, ocean processes, clouds and storms can be searched through the EOS home page:

http://spso2.gsfc.nasa.gov/spso_homepage.html
Slides and images intended specifically for educational purposes can be reached via

http://eospso.gsfc.nasa.gov/eos_edu.pack/toc.html

Charles Sturt University has set up a site assembling links to a variety of sites pertinent to weather and global monitoring, grouped by topic and region; see

http://www.csu.edu.au/weather/weatherl.html

The Desert Research Institute engages in a wide variety of geologic, ecologic, and archeologic studies of deserts; their comprehensive web site is located at

http://www.dri.edu

The International Arid Lands Consortium homepage offers a variety of links to information about drylands; see

http://ag.arizona.edu/OALS/IALC/Home.html

The Smithsonian National Air and Space Museum has an online exhibit, "Bright edges of the world—the world's evolving drylands," at

http://drylands.nasm.edu:1995

The British Antarctic Survey Ice and Climate Division presents a variety of information on snow and ice studies and related climate issues at

http://www.nbs.ac.uk/public/icd/

INSTAAR, the Institute for Arctic and Alpine Research at the University of Colorado, offers glacier-related information through

http://ekman.colorado.edu/glaciology/

Section 4

Resources

Aerial view of surface mining for clay, Devon, England. © Jason Hawkes/Tony Stone Images

In a general sense, *resources* are all those things that are necessary or important to human life and civilization, that have some value to individuals and/or to society. Implicit in the term is the availability of a supply that can be drawn on as required in the future. The requirement that resources have some value to people means that what constitutes a resource may change with time or social context. Many of the fuels, building materials, metals, and other substances important to modern technological civilization (which are thus resources in a modern context) were of no use to the earliest cave dwellers. In the next five chapters we will survey earth resources, that subset of all resources involved in or formed by geologic processes.

In the early 1970s, the Arab oil embargo precipitated sharp increases in the prices of gasoline and heating oil, long lines at gas stations, and a lot of talk about an "energy crisis." Ten years later, news reports were full of the "oil glut" and of widespread layoffs among petroleum industry employees. Were these developments completely inconsistent? What had happened in the interim?

There is more and more discussion nowadays of "mining" water, and some warn of impending water shortages. Yet over 70 percent of the earth's surface is covered by water. Why, then, is there concern about water resources?

219

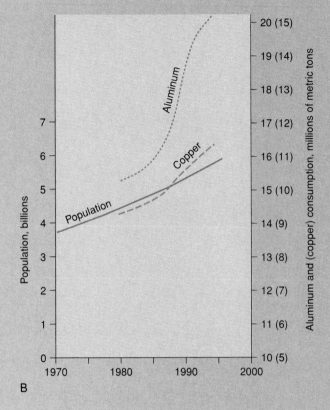

Figure R.1 Comparison of rates of population growth and energy and mineral-resource consumption. (*A*) U. S. energy consumption has been rising faster than the population, except during the Great Depression and the Arab oil embargo of the early 1970s. (*B*) Worldwide consumption of copper ore (dashed line) and aluminum (dotted line) is also increasing faster than population (solid line), and this pattern is true for many industrial materials.   Source: (*A*) Data from Earl Cook, "The Flow of Energy in an Industrialized Society," *Scientific American*, September 1971. (*B*) Data from World Resources Institute.

The world hardly seems to be running out of the surface accumulations of rock and mineral fragments and other debris that make up soil. However, in many places, soil is eroding far faster than new soil is being produced by natural processes, and the fertility of the remaining soil is decreasing. This may have serious implications for the world's food supply.

The chapters that follow describe the various processes by which many mineral and fuel deposits and soils form and the pathways by which water moves through the environment. Certain resources are **renewable** (replaceable on a human time scale), while others are not. We will look at the rates at which resources are being consumed, and for those in finite supply, how long they may be expected to last. In some cases, it may become necessary to look to different resources in the future from those that have been used in the past. We will also consider some of the potential adverse environmental impacts of the use of our current and possible future resources.

Resources, People, and Standards of Living

Resource supplies have become a pressing problem, in part because of the surge in the world's population discussed in chapter 1. The more people on the earth, the more water consumed, the more fuel burned, the more minerals used, and so on. Yet population growth alone is not the whole story. If the rate of population growth is compared with the rate of energy consumption or use of many mineral resources as a function of time (figure R.1), it is apparent that the rates of resource use are increasing even faster than the population! Not only are there more and more people, but *per capita* resource consumption is rising.

This is mainly a result of elevated standards of living, for both people in underdeveloped, or developing, countries and those in developed ones. Most societies continue to develop increasingly sophisticated technologies and greater use of machinery, and to have lifestyles that involve increasing quantities of manufactured goods

(home furnishings, clothes, cars, and so on). Therefore, each person tends to account for the consumption of more and more energy and mineral resources. The connection between per capita energy consumption and standard of living can be seen in figure R.2. In it, energy consumption is plotted against gross national product (GNP), a measure of the value of all goods and services produced. In a general way, per capita GNP reflects the level of technological development and standard of living. Increased per capita consumption, then, together with a fast-growing world population, makes questions of resource supply even more pressing.

Projection of Resource Supply and Demand

Projecting the lengths of time that existing supplies of various resources are likely to last is a complex task, requiring accurate assessment of the available supplies and an estimate of how fast they will be used. With regard to the first part of the task, certain terms need to be defined.

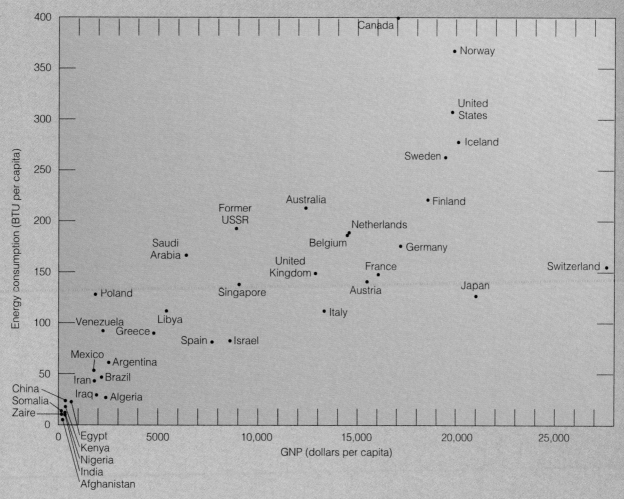

Figure R.2 There is a variable but generally positive correlation between GNP and energy consumption: The more energy consumed, the higher the value of goods and services produced, and generally, the higher the level of technological development as well.
Source: Data from *1992 Information Please Environmental Almanac*, World Resources Institute.

Up to this point, the phrase *earth resources* has been used informally to describe any useful or valuable geologic materials. In discussing the supply/demand question, however, distinctions are made between *reserves* and various kinds of *resources*, as summarized in figure R.3. The **reserves** are that quantity of a given material that has been found and that can be recovered economically with existing technology. Usually, the term is used only for material not already consumed, as distinguished from **cumulative reserves,** which include the quantity of the material already used up in addition to the remaining (unused) reserves. The reserves represent the most conservative estimate of how much of a given metal,

mineral, or fuel remains unused. Beyond that, several additional categories of resources can be identified.

Those deposits that have already been found but that cannot presently be profitably exploited are the **subeconomic resources** (also known as **conditional resources**). Some may be exploitable with existing technology but contain ore that is too low-grade or fuel that is too dispersed to produce a profit at current prices. Others may require further advances in technology before they can be exploited. Still, these deposits have at least been located, and the quantity of material they represent can be estimated fairly well.

Then there are the undiscovered resources. These are often subdivided

into **hypothetical resources,** additional deposits expected to be found in areas in which some deposits of the material of interest have already been found, and the **speculative resources,** those deposits that might be found in explored or unexplored regions where deposits of the material are not already known to occur. Estimates of undiscovered resources are extremely rough by nature, and it would be unwise to count too heavily on deposits that have not even been found yet, especially for the near future.

Consequently, the reserves are the amounts of materials normally used to make supply projections. Even those figures may be somewhat imprecise, especially on a worldwide basis. Many mining and energy companies and

Figure R.3 Different categories of reserves and resources.

some nations prefer to reveal as little about their unexploited assets as possible. In some countries, firms are taxed, in part, on the size of their remaining reserves, so their estimates may be on the low side. Still, over past decades, scientists have combined published information on reserves and mineral production data with basic geologic knowledge to make what are believed to be reasonably good estimates of worldwide reserves. Estimates of U.S. reserves are presumably more accurate.

Predictions of how long supplies of various resources will last depend very strongly on projections of future demand. The subject of exponential growth was explored in chapter 1. For many mineral and energy resources, demand has been growing exponentially over decades, which in some cases leads to startlingly short projections for the length of time reserves of certain materials will last. Moreover, any such calculations done on a global scale necessarily ignore significant political realities arising

from the fact that geologic resources are hardly uniformly distributed around the world.

Economics influence both sides of the problem, for costs of materials affect both consumer demand for, and profitability of exploitation or extraction of, particular minerals or fuels. The various aspects of supply/demand problems are explored more fully with respect to specific materials in the chapters that follow.

Chapter 10

Water as a Resource

Some aspects of surface-water flow, including the hydrologic cycle, were already considered in chapter 6. We now take a larger view of water on earth, to consider water as a resource. The importance of water availability for domestic use, agriculture, and industry is apparent. What may be less obvious is that water, or the lack of it, may control the extent to which we can develop certain other resources, such as fossil fuels (see chapter 13). To begin, we will look at the earth's water supply and its distribution in nature. This involves considering not only surface water but ground water and ice.

Avalanche Creek Gorge at Glacier National Park, Montana.
© Doug Sherman/Geofile

Table 10.1 The Water in the Hydrosphere

Reservoir	Percentage of total water*	Percentage of fresh water†	Percentage of unfrozen fresh water
oceans	97.54	—	—
ice	1.81	73.9	—
ground water	0.63	25.7	98.4
lakes and streams			
salt	0.007	—	—
fresh	0.009	0.36	1.4
atmosphere	0.001	0.04	0.2

Source: Data from J. R. Mather, *Water Resources*, 1984, John Wiley & Sons, Inc., New York.

*These figures account for over 99.9 percent of the water. Some water is also held in organisms (the biosphere).

†This assumes that all ground water is more or less fresh, since it is not all readily accessible to be tested and classified.

The Global Water Budget

Table 10.1 shows how the water in the hydrosphere is distributed. Several points emerge immediately from the data. One is that there is, relatively speaking, little fresh liquid water on the earth. Most of the fresh water is locked up as ice, mainly in the large polar ice caps. Even the ground water beneath continental surfaces is not all fresh. These facts underscore the need for restraint in our use of fresh water. From the long-term geologic perspective, water is a renewable resource, but local supplies may be inadequate in the short term. Given the relative volumes of fresh surface water and of ground water, it is clear why one might consider ground water as a source of supply. The geologic setting in which it occurs influences not only how much is locally available, but how accessible it is. The critical rock properties in this connection are *porosity* and *permeability*.

Fluid Storage and Movement: Porosity and Permeability

Porosity and permeability involve the ability of rocks or other mineral materials (sediments, soils) to contain fluids and to allow fluids to pass through them. **Porosity** is the proportion of void space in the material—holes or cracks, unfilled by solid material, within or between individual mineral grains. Porosity may be expressed either as a percentage (for example, 1.5 percent) or as an equivalent decimal fraction (0.015). The pore spaces may be occupied by gas, liquid, or a combination of the two. **Permeability** is a measure of how readily fluids pass through the

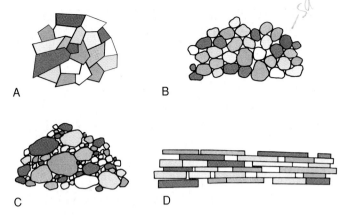

Figure 10.1 Porosity and permeability vary with grain shapes and the way grains fit together. (*A*) Low porosity in an igneous rock. (*B*) A sandstone made of rounded grains similar in size has more pore space. (*C*) Poorly sorted sediment: fine grains fill pores between coarse ones. (*D*) Packing of plates of clay in a shale may result in high porosity but low permeability. In all cases, note that fracturing or weathering may increase porosity and permeability beyond that resulting from grain shapes and packing.

material and is related to the extent to which pores or cracks are interconnected. Porosity and permeability of geologic materials are both influenced by the shapes of mineral grains or rock fragments in the material, the range of grain sizes present, and the way in which the grains fit together (figure 10.1).

Igneous and metamorphic rocks consist of tightly interlocking crystals, and they usually have low porosity and permeability unless they have been broken up by fracturing or weathering. The same is true of many of the chemical sedimentary rocks, unless cavities have been produced in them by slow dissolution. Clastic sediments, however, accumulate as layers of loosely piled particles. Even after compaction or cementation into rock, they may contain more open pore space. Well-rounded equidimensional grains of similar size can produce sediments of quite high porosity and permeability. (This can be illustrated by pouring water through a pile of marbles.) Many sandstones have these characteristics. (Consider, for example, how rapidly water drains into beach sands.) In materials containing a wide range of grain sizes, finer materials can fill the gaps between coarser grains. Porosity can thereby be reduced, though permeability may remain high. Clastic sediments consisting predominantly of flat, platelike grains, such as clay minerals or micas, may be porous, but because the flat grains can be packed closely together parallel to the plates, these sediments may not be very permeable, especially in the direction perpendicular to the plates. Shale is a rock commonly made from such sediments. The low permeability of clays can readily be demonstrated by pouring water onto a slab of artists' clay.

Table 10.2 gives some representative values of porosity and permeability. These properties are relevant not only to

Table 10.2	Representative Porosities and Permeabilities of Geological Materials	
Material	Porosity (%)	Permeability (m/day)
Unconsolidated		
clay	45–55	less than 0.01
fine sand	30–52	0.01–10
gravel	25–40	1,000–10,000
glacial till	25–45	0.001–10
Consolidated (rock)		
sandstone and conglomerate	5–30	0.3–3
limestone (crystalline, unfractured)	1–10	0.00003–0.1
granite (unweathered)	less than 1–5	0.0003–0.003
volcanic rock	1–30; mostly less than 10	0.0003–3; depends on presence or absence of fractures or interconnected gas bubbles

Source: Data from *Water in Environmental Planning*, by T. Dunne and L. B. Leopold. Copyright © 1978 W. H. Freeman and Company.

discussions of groundwater availability and use, but also to issues such as stream flooding, petroleum resources, water pollution, and waste disposal.

Subsurface Waters

If soil on which precipitation falls is sufficiently permeable, infiltration occurs, as noted in chapter 6. Gravity continues to draw the water downward until an impermeable rock or soil layer is reached, and the water begins to accumulate above it. Immediately above the impermeable material is a zone of rock or soil that is water-saturated, in which water fills all the accessible pore space, called the **zone of saturation,** or **phreatic zone. Ground water** is the water in the zone of saturation. Usually, ground water is found, at most, a few kilometers into the crust. In the deep crust and below, pressures on rocks are so great that compression closes up any pores that water might fill. Above the saturated zone is rock or soil in which the pore spaces are filled partly with water, partly with air: the **zone of aeration,** or **vadose zone.** The water in unsaturated soil layers is **soil moisture,** and is often an important factor in agricultural productivity. All of the water occupying pore space below the ground surface is, logically, called *subsurface water,* a term that thus includes

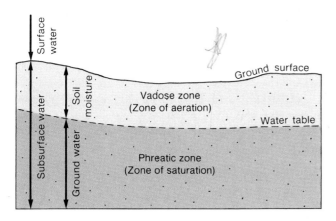

Figure 10.2 Nomenclature of surface and subsurface waters. Ground water is water in the zone of saturation, below the water table.

Figure 10.3 In this natural system, the stream drains water from the region—not only surface runoff, but ground water from adjacent saturated rocks and soil that slowly drains into the stream. Such slow groundwater influx explains why many streams can keep flowing even when there has been no rain for weeks.

both ground water and soil moisture. The **water table** is defined as the top of the zone of saturation, where the saturated zone is not confined by overlying impermeable rocks. (See also the discussion of aquifer geometry in the next section of this chapter.) These relationships are illustrated in figure 10.2.

The water table is not always below the ground surface. Where the water table locally intersects the ground surface, the result may be a lake, stream, spring, or wetland; the water's surface is the water table. The water table below ground is not flat like a tabletop, either. It may undulate with the surface topography and with the changing distribution of permeable and impermeable rocks underground. (See figure 10.3.) The height of the water table varies, too. It is highest when the ratio of input water to water removed is greatest, typically in the spring, when rain is heavy or snow and ice accumulations melt. In dry seasons, or when local human use of ground water is intensive, the water table drops, and the amount of available ground water remaining decreases. Ground water can flow laterally through permeable soil and rock, from higher elevations to lower, from areas of higher pressure or potential (see below) to lower, from areas of abundant infiltration to drier ones, or from areas of little groundwater use toward areas of heavy use. The processes of infiltration and migration or percolation through which ground water is replaced are collectively called **recharge.**

If ground water drawn from a well is to be used as a source of water supply, the porosity and permeability of the surrounding rocks are critical. The porosity controls the total amount of water available. In most places, there are not vast pools of open water underground to be tapped, just the water dispersed in the rocks' pore spaces. This pore water amounts, in most cases, to a few percent of the rocks' volume at best, although some sands and gravels may occasionally have porosities as high as 30 percent. The permeability, in turn, governs both the rate at which water can be withdrawn and

the rate at which recharge can occur. All the pore water in the world would be of no practical use if it were so tightly locked in the rocks that it could not be pumped out. A rock that holds enough water and transmits it rapidly enough to be useful as a source of water is an **aquifer.** Many of the best aquifers are sandstones or other coarse, clastic sedimentary rocks, but any other type of rock may serve if it is sufficiently porous and permeable—a porous limestone, fractured basalt, or weathered granite, for instance. One can describe rocks of very low and low permeability, respectively, as *aquicludes* or *aquitards*. An **aquitard** is a rock that may store a considerable quantity of water, but in which water flow is slowed, or retarded. Shales are common aquitards. Historically, the term *aquiclude* was used by describe an extreme aquitard, a rock that is essentially impermeable on a human timescale; but virtually no rock would be impermeable indefinitely, and the term aquiclude has fallen into disuse.

Aquifer Geometry

Confined and Unconfined Aquifers

The behavior of ground water is controlled to some extent by the geology and geometry of the particular aquifer in which it is found. When the aquifer is directly overlain only by permeable rocks and soil, it is described as an **unconfined aquifer** (figure 10.4). An unconfined aquifer may be recharged by infiltration over the whole area underlain by that aquifer, because there is nothing to stop the downward flow of water from surface to aquifer. If a well is drilled into an unconfined aquifer, the water will rise in the well to the same height as the water table in the aquifer. The water must be actively pumped up to the ground surface.

A **confined aquifer** is bounded above and below by low-permeability rocks. Water in a confined aquifer may be under considerable pressure from the adjacent rocks, or as a consequence of lateral differences in elevation within the aquifer. The confining layers prevent the free flow of water to relieve this pressure. At some places within the aquifer, the water level in the saturated zone may be held above or below the level it would assume if the aquifer were unconfined and the water allowed to spread freely (figure 10.5).

If a well is drilled into a confined aquifer, the water can rise above its level in the aquifer because of this extra hydrostatic (fluid) pressure. This is called an **artesian system.** The water in an artesian system may or may not rise all the way to the ground surface; some pumping may still be necessary to bring it to the surface for use. In such a system, rather than describing the height of the water table, geologists refer to the height of the **potentiometric surface,** which represents the height to which the water's pressure would raise the water if the water were unconfined. This level will be somewhat higher than the top of the confined aquifer where its rocks are saturated, and it may be above the ground surface, as shown in figure 10.5.

Advertising claims notwithstanding, this is all that "artesian water" means—artesian water is no different chemically from, and no purer, better-tasting, or more wholesome than, any other ground water; it is just under natural pressure. Water towers create the same effect artificially: After water has been pumped into a high tower, gravity increases the fluid pressure in the water-delivery system so that water rises up in the pipes of individual users' homes and businesses without the need for many pumping stations. The same phenomenon can be demonstrated very simply by holding up a water-filled length of rubber tubing or hose at a steep angle. The water is confined by the hose, and the water in the low end is under pressure from the weight of the water above it. Poke a small hole in the top side of the lower end of the hose, and a stream of water will shoot up to the level of the water in the hose.

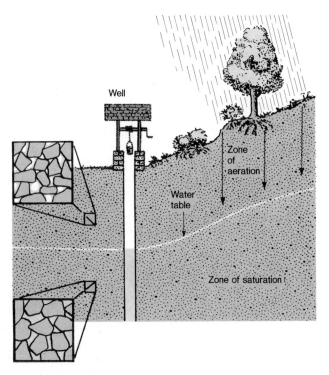

Figure 10.4 An unconfined aquifer. Water rises in an unpumped well just to the height of the water table.
From Charles C. Plummer and David McGeary, *Physical Geology*, 5th ed. Copyright © 1991 McGraw-Hill Companies, Inc., Dubuque, Iowa. All Rights Reserved. Reprinted by permission.

Other Factors in Water Availability

More complex local geologic conditions can make it difficult to determine the availability of ground water without thorough study. For example, locally occurring lenses or patches of relatively impermeable rocks within otherwise permeable ones may result in a *perched water table* (figure 10.6). Immediately above the aquitard is a local saturated zone, far above the true regional water table. Someone drilling a well above this area could be deceived about the apparent depth to the water table and might also find that the perched saturated zone contains very little total water. The quantity of water available would be especially sensitive to local precipitation levels and fluctuations. Using ground water over the longer term might well require drilling down to the regional water table, at a much greater cost.

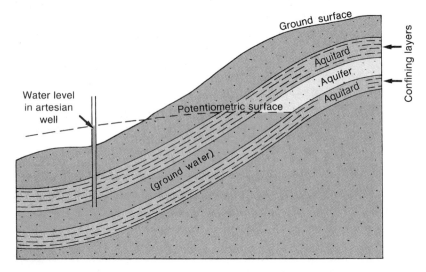

Figure 10.5 Natural internal pressure in a confined aquifer system creates artesian conditions, in which water may rise above the apparent (confined) local water table. The aquifer here is a sandstone; the confining layers, shale.

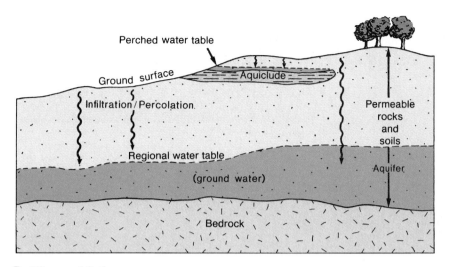

Figure 10.6 A perched water table may create the illusion of a shallow regional water table.

As noted earlier, ground water flows (albeit often slowly). Both flow paths (directions) and locations of recharge zones may have to be identified in assessing water availability. For example, if water is being consumed very close to the recharge area, and consumption rate exceeds recharge rate, the stored water may be exhausted rather quickly. If the point of extraction is far down the flow path from the recharge zone, a larger reserve of stored water may be available to draw upon. Like stream systems, aquifer systems also have divides from which ground water flows in different directions, which determine how large a section of the aquifer system can be tapped at any point. Groundwater divides may also separate polluted from unpolluted waters within the same aquifer system.

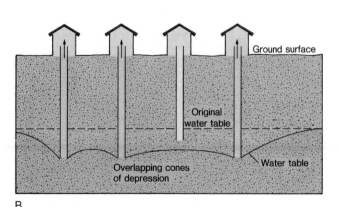

Figure 10.7 The formation and effect of cones of depression. (*A*) Lowering of the water table in a cone of depression around a pumped well in an unconfined aquifer. (*B*) Overlapping cones of depression lead to net lowering of the water table; shallower wells may run dry.

Consequences of Groundwater Withdrawal

Lowering the Water Table

When ground water must be pumped from an aquifer, the rate at which water flows in from surrounding rock to replace that which is extracted is generally slower than the rate at which water is taken out. In an unconfined aquifer, the result is a circular lowering of the water table immediately around the well, which is called a **cone of depression** (figure 10.7A). A similar feature is produced on the surface of a liquid in a drinking glass when it is sipped hard through a straw. When there are many closely spaced wells, the cones of depression of adjacent wells may overlap, further lowering the water table between wells (figure 10.7B). If over a period of time, groundwater withdrawal rates consistently exceed recharge rates, the regional water table may drop. A clue that this is happening is the need for wells throughout a region to be drilled deeper periodically to keep the water flowing. Cones of depression can likewise develop in potentiometric surfaces. When artesian ground water is withdrawn at a rate exceeding the recharge rate, the potentiometric surface can be lowered. (See box 10.1.)

These should be warning signs, for the process of deepening wells to reach water cannot continue indefinitely, nor can a potentiometric surface be lowered without limit. In many areas, impermeable rocks are reached at very shallow depths. In an individual aquifer system, the lower confining layer may be far above the bedrock depth. Therefore, there is a "bottom" to the groundwater supply, though without drilling, it is not always possible to know just where it is. Groundwater flow rates are highly variable (recall table 10.2), but in many aquifers, they are of the order of only meters or

Box 10.1 The Great Depression— in Ground Water

In areas of the dry south and southwest, where groundwater use is heavy and withdrawal exceeds recharge, water tables have dropped considerably (see also box 10.2). What is more surprising is that the effects of excessive pumping can be seen even in areas widely regarded as adequately wet. Consider, as an example, northern Illinois.

Sandstones of the Cambro-Ordovician aquifer system, confined over much of their areal extent, have long been used as the principal or sole source of municipal water in many parts of northern Illinois. When the height of the potentiometric surface for the confined aquifer is mapped, as in figure 1, a dramatic drop is seen in the Chicago metropolitan area, from more than 750 feet (250 meters) above sea level to, locally, more than 100 feet (30+ meters) below sea level. What the contours really represent is a giant cone of depression in the potentiometric surface.

This cone of depression is due to yet another instance of groundwater withdrawal rates far exceeding recharge rates. Over just a decade, water levels in some area wells dropped more than 30 meters. Close to Lake Michigan, the situation can be alleviated by using proportionately more lake water. However, nearby communities and rural homesteads lacking ready access to the lake waters are finding it increasingly difficult and expensive to keep the ground waters flowing. Even if all groundwater withdrawal ceased for a time—an unlikely possibility—centuries of recharge at least would be required to restore the water levels.

Figure 1 Potentiometric surface of the principal aquifer system in northern Illinois. Contours are in feet above sea level. The shaded area denotes where the potentiometric surface is below sea level. The city of Chicago is located in Cook County.   From R. T. Sasman, et al., "Water Level Decline and Pumpage in Deep Wells in the Chicago Region, 1966–1971," *Illinois State Water Survey Circular 113*, 1973. Reprinted by permission.

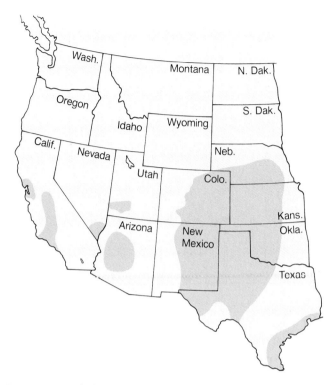

Figure 10.8 Areas of the arid western United States in which ground water is clearly being "mined," with withdrawal rates exceeding recharge.   Source: U.S. Water Resources Council.

Figure 10.9 Excessive groundwater withdrawal and associated land subsidence have contributed to the sinking of Venice and slow drowning of architectural and artistic treasures.   © David H. Endersbee/Tony Stone Images

tens of meters per year. Recharge of significant amounts of ground water, especially to confined aquifers with limited recharge areas, can thus require decades or centuries.

Where ground water is being depleted by too much withdrawal too fast, one can speak of "mining" ground water (figure 10.8). The idea is not necessarily that the water will never be recharged but that the rate is so slow on the human timescale as to be insignificant. From the human point of view, we may indeed use up ground water in some heavy-use areas. Also, as we will see in the next section, human activities may themselves reduce natural recharge, so ground water consumed may not be replaced, even slowly.

Compaction and Surface Subsidence

Lowering of the water table may have secondary consequences. The aquifer rocks, no longer saturated with water, may become compacted from the weight of overlying rocks. This decreases their porosity, permanently reducing their water-holding capacity, and may also decrease their permeability. At the same time, as the rocks below compact and settle, the ground surface itself may subside. Where water depletion is extreme, the surface subsidence may be several meters. Lowering of the water table/potentiometric surface also may contribute to *sinkhole* formation, as described in a later section.

At high elevations or in inland areas, this subsidence causes only structural problems as building foundations are disrupted. In low-elevation coastal regions, the subsidence may lead to extensive flooding, as well as to increased rates of coastal erosion. The city of Venice, Italy, is in one such slowly drowning coastal area (figure 10.9). Many of its historical, architectural, and artistic treasures are threatened by the combined effects of the gradual rise in worldwide sea levels, the tectonic sinking of the Adriatic coast, and surface subsidence from extensive groundwater withdrawal. Drastic and expensive engineering efforts will be needed to save them. The groundwater withdrawal began about 1950. Since 1969, no new wells have been drilled, groundwater use has declined, and subsidence from this cause has apparently stopped, but the city remains awash. Closer to home, the Houston/Galveston Bay area has suffered subsidence of up to several meters from a combination of groundwater withdrawal and, locally, petroleum extraction (figure 10.10A). Some 80 sq km (over 30 sq mi.) are permanently flooded. In the San Joaquin Valley in California, half a century of groundwater extraction, much of it for irrigation, produced as much as nine meters of surface subsidence (figure 10.10B).

In such areas, simple solutions, such as pumping water back underground, are unlikely to work. The rocks may have been permanently compacted: This is the case in Venice, where subsidence may have been halted but rebound is not expected because clayey rocks in the aquifer system are irreversibly compacted, and compaction has occurred in the San Joaquin Valley too. Also, what water is to be employed? If groundwater use is heavy, it may well be

A

Figure 10.10 (*A*) Surface subsidence near Galveston Bay, Texas. Faulting is a consequence of the subsidence. (Consider that the average height of an adult is between five and six feet.) (*B*) Subsidence of as much as 9 meters (almost 30 feet) occurred between 1925 and 1977 in the San Joaquin Valley, California, as a consequence of groundwater withdrawal. Signs indicate former ground surface elevation. Photograph was taken in December 1976.
(*A*) Source: U.S. Geological Survey, *1982 Annual Report.* (*B*) Thomas L. Holzer (ed.), Man-Induced Land Subsidence, 1984, Geological Society of America.

B

because there simply is no great supply of fresh surface water. Pumping in salt water, in a coastal area, will in time make the rest of the water in the aquifer salty, too. And, presumably, a supply of fresh water is still needed for local use, so groundwater withdrawal often cannot be easily or conveniently curtailed.

Saltwater Intrusion

A further problem arising from groundwater use in coastal regions, aside from the possibility of surface subsidence, is **saltwater intrusion** (figure 10.11). When rain falls into the ocean, the fresh water promptly mixes with the salt water. However, fresh water falling on land does not mix so readily with saline ground water at depth because water in the pore spaces in rock or soil is not vigorously churned by currents or wave action. Fresh water is also less dense than salt water. So the fresh water accumulates in a lens, which floats

Figure 10.11 Saltwater intrusion in a coastal zone. If groundwater withdrawal exceeds recharge, the lens of fresh water thins, and salt water flows into more of the aquifer system from below. "Upconing" of saline water also occurs below a cone of depression. Similar effects may occur in an inland setting where water is drawn from a fresh groundwater zone underlain by more saline waters.

above the denser salt water. If water use approximately equals the rate of recharge, the freshwater lens stays about the same thickness.

However, if consumption of fresh ground water is more rapid, the freshwater lens thins, and the denser saline ground water, laden with dissolved sodium chloride, moves up to fill in pores emptied by removal of fresh water. *Upconing* of salt water below cones of depression in the freshwater lens may also occur. Wells that had been tapping the freshwater lens may begin pumping unwanted salt water instead, as the limited freshwater supply gradually decreases. Salt-water intrusion destroyed useful aquifers beneath Brooklyn, New York, in the 1930s and is a serious problem in many coastal areas of the southeastern and Gulf coastal states and in some densely populated parts of California.

Other Impacts of Urbanization on Groundwater Systems

Obviously, an increasing concentration of people means an increased demand for water. We saw in chapter 6 that urbanization may involve extensive modification of surface-water runoff patterns and stream channels. Insofar as it modifies surface runoff and the ratio of runoff to infiltration, urbanization also influences groundwater hydrology.

Loss of Recharge

Impermeable cover—buildings, asphalt and concrete roads, sidewalks, parking lots, airport runways—over one part of a broad area underlain by an unconfined aquifer has relatively little impact on that aquifer's recharge. Infiltration will continue over most of the region. In the case of a confined aquifer, however, the available recharge area may be very limited, since the overlying confining layer prevents direct downward infiltration in most places (figure 10.12). If

impermeable cover is built over the recharge area of a confined aquifer, then, recharge can be considerably reduced, thus aggravating the water-supply situation.

Filling in wetlands is a common way to provide more land for construction. This practice, too, can interfere with recharge, especially if surface runoff is rapid elsewhere in the area. The stagnant swamp in a recharge area, holding water for long periods, can be a major source of infiltration and recharge. Filling it in so water no longer accumulates there, and worse yet, topping the fill with impermeable cover, again may greatly reduce local groundwater recharge.

Artificial Recharge

In steeply sloping areas or those with low-permeability soils, well-planned construction that includes artificial recharge basins can aid in increasing groundwater recharge (figure 10.13). The basin acts similarly to a flood-control retention pond (chapter 6), in that it is designed to catch some of the surface runoff during high-runoff events (heavy rain or snowmelt). Trapping the water allows more time for infiltration and thus more recharge in an area from which the fresh water might otherwise be quickly lost to streams and carried away. Recharge basins are a partial solution to the problem of areas where groundwater use exceeds natural recharge rate, but, of course, they are only effective where there is surface runoff to catch.

Other Features Involving Subsurface Water

Sinkholes

An abundance of surface and near-surface water may indicate an ample water supply, but it also means more water available to dissolve rocks. Most rocks are not very soluble, so this is not a concern in all areas. A few rock types,

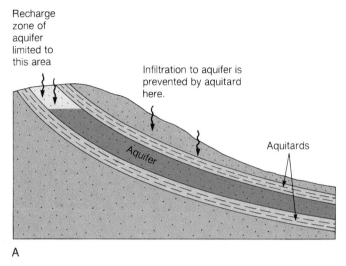

Recharge zone of aquifer limited to this area

Infiltration to aquifer is prevented by aquitard here.

Aquifer

Aquitards

A

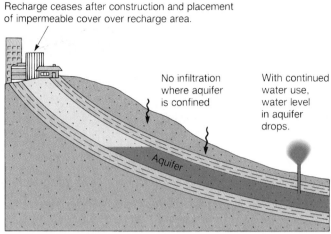

Recharge ceases after construction and placement of impermeable cover over recharge area.

No infiltration where aquifer is confined

With continued water use, water level in aquifer drops.

Aquifer

B

Figure 10.12   Recharge to a confined aquifer. (*A*) The recharge area of this confined aquifer is limited to the area where permeable rocks intersect the surface. (*B*) Recharge to the confined aquifer may be reduced by placement of impermeable cover over the limited recharge area.

Recharge basin retards surface runoff.

Water table

Increased infiltration under recharge basin

Figure 10.13   Artificial recharge basins can aid recharge by slowing surface runoff.

however, are extremely soluble. The most common of these is limestone. Over long periods of time, underground water may dissolve large volumes of limestone, slowly enlarging underground caverns and eroding support for the land above. There may be no obvious evidence at the surface of what is taking place until the ground collapses abruptly into the void, producing a **sinkhole** (figure 10.14).

The collapse of a sinkhole may be triggered by a drop in the water table as a result of drought or water use that leaves rocks and soil that were previously buoyed up by water pressure unsupported. Failure may also be caused by the rapid input of large quantities of water from heavy rains that wash overlying soil down into the cavern, or by an increase in subsurface water flow rates.

Sinkholes come in many sizes. The larger ones are quite capable of swallowing many houses at a time: They may be over 50 meters deep and cover several tens of acres. If one occurs in a developed area, a single sinkhole can cause millions of dollars in property damage. Sudden sinkhole collapses beneath bridges, roads, and railways have occasionally caused accidents and even deaths.

Figure 10.14   Sinkhole at Winter Park, Florida. Collapse on May 8–9, 1981 was caused in part by drought. The sinkhole was over 100 meters across and 30 meters deep; estimated damage, over $2 million.   Photograph courtesy of USGS Photo Library, Denver, CO.

Karst

Sinkholes are rarely isolated phenomena. Limestone most often is formed from chemical sediments deposited in shallow seas, and limestone beds cover broad areas. The same is true of gypsum beds, and gypsum, too, is quite soluble. Therefore, where there is one sinkhole in a region underlain by such soluble rocks, there are likely to be others. Thousands have formed in Alabama alone. An abundance of sinkholes in an area is a strong hint that more can be expected.

Figure 10.15 Telltale karst topography, showing many round lakes formed in sinkholes. (Red is false color, and corresponds to green vegetation.)   Photograph © NASA.

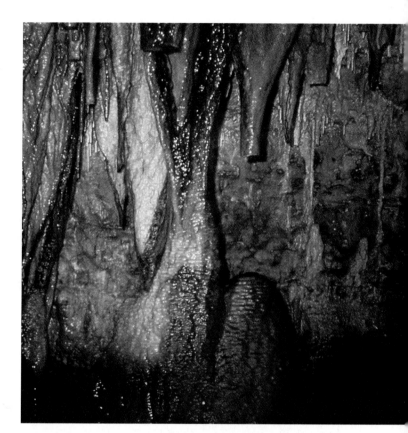

Figure 10.16 In a cave produced by subsurface solution of limestone, later precipitation from dripping water produces stalactites and stalagmites. Cave of the Mounds, Wisconsin.

The situation may sometimes be easily recognized through aerial or satellite photography, which can show a region to be pockmarked with the circular lakes or holes characteristic of this solution-dominated **karst topography** (figure 10.15). Clearly, in such an area, the subsurface situation should be investigated before buying or building a home or business. Circular patterns of cracks on the ground or conical depressions in the ground surface may be early signs of trouble developing below.

Even where surface manifestations of solution processes are few, there may be extensive development of solution channels, or even caves, at depth (figure 10.16). Pervasive channels and large underground voids allow rapid drainage and infiltration of water, and underground streams occasionally develop; surface streams are uncommon in karst terranes.

Water Quality

As noted earlier, most of the water in the hydrosphere is in the very salty oceans, and almost all of the remainder is tied up in ice. That leaves relatively little surface or subsurface water for potential freshwater sources. Moreover, much of the water on and in the continents is not strictly fresh. Even rainwater, long the standard for "pure" water, contains dissolved chemicals of various kinds, especially in industrialized areas with substantial air pollution. Once precipitation reaches the ground, it reacts with soil, rock, and organic debris, dissolving still more chemicals naturally, aside from any pollution generated by human activities. Water quality thus must be a consideration when evaluating water supplies.

Pollution of both surface water and ground water, and some additional measures of water quality used particularly in describing pollution, will be discussed in chapter 16. Here, we focus on aspects of natural water quality.

Measures of Water Quality

Water quality may be described in a variety of ways. A common approach is to express the amount of a dissolved chemical substance present as a concentration in parts per million (ppm) or, for very dilute substances, parts per billion (ppb). These units are analogous to weight percentages (which are really "parts per hundred") but are used for lower (more dilute) concentrations. For example, if water contains 1 weight percent salt, it contains one gram of salt per hundred grams of water, or one ton of salt per hundred tons of water, or whatever unit one wants to use. Likewise, if the water contains only 1 ppm salt, it contains one gram of salt per million grams of water (about 250 gallons), and so on.

Another way to express overall water quality is in terms of *total dissolved solids* (TDS), the sum of the concentrations of all dissolved solid chemicals in the water. How low a level of TDS is required or acceptable varies with the application. Standards might specify a maximum of 500 or 1,000 ppm TDS for drinking water; 2,000 ppm TDS might be acceptable for watering livestock; industrial applications where water chemistry is important (in pharmaceuticals or textiles, for instance) might need water even purer than normal drinking water.

Yet describing water in terms of total content of dissolved solids does not present the whole picture: At least as important as the quantities of impurities present is *what* those impurities are. If the main dissolved component is calcite (calcium carbonate) from a limestone aquifer, the water may taste fine and be perfectly wholesome with well over 1,000 ppm TDS in it. If iron or sulfur is the dissolved substance, even a few parts per million may be enough to make the water taste bad. Many synthetic chemicals that have leaked into water through improper waste disposal are toxic even at concentrations of 1 ppb or less.

Other parameters also may be relevant in describing water quality. One is pH, which is a measure of the acidity or alkalinity of the water. The pH of water is inversely related to acidity: the lower the pH, the more acid the water. (Water that is neither acid nor alkaline has a pH of 7.) For health reasons, concentrations of certain bacteria may also be monitored in drinking-water supplies.

A water-quality concern that has only recently drawn close attention is the presence of naturally occurring radioactive elements that may present a radiation hazard to the water consumer. Uranium, which can be found in most rocks, including those serving commonly as aquifers, decays through a series of steps. Several of the intermediate decay products pose special hazards. One—radium—behaves chemically much like calcium and therefore tends to be concentrated in the body in bones and teeth. Another—radon—is a chemically inert gas but is radioactive itself and decays to other radioactive elements in turn. Radon leaking into indoor air from water supplies contributes to indoor air pollution (see chapter 17). Because of its chemical similarity to calcium, radium can be removed from water by passage through a water softener, but even in hard-water areas, not all water consumed is "softened."

Hard Water

Aside from the issue of health, water quality may be of concern because of the particular ways certain dissolved substances alter water properties. In areas where water supplies have passed through soluble carbonate rocks, like limestone, the water may be described as "hard." **Hard water** simply contains substantial amounts of dissolved calcium and magnesium. When calcium and magnesium concentrations reach or exceed the range of 80 to 100 ppm, the hardness may become objectionable.

Table 10.3   Concentrations of Some Dissolved Constituents in Rain, River Water, and Seawater

Constituent	Concentration (ppm)						
	Rainwater			River water			Average seawater
	1-year avg., mid-Atlantic states	Single storm, California	1-year avg., inland U.S.	Amazon River	Mississippi River	World avg. (est.)	
silica (SiO_2)	—	0.3	—	7.0	6.7	13	6.4
calcium (Ca)	0.65	0.8	1.41	4.3	42	15	400
sodium (Na)	0.56	9.4	0.42	1.8	25	6.3	10,500
potassium (K)	0.11	0.0	—	—	2.9	2.3	380
magnesium (Mg)	0.14	1.2	—	1.1	12	4.1	1,350
chloride (Cl)	0.57	17	0.22	1.9	30	7.8	19,000
fluoride (F)	—	—	—	0.2	0.2	—	1.3
sulfate (SO_4)	2.18	7.6	2.14	3.0	56	11	2,700
bicarbonate (HCO_3)	—	4	—	19	132	58	142
nitrate (NO_3)	0.62	0	—	0.1	2.4	1	0.5

Source: J. D. Hem, *Study and Interpretation of the Chemical Characteristics of Natural Water*, U.S. Geological Survey Water-Supply Paper 1473, 1970, pp. 11, 12, and 50.

Perhaps the most irritating routine problem with hard water is the way it reacts with soap, preventing the soap from lathering properly, causing bathtubs to develop rings and laundered clothes to retain a gray soap scum. Hard water or water otherwise high in dissolved minerals may also leave mineral deposits in plumbing and in appliances, such as coffeepots and steam irons. Primarily for these reasons, many people in hard-water areas use water softeners, which remove calcium, magnesium, and certain other ions from water in exchange for added sodium ions. The sodium ions are replenished from the salt (sodium chloride) supply in the water softener. (While softened water containing sodium ions in moderate concentration is unobjectionable in taste or household use, it may be of concern to those on diets involving restricted sodium intake.) The "active ingredient" in water softeners is a group of hydrous silicate minerals known as *zeolites*. The zeolites have an unusual capacity for *ion exchange*, a process in which ions loosely bound in the crystal structure can be exchanged for other ions in solution.

Overall, groundwater quality is highly variable. It may be nearly as pure as rainwater or saltier than the oceans. Some representative analyses of different waters in the hydrosphere are shown in table 10.3 for reference.

Water Use, Water Supply

General U.S. Water Use

Inspection of the U.S. water budget overall would suggest that ample water is available for use (figure 10.17). Some 4,200 billion gallons of precipitation fall on this country each day; subtracting 2,750 billion gallons per day lost to evapotranspiration still leaves a net of 1,400 billion gallons per day for streamflow and groundwater recharge. Water-supply problems arise, in part, because the areas of greatest water availability do not always coincide with the areas of concentrated population or greatest demand, and also because a portion of the added fresh water quickly becomes polluted by mixing with impure or contaminated water.

People in the United States use a large amount of water. Biologically, humans require about a gallon of water a day per person, or, in the United States, about 250 million gallons per day for the country. Yet, Americans divert, or "withdraw," about 400 *billion* gallons of water each day—about 1,800 gallons per person—for cooking, washing, and other household uses, for industrial processes and power generation, and for livestock and irrigation, a wide range of *"offstream"* water uses. Another several trillion gallons of

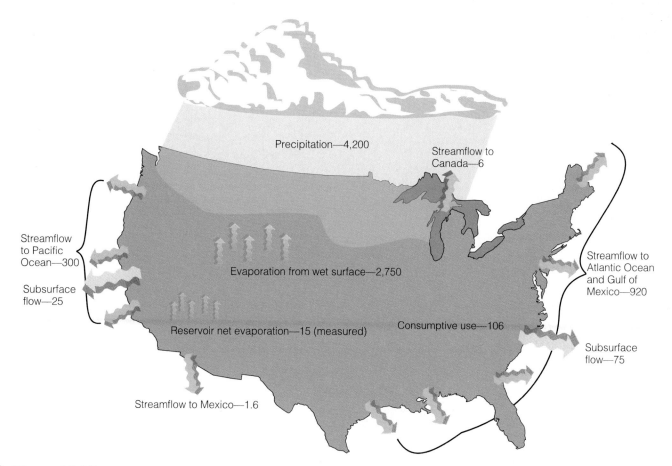

Figure 10.17 In terms of gross water flow, U.S. water budget seems ample. Figures are in billions of gallons per day.   Source: Data from *The Nation's Water Resources 1975–2000*, Vol. 1, U.S. Water Resources Council.

water are used each day to power hydroelectric plants (*"in-stream"* use). Of the total water withdrawn, more than 100 billion gallons per day are *consumed,* meaning that the water is not returned as wastewater. Most of the consumed water is lost to evaporation; some is lost in transport (for example, through piping systems).

Surface Water versus Ground Water as Supply

It might seem easier to use surface waters rather than sub-surface waters for water supplies. Why, then, worry about using ground water at all? One basic reason is that, in many dry areas, there is little or no surface water available, while there may be a substantial supply of water deep underground. Tapping the latter supply allows us to live and farm in otherwise uninhabitable areas.

Then, too, streamflow varies seasonally. During dry seasons, the water supply may be inadequate. Dams and reservoirs allow a reserve of water to be accumulated during wet seasons for use in dry times, but we have already seen some of the negative consequences of reservoir construction (chapter 6). Furthermore, if a region is so dry at some times that dams and reservoirs are necessary, then the rate of water evaporation from the broad, still surface of the reservoir may itself represent a considerable water loss and aggravate the water-supply problem.

Precipitation (the prime source of abundant surface runoff) varies widely geographically (figure 10.18), as does population density. In many areas, the concentration of people far exceeds what can be supported by available local surface waters, even during the wettest season.

Also, streams and large lakes have historically been used as disposal sites for untreated wastewater and sewage, which makes the surface waters decidedly less appealing as drinking waters. A lake, in particular, may remain polluted for decades after the input of pollutants has stopped if there is no place for those pollutants to go or if there is a limited input of fresh water to flush them out. Ground water from the saturated zone, on the other hand, has passed through the rock of an aquifer and has been naturally filtered to remove some impurities—soil or sediment particles and even the larger bacteria—although it can still contain many dissolved chemicals.

Finally, ground water is by far the largest reservoir of unfrozen fresh water. For a variety of reasons, then, underground waters may be preferred as a supplementary or even sole water source.

Regional Variations in Water Use

Water withdrawal varies greatly regionally (figure 10.19), as does water consumption. These two quantities are not necessarily directly related, for the fraction of water withdrawn that is actually consumed depends, in part, on the principal

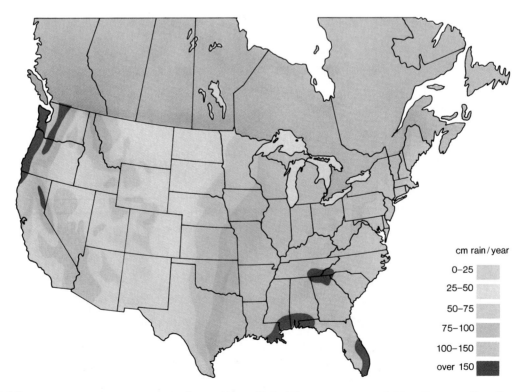

cm rain/year

0–25
25–50
50–75
75–100
100–150
over 150

Figure 10.18 Average annual precipitation in the contiguous United States.   Source: U.S. Water Resources Council.

purposes for which it is withdrawn, as will be described below. This, in turn, influences the extent to which local water use may deplete groundwater supplies. Of the fresh water withdrawn, only about 25 percent is ground water; but many of the areas of heaviest groundwater use rely on that water for irrigation, which consumes a large fraction of the water withdrawn.

Aside from hydropower generation, several principal categories of water use can be identified: municipal (public) supplies (home use and some industrial use in urban and suburban areas), rural use (supplying domestic needs for rural homes and watering livestock), irrigation (a form of rural use, too, but worthy of special consideration for reasons to be noted shortly), self-supplied industrial use (use other than for electricity generation, by industries for which water supplies are separate from municipal sources), and thermo-electric power generation (using fuel—coal, nuclear power, etc.—to generate heat). The quantities withdrawn for and consumed by each of these categories of users are summarized in figure 10.20. The sources and disposition of water for the four major water uses are shown in figure 10.21.

A point that quickly becomes apparent from figure 10.21 is that while power generation may be the major water *user*, agriculture is the big water *consumer*. Power generators and industrial users account for more than half the water withdrawn, but nearly all their wastewater is returned as liquid water at or near the point in the hydrologic cycle from which it was taken. Most of these users are diverting surface waters and dumping wastewaters (suitably treated, one hopes!) back into the same lake or stream. Together, industrial users and power generation *consume* only about 10 billion gallons per day, or 10 percent of the total.

A much higher fraction of irrigation water is consumed: lost to evaporation, lost through transpiration from plants, or lost

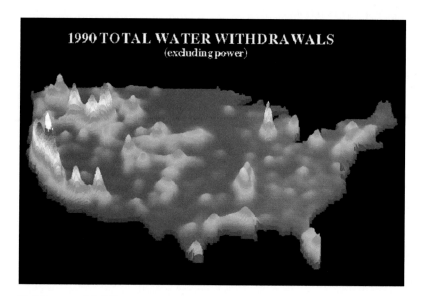

Figure 10.19 U.S. regional variations in water use. Compare with figure 10.18.   Source: U.S. Geological Survey Water Resources Division.

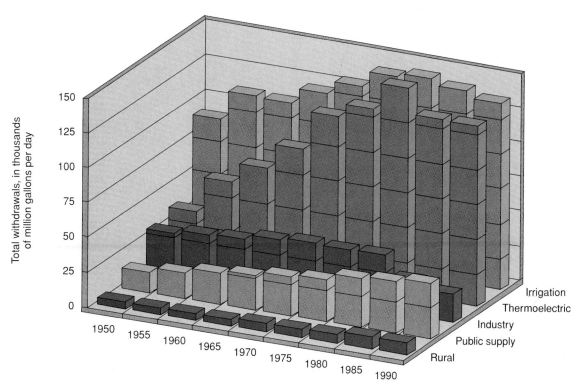

Figure 10.20 Water withdrawal by sector, 1950–1990.   Source: U.S. Geological Survey National Circular 1081.

SOURCE

Surface water saline 33%

Surface water fresh 67%

DISPOSITION

Consumptive use fresh 2%

Return flow 98%

195,000 million gallons per day
THERMOELECTRIC POWER source and disposition, 1990

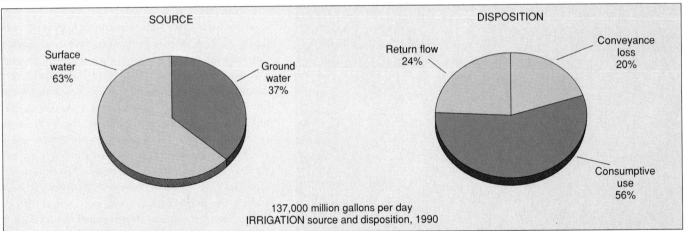

SOURCE

Surface water 63%

Ground water 37%

DISPOSITION

Return flow 24%

Conveyance loss 20%

Consumptive use 56%

137,000 million gallons per day
IRRIGATION source and disposition, 1990

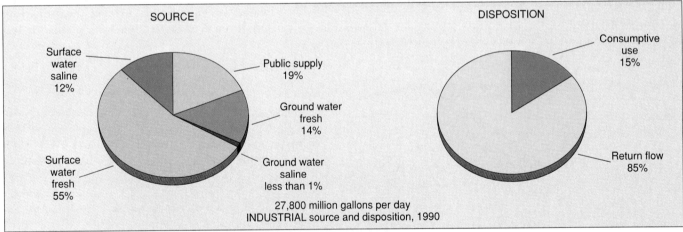

SOURCE

Surface water saline 12%

Public supply 19%

Ground water fresh 14%

Ground water saline less than 1%

Surface water fresh 55%

DISPOSITION

Consumptive use 15%

Return flow 85%

27,800 million gallons per day
INDUSTRIAL source and disposition, 1990

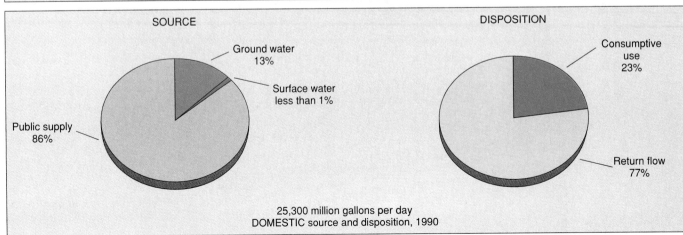

SOURCE

Ground water 13%

Surface water less than 1%

Public supply 86%

DISPOSITION

Consumptive use 23%

Return flow 77%

25,300 million gallons per day
DOMESTIC source and disposition, 1990

Figure 10.21 Sources and disposition of water withdrawn/diverted for the four major uses. Source: U.S. Geological Survey Circular 1081.

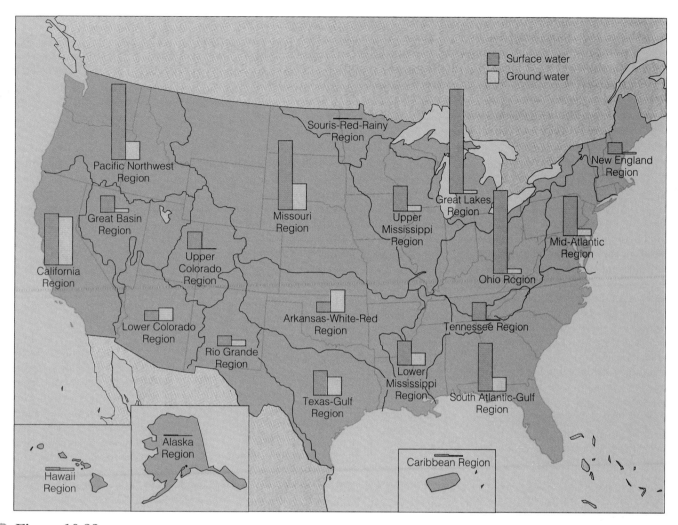

Figure 10.22 Withdrawal of fresh surface water and fresh ground water, by water-resource region.   Source: Data from U.S. Water Resources Council.

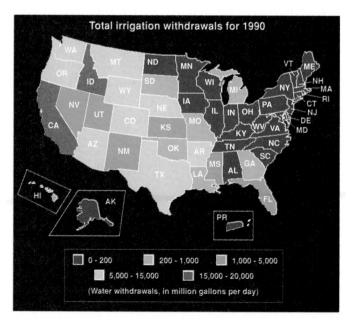

Figure 10.23 Water use for irrigation, by state.
Source: U.S. Geological Survey Water Resources Division.

because of leakage from ditches and pipes. Moreover, close to 40 percent of the water used for irrigation is ground water. Most of the water lost to evaporation drifts out of the area to come down as rain or snow somewhere far removed from the irrigation site. It then does not contribute to the recharge of aquifers or to runoff to the streams from which the water was drawn.

The regional implications of all this can be examined through figures 10.22 and 10.23. The former indicates the relative volumes of fresh surface and ground waters withdrawn, by water-resource region, and reflects the comparatively greater dependence of the west on ground water. The latter figure shows that, with few exceptions, states showing high groundwater use and high overall water withdrawal use their water extensively for irrigation. Where irrigation use of water is heavy, water tables have dropped by tens of meters and streams have been drained nearly dry, while we have become increasingly dependent on the crops. Box 10.2 emphasizes the magnitude of the problem in certain parts of the United States.

Box 10.2 The High Plains (Ogallala) Aquifer System

In the metropolitan Chicago area and environs, highlighted in box 10.1, there exist both moderate rainfall to support agriculture and alternatives to dwindling groundwater supplies for municipal water users. Where ground water is the only substantial actual and potential water source, however, the danger of mining that ground water is intensified.

The Ogallala Formation, a sedimentary aquifer, underlies most of Nebraska and sizeable portions of Colorado, Kansas, and the Texas and Oklahoma panhandles (figure 1). The most productive units of the aquifer are sandstones and gravels. The area under which the Ogallala lies is one of the largest and most important agricultural regions in the United States. It accounts for about 25 percent of U.S. feed-grain exports and 40 percent of wheat, flour, and cotton exports. More than 14 million acres of land are irrigated with water pumped from the Ogallala. Yields on irrigated land may be triple the yields on similar land cultivated by dry farming (no irrigation).

The Ogallala's water was, for the most part, stored during the retreat of the Pleistocene continental ice sheets. Present recharge is negligible over most of the region. The original groundwater reserve in the Ogallala is estimated to have been approximately 2 billion acre-feet. (One acre-foot is the amount of water required to cover an area of one acre to a depth of one foot; it is more than 300,000 gallons.) But each year, farmers draw from the Ogallala more water than the entire flow of the Colorado River: In 1990, an estimated 15.6 million acre-feet of Ogallala water were used, to irrigate about 14 million acres of agricultural land.

Soon after development in the plains began, around 1940, the water table began to decline noticeably. In areas of heavy use—Texas, southwestern Kansas, the Oklahoma panhandle—it had dropped over 100 feet by 1980. That observation, together with simple arithmetic comparing the total water reserve with current withdrawal rates, suggests that depletion within a matter of decades is possible. What then?

Reversion to dry farming, where possible at all, would greatly diminish yields. Reduced vigor of vegetation could lead to a partial return to pre-irrigation, Dust-Bowl-type conditions. Alternative local sources of municipal water are in many places not at all apparent. Planners in Texas and Oklahoma have advanced ambitious water-transport schemes as solutions. Texas's various proposed alternatives would basically all involve transferring water now draining into the Mississippi River from northeastern Texas across the state to the High Plains of the panhandle. Oklahoma's Comprehensive Water Plan would draw on the Red River and Arkansas River basins (see figure 10.22). Each such scheme would cost billions of dollars, perhaps tens of billions. Water-transport systems of this scale could take a decade or longer to complete from the time the public commits itself to the projects. Before the projects are completed, however, acute water shortages can be expected in the areas of most urgent need. Even if and when the transport networks are finished, the cost of the water might be ten times what farmers in the region can comfortably afford to pay if their products are to remain fully competitive in the marketplace. There seem to be no easy solutions. Perhaps because of this, progress toward any solution has been slow. Meanwhile, the draining of the Ogallala Formation continues day by day.

It is noteworthy that there has been little incentive in the short term for the exercise of restraint, and that this is due in large measure to federal policies. Price supports encourage the growing of crops like cotton that require irrigation in the southern part of the region. The government cost-shares in soil-conservation programs and provides for crop-disaster payments, thus giving compensation for the natural consequences of water depletion. And, in fact, federal tax policy provides for groundwater depletion allowances (tax breaks) for High Plains area farmers using pumped ground water, with larger breaks for heavier groundwater use. The results of all these policies aren't surprising.

Still, water-resource professionals are convincing more and more of those dependent on the Ogallala to conserve. In 1988, too, the U.S. Geological Survey initiated a continuing program of monitoring water-level changes in the aquifer system. Federal, state, and local water-resource specialists are working with farmers and communities to slow water depletion. Aside from increasing irrigation efficiency, some Ogallala users are trying more creative measures: For example, some farmers in Kansas and Nebraska pump fresh ground water from their land for local towns, which return their treated wastewater to irrigate the fields. Over the longer term, further emphasis on conservation, plus modifications of existing laws that promote groundwater use, could substantially extend the useful life of the Ogallala aquifer.

Figure 1   Changes in water levels in the Ogallala Formation, 1980–1994. Only in a few places, such as along the Platte River, has the water table risen significantly, in response to short-term precipitation increases; the overall long-term trend is definitely down.   Source: U. S. Geological Survey.

Table 10.4   International Water Disputes

River	Countries in dispute	Issues
Nile	Egypt, Ethiopia, Sudan	Siltation, flooding, water flow/diversion
Euphrates, Tigris	Iraq, Syria, Turkey	Reduced water flow, salinization
Jordan, Yarmuk, Litani, West Bank aquifer	Israel, Jordan, Syria, Lebanon	Water flow/diversion
Indus, Sutlei	India, Pakistan	Irrigation
Ganges-Brahmaputra	Bangladesh, India	Siltation, flooding, water flow
Salween	Myanmar, China	Siltation, flooding
Mekong	Cambodia, Laos, Thailand, Vietnam	Water flow, flooding
Paraná	Argentina, Brazil	Dam, land inundation
Lauca	Bolivia, Chile	Dam, salinization
Rio Grande, Colorado	Mexico, United States	Salinization, water flow, agrochemical pollution
Rhine	France, Netherlands, Switzerland, Germany	Industrial pollution
Maas, Schelde	Belgium, Netherlands	Salinization, industrial pollution
Elbe	Czechoslovakia, Germany	Industrial pollution
Szamos	Hungary, Romania	Industrial pollution

Note that in most cases, central issues to the disputes involve adequacy and/or quality of water supply.

From: Michael Renner, *National Security: The Economic and Environmental Dimensions,* Worldwatch Paper 89, p. 32. Copyright © 1989 Worldwatch Institute, Washington, D.C. Reprinted by permission.

Nor is water supply an issue only in the western U.S., by any means. It is an international political matter as well (table 10.4). In most of the cases cited, the central issues relate either to water availability, or to the quality of the available water. One country's water-use preferences and practices can severely disadvantage another, especially if they share a common water source.

Extending the Water Supply

Conservation

The most basic approach to improving the U.S. water-supply situation is conservation. Water is wasted in home use every day—by long showers, inefficient plumbing, insistence on lush, green lawns even in the heat of summer, and in dozens of other ways. Raising livestock for meat requires far more water per pound of protein than growing vegetables for protein. Still, municipal and rural water uses (excluding irrigation) together account for only about 10 percent of total U.S. water consumption.

The big water drain is plainly irrigation, and that use must be moderated if the depletion rate of water supplies is to be reduced appreciably. For example, the raising of crops that require a great deal of water could be shifted, in some cases at least, to areas where natural rainfall is adequate to support them. Irrigation methods can also be made more efficient so that far less water is lost by evaporation. This can

be done, for instance, by drip irrigation. Instead of running irrigation water in open ditches from which evaporation loss is high, the water can be distributed via pipes with tiny holes from which water seeps slowly into the ground at a rate more closely approaching that at which plants use it. However, the more efficient methods are often considerably more expensive, too.

Domestic use can be reduced in a variety of ways. For example, lawns can be watered morning or evening when evaporation is less rapid than at midday; or one can forgo traditional lawns altogether in favor of ground covers that don't need watering. Stormwater can be directed into recharge basins rather than dumped into sewer systems. Increasingly, municipalities in dry areas are looking to recycling their wastewater (discussed further in chapter 15).

Interbasin Water Transfer

In the short term, conservation alone will not resolve the imbalance between demand and supply. New sources of supply are needed. Part of the supply problem, of course, is purely local. For example, people persist in settling and farming in areas that may not be especially well supplied with fresh water (see, for example, box 10.3), while other areas with abundant water go undeveloped. If the people cannot be persuaded to be more practical, perhaps the water can be redirected. This is the idea behind interbasin transfers—moving surface waters from one stream system's drainage basin to another's where demand is higher.

California pioneered the idea with the Los Angeles Aqueduct. The aqueduct was completed in 1913 and carried

Box 10.3 A Crisis of Surface-Water Supply: The Colorado River Basin

The stream system of the Colorado River's drainage basin drains portions of seven western states (figure 1). Many of these states have extremely dry climates, and it was recognized decades ago that some agreement would have to be reached about which region was entitled to how much of that water. Intense negotiations during the early 1900s led in 1922 to adoption of the Colorado River Compact, which apportioned 7.5 million acre-feet of water per year each to the Upper Basin (Colorado, New Mexico, Utah, and Wyoming) and the Lower Basin (Arizona, California, and Nevada). The Lower Basin was also to be allowed to increase its consumption by 1 million acre-feet per year. No specific provision was made for Mexico, into which the river ultimately flows.

Rapid development occurred throughout the region. Huge dams impounded enormous reservoirs of water for irrigation or for transport out of the basin. In 1944, Mexico was awarded by treaty some 1.5 million acre-feet of water per year, but with no stipulation concerning water quality. Mexico's share was to come from the surplus above the allocations within the United States or, if that was inadequate, equally from the Upper and Lower Basins. In 1948, the Upper Basin states entered into a compact to apportion their share among them.

Heavy water use has led to a reduction in both water flow and water quality in the Colorado River. The reduced flow results not only from diversion of the water for use but also from large evaporation losses from the numerous reservoirs in the system. The reduced water quality is partly a consequence of that same evaporation, concentrating dissolved minerals, and of selective removal of fresh water. Also, many of the streams flow through soluble rocks, which then are partially dissolved, increasing the dissolved mineral load. By 1961, the water delivered to Mexico contained up to 2,700 ppm TDS, and partial crop failures resulted from use of such saline water for irrigation. In response to protests from the

Mexican government, the United States agreed in 1974 to build a desalination plant at the U.S.-Mexican border to reduce the salinity of Mexico's small share of the Colorado.

It has now been shown that sufficient water flow simply does not exist in the Colorado River basin even to supply the full allocations to the U.S. users. The streamflow measurements made in the early 1900s were grossly inaccurate and apparently also were made during an unusually wet period. Current estimates put the total flow at only 14 million acre-feet, and of that, an estimated 2 million acre-feet are lost to evaporation, especially from reservoirs. The 1922 agreement and Mexican treaties still work tolerably well only because the Upper Basin states are using little more than half their allotted water. The dry Lower Basin continues to need ever more water as its population grows. Arizona has already gone to court against California over the issue of which state is entitled to how much water; the consequence is that California has been compelled to *reduce* its use of Colorado River water. And California has grown very dependent on that water: The Colorado River Aqueduct, completed in 1941, alone carries over 1.2 million acre-feet of water per year from Lake Havasu to supply Los Angeles and San Diego, and California's 4.4 million acre-feet is already the lion's share of the Lower Basin's allocation.

Moreover, the possibility of greatly increased future water needs looms for another reason: energy. The western states contain a variety of energy resources, conventional and new. Chief among these are coal and oil shale. Extraction of fuel from oil shale is, as we will see in chapter 13, a water-intensive

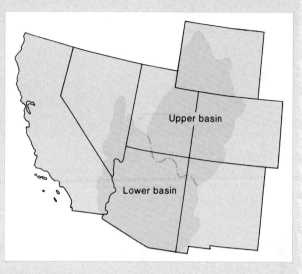

Figure 1 The Colorado River basin. Source: U.S. Geological Survey Water Supply Circular 1001.

process. In addition, both the coal and oil shale are very likely to be strip-mined, and current laws require reclamation of strip-mined land. Abundant water will be needed to reestablish vegetation on the reclaimed land; where rainfall is insufficient, river water will be in demand. Already, energy interests have water-rights claims to more than 1 million acre-feet of water in the Colorado oil shale area alone. The streamflow there, as in many other parts of the Colorado River basin, is already over-appropriated. The crunch will get worse if more claims are fully utilized and additional claims filed with the development of these energy resources.

Projections made in 1978 by the U.S. Water Resources Council anticipated a 50 percent increase in municipal water consumption in the lower Colorado basin from 1975 to 2000, and a doubling of industrial consumption there over the same period, but water consumption for extraction of minerals and fuels was projected nearly to double in the lower basin, and to triple in the upper basin, partly in reflection of development of alternative energy sources. In fact, the unconventional fuels have so far remained uneconomic to exploit; but the potential (and associated potential water demand) is still there.

Figure 10.24 Diversion of surface water that formerly flowed into Mono Lake, California, has caused lake level to drop, exposing sedimentary formations built underwater by reaction of lake water with spring water seeping up from below.

Figure 10.25 Methods of desalination. (*A*) Filtration (simplified schematic): Dissolved and suspended material is screened out by very fine filters. (*B*) Distillation: As water is heated, pure water is evaporated, then recondensed for use. Dissolved and suspended materials stay behind.

nearly 150 million gallons of water per day from the eastern slopes of the Sierra Nevada to Los Angeles. In 1958, the system was expanded to bring water from northern California to the southern part of the state. More and larger projects have been undertaken since. Bringing water from the Colorado River to southern coastal California, for example, required construction of over 300 kilometers (200 miles) of tunnels and canals. Other water projects have transported water over whole mountain ranges. It should be emphasized, too, that such projects are not confined to the drier west: for example, New York City draws on several reservoirs in upstate New York. If population density or other sources of water demand are high, a local supply shortfall can occur even in an area thought of as quite moist.

Dozens of interbasin transfers of surface water have been proposed. Political problems are common even when the transfer involves diverting water from one part of a single state to another: In 1982, it was proposed to expand the aqueduct system to carry water from northern California to the south; 60 percent of the voters in the southern part of the state were in favor, but 90 percent of those in the north were opposed, and the proposition lost. The opposition often increases when transfers among several states are considered: recently, for instance, officials in states around the Great Lakes objected to the idea of diverting some lake water to states in the southern and southwestern United States. The problems may be far greater when transfers between nations are involved. Various proposals have been made to transfer water from little-developed areas of Canada to high-demand areas in the United States and Mexico. Such proposals, which could involve transporting water over distances of thousands of kilometers, are not only expensive (one such scheme, the North American Water and Power Alliance, had a projected price of $100 billion); they also presume a continued willingness on the

part of other nations to share their water. Sometimes, too, the diversion in turn causes problems in the region from which the water is drawn (figure 10.24).

Desalination

Another alternative for extending the water supply is to improve the quality of waters not now used, purifying them sufficiently to make them usable. Desalination of seawater, in particular, would allow parched coastal regions to tap the vast ocean reservoirs. Also, some ground waters are not presently used for water supplies because they contain excessive concentrations of dissolved materials. There are two basic methods used to purify water of dissolved minerals: filtration and distillation (figure 10.25).

In a filtration system, the water is passed through fine filters or membranes to screen out dissolved impurities. An advantage of this method is that it can rapidly filter great quantities of water. A large municipal filtration operation may produce several billion gallons of purified water per day. A disadvantage is that the method works best on water not containing very high levels of dissolved minerals. Pumping anything as salty as seawater through the system quickly clogs the filters. This method, then, is most useful for cleaning up only moderately saline ground waters or lake or stream water.

Distillation involves heating or boiling water full of dissolved minerals. The water vapor driven off is pure water, while the minerals stay behind in what remains of the liquid. Because this is true regardless of how concentrated the dissolved minerals are, the method works fine on seawater as well as on less saline waters.

A difficulty, however, is the nature of the necessary heat source. Furnaces fired by coal, gas, or other fuels can be used, but any fuel may be costly in large quantity, and as we will see in chapter 13, many conventional fuels are becoming scarce. The sun is an alternative possible heat source. Sunlight is free and inexhaustible, and some solar desalination facilities already exist. Their efficiency is limited by the fact that solar heat is low-intensity heat. If a large quantity of desalinated water is required rapidly, the water to be heated must be spread out shallowly over a large area, or the rate of water output will be slow. A large city might need a solar desalination facility covering thousands of square kilometers to provide adequate water, and construction on such a scale would be prohibitively expensive even if the space were available.

Desalinated water may be five to ten times more costly to deliver than water pumped straight from a stream or aquifer. For most homeowners, the water bill is a relatively minor expense, so a jump in water costs, if necessitated by the use of desalinated water, would not be a great hardship. Water for irrigation, however, must be both plentiful and cheap if the farmer is to compete with others here and abroad who need not irrigate and if the cost of food production is to be held down. Desalinated water in most areas is prohibitively expensive for irrigation use. Unless ways can be found to reduce drastically the cost of desalinated water, agriculture will continue to drain limited and dwindling surface and groundwater supplies. Water from the ocean will not be a viable alternative for large-scale irrigation for some time.

Overview

Summary

Most of the water in the hydrosphere at any given time is in the oceans; most of the remaining fresh water is stored in ice sheets. Relatively little water is found in lakes and streams. A portion of the fresher waters on the continents—both surface water and ground water—is diverted for human use. The availability of ground water is influenced by such factors as presence of suitable aquifers, water quality, and rate of recharge relative to rate of water use.

In the United States, offstream water use (withdrawal) amounts to about 400 billion gallons per day, more than half of which is for industrial use and thermoelectric power generation. However, these industrial applications generally consume very little water. Of the roughly 100 billion gallons consumed each day in the United States, approximately 80 percent is water loss associated with irrigation. While three-fourths of the water *withdrawn* for use is surface water, close to half of the water *consumed* is ground water, which, in many cases, is being consumed faster than recharge can replace it. Adverse consequences of such rapid consumption of ground water include lowering water tables, surface subsidence, and, in coastal areas, saltwater intrusion. Conservation, interbasin transfers of surface water, and desalination are possible ways to extend the water supplies of high-demand regions. Desalinated water is presently too expensive to use for irrigation on a large scale, which means that it is unlikely to have a significant impact on the largest consumptive water use for some time to come.

Terms to Remember

Exercises

Questions for Review

1. Explain the importance of porosity and of permeability to groundwater availability for use.

2. Define the following terms: *ground water, water table, and potentiometric surface.*

3. How is an artesian aquifer system formed?

4. Explain how sinkholes develop. What name is given to a terrane in which sinkholes are common?

5. Explain three parameters used to describe ground-water quality.

6. What is *hard water,* and why is it often considered un-desirable?

7. Describe two possible consequences of groundwater withdrawal exceeding recharge.

8. Explain the process of saltwater intrusion.

9. In what ways may urbanization affect groundwater recharge?

10. Industry is the big water *user,* but agriculture is the big water *consumer.* Explain.

11. Compare and contrast filtration and distillation as de-salination methods, noting advantages and drawbacks of each.

12. What factor presently limits the potential of desalina-tion to alleviate agricultural water shortages?

For Further Thought

1. Where does your water come from? What is its qual-ity? How is it treated, if at all, before it is used? Is a long-term shortage likely in your area? If so, what plans are being made to avert it?

2. Compare figures 10.18 and 10.19. Speculate on the reasons for some of the extremes of water withdrawal shown in figure 10.19, and consider the likelihood that this withdrawal is predominantly *consumption* in the highest-use states.

3. Some activities are commonly cited as great water-wasters. One is letting the tap run while brushing one's teeth. Try plugging the drain, then doing this; next, measure the volume of water accumulated during that single toothbrushing (a measuring cup and bucket may be helpful). Consider the implications for water use, bearing in mind the 255 million persons in the U.S. Read your home water meter before and after watering the lawn or washing the car, and make simi-lar projections.

4. Select an area, or country outside the United States, and investigate current precipitation and water use. See NetNotes for potential Internet sites at which to start.

Suggested Readings/References

Ballard, S. C., Devine, M. D., et al. 1982. *Water and Western energy: Impacts, issues, and choices.* Boulder, Colo.: Westview Press.

Berghinz, D. 1982. Venice is sinking into the sea. *Civil Engineering* 41, 67–71.

Dolan, R., and Goodell, H. G. 1986. Sinking cities. *American Scientist* 74 (January–February): 38–47.

Dunne, T., and Leopold, L. B. 1978. *Water in environmental planning.* San Francisco: W. H. Freeman.

Estimated use of water in the United States in 1990. 1996. U.S. Geological Survey National Circular 1081. [Online] http://h2o.usgs.gov/public/watuse/wucircular2.html

Goodman, A. S. 1984. *Principles of water resources planning.* Englewood Cliffs, N.J.: Prentice-Hall.

Graves, W., ed. 1993. *Water.* Washington, D.C.: National Geographic Society.

Griggs, G. B., and Gilchrist, J. A. 1983. *Geologic hazards, resources, and environmental planning.* Belmont, Calif.: Wadsworth.

Helweg, O. J. 1985. *Water resources planning and management.* New York: John Wiley & Sons.

Heppenheimer, T. A. 1991. The man who made Los Angeles possible. *Invention and Technology* (Spring/Summer): 11–18.

Howe, C. W., and Easter, K. W. 1971. *Interbasin transfers of water, economic issues and impacts.* Baltimore, Md.: Johns Hopkins University Press.

Mather, J. R. 1984. *Water resources.* New York: John Wiley & Sons.

Postel, S. L., Daily, G. C., and Ehrlich, P. R. 1996. Human appropriation of renewable fresh water. *Science* 271 (9 February): 785–88.

Sasman, R. T., et al. 1973. *Water level decline and pumpage in deep wells in the Chicago region, 1966–1971.* Illinois State Water Survey Circular 113.

Seckler, D., ed. 1971. *California water, a study in resource management.* Berkeley, Calif.: University of California Press.

Sheridan, D. 1981. *Desertification of the United States.* Washington, D.C.: Council on Environmental Quality.

Smith, Z. A. 1988. *Groundwater in the west.* New York: Academic Press.

U.S. Water Resources Council. 1978. *The nation's water resources 1975–2000.* Washington, D.C.: U.S. Government Printing Office.

Ward, C. H., Giger, W., and McCarty, P. L., eds. 1985. *Groundwater quality.* New York: John Wiley & Sons.

World Resources Institute. 1994. *World resources 1994–95* (Chapter 5, "Water"). New York: Oxford University Press.

World Resources Institute, United Nations Environment Programme, United Nations Development Programme, and The World Bank. 1996. *World resources 1996–97* (Chapter 13, "Water and fisheries"). New York: Oxford University Press.

Zwingle, E. 1993. Ogallala aquifer: Wellspring of the High Plains. *National Geographic* 184 (March): 80–109.

The U.S Geological Survey Water Resources Division home page is a gateway to a variety of water-resource information, at
http://h2o.usgs.gov/
"National Water Conditions"/current streamflow data for the U.S. and southern Canada, provided by the USGS in collaboration with the Climate Information Branch of Environment Canada, is found at
http://h2o.usgs.gov/nwc/
The USGS has a National Water-Use Information Program offering information to the public; its home page is at
http://h2o.usgs.gov/public/watuse/wufactsheet.html

Information on recent developments in the Ogallala (High Plains) aquifer system can be found at
http://h2o.usgs.gov/public/wid/FS_215-95/ FS_215-95.html

The World Resources Institute's water-resource information can be accessed via
http://www.wri.org/wri/enved/trends/wtr-home.html

The Climate Prediction Center of the National Centers for Environmental Prediction engages in short-term climate predictions, which involves considering historic climate trends. Historic precipitation data, U.S. and foreign, can be accessed via links at
http://nic.fb4.noaa.gov/

NASA, in collaboration with the National Space Development Agency of Japan, is developing better methods of monitoring rainfall in tropical and subtropical regions, which has both streamflow and water-resource implications. The Tropical Rainfall Measuring Mission site is at
http://trmm.gsfc.nasa.gov/trmm.office/trmm_overview.html

Chapter 11

Soil as a Resource

oil does not, at first glance, strike many people as a resource requiring special care for its preservation. In most places, even where soil erosion is active, a substantial quantity seems to remain underfoot. Associated problems, such as loss of soil fertility and sediment pollution of surface waters, are even less obvious to the untutored eye and may be too subtle to be noticed readily. Nevertheless, soil is an essential resource on which we depend for the production of the major portion of our food. Soils vary in their suitability not only for agriculture but also for construction and other purposes. Unfortunately, soil erosion is a significant and expensive problem in an increasing number of places as human activities disturb more and more land. In this chapter, we will examine the nature of soil, its formation and properties, aspects of the soil erosion problem, and some strategies for reducing erosion.

Soybeans grown using minimum-tillage methods to minimize soil erosion, Lake County, Illinois. © Doug Sherman/Geofile

Soil Formation

Soil is defined in different ways for different purposes. Engineering geologists define soil very broadly to include all unconsolidated material overlying bedrock. Soil scientists restrict the term *soil* to those materials capable of supporting plant growth, and distinguish it from *regolith*, which encompasses all unconsolidated material at the surface, fertile or not. Conventionally, the term *soil* implies little transportation away from the site at which the soil formed, while the term *sediment* indicates matter that has been transported and redeposited by wind, water, or ice.

Soil is produced by *weathering*, a term that encompasses a variety of chemical, physical, and biological processes acting to break down rocks. It may be formed directly from bedrock, or from further breakdown of transported sediment such as glacial till. The relative importance of the different kinds of weathering processes is largely determined by climate. Climate, topography, the composition of the material from which the soil is formed, and time govern a soil's final composition.

Soil-Forming Processes: Weathering

Mechanical weathering is the physical breakup of rocks without changes in the rocks' composition. In a cold climate, with temperatures that fluctuate above and below freezing, water in cracks repeatedly freezes and expands, forcing rocks apart, then thaws and contracts or flows away. Crystallizing salts in cracks may have the same wedging effect. Whatever the cause, the principal effect of mechanical weathering is the breakup of large chunks of rock into smaller ones. In the process, the total exposed surface area of the particles is increased (figure 11.1).

Chemical weathering involves the breakdown of minerals by chemical reaction with water, with other chemicals dissolved in water, or with gases in the air. Minerals differ in the kinds of chemical reactions they undergo. Calcite (calcium carbonate) tends to dissolve completely, leaving no other minerals behind in its place. Calcite dissolves rather slowly in plain water but more rapidly in acidic water. Many natural waters are slightly acidic; acid rainfall (see chapters 13 and 17) or acid runoff from coal strip mines (chapter 13) is more so and causes more rapid dissolution. This is, in fact, becoming a serious problem where limestone and its metamorphic equivalent—marble—are widely used for outdoor sculptures and building stone. Calcite dissolution is gradually destroying delicate sculptural features and eating away at the very fabric of many buildings in urban areas and where acid rain is common.

Silicates tend to be somewhat less susceptible to chemical weathering and leave other minerals behind when they are attacked. Feldspars principally weather into clay minerals. Ferromagnesian silicates leave behind insoluble iron oxides and hydroxides and some clays, with other chemical components being dissolved away. Those residual iron compounds are responsible for the reddish or yellowish colors of

6 sides 1 cube

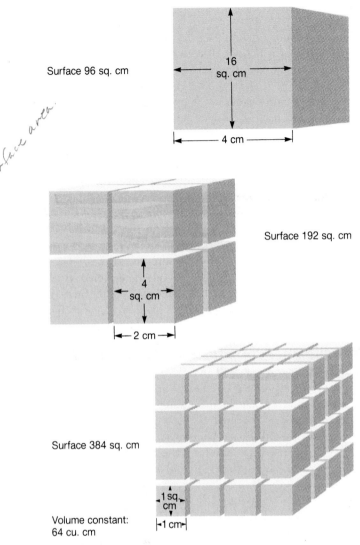

Surface 96 sq. cm

16 sq. cm

4 cm

surface area

Surface 192 sq. cm

4 sq. cm

2 cm

Surface 384 sq. cm

1 sq. cm

1 cm

Volume constant: 64 cu. cm

Figure 11.1 Mechanical breakup increases surface area and surface-to-surface volume ratio.

Table 11.1 *Some Chemical Weathering Reactions*

Solution of calcite (no solid residue)
$CaCO_3 + 2\ H^+ = Ca^{2+} + H_2O + CO_2$ (gas)

Breakdown of ferromagnesians (possible mineral residues include iron compounds and clays)
$FeMgSiO_4$ (olivine) $+ 2\ H^+ = Mg^{2+} + Fe(OH)_2 + SiO_2^*$
$2\ KMg_2FeAlSi_3O_{10}(OH)_2$ (biotite) $+ 10\ H^+ + 1/2\ O_2$ (gas)
$= 2\ Fe(OH)_3 + Al_2Si_2O_5(OH)_4$ (kaolinite, a clay) $+$
$4\ SiO_2^* + 2\ K^+ + 4\ Mg^{2+} + 2\ H_2O$

Breakdown of feldspar (clay is the common residue)
$2\ NaAlSi_3O_8$ (sodium feldspar) $+ 2\ H^+ + H_2O$
$= Al_2Si_2O_5(OH)_4 + 4\ SiO_2^* + 2\ Na^+$

Solution of pyrite (making dissolved sulfuric acid, H_2SO_4)
$2\ FeS_2 + 5\ H_2O + \frac{15}{2}\ O_2$ (gas) $= 4\ H_2SO_4 + Fe_2O_3 \cdot H_2O$

Notes: Hundreds of possible reactions could be written; the above are only examples of the kinds of processes involved.

All ions (charged species) are dissolved in solution; all other substances, except water, are solid unless specified otherwise.

Commonly, the source of the H^+ ions for solution of calcite and weathering of silicates is carbonic acid, H_2CO_3, formed by solution of atmospheric CO_2.

*Silica is commonly removed in solution.

many soils. In most climates, quartz is extremely resistant to chemical weathering, dissolving only slightly. Representative weathering reactions are shown in table 11.1.

The susceptibility of many silicates to chemical weathering can be inferred from the conditions under which the silicates formed. Given several silicates that have crystallized from the same magma, those that formed at the highest temperatures tend to be the least stable, or most easily weathered, at the low temperatures of soil formation at the earth's surface, and vice versa. A rock's tendency to weather chemically is determined by its mineral composition. For example, a gabbro (the coarsely crystalline equivalent of basalt), formed at high temperatures and rich in ferromagnesian minerals, generally weathers more readily than a granite rich in quartz and low-temperature feldspars.

Climate plays a major role in the intensity of chemical weathering. Most of the relevant chemical reactions involve water. All else being equal, then, the more water, the more chemical weathering. Also, most chemical reactions proceed more rapidly at high temperatures than at low ones. Therefore, warm climates are more conducive to chemical weathering than cold ones.

The rates of chemical and mechanical weathering are interrelated. Chemical weathering may speed up the mechanical breakup of rocks if the minerals being dissolved are holding the rock together by cementing the mineral grains, as in some sedimentary rocks. Increased mechanical weathering may, in turn, accelerate chemical weathering through the increase in exposed surface area, because it is only at grain surfaces that minerals, air, and water interact. The higher the ratio of surface area to volume—that is, the smaller the particles—the more rapid the chemical weathering. The vulnerability of surfaces to attack is also shown by the tendency of angular fragments to become rounded by weathering (figure 11.2).

As will be seen below, the organic component of soil is particularly important to its fertility, and influences other physical properties as well. Biological processes are also important to the formation of soil. Biological weathering effects can be either mechanical or chemical. Among the mechanical effects is the action of tree roots in working into cracks to split rocks apart (figure 11.3). Chemically, many organisms produce compounds that may react with and dissolve or break down minerals. Plants, animals, and microorganisms develop more abundantly and in greater variety in warm,

A

C

D

B

Figure 11.7 Though we may not think much about it, looking down from above, soils in cross section vary widely in color, texture, structure, and mineralogy; there are corresponding variations in suitability for construction and agriculture, ease of drainage, and other properties. (*A*) Soil is developed from loess; (*B*) shows a black, organic-matter-rich layer; (*C*) is a podsol, typical of soil found underlying forests in temperate climates; and (*D*) is a marshland soil.   Photos by Institute of Soil Science, Dept. of Agriculture, Göttingen, Germany

The highly leached character of lateritic soils is one reason. Even where the vegetation is dense, the soil itself has few soluble nutrients left in it. A forest holds a huge reserve of nutrients, but there is no corresponding reserve in the soil. Further growth is supported by the decay of earlier vegetation. As one plant dies and decomposes, the nutrients it contained are quickly taken up by other plants or leached away. If the forest is cleared to plant crops, most of the nutrients are cleared away with it, leaving little in the soil to nourish the crops. Many natives of tropical climates practice a slash-and-burn agriculture, cutting and burning the jungle to clear the land. Some of the nutrients in the burned vegetation settle into the topsoil temporarily, but relentless leaching by the warm rains makes the soil nutrient-poor and infertile within a few growing seasons. Nutrients could, in principle, be added through synthetic chemical fertilizers. However, many of the nations in regions of lateritic soil are among the poorer developing countries, and vast expenditures for fertilizer are simply impractical.

Even with fertilizers available, a second problem with lateritic soils remains. A clue to the problem is found in the term *laterite* itself, which is derived from the Latin for "brick." A lush rain forest shields lateritic soil from the drying and baking effects of the sun, while vigorous root action helps to keep the

Figure 11.8 Lateritic soil, like this in North Queensland, Australia, is characteristic of tropical climates. Bright red color is produced by residual iron oxides in soil.

soil well broken up. When its vegetative cover is cleared and it is exposed to the baking tropical sun, lateritic soil can quickly harden to a solid, bricklike consistency that resists infiltration by water or penetration by crops' roots. What crops can be grown provide little protection for the soil and do not slow the hardening process very much. Within five years or less, a freshly cleared field may become completely uncultivatable. The tendency of laterite to bake into brick has been used to advantage in some tropical regions, where blocks of hardened laterite have served as building materials. Still, the problem of how to maintain crops remains. Often, the only recourse is to abandon each field after a few years and clear a replacement. This results, over time, in the destruction of vast tracts of rain forest where only a moderate expanse of farmland is needed. Also, once the soil has hardened and efforts to farm it have been abandoned, it may revegetate only very slowly, if at all.

In Indochina, in what is now Cambodian jungle, are the remains of the Khmer civilization that flourished from the ninth to the sixteenth centuries. There is no clear evidence of why that civilization vanished. A major reason may have been the difficulty of agriculture in the region's lateritic soil. (It is clear that the phenomenon of laterite solidification was occurring at that time: Chunks of hardened laterite were used to construct temples like those at Angkor Wat.) Perhaps the Mayas, too, moved north into Mexico to escape the problems of lateritic soils. In Sierra Leone in west Africa, increased clearing of forests for firewood, followed by deterioration of the exposed soil, has reduced the estimated carrying capacity

of the land to 25 persons per square kilometer; the actual population is already over 38 persons per square kilometer. Part of the concern over deforestation associated with harvesting of hardwoods in South America relates to loss of habitat and possible extinction of unique species (loss of biodiversity), but in part it relates to the fact that what is left behind is barren lateritic soil. Today, some countries with lateritic soil achieve agricultural success only because frequent floods deposit fresh, nutrient-rich soil over the depleted laterite. The floods, however, cause enough problems that flood-control efforts are in progress or under serious consideration in many of these areas of Africa and Asia. Unfortunately, successful flood control could create an agricultural disaster for such regions.

Deterioration of lateritic soil is not the only form of soil degradation worldwide. Desertification, erosion, and contamination from pollution all contribute to reduced soil quality, fertility, and productivity. The problems are widespread (figure 11.9A); the contributing causes differ regionally (figure 11.9B).

Soil Erosion

Weathering is the breakdown of rock or mineral materials in place; *erosion* involves physical removal of material from one place to another. Soil erosion is caused by the action of water and wind. Rain striking the ground helps to break soil particles loose (figure 11.10). Surface runoff and wind together carry away loosened soil. The faster the wind and water travel, the larger the particles and the greater the load

A

B

Figure 11.2 (*A*) Angular fragments are rounded by weathering. Corners are attacked on three surfaces, edges on two. In time, fragments become rounded as a result. (*B*) Fractured Hawaiian basalt weathers into rounded chunks.

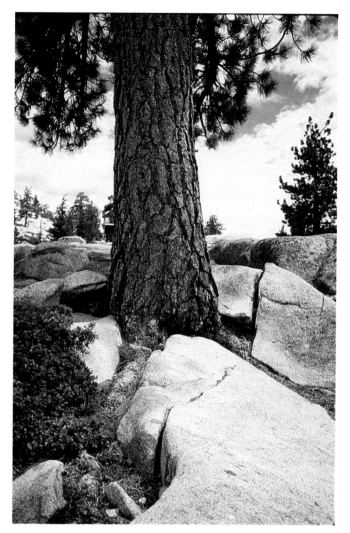

Figure 11.3 Tree roots working into cracks in this granite break up the rock mass.

wet climates. Mechanical weathering is generally the dominant process only in areas where climatic conditions have limited the impact of chemical weathering and biological effects—that is, in cold or dry areas.

Finally, airborne chemicals and sediments may *add* components to soil—e.g., sulfate from acid rain, salts from sea spray in coastal areas, clay minerals and other fine-grained minerals blown on the wind.

Soil Profiles, Soil Horizons

The result of mechanical, chemical, and biological weathering, together with the accumulation of decaying remains from organisms living on the land and any input from the atmosphere, is the formation of a blanket of soil between bedrock and atmosphere. A cross section of this soil blanket usually reveals a series of zones of different colors, compositions, and physical properties. The number of recognizable zones and the thickness of each vary. A basic, generalized soil profile as developed directly over bedrock is shown in

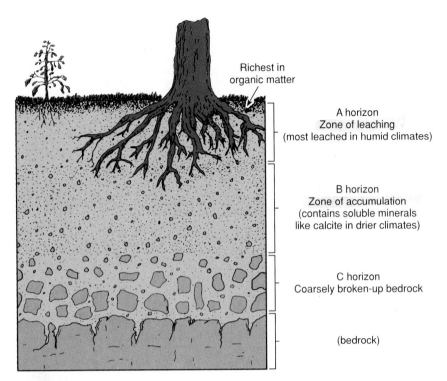

Richest in
organic matter

A horizon
Zone of leaching
(most leached in humid climates)

B horizon
Zone of accumulation
(contains soluble minerals
like calcite in drier climates)

C horizon
Coarsely broken-up bedrock

(bedrock)

Figure 11.4 A generalized soil profile. Individual horizons can vary in thickness. Some may be locally absent, or additional horizons or subhorizons may be identifiable.

figure 11.4. At the top is the **A horizon.** It consists of the most intensively weathered rock material, being the zone most exposed to surface processes. It also contains most of the organic remains. Unless the local water table is exceptionally high, precipitation infiltrates down through the A horizon. In so doing, the water may dissolve soluble minerals and carry them away with it. This process is known as **leaching,** and the A horizon is also called the **zone of leaching.**

Particularly in drier climates, many of the minerals leached from the A horizon accumulate in the layer below, the **B horizon,** also known as the **zone of accumulation** (or **zone of deposition**). Soil in the B horizon has been somewhat protected from surface processes. Organic matter from the surface has also been less well mixed into the B horizon. Below the B horizon is a zone consisting principally of very coarsely broken-up bedrock and little else. This is the **C horizon,** which does not resemble our usual idea of soil at all. Sometimes, the bedrock or parent material itself is called the R horizon. Similar zonation—though without bedrock at the base of the soil profile—is found in soils developed on transported sediment.

The boundaries between adjacent soil horizons may be sharp or indistinct. In some instances, one horizon may be divided into several recognizable subhorizons—for example, the top of the A horizon might consist of a layer of topsoil particularly rich in organic matter. This layer may be so

organic-matter-rich that a distinct "O horizon" can be designated. Subhorizons also may exist that are gradational between A and B or B and C. It is also possible for one or more horizons locally to be absent from the soil profile. Some of the diversity of soil types is illustrated in the next section.

All variations in the soil profile arise from the different mix of soil-forming processes and starting materials found from place to place. The overall total thickness of soil is partly a function of the local rate of soil formation and partly a function of the rate of soil erosion. The latter reflects the work of wind and water, the topography, and often the extent and kinds of human activities.

Chemical and Physical Properties of Soils

Mechanical weathering merely breaks up rock without changing its composition. In the absence of pollution, wind and rain-water rarely add many chemicals, and runoff water may carry away some leached chemicals in solution. Thus, chemical weathering tends to involve a net subtraction of elements from rock or soil. One control on a soil's composition, then, is the composition of the material from which it is formed. If the bedrock or parent sediment is low in certain critical plant nutrients, the soil produced from it will also be low in those nutrients, and chemical fertilizers may be needed to grow particular crops on that soil even if the soil has never been farmed before. The extent and balance of weathering processes involved in soil formation then determine the extent of further depletion of the soil in various elements. The weathering processes also influence the mineralogy of the soil, the compounds in which its elements occur. The physical properties of the soil are affected by its mineralogy, the texture of the mineral grains (coarse or fine, well or poorly sorted, rounded or angular, and so on), and any organic matter present.

Color, Texture, and Structure of Soils

Soil *color* tends to reflect compositional characteristics. Soils rich in organic matter tend to be black or brown, while those poor in organic matter are paler in color, often white or gray. When iron is present and has been oxidized by reaction with oxygen in air or water, it adds a yellow or red color (figure 11.5). (Rust on iron or steel is also produced by oxidation of iron.)

Soil *texture* is related to the sizes of fragments in the soil. The U.S. Department of Agriculture recognizes three size components: sand (grain diameters 2–0.05 mm), silt

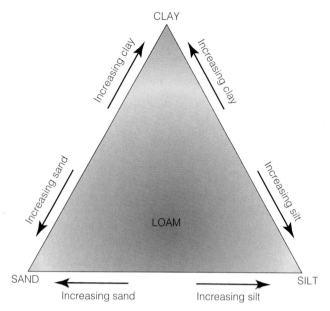

Figure 11.6 Soil-texture terminology reflects particle size, not mineralogy. This diagram reflects simplified U.S. Department of Agriculture classification scheme.

drain, by virtue of their much lower permeability, a characteristic shared by their lithified counterpart, shale.

Soil *structure* relates to the soil's tendency to form lumps or clods of soil particles. These clumps are technically called *peds*, their name deriving from the Latin root *pedo-*, meaning "soil." A soil that clumps readily may be more resistant to erosion; some of the fine-grained soils developed on fine loess in the U.S. and in China may erode very readily. On the other hand, soil consisting of very large peds with large cracks between them may be a poor growing medium for small plants with fine roots. Abundant organic matter may promote the aggregation of soil particles into crumblike peds especially conducive to good plant growth; mechanical weathering can break up larger clumps into smaller, just as it breaks up rock fragments.

Soil Classification

Early attempts at soil classification emphasized compositional differences among soils and thus principally reflected the effects of chemical weathering. The resultant classification into two broad categories of soil was basically a climatic one. The **pedalfer** soils were seen as characteristic of more humid regions. Where the climate is wetter, there is naturally more extensive leaching of the soil, and what remains is enriched in the less-soluble oxides and hydroxides of aluminum and iron, with clays accumulated in the B horizon. The term *pedalfer* comes from the prefix *pedo-* and the Latin words for aluminum (*alumium*) and iron (*ferrum*). In North America, pedalfer-type soils are found in higher-rainfall areas like the eastern United States and most of Canada. Pedalfer soils are typically acidic. Where the climate is drier,

Figure 11.5 Extensive weathering of iron-rich Hawaiian basalt, enriched with decaying vegetation, produces red-brown soil.

(0.05–0.002 mm), and clay (less than .002 mm). Soils are named on the basis of the dominant grain size(s) present (figure 11.6). An additional term, *loam*, describes a soil that is a mixture of all three particle sizes in similar proportions (10–30 percent clay, the balance nearly equal amounts of sand and silt). Soils of intermediate particle-size distribution have mixed names: for example, "silty clay" would be a soil consisting of about half silt-sized particles, half clay-sized particles. The significance of soil texture is primarily the way it influences drainage. Sandy soils, with high permeability (like many sandstone aquifers), drain quickly. In an agricultural setting, this may mean a need for frequent watering or irrigation. In a flood-control context, it means relatively rapid infiltration. Clay-rich soils, by contrast, may hold a great deal of water but be relatively slow to

Table 11.2 Soil Orders of the Seventh Approximation, with General Descriptions of Terms

1.	Entisols	soils without layering, except perhaps a plowed layer
2.	Vertisols	soils with upper layers mixed or inverted because they contain expandable clays (clays that swell when wet, contract and crack when dry)
3.	Inceptisols	very young soils with weakly developed soil layers, and not much leaching or mineral alteration
4.	Aridosols	soils of deserts and semiarid regions, and related saline or alkaline soils
5.	Mollisols	grassland soils, mostly rich in calcium; also, forest soils developed on calcium-rich parent materials; characterized by a thick surface layer rich in organic material
6.	Spodosols	soils with a light, ashy gray A horizon, and a B horizon containing organic matter and clay leached from the A horizon
7.	Alfisols	include most other acid soils with clay-enriched subsoils
8.	Ultisols	similar to alfisols but with weathering more advanced; include some lateritic soils
9.	Oxisols	still more weathered than the ultisols; includes most laterites
10.	Histosols	bog-type soils

Source: Data from *Geology of Soils: Their Evolution, Classification and Uses,* by C. B. Hunt, p. 181. Copyright © 1972 W. H. Freeman and Company.

such as in the western and especially the southwestern United States, leaching is much less extensive. Even quite soluble compounds like calcium carbonate remain in the soil, especially in the B horizon. From this observation came the term **pedocal** for the soil of a dry climate. The presence of calcium carbonate makes pedocal soils more alkaline.

One problem with this simple classification scheme is that, to be strictly applied, the soils it describes must have formed over suitable bedrock. For example, a rock poor in iron and aluminum, such as a pure limestone or quartz sandstone, does not leave an iron- and aluminum-rich residue, no matter how extensively leached or intensely weathered. Still, the terms *pedalfer* and *pedocal* can be used generally to indicate, respectively, more and less extensively leached soils.

Modern soil classification has become considerably more sophisticated and complex. Various schemes may take into account characteristics of present soil composition and texture, the type of bedrock on which the soil was formed, the present climate of the region, the degree of "maturity" of the soil, and the extent of development of the different soil horizons. Different countries have adopted different schemes. The new United Nations Education, Scientific, and Cultural Organization (UNESCO) world map uses 110 different soil map units. This is, in fact, a small number compared to the number of distinctions made under other classification schemes.

The U.S. comprehensive soil classification, known as the Seventh Approximation, has ten major categories (orders), which are subdivided through five more levels of classification into a total of some twelve thousand soil series. Even soil scientists must use reference books to keep all of the distinctions straight. A short summary of the ten orders is presented for reference in table 11.2. Some of the orders are characterized by a particular environment of formation and

the distinctive soil properties resulting from it—for example, the Histosols, which are bog soils. Others are characterized principally by physical properties—for example, the Entisols, which lack horizon zonation, and the Vertisols, in which the upper layers are mixed because these soils contain expansive clays. The practical significance of the Vertisols lies in the considerable engineering problems posed by expansive clays (see chapter 19). Most of the Oxisols and some Ultisols, on the other hand, are soils of a type that has serious implications for agriculture, especially in much of the Third World: lateritic soils, which will be discussed below. Figure 11.7 illustrates some of the variety possible among soils.

Soils and Human Activities

Soils and their characteristics are not only critical to agricultural activities but their properties influence the stability of construction projects. In the area of soil erosion, human activities may aggravate or moderate the problems—and sometimes may in turn be influenced by the erosion. This section samples several relationships between soil and human activities.

Lateritic Soil

Lateritic soil is common to many less-developed nations and poses special agricultural challenges. A **laterite** may be regarded as an extreme kind of pedalfer. Lateritic soils develop in tropical climates with high temperatures and heavy rainfall, so they are severely leached. Even quartz may have been dissolved out of the soil under these conditions. Lateritic soil (figure 11.8) may contain very little besides the insoluble aluminum and iron compounds. Soils of the lush tropical rain forests are commonly lateritic, which seems to suggest that lateritic soils have great farmland potential. Surprisingly, however, the opposite is true, for two reasons.

Figure 11.9 Soil degradation of various kinds is a global problem. (*A*) Six continents are significantly affected; on four (Africa, Australia, Europe, and South America), little land is considered stably vegetated. (*B*) The causes of soil degradation, however, vary around the world.   Source: World Resources Institute.

Figure 11.10 The impact of a single raindrop loosens many soil particles.   Photograph courtesy of USDA Soil Conservation Service.

they move. Therefore, high winds cause more erosion than calmer ones, and fast-flowing surface runoff moves more soil than slow runoff. This in turn suggests that, all else being equal, steep and unobstructed slopes are more susceptible to erosion by water, for surface runoff flows more rapidly over them. Flat, exposed land is correspondingly more vulnerable to wind erosion. The physical properties of the soil also influence its vulnerability to erosion.

Rates of soil erosion can be estimated in a variety of ways. Over a large area, erosion due to surface runoff may be judged by estimating the sediment load of streams draining the area. On small plots, runoff can be collected and its sediment load measured. Because wind is harder to monitor comprehensively, especially over a range of altitudes, the extent of wind erosion is more difficult to estimate. Generally, it is much less significant than erosion through surface-water runoff, except under drought conditions (for example, during the Dust Bowl era; see box 11.1). Controlled laboratory experiments can be used to simulate wind and water erosion and measure their impact.

If all the soil in an area is lost, farming clearly becomes impossible. Even while some remains, however, the present rapid erosion rates are cause for concern. The topsoil, with its higher content of organic matter and nutrients, is especially fertile and suitable for agriculture, and it is the topsoil that is lost first as the soil erodes. Estimates indicate that conserving the nutrients now being lost could save U.S. farmers $20 billion annually in fertilizer costs. In addition, the organic-

matter-rich topsoil usually has the best structure for agriculture—it is more permeable, more readily infiltrated by water, and retains moisture better than deeper soil layers.

Soil erosion from cropland leads to reduced crop quality and reduced agricultural income. Six inches of topsoil loss in western Tennessee has reduced corn yields by 42 percent, and such relationships are not uncommon. Even when the nutrients required for adequate crop growth are added through fertilizers, other chemicals that contribute to the nutritional quality of the food grown may be lacking; in other words, the food itself may be less healthful. Also, the soil eroded from one place is deposited, sooner or later, somewhere else. If a large quantity is moved to other farmland, the crops on which it is deposited may be stunted or destroyed, although small additions of fresh topsoil may enrich cropland and make it more productive.

A subtle consequence of soil erosion in some places has been increased persistence of toxic residues of herbicides and pesticides in the soil. The loss of nutrients and organic matter through topsoil erosion may decrease the activity of soil microorganisms that normally speed the breakdown of many toxic agricultural chemicals. Many of these chemicals, which contribute significantly to water pollution, also pollute soils.

Another major soil erosion problem is sediment pollution. In the United States, about 750 million tons per year of eroded sediment end up in lakes and streams. This decreases

Box 11.1 The Dust Bowl Dust Storms—A One-Time Event?

The Dust Bowl area proper, although never exactly defined, comprised close to 100 million acres of southeastern Colorado, northeastern New Mexico, western Kansas, and the Texas and Oklahoma panhandles. The farming crisis there during the 1930s resulted from an unfortunate combination of factors: clearing or close grazing of natural vegetation, drought (rainfall less than 50 centimeters [20 inches] per year for several years), sustained winds (averaging more than 15 kilometers/hour [10 miles/hour], with velocities ranging much higher during storms), and poor farming practices, including widespread disregard of wind erosion as a potential problem. There had been droughts in the area previously. But in the late 1800s and the early decades of this century, mechanization of farming made possible the rapid expansion of cultivated acreage, from about 12 million acres in 1879 to over 100 million acres by 1929, which, in turn, greatly increased the size of the area threatened by adverse conditions.

The action of the wind was most dramatically illustrated during the fierce dust storms that began in 1932. The storms were described as "black blizzards" that blotted out the sun (figure 1). Black rain fell in New York, black snow in Vermont, as windblown dust moved eastward. In May 1934, one thirty-six-hour storm whipped up a dust cloud more than 2,000 kilometers (1,250 miles) long. People choked on the dust, some dying of suffocation or of a "dust pneumonia" similar to the silicosis miners develop from breathing rock dust.

As the dust from Kansas and Oklahoma settled on their desks, politicians in Washington, D.C., realized that something had to be done. In April 1935, a permanent Soil Conservation Service was established to advise farmers on land use, drainage, erosion control, and other matters. (More recently, the Environmental Protection Agency has become involved, principally from the standpoint of reducing sediment pollution of water, and has begun establishing target levels of reduced soil loss toward which farmers should work.)

By the late 1930s, concerted efforts by individuals and state and federal government agencies to improve farming practices to reduce wind erosion—together with a return to more normal rainfall—had considerably reduced the problems. However, drought struck again in the 1950s, and tens of millions of cropland acres were damaged by wind erosion in 1954. In a dry period in the winter of 1965, high winds raised dust 10,000 meters (about 6 miles) into the air and carried some of it east as far as Pennsylvania. Millions of acres more were damaged in the mid-1970s. In recent years, the extent of wind damage has actually been limited somewhat by the widespread use of irrigation to maintain crops in dry areas and dry times. But, as was noted in chapter 10, some of the important sources of that irrigation water are rapidly being depleted. Future spells of combined drought and wind may yet produce more scenes like those in figures 1 and 2.

This is not a concern unique to the United States. Major dust storms have increased tenfold in frequency worldwide in less than two centuries. The increase in atmospheric dust is attributed primarily to deforestation and ever-expanding cultivation, which are exposing ever more soil to wind erosion. If greenhouse-effect heating becomes appreciable, some areas now marginally suitable for agriculture will become too dry, vegetative cover will die, and wind erosion will increase further. There will be more tragedies like the recent drought-aggravated famine in Ethiopia. Increased soil erosion by wind would not merely be inconvenient or unpleasant; it could have serious implications for soil fertility and, hence, future world food production.

Figure 1 A 1930s dust storm in the Dust Bowl.
Photograph courtesy of USDA Soil Conservation Service.

Figure 2 Barren fields buried under windblown soil.
Photograph courtesy of USDA Soil Conservation Service.

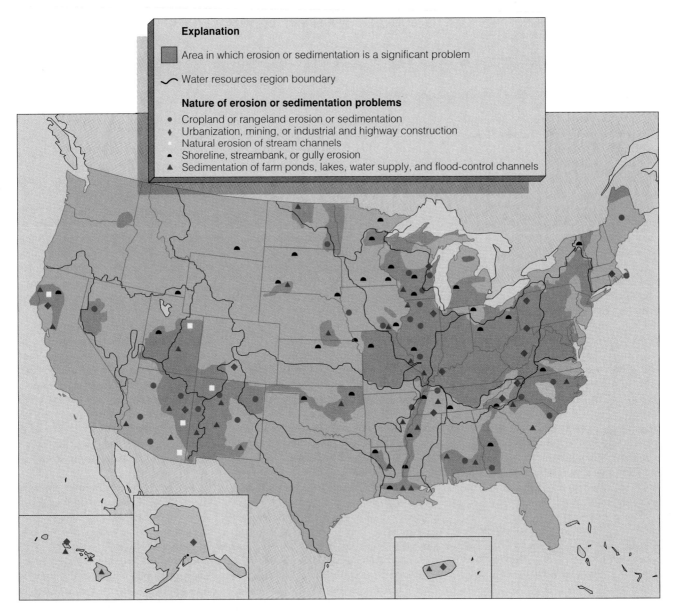

Figure 11.11 Erosion and associated sediment deposition in stream channels is a common problem in many parts of the United States. Source: After U.S. Water Resources Council, *The Nation's Water Resources 1975–2000*, Vol. 1, p. 71.

the water quality and may harm wildlife. The problem is still more acute when those sediments contain toxic chemical residues, as from agricultural herbicides and pesticides: The sediment is then both a physical and a potential chemical pollutant. A secondary consequence of this sediment load is the infilling of stream channels and reservoirs, restricting navigation and decreasing the volume of reservoirs and thus their usefulness for their intended purposes, whether for water supply, hydropower, or flood control. Tens of millions of dollars are spent each year to dredge and remove unwanted sediment deposits. The problems are widespread; see figure 11.11.

In recent years, there has been an increase in several kinds of activities that may increase soil erosion in places other than farms and cities. One is strip mining, which leaves behind readily erodable piles of soil and broken rock, at least until the land is reclaimed. Another is the use

of off-road recreational vehicles (ORVs). This last is a special problem in dry areas where vegetation is not very vigorous or easily reestablished. Fragile plants can easily be destroyed by a passing ORV, leaving the land bare and vulnerable to intensified erosion (figure 11.12).

Figure 11.13 shows that erosion during active urbanization (highway and building construction and so forth) is considerably more severe than erosion on any sort of undeveloped land. Erosion during highway construction in the Washington, D.C., area, for example, was measured at rates up to 75 tons/acre/year. However, the total U.S. area undergoing active development is only about 1.5 million acres each year, as compared to nearly 400 million devoted to cropland. Also, construction disturbance is of relatively short duration. Once the land is covered by buildings, pavement, or established landscaping, erosion rates typically drop to below

A

B

Figure 11.12 Erosion by off-road vehicles is not necessarily a new problem. (*A*) Fragile vegetation on sand dunes is easily disrupted by ORV traffic, Oregon Dunes National Recreation Area. (*B*) Oregon Trail Ruts were carved in soft sedimentary rocks by the passage of many westward-bound wagon trains.

even natural, predevelopment levels. (Whether this represents an optimum land use is another issue.) The problems of erosion during construction, therefore, may be very intensive but are typically localized and short-lived.

Soil Erosion versus Soil Formation

Estimates of the total amount of soil erosion in the United States vary widely, but estimates by the U.S. Department of Agriculture Natural Resources Conservation Service put the figure at several billion tons per year. To a great extent, human activities, such as construction and especially farming (figure 11.14), account for the magnitude of soil loss. On cropland, average erosion losses from wind and water together were estimated at 5.6 tons per acre per year. In very round numbers, that is about a 0.04 centimeter (0.02 inch) average thickness of soil removed each year. The rate of erosion from land under construction might be triple that (figure 11.14A). Erosion rates from other intensely disturbed lands, such as unreclaimed strip mines, might be at least as high as those for construction sites.

A

Figure 11.13 The impact of urbanization and other human activities on soil erosion rates. Construction may cause even more intensive erosion than farming, but far more land is affected by farming, and for a longer time. In this study of the Washington, D.C. area, erosion rates rose sharply during a highway construction boom around 1960; once construction was completed, erosion became negligible.   Reprinted from "A Cycle of Sedimentation and Erosion in Urban River Channels," M. G. Wolman from *Geografiska Annaler,* Vol. 49, series A, figure 1, by permission of Scandinavian University Press.

How this compares with the rate of soil *formation* is difficult to generalize because the rate of soil formation is so sensitive to climate, the nature of the parent rock material, and other factors. Upper limits on soil formation rates, however, can be determined by looking at soils in glaciated areas of the northern United States last scraped clean by the glaciers tens of thousands of years ago. Where the parent material was glacial till, which should weather more easily than solid rock, perhaps 1 meter (about 3 feet) of soil has formed in 15,000 years in the temperate, fairly humid climate of the upper Midwest. The corresponding average rate of 0.006 centimeters (0.002 inches) of soil formed per year is an order of magnitude less than the average cropland erosion rate. Furthermore, soil formation in many areas and over more resistant bedrocks is slower still. In some places in the Midwest, virtually no soil has formed on glaciated bedrocks, and in the drier southwest, soil formation is also likely to be very much slower. It seems clear that erosion is stripping away farmland far faster than the soil is being replaced.

Strategies for Reducing Erosion

The wide variety of approaches for reducing erosion on farmland basically involve either reducing the velocity of an eroding agent or protecting the soil from its effects. Under the latter heading come such practices as leaving stubble in the fields after a crop has been harvested and planting cover crops in the off-season between cash crops. In either case, the plants' roots help to hold the soil in place, and the plants themselves, to some extent, shield the soil from wind and rain.

The lower the wind or water runoff velocity, the less material carried. Wind can be slowed down by planting hedges or rows of trees as windbreaks along field borders in

B

Figure 11.14 Examples of water erosion of soil. (*A*) Severe erosion of steep bank exposed during construction. (*B*) Gullying caused by surface runoff over farmland.   Photograph (*A*) © The McGraw-Hill Companies, Inc./Bob Coyle, photographer; (*B*) courtesy of USDA Soil Conservation Service.

rows perpendicular to the dominant wind direction (figure 11.15A) or by erecting low fences, like snow fences, similarly arrayed. This does not altogether stop soil movement, as shown by the ridges of soil that sometimes pile up along the windbreaks. However, it does reduce the distance over which soil is transported, and some of the soil caught along the windbreaks can be redistributed over the fields. Strip cropping (figure 11.15B), alternating crops of different heights, slows near-ground wind by making the land surface more irregular.

Surface runoff may be slowed on moderate slopes by contour plowing (figure 11.16A). Plowing rows parallel to the contours of the hill, perpendicular to the direction of water flow, creates a ridged land surface so that water does not rush downhill as readily. Other slopes may require terracing (figure 11.16B), whereby a single slope is terraced into a series of shallower slopes, or even steps that slant backward into the hill. Again, surface runoff makes its way down the slope more slowly, if at all, and therefore carries

A

B

Figure 11.15 Creating near-ground irregularities decreases wind velocity. (A) Windbreaks retard wind erosion by decreasing wind velocity. (B) Strip cropping also breaks up the surface topography, disrupting sweeping winds.   (A) Photograph © John D. Cunningham/Visuals Unlimited. (B) Photograph courtesy of USDA Soil Conservation Service.

far less sediment with it. Terracing has, in fact, been practiced since ancient times. Both terracing and contour plowing, by slowing surface runoff, increase infiltration and enhance water conservation as well as soil conservation. Increased water retention from contour plowing in Texas increased cotton yields by 25 percent. In places, terracing and contouring are used in conjunction (figure 11.16C).

A

B

C

Figure 11.16 Decreasing slope of land, or breaking up the slope, decreases erosion by surface runoff. (A) Contour plowing/planting impedes water flow. (B) Terracing creates "steps" of shallower slope from one long, steep slope. (C) Terraces, too, may be contoured to fit topography, as here in the Great Plains; notice the foreground particularly.   (A) and (B) photographs courtesy of USDA Soil Conservation Service.

Box 11.2 Soil Erosion, and Feeding a Hungry World

Where cropland soil erosion is seen as a threat to productivity as well as a sediment-pollution problem, one widely advocated strategy is minimum-tillage farming, in which the land is not plowed separately before planting or after harvest. Instead, in a single step, the land is plowed and planted and any needed fertilizers and pesticides applied. Minimum-tillage agriculture leaves more residues from past crops in the soil between plantings and does not lay the soil bare before planting as do conventional methods.

Associated with minimum tillage are several considerable advantages. Both wind and water erosion are reduced because of the stabilizing effects of past crop residues. The preserved plant residues improve infiltration and help to retain water in the soil. The one-pass planting method results in reduced labor and energy costs. And the extra plant material in the soil gives the soil a firmer, less muddy texture through the growing season and makes harvesting easier if the harvest season is wet.

Unfortunately, the method has significant disadvantages as well. More money—up to twice as much—may need to be spent on herbicides and pesticides, since the plant residues include weed seeds and also harbor insects. In addition, higher levels of these toxic agricultural chemicals may contribute to increased water pollution and lead to the development of resistant strains of weeds and insect pests (see chapter 16). Residues of the

A

Figure 1 (*A*) Circular pattern on farmland reflects irrigation by the circling piping systems used in the midwestern United States. (*B*) False-color Landsat photograph of the Imperial Valley of California; what appears as red in this image is actually green vegetation. The intensive agricultural development indicated by the checkerboard pattern of cultivated fields is largely the result of irrigation; land is dry and barren outside irrigated area. The U.S./Mexican border (lower right) is sharply defined as a result of differences in land use and agricultural practices in the two countries. Photograph (*B*) © NASA.

herbicides left in the soil may also damage subsequent crops. Leaving old plant stubble on fields in the spring keeps the soil both cooler and wetter longer, which can delay spring planting. Moreover, on some flat, poorly drained farmland, such as is found over much of Illinois, Indiana,

A major obstacle to erosion control on farmland, of course, is money. Many of the recommended measures—for example, terracing and planting cover crops—can be expensive, especially on a large scale. Even though the long-term benefits of reduced erosion are obvious, and may involve long-term savings or increased income, the effort may not be made if short-term costs seem too high. The benefits of a particular soil conservation measure may be counterbalanced by substantial drawbacks, such as reduced crop yields (see, for example, box 11.2). Also, if strict erosion-control standards are imposed selectively on farmers of one state (as by a state environmental agency, for instance), those farmers

B

Those who see this as a minor problem point out that, soil erosion or no, any associated productivity losses are apparently being more than adequately compensated. For the past three decades, world production of cereal crops has increased faster than population. It can be projected that in 100 years, there could be enough food for a stable world population of 10 billion persons.

These productivity gains, however, are neither risk- nor cost-free. Much of the increase is attributed to increased irrigation, with irrigated acreage worldwide tripling since 1950 (figure 1). Many of the irrigation schemes are "mining" ground water or making major alterations in the hydrology of large-scale drainage systems. New high-productivity crop varieties are being developed; but as farmers plant vast tracts to a single genetic strain, they leave themselves vulnerable should that strain prove particularly susceptible to some pest or disease. Energy use in agriculture worldwide has increased sixfold since 1950, while supplies of fuels commonly used in agriculture are dwindling. Fertilizer use has tripled in less than three decades, and pesticide use is also higher; this puts increased stress on the environment in the form of pollution, as will be seen in chapter 16.

The need to balance many costs and benefits is not unique to agricultural issues, but such issues do provide a good example of some of the complexities of cost-benefit analysis.

and Ohio, the increased moisture retention associated with minimum-tillage agriculture may actually reduce crop yields.

Similar dilemmas arise in connection with other aspects of the world food problem. Analysts actually differ in the extent to which they even perceive soil erosion as a threat to the world's food supply. Certainly as the world's population grows, the

amount of land potentially at risk from erosion grows also: from 1950 to 1985, the area planted to cereal crops worldwide (two-thirds of global cropland) increased by about 20 percent. Productivity declines related to soil erosion are well documented in the United States; while similar data for other nations are scant, one can anticipate similar problems globally.

are at a competitive disadvantage with respect to farmers elsewhere who are not faced with equivalent expenses. Even meeting the same standards does not cost all farmers everywhere equally. Increasingly, the government is becoming involved in cost-sharing for erosion-reduction programs so that the financial burden does not fall too heavily on individual farmers. Since the Dust Bowl era, over $20 billion in federal funds have been spent on soil conservation efforts.

Tangible progress has been made. The USDA Natural Resources Conservation Service reports that from 1982 to 1992, average soil erosion losses on cropland were reduced from 4.1 to 3.1 tons per acre per year for erosion

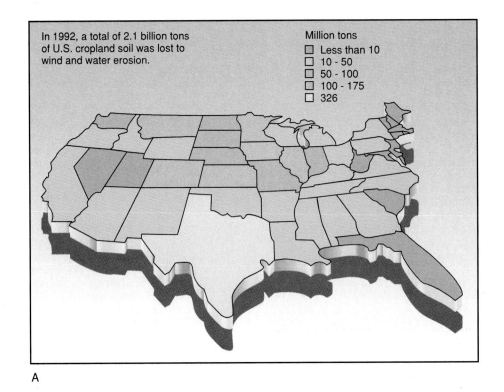

In 1992, a total of 2.1 billion tons of U.S. cropland soil was lost to wind and water erosion.

Million tons
☐ Less than 10
☐ 10 - 50
☐ 50 - 100
☐ 100 - 175
☐ 326

A

The top 5 states in soil loss on cropland, 1992:

By WIND

Texas	255
Minnesota	121
Montana	91
Colorado	90
Kansas	55

By WATER

Iowa	134
Illinois	104
Missouri	74
Texas	71
Nebraska	64

Million tons/year

B

Figure 11.17 Soil erosion in the U.S. has been reduced substantially, but remains a significant problem on cropland. It is particularly acute in a limited number of states (*A*); whether wind or water is the more serious threat varies by location (*B*). Source: *National Resources Inventory April 1995*, USDA Natural Resources Conservation Service.

loss by runoff water, and from 3.3 to 2.5 tons per acre per year for wind erosion. Still, that meant a total loss of 2.1 billion tons of soil from cropland in 1992 (figure 11.17A), 40 percent of it from just five states (Texas, Minnesota, Iowa, Montana, and Kansas); the losses are concentrated in states with significant agricultural activity, naturally enough, but local climate, soil types, and topography influence vulnerability to erosion, and which kind predominates (figure 11.17B). Clearly, there is room for further improvements.

Other strategies can minimize erosion in nonfarm areas. For example, in areas where vegetation is sparse, rainfall limited, and erosion a potential problem, ORVs should be restricted to prescribed trails. In the case of urban construction projects, one reason for the severity of the resulting erosion is that it is common practice to clear a whole area, such as an entire housing-project site, at the beginning of the work, even though only a small portion of the site is actively worked at any given time. Clearing the land in stages, as needed, minimizes the length of time the soil is exposed and reduces urban erosion. Mulch or temporary ground covers can protect soil that must be left exposed for longer periods. Stricter mining regulations requiring reclamation of strip-mined land (see chapter 12) are already significantly reducing soil erosion and related problems in these areas.

Where erosion cannot be eliminated, sediment pollution may nevertheless be reduced. "Soil fences" that act somewhat like snow fences, and hay-lined soil traps through which runoff water must flow, may reduce soil runoff from construction sites. On either construction sites or farmland, surface runoff water may be trapped and held in ponds to allow suspended sediment to settle out in the still ponds before the clarified water is released; such settling ponds are illustrated in chapter 16.

Overview

Summary

Soil forms from the breakdown of rock materials through mechanical, chemical, and biological weathering. Weathering rates are closely related to climate. The character of the soil reflects the nature of the parent material and the kinds and intensities of weathering processes that formed it. The wetter the climate, the more leached the soil. The lateritic soil of tropical climates is particularly unsuitable for agriculture: Not only is it highly leached of nutrients, but when exposed, it may harden to a bricklike solidity. The texture of soil is a major determinant of its drainage characteristics; soil structure is related to its suitability for agriculture.

Soil erosion by wind and water is a natural part of the rock cycle. Where accelerated by human activity, however, it can also be a serious problem, especially on farmland or, locally, in areas subject to construction or strip mining. Erosion rates far exceed inferred rates of soil formation in many places. A secondary problem is the resultant sediment pollution of lakes and streams. Strategies to reduce soil erosion on farmland include terracing, contour plowing, planting or erecting windbreaks, the use of cover crops, and minimum-tillage farming. Elsewhere, restriction of ORVs, more selective clearing of land during construction, use of sediment traps and settling ponds, and careful reclamation of strip-mined areas could all help to minimize soil erosion.

Terms to Remember

A horizon 254	pedalfer 255
B horizon 254	pedocal 256
chemical weathering 251	soil 251
C horizon 254	zone of
laterite 256	accumulation 254
leaching 254	zone of deposition 254
mechanical weathering 251	zone of leaching 254

Exercises

Questions for Review

1. Briefly explain how the rate of chemical weathering is related to (a) the amount of precipitation, (b) the temperature, and (c) the amount of mechanical weathering.
2. Sketch a generalized soil profile and indicate the A horizon, B horizon, C horizon, zone of leaching, and zone of accumulation. Is such a profile always present?
3. How do pedalfer and pedocal types of soil differ, and in what kind of climate is each more common?
4. The lateritic soil of the tropical jungle is poor soil for cultivation. Explain.

5. Soil erosion during active urbanization is far more rapid than it is on cultivated farmland, and yet the majority of soil-conservation efforts are concentrated on farmland. Why?
6. Cite and briefly describe three strategies for reducing cropland erosion.

For Further Thought

1. Choose any major agricultural region of the United States and investigate the severity of any soil erosion problems there. What efforts, if any, are being made to control that erosion, and what is the cost of those efforts?
2. Visit the site of an active construction project; examine it for evidence of erosion, and erosion-control activities.
3. How does your state fare with respect to soil erosion? What factors can you identify that contribute to the extent—or the lack—of soil erosion? (The National Resources Inventory of April 1995 may be a place to start, the Natural Resources Conservation Service or your local geological survey an additional resource.)

Suggested Readings/References

Beasley, R. P. 1971. *Erosion and sediment pollution control.* Ames, Iowa: Iowa State University Press.

Birkeland, P. W. 1984. *Soils and geomorphology.* New York: Oxford University Press.

Blaxter, Sir K. 1986. *People, food, and resources.* Cambridge, England: Cambridge University Press.

Bridges, E. M. 1978. *World soils.* 2d ed. Cambridge, England: Cambridge University Press.

Buol, S. W., Hole, F. D., and McCracken, R. J. 1980. *Soil genesis and classification.* 2d ed. Ames, Iowa: Iowa State University Press.

Crosson, P. R., and Brubaker, S. 1982. *Resource and environmental effects of U.S. agriculture.* Washington, D.C.: Resources for the Future.

Crosson, P. R., and Rosenberg, N. J. 1989. Strategies for Agriculture. *Scientific American* 261 (September): 128–35.

Harlin, J. M., and Berardi, G. M. 1987. *Agricultural soil loss: Processes, policies, and prospects.* Boulder, Colo.: Westview Press.

Hunt, C. B. 1972. *Geology of soils: Their evolution, classification and uses.* San Francisco: W. H. Freeman.

Lal, R. 1987. Managing the soils of sub-Saharan Africa. *Science* 236: 1069–76.

Morgan, R. P. C. 1979. *Soil erosion.* London: Longman Group.

Pimentel, D., et al. 1995. Environmental and economic costs of soil erosion and conservation benefits. *Science* (24 February): 1117–23.

Sposito, G. 1989. *The chemistry of soils.* New York: Oxford University Press.

Thompson, L. M., and Troeh, F. R. 1978. *Soils and soil fertility.* 4th ed. New York: McGraw-Hill.

Troeh, F. R., Hobbs, J. A., and Donahue, R. L. 1980. *Soil and water conservation.* Englewood Cliffs, N.J.: Prentice-Hall.

U.S. Council on Environmental Quality. 1979. *Environmental quality.* 10th annual report. Washington, D.C.: U.S. Government Printing Office.

U.S. Department of Agriculture Natural Resources Conservation Service. 1995. *National resources inventory: Graphic highlights of natural resource trends in the United States between 1982 and 1992.* [Online] http://www.nhq.nrcs.usda.gov/nri_sumbook.html

World Resources Institute. 1994. *World Resources 1994–95* (Chapter 6, "Food and Agriculture.") New York: Oxford University Press.

World Resources Institute, United Nations Environment Programme, United Nations Development Programme, and The World Bank. 1996. *World Resources 1996–97* (Chapter 9, "Forests and Land Cover"; Chapter 10, "Food and Agriculture.") New York: Oxford University Press.

The U.S. Department of Agriculture Natural Resources Conservation Service (formerly USDA Soil Conservation Service) is located at

http://www.nhq.nrcs.usda.gov/

The Canadian Soil Information System, through which one can access *The Soil Landscapes of Canada* and illustrations from it, is found at

http://res.agr.ca/PUB/CANSIS/_overview.html

A wider selection of soil profiles from which those of figure 11.7 were taken can be viewed at a site maintained by the Institute of Soil Science (Germany). At the time of this writing, the captions were in German, but in many cases, the English equivalents are recognizable. See

http://vendigo.uni-soilsci.gwdg.de/soilidbr.html

Information about 20-year trends in soils and in the areas of food and agriculture is available at the World Resources Institute Environmental Education Project Web site

http://www.wri.org/wri/enved/trends.html

The USDA/Agricultural Research Service National Soil Erosion Lab site, which includes links to a number of other USDA-ARS research labs (e.g., Great Plains; Sedimentation; Soil Dynamics; Wind Erosion) is at

http://soils.ecn.purdue.edu:20002/~wepp/nserl.html

For a space soil mystery, consider the early Martian-lander experiments, which showed oxidizing actions in soil samples, originally interpreted as evidence of life; see "The Mars soil oxidant: A science puzzle" at

http://pokey.arc.nasa.gov/division/rt_reports/freund.html

For the official descriptions of the soil series of the Seventh Approximation

http://www.statlab.iastate.edu/soils/osd

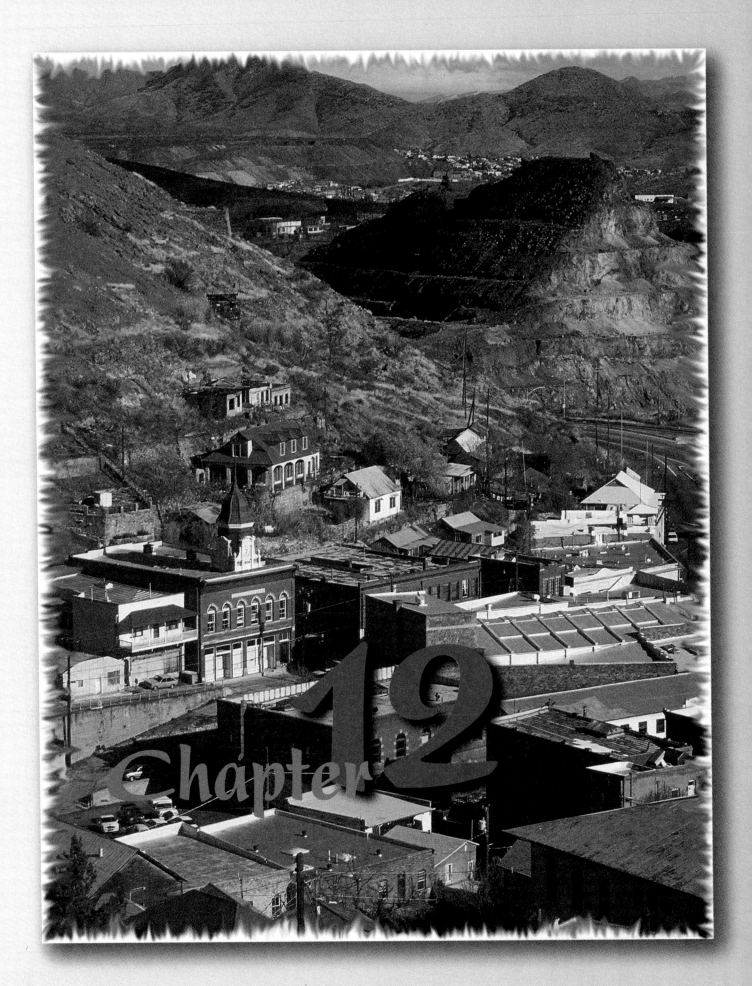

Chapter 12

Mineral Resources

The bulk of the earth's crust is composed of fewer than a dozen elements. In fact, eight chemical elements make up more than 98 percent of the crust. Many of the elements *not* found in abundance in the earth's crust, including industrial and precious metals, essential components of chemical fertilizers, and elements like uranium that serve as energy sources, are vitally important to society. Some of these are found in very minute amounts in the average rock of the continental crust: copper, 0.006 percent; tin, 2 ppm; gold, 4 ppb. Clearly, then, many useful elements must be mined from very atypical rocks. This chapter begins with a look at the occurrences of a variety of rock and mineral resources and then examines the U.S. and world supply-and-demand picture.

Historically, mining towns evolved right next to the mines, which often dominated the landscape; copper mining, Bisbee, Arizona. © Renee Lynn/Photo Researchers, Inc.

Ore Deposits

Definition

An **ore** is a rock in which a valuable or useful metal occurs at a concentration sufficiently high, relative to average rocks, to make it economically worth mining. A given ore deposit may be described in terms of the enrichment or **concentration factor** of a metal of interest:

Concentration factor of a given metal in an ore =

$$\frac{\text{Concentration of the metal in the ore deposit}}{\text{Concentration of the metal in average continental crust}}$$

The higher the concentration factor, the richer the ore, and the less of it needs to be mined to extract a given amount of the metal.

In general, the minimum concentration factor required for profitable mining is inversely proportional to the average crustal concentration: If ordinary rocks are already fairly rich in the metal, it need not be concentrated much further to make mining it economic, and vice versa. Metals like iron or aluminum, which make up about 6 percent and 8 percent of average continental crust, respectively, need only be concentrated by a factor of 4 to 5 times for mining to be profitable. Copper must be enriched about 100 times relative to average rock, while mercury (average concentration 80 ppb) must be enriched to about 25,000 times its average concentration before the ore is rich enough to mine profitably. The exceptions to this rule are those few extremely valuable elements, such as gold, that are so high-priced that a small quantity justifies a considerable amount of mining. At several hundred dollars an ounce, gold need only be found in concentrations a few thousand times its average 4 ppb to be worth mining. This inverse relationship between average concentration and the concentration factor required for profitable mining further suggests that economic ore deposits of the relatively abundant metals (such as iron and aluminum) might be more plentiful than economic deposits of rarer metals, and this is indeed commonly the case.

The value of the mineral or metal extracted and its concentration in a particular deposit, then, are major factors determining the profitability of mining a specific deposit. The economics are naturally sensitive to world demand. If demand climbs and prices rise in response, additional, not-so-rich ore deposits may be opened up; a fall in price causes economically marginal mines to close. The discovery of a new, rich deposit of a given ore may make other mines with poorer ores noncompetitive and thus uneconomic in the changing market. The practicality of mining a specific ore body may also depend on the mineral(s) in which a metal of interest is found, because this affects the cost to extract the pure metal. Three iron deposits containing equal concentrations of iron are not equally economic if the iron is mainly in

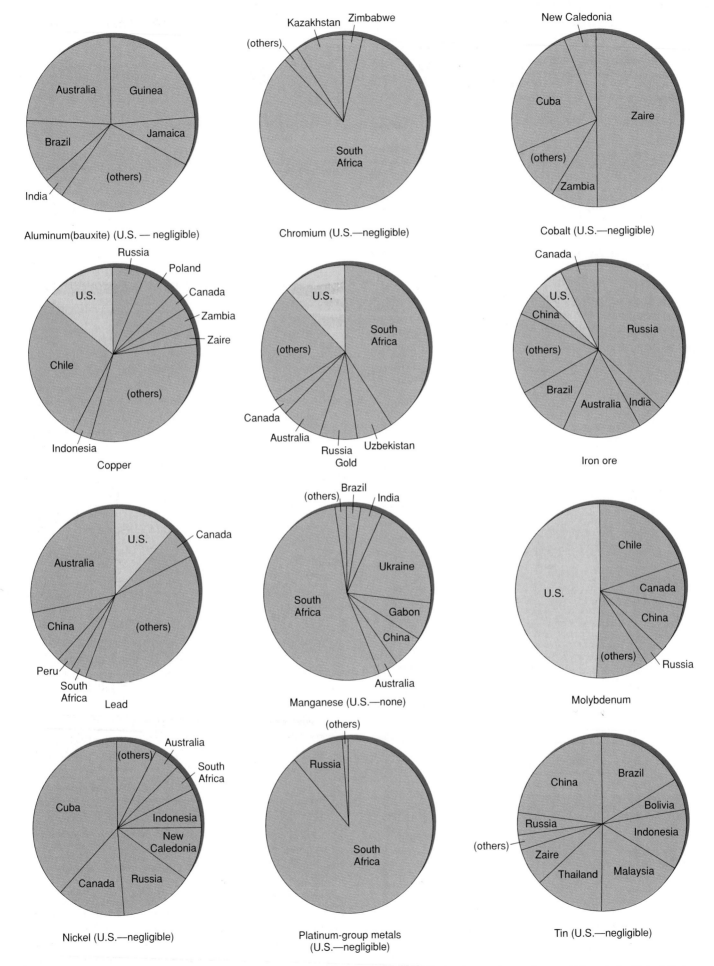

Figure 12.1 Proportions of world reserves of some nonfuel minerals controlled by various major producers. Although the United States is a major consumer of most metals, it is a major producer of very few. Source: Data from *Mineral Commodity Summaries 1996*, U.S. Bureau of Mines.

oxides in one, in silicates in another, and in sulfides in the third. The relative densities of the iron-bearing minerals, compared to other minerals in the rock, will affect how readily those iron-rich minerals can be separated from the rest; the nature and strengths of the chemical bonds in the iron-rich phases will influence the energy costs to break them down to extract the iron, and the complexity of chemical processes required; the concentration of iron in the iron-rich minerals (rather than in the rock as a whole) will determine the yield of iron from those minerals, after they are separated. These factors and others influence the economics of mining.

Distribution

By definition, ores are somewhat unusual rocks. Therefore, it is not surprising that the known economic mineral deposits are very unevenly distributed around the world. (See, for example, figure 12.1.) The United States controls about 50% of the world's known molybdenum deposits and about 15% of the lead. But although the United States is the major world consumer of aluminum, using about 30% of the total produced, it has virtually no domestic aluminum ore deposits; Australia and Guinea each control one-fourth of the world's aluminum ore. Zaire accounts for approximately half the known economic cobalt deposits, Cuba for nearly 40% of the nickel, Chile for almost 30% of the copper. South Africa controls nearly half the world's known gold ore, 75% of the chromium, and almost 90% of the platinum-group metals. These great disparities in mineral wealth among nations are relevant to later discussions of world supply/demand projections.

Types of Mineral Deposits

Deposits of economically valuable rocks and minerals form in a variety of ways. This section does not attempt to review them all but describes some of the more important processes involved.

Igneous Rocks and Magmatic Deposits

Magmatic activity gives rise to several different kinds of deposits. Certain igneous rocks, just by virtue of their compositions, contain high concentrations of useful silicate or other minerals. The deposits may be especially valuable if the rocks are coarse-grained, so that the mineral(s) of interest occur as large, easily separated crystals. **Pegmatite** is the term given to unusually coarse-grained igneous intrusions (figure 12.2). In some pegmatites, single crystals may be over 10 meters (30 feet) long. Feldspars, which provide raw materials for the ceramics industry, are common in pegmatites. Many pegmatites also are enriched in uncommon elements. Examples of rarer minerals mined from pegmatites include tourmaline (used both as a gemstone and for crystals in radio equipment) and beryl (mined for the metal beryllium when crystals are of poor quality, or used as the gemstones aquamarine and emerald when found as clear, well-colored crystals).

Figure 12.2 Pegmatite, a very coarse-grained plutonic rock, may yield large, valuable crystals.

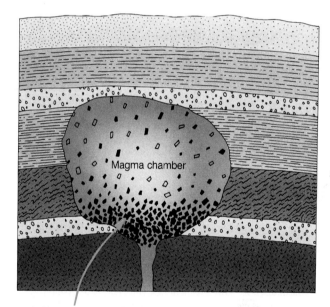

A dense mineral like chromite or magnetite may settle out of a crystallizing magma to be concentrated at the bottom of the chamber.

Figure 12.3 Formation of magmatic ore deposit by gravitational settling of a dense mineral during crystallization.

Other useful minerals may be concentrated within a cooling magma chamber by gravity. If they are more or less dense than the magma, they may sink or float as they crystallize, instead of remaining suspended in the solidifying silicate mush, and accumulate in thick layers that are easily mined (figure 12.3). Chromite, an oxide of the metal chromium, and magnetite, one of the iron oxides, are both

A

B

C

Figure 12.4 Hydrothermal ore deposits form in various ways. (*A*) Ore deposition in veins around a magma chamber. Less-concentrated ore is deposited disseminated through the rock as fluids seep outward from magma. (*B*) Hydrothermal sulfides permeate this basaltic rock. (*C*) Sulfides can be deposited by hydrothermal circulation around a spreading ridge.   Photograph (*B*) by W. R. Normark, USGS Photo Library, Denver, CO.

quite dense. In a magma of suitable bulk composition, rich concentrations of these minerals may form in the lower part of the magma chamber during crystallization of the melt.

The dense precious metals, such as gold and platinum, may also be concentrated during magmatic crystallization. However, these metals are valuable enough that, even where they are disseminated throughout an igneous rock body, they may be worth mining if their concentrations are sufficiently high. This is also true of diamonds. One reason for the rarity of diamonds is that they must be formed at extremely high pres-

sures, such as are found within the mantle, and then brought rapidly up into the crust. They are mined primarily from igneous rocks called **kimberlites,** which occur as pipelike intrusive bodies that must have originated in the mantle. Even where only a few gem-quality diamonds are scattered within many tons of kimberlite, their high value makes the mining profitable.

Hydrothermal Ores

Not all mineral deposits related to igneous activity form within igneous rock bodies. Magmas have water and other

Figure 12.5 Sulfur deposition around a fumarole.
Photograph by J. C. Ratté, USGS Photo Library, Denver, CO.

fluids dissolved in or associated with them. Particularly during the later stages of crystallization, the fluids may escape from the cooling magma, seeping through cracks and pores in the surrounding rocks, carrying with them dissolved salts, gases, and metals. These warm fluids can leach additional metals from the rocks through which they pass. In time, the fluids cool and deposit their dissolved minerals, creating a **hydrothermal** (literally, "hot water") ore deposit (figure 12.4).

The particular minerals deposited vary with the composition of the hydrothermal fluids, but worldwide, a great variety of metals occur in hydrothermal deposits: copper, lead, zinc, gold, silver, platinum, uranium, and others. Because sulfur is a common constituent of magmatic gases and fluids (figure 12.5), the ore minerals are frequently sulfides. For example, the lead in lead ore deposits is found in galena (PbS); zinc, in sphalerite (ZnS); and copper, in a variety of copper and copper-iron sulfides ($CuFeS_2$, CuS, Cu_2S, and others).

The hydrothermal fluids need not all originate within the magma. Sometimes, circulating subsurface waters are heated sufficiently by a nearby cooling magma to dissolve, concentrate, and redeposit valuable metals in a hydrothermal ore deposit. Or the fluid involved may be a mix of magmatic and nonmagmatic fluids. Indeed, some hydrothermal deposits are found far from any obvious magma source, and those fluids might have been of metamorphic origin.

Relationship to Plate Margins

The relationship between magmatic or metamorphic activity and formation of many ore deposits suggests that hydrothermal and igneous-rock deposits should be especially common in regions of great magmatic activity—that is, plate boundaries. This is generally true. A striking example of this link can be seen in figure 12.6A, which shows the locations of a certain type of copper and molybdenum deposit of igneous origin in North and South America. Comparison with figure 3.6 shows that these locations correspond closely to present or recent subduction zones. The same plate-tectonic influence can be seen in figure 12.6B: The precious-metal deposits in the western United States are predominantly hydrothermal deposits related to magmatic activity associated with plate collision and subduction along the western margin of the continent.

Scientists using submersible research vessels to investigate underwater spreading ridges have discovered hydrothermal fluids gushing from the sea floor at places along these ridges, depositing sulfides as they cool (figure 12.7). Many of these hydrothermal vents have been nicknamed [Iron sulfide] "black smokers" for the dark cloud of suspended sulfide minerals issuing from them. Similar activity probably accounts for the metal-rich muds at the bottom of the Red Sea, which is also a spreading rift zone and the site of a developing ocean basin (see figure 12.7B).

Sedimentary Deposits

Sedimentary processes can also produce economic mineral deposits. Some such ores have been deposited directly as chemical sedimentary rocks. Layered sedimentary iron ores, called **banded iron formation,** are an example (figure 12.8). Iron-rich layers (predominantly hematite or magnetite) alternate with silicate- or carbonate-rich layers. These large deposits, which may extend for tens of kilometers, are, for the most part, very ancient. Their formation is believed to be related to the development of the earth's atmosphere. As pointed out in chapter 1, the primitive atmosphere is believed to have lacked free oxygen. Under those conditions, iron from the weathering of continental rocks would have been very soluble in the oceans. As photosynthetic organisms began producing oxygen, that oxygen would have reacted with the dissolved iron and caused it to precipitate. If the majority of large iron ore deposits formed in this way, during a time long past when the earth's surface chemistry was very different from what it is now, it follows that similar deposits are unlikely to form now or in the future.

A

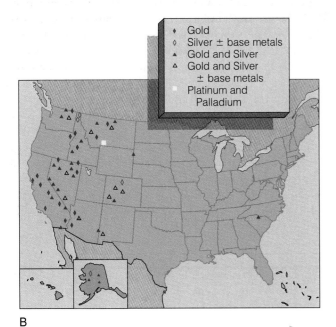

B

Figure 12.6 (*A*) Distribution of copper and molybdenum deposits. (Deposit in eastern Canada relates to an ancient subduction zone, rather than modern subduction.) (*B*) Major precious-metal-producing areas in the United States.   (*A*) Source: Data from M. L. Jensen and A. M. Bateman, *Economic Mineral Deposits*, 3d ed. Copyright © 1981 John Wiley & Sons, Inc., New York. (*B*) Source: Data from *Mineral Commodity Summaries 1993*, U.S. Bureau of Mines.

A

B

Figure 12.7 Effects of hydrothermal activity along the East Pacific Rise. (*A*) An active "black smoker chimney" vent. The black "smoke" is mainly fine-grained iron sulfide. (*B*) Sulfides are also being deposited in Red Sea muds.   (*A*) Photograph by W. R. Normark, USGS Photo Library, Denver, CO.

Figure 12.8 Banded iron formation; this sample, displayed outside the Smithsonian National Museum of Natural History in Washington, D.C., has been deformed by metamorphism.

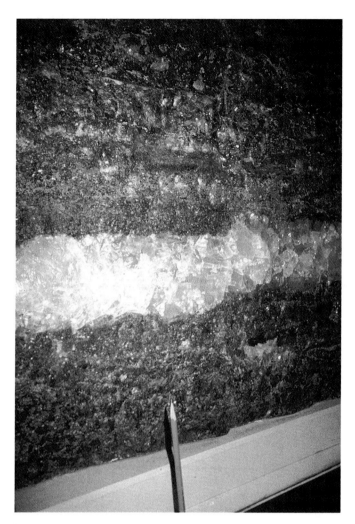

Figure 12.9   Rock salt from an evaporite deposit.

Other sedimentary mineral deposits can form from seawater, which contains a variety of dissolved salts and other chemicals. When a body of seawater trapped in a shallow sea dries up, it deposits these minerals in **evaporite** deposits (figure 12.9). Some evaporites may be hundreds of meters thick. Ordinary table salt, known mineralogically as halite, is one mineral commonly mined from evaporite deposits. Others include gypsum and salts of the metals potassium and magnesium.

Other Low-Temperature Ore-Forming Processes

Streams also play a role in the formation of mineral deposits. They are rarely the sites of primary formation of ore minerals. However, as noted in chapter 6, streams often deposit sediments well sorted by size and density. The sorting action can effectively concentrate certain weathering-resistant, dense minerals in places along the stream channel. The currents of a coastal environment can also cause sediment sorting and selective concentration of minerals. Such deposits, mechanically concentrated by water, are called **placers.** The

minerals of interest are typically weathered out of local rocks, then transported, sorted, and concentrated while other minerals are dissolved or swept away (figure 12.10). Gold, diamonds, and tin oxide are examples of minerals that have been mined from the sands and gravels of placer deposits. (Indeed, the ability to separate high-density gold from other minerals in water by gravity is the underlying principle of gold panning.)

Even weathering alone can produce useful ores by leaching away unwanted minerals, leaving a residue enriched in some valuable metal. As described in chapter 11, the extreme leaching of tropical climates gives rise to lateritic soils from which nearly everything has been leached except for aluminum and iron compounds. Many of the aluminum deposits presently being mined were enriched in aluminum to the point of being economic by lateritic weathering. The iron content of the laterites is generally not as high as that of the richest sedimentary iron ores described earlier. However, as the richest of those ore deposits are mined out, it may become profitable to mine laterites for iron as well as for aluminum.

Metamorphic Deposits

The mineralogical changes caused by the heat or pressure of metamorphism also can produce economic mineral deposits. Graphite, used in "lead" pencils, in batteries, as a lubricant, and for many applications where its very high melting point is vital, is usually mined from metamorphic deposits. Graphite consists of carbon, and one way in which it can be formed is by the metamorphism of coal which, as will be seen in chapter 13, is also carbon-rich. Asbestos is not a single mineral but a general term applied to a group of fibrous silicates that are formed by the metamorphism of igneous rocks rich in ferromagnesian minerals, with the addition of water. The asbestos minerals are used less in insulation now but are still valuable for their heat- and fire-resistant properties, especially because their unusual fibrous character allows them to be woven into cloth.

Mineral and Rock Resources—Examples

Dozens of minerals and rocks have some economic value. In this section, we survey a sampling of these resources, noting their occurrences and principal applications. Figure 12.11 represents per capita consumption of a number of these materials in the United States.

Metals

The term *mineral resources* usually brings metals to mind first. Overwhelmingly, the most heavily used metal is iron. Fortunately, it is also one of the most common metals. Nearly all iron ore is used for the manufacture of iron and especially steel products. It is mined principally from the ancient sedimentary deposits described in the previous section but also from some laterites and from concentrations of magnetite in some igneous bodies.

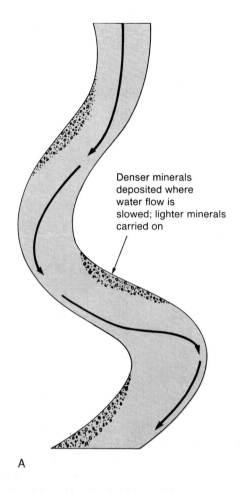

Denser minerals deposited where water flow is slowed; lighter minerals carried on

A

Crushed stone
4,200 kg

Sand and gravel
2,900 kg

Salt
180 kg

Gypsum
94 kg

Phosphate
170 kg

Potash
22 kg

Iron ore
260 kg

Aluminum
22 kg

Zinc
5 kg

Lead
5 kg

Copper
9 kg

Figure 12.11   Per capita consumption of mineral and rock resources in the United States, 1992, based on population of approximately 250 million persons. (1 kg = 2.2 lb.)

B

Figure 12.10   (*A*) Formation of a placer deposit. (*B*) Since placer deposits typically occur as loose gravels, mining them can be relatively easy; here, water flowing from a giant nozzle washes gold-bearing gravels to sluices, in the Yukon, Canada. Photograph © Steve McCutcheon/Visuals Unlimited.

Aluminum is another relatively common metal, and it is the second most widely used. Its light weight, coupled with its strength, make it particularly useful in the transportation and construction industries; it is also widely used in packaging, especially for beverage cans. Aluminum is the third most common element in the crust, but there it is most often found in silicates, from which it is extremely difficult to extract. Most of what is mined commercially is bauxite, an aluminum-rich laterite in which the aluminum is found as a hydroxide. Even in this form, the extraction of the aluminum is somewhat difficult and energy-intensive: 3 to 4 percent of the electricity consumed in the United States is used just in the production of aluminum metal from aluminum ore.

Many less-common but also important metals, including copper, lead, zinc, nickel, cobalt, and others, are found in sulfide ore deposits. Sulfides occur frequently in hydrothermal deposits and may also be concentrated in igneous rocks. Copper, lead, and zinc may also be found in sedimentary ores; some laterites are moderately rich in nickel and cobalt. Clearly, these metals may be concentrated into economically valuable deposits in a variety of ways. Copper is primarily used for electrical applications because it is an excellent conductor of electricity; it is also used in the construction and transportation industries. An important use of lead is in batteries; among its many other applications, it is a component of many solders and is used in paints and ceramics. The zinc coating on steel cans (misnamed "tin cans") keeps the cans from rusting, and zinc is also used in the manufacture of brass and other alloys.

The so-called precious metals—gold, silver, and platinum—seem to have a special romance about them that most metals lack, but they also have some unique practical uses. Gold is used not only for jewelry, in the arts, and in commerce, but also in the electronics industry and in dentistry. It is particularly valued for its resistance to tarnishing. Silver's principal use, accounting for nearly half the silver consumed in the United States, is for photographic materials (for example, film), and the next broadest applications are in electronics. Platinum is an excellent *catalyst*, a substance that promotes chemical reactions. Currently, close to half the platinum used in the United States goes into automobile emissions-control systems, with the rest finding important applications in the petroleum and chemical industries, in electronics, and in medicine, among other areas. All of the precious metals can be found as native metals, most frequently in igneous or hydrothermal ore deposits. Silver also commonly forms sulfide minerals. Most gold and silver production in this country is, in fact, a by-product of mining ores of more abundant metals like copper, lead, and zinc, with which small amounts of precious metals may be associated.

Nonmetallic Minerals

Another by-product of mining sulfides is the nonmetal sulfur. Sulfur may also be recovered from petroleum during refining, from volcanic deposits (sulfur is sometimes precipitated as pure native sulfur from fumes escaping from volcanic vents), and from evaporites. The primary use of sulfur (90% of it in this country) is for the manufacture of sulfuric acid for industrial purposes.

Several important minerals are recovered from evaporite deposits. The most abundant is halite, or rock salt, used principally as a source of the sodium and chlorine of which it is composed, and secondarily for road salt, either directly or through the production of other salts from it. Halite has many lower-volume applications, including, of course, seasoning food as table salt. Gypsum, essential to the manufacture of plaster, portland cement, and wallboard for construction, is another evaporite mineral. Others include phosphate rock and potassium-rich potash, key ingredients of the synthetic fertilizers on which much of U.S. agriculture depends.

As noted in chapter 2, clay is not a single mineral, but a group of layered hydrous silicates that are formed at low temperature, commonly by weathering, and that are abundant in sedimentary deposits in the United States. The diversity of clay minerals leads to a variety of applications, from fine ceramics to the making of clay piping and other construction materials, the processing of iron ore, and drilling for oil. In the last case, the clay is mixed with water and sometimes other minerals to make "drilling mud," which lubricates the drill bit; the low permeability of clays also enables them to seal off porous rock layers encountered during drilling.

Rock Resources

When it comes to overall mass, the quantities used of the various metals and minerals mentioned previously shrink to insignificance beside the amount of rock, sand, and gravel used. In the United States in 1995, nearly 900 million tons of sand and gravel were used in construction, especially in making cement and concrete; another 1.3 billion tons of crushed rock were consumed for fill and other applications (such as limestone for making cement). That corresponds to several tons of rock products for every person in the country. In addition, more than 28 million tons of quartz-rich sand were used in industry, particularly for glassmaking (a pure quartz sand is nearly all silica, SiO_2, the main ingredient in glass) and for abrasives. Also, more than 1.3 million tons of dimension and facing stone were used—slate for flagstones, and various other attractive and/or durable rocks, such as marble, granite, sandstone, and limestone, for monuments and building facings.

Mineral Supply and Demand

Assessments of the adequacy of known supplies of various rock and mineral resources naturally require information both on those supplies and on the rates at which they are, and will be, consumed. We have already examined some of the variables on the supply side of the calculation, including geologic and economic factors. Large uncertainties also exist on the demand side.

For example, for most metals, growth in worldwide demand ranged from just over 2 percent per year to nearly 10 percent per year from World War II to the mid-1970s. More recently, world economic conditions restrained the rising demand, at least temporarily. Over the period from 1976 to 1982, demand for most minerals grew much more slowly than it had in prior decades, and for a few materials, demand actually declined. As the global economy has rebounded, demand has tended to increase. Improvements in the standard of living for billions of people living in less-developed countries also increase demand pressures, as is illustrated in figure 12.12.

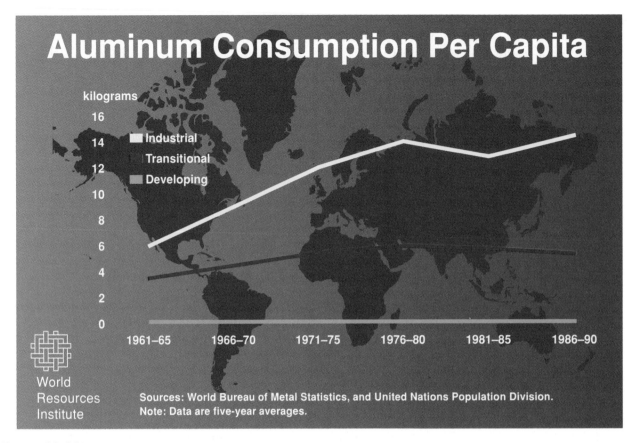

Figure 12.12   Per capita aluminum consumption in more and less developed countries.   Source: World Resources Institute.

U.S. Mineral Production and Consumption

All told, the United States consumes huge quantities of mineral and rock materials, relative to both its population and its production of most minerals. Per capita consumption of many materials was shown in figure 12.11. Table 12.1 gives a global perspective by showing 1995 U.S. primary production and consumption data and the percentages of world totals that these figures represent. Keep in mind when looking at the consumption figures that fewer than 5 percent of the people in the world live in the United States.

The relationship of domestic consumption to domestic production can be complex. "Primary production" is production from new ore, excluding production from recycling or from partially processed ore imported from elsewhere; total production includes the latter kinds of production as well. (This accounts for differences in figures for metals in the last two columns of table 12.1.) Sometimes, the United States even imports materials of which it has ample domestic supplies, so as to conserve what it has, or to stay on good trade terms with countries having other, more critical commodities that the United States also wishes to import. For example, in the case of molybde-

num, in 1995, U.S. primary production was 59,000 metric tons, against consumption of 29,000 tons—yet we imported 10,000 tons. Flow into and out of government stockpiles can also be significant: For mercury in 1992, total production was 160 metric tons, consumption about 500 tons; but we imported only 100 tons and exported 400, reflecting releases of about 500 tons from the National Defense Stockpile and U.S. Department of Energy.

Of course, where production falls short of consumption, importing is necessary to meet domestic demand. Moreover, low domestic production may not necessarily mean that the United States is conserving its own resources, but that it has none to produce or, at least, no economic deposits. This is true, for example, of chromium, an essential ingredient of stainless steel, and manganese, which is presently irreplaceable in the manufacture of steel. Currently, for materials such as these, the United States is almost wholly dependent on imports (figure 12.13).

Leaving aside the unanswerable political question of how long any country can count on an uninterrupted flow of necessary mineral imports, how well supplied is the world, overall, with minerals? How imminent are the shortages? We will address these questions in the next section.

Table 12.1 U.S. Production and Consumption of Rock and Mineral Resources, 1995[†]

	Material	U.S. primary production	U.S. primary production as % of world	U.S. consumption	U.S. consumption as % of world[‡]	U.S. production as % of U.S. consumption[■]	
						PRIMARY	TOTAL
METALS	aluminum[x]	3,350	17.4	6,500	33.7	52	76
	chromium	0	0	537	5.1	0	22
	cobalt	0	0	8.4	43.1	0	18
	copper	1,890	19.3	2,570	28.1	74	89
	iron ore	62,000	6.2	75,300	7.5	82	82
	lead	390	13.9	1,450	51.8	27	86
	manganese	0	0	710	9.7	0	0
	nickel	1.6	0.01	159	17.3	1	6
	tin	negligible	0	47.0	26.1	0	25
	zinc	600	8.5	1,460	20.6	41	74
	gold	320	14.5	75	3.4	427	493
	silver	1,500	10.7	ø	ø	ø	ø
	platinum group	8.0	3.5	ø	ø	ø	ø
NONMETALS	clays	46,600	*	42,100	*	111	111
	gypsum	17,300	16.8	25,900	25.1	68	70
	phosphate	45,500	33.2	43,600	31.8	104	104
	potash	1,460	5.6	5,670	21.6	26	26
	salt	43,300	23.4	49,500	26.8	87	87
	sulfur	11,800	22.7	13,900	26.7	85	85
CONSTRUCTION	sand and gravel, construction	879,000	*	878,000	*	100	100
	stone, crushed	1,280,000	*	1,280,000	*	100	100
	stone, dimension and facing	1,320	*	**	*	**	**

Source: *Mineral Commodity Summaries 1996*, U.S. Bureau of Mines.

[†]All production and consumption figures in thousands of metric tons, except for gold, silver, and platinum-group metals, for which figures are in metric tons, and construction materials, for which figures are in thousands of tons.

[‡]Assumes that overall, world production approximates world consumption. This may not be accurate if, for example, recycling is extensive.

[■]"Primary" and "Total" headings refer to U.S. production; total production includes production from recycled scrap.

* World data not available

øConsumption data not available

**Consumption not available in tonnage. However, $470 million worth of dimension stone was imported (value of domestic production, $220 million, of which $51 million worth was exported).

[x]Partly from bauxite imports

World Mineral Supply and Demand

Certain assumptions must underlie any projections of "lifetimes" of mineral reserves. One is future demand. For reasons already outlined, demand can fluctuate both up and down. Furthermore, not all commodities follow the same trends. Making the assumption that world production annually approximates demand, at least for "new" mine production, we find that just over one recent decade, annual demand for iron ore remained approximately constant; for copper and zinc, it increased by 10% and 20%, respectively; for lead, it declined by 7%. Rather than attempt to make precise demand estimates for the future, the following discussion assumes demand equal to primary production at constant 1995 levels.

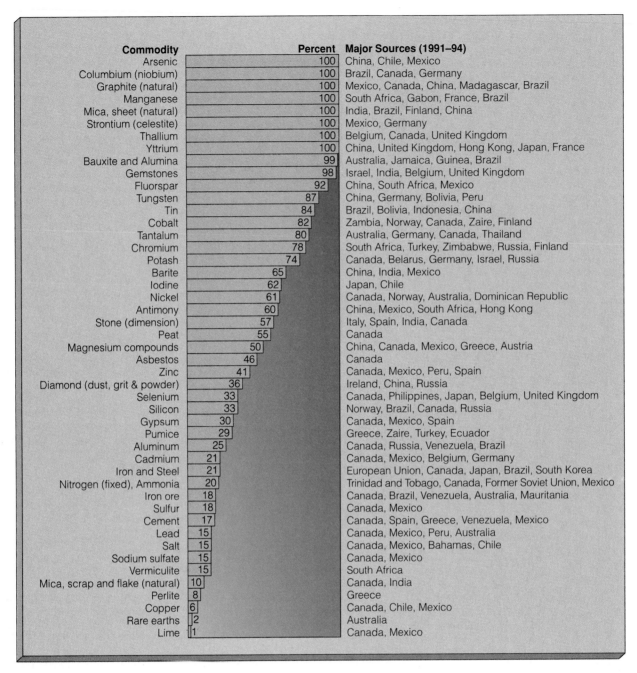

Figure 12.13 Proportions of U.S. mineral needs supplied by imports as a percentage of apparent consumption. Principal sources of imports indicated at right.   Source: *U.S. Commodity Summaries 1996,* U.S. Bureau of Mines.

Table 12.2 shows projections of the lifetimes of selected world mineral reserves, assuming that constant demand. Note that, for most of the metals, the present reserves are projected to last only a few decades, even at these constant consumption levels. Consider the implications if consumption resumes the rapid growth rates of the mid-twentieth century.

Such projections also presume unrestricted distribution of minerals so that the minerals can be used as needed and where needed, regardless of political boundaries or such economic factors as nations' purchasing power. It is interesting to compare global projections with corresponding data for the United States only (table 12.3). Even at constant 1995 consumption rates, the United States has less than half a century's worth of reserves of most of these materials. Of some metals, including chromium, manganese, and platinum, the United States has virtually no reserves at all.

Some relief can be anticipated from the economic component of the definition of reserves. For example, as currently identified reserves are depleted, the law of supply

Table 12.2 World Production and Reserves Statistics, 1995*

Material	Production	Reserves	Projected lifetime of reserves (years)	Estimated resources
bauxite	109,000	23,000,000	211	55,000,000–75,000,000
chromium	10,600	3,700,000	349	11,000,000
cobalt	19.5	4,000	205	11,000
copper	9,800	310,000	32	2,300,000†
iron ore	1,000,000	150,000,000	150	over 800,000,000
lead	2,800	68,000	24	1,500,000
manganese	7,300	680,000	93	**
nickel	920	47,000	51	130,000
tin	180	7,000	39	**
zinc	7,070	140,000	20	1,800,000
gold	2,200	44,000	20	75,000
silver	14,000	280,000	20	n.a.
platinum group	230	56,000	243	100,000
gypsum	103,000	**	**	**
phosphate	137,000	11,000,000	80	**
potash	26,200	8,400,000	320	250,000,000
sulfur	52,000	1,400,000	27	5,000,000

Source: *Mineral Commodity Summaries 1996*, U.S. Bureau of Mines.

Reserves include only currently economic deposits.

*All production, reserve, and resource figures in thousands of metric tons, except for gold, silver, and platinum-group metals, for which figures are in metric tons.

**Resources and reserves estimated as "large" but not quantified.

†Includes 0.7 billion tons copper estimated to occur in manganese nodules; "extensive" resources of nickel are also projected to occur in these nodules.

and demand will drive up minerals' prices. This, in turn, will make some of what are presently subeconomic resources profitable to mine; some of those deposits will effectively be reclassified as reserves. (See, for example, box 12.1.) Improvements in mineral-processing technology may accomplish the same result. In addition, continued exploration can be expected to locate additional economic deposits, particularly as exploration methods become more sophisticated.

The technological advances necessary for the development of some of the subeconomic resources, however, may not be achieved rapidly enough to help substantially. Also, if the move is toward developing lower- and lower-grade ore, then by definition, more and more rock will have to be processed and more and more land disturbed to extract a given quantity of mineral or metal. Some of the consequences of mining activities, which will have increasing impact as more mines are developed, are discussed later in this chapter.

What other prospects exist for averting future mineral-resource shortages? Some possibilities are discussed in the next section.

Minerals for the Future: Some Options Considered

One way to extend nonrenewable resources, in theory, would be to reduce consumption rates, or at least hold them steady. In practice, this seems unlikely to occur, even in the developed nations where resource consumption is already high.

The United States has the world's highest rate of car ownership: In 1993, nearly 60% of U.S. households owned two or more cars, and 20%, three or more, for total vehicle ownership of 561 passenger cars per 1,000 people. (The figure was 366 per 1,000 people for all the rest of the member nations of the Organisation for Economic Cooperation and Development together, 72 for central and eastern Europe, 29 for east Asia, 14 for Africa.) Many factors—trends toward smaller households, necessity, more leisure for travel, desires for convenience or status—put upward pressure even on those U.S. car-ownership numbers. Or consider the impact of new technologies: for example, microwave ovens, personal computers, or cellular phones (increasing in number so fast that some telephone companies in urban areas are scrambling to create new area codes to accommodate them).

Table 12.3
Projected Lifetimes of U.S. Mineral Reserves (assuming complete reliance on domestic reserves)*

Material	Reserves	Projected lifetime (years)
bauxite†	20,000	5
chromium	0	0
cobalt	0	0
copper	45,000	17.5
iron ore	16,000,000	205
lead	8,000	5.5
manganese	0	0
nickel	23	0.1
tin	20	0.4
zinc	16,000	11
gold	5,400	72
silver	31,000	**
platinum group	250	**
gypsum	700,000	27
phosphate	1,200,000	27
potash	76,000	13.4
sulfur	140,000	10

Source: *Mineral Commodity Summaries 1996*, U.S. Bureau of Mines.

*Reserves in thousands of metric tons, except for gold, silver, and platinum-group metals, for which figures are in metric tons.

**Accurate consumption data not available.

†Note that bauxite consumption is only a partial measure of total aluminum consumption; additional aluminum is consumed as refined aluminum metal, of which there are no reserves.

minerals, in either traditional or nontraditional areas, or we must conserve our minerals better so that they will last longer.

New Methods in Mineral Exploration

Most of the "easy ores," or near-surface mineral deposits in readily accessible places, have probably already been discovered. Fortunately, however, a variety of new methods are being applied to the search. Geophysics provides some assistance. Rocks and minerals vary in density, and in magnetic and electrical properties, so changes in rock types or distribution below the earth's surface can cause small variations in the gravitational or magnetic field measured at the surface, as well as in the electrical conductivity of the rocks. Some ore deposits may be detected in this way. As a simple example, because many iron deposits are strongly magnetic, large magnetic anomalies may indicate the presence and the extent of a subsurface body of iron ore. Radioactivity is another readily detected property; uranium deposits may be located with the aid of a Geiger counter.

Geochemical prospecting is another idea that is gaining acceptance. It may take a variety of forms. Some studies are based on the recognition that soils reflect, to a degree, the chemistry of the materials from which they formed. The soil over a copper-rich ore body, for instance, may itself be enriched in copper relative to surrounding soils. Surveys of soil chemistry over a wide area can help to pinpoint likely spots to drill in search of ores. Occasionally, plants can be sampled instead of soil: Certain plants tend to concentrate particular metals, making the plants very sensitive indicators of locally high concentrations of those metals. Even soil gases can supply clues to ore deposits. Mercury is a very volatile metal, and high concentrations of mercury vapor have been found in gases filling soil pore spaces over mercury ore bodies. Geochemical prospecting has also been used in petroleum exploration.

Remote sensing methods are becoming increasingly sophisticated and valuable in mineral exploration. These methods rely on detection, recording, and analysis of wave-transmitted energy, such as visible light and infrared radiation, rather than on direct physical contact and sampling. Aerial photography is one example, satellite imagery another. Remote sensing, especially using satellites, is a quick and efficient way to scan broad areas, to examine regions having such rugged topography or hostile climate that they cannot easily be explored on foot or with surface-based vehicles, and to view areas to which ground access is limited for political reasons. Probably the best known and most comprehensive earth satellite imaging system is the one initiated

In many cases, the new devices are additions to, rather than replacements for, other devices already owned (conventional ovens, typewriters, stationary in-home phones). All represent consumption of materials (as well as energy, to manufacture and often to use). While it is perhaps perfectly natural for people to want to acquire goods that increase comfort, convenience, efficiency, or perceived quality of life, the result may be increases in already high resource-consumption rates (figure 12.14): In the U.S., from 1950 to 1990, the population grew by about 65%, while materials consumption increased twice as fast, by more than 130%. Most of the increase was in plastics and industrial minerals.

Furthermore, even if the technologically advanced nations were to restrain their appetite for minerals, they represent a minority of the world's people. If the less-developed nations are to achieve a standard of living for their citizens that is even remotely comparable to that in the most industrialized nations, vast quantities of minerals and energy will be needed. Moreover, as pointed out in chapter 1, those same less-developed countries represent the fastest-growing segment of the global population.

If demand cannot realistically be cut, supplies must be increased or extended. We must develop new sources of

Box 12.1 Additional Reflections on Mineral Economics and Supplies: Copper as an Example

The distinction between reserves and resources is, in part, an economic one. As a metal's price goes up, additional deposits become profitable to mine. However, the relationship is not a direct one; that is, a doubling in price does not double the reserves. This can be illustrated using U.S. Department of the Interior estimates of U.S. and world copper reserves (in millions of tons), based on various copper prices (Banks, 1974):

	50¢/lb	60¢/lb	70¢/lb	80¢/lb
World	268	301	329	365
United States	73	85	85	90

There is also a time-lag factor to consider. Time is required to bring each new mine into production. For copper mines, there is presently a ten- to fifteen-year lag between the identification of the deposit and the start of production. A sudden price increase may not result in a rush to open new mines if producers believe that the price rise is likely to be only temporary because, by the time the metal is actually extracted, the price may have fallen again.

Technological improvements, however, can make additional deposits profitable even in the absence of a price rise. At the turn of the century, the lowest-grade copper ore that could profitably be mined was about 3 percent copper. By the early 1970s, at which time the price of copper was actually lower in inflation-adjusted dollars, deposits as poor as 0.2 percent copper could be mined economically. On the other hand, technological advances take time to develop, and rising world consumption rates may not allow for this. Whether further technological developments can keep pace with future demand is unclear.

World commodity prices, too, are sensitive to a large number of factors too numerous and complex to explore fully here. Governments restrict exports or nationalize foreign-owned mining operations; they form cartels to control prices, or subsidize prices to increase exports. From nation to nation, production costs fluctuate with labor costs, pollution controls, costs of raw materials and land, and so on. In

the early 1970s, the price of copper rose more than 50% in response to a combination of factors, including increased demand by industrialized countries and panic buying by those anticipating shortages resulting from that demand. But by 1975, the price began to plunge. Only the nations with very low production costs, like Chile, could then sell copper at a profit; most U.S. mines closed. If U.S. imports were curtailed, there would be at least short-term shortages until U.S. copper mines could be brought into production again. The rapidity of price fluctuations, set against long mining startup lead times, also presents a challenge to mining-company planners: The price of copper rose 10% from 1988 to 1989; from 1989 to 1992, it declined 18%, then plunged 16% more in 1993, following which it soared 52% in the next two years.

Complexities such as these make mineral reserve estimates and "lifetime" projections extremely difficult.

---

in 1972, known as Landsat until sold by the government to private interests, and renamed Eosat.

Landsat satellites orbit the earth in such a way that images can be made of each part of the earth. Each orbit is slightly offset from the previous one, with the areas viewed on one orbit overlapping the scenes of the previous orbit. Each satellite makes fourteen orbits each day; complete coverage of the earth takes eighteen days.

The sensors in the Landsat satellites do not detect all wavelengths of energy reflected from the surface. They do not take photographs in the conventional sense. They are particularly sensitive to selected green and red wavelengths in the visible light spectrum and to a portion of the infrared (invisible heat radiation, with wavelengths somewhat longer than those of red light). These wavelengths were chosen because plants reflect light most strongly in the green and the infrared. Different plants, rocks, and soils reflect different proportions of radiation of different wavelengths. Even the same feature may produce a somewhat different image under different con-

ditions. Wet soil differs from dry; sediment-laden water looks different from clear; a given plant variety may reflect a different radiation spectrum depending on what trace elements it has concentrated from the underlying soil or how vigorously it is growing. Landsat images can be powerful mapping tools. The smallest features that can be distinguished in a Landsat image are about 80 meters (250 feet) in size, which indicates the quality of the resolution. Multiple images can be joined into mosaics covering whole countries or continents.

Basic geologic mapping, identification of geologic structures, and resource exploration are only some of the applications of Landsat imagery (figure 12.15). Such imagery is especially useful when some *ground truth* can be obtained—information gathered by direct surface examination (best done at the time of imaging, if vegetation is involved). Alternatively, Landsat images of inaccessible regions can be compared with images of other regions that have been mapped and sampled directly, and the similarities in imagery characteristics used to infer the actual geology or vegetation.

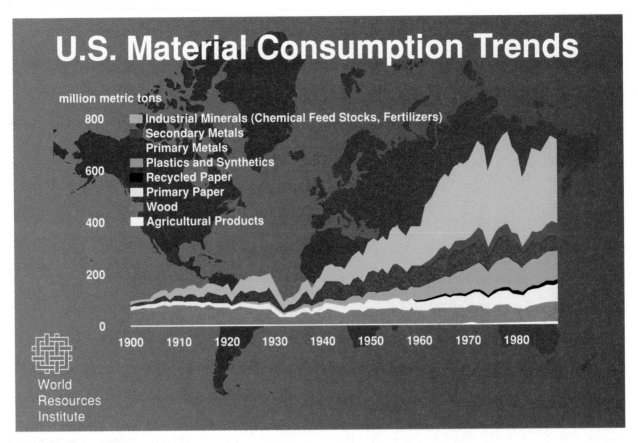

Figure 12.14 Trends in U.S. material consumption, 1900–1990.   Source: World Resources Institute.

A

B

Figure 12.15 Landsat satellite images may reveal details of the geology that will aid in mineral exploration. Vegetation, also sensitive to geology, may enhance the image. (*A*) View of South Africa, dry season. (*B*) Same view, rainy season. Recognizable geologic features include a granite pluton (round feature at top left) and folded layers of sedimentary rock (below).   Photographs © NASA.

Finally, advances in geologic understanding also play a role in mineral exploration. Development of plate-tectonic theory has helped geologists to recognize the association between particular types of plate boundaries and the occurrence of certain kinds of mineral deposits. This recognition may direct the search for new ore deposits. For example, because geologists know that molybdenum deposits are often found over existing subduction zones, they can logically explore for more molybdenum deposits in other present or past subduction zones. Also, the realization that many of the continents were once united can suggest likely areas to look for more ores (figure 12.16). If mineral deposits of a particular kind are known to occur near the margin of one continent, similar deposits might be found at the corresponding edge of another continent once connected to it. If a mountain belt rich in some kind of ore seems to have a counterpart range on another continent once linked to it, the same kind of ore may occur in the matching mountain belt. Such reasoning is partly responsible for the supposition that economically valuable ore deposits may exist on Antarctica. (Some of the legal/political consequences of such speculations are explored in chapter 18, and environmental consequences are discussed in box 12.2.)

Marine Mineral Resources

Increased consideration is also being given to seeking mineral wealth in unconventional places. In particular, the sea may provide partial solutions to some mineral shortages. Seawater itself contains dissolved in it virtually every chemical element. However, most of what it contains is dissolved halite, or sodium chloride. Most needed metals occur in far lower concentrations. Vast volumes of seawater would have to be processed to extract small quantities of such metals as copper and gold. The costs would be very high, especially since existing technology is not adequate to extract, selectively and efficiently, a few specific metals of interest. Several types of underwater mineral deposits have greater potential.

During the last Ice Age, when a great deal of water was locked in ice sheets, worldwide sea levels were lower than at present (recall figure 9.28). Much of the area of the now-submerged continental shelves was dry land, and streams running off the continents flowed out across the exposed shelves to reach the sea. Where the streams' drainage basins included appropriate source rocks, these streams might have produced valuable placer deposits. With the melting of the ice and rising sea levels, any placers on the continental shelves were submerged. Finding and mining these deposits may be costly and difficult, but as land deposits are exhausted, placer deposits on the continental shelves may well be worth seeking. Already, more than 10 percent of world tin production is from such marine placers; potentially significant amounts of titanium, zirconium, and chromium could also be mined from them if the economics were right.

The hydrothermal ore deposits forming along some seafloor spreading ridges are another possible source of needed metals. In many places, the quantity of material being deposited

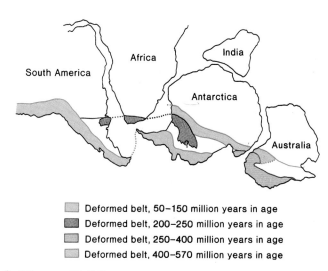

Deformed belt, 50–150 million years in age
Deformed belt, 200–250 million years in age
Deformed belt, 250–400 million years in age
Deformed belt, 400–570 million years in age

Figure 12.16 Pre-continental-drift reassembly of landmasses suggests locations of possible ore deposits in unexplored regions by extrapolation from known deposits on other continents.   Source: Data from C. Craddock, et al., *Geological Maps of Antarctica,* Antarctic Map Folio Series, Folio 12, copyright 1970 by the American Geographical Society.

is too small, and the depth of water above would make recovery prohibitively expensive. However, the metal-rich muds of the Red Sea contain sufficient concentrations of such metals as copper, lead, and zinc that some exploratory dredging is underway, and several companies are interested in the possibility of mining those sediments. Along a section of the Juan de Fuca ridge, off the coast of Oregon and Washington, hundreds of thousands of tons of zinc- and silver-rich sulfides may already have been deposited, and the hydrothermal activity continues.

Perhaps the most widespread undersea mineral resource, as well as the most frequently and seriously discussed as a practical resource for the near future, is **manganese nodules.** These are lumps up to about 10 centimeters in diameter, composed mostly of manganese minerals. They also contain lesser but potentially valuable amounts of copper, nickel, cobalt, platinum, and other metals. Indeed, the value of the minor metals may be a greater motive for mining manganese nodules than the manganese itself. The nodules are found over much of the deep-sea floor, in regions where sedimentation rates are slow enough not to bury them (figure 12.17A). At present, the costs of recovering these nodules would be high compared to the costs of mining the same metals on land, and the technical problems associated with recovering the nodules from beneath several kilometers of seawater remain to be worked out. Still, manganese nodules represent a sufficiently large metal resource that they have become a subject of International Law of the Sea conferences (see chapter 18). With any ocean resources outside territorial waters—manganese nodules, hydrothermal deposits at spreading ridges, fish, whales, or whatever—questions of ownership, or of who has the right to exploit these resources, inevitably lead to heated debate.

Box 12.2 Antarctica and the Question of Environmental Impact of Mineral Resource Development

Antarctica, coldest and windiest continent on earth, covers 13.5 million square kilometers (5 million square miles). Most of it is ice-covered year-round, and in winter it is isolated by pack ice extending up to 1,400 kilometers (875 miles) out beyond the continent. Some of its soils are believed to be unique in the world. Life on the continent is limited to microorganisms, lichens, mosses, algae and fungi, and a few arthropods (many-legged animals, including insects). However, the surrounding southern oceans teem with life, including birds, seals, whales, squid, and some fish, along with the simpler life-forms on which they feed. The continent provides breeding grounds to many of the birds and seals.

The thick ice cover in most areas has so far restricted geologic explo-ration of Antarctica. Limited coal beds and iron ore deposits have been found; there is some evidence of copper mineralization and small concentrations of other economic minerals, including graphite, beryl, and mica. Extrapolation from mineral-producing regions of other southern continents suggests that oil and natural gas, chromite, and platinum-group metals might also be found.

The *types* of impacts to be expected from mineral or petroleum exploitation in Antarctica are the same as the impacts of corresponding activities on or offshore from other continents, but the *seriousness* of these impacts may be greater. In such a hostile climate, many life-forms that already exist at the limits of survival could be destroyed by additional environmental stress. There is limited open land available for breeding grounds; if some of it is occupied and/or disturbed by mining or mineral-processing operations, animal populations could suffer. Water pollution could considerably diminish the abundance of life in and the productivity of the southern oceans. Because chemical weathering processes are so slow in this environment, soils once disturbed may not readily be regenerated.

The world's need for minerals is not yet sufficiently desperate to encourage active mineral exploitation under such difficult (and expensive) conditions. When and if such exploitation begins, particular care must be taken to minimize the adverse impacts on this distinctive environment and its diverse life-forms. Such concerns are now recognized by international treaty, as described in chapter 18.

Conservation of Mineral Resources

Overall need for resources is likely to increase, but for selected individual materials that are particularly scarce, perhaps demand can be moderated. One way is by making substitutions. One might, for certain applications, replace a very rare metal with a more abundant one. The extent to which this is likely to succeed is limited, since reserves of most metals are quite limited, as is clear from tables 12.2 and 12.3. Substituting one metal for another reduces demand for the second while increasing demand for the first; the focus of the shortage problem shifts, but the problem persists. Nonmetals are replacing metals in other applications. Unfortunately, many of these nonmetals are plastics or other materials derived from petroleum, which has simply contributed to petroleum consumption (figure 12.18). As we will see in chapter 13, supplies of that resource are limited, too. For some applications, ceramics or fiber products can be substituted; however, sometimes the physical, electrical, or other properties required for the particular application demand the use of metals.

The most effective way to extend mineral reserves, for some metals at least, may be through recycling. Overall use of the metals may increase, but if ways can be found to reuse a significant fraction of these metals repeatedly, proportionately less new metal will need to be extracted from mines, reducing—or at least moderating increases in—primary production. Some metals are already extensively recycled, at least in the United States (table 12.4 lists a few examples). Worldwide, recycling is less widely practiced, in part because the less-industrialized countries have accumulated fewer manufactured products from which materials can be retrieved. Still, given the limited U.S. reserves of many metals, recycling can be a very important means of extending those reserves. Additional benefits of recycling include a reduction in the volume of the waste-disposal problem and a decrease in the extent to which more land must be disturbed by new mining activities.

A

Figure 12.17 Manganese nodules. (*A*) Manganese nodule distribution on the sea floor. (*B*) Manganese nodules off the Marshall Islands.   (*A*) Data from G.R. Heath, "Manganese Nodules: Unanswered Questions," *Oceanus* 25, Vol. 25, No. 3, pp. 37–41. 1982. Copyright © Woods Hole Oceanographic Institute. (*B*) Photograph by K.O. Emery, USGS Photo Library, Denver, CO.

B

Figure 12.18 Growth in consumption of nonfuel organic (mainly plastic) and inorganic (mineral) materials. Major plastics and polymers shown with date of development. As more varied plastics were developed, they became a larger share of resource consumption — but mineral consumption continued to rise also.   Source: U.S. Bureau of Mines.

Unfortunately, not all materials lend themselves equally well to recycling. Among those that work out best are metals that are used in pure form in sizeable pieces—copper in pipes and wiring, lead in batteries, and aluminum in beverage cans. The individual metals are relatively easy to recover and require minimal effort to purify for reuse. Recycling aluminum has additional appeal from an energy standpoint: It takes only one-twentieth as much energy to produce aluminum by recycling old scrap as it does to extract new aluminum from aluminum ore, which results in the dual benefits of saving energy and cutting costs.

Where different materials are intermingled in complex manufactured objects, it is more difficult and costly to extract individual metals. Consider trying to separate the various met-als from a refrigerator, a lawn mower, or a television set. The effort required to do that, even if it were technically possible, would make most of the materials recovered far too costly and thus noncompetitive with new production. Only in a few rare cases are metals valuable enough that the recycling effort may be worthwhile; for example, there is interest in recovering the platinum from catalytic converters in exhaust systems.

Alloys present special problems. The United States uses tens of millions of tons of steel each year, and some of it is indeed recycled scrap. However, steel is not a single chemical substance. It is iron alloyed with one or more other elements, and, often, the composition is very specific to the application. Chromium is added to make stainless steel; titanium, molybdenum, or tungsten steels are high-strength steels; other alloys are used when other properties are important. If the composition of the alloy is critical to a particular application, clearly one cannot just toss any old steel scrap into the recycling pot. Each type of steel would have to be recycled individually. Facilities for separate collection and recycling of many different compositions of steel are rare.

Some materials are not used in discrete objects at all. The potash and phosphorus in fertilizers are strewn across the land and cannot be recovered. Road salt washes off streets and into the soil and storm sewers. Lead in gasoline, now essentially phased out, was emitted with exhaust fumes into the atmosphere. These things clearly cannot be recovered and reused.

For the foregoing and other reasons, it is unrealistic to expect that all minerals can ever be wholly recycled. However, more than half of the U.S. consumption of certain metals already is being recovered from old scrap. Where this is possible, the benefits are substantial.

Table 12.4	Metal Recycling in the United States, 1986–95 (recycled scrap as percentage of consumption)			
Metal	**1986**	**1989**	**1992**	**1995**
aluminum	15	22	30	25
chromium	21	21	26	22
cobalt	15	14	25	22
copper	22	25	42	39
lead	50	60	62	57
manganese	0	0	0	0
nickel	25	34	30	34
zinc	7	10	29	25

Source: *Mineral Commodity Summaries 1990, Mineral Commodity Summaries 1993, and Mineral Commodity Summaries, 1996,* U.S. Bureau of Mines.

Figure 12.19 Even a partial map of sites of mining for industrial rock and mineral resources shows that these activities are widespread in the U.S. Still more places are affected by mining for metal ores. Source: *Mineral Commodity Summaries 1996,* U.S. Bureau of Mines.

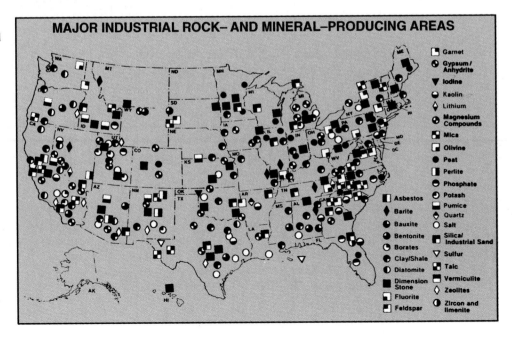

MAJOR INDUSTRIAL ROCK– AND MINERAL–PRODUCING AREAS

Garnet
Gypsum/Anhydrite
Iodine
Kaolin
Lithium
Magnesium Compounds
Mica
Olivine
Peat
Perlite
Phosphate
Potash
Pumice
Quartz
Salt
Silica/Industrial Sand
Sulfur
Talc
Vermiculite
Zeolites
Zircon and Ilmenite

Asbestos
Barite
Bauxite
Bentonite
Borates
Clay/Shale
Diatomite
Dimension Stone
Fluorite
Feldspar

Impacts of Mining Activities

Mining and mineral-processing activities can modify the environment in a variety of ways, and few areas in the U.S. are untouched by mining activities (figure 12.19). Most obvious is the presence of the mine itself. Both underground mines and surface mines have their own sets of associated impacts. Though safety is improving, the hazards of mining activities to the miners should also not be overlooked: According to the National Safety Council, mining is the most dangerous occupation in the United States, with 43 deaths per 100,000 workers per year in 1989 (followed by agriculture at 40 and construction at 32 deaths per 100,000). Aspects of these hazards will be noted in the next chapter.

Underground Mines

Underground mines are generally much less apparent than surface mines. They disturb a relatively small area of the land's surface close to the principal shaft(s). Waste rock dug out of the mine may be piled close to the mine's entrance, but in most underground mines, the tunnels follow the ore body as closely as possible to minimize the amount of non-ore rock to be removed and, thus, mining costs. When mining activities are complete, the shafts can be sealed, and the area often returns very nearly to its pre-mining condition. However, near-surface underground mines occasionally have collapsed years after abandonment, when supporting timbers have rotted away or ground water has enlarged the underground cavities through solution (figure 12.20). In some cases, the collapse occurred so long after the mining had ended that the existence of the old mines had been entirely forgotten.

Surface Mines

Surface-mining activities consist of either open-pit mining (including quarrying) or strip-mining. Open-pit mining is practical when a large, three-dimensional ore body is located near the surface (figure 12.21A). Most of the material in the pit is the valuable commodity and is extracted for processing. Thus, this procedure permanently changes the topography, leaving a large hole in its wake. The exposed rock may begin to weather and, depending on the nature of the ore body, may release pollutants into surface runoff water.

Strip-mining, more often used to extract coal than mineral resources, is practiced most commonly when the material of interest occurs in a layer near and approximately parallel to the surface (figure 12.21B). Overlying vegetation, soil, and rock are stripped off, the coal or other material is removed, and the waste rock and soil is dumped back as a series of **spoil banks** (figure 12.22A). Until recently, that was all that was done. The broken-up material of the spoil banks, with its high surface area, is very susceptible to both erosion and chemical weathering. Chemical and sediment pollution of runoff from spoil banks was common. Vegetation reappeared on the steep, unstable slopes only gradually, if at all (figure 12.22B). Now, much stricter laws govern strip-mining activities, and reclamation of new strip mines is required. Reclamation usually involves regrading the area to level the spoil banks and to provide a more gently sloping land surface (figure 12.22C); restoring the soil; replanting grass, shrubs, or other vegetation; and, where necessary, fertilizing and/or watering the area to help establish vegetation. The result, when successful, can be land on which evidence of mining activities has essentially been obliterated (figure 12.22D).

A

B

▌ **Figure 12.20** Subsurface mining activities sometimes affect the earth's surface. (*A*) Collapse of land surface over old abandoned copper mine in Arizona. Note roads, trees, and houses for scale. (*B*) Subsidence pits and troughs over abandoned underground coal mines in Sheridan County, Wyoming.   Photograph (*A*) by H. E. Malde, photograph (*B*) by C. R. Dunrud, both courtesy USGS Photo Library, Denver, CO.

Figure 12.21 Surface mining. (*A*) The world's largest open-pit mine: Bingham Canyon, Utah. Over $6 billion worth of minerals have been extracted from this one mine over the decades of it operation. (*B*) Strip-mining for manganese in South Africa. Photograph (*A*) courtesy of Kennecott. Photograph (*B*) courtesy of USGS Photo Library, Denver, CO.

Figure 12.22 Strip-mining and land reclamation. (*A*) Spoil banks, Rainbow coal strip mine, Sweetwater County, Wyoming. (*B*) Grading of spoils at Indian Head mine. (*C*) Indian Head coal strip mine, Mercer County, North Dakota; revegetation of ungraded spoils is slow. (*D*) Reclaimed portion of Indian Head mine one year after seeding.   Photographs by H. E. Malde, USGS Photo Library, Denver, CO.

Box 12.3 U.S. National Policy Response to Potential Mineral Shortages: Strategic Mineral Stockpiling

Leaving aside the overall adequacy of world mineral resources to meet world demands, each country that relies significantly on imports of mineral commodities must consider the implications of possible disruptions of those imports.

Not until after World War II was there widespread recognition among high-level U.S. government officials of U.S. vulnerability to loss of imported mineral supplies. The official response was the establishment of a peacetime stockpiling program and encouragement of more comprehensive mapping programs to identify and delineate domestic mineral deposits. Unfortunately, Congress failed to fund either of these programs fully. The need for such programs was again underscored in the early 1950s by events of the Cold War and Korean War: For example, the cutoff of tungsten from mainland China led to a critical shortage of that metal in the United States. Under President Eisenhower, high priority was again given to mineral resources as related to national security, and to stockpiling in particular.

However, after President Eisenhower's administration, stockpile purchases stopped. Under the next several administrations, both Democratic and Republican, there were major sales of stockpiled mineral materials, in part to reduce budget deficits, and in part to help keep prices down and reduce inflationary pressures. Events of the early 1970s combined to swing the pendulum back: The Arab oil embargo quadrupled the price of oil and tightened immediate availability; political/military disruption in Zaire doubled the price of cobalt; various other global events contributed to doubling of the prices of tin and zinc and to an increase of over 50 percent in the price of copper. National attention was once more forcibly focused on mineral resources.

A reexamination of the strategic importance of minerals led to the Strategic and Critical Minerals Stockpiling Revision Act of 1979, which provides for emergency stockpiles of some ninety-three materials for national defense purposes, sufficient to sustain the United States for at least three years in case of national emergency. Over 80 percent of the identified strategic materials are mineral materials. The National Materials and Minerals Policy, Research, and Development Act of 1980 further directed the federal government to develop policies to strengthen the nation's position with respect to such materials. A variety of activities have been planned or undertaken as part of this program, including conducting inventories of federal lands to assess their mineral potential, collecting data on mineral supply and demand trends, encouraging private mineral research and development projects with wide application to materials and productivity problems, and resuming stockpiling once again. None of these activities, however, will instantly eliminate substantial U.S. dependence on mineral imports (indeed, stockpiling will require short-term increases in some imports), nor do most of these actions address the larger issue of the ultimate limits of mineral resource supplies.

Naturally, reclamation is much more costly to the mining company than just leaving spoil banks behind. Costs can reach several thousand dollars per acre. Also, even when conscientious reclamation efforts are undertaken, the consequences may not be as anticipated. For example, the inevitable changes in topography after mining and reclamation may alter the drainage patterns of regional streams and change the proportion of runoff flowing to each. In dry parts of the western United States, where every drop of water in some streams is already spoken for, the consequences may be serious for landowners downstream. In such areas, the water needed to sustain the replanted vegetation may also be in very short supply.

Mineral Processing

Mineral processing to extract a specific metal from an ore can also cause serious environmental problems. Processing generally involves finely crushing or grinding the ore. The fine waste materials, or **tailings,** that are left over may end up heaped around the processing plant to weather and wash away, much like spoil banks. Generally, traces of the ore are also left behind in the tailings. Depending on the nature of the ore, rapid weathering of the tailings may leach out harmful elements, such as mercury, arsenic, cadmium, and uranium, to contaminate surface and ground waters. Uncontrolled usage of uranium-bearing tailings as fill and in concrete in Grand Junction, Colorado, has resulted in radioactive buildings that must be extensively torn up and rebuilt before they are safe for occupation.

The chemicals used in processing are often hazardous also. For example, cyanide is commonly used to extract gold from its ore. Water pollution from poorly-controlled runoff of cyanide solutions can be a significant problem. Smelting to extract metals from ores may, depending on the ores involved and on emission controls, release arsenic, lead, mercury, and other potentially toxic elements along with exhaust gases and ash. Sulfide-ore processing also releases the sulfur oxide gases that are implicated in the production of acid rain (chapter 17). These same gases, at high concentrations, can destroy nearby vegetation: Earlier in this century, one large smelter in British Columbia destroyed all conifers within *19 kilometers* (about 12 miles) of the operation. More stringent pollution controls are curbing such problems today. Still, mineral processing can be a huge potential source of pollutants if not carefully executed. Worldwide, not all nations are equally sensitive to the hazards.

Overview

Summary

Economically valuable mineral deposits occur in a variety of specialized geologic settings—igneous, sedimentary, and metamorphic. Both occurrence of and demand for minerals are very unevenly distributed worldwide. Projections for mineral use, even with conservative estimates of consumption levels, suggest that present reserves of most metals and other minerals could be exhausted within decades. As shortages arise, some currently subeconomic resources will be added to the reserves, but the quantities are unlikely to be sufficient to extend supplies greatly. Other strategies for averting further mineral shortages include applying new exploration methods to find more ores, looking to undersea mineral deposits not exploited in the past, and recycling metals to reduce the demand for newly mined material. Decreasing mining activity would minimize the negative impacts of mining, such as disturbance of the land surface and the release of harmful chemicals through accelerated weathering of pulverized rock or as a consequence of mineral-processing activities.

Terms to Remember

banded iron formation 277	ore 273
concentration factor 273	pegmatite 275
evaporite 279	placer 279
hydrothermal 277	remote sensing 286
kimberlites 276	spoil banks 293
manganese nodules 289	tailings 295

Exercises

Questions for Review

1. Explain how economics and concentration factor relate to the definition of an ore.
2. Describe two examples of magmatic ore deposits.
3. Hydrothermal ore deposits tend to be associated with plate boundaries. Why?
4. What is an evaporite? Give an example of a common evaporite mineral.
5. Explain how stream action may lead to the formation of placer deposits. Why is there interest in exploring for placers on the continental shelves?
6. As mineral reserves are exhausted, some resources may be reclassified as reserves. Explain.
7. How might plate-tectonic theory contribute to the search for new ore deposits?
8. What metallic mineral resource is found over much of the deep-sea floor, and what political problem arises in connection with it?
9. Why are aluminum and lead comparatively easy to recycle, while steel is less so?
10. Describe one hazard associated with underground mining.
11. What steps are involved in strip-mine reclamation? Can land always be fully restored to its pre-mining condition with sincere effort? Explain.
12. Why are tailings from mineral processing a potential environmental concern?

For Further Thought

1. Canada is the world's largest exporter of minerals and mineral products; it is the world's largest producer of uranium (28%), zinc (15%), and potash (34%), and among the world's top five producers of gold and of lead. Choose any mineral of which Canada is a major producer, and investigate its abundance and geologic occurrence. Do similar deposits occur in the U.S.? If so, how do they compare with the Canadian deposits? (See NetNotes for a start on Canadian data.)
2. Choose an area that has been subjected to extensive surface mining (examples include the coal country of eastern Montana or Appalachia and the iron ranges of upper Michigan). Investigate the history of mining activities in this area and of legislation relating to mine reclamation. Assess the impact of the legislation.

3. Select one metallic mineral resource and investigate its occurrence, distribution, consumption, and reserves. Evaluate the impact of its customary mining and extraction methods. (How and where is it mined? How much energy is used in processing it? What special pollution problems, if any, are associated with the processing?) Can this metal be recycled, and if so, from what sorts of material?

Suggested Readings/References

Agnew, A. F., ed. 1983. *International minerals: A national perspective.* Boulder, Colo.: Westview Press.

Banks, F. E. 1974. *The world copper market: An economic analysis.* Cambridge, Mass.: Ballinger Publishing.

Borgese, E. M. 1985. *The mines of Neptune: Metals and minerals from the sea.* New York: H. M. Abrams.

Broadus, J. M. 1987. Seabed materials. *Science* 235: 853–60.

Craig, J. R., Vaughan, D. J., and Skinner, B. J. 1988. *Resources of the Earth.* Englewood Cliffs, N.J.: Prentice-Hall.

Davis, L. L. 1992. Construction—the essential role of minerals. *Minerals Today* (April): 12–17.

Deep ocean mining. 1982. *Oceanus* 25(3). (This whole issue is devoted to various aspects of deep-sea mineral resources and the legal and technical problems of mining them.)

Fergusson, J. E. 1982. *Inorganic chemistry and the earth.* New York: Pergamon Press.

Heath, G. R. 1982. Manganese nodules: Unanswered questions. *Oceanus* 25(3): 37–41.

Hodges, C. A. 1995. Mineral resources, environmental issues, and land use. *Science* (2 June): 1305–12.

Institute for Polar Studies. 1977. *A framework for assessing environmental impacts of possible Antarctic mineral development.* Columbus, Ohio: Institute for Polar Studies, Ohio State University.

Jensen, M. L., and Bateman, A. M. 1981. *Economic mineral deposits.* 3d ed. New York: John Wiley & Sons.

Jordan, A. A., and Kilmarx, R. A. 1979. *Strategic mineral dependence: The stockpile dilemma.* Beverly Hills, Calif.: Sage Publications.

Leontief, W., Koo, J. C. M., Nasar, S., and Sohn, I. 1983. *The future of nonfuel minerals in the U.S. and world economy.* Lexington, Mass.: D. C. Heath.

Lumsden, G. I. (ed.). 1992. *Geology and the environment in western Europe.* Oxford: Clarendon Press.

McGregor, B. A., and Offield, T. W. 1987. *The Exclusive Economic Zone: An exciting new frontier.* U.S. Geological Survey.

Rahn, P. H. 1986. *Engineering geology.* New York: Elsevier.

Rona, P. A. 1986. Mineral deposits from seafloor hot springs. *Scientific American* 254 (January): 84–92.

Sawkins, F. J. 1984. *Metal deposits in relation to plate tectonics.* New York: Springer-Verlag.

Shusterich, K. M. 1982. *Resource management and the oceans.* Boulder, Colo.: Westview Press.

U.S. Bureau of Mines. 1996. *Mineral commodity summaries* 1996. (published annually)

———. *Minerals yearbook.* (published annually until the demise of the Bureau of Mines)

Whitmore, F. C., Jr., and Williams, M. E., eds. 1982. *Resources for the twenty-first century.* U.S. Geological Survey Professional Paper 1193.

Zumberge, J. H. 1979. Mineral resources and geopolitics in Antarctica. *American Scientist* 67 (January/February): 68–76.

The functions of the former U.S. Bureau of Mines, longtime source of mineral-resource information, have been folded into the Division of Energy and Mineral Resources of the U.S. Geological Survey. The USGS Mineral Resource Surveys Program maintains a site with numerous links to U.S. and international resource data at

    http://minerals.er.usgs.gov/

The Office of Surface Mining is found at

    http://www.osmre.gov/

The U.S. Minerals Management Service (MMS) oversees offshore mineral and fuel leasing arrangements, with special emphasis on the Gulf of Mexico, the Pacific Basin, and Alaska; see

    http://www.mms.gov/
    For Environmental Studies Program Information System (ESPIS) reports of MMS, filed from 1973 to the present; see

    http://www.mms.gov/espis/

Natural Resources Canada (NRC), which deals with a broad range of resources, including minerals, energy, and forestry, has its home page at

    http://www.emr.ca

The Minerals and Mining Statistics division of NRC is at

    http://www.nrcan.gc.ca/ms/efab/mmsd/
    (At the time of writing, this site was under construction and in the process of being expanded substantially.)

Some data on both U.S. and global trends in mineral-resource consumption, from the World Resources Institute, can be accessed via

    http://www.wri.org/wri/enved/trends/con-home.html

The United Nations Division for Sustainable Development's Committee on Natural Resources is found at

    http://www.un.org/dpcsd/dsd/natural.html

Chapter **13**

Energy Resources— Fossil Fuels

The earliest, and, at first, only energy source used by primitive people was the food they ate. Then, wood-fueled fires produced energy for cooking, heat, light, and protection from predators. As hunting societies became agricultural, people used energy provided by animals—horsepower and ox power. The world's total energy demands were quite small and could readily be met by these sources.

Gradually, animals' labor was replaced by that of machines, new and more sophisticated technologies were developed, and demand for manufactured goods rose. These factors all greatly increased the need for energy and spurred the search for new energy sources (figure 13.1). It was first wood, then primarily the fossil fuels that eventually met those needs, and they continue to dominate our energy consumption (figure 13.2). They also pose certain environmental problems. All contribute to the carbon-dioxide pollution in the atmosphere, and several can present other significant hazards, such as toxic spills and land subsidence (oil), or sulfur pollution, acid runoff, and mine collapse (coal).

The term *fossil* refers to any remains or evidence of ancient life. The **fossil fuels,** then, are those energy sources that formed from the remains of once-living organisms. These include oil, natural gas, coal, and fuels derived from oil shale and tar sand. The differences in the physical properties among the various fossil fuels arise from differences in the starting materials from which the fuels formed and in what happened to those materials after the organisms died and were buried within the earth.

Figure 13.1 As societies evolve from primitive to agricultural to industrialized, the daily per capita consumption of energy increases sharply. The body needs, on average, about 2,000 kilocalories per day to live (the "kilocalorie," a measure of energy, is the food calorie that dieters count). However, in the United States in 1989, the per capita energy consumption was 224,000 kilocalories per day: 36 percent for residential/commercial uses, 36 percent for industry, 27 percent for transportation. Of the total, about 36 percent was consumed as electricity. Note that, as societies advance, even the energy associated with the food individuals eat increases. Members of modern societies do not consume 10,000 kilocalories of food energy directly, but behind the 2,000 or so actually consumed are large additional energy expenditures associated with agriculture, food processing, and preservation, and even the conversion of plants to meat by animals, which in turn are consumed as food. The shaded area in the far left bar of the graph is electricity usage.   Source: Data from Earl Cook, "The Flow of Energy in an Industrialized Society," *Scientific American,* © 1971.

Even to find a single economic petroleum deposit takes many wells; to extract the oil, many more. Photo Disc/U.S. Travel & Landmark.

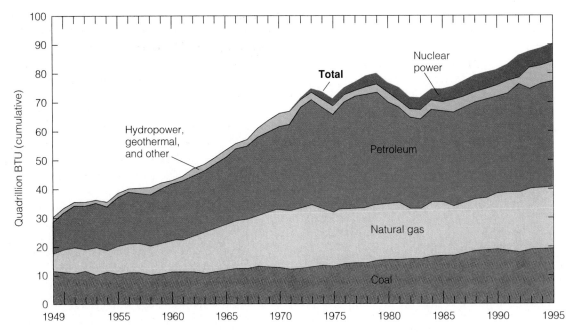

Figure 13.2 U.S. energy consumption 1949–1995. Note the consistent rise except following the Arab oil embargo of the early 1970s, and in the early 1980s, when high prices and a sluggish economy combined to depress energy consumption.   Source: U.S. Energy Information Administration, *Annual Energy Review 1995*, Department of Energy.

Oil and Natural Gas

Oil, or petroleum, is not a single chemical compound. Petroleum comprises a variety of liquid hydrocarbon compounds (compounds made up of different proportions of the elements carbon and hydrogen). There are also gaseous hydrocarbons (**natural gas**), of which the compound methane (CH_4) is the most common. How organic matter is transformed into liquid and gaseous hydrocarbons is not fully understood, but the main features of the process are outlined in the following section.

Formation of Oil and Gas Deposits

The production of a large deposit of any fossil fuel requires a large initial accumulation of organic matter, which is rich in carbon and hydrogen. Another requirement is that the organic debris be buried quickly to protect it from the air, so that decay by biological means or reaction with oxygen will not destroy it.

Microscopic life is abundant over much of the oceans. When these organisms die, their remains can settle to the sea floor. There are also underwater areas near shore, such as on many continental shelves, where sediments derived from continental erosion accumulate rapidly. In such a setting, the starting requirements for the formation of oil are satisfied: There is an abundance of organic matter rapidly buried by sediment. Oil and natural gas are believed to form from such accumulated marine microorganisms. Continental oil fields reflect the presence of marine sedimentary rocks below the surface.

As burial continues, the organic matter begins to change. Pressures increase with the weight of the overlying sediment or rock; temperatures increase with depth in the earth; and slowly, over long periods of time, chemical reactions take place. These reactions break down the large, complex organic molecules into simpler, smaller hydrocarbon molecules. The nature of the hydrocarbons changes with time and continued heat and pressure. In the early stages of petroleum formation, the deposit may consist mainly of larger hydrocarbon molecules ("heavy" hydrocarbons), which have the thick, nearly solid consistency of asphalt. As the petroleum matures, and as the breakdown of large molecules continues, successively "lighter" hydrocarbons are produced. Thick liquids give way to thinner ones, from which are derived lubricating oils, heating oils, and gasoline. In the final stages, most or all of the petroleum is further broken down into very simple, light, gaseous molecules—natural gas. Most of the maturation process occurs in the temperature range of 50° to 100° C (approximately 120° to 210° F). Above these temperatures, the remaining hydrocarbon is almost wholly methane; with further temperature increases, methane can be broken down and destroyed in turn (see figure 13.3).

A given oil field yields crude oil containing a distinctive mix of hydrocarbon compounds, depending on the history of the material. The refining process separates the different types of hydrocarbons for different uses (see table 13.1). Some of the heavier hydrocarbons may also be broken up during refining into smaller, lighter molecules through a process called *cracking,* which allows some of the lighter compounds such as gasoline to be produced as needed from heavier components of crude oil.

Oil and Gas Migration

Once the solid organic matter is converted to liquids and gases, the hydrocarbons can migrate out of the rocks in which they formed. Such migration is necessary if the oil or gas is to be collected into an economically valuable and practically usable deposit. The majority of petroleum source rocks are fine-grained clastic sedimentary rocks of low permeability, from which it would be difficult to extract large quantities of oil or gas quickly. Despite the low permeabilities, oil and gas are able to migrate out of their source rocks and through more permeable rocks over the long spans of geologic time. The pores, holes, and cracks in rocks in which fluids can be trapped are commonly full of water. Most oils

and all natural gases are less dense than water, so they tend to rise as well as to migrate laterally through the water-filled pores of permeable rocks.

Unless stopped by impermeable rocks, oil and gas may keep rising right up to the earth's surface. At many known oil and gas seeps, these substances escape into the air or the oceans or flow out onto the ground. These natural seeps, which are one of nature's own pollution sources, are not very efficient sources of hydrocarbons for fuel if compared with present drilling and extraction processes, although asphalt from seeps in the Middle East was used in construction five thousand years ago.

Commercially, the most valuable deposits are those in which a large quantity of oil and/or gas has been concentrated and confined by impermeable rocks in structures (see figure 13.4). The reservoir rocks in which the oil or gas has accumulated should be relatively porous if a large quantity of petroleum is to be found in a small volume of rock, and should also be relatively permeable so that the oil or gas flows out readily once a well is drilled into the reservoir. If the reservoir rocks are not naturally very permeable, it may be possible to fracture them artificially with explosives or with water or gas under high pressure to increase the rate at which oil or gas flows through them.

The Time Factor

The amount of time required for oil and gas to form is not known precisely. Since virtually no petroleum is found in rocks younger than 1 to 2 million years old, geologists infer that the process is comparatively slow. Even if it took only a few tens of thousands of years (a geologically short period), the world's oil and gas is being used up far faster than significant new supplies could be produced. Therefore, oil and natural gas are among the **nonrenewable** energy sources. We have an essentially finite supply with which to work.

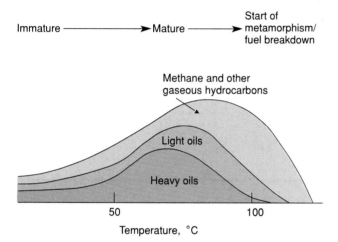

Figure 13.3 The process of petroleum maturation in simplified form.   Source: Simplified from J. Watson, *Geology and Man*. 1983 Allen and Unwin, Inc.

Table 13.1 Fuels Derived from Liquid Petroleum and Gas

	Material	Principal uses
Heavier hydrocarbons	waxes (for example, paraffin)	candles
	heavy (residual) oils	heavy fuel oils for ships, power plants, and industrial boilers
	medium oils	kerosene, diesel fuels, aviation (jet) fuels, power plants, and domestic and industrial boilers
	light oils	gasoline, benzene, and aviation fuels for propeller-driven aircraft
	"bottled gas" (mainly butane, C_4H_{10})	primarily domestic use
Lighter hydrocarbons	natural gas (mostly methane, CH_4)	domestic/industrial use and power plants

Source: Data from J. Watson, *Geology and Man*, copyright © 1983 Allen and Unwin, Inc.

Figure 13.4 Types of petroleum traps. (*A*) A simple fold trap. (*B*) Petroleum accumulated in a fossilized ancient coral reef. (*C*) A fault trap. (*D*) Petroleum trapped against an impermeable salt dome, which has risen up from a buried evaporite deposit.

Supply and Demand for Oil and Natural Gas

As with minerals, the most conservative estimate of the supply of an energy source is the amount of known reserves, "proven" accumulations that can be produced economically with existing technology. A more optimistic estimate is total resources, which include reserves, plus known accumulations that are technologically impractical or too expensive to tap at present, plus some quantity of the substance that is expected to be found and extractable. As with minerals, estimates of energy reserves are thus sensitive both to price fluctuations and to technological advances.

Oil

Oil is commonly discussed in units of barrels (1 barrel = 42 gallons). Worldwide, over 400 billion barrels of oil have been consumed, and the estimated remaining reserves are somewhat over 1 trillion barrels (figure 13.5). That does not sound too ominous until one realizes that more than half of the consumption has occurred in the last two decades or so. Also, global demand will continue to increase as more countries advance technologically.

Oil supply and demand are very unevenly distributed around the world. Some low-population, low-technology, oil-rich countries (Libya, for example) may be producing fifty or a hundred times as much oil as they themselves con-

sume. At the other extreme are countries like Japan, highly industrialized and energy-hungry, with no petroleum reserves at all. The United States alone consumes over 25 percent of the oil used worldwide. As with mineral resources, the size of petroleum resources of a given nation are not correlated with geographic size. Many have speculated that the quick international reaction to the takeover of Kuwait was motivated as much by concern for its 10 percent share of oil reserves as by humanitarian motives.

U.S. Oil Supplies

The United States initially had perhaps 10 percent of all the world's oil resources. The total U.S. oil resources, including what has been consumed to date, are estimated to have been not much above 200 billion barrels. Already more than half of that has been produced and consumed; remaining U.S. *re-sources* amount to about 80 billion barrels. Out of that, present proven *reserves* are under 30 billion barrels. For more than a decade, the United States has been using up about as much domestic oil each year as the amount of new domestic reserves discovered, or proven, and in many years somewhat more, so the net U.S. oil reserves have generally been decreasing year by year (see table 13.2). Domestic production has likewise been declining (figure 13.6).

Furthermore, the United States has for many years relied heavily on imported oil to meet part of its energy needs. Without it, U.S. reserves would dwindle more quickly. The

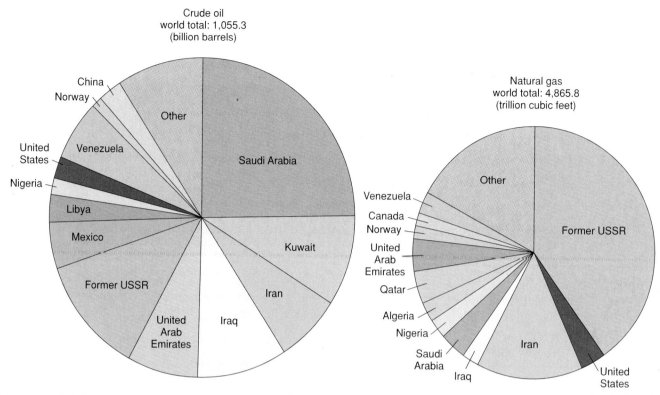

Figure 13.5 Estimated proven world reserves of crude oil and natural gas, January 1996.   Source: Averages of estimates of *Oil and Gas Journal* and *World Oil*, as summarized in *Annual Energy Review 1995*. U.S. Energy Information Administration.

Table 13.2 Proven U.S. Reserves of Crude Oil and Natural Gas, 1980–1994

Year	Crude oil (billions of barrels)	Natural gas (trillions of cu. ft.)
1980	29.8	199.0
1981	29.4	201.7
1982	27.9	201.5
1983	27.7	200.2
1984	28.4	197.5
1985	28.4	193.4
1986	26.9	191.6
1987	27.3	187.2
1988	26.8	168.0
1989	26.5	167.1
1990	26.3	169.3
1991	24.7	167.1
1992	23.7	165.0
1993	23.0	162.4
1994	22.5	163.8

Source: Data from *Annual Energy Review 1995*, U.S. Energy Information Administration, Department of Energy.

United States consumes over 6 billion barrels of oil per year, to supply about 40% of all the energy used. At times, half the oil consumed has been imported from other countries. That percentage fluctuated between 29% and 50% from 1985 to 1995; an estimated 49% of petroleum consumed in the U.S. in 1995 was imported. Simple arithmetic demonstrates that the rate of oil consumption in the United States is very high compared even to the estimated total U.S. resources available. Some of those resources have yet to be found, too, and will require time to discover and develop. Both U.S. and world oil supplies are likely to be nearly exhausted within decades (figure 13.7), especially considering the possible acceleration of world energy demands. On occasion, explorationists do find the rare, very large concentrations of petroleum—the deposits on Alaska's North Slope and beneath Europe's North Sea are examples. Yet even these make only a modest difference in the long-term picture. The Alaskan oil, for instance, represented reserves of only about 10 billion barrels, a great deal for a single region, but less than two years' worth of U.S. consumption.

Natural Gas

The supply/demand picture for conventional natural gas is similar to that for oil (see figures 13.2 and 13.6). Natural gas presently supplies close to 25 percent of the energy used in the United States. The United States has proven natural gas reserves of under 200 trillion cubic feet. However,

Figure 13.6 U.S. energy production since the mid-twentieth century, by energy source. Note that domestic production began to decline despite sharp rises in fossil-fuel prices. (Natural gas plant liquids (NGPL) are liquid fuels produced during refining/processing of natural gas.)   Source: *Annual Energy Review 1995*, U.S. Energy Information Administration, Department of Energy.

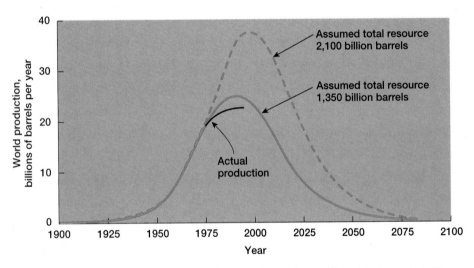

Figure 13.7 When this classic projection by M. King Hubbert was first widely published in the early 1970s, many people were shocked to discover that the total supply of oil appeared so limited. In fact, the total resource estimates had long existed in the geologic literature. The supply problem was made acute by the rapid rise in consumption starting in the 1950s. Interestingly, the basic picture has not changed appreciably in decades. Hubbert accurately predicted the slower rise in production of oil after 1975, which to a great extent is the result of depletion of reserves, especially easily accessible ones.   Source: Data from M. King Hubbert, "The Energy Resources of the Earth," *Scientific American*, Sept. 1971, p. 69.

roughly 20 trillion cubic feet are consumed per year, and each year, less is found in new domestic reserves than the quantity consumed. The net result is, again, declining reserves (table 13.2). The U.S. supply of conventional natural gas is thus expected to be used up in a matter of decades.

Future Prospects

Many people tend to assume that, as oil and natural gas supplies dwindle and prices rise, there will be increased exploration and discovery of new reserves, so that we will not really run out of these fuels. Most experts in the petroleum industry disagree.

While decreasing reserves have prompted exploration of more areas, most regions not yet explored have been neglected precisely because they are unlikely to yield appreciable amounts of petroleum. The high temperatures involved in the formation of igneous and most metamorphic rocks would destroy organic matter, so oil would not have formed or been preserved in these rocks. Nor do these rock types tend to be very porous or permeable, so they generally make poor reservoir rocks as well, unless fractured. The large portions of the continents underlain predominantly by igneous and metamorphic rocks are, for the most part, very unpromising places to look for oil. Some such areas are now being explored, but with limited success. Exploratory wells drilled off the east coast of the United States have likewise come up dry.

Despite a quadrupling in oil prices between 1970 and 1980 (*after* adjustment for inflation), U.S. proven reserves continued to decline. This is further evidence that higher prices do not automatically lead to proportionate, or even appreciable, increases in fuel supplies. Also, many major oil companies have been branching out into other energy sources beyond oil and natural gas, which indicates their expectation of a shift away from petroleum in the future.

Enhanced Oil Recovery

A few techniques are being developed to increase petroleum production from known deposits. An oil well initially yields its oil with minimal pumping, or even "gushes" on its own, because the oil and any associated gas are under pressure from overlying rocks, or perhaps because the oil is confined like water in an artesian system. This is *primary recovery*. When flow falls off, water may be pumped into the reservoir, filling empty pores and buoying up more oil to the well (*secondary recovery*). Primary and secondary recovery together extract an average of one-third of the oil in a given trap, though the figure varies greatly with the physical properties of the oil and host rocks in a given oil field. On average, then, two-thirds of the oil in each deposit has historically been left in the ground. Thus, *enhanced recovery* methods have attracted much interest.

Enhanced recovery comprises a variety of methods. Permeability of rocks may be increased by deliberate fracturing, using explosives or even water under very high pressure. Carbon dioxide gas under pressure can be used to force out more oil. Hot water or steam may be pumped underground to warm thick, viscous oils so that they flow more easily and can be extracted more completely. There have also been experiments using detergents or other substances to break up the oil.

All of these methods add to the cost of oil extraction. That they are now being used at all is largely a consequence of the dramatic increase in oil prices in the last decade or two. Researchers in the petroleum industry believe that, from a technological standpoint, up to an additional 40 percent of the oil initially in a reservoir might be extractable by enhanced-recovery methods. This would substantially increase oil reserves. A further positive feature of enhanced-recovery methods is that they can be applied to old oil fields that have already been discovered, developed by conventional methods, and abandoned, as well as to new discoveries. In other words, numerous areas in which these techniques could be used (if the economics were right) are already known, not waiting to be found. It should, however, be kept in mind that enhanced recovery may involve increases in problems such as ground subsidence or groundwater pollution that arise also with conventional recovery methods. Even conventional drilling uses large volumes of drilling mud to cool and lubricate the drill bits. The drilling muds may become contaminated with oil and must be handled carefully to avoid polluting surface or ground water. When water is more extensively used in fracturing rock or warming the oil, the potential for pollution is correspondingly greater.

Geopressurized Natural Gas and Other Alternate Gas Sources

Some evidence suggests that additional natural gas reserves exist at very great depths in the earth. Thousands of meters below the surface, conditions are sufficiently hot that any oil would have been broken down to natural gas (recall figure 13.3). This natural gas, under the tremendous pressures exerted by the overlying rock, may be dissolved in the water filling the pore spaces in the rock, much as carbon dioxide is dissolved in a bottle of soda. Pumping this water to the surface is like taking off the bottle cap: The pressure is released, and the gas bubbles out. Enormous quantities of natural gas might exist in such **geopressurized zones;** recent estimates of the amount of potentially recoverable gas in this form range from 150 to 2,000 trillion cubic feet.

Special technological considerations are involved in developing these deposits. It is difficult and very expensive to drill to these deep, high-pressure accumulations. Also, many of the fluids in which the gas is dissolved are very saline brines that cannot be casually disposed of at the surface without damage to the quality of surface water; the most effective means of disposal is to pump the brines back into the ground, but this is costly. On the plus side, however, the hot fluids themselves may represent an additional supplementary geothermal energy source (see the discussion of geothermal energy in chapter 14). Geopressurized natural gas may well become an important supplementary energy source in the future, though its total potential and the economics of its development are poorly known. It is unlikely that enough can be found and produced soon enough to solve near-future energy problems.

Still more recently, the significance of **methane hydrates** has been increasingly discussed. Gas (usually methane) hydrates are crystalline solids of gas and water molecules (figure 13.8). These hydrates have been found to be abundant in arctic regions and in marine sediments. The U.S. Geological Survey has estimated that the amount of carbon found in gas hydrates may be twice that in all known fossil fuels on earth (figure 13.9). Methane in methane hydrate thus represents a huge potential natural gas resource. Two recent studies of the coast of the Carolinas suggest methane resources of over 1,300 trillion cubic feet there alone. Just how to tap methane hydrates for their energy is not yet clear.

Beyond that, new questions are being raised about methane hydrates and the greenhouse effect. In arctic regions, the hydrates are stabilized by the low temperature. If these arctic sediments were warmed enough to break down some of the hydrates, methane gas would be released into the atmosphere, which would further increase greenhouse-effect

Figure 13.8 Irregular chunks of gas hydrate from sediments in the Sea of Okhotsk. Scale is in cm.   Photograph courtesy U.S. Geological Survey.

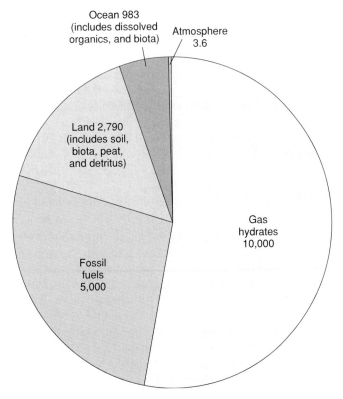

Figure 13.9 The amount of carbon now believed to exist in gas hydrates exceeds that in all known fossil fuels and other organic-carbon reservoirs combined.   Data from U.S. Geological Survey.

heating. Considering that methane is a far more efficient greenhouse gas than is CO_2, and that the amount of methane locked in hydrates now is estimated at about 3,000 times the methane currently in the atmosphere, the potential impact is large, and represents yet another source of uncertainty in global-climate modeling.

In the meantime, scientists are studying the feasibility of using fracturing techniques to initiate enhanced gas recovery from "tight" (low-permeability) sandstones in the Rocky Mountains and gas-bearing shales in the Appalachians. These projects are still in the experimental stages. As with geopressurized gas, the quantity of recoverable gas in these rocks is uncertain. Recent estimates, however, range from 60 to 840 trillion cubic feet. A factor that limits their current attractiveness is lack of price competitiveness, a problem they share with fuels derived from oil shale and tar sand, as will be seen later in the chapter.

Conservation

Conservation of oil and gas is, potentially, a very important way to stretch remaining supplies. More and more individuals, businesses, and governments in industrialized societies are practicing energy conservation out of personal concern for the environment, from fear of running out of nonrenewable fuels, or simply for basic economic reasons when energy costs have soared. (See box 13.1.) The combined effects of conservation and economic recession can be seen in the flattening of the U.S. energy-consumption curve in figure 13.2 during the mid-1970s to mid-1980s. However, beginning about 1990, energy consumption resumed rising steadily, reaching an all-time high in 1995. Some of the factors in 1995's record energy consumption were climatic—unusual heat in July and August, unusual cold in November and December—but other consumption increases were discretionary. Low gasoline prices, for example, made fuel-efficient cars and other fuel-conservation methods such as carpooling less urgent priorities for some consumers.

Conservation can buy some much-needed time to develop alternative energy sources. However, world energy consumption will probably not decrease significantly in the near future. Even if industrialized countries consistently adopt more energy-efficient practices, demand in the many nonindustrialized countries is expected to continue to rise. Many technologically less developed countries view industry and technology as keys to better, more prosperous lives for their people. They resent being told to conserve energy by nations whose past voracious energy consumption has largely led to the present squeeze.

From time to time, poor global economic conditions may depress energy use in the industrialized countries, as has happened over the past several years worldwide. Such events may lead to talk of an "oil glut" on world markets and to temporarily falling prices, but this surplus really

Box 13.1 An Aside on Some Economics of Drilling for Petroleum Fuels

The news media directed much attention in the 1980s to "windfall profits" made by oil companies. To such accusations the companies reply that the vast sums involved are needed for further exploration.

It is true, given rapid increases in the price of oil in the early 1980s, that companies were selling their oil for far more than it had cost to discover, drill for, and produce. A company selling off its "old" oil and planning no further exploration or drilling could indeed have pocketed a huge return.

On the other hand, the oil companies presumably intend to stay in the oil business as long as there is oil left. What oil remains is becoming more difficult to find and more expensive to drill for. It is widely believed that the "easy oil"—large accumulations located at shallow depths that can be tapped by moderate drilling from land-based drilling rigs—has already been found. The oil now being discovered is buried deeper in the earth. The first U.S. oil well was less than 15 meters deep; the average exploratory well today is nearly 1,800 meters deep, and the deepest gas wells now reach down nearly 10,000 meters (over 30,000 feet). Much of the new oil is being found offshore, but offshore drilling is far more expensive—perhaps fifteen to twenty times as expensive as land-based drilling—and raises concerns about environmental impacts on marine life. A single offshore well now may cost several million dollars to drill. The *average* cost per well drilled, whether it is a productive well or a dry hole, is close to $300,000.

Even with the best efforts of geologists and geophysicists directing the drilling, the rate at which wells in new areas find oil and gas is embarrassingly low. For every ten exploratory wells drilled, only about two find oil or gas, and fewer than one yields a commercially valuable deposit. Once a new find is made, the oil company must drill multiple wells to produce it.

Geologists have already extensively explored the most geologically promising areas. Further exploration will move into areas long regarded as less likely to be fruitful, previously left unexplored for just that reason. As M. King Hubbert predicted decades ago, the rate of discovery of new oil as a function of drilling activity has declined with time (see figure 1), despite advances in exploration technology and increased understanding of petroleum geology. That, too, means higher cost for new oil.

The supply problem has been compounded recently by a plummeting in the price of oil. It peaked in 1981, slipped about 25 percent over the period 1982–85, and then dropped precipitously, by about half, during 1986. In inflation-adjusted dollars, the average domestic crude-oil price in 1986 was only about one-third what it had been five years before.

With such a drastic decline in revenue from oil sales, companies were forced to curtail sharply their expenditures for exploration. The number of exploratory wells drilled dropped from 17,500 in 1981 to 3,920 in 1991; the success rate also declined, from nearly 30 percent to 23 percent, over the same period. Without exploration, new reserves go unfound and unproven. Despite widespread concern, the long-term effects of the Mideast

Figure 1   U.S. oil discoveries, projected and actual.   Source: U.S. Geological Survey Professional Paper 1193.

crisis of 1990–91 were negligible: prices rose about 20 percent in 1990 relative to 1989, but then dropped in 1991 to below 1989 prices (after adjustment for inflation). In the mid-1990s, oil prices, and consequently exploration activities, remain at depressed levels, with obvious implications for oil reserves.

Government policies also can greatly influence the economics of particular fuels. The National Gas Policy Act of 1978 introduced phased-in deregulation of natural gas prices, at least for "new gas" brought into production after 1977 and for expensive-to-produce "deep gas" from wells deeper than 5,000 meters. Deregulation has made exploration for new natural gas fields much more attractive and, consequently, has spurred considerable activity in that area. As table 13.2 indicates, however, the result has not been a consistent upsurge in natural gas reserves, providing further evidence that economics can do only so much when the basic constraints are those of geologic availability.

represents just a short-term temporary excess of current production over current demand. Assuming that world demand for energy stays at least as high as it is now, we face a crisis of dwindling petroleum and natural gas supplies over the next few decades. In the meantime, heavy reliance on oil and gas continues to have some serious environmental consequences, notably oil spills.

Oil Spills

Natural oil seeps are not unknown. In fact, it is estimated that oil rising up through permeable rocks escapes into the ocean at the rate of 600,000 tons per year. Tankers that flush out their holds at sea continually add to the oil pollution of the oceans and, collectively, are a significant source of such pollution. The oil spills about which most people have been concerned, however, are the large, sudden, catastrophic spills that occur in two principal ways: from accidents during drilling of offshore oil wells and from wrecks of oil tankers at sea. Oil spills represent the largest negative impacts from the extraction and transportation of petroleum, although as a source of water pollution, they are less significant volumetrically than petroleum pollution from careless disposal of used oil. And although we tend to hear only about the occasional massive, disastrous spill, there are many more that go unreported in the media: The U.S. Coast Guard reports about *ten thousand* spills in and around U.S. waters each year, totaling 15 to 25 million gallons of oil annually.

Drilling accidents have been a growing concern as more areas of the continental shelves are opened to drilling. Normally, the drill hole is lined with a steel casing to prevent lateral leakage of the oil, but on occasion, the oil finds an escape route before the casing is complete. This is what happened in Santa Barbara in 1979, when a spill produced a 200-square-kilometer (about 80-square-mile) oil slick. Alternatively, drillers may unexpectedly hit a high-pressure pocket that causes a blowout. This was the cause of a spill in the Gulf of Mexico in 1979 that released millions of gallons of oil.

Tanker disasters are potentially becoming larger all the time. The largest supertankers now being built are longer than the Empire State Building is high (over 335 meters), and can carry nearly 2 million barrels of oil (about 80 million gallons). The largest single marine spill ever resulted from the wreck of the *Amoco Cadiz* near Portsall, France, in 1978; the bill for cleaning up what could be recovered of the 1.6 million barrels spilled was more than $50 million, and at that, negative environmental impacts were still detectable seven years later.

In 1991, the war in the Persian Gulf demonstrated yet another possible cause of major oil spills: destruction of major pipelines and refinery facilities.

When an oil spill occurs, the oil, being less dense than water, floats. The lightest, most volatile hydrocarbons start to evaporate initially, decreasing the volume of the spill somewhat but polluting the air. Then a slow decomposition process sets in, due to sunlight and bacterial action. After several months, the mass may be reduced to about 15 percent of the starting quantity, and what is left is mainly thick asphalt lumps. These can persist for many months more. Oil is toxic to marine life, causes water birds to drown when it soaks their feathers, may decimate fish and shellfish populations, and can severely damage the economies of beach resort areas.

A variety of approaches to damage control following an oil spill have been tried. In calm seas, if a spill is small, it may be contained by floating barriers and picked up by specially designed "skimmer ships" that can skim up to fifty barrels of oil per hour off the water surface. Some attempts have been made to soak up oil spills with peat moss, wood shavings, and even chicken feathers. Large spills or spills in rough seas are a greater problem. When the tanker *Torrey Canyon* broke up off Land's End, England, in 1967, several strategies were tried, none very successfully. First, an attempt was made to burn off the spill. By the time this was tried, the more flammable, volatile compounds had evaporated, so aviation fuel was poured over the spill, and bombs were dropped to ignite it! This did not really work very well, and in any event, it would have produced a lot of air pollution. Some French workers used ground chalk to absorb and sink the oil. Sinking agents like chalk, sand, clay, and ash can be effective in removing an oil spill from the sea surface, but the oil is no healthier for marine life on the ocean bottom. Furthermore, the oil may separate out again later and resurface. The British mixed some 2 million gallons of detergent with part of the spill, hoping to break up the spill so that decomposition would work more rapidly. The detergents, in turn, turned out to be toxic to some organisms, too.

Such experiences, in which many cleanup methods are tried but none is notably effective, are typical. Perhaps the best, most environmentally benign prospect for future oil spills is the development of "oil-hungry" microorganisms that will eat the spill for food and thus get rid of it. Scientists are currently trying to develop suitable bacterial strains. For the time being, no good solution exists to the problems posed by a major oil spill. Preventive measures can be increased—for instance, double-hulled tankers and tugboat escorts are now mandatory for oil tankers in Prince William Sound, Alaska—but as long as demand for petroleum remains high, some accidents are, perhaps, inevitable. See also box 13.2, pp. 310–311.

Coal

Prior to the discovery and widespread exploitation of oil and natural gas, wood was the most commonly used fossil fuel, followed by coal as the industrial age began in earnest last century. However, coal was bulky, cumbersome, and dirty to handle and to burn, so it fell somewhat out of favor, particularly for home use, when the liquid and gaseous fossil fuels became available. Currently, though it supplies only about 20 percent of total U.S. energy needs, coal plays a key role in the energy picture, for coal-fired power plants account for more than half of U.S. electric power generation. Now that oil and gas supplies are dwindling, many energy users are looking again at the potential of coal for other applications also.

Formation of Coal Deposits

Coal is formed not from marine organisms, but from the remains of land plants. A swampy setting, in which plant growth is lush and where there is water to cover fallen trees, dead leaves, and other debris, is especially favorable to the initial stages of coal formation. The process requires **anaerobic** conditions, in which oxygen is absent or nearly so, since reaction with oxygen destroys the organic matter. (A pile of dead leaves in the street does not turn into fuel, and in fact, over time, leaves very little solid residue.)

The first combustible product formed under suitable conditions is *peat*. Peat can form at the earth's surface, and there are places on earth where peat can be seen forming today. Further burial, with more heat, pressure, and time, gradually dehydrates the organic matter and transforms the spongy peat into soft brown coal (**lignite**) and then to the harder coals (**bituminous** and **anthracite**). (See figure 13.10.) As the coals become harder, their carbon content increases, and so does the amount of heat released by burning a given weight of coal. The hardest, high-carbon coals (especially anthracite), then, are the most desirable as fuels because of their potential energy yield. As with oil, however, the heat to which coals can be subjected is limited: overly high temperatures lead to metamorphism of coal into graphite.

The higher-grade coals, like oil, apparently require formation periods that are long compared to the rate at which coal is being used. Consequently, coal too can be regarded as a nonrenewable resource. However, the world supply of coal represents an energy resource far larger than that of petroleum, and this is also true with regard to the U.S. coal supply (figure 13.11).

Coal Reserves and Resources

Coal resource estimates are subject to fewer uncertainties than corresponding estimates for oil and gas. Coal is a solid, so it does not migrate. It is found, therefore, in the sedimentary rocks in which it formed; one need not seek it in igneous or metamorphic rocks. It occurs in well-defined beds that are easier to map than underground oil and gas concentrations. And because it formed from land plants, which did not

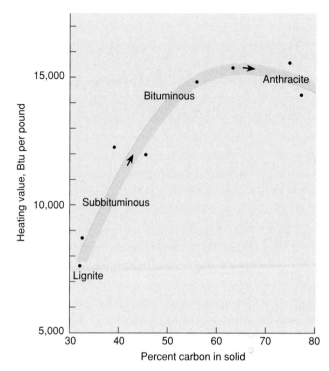

Figure 13.10 Change in character of coal with increasing application of heat and pressure. There is a general trend toward higher carbon content and higher heat value with increasing grade, though heat value declines somewhat as coal tends toward graphite.

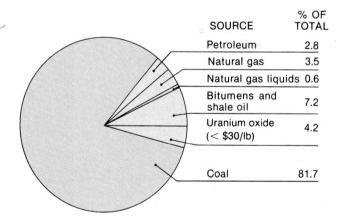

SOURCE	% OF TOTAL
Petroleum	2.8
Natural gas	3.5
Natural gas liquids	0.6
Bitumens and shale oil	7.2
Uranium oxide (< $30/lb)	4.2
Coal	81.7

Figure 13.11 Comparison of the total recoverable U.S. energy reserves from various fossil fuels and uranium. The key word is *recoverable*, which reflects the limitations of present technology.   Source: *The Coal Data Book*, 1980, The President's Commission on Coal.

become widespread until 400 million years ago, one need not look for coal in more ancient rocks.

The estimated world reserve of coal is about one trillion tons; total resources are estimated at over 10 trillion tons. The United States is particularly well supplied with

Box 13.2 Diary of an Oil Spill — The Exxon Valdez

Oil spills from tankers amount to only about 5 percent of the petroleum released into the environment, but they are of special concern because they may have a concentrated local impact.

The port of Valdez lies at the southern end of the Trans-Alaska Pipeline (figure 1). Through that pipeline flow 1.6–1.7 million barrels of oil a day. From Valdez, the oil is transported by tanker to refinery facilities elsewhere. Prince William Sound, in which Valdez is located, is home to a great variety of wildlife, from large mammals (whales, dolphins, sea otters, seals, and sea lions) to birds (bald eagles, cormorants, Canada geese, and many others) to shellfish (crabs, shrimp, clams, oysters) and fish (of which salmon and herring are particularly important economically). Most of the time, humans and animals coexist uneventfully (figure 2).

Early in the morning of 23 March 1989, the tanker *Exxon Valdez,* loaded with 1.2 million barrels (50 million gallons) of crude oil from Alaska's North Slope, ran aground on Bligh Island in Prince William Sound. The grounding may have been due to human error. The result was the worst oil spill yet in U.S. waters. Response to the spill was delayed for ten to twelve hours after the accident, in part because an oil-containment barge happened to be disabled. Within the first twelve hours, an estimated 10.2 million gallons of oil escaped, eventually to spread over more than 900 square miles of water. The various cleanup efforts cost Exxon an estimated $2.5 billion.

Skimmers recovered relatively little of the oil, as is typical: Three weeks after the accident, only about 5 percent

Figure 1   The oil tanker port at Valdez, Prince William Sound, Alaska, southern end of the Trans-Alaska Pipeline. (Note classic glacial valley in background.)

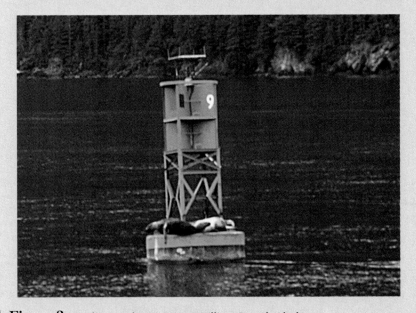

Figure 2   Sea lions on buoy, Prince William Sound, Alaska.

of the oil had been recovered. Chemical dispersants were not very successful (perhaps because their use was delayed by concerns over *their* impact on the environment), nor were attempts to burn the spill. Four days after the accident, the oil had emulsified by interaction with the water into a thick, mousse-like substance, difficult to handle, too thick to skim, impossible to break up or burn effectively. Oil that had washed up on shore had coated many miles of beaches and rocky coast (figure 3).

Thousands of birds and marine mammals died or were sickened by the oil. The annual herring season in Valdez was cancelled, and salmon hatcheries were threatened. When other methods failed, it seemed that only slow degradation over time would get rid of the oil. Given the size of the spill and the cold Alaskan temperatures, prospects seemed grim; microorganisms that might assist in the breakdown grow and multiply slowly in the cold. It appeared highly likely that the spill would be very persistent.

So, in a $10-million experiment, Exxon scientists sprayed miles of beaches in the Sound with a fertilizer solution designed to stimulate the growth of naturally occurring microorganisms that are known to "eat" oil, thereby contributing to its decomposition. The experiment seemed to succeed. Within two weeks, treated beaches were markedly cleaner than untreated ones. At first, there was concern that the chemicals were merely breaking up the oil mechanically, but further tests coupled with laboratory experiments confirmed that microbe-assisted chemical breakdown was occurring. Five months later, the treated beaches still showed higher levels of those microorganisms than did untreated beaches. The "cure" isn't perfect. The fertilizer solution runs off rocky shores, so this treatment won't work on rocky stretches of the oil-contaminated Alaskan coast; nor is it as effective once the oil has seeped deep into beach sands. But microorganisms may indeed be the future treatment of choice.

Analysis of the success of the *Valdez* cleanup is ongoing. The rough, cobbled, irregular shore is difficult to sample adequately, but so far, follow-up studies have failed to show statistically significant differences in oil removal from sediments on nutrient-treated and untreated beaches.

Examination of the long-term effects of another strategy, hot-water washing of beaches, has suggested that it may actually do more harm than good. The heat can kill organisms that had survived the oil, the pressure of the hot-water blast can drive oil deeper into the sediments where degradation is slower, and the water treatment may wash oil onto beaches and into tidepools that were unaffected by the original spill.

Figure 3 Beach cleanup workers in oil-splattered raingear mop oil from beach on La Touche Island, Prince William Sound. Alaska State Archive slide, courtesy Oil Spill Public Information Center.

From 1989 to 1991, worldwide, there were five spills of over 5 million gallons apiece, and numerous smaller spills. Clearly there is a need for more research on, and more answers about, how to deal with spills of this magnitude.

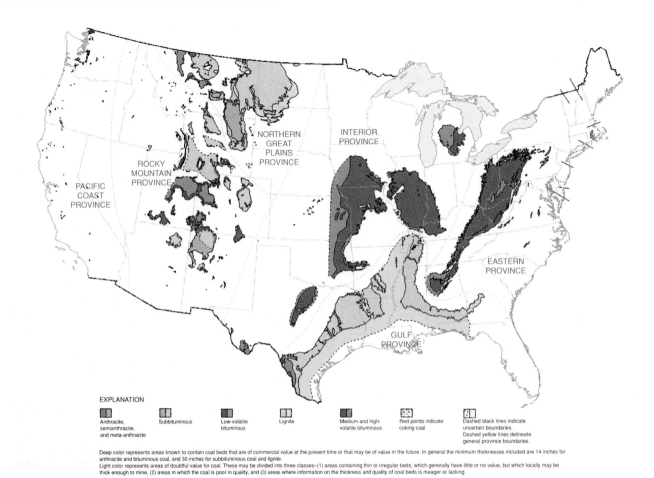

Figure 13.12 Distribution of U.S. coal fields.   Source: U.S. Geological Survey.

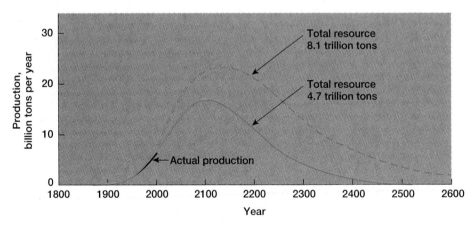

Figure 13.13 Projected world coal production, assuming coal as the principal substitute for petroleum. Note what a small fraction of coal has been consumed, and compare that with oil consumption (figure 13.7).   Source: Data from M. King Hubbert, "The Energy Resources of the Earth," *Scientific American*, Sept. 1971, p. 69. Actual production data to 1995 from *Annual Energy Review 1995*, U.S. Energy Information Administration.

coal (figure 13.12), possessing nearly 30 percent of the world's reserves, over 290 billion tons of recoverable coal. Total U.S. coal resources may approach ten times that amount, and most of that coal is yet unused and unmined. At present, coal provides about 25 percent of the energy used in the United States. While the United States has consumed probably half of its total petroleum resources, it has used up only a few percent of its coal. Even if only the reserves are counted, the U.S. coal supply could satisfy U.S. energy needs for more than two hundred years at current levels of energy use, if coal could be used for all purposes. As a supplement to other energy sources, coal could last many centuries (figure 13.13). Unfortunately, heavy reliance on coal has drawbacks.

Limitations on Coal Use

A primary limitation of coal use is that solid coal is simply not as versatile a substance as petroleum or natural gas. A major shortcoming is that it cannot be used directly in most forms of modern transportation, such as automobiles and airplanes. (It was, of course, used to power trains in the past, but coal-fired locomotives quickly fell out of favor when cleaner fuels became available.) Coal is also a dirty and inconvenient fuel for home heating, which is why it was abandoned in favor of oil or natural gas. Therefore, given present technology, coal simply cannot be adopted as a substitute for oil in all applications.

Coal can be converted to liquid or gaseous hydrocarbon fuels—gasoline or natural gas—by causing the coal to react with steam or with hydrogen gas at high temperatures. The conversion processes are called **gasification** (when the product is gaseous) and **liquefaction** (when the product is liquid fuel). Both processes are intended to transform coal into a cleaner-burning, more versatile fuel, thus expanding its range of possible applications.

Gasification

Commercial coal gasification has existed on some scale for 150 years. Many U.S. cities used the method before inexpensive natural gas became widely available following World War II. Europeans continued to develop the technologies into the 1950s. At present, only experimental coal-gasification plants are operating in the United States.

Current gasification processes yield a gas that is a mixture of carbon monoxide and hydrogen with a little methane. The heat derived from burning this mix is only 15 to 30 percent of what can be obtained from burning an equivalent volume of natural gas. Because the low heat value makes it uneconomic to transport this gas over long distances, it is typically burned only where produced. Technology exists to produce high-quality gas from coal, equivalent in heat content to natural gas, but it is presently uneconomical when compared to natural gas. Research into improved technologies continues.

Pilot projects are also underway to study *in situ* underground coal gasification, through which coal would be gasified in place and the gas extracted directly. The expected environmental benefits of not actually mining the coal include reduced land disturbance, water use, air pollution, and solid waste produced at the surface. Potential drawbacks include possible groundwater pollution and surface subsidence over gasified coal beds. Underground gasification may provide a means of using coals that are too thin to mine economically or that would require excessive land disruption to extract.

Liquefaction

Like gasification, the practice of generating liquid fuel from coal has a longer history than many people realize. The Germans used it to produce gasoline from their abundant coal during World War II; South Africa has a liquefaction plant in operation now, producing gasoline and fuel oil. A variety of technologies exist, and several noncommercial pilot plants have been in operation in the United States. Their liquid fuel products are not yet economically competitive with conventional petroleum. In the absence of substantial government subsidies, coal liquefaction will probably not become commercially feasible before the late 1990s, if then. However, technological advances made during the 1980s have slashed costs by 60 percent and greatly improved yields (to about 70 percent of original coal converted to liquid fuel). Such developments are encouraging, given our dependence on liquid petroleum and relative reserves of oil and coal.

Environmental Impacts of Coal Use

A major problem posed by coal is the pollution associated with its mining and use. Like all fossil fuels, coal produces carbon dioxide (CO_2) when burned. In fact, it produces significantly more carbon dioxide per unit energy released than oil or natural gas. (It is the burning of the carbon component [to make CO_2] that is the energy-producing reaction when coal is burned; when hydrocarbons are burned, the reactions produce partly CO_2, partly H_2O as their principal by-products, the proportions varying with the ratio of carbon to hydrogen in the particular hydrocarbon(s).) The potentially harmful effects of carbon dioxide and greenhouse-effect heating were discussed in chapter 9. The additional pollutant that is of special concern with coal is sulfur.

Sulfur in Coal

The sulfur content of coal can be more than 3 percent, some in the form of the iron sulfide mineral pyrite (FeS_2), some bound in the organic matter of the coal itself. When the sulfur is burned along with the coal, sulfur gases, notably sulfur dioxide (SO_2), are produced. These gases are poisonous and are extremely irritating to eyes and lungs. The gases also react with water in the atmosphere to produce sulfuric acid, a very strong acid. This acid then falls to earth as acid rainfall. Acid rain falling into streams and lakes can kill fish and other aquatic life. It can acidify soil, stunting plant growth. It can dissolve rock; in areas where acid rainfall is a severe problem, buildings and monuments are visibly corroding away because of the acid. The geology of acid rain is explored further in chapter 17.

Oil can contain appreciable sulfur derived from organic matter, too, but most of that sulfur can be removed during the refining process, so that burning oil releases only about one-tenth the sulfur gases of burning coal. Some of the sulfur can be removed from coal prior to burning, but the process is expensive and only partially effective, especially with organic sulfur. Alternatively, sulfur gases can be trapped by special devices ("scrubbers") in exhaust stacks, but again, the process is expensive (in terms of both money and energy) and not totally efficient. From the standpoint of

Table 13.3 Regional Differences in U.S. Coals

Region (% of total)	Coal category (% of total)		Sulfur content* (% of total)		Potential recovery method (% of total)	
Appalachia 22.5%	lignite	1	high	21	surface	18
	bituminous	92	medium	39	deep	82
	anthracite	7	low	28		
			unknown	12		
Midwest 28.2%	lignite	10	high	61	surface	30
	bituminous	90	medium	14	deep	70
	anthracite	—	low	2		
			unknown	23		
West 49.3%	lignite	13	high	1	surface	40
	bituminous	87	medium	16	deep	60
	anthracite	—	low	77		
			unknown	6		

Sources: *The Coal Data Book*, 1980, The President's Commission on Coal; *Annual Energy Review 1986*, Energy Information Administration, U.S. Department of Energy.

*High-sulfur, over 3 percent; medium, 1.1 to 3 percent; low-sulfur, 1 percent or less.

environmental quality, then, low-sulfur coal (1 percent sulfur or less) is more desirable than high-sulfur coal for two reasons: first, it poses less of a threat to air quality, and second, if the sulfur must be removed before or after burning, there is less of it to remove, so stricter emissions standards can be met more cheaply. On the other hand, much of the low-sulfur coal in the United States, especially western coal, is also lower-grade coal (see table 13.3), which means that more of it must be burned to yield the same amount of energy. This dilemma is presently unresolved.

Ash

Coal use also produces a great deal of solid waste. The ash residue left after coal is burned typically ranges from 5 to 20 percent of the original volume. The ash, which consists mostly of noncombustible silicate minerals, also contains toxic metals. If released with waste gases, the ash fouls the air. If captured by scrubbers or otherwise confined within the combustion chamber, this ash still must be disposed of. If exposed at the surface, the fine ash, with its proportionately high surface area, may weather very rapidly, and the toxic metals can be leached from it, thus posing a water-pollution threat. Uncontrolled erosion of the ash could likewise cause sediment pollution. The magnitude of this waste-disposal problem should not be underestimated. A single coal-fired electric power plant can produce over a million tons of solid waste a year. There is no obvious safe place to dump all that waste.

Coal-Mining Hazards and Environmental Impacts

Coal mining poses further problems. Underground mining of coal is notoriously dangerous, as well as expensive. Mines

can collapse; miners may contract black lung disease from breathing the dust; there is always danger of explosion from pockets of natural gas that occur in many coal seams. (See also box 13.3.) A tragic coal-mine fire in Utah in December 1984 was a reminder of the seriousness of the fire hazards. Recent evidence further suggests that coal miners are exposed to increased cancer risks from breathing the radioactive gas radon, which is produced by natural decay of uranium in rocks surrounding the coal seam (see chapter 17). The rising costs of providing a safer working environment for underground coal miners, together with the greater technical difficulty (and thus lower productivity) of underground mines, are largely responsible for a steady shift in coal-mining methods, from about 20 percent surface mining around 1950 to over 60 percent surface mining by 1980. Surface mining continues to dominate, as demand for low-sulfur western coal increases (figure 13.14). The positive side of this shift in methods is that it has been accompanied by a sharp decline in U.S. coal-mine fatalities (figure 13.15); strip-mining of coal is far safer for the miners.

A significant portion of U.S. coal, particularly in the west, occurs in thick beds very close to the surface. It is relatively cheap to strip off the vegetation and soil and dig out the coal at the surface, thus making strip-minable coal very attractive economically in spite of land reclamation costs.

Still, strip-mining presents its own problems. Strip-mining in general was discussed in chapter 12 (see also figure 12.22). A particular problem with strip-mining coal involves, again, the sulfur in the coal. Not every bit of coal is extracted from the surrounding rock. Some coal and its associated sulfur are left behind in the waste rock in spoil banks. Pyrite is also common in the shales with which coal often is

Box 13.3 Caution— Fossil Fuels Below!

The potential evils of strip mines have received wide publicity. Less well-known, but often equally serious, are the possible consequences of underground extraction of fossil fuels.

Withdrawal of oil, like that of water, can cause surface subsidence, resulting in building failure and, in low-lying coastal areas, flooding. Subsidence of up to several meters has been observed in more than thirty oil fields in California and Texas, and the flooding of low-lying coastal properties at Long Beach, California, has been widely publicized. The pumping of water associated with secondary oil recovery often serves the additional purpose of "propping up" the land surface as the petroleum is withdrawn.

Surface subsidence over old, shallow underground coal mines also is not uncommon (recall figure 12.20B). In fact, it is much more common over coal mines than other types of mines, partly because the sedimentary rocks with which coal beds are interlayered may be weaker than the igneous or metamorphic rocks that may host ore deposits, partly as a consequence of coal-mining methods that maximize coal extracted, often leaving only "pillars" of coal to support the overlying rock layers. Indeed, sometimes even the pillars are mined out as the last stage of mining, and the overlying rock slowly sinks under its own weight. This subsidence typically takes the form of pits or trenches, which may be deeper than the thickness of coal mined. Surface failure may lag years or decades behind mining, developing as supporting structures decay and subsurface waters weaken rocks through weathering. Areas so affected are particularly common in the northwestern and north-central United States.

As water and oxygen seep into abandoned coal mines, spontaneous combustion may ignite the remaining coal (figure 1). In the mid-1980s, the U.S. Bureau of Mines estimated that more than 250 uncontrolled mine fires were presently burning in seventeen states. The burning can lead to more settling, more subsidence pits, and increased air supply to the fire. Carbon monoxide and noxious sulfur gases rise to the surface. Such fires are not easily extinguished. Pouring in water from the surface has been known to *increase* some mine fires' intensity. The flames can, in principle, be suffocated by blocking all pits, shafts, and other openings through which air is reaching the flames, but the geometry of the mine workings and cracks above may be so complex that locating and sealing every airway is impossible. As a result, such fires can and do burn for years. An underground coal-mine fire that may initially have been ignited by a trash fire at a dump has been burning beneath Centralia, Pennsylvania, for a quarter of a century. The fire has traveled several kilo-meters underground in that time, and, periodically, people and property have dropped into new collapse pits at the surface. In 1983, conceding the futility of repeated fire-control efforts, the federal government approved a $42 million plan to buy out owners of threatened properties—not all of whom wanted to leave their homes.

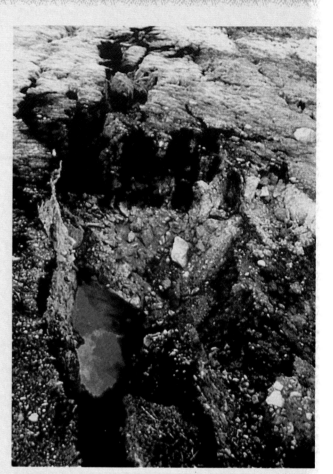

Figure 1   Fire out of control in abandoned underground coal mine in Sheridan County, Wyoming.   Photograph by C. R. Dunrud, USGS Photo Library, Denver, CO.

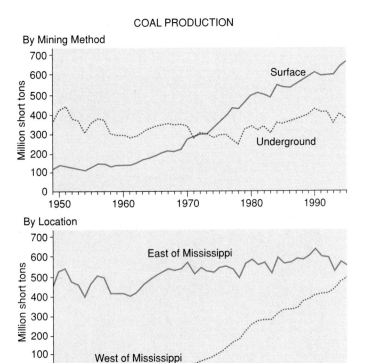

Figure 13.14 Rise in proportion of surface vs. underground mining parallels relative increase in production of western coal.   Source: Data from *Annual Energy Review 1995*, U.S. Energy Information Administration, Department of Energy.

interlayered, and these pyritic shales form part of the spoil banks. The sulfur in the spoils can react with water and air to produce runoff water containing sulfuric acid:

$$2\ FeS_2 + 5\ H_2O + \tfrac{15}{2}\ O_2\ =\ 4\ H_2SO_4 + Fe_2O_3 \cdot H_2O$$

pyrite   water   oxygen   sulfuric   hydrated iron
                                        acid   oxide

Because plants grow poorly in very acid conditions, this acid slows revegetation of the area by stunting plant growth or even preventing it altogether (figure 13.16). The acid runoff can also pollute area ground and surface waters, killing aquatic plants and animals in lakes and streams and contaminating the water supply. Very acidic water also can be particularly effective at leaching or dissolving some toxic elements from soils, further contributing to water pollution. Coal strip mines, like others, can be reclaimed, but in addition to regrading and replanting the land, it is frequently necessary to replace the original topsoil (if it was saved) or to bring in fresh soil so that sulfur-rich rocks are not left exposed at the surface, where weathering is especially rapid. (Even underground coal mines may have associated acid drainage, but water circulation there is generally more restricted.)

Coal-mine reclamation can be very effective, but it can also be slow and/or expensive. The ease of reclamation, as with any surface-mining operation, varies with the chemistry of spoils or soil and how it influences water chemistry, with

the amount of available precipitation, and with the nature of local vegetation (how hardy it is and how readily it spreads). The question of water availability is of special concern in the western United States: 50 percent of U.S. coal reserves are found there, 40 percent of these reserves would be surface-mined, and most regions underlain by western coal are very dry (figure 13.17).

When coal surface mines are reclaimed, efforts are commonly made to restore original topography—although if a thick coal seam has been extracted, this may not be easy. The artificial slopes of the restored landscape may not be altogether stable, and as natural erosional processes begin to sculpture the land surface, natural slope adjustments in the form of slumps and slides may occur. In dry areas, thought must be given to how the reclamation will modify drainage patterns (surface and subsurface). In wetter areas, the runoff may itself be a problem, gullying slopes and contributing to sediment pollution of surface waters until new vegetation is established; note gullying in figure 13.16.

Sometimes, coal-mine reclamation can even take a creative turn. In late 1984, reclamation efforts were begun on a coal strip mine in north-central Illinois, abandoned and barren since the 1940s. Instead of merely creating a normal landscape, the restorers are arranging some of the soil into mounds shaped like giant, abstract animal forms, reminiscent of the effigies built by the Mound Builder Indians in the region more than one thousand years ago. When the project is complete, the area will become a park.

Oil Shale

Oil shale is very poorly named. The rock, while always sedimentary, need not be a shale, and the hydrocarbon in it is not oil! The potential fuel in oil shale is a sometimes-waxy solid called **kerogen,** which is formed from the remains of plants, algae, and bacteria. Kerogen is not a single compound; the term is a general one describing organic matter that is not readily dissolved either in water or in organic solvents. These physical properties dictate that the rock must be crushed and heated to distill out the hydrocarbon as "shale oil," which then is refined somewhat as crude oil is to produce various liquid petroleum products.

The United States has about two-thirds of the world's known supply of oil shale (see locations, figure 13.18), for a total estimated resource of 2 to 5 trillion barrels of shale oil. For a number of reasons, the United States is not yet using this apparently vast resource to any significant extent, and it may not be able to do so in the near future.

One reason for this is that much of the kerogen is so widely dispersed through the oil shale that huge volumes of rock must be processed to obtain moderate amounts of shale oil. Even the richest oil shale yields only about three barrels of shale oil per ton of rock processed. The cost is not presently competitive with that of conventional petroleum, even at current oil prices. Nor have large-scale processing

Figure 13.15 Drop in coal-mine fatalities reflects partly safer mining practices, partly the shift from underground mining to strip-mining.   Source: U.S. Bureau of Mines, *Minerals Today*, Oct. 1991, p. 19.

Figure 13.16 Abandoned coal strip mine, Fulton County, Illinois. There is no lack of moisture, but plants will not thrive in acid soil.   Photograph by Arthur Greenberg, courtesy of EPA/National Archives.

Figure 13.17 Rainbow coal strip mine, Sweetwater County, Wyoming.   Photograph by H. E. Malde, USGS Photo Library, Denver, CO.

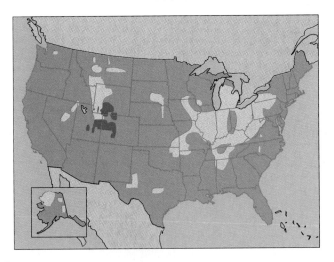

Figure 13.18 Distribution of U.S. oil shale deposits. The richest of these deposits is the Green River formation (blue). Source: Data from J. W. Smith, "Synfuels: Oil Shale and Tar Sands," in *Perspectives on Energy*, 3d ed., ed. by Lon C. Ruedisili and Morris W. Firebaugh, 1982, Oxford University Press, NY.

Figure 13.19 Oil shale in outcrop: the Green River formation in Colorado. Note the sparse vegetation in this area. Photograph by R. L. Elderkin, Jr., USGS Photo Library, Denver, CO.

facilities yet been built—present plants are strictly experimental. More efficient and cheaper processing technologies are needed.

Another problem is that a large part of the oil shale is located at or near the surface. At present, the economical way to mine it, therefore, appears to be surface- or strip-mining with its attendant land disturbance. Most of the richest oil shale is located in areas of the western United States that are already chronically short of water (figure 13.19). The dry conditions would make revegetation of the land after strip-mining especially difficult. Some consideration has been given to *in situ* heating and extraction of the kerogen without mining to minimize land disruption (figure 13.20), but the necessary technology is poorly developed, and the process leaves much of the fuel behind.

The water shortage presents a further problem. Current processing technologies require large amounts of water—on the order of three barrels of water per barrel of shale oil produced. Just as water for reclamation is in short supply in the west, so, too, is the water to process the oil shale.

Finally, since the volume of the rock actually increases during processing, it is possible to end up with a 20 to 30 percent larger volume of waste rock to dispose of than the original volume of rock mined. Aside from any problems of accelerated weathering of the crushed material, there remains the basic question of where to put it. Because it will not all fit back into the space mined, the topography will inevitably be altered.

Scientists and economists differ widely in the extent to which they see oil shale as a promising alternative to conventional oil and gas. Certainly, the water-shortage, waste-disposal, and land-reclamation problems will have to be solved before shale oil can be used on a large scale. Also, the lower oil prices fall, the less competitive oil shale becomes. It is unlikely that those problems can be solved within the next few decades, although over the longer term, oil shale may become an important resource.

Tar Sand

Tar sands are sedimentary rocks containing a very thick, semisolid, tarlike petroleum. The heavy petroleum in tar sands is believed to be formed in the same way and from the same materials as lighter oils. Tar-sand deposits may represent very immature petroleum deposits, in which the breakdown of large molecules has not progressed to the production of the lighter liquid and gaseous hydrocarbons. Alternatively, the lighter compounds may have migrated away, leaving this dense material behind. Either way, the tar is too thick to flow out of the rock. Like oil shale, tar sand presently must be mined, crushed, and heated to extract the petroleum, which can then be refined into various fuels.

Many of the environmental problems associated with oil shale likewise apply to tar sand. Because the tar is disseminated through the rock, large volumes of rock must be mined and processed to extract significant amounts of petroleum. Many tar sands are near-surface deposits, so the mining method used is commonly strip-mining. The processing requires a great deal of water, and the amount of waste rock after processing may be larger than the original volume of tar sand. The negative impact of tar-sand production will naturally increase as production increases.

Figure 13.20 Schematic diagram of *in situ* recovery of "shale oil." "Retorting" is the process by which shale oil is distilled out of the rock by application of heat.   Source: After R. L.Elderkin, Jr., U.S. Geological Survey.

Figure 13.21 Locations of unconventional petroleum deposits in western Canada.   After R. M. Procter, et al., 1984, *Oil and Natural Gas Resources of Canada*, Energy, Mines and Resources Canada. Reproduced with permission of the Minister of Supply and Services Canada, 1994.

In Canada, the extensive Athabasca tar sand (figure 13.21) may contain several hundred billion barrels of petroleum. Some mining activities are already underway in the province of Alberta. Canadians hoped to meet about one-third of their oil needs with petroleum from tar sand by 1990; indeed, Canadian supplies of "unconventional" petroleum resources far exceed that nation's "conventional" oil deposits (figure 13.22). Actual development has lagged somewhat behind aspirations, in part because of environmental concerns. Two open-pit mining operations and associated refinery facilities are functioning. To reduce the impact of mining, pilot projects involving *in situ* extraction of the "tar" are underway. One, nicknamed the "huff-and-puff" procedure, involves injection of hot steam for several weeks to warm the oil, after which the less-viscous fluid can be pumped out for a month or so before it chills and stiffens and reheating is necessary. The United States, however, has virtually no tar-sand resources (figure 13.23), so it cannot look to tar sand to solve its domestic energy problems even if the environmental difficulties could be overcome.

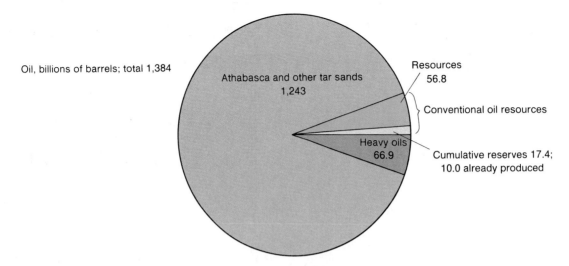

Oil, billions of barrels; total 1,384

Athabasca and other tar sands
1,243

Resources
56.8

Conventional oil resources

Heavy oils
66.9

Cumulative reserves 17.4;
10.0 already produced

Figure 13.22 Canadian petroleum resources. Note the significance of the tar-sand deposits. Source: Data from R. M. Procter, et al., 1984, *Oil and Natural Gas Resources of Canada*, Energy, Mines and Resources Canada.

Figure 13.23 Distribution of U.S. tar sands. Source: Data from J. W. Smith, "Synfuels, Oil Shale and Tar Sands," in *Perspectives on Energy*, 3d ed., ed. by Lon C. Ruedisili and Morris W. Firebaugh, 1982, Oxford University Press, NY.

Overview

Summary

The United States now relies on fossil fuels for almost 90% of its energy: About 45% comes from oil, and 20 to 25% each from natural gas and coal. All the fossil fuels are non-renewable energy sources. Known oil and gas supplies have been seriously depleted and are likely to be exhausted within a few decades. Remaining coal supplies are much larger, but many hazards and unsolved environmental problems are associated with possible heavy reliance on coal in the future. Technical, environmental, and economic problems also limit the potential of other fossil fuels— petroleum derived from oil shale or tar sand—to become major energy sources. It seems likely that other energy alternatives will be essential in the future.

Terms to Remember

anaerobic 309
anthracite 309
bituminous 309
fossil fuel 299
gasification 313
geopressurized zones 305
kerogen 316
lignite 309

liquefaction 313
methane hydrate 305
natural gas 300
nonrenewable 301
oil 300
oil shale 316
tar sand 318

Exercises

Questions for Review

1. A society's level of technological development strongly influences its per capita energy consumption. Explain.
2. What are fossil fuels?
3. Briefly describe how oil and gas deposits form and mature.
4. Compare and contrast past and projected U.S. consumption of petroleum and coal.
5. What is enhanced recovery, and why is it of interest? Give at least two examples of the method.
6. Explain the nature of geopressurized natural gas resources, and note at least one obstacle to their exploitation.
7. What are methane hydrates, and where are they found? Why are they both a substantial potential energy resource, and a concern with respect to global climate change?
7. What is coal? Why is it a less versatile fuel than oil, and how might its versatility be increased?
8. What air-pollution problems are associated particularly with coal, relative to other fossil fuels?
9. List and describe at least three potential negative environmental impacts of coal mining.
10. What are oil shales and tar sands?
11. Oil shale and tar sand share several drawbacks. Cite and explain several of these.

For Further Thought

1. Select a particular region or major city and investigate its energy consumption. Identify the principal energy sources and the proportion that each contributes; see whether these proportions are significantly different now from what they were ten and twenty-five years ago. To what extent have the types and quantities of energy used been sensitive to economic factors?
2. If possible, visit an area that has been or is being drilled for oil or mined for coal. Observe any visible signs of negative effects on the environment, and note any efforts being made to minimize such impacts.
3. Compare the distribution of precipitation given in figure 10.18 with the locations of principal U.S. oil-shale and tar-sand deposits (figures 13.18 and 13.23). Consider the implications for the economics of developing these fuel sources. Where *would* you get the water, economics aside?
4. Check the current status of the *Exxon Valdez* cleanup analysis. How many oil spills of similar size have occurred since 1989, where have they been, and what cleanup strategies have been tried? (See NetNotes for some sources.)

Suggested Readings/References

Cook, E. 1971. The flow of energy in an industrial society. In *Planet earth*, 15–25. San Francisco: W. H. Freeman.

Dick, R. A., and Wimpfen, S. P. 1980. Oil mining. *Scientific American* 243 (April): 182–88.

Dunrud, C. R., and Osterwald, F. W. 1980. *Effects of coal mine subsidence in the Sheridan, Wyoming, area.* U.S. Geological Survey Professional Paper 1164.

Earle, S. A. 1992. Assessing the damage one year later. *National Geographic* 179 (February): 122–34. (Pertains to the oil spills related to the Persian Gulf War and invasion of Kuwait.)

Gray, R. E., and Bruhn, R. W. 1984. Coal mine subsidence—eastern United States. Geological Society of America. *Reviews in Engineering Geology VI*, 123–49.

Hodgson, B. 1990. Can the wilderness heal? *National Geographic* 177 (January): 4–43. (Pertains to the *Exxon Valdez* spill.)

Holloway, M. 1991. Soiled shores. *Scientific American* 265 (October): 100–16.

Hubbert, M. King. 1971. The energy resources of the earth. In *Planet earth*, 5–14. San Francisco: W. H. Freeman.

Lucier, P. 1991. Petroleum: What is it good for? *Invention & Technology* (Fall): 56–63.

Lumpkin, R. E. 1988. Recent progress in the direct liquefaction of coal. *Science* 239: 873–77.

Macauley, G., Snowdon, L. R., and Ball, F. D. 1985. *Geochemistry and geological factors governing exploitation of selected Canadian oil-shale deposits.* Geological Society of Canada, Paper 85–13.

Marshall, E. 1989. Valdez: The predicted oil spill. *Science* 244 (7 April): 20–21.

Masters, C. D., Root, D. H., and Attanasi, E. D. 1991. Resource constraints in petroleum production potential. *Science* 253 (12 July): 146–52.

National Academy of Sciences. 1979. *Energy in transition, 1985–2010.* San Francisco: W. H. Freeman.

National Geographic Society. 1981. *Special report on energy.* Washington, D.C.: National Geographic Society.

Newman, S. O., and Smith, A. O. 1991. Safer coal mines: Decades of progress. *Minerals Today* (October): 6–15.

Oil Substitution Task Force, World Energy Conservation Commission. 1983. *Oil substitution: World outlook to 2020.* Oxford: Oxford University Press.

President's Commission on Coal. 1980. *Coal data book.* Washington, D.C.: President's Commission on Coal.

Procter, R. M., Taylor, G. C., and Wade, J. A. 1984. *Oil and natural gas resources of Canada 1983.* Geological Society of Canada, Paper 83–31.

Rowand, A. 1984. Synfuels: Stayin' alive? *Science News* 126: 74–75.

Ruedisili, L. C., and Firebaugh, M. W., eds. 1982. *Perspectives on energy.* 3d ed. New York: Oxford University Press.

Scientific American. 1980. *Energy.* San Francisco: W. H. Freeman. (A collection of offprints from *Scientific American*, 1970–79.)

U.S. Department of Energy, Energy Information Administration. 1996. *Annual Energy Review 1995.* Washington, D.C.: U.S. Department of Energy. See also NetNotes.

Wadi, M. K., ed. 1975. *Practices and problems of land reclamation in western North America.* Grand Forks, N.D.: University of North Dakota Press.

Watson, J. 1983. *Geology and man.* Boston: Allen and Unwin.

World Resources Institute, United Nations Environment Programme, United Nations Development Programme, and The World Bank. 1996. *World Resources 1996–97.* (Chapter 12, "Energy and Materials.") New York: Oxford University Press.

The U.S. Department of Energy's Energy Information Administration now publishes *Annual Energy Renew* online. This is a comprehensive compilation of U.S. and international energy information. As of this writing, the newest edition (1995) could be found at

http://www.eia.doe.gov/emeu/aer95/pubpdf/aer95pdf.html

The general Energy Information Administration site is located at

http://www.eia.doe.gov/

The Office of Surface Mining, which provides information specifically about coal mining, can be reached at

http://www.osmre.gov/

Information on energy consumption in Canada, including fossil-fuel consumption, can be obtained via

http://199.212.18.12/folio.pgi/text.nfo/query=/doc/ {1981,0,0,0}?

In 1995, the U.S. Geological Survey reexamined U.S. fossil-fuel resources; the results are available via

http://www-gd.cr.usgs.gov/resass/resass_summary.html

General information about energy is accessible from USGS via

http://www.usgs.gov/themes/energy.html

The USGS Energy Resource Surveys Program home page is at

http://energy.usgs.gov/

Information on world energy trends from World Resources Institute can be found at

http://www.wri.org/wri/enved/trends/en-home.html

The Oil Spill Public Information Center, in Alaska, was established by the U.S. Department of Justice to provide information about the *Valdez* spill. Their Internet site, established in April 1995, has links to information and slides concerning the spill. See

http://www.alaska.net:80/~ospic/

For more general information about oil spills and other sources of oil in the marine environment, the International Tanker Owners' Pollution Federation maintains a page (under construction at time of writing) at

http://www.itopf.com/unfinish.html

The e-mail addresses of contact persons within Natural Resources Canada who deal with a variety of energy issues can be found at

http://www.NRCan.gc.ca/htbin-post/x500services

Chapter 14

Energy Resources— Alternative Sources

Figure 14.1 World energy production by source, 1993. Source: Data from *World Resources 1996–97*.

Figure 14.2 Regional variations in energy consumption in the United States, 1991. Source: Data from *State Energy Data Report, Consumption Estimates, 1960–1990* and *Annual Energy Review 1991*, U.S. Energy Information Administration.

I n the discussion of nonrenewable fossil fuels in chapter 13, the point was made that, with the conspicuous exception of coal, most of the U.S. supply of recoverable fossil fuels could be exhausted within decades. It was further pointed out that coal is not the most environmentally benign of energy sources. Alternative energy sources for the future are thus needed, both to supply essential energy and to spare the environment as much disruption as possible. And this is true worldwide, for globally, too, fossil fuels are the primary energy source (figure 14.1).

The extent to which alternative energy sources are required, and how soon they will be needed, is directly related to future world energy demand, which is difficult to predict precisely. In general, consumption can be expected to rise as population increases and as standards of living improve. However, the correlation between energy consumption and standard of living is perhaps even less direct than the correlation between mineral consumption and living standards.

By way of a simplified example, consider the addition of central heating to a home. The furnace is likely to contain about the same quantity of material, regardless of its type or efficiency; the drain on mineral reserves is approximately fixed. The very addition of that furnace represents a certain jump in both mineral and energy consumption. However, depending on the efficiency of the unit, the "tightness" of the home's insulation, and the climate, the amount of fuel the furnace must consume to maintain a certain level of heat in the home will vary enormously. The best-insulated modern homes may require as little as one-tenth the energy for heating as the average U.S. home. Also, the quantity of material used to manufacture the furnace will not change whether the

homeowner maintains the temperature at 60° F or 80° F, but the amount of energy consumed plainly will. And, of course, the climate in which the home is built influences its heating costs. Thus there are considerable regional variations in energy consumption in the United States (figure 14.2) not only because of varying population density and degree of industrialization, but also because of climatic variations. Both cold and heat, taken to extremes, cause increased energy demand—for heating and air conditioning respectively.

Solar Two power tower and reflectors at Dagget, California; molten salt is used to transfer the heat collected. © Doug Sherman/Geofile

Table 14.1 Comparative Fuel Efficiency in Transportation

Mode of transportation	Average fuel economy (mpg)	Fuel use per passenger-mile (gallons)
motorcycle	50.0	
rider only		0.020
passenger car	19.2	
driver only		0.052
driver + 3		
passengers		0.013
bus	6.0	
driver only		0.167
driver + 30		
passengers		0.005

Source: *Annual Energy Review 1983, 1988,* and *1991,* Energy Information Administration, Department of Energy.

Note: Given these typical mileage ratings, it is clear that moving people by the busload is by far the most fuel-efficient approach among the listed options. Passenger-car fuel efficiency has been increasing over the past two decades, but still averaged only 20.92 mpg in 1990.

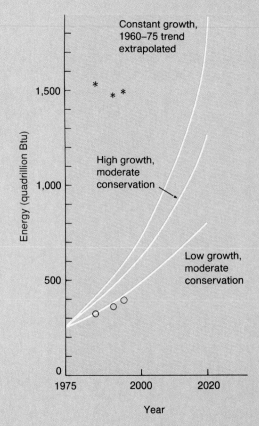

Figure 14.3 Projections of future world energy demand made by World Energy Conference. Estimates vary widely, depending on the assumptions made. Circles indicate actual world energy consumption for 1985, 1991, and 1995; asterisks show what consumption would have been if the world's entire population were consuming energy at the per capita rate of the United States in those same years.

In the area of transportation, similar variability arises. As standards of living rise to the point that motorized transportation becomes commonplace, consumption of both materials and energy rises. However, the amount of energy used depends heavily on the mode of transportation chosen and its fuel efficiency (see table 14.1).

In deciding just how much energy various alternative sources must supply, assumptions must be made not only about the rates of increase in standards of living (growth of GNP) but also about the degree of energy efficiency with which the growth is achieved. Other factors that influence both demand and projections include prices of various forms of energy (which recent history suggests is difficult to assess years, let alone decades, into the future) and public policy decisions (to pursue or not to pursue nuclear power, for example) in those nations that are major energy consumers.

Figure 14.3 shows how the assumptions collectively make a very large difference in projected demand. Rather than try to guess which scenario is most probable and thus attempt to make a quantitative determination of the expected shortfall of oil and gas, we will survey in this chapter some of the most promising alternative energy sources and assess the potential of each to supply substantial amounts of energy in the near and more distant future.

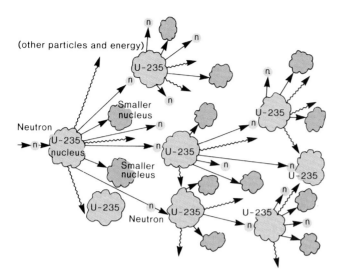

Figure 14.4 (Schematic) Nuclear fission and chain reaction involving uranium-235. Neutron capture by uranium-235 causes fission into two smaller nuclei plus additional neutrons, other subatomic particles, and energy. Released neutrons, in turn, cause fission in other uranium-235 nuclei.

Nuclear Power—Fission

Fission—Basic Principles

The phrase *nuclear power* actually comprises two different types of processes with different advantages and limitations. **Fission** is the splitting apart of atomic nuclei into smaller ones, with the release of energy. **Fusion** is the combining of smaller nuclei into larger ones, also releasing energy. Currently, only one of these processes is commercially feasible: fission. Fission basics are outlined in figure 14.4. (For a review of the terminology related to atoms, see chapter 2.) Very few isotopes (some 20 out of more than 250 naturally occurring isotopes) can undergo fission spontaneously, and do so in nature. Some additional nuclei can be induced to split apart, and the naturally fissionable nuclei can be made to split up more rapidly, thus increasing the rate of energy release. The fissionable nucleus of most interest in modern nuclear power reactors is the isotope of uranium with 92 protons and 143 neutrons, uranium-235.

A uranium-235 nucleus can be induced to undergo fission by firing another neutron into the nucleus. The nucleus splits into two lighter nuclei (not always the same two) and releases additional neutrons as well as energy. Some of the newly released neutrons can induce fission in other nearby uranium-235 nuclei, which, in turn, release more neutrons and more energy in breaking up, and so the process continues in a **chain reaction.** A controlled chain reaction, with a continuous, moderate release of energy, is the basis for fission-powered reactors (figure 14.5). The energy released heats cooling water that circulates through the reactor's core. The heat removed from the core is transferred through

Containment building

Figure 14.5 Schematic diagram of conventional nuclear fission reactor. Heat is generated by chain reaction; withdrawing or inserting control rods between fuel elements varies rate of reaction, and thus release of heat energy. Cooling water also serves to extract heat for use. Heat is transferred to power loop via heat exchanger, so the cooling water, which may contain radioactive contaminants, is isolated from the power-generating equipment.

a heat exchanger to a second water loop in which steam is produced. The steam, in turn, is used to run turbines to produce electricity.

This scheme is somewhat complicated by the fact that a chain reaction is not sustained by ordinary uranium. Only 0.7 percent of natural uranium is uranium-235. The material must be processed to increase the concentration of this isotope to several percent of the total to produce reactor-grade uranium. As the reactor operates, the uranium-235 atoms are split and destroyed, so that, in time, the fuel is so depleted (spent) in this isotope that it must be replaced with fresh enriched uranium.

The Geology of Uranium Deposits

Worldwide, 95 percent of known uranium reserves are found in sedimentary or metasedimentary rocks. In the United States, the great majority of deposits are found in sandstone. They were formed by weathering of uranium source rocks, followed by uranium migration in and deposition by ground water.

Minor amounts of uranium are present in many crustal rocks. Granitic rocks and carbonates may be particularly rich in uranium (meaning that its concentration in these may be in the range of ppm to tens of ppm; in most rocks, it is even lower). In granites, the uranium is concentrated in the late stages of magma crystallization; its large atoms fit relatively poorly in most early-crystallizing silicate minerals. Uranium is concentrated in carbonate rocks during precipitation of the carbonates from water, including seawater. As uranium-bearing rocks weather under near-surface conditions, that uranium goes into solution: uranium is particularly soluble in an oxygen-rich environment. The uranium-bearing solutions then infiltrate and join the groundwater system. As they percolate through permeable rocks, such as sandstone, they may encounter chemically reducing conditions, created by some factor such as an abundance of carbon-rich organic matter, or sulfide minerals, in shales bounding the sandstone. Under reducing conditions,

solubility of uranium is much lower. The dissolved uranium is then precipitated, and concentrated, in these reducing zones. Over time, as great quantities of uranium-bearing ground water percolate slowly through such a zone, a large deposit of precipitated uranium ore may form.

Limitations of Uranium Supply

World estimates of available uranium are somewhat difficult to obtain, partly because the strategic importance of uranium leads to some secrecy. For the United States, the estimates are strongly sensitive to the price obtainable for the processed "yellowcake"—uranium oxide (U_3O_8)—which has oscillated considerably. Table 14.2 summarizes present reserve and resource estimates at different price levels. With the type of nuclear reactor currently in commercial operation in the United States, nuclear electricity-generating capacity probably could not be increased to much more than four times present levels by the year 2010 without serious fuel shortages. This increase in production would supply less than 15 percent of total U.S. energy needs. In other words, the rare isotope uranium-235 is in such short supply that the United States could use it up within several decades, assuming no improvements in reactor technology and no significant uranium imports. One might well ask, then, why all the debate about nuclear power, and why consider constructing more power plants (figure 14.6)?

Table 14.2 U.S. Uranium Reserve and Resource Estimates

Recoverable costs	Reserves (million pounds of U_3O_8)	Resources, including reserves (million pounds)
$30/lb U_3O_8	290	3,780
$50/lb U_3O_8	947	6,487
$100/lb U_3O_8	1,493	9,823

Source: Data from *Annual Energy Review 1991* and *1995*, Energy Information Administration, U.S. Department of Energy.

Actual uranium prices (in constant 1983 dollars) rose from under $20/lb U_3O_8 in the early 1970s, to over $70/lb on the spot market in 1976, then declined sharply to $15 to $20/lb in 1985, and have remained depressed since. By the early 1990s, uranium prices were so low—partly as a result of reduced demand for power-plant fuel—that domestic production had virtually ceased. In 1995, with prices averaging about $11/lb U_3O_8, only about 6 million pounds were produced in the U.S., while 32 million pounds were imported.

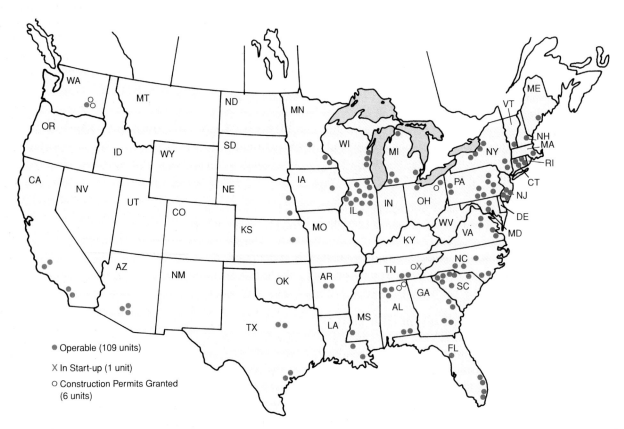

• Operable (109 units)

X In Start-up (1 unit)

○ Construction Permits Granted (6 units)

Figure 14.6 Distribution of U.S. nuclear power plants. In December 1995, there were 109 operable plants, one in startup, and 6 for which construction permits had been granted. (Due to space limitations, symbols do not represent actual locations; the number of plants in each state *is* accurate.)   Source: Data from *Annual Energy Review 1995*, U.S. Energy Information Administration.

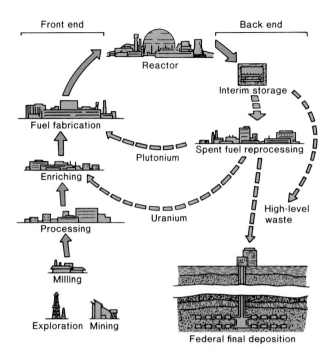

Front end | Back end

Reactor

Interim storage

Fuel fabrication

Plutonium

Spent fuel reprocessing

Enriching

Uranium

High-level waste

Processing

Milling

Exploration   Mining

Federal final deposition

➡ Fuel cycle as it operates currently

▢➡ Fuel cycle as it would operate with spent fuel reprocessing and federal waste storage

Figure 14.7 The nuclear fuel cycle, as it presently operates and as it would function with fuel reprocessing.   Source: U.S. Energy Information Administration, Department of Energy.

Extending the Nuclear Fuel Supply

Uranium-235 is not the only possible fission-reactor fuel, although it is the most plentiful naturally occurring one. When an atom of the far more abundant uranium-238 absorbs a neutron, it is converted into plutonium-239, which is, in turn, fissionable. Uranium-238 makes up 99.3% of natural uranium and over 90% of reactor-grade enriched uranium. During the chain reaction inside the reactor, as freed neutrons move about, some are captured by uranium-238 atoms, making plutonium. Spent fuel could be *reprocessed* to extract this plutonium, which could be purified into fuel for future reactors, as well as to re-enrich the remaining uranium in uranium-235. How reprocessing would alter the nuclear fuel cycle is shown in figure 14.7. Fuel reprocessing with recovery of both plutonium and uranium could reduce the demand for "new" enriched uranium by an estimated 15%.

A **breeder reactor** can maximize production of new fuel. Breeder reactors produce useful energy during operation, just as conventional "burners" using up uranium-235 do, by fission in a sustained chain reaction within the reactor core. In addition, they are designed so that surplus neutrons not required to sustain the chain reaction are used to produce more fissionable fuels from suitable materials, like plutonium-239 from uranium-238, or thorium-233, which is

derived from the common isotope thorium-232. Breeder reactors could synthesize more nuclear fuel for future use than they would actively consume while generating power.

Breeder-reactor technology is more complex than that of conventional water-cooled burners. The core coolant is liquid metallic sodium; the reactor operates at much higher temperatures. The costs would be substantially higher than for burner reactors, perhaps close to $10 billion per reactor by the time any breeders could be completed in the United States. The breeding process is slow, too: The break-even point after which fuel produced would exceed fuel consumed might be several decades after initial operation. If the nuclear-fission option is to be pursued vigorously into the next century, the reprocessing of spent fuels and the use of breeder reactors are essential. Yet at present, reprocessing is minimal in the United States, and in early 1985, Congress cancelled funding for the experimental Clinch River breeder reactor, in part because of high estimated costs. No commercial breeders currently exist in the United States, and only a handful of breeder reactors are operating in the world. Even conventional reactors are falling out of favor in the United States (see box 14.1). There are several reasons for this declining enthusiasm.

Concerns Related to Nuclear Reactor Safety

A major concern regarding the use of fission power is reactor safety. In normal operation, nuclear power plants release very minor amounts of radiation, which are believed to be harmless. (A general discussion of radiation and its hazards is presented in chapter 15.) The small but finite risk of damage to nuclear reactors through accident or deliberate sabotage is more worrisome to many.

One of the most serious possibilities is a so-called loss-of-coolant event, in which the flow of cooling water to the reactor core would be interrupted. Resultant overheating of the core might lead to **core meltdown,** in which the fuel and core materials would deteriorate into a molten mass that might or might not melt its way out of the containment building and thus release high levels of radiation into the environment (opinion is divided on this possibility). A partial loss of coolant, with 35 to 45 percent meltdown, occurred at Three Mile Island in 1979.

No known full-scale loss-of-coolant accident with complete core meltdown has occurred in the United States. A comprehensive study commissioned by the Nuclear Regulatory Commission in 1975 (the Rassmussen Report) projected the risk of such a core meltdown at about one in a million per reactor per year. This risk factor certainly is small, particularly considering other risks of daily life, but it is also only an educated guess. (Through 1995, cumulative total collective nuclear-reactor operation globally was about 7,700 reactor-years.)

No matter how far awry the operation of a commercial power plant might go, and even if there were a complete loss of coolant, the reactor could *not* explode like an atomic bomb.

Box 14.1 Nuclear Fission Power: Fallen Idol?

In the early 1970s, it was predicted that 25% of U.S. energy would be supplied by nuclear power by the year 2000. At the end of 1995, there were 109 nuclear power plants (figure 1) operating in the United States (out of 437 worldwide), with 1 in startup and 6 in some phase of construction. However, during the last fifteen years, nuclear plant cancellations have far exceeded new orders; since 1978, there have been no new orders at all, and several utilities have decided to power their new generating plants with coal, not nuclear fission. The 109 nuclear plants produced only 22% of the electricity consumed in the United States in 1995, or about 8% of the total energy consumed. Worldwide, reliance on nuclear power varies widely, accounting for only 1 or 2% of electricity production in such countries as Brazil, India, and China, but over 30% in much of Europe, and in some countries (Belgium, France, Lithuania), over 50% (figure 2).

Nuclear plants do have lower fueling and operating costs than coal-fired plants. The small quantity of fuel required for their operation can also be more easily stockpiled against interruption by strikes or transportation delays. However, nuclear plants have become more costly to build than coal-fired plants and now require longer to plan, construct, and license (nine to twelve years for U.S. nuclear plants; six to ten for coal). Especially in times of high interest rates, these economic and time

Figure 1 Three Mile Island near Harrisburg, Pennsylvania; damaged reactor remains shut down, while others are still operative. © Doug Sherman/Geofile.

factors loom large. Those, together with increasingly vocal public opposition to nuclear power plants (itself a contributing factor in many licensing delays), have made the coal option more appealing to many utilities. Reduced growth in energy demand has further diminished the need for new generating facilities of any kind.

Nuclear-plant cancellation is not without its costs. Since the first such cancellations in 1972, orders for more than one hundred nuclear power plants have been cancelled. Where construction has not yet begun, abandonment costs are typically less than $100 million. Where a partially built plant is cancelled, the costs are much higher. The widely publicized cancel-

lation of Washington Public Power Supply System Units 4 and 5, which were, respectively, 25 percent and 17 percent complete at the time of cancellation, involved abandonment costs of over $2.2 billion. Abandonment costs of nuclear plants are typically borne by investors, utility users, and taxpayers, generally in proportions specified under local law.

As existing nuclear plants have aged, yet another issue peculiar to fission power has emerged: the fact that nuclear fission plants have a finite lifetime. Thereafter, they must be *decommissioned,* as described in the text, adding to both cost and waste-disposal problems.

Bomb-grade fuels must be much more highly enriched in the fissionable isotope uranium-235 for the reaction to be that intensive and rapid. However, an ordinary explosion originating within the reactor (or by saboteurs' use of conventional explosives) could rupture both the containment building and reac-

tor core and thus release large amounts of radioactive material. The serious accident at Chernobyl reinforced reservations about reactor safety in many people's minds (box 14.2).

Plant siting is another problem. Siting nuclear plants close to urban areas puts more people potentially at risk

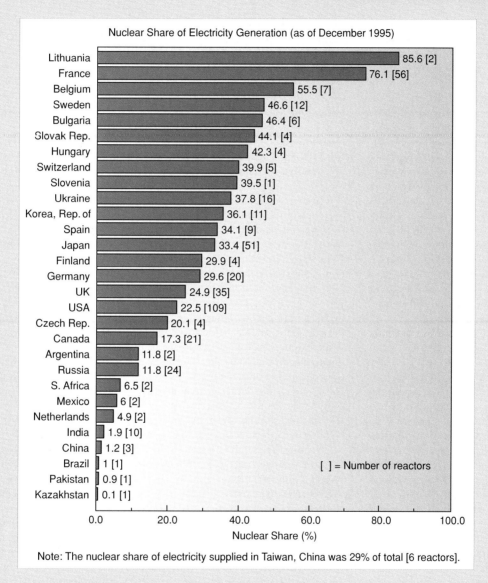

Figure 2 Proportion of electricity generated by nuclear fission varies greatly by country, and not simply in proportion to numbers of reactors: where electricity consumption is moderate, a few reactors can account for a large share.   Source: Data from International Atomic Energy Agency.

in case of accident; placing the plants far from population centers where energy is needed means more transmission loss of electricity (which already claims nearly 10 percent of electricity generated). Proximity to water is often important for cooling purposes but makes water pollution in case of mishap more likely. There are also concerns about the structural integrity of nuclear plants located close to fault zones— the Diablo Canyon station in California, the Humboldt Bay nuclear plant, and others; several proposed reactor sites have been rejected when investigation revealed nearby faults.

Box 14.2 Crisis at Chernobyl—and After

On 26 April 1986, an accident occurring at the nuclear power plant at Chernobyl, in the former USSR, dwarfed any U.S. reactor accident in scale and regional impact and reinforced many fears about the safety of the nuclear industry. The reactor core overheated; an explosion and core meltdown occurred. The initial overheating of the core resulted from a still-unexplained power surge to 50 percent of reactor capacity from standby status. A runaway chain reaction ensued. What then caused the explosion is not clear. Possibly, leaks allowed water and steam to hit the hot graphite of the reactor core, producing hydrogen gas, which is highly explosive. The explosion, in turn, may have disrupted emergency cooling systems and also prevented insertion of control rods to slow the reaction.

Quantities of radioactive material escaped from the reactor building, drifting with atmospheric circulation over Scandinavia and eastern Europe (figure 1). The official report was that thirty-one deaths were directly attributable to the accident; 203 persons were hospitalized for acute radiation sickness; 135,000 people were evacuated. Downwind, preventive measures to minimize possible health impacts were suggested: For example, some people consumed large doses of natural iodine to try to saturate the thyroid gland (which concentrates iodine) and thus prevent its absorption of the radioactive iodine in fallout from the reactor. The republics most severely affected were Ukraine, Russia, and Byelorussia.

Some of the long-range health effects of the accident will not be known for decades, partly because some of these consequences develop long after radiation exposure and partly because the effects of radiation are not always distinguishable from the effects of other agents (see chapter 15). In 1990–91, at the request of the Soviet government, teams of specialists from many nations converged on those three republics to make a comprehensive evaluation of the shorter-term consequences of the accident. For the most part, residual radiation levels in water and soil were below acceptable limits. Studies of health effects were more difficult, in part because many of those in the contaminated area had moved in the four to five years following the accident; local clinic records were imperfect; and resources did not permit a thorough health evaluation of all persons exposed to the radiation from the accident. Still, a number of persons in contaminated and uncontaminated (control) areas were surveyed; and no statistically significant differences between the two groups in health disorders attributable to radiation exposure were found. On the other hand, psychological problems in the contaminated area were considerable—anxiety and stress, partly related to uncertainty; desire to relocate among those who could not, fear of relocation among those who had lived all their lives in the same settlements and didn't wish to move. General anxiety levels were heightened by political changes in the former Soviet Union.

Later studies have begun to document increases in medical problems attributable to radiation—thyroid cancer in children, chromosome damage—near the reactor site. Estimates vary widely, but put the increased lifetime risk of contracting radiogenic cancer at 0 to 0.02 percent in Europe, 0.003 percent in the Northern Hemisphere overall. Over the long term, total cancer cases resulting from the Chernobyl incident are projected at 5,000 to 100,000.

Can a Chernobyl-style accident happen at a commercial nuclear reactor in the United States? Basically, no, because the design of these reactors is fundamentally different from the design of the Chernobyl reactor (though a few research reactors in this country have a core of Chernobyl type). In any reactor, some material must serve as a *moderator*, to slow the neutrons streaming through the core enough that they can interact with nuclei to sustain the chain reaction. In the Chernobyl reactor, the moderator is graphite; the core is built largely of graphite blocks. In commercial U.S. reactors (and most reactors in western nations), the moderator is water. Should the core of a water-moderated reactor overheat, the water vaporizes. Steam is a poor moderator, so the rate of chain reaction slows down. (Emergency cooling of the core is still needed, but as the reaction slows, so does the rate of further heat production.) However, graphite remains solid up to extremely high temperatures, and hot graphite and cold graphite are both effective moderators. In a graphite reactor, therefore, a runaway chain reaction is not stopped by changes in the moderator. At Chernobyl, as the graphite continued to overheat, it eventually began to burn, turning the reactor core into the equivalent of a giant block of radioactive charcoal. (To put it out, workers eventually had to smother it in concrete.) The fuel rods in the burning core ruptured and melted from the extreme heat. The core meltdown, together with the explosion, led to the release of high levels of radioactive materials into the environment. The extent of that release was also far greater for the Chernobyl reactor than it would have been for a U.S. reactor suffering a similar accident because the containment structures of U.S. reactors are much more extensive and effective. The accident focused international attention sharply on the issue of reactor safety and also on the reactor-operator interface (operator error may have contributed to the accident). Out of this disaster have come a number of constructive suggestions for the improvement of reactor safety, both in design and in the operation of existing reactors. The incident also emphasized that, like air pollution, radioactive fallout ignores international boundaries.

Scale in km

0 100 200   400   600   800   1000   1200

Figure 1 A cloud of radioactive iodine (iodine-131) drifts and spreads out from the Chernobyl reactor site in the week after the accident. For comparison, maximum *annual* radiation emission permitted at the boundary of a nuclear facility in the United States is 25 rem. (A *rem* is a unit of radiation exposure that takes into account both the energy and nature of the radiation, so that different types of radiation can be compared in terms of their effects on the human body.)
(*A*) *Day 2* Time: 28 April 1986; 0000 GMT (accident began April 26; 1:23 GMT). (*B*) *Day 4* Time: 30 April 1986; 0000 GMT. (*C*) *Day 6* Time: 2 May 1986; 0000 GMT.
Source: Data from Mark A. Fischetti, "The Puzzle of Chernobyl" in *IEEE Spectrum* 23, July 1986, pp. 38–39. Copyright © 1986 IEEE.

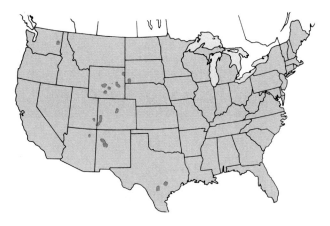

Figure 14.8 U.S. uranium deposits.   Source: U.S. Energy Information Administration, Department of Energy.

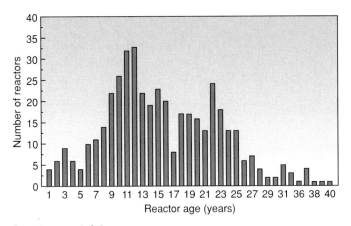

Figure 14.9 Age distribution of nuclear reactors worldwide. Few face decommissioning yet, but many may in the next decade or two.   Source: Data from the International Atomic Energy Agency.

Concerns Related to Fuel Handling

The mining and processing of uranium ore are operations that affect relatively few places in the United States (figure 14.8). Nevertheless, they pose hazards because of uranium's natural radioactivity. Miners exposed to the higher radiation levels in uranium mines have experienced higher occurrence rates of some types of cancer. Carelessly handled tailings from processing plants have exposed others to radiation hazards; recall the case of Grand Junction, Colorado, mentioned in chapter 12, where radioactive tailings were unknowingly mixed into concrete used in construction.

The use of reprocessing or breeder reactors to produce and recover plutonium to extend the supply of fissionable fuel poses special problems. Plutonium itself is both radioactive and chemically toxic. Of greater concern to many people is that, as a readily fissionable material, it can also be used to make nuclear weapons. Extensive handling, transport, and use of plutonium would pose a significant security problem and would therefore require very tight inventory control to prevent the material from falling into hostile hands.

Radioactive Wastes

The radioactive wastes from the production of fission power are another concern. Radiation hazards and radioactive-waste disposal are considered within the broader general context of waste disposal in chapter 15. Here, two aspects of the problem are highlighted. First, radioactive materials cannot be treated by chemical reaction, heating, and so on to make them nonradioactive. In this respect, they differ from many toxic chemical wastes that can be broken down by appropriate treatment. Second, there has been sufficient indecision about the best method of radioactive-waste disposal and the appropriate site(s) for it; none of the radioactive wastes generated anywhere in the world have been disposed of permanently to date. Currently, the wastes are in temporary storage while various disposal methods are being explored, and many of the temporary waste-holding sites are

filled almost to capacity. Clearly, acceptable waste-disposal methods must be identified and adopted, if only to dispose of wastes already accumulated.

Nuclear plants have another unique waste problem. The bombardment of the reactor core and structure by neutrons and other atomic debris from the fission process converts some of the structural materials to radioactive ones and changes the physical properties of others, weakening the structure. At some point, then, the plant must be **decommissioned**—taken out of operation, broken down, and the most radioactive parts delivered to radioactive-waste disposal sites. In late 1982, an electricity-generating plant at Shippingport, Pennsylvania, became the first commercial U.S. fission power plant to face decommissioning, after twenty-five years of operation. Typical costs of demolition and disposal are over $100 million per reactor in the United States, and the process may take a decade or more. In other countries, cost estimates run as high as $500 million per reactor, and dozens of reactors are being, or have been, decommissioned worldwide. The U.S. Energy Department estimates that another twenty commercial reactors will require decommissioning by the year 2000; a total of eight, including Shippingport, had been retired by the close of 1989. Historically, the nominal lifetime allowed by regulatory agencies was about 30 years; currently, it is about 40 years; with rigorous plant monitoring and replacement of some key components, it may be possible to operate a fission-powered electricity-generating plant for 70 years. Still, sooner or later, decommissioning will take its fiscal and waste-disposal toll. In the U.S. and worldwide, reactors are aging (figure 14.9).

Concluding Observations

Different people weigh the pros and cons of nuclear fission power in different ways. For many, the uncertainties and potential risks outweigh the benefits. For others, the problems associated with using coal (the most readily available and likely alternative) appear at least as great (see box 14.3).

Box 14.3 Risk Assessment, Risk Projection

Much of the debate about energy sources focuses on hazards: what kinds, how many, how large. Many of the risks associated with various energy sources may not be immediately obvious. No energy source is risk-free, but, then, neither is living. The question is, what constitutes "acceptable" risk (and, acceptable to whom)? Underground miners of both coal and uranium face increased health risks, but the power consumer does not see those risks directly. Those living downstream from a dam are at risk; so are those downwind from a nuclear or fossil-fuel plant. Table 1 presents data on the number of accidental deaths to be expected just from normal operations each year.

A power plant of 1-billion-watt capacity would serve the electricity needs of about 1 million people. By way of comparison, in 1974, per million people, there were 220 deaths from motor vehicle accidents, 50 from falls, 30 from burns, and 10 from accidents involving firearms.

A further consideration when comparing risks may be how well-defined the risks are. That is, estimates of risks from fires, automobile accidents, and so on are actuarial, based on considerable actual past experience and hard data. By contrast, projections of risks associated with extensive use of fission power are based on limited past experience, which does not even include the worst possible events, plus educated guesses about future events based, in part, on hard-to-quantify parameters like probability of human errors. Since the accidents at Three Mile Island and particularly Chernobyl, and given the numerous other less serious "events" at other commercial reactors, many have begun to speculate that the risk estimates in the Rassmussen Report mentioned earlier are too low. There is, however, no way to confirm or refute this suspicion directly in the absence of more data.

Table 1 Accidental Deaths per Billion-Watt Power Plant per Year

Energy source	Extraction	Processing/transport	Power plant	Total
coal		2.32*	0.01	
if mined underground	1.7			4.0
if strip-mined	0.3			2.6
oil	0.2	0.13	0.01	0.4
natural gas	0.16	0.03	0.01	0.2
uranium	0.2	0.011	0.01	0.2

Source: Data from National Academy of Sciences, "NRC Report" in *Energy in Transition 1985–2010*, p. 429. Copyright © W. H. Freeman and Company.
*Includes coal-train accidents projected from average train accidents; may be high.

The use of nuclear fission power would, in any case, be restricted to the generation of electricity; thus, the extent to which it can supply U.S. energy needs is limited. In the United States, electricity production accounts for about 35 percent of total energy consumed, though its proportion of the total is growing with time. Even with wholehearted commitment to nuclear fission power, then, it could not realistically become the principal energy source of the United States for many decades. The most enthusiastic recent estimates of its potential project that nuclear fission power might supply about a third of U.S. electricity, or approximately 10 percent of total energy used, by the year 2020, and that assumes more active pursuit of that option.

Nuclear Power—Fusion

Nuclear *fusion* is the opposite of fission. As noted earlier, fusion is the process by which two or more smaller atomic nuclei combine to form a larger one, with an accompanying release of energy. It is the process by which the sun generates

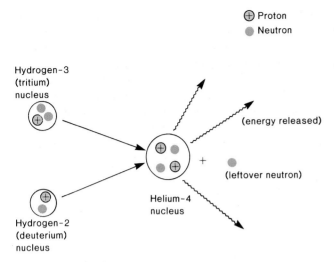

⊕ Proton
● Neutron

Hydrogen-3
(tritium)
nucleus

(energy released)

+

(leftover neutron)

Helium-4
nucleus

Hydrogen-2
(deuterium)
nucleus

Figure 14.10 Schematic diagram of the process of nuclear fusion.

its vast amounts of energy. In the sun, simple hydrogen nuclei containing one proton are fused to produce helium. For technical reasons, fusion of the heavier hydrogen isotopes deuterium (nucleus containing one proton and one neutron) and tritium (one proton and two neutrons) would be easier to achieve on earth. The relevant fusion reaction is diagrammed in figure 14.10.

Hydrogen is plentiful since it is a component of water; the oceans contain, in effect, a huge reserve of hydrogen, an essentially inexhaustible supply of the necessary fuel for fusion. This is true even considering the scarcity of deuterium, which is only 0.015 percent of natural hydrogen. (Tritium is rarer still and would probably have to be produced from the comparatively rare metal lithium. However, the magnitude of energy release from fusion reactions is such that this is not a serious constraint.) The principal product of the projected fusion reactions—helium—is a nontoxic, chemically inert, harmless gas. There could be some mildly radioactive light-isotope by-products of fusion reactors, but they would be much less hazardous than many of the products of fission reactors.

Since fusion is a far "cleaner" form of nuclear power than fission, why not use it? The principal reason is technology, or lack of it. To bring about a fusion reaction, the reacting nuclei must be brought very close together at extremely high temperatures (millions of degrees at least). The natural tendency of hot gases is to expand, not come together, and no known physical material could withstand such temperatures to contain the reacting nuclei. The techniques being tested in laboratory fusion experiments are elaborate and complex, either involving containment of the fusing materials with strong magnetic fields or using lasers to heat frozen pellets of the reactants very rapidly. At best, the experimenters have been able to achieve the necessary conditions for fractions of a second, and the energy required to bring about the fusion reactions has exceeded the energy released

thereby. Certainly, the present technology does not permit the construction of commercial fusion reactors in which controlled, sustained fusion could be used as a source of energy.

Scientists in the field estimate that several decades of intensive, expensive research will be needed before fusion can become a commercial reality. Even then, it will be costly. Some projections suggest that a commercial fusion electricity-generating plant could cost tens of billions of dollars. Nevertheless, the abundance of the fuel supply and the relative cleanness of fusion (as compared to both fission and fossil-fuel power generation) make it an attractive prospect, at least for the twenty-first century. Fusion, however, is a means of generating electricity in stationary power plants only. This fact limits its potential contribution toward satisfying total energy needs.

It should be noted that in 1989, two scientists announced what would have been an immensely important technical breakthrough: "cold fusion," or fusion at room temperature in a beaker. The experimental setup involved palladium electrodes in a beaker full of "heavy" (deuterium-enriched) water. The idea was that the electric current would split the water molecules, and deuterium ions from the water would be driven into one of the electrodes, where some would fuse. Evidence cited to support the occurrence of cold fusion included heat in excess of what could be accounted for by chemical processes and current flow in the apparatus, and the presence of tritium and helium (expected products of deuterium fusion) afterwards. If fusion could be achieved under such simple, inexpensive conditions, the principal obstacle to use of fusion power would be removed, and this vast resource would become readily and cheaply available. Hundreds of scientists rushed to replicate the experiments. Unfortunately, the overwhelming majority failed to reproduce the anomalies reported; some proposed that any helium found was simply a contaminant, from the atmosphere or introduced in the electrodes. The consensus is that whatever is going on in such apparatus is not "cold fusion," and commercial fusion power is still a long way off.

Solar Energy

The earth intercepts only a small fraction of the energy radiated by the sun. Much of that energy is reflected or dissipated in the atmosphere. Even so, the total solar energy reaching the earth's surface far exceeds the energy needs of the world at present and for the foreseeable future. The sun can be expected to go on shining for approximately 5 billion years—in other words, the resource is inexhaustible, which contrasts sharply with nonrenewable sources like uranium or fossil fuels. Sunlight falls on the earth without any mining, drilling, pumping, or disruption of the land. Sunshine is free; it is not under the control of any company or cartel, and it is not subject to embargo or other political disruption of supply. The *use* of solar energy is essentially pollution-free, at least in the sense that absorption of sunlight for heat or

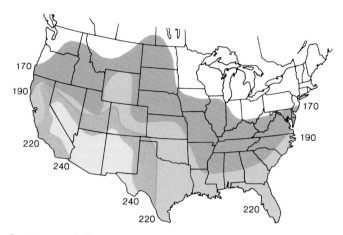

Figure 14.11 Distribution of solar energy over the continental United States. Maximum insolation occurs over the southwestern region. Numbers are watts per square meter; lines connect points of equal insolation.   Source: Data from *Energy Strategies: Toward a Solar Future*, ed. by S.W. Kendall and S.J. Nadis, Union of Concerned Scientists, 1980, Ballinger Publishing Co.

operation of a solar cell for electricity are very "clean" processes (though solar electricity has other environmental costs; see below). It produces no hazardous solid wastes, air or water pollution, or noise. For most current applications, solar energy is used where it falls, thereby avoiding transmission losses. All these features make solar energy an attractive option for the future. Several practical limitations on its use also exist, however, particularly in the short term.

The solar energy reaching the earth is dissipated in various ways. Some is reflected into space; some heats the atmosphere, land, and oceans, driving ocean currents and winds. The sun supplies the energy needed to cause evaporation and thus to keep the hydrologic cycle going; it allows green plants to generate food by photosynthesis. More than ample energy is still left over to provide for all human energy needs, in principle. However, the energy is distributed over the whole surface of the earth. In other words, it is a very dispersed resource: Where large quantities of solar energy are used, solar energy collectors must cover a wide area. Sunlight is also variable in intensity, both from region to region (figure 14.11) and from day to day as weather conditions change. The two areas in which solar energy can make the greatest immediate contribution are in space heating and in the generation of electricity, uses that together account for about two-thirds of U.S. energy consumption.

Solar Heating

Solar space heating typically combines direct use of sunlight for warmth with some provision for collecting and storing additional heat to draw on when the sun is not shining. The simplest approach is **passive-solar heating,** which does not require mechanical assistance. The building design should allow the maximum amount of light to stream in through south and west windows during the cooler months. This heats the materials inside the house, including the structure itself, and the radiating heat warms indoor air. Media used specifically for storing heat include water—in barrels, tanks, even indoor swimming pools—and the rock, brick, concrete, or other dense solids used in the building's construction (figure 14.12A). These supply a thermal mass that radiates heat back when needed, in times of less (or no) sunshine. Additional common features of passive-solar design include broad eaves to block sunshine during hotter months (feasible because the sun is higher in the sky during summer than during winter) and drapes or shutters to help insulate window areas during long winter nights (figure 14.12B). Other possible variations on passive-solar heating systems are too numerous to describe here.

Active-solar heating systems usually involve the mechanical circulation of solar-heated water (figure 14.12C). The flat solar collectors are water-filled shallow boxes with a glass surface to admit sunlight and a dark lining to absorb sunlight and help heat the water. The warmed water is circulated either directly into a storage tank or into a heat exchanger through which a tank of water is heated. The solar-heated water can provide both space heat and a hot-water supply. If a building already uses conventional hot-water heat, incorporation of solar collectors is not necessarily extremely expensive (especially considering the free "fuel" to be used). With the solar collectors mounted on the roof, an active-solar system does not require the commitment of any additional land to the heating system—another positive feature. The method can be as practical for urban row houses or office buildings as for widely spaced country homes.

While solar heating may be adequate by itself in mild sunny climates, in areas subject to prolonged spells of cloudiness or extreme cold, a conventional backup heating system is almost always needed. In the latter areas, then, solar energy can greatly reduce but not wholly eliminate the need for some consumption of conventional fuels. It has been estimated that, in the United States, 40 to 90 percent of most homes' heating requirements could be supplied by passive-solar heating systems, depending on location. It is usually more economical to design and build passive-solar technology features into a new structure initially than to incorporate them into an existing building later (retrofit). (A few 100-percent-solar homes have been built, even in such severe climates as northern Canada. However, the variety of solar-powered devices, extra insulation, and other features required can add tens of thousands of dollars to the cost of such homes, especially when an existing home is retrofit, pricing this option out of the reach of many home buyers. Moreover, superinsulation can aggravate indoor air pollution; see chapter 17.)

Solar Electricity

Direct production of electricity using sunlight is accomplished through **photovoltaic cells,** also called simply "solar cells" (see figure 14.13). They have no moving parts and,

A

B

C

Figure 14.12 (*A*) Basics of passive-solar heating with water or structural materials as thermal reservoir: Sunlight streams into greenhouse with glass roof and walls, heat is stored for nights and cloudy days. (*B*) Design features of home and landscaping can optimize use of sun in colder weather, provide protection from it in summer. (*C*) A common type of active-solar heating system with a pump to circulate the water between the collector and the heat exchanger/storage tank.

Figure 14.13 Schematic diagram of a photovoltaic (solar) cell for the generation of electricity.

like solar heating systems, do not emit pollutants during operation. For many years, they have been the principal power source for satellites and for a few remote areas on earth difficult to reach with power lines (figure 14.14). A major limitation on solar-cell use has historically been cost, which is several times higher per unit of power-generating capacity than for either fossil-fuel or nuclear-powered generating plants. This has restricted the appeal of home-generated solar electricity. The high cost is partly a matter of technology (present solar cells are not very efficient, though they are being improved) and partly one of scale (the industry is not large enough to enjoy the economies of mass production). Major expansion of the use of solar cells also would require considerable growth of the semiconductor industry; at the moment, the industry's capacity is limited.

Currently, low solar-cell efficiency and the diffuse character of sunlight combine to make photovoltaic conversion an inadequate option for energy-intensive applications, such as many industrial and manufacturing operations. Even in the areas of strongest sunlight in the United States, incident radiation is of the order of 250 watts per square meter. Operating with commercial solar cells of about 20 percent efficiency means power generation of only 50 watts per square meter. In other words, to keep one 100-watt light bulb burning would require at least 2 square meters of collectors (with the sun always shining). A 100-megawatt power plant would require 2 square kilometers of collectors (nearly one square mile!). This represents a large commitment of both land and the mineral resources from which the collectors are made. Even solar-cell technologies currently in the experimental stages promise efficiencies of little more than 30 percent. An alternative approach is to use a bank of precisely oriented mirrors to focus sunlight on a boiler, and generate steam to run a turbine to generate power; but a large expanse of mirrors would still be needed.

Storing solar electricity is also a more complex matter than storing heat. For individual homeowners, batteries may

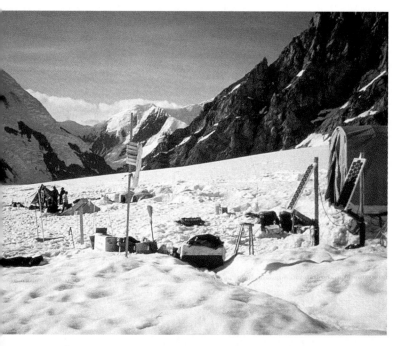

Figure 14.14 High in the mountains of Denali National Park, at the base camp that is the takeoff point for expeditions to climb Denali (Mount McKinley), solar cells (right) power vital communications equipment.

suffice, but no wholly practical scheme for large-scale storage has been devised. Some of the proposals are shown in figure 14.15.

For the time being, it appears that solar-generated electricity could eventually supply perhaps 10 to 15 percent of U.S. electricity needs. Major improvements in efficiency of generation and in storage technology are needed before the sun can be the principal source of electricity. It may be still longer before solar electricity can contribute to the transportation energy budget.

Potential Environmental Impacts of Large-Scale Commitment to Solar Electricity

The use of solar energy is, as noted earlier, environmentally benign. Unfortunately, the construction of solar facilities is not.

The most efficient photovoltaic cells use potentially toxic materials, such as gallium and arsenic. These toxic materials already present health hazards in both the mining and manufacturing phases of the semiconductor industry. The risks will be greatly magnified if that industry is expanded to the power-plant scale outlined earlier.

For the collector array alone, a 100-megawatt solar electric plant would use at least an estimated 30,000 to 40,000 tons of steel, 5,000 tons of glass, and 200,000 tons of concrete. A nuclear plant, by contrast, would require about 5,000 tons of steel and 50,000 tons of concrete; a coal-fired plant, still less.

Figure 14.15 Some possible schemes for storing the energy of solar-generated electricity. (*A*) Use solar electricity to break up water molecules into hydrogen and oxygen; recombine them later (burn the hydrogen) to release energy. (*B*) Use solar energy to pump water up in elevation; when the energy is needed, let the water fall back and use it to generate hydropower.

Siting such a sizable array requires a substantial commitment and disturbance of land. Its presence may alter patterns of evaporation and surface runoff. These considerations could be especially critical in desert areas, which are most favorable sites for such facilities from the standpoint of intensity and constancy of incident sunlight. Construction could also disturb desert-pavement surfaces and accelerate erosion.

Solar Energy Summary

Given all these considerations, space heating seems to be the area in which solar energy can make the most immediate significant contribution with the minimum negative environmental impact. With solar space heating, the commitment of

Hot water, steam
tapped for energy

Natural recharge by cool water
infiltrating from surface through
fractures, etc.

Cold water returned

Impermeable
rocks

Permeable rocks

Circulating warmed water

Cooling magma

Figure 14.16 Geothermal energy is utilized by tapping circulating warmed ground water.

Figure 14.17 One of the many thermal features in Yellowstone National Park: Riverside geyser.

materials is comparable to that required by conventional technologies, and the availability of abundant free energy and the lack of pollution during operation constitute substantial benefits.

At present, however, even this use of solar energy is minimal. Solar energy accounted for only 1 percent of energy derived from *renewable* sources in the U.S. in 1995, and renewable sources, in turn, together produced just 8 percent of total energy consumed.

Geothermal Power

The earth contains a great deal of heat, most of it left over from earth's early history, some continually generated by decay of radioactive elements in the earth. Slowly, this heat is radiating away and the earth is cooling down, but under normal circumstances, the rate of heat escape at the earth's surface is so slow that we do not even notice it and certainly cannot use it. If the heat escaping at the earth's surface were collected over an average square meter for a year, it would be sufficient only to heat about 2 gallons of water to the boiling point, though local heat flow can be substantially higher, for example in young volcanic areas.

The Geothermal Resource

Magma rising into the crust from the mantle brings unusually hot material nearer the surface. Heat from the cooling magma heats any ground water circulating nearby (figure 14.16). This is the basis for generating **geothermal energy.** The magma-warmed waters may escape at the surface in geysers and hot springs, signalling the existence of the shallow heat source below (figure 14.17). More subtle evidence of the presence of hot rock at depth comes from sensitive measurements of heat flow at the surface, the rate at which heat is being conducted out of the ever-cooling earth: High heat flow signals unusually high temperatures at shallow depths. High heat flow and recent (or even current) magmatic activity go together and, in turn, are most often associ-

ated with plate boundaries. Therefore, most areas in which geothermal energy is being tapped extensively are along or near plate boundaries (figure 14.18).

Applications of Geothermal Energy

Exactly how the geothermal energy is used depends largely on how hot the system is. In some places, the ground water is warmed, but not enough to turn to steam. Still, the water temperatures may be comparable to what a home heating unit produces (50 to 90° C, or about 120 to 180° F). Such warm waters can be circulated directly through homes to heat them. This is being done in Iceland and in parts of the former Soviet Union.

Other geothermal areas may be so hot that the water is turned to steam. The steam can be used, like any boiler-generated steam produced using conventional fuels, to run electric generators. The largest U.S. geothermal-electricity operation is The Geysers, in California (figure 14.19), which has operated since 1960 and now has a generating capacity of close to 2 billion watts. In 1989, The Geysers and six smaller geothermal areas together generated close to 10 billion kilowatt-hours of electricity (0.3% of total energy consumed) in this country. By 1995, utilities were generating only about 5 billion kilowatt-hours of geothermal electricity, while non-utility power producers generated about twice that much. Other steam systems are being used in Larderello, Italy, and in Japan, Mexico, the Philippines, and elsewhere. Altogether, there are about forty sites worldwide where geothermal electricity is actively being developed.

Figure 14.18 Geothermal power plants worldwide.   Source: Figure prepared by L. J. Patrick Muffler and Ellen Lougee, U.S. Geological Survey; plate-tectonic boundaries supplied by Charles DeMets, University of Wisconsin at Madison.

A

B

Figure 14.19 The Geysers, California. (*A*) Close-up of power plant. (*B*) Aerial view of The Geysers geothermal area.   Photographs by R. E. Wallace, USGS Photo Library, Denver, CO.

Figure 14.20 Mammoth Terraces in Yellowstone National Park have been built from centuries of deposition of minerals dissolved in circulating geothermal waters that have seeped out at the surface.

Figure 14.21 Decrease in productivity of a geothermal field due to chilling of the aquifer and adjacent rocks limits the lifetime of the average geothermal field to about thirty years.

Environmental Considerations of Geothermal Power

Where most feasible, geothermal power is quite competitive economically with conventional methods of generating electricity. The use of geothermal steam is also largely pollution-free. Some sulfur gases derived from the magmatic heat source may be mixed with the steam, but these certainly pose no more serious a pollution problem than sulfur from coal burning. Moreover, there are no ash, radioactive-waste, or carbon-dioxide problems as with other fuels. Warm geothermal *waters* may be a somewhat larger problem. They frequently contain large quantities of dissolved chemicals that can clog or corrode pipes (a potentially significant problem that will increase operational costs), or may pollute local ground or surface waters if allowed to run off freely (figure 14.20). Sometimes, there are surface subsidence problems, as at Wairakei (New Zealand), where subsidence of up to 0.4 meters per year has been measured. Subsidence problems may be addressed by reinjecting water, but the long-term success of this strategy is unproven.

Limitations on Geothermal Power

While the environmental difficulties associated with geothermal power are relatively small, three other limitations severely restrict its potential. First, each geothermal field can only be used for a period of time—a few decades, on average—before the rate of heat extraction is seriously reduced. This is because rocks conduct heat very poorly. That characteristic can easily be demonstrated by turning over a flat, sunbaked rock on a bright but cool day. The surface exposed to the sun may be hot to the touch, but the heat will not have traveled far into the rock; the underside stays cool for some time. Conversely, as hot water or steam is withdrawn from a geothermal field, it is replaced by cooler water

that must be heated before use. Initially, the heating can be rapid, but in time the permeable rocks become chilled to such an extent that water circulating through them heats too slowly or too little to be useful (figure 14.21). The heat of the magma has not been exhausted, but its transmittal into the permeable rocks is slow. Some time must then elapse before the permeable rocks are sufficiently reheated to resume normal operations. Steam pressure at The Geysers has declined rapidly in recent years, forcing the idling of some of its generating capacity: By 1991, despite a capacity of 2 billion watts, electricity production was only $1\frac{1}{2}$ billion watts, and power generation had declined to half its 1987 peak within a decade.

A second limitation of geothermal power is that not only are geothermal power plants stationary, but so is the resource itself. Oil, coal, or other fuels can be moved to power-hungry population centers. Geothermal power plants must be put where the hot rocks are, and long-distance transmission of the power they generate is not technically practical. Most large cities are far removed from major geothermal resources. Also, of course, geothermal power cannot contribute to such energy uses as transportation.

The total number of sites suitable for geothermal power generation is the third limitation. Clearly, plate boundaries cover only a small part of the earth's surface, and many of them are inaccessible (seafloor spreading ridges, for instance). Not all have abundant circulating subsurface water in the area, either. Even accessible regions that do have adequate subsurface water may not be exploited. Yellowstone National Park has the highest concentration of thermal features of any single geothermal area in the world, but because of its scenic value and uniqueness, the decision was made years ago not to build geothermal power plants there.

Alternative Geothermal Sources

Many areas away from plate boundaries have heat flow somewhat above the normal level, and rocks in which temperatures increase with depth more rapidly than in the

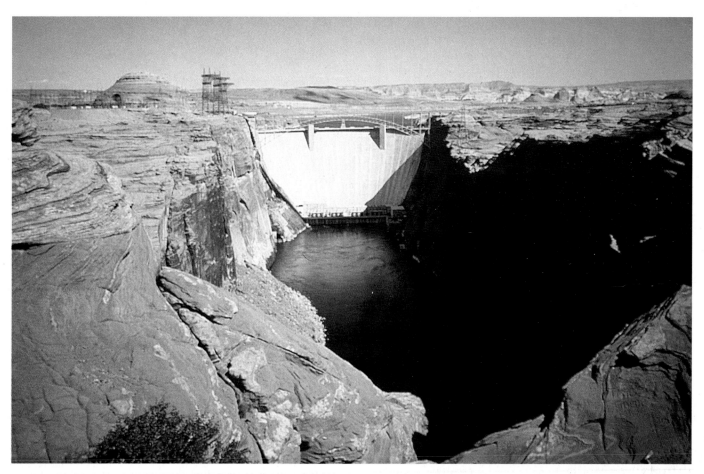

Figure 14.22 Glen Canyon Dam hydroelectric project. In dry surroundings such as these, evaporation losses from reservoirs can be high.

average continental crust. Even in ordinary crust, rock temperature increases with depth at a rate of 30° C/kilometer (about 85° F/mile). Where thermal gradients are at least 40° C/kilometer, even in the absence of much subsurface water, the region can be regarded as a potential geothermal resource of the **hot-dry-rock** type. Deep drilling to reach usefully high temperatures must be combined with induced circulation of water pumped in from the surface to make use of these hot rocks. The amount of heat extractable from hot-dry-rock fields is estimated at more than ten times that of natural hot-water and steam geothermal fields, just because the former are much more extensive. Most of the regions identified as possible hot-dry-rock geothermal fields in the United States are in thinly populated western states with restricted water supplies. There is, therefore, a large degree of uncertainty about how much of an energy contribution they may ultimately make. Certainly, hot-dry-rock geothermal energy will be less economical than that of the hot-water or steam fields where circulating water is already present. Although some experimentation with hot-dry-rock fields is underway— Los Alamos National Laboratory has an experimental project in operation in the Jemez Mountains of New Mexico—commercial development in such areas is not expected before the end of the century.

The geopressurized natural gas zones described in chapter 13 represent a third possible type of geothermal resource. As hot, gas-filled fluids are extracted for the gas, the heat from the hot water might, in theory, also be used to generate electricity. Substantial technical problems probably will be associated with developing this resource, as noted in chapter 13. Moreover, it probably would be economically infeasible to pump the spent water back to such depths for recycling. No geopressurized zones are presently being developed for geothermal power, even on an experimental basis.

Summary of Geothermal Potential

For the near term and well into the next century, geothermal energy probably will be a major energy source only in the handful of areas best suited to its production, supplying them with heat and/or electricity. Its overall contribution to either U.S. or world energy consumption in the foreseeable future is likely to be minor.

Hydropower

The energy of falling or flowing water has been used for centuries. It is now used primarily to generate electricity (figure 14.22). Hydroelectric power has consistently supplied a small percentage of U.S. energy needs for several decades; it currently provides about 4% of U.S. energy (about 10% of U.S. electricity). The principal requirements for the generation of substantial amounts of hydroelectric power are a large volume of water and the rapid movement of that water. Nowadays, commercial generation of hydropower typically involves damming up a high-discharge stream, impounding a large volume of water, and releasing it as desired, rather than operating subject to great seasonal variations in discharge.

Hydropower is a very clean energy source. The water is not polluted as it flows through the generating equipment. No chemicals are added to it, nor are any dissolved or airborne pollutants produced. The water itself is not consumed during power generation; it merely passes through the generating equipment. Hydropower is renewable as long as the streams continue to flow. Its economic competitiveness with other sources is demonstrated by the fact that nearly one-third of U.S. electricity-generating plants are hydropower plants; also, worldwide, about 6 percent of all energy consumed is hydropower.

The Federal Power Commission has estimated that the potential energy to be derived from hydropower in the United States, if it were tapped in every possible location, is about triple current hydropower use. In principle, then, hydropower could supply one-third to one-half of U.S. electricity if consumption remained near present levels. However, ultimate development may not occur on such a scale.

We have already considered, in chapters 6 and 10, some of the problems posed by dam construction, including silting-up of reservoirs, habitat destruction, water loss by evaporation, and even, sometimes, earthquakes. Evaluation of the risks of various energy sources must also consider the possibility of dam failure. There are over a thousand dams in the United States (not all constructed for hydropower generation). Several dozen have failed within this century. Aside from age and poor design or construction, the reasons for these failures may include geology itself. Fault zones often occur as topographic lows, and streams thus frequently flow along fault zones. It follows that a dam built across such a stream is built across a fault zone, which may be active or may be reactivated by filling the reservoir. Not all otherwise-suitable sites, then, are safe for hydropower dams.

Other sites with considerable power potential may not be available or appropriate for development. Construction might destroy a unique wildlife habitat or threaten an endangered species. It might deface a scenic area or alter its natural character; suggestions for additional power dams along the Colorado River that would have involved backup of reservoir water into the Grand Canyon were met by vigorous protests. Many potential sites—in Alaska, for instance—are just too remote from population centers to be practical unless power transmission efficiency is improved.

An alternative to development of many new hydropower sites would be to add hydroelectric-generating facilities to dams already in place for flood control, recreational

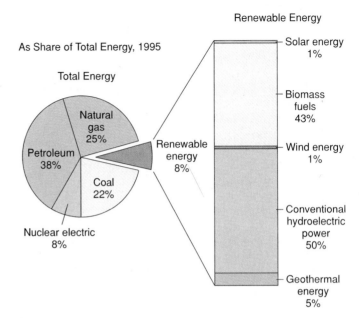

Figure 14.23 Hydropower is currently the dominant renewable energy source in the U.S., though renewable sources still provide a relatively small proportion of energy used. Source: *Annual Energy Review 1995*, U.S. Energy Information Administration.

purposes, and so on. Although the release of impounded water for power generation alters streamflow patterns, it is likely to have far less negative impact than either the original dam construction or the creation of new dam/reservoir complexes. Still, it is clear that flood control and power generation are somewhat conflicting aims: the former requires considerable reserve storage capacity, while the latter is enhanced by impounding the maximum volume of water.

Like geothermal power, conventional hydropower is also limited by the stationary nature of the resource. In addition, hydropower is more susceptible to natural disruptions than other sources considered so far. Just as torrential precipitation and one-hundred-year floods are rare, so, too, are prolonged droughts—but they do happen, as residents of the Midwest were reminded in 1988. In fact, hydropower generation declined by about 25 percent from 1986 to 1988, largely as a consequence of drought. Heavier reliance on hydropower would leave many energy consumers vulnerable to interruption of service in times of extreme weather.

For various reasons, then, it is unlikely that numerous additional hydroelectric power plants will be developed. This clean, cheap, renewable energy source can continue indefinitely to make a modest contribution to energy use, but it cannot be expected to supply much more energy in the future than it does now. Still, in 1995, hydropower was the dominant renewable energy source used in the U.S. (figure 14.23).

Tidal Power

All large bodies of standing water on the earth, including the oceans and large lakes like the Great Lakes, show tides. Why not also harness this moving water as an energy

(bay)   (open water)

As the tide rises outside restricted bay, water rushes in through turbines, generating electricity.

When tide is out, water confined in bay again moves through turbines to generate power.

Figure 14.24 Tidal-power generation uses flowing water to generate electricity, as with conventional hydropower.

Wind Resource in the United States

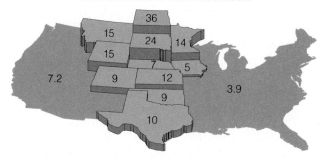

Figure 14.25 Numbers indicate the percentage of 1990 electricity demand that the corresponding region could theoretically supply with full utilization of current wind-power technology.   Source: After Jon G. McGowan, "Tilting Toward Windmills," *Technology Review,* July 1993, p. 41, MIT.

source? Unfortunately, the energy represented by tides is too dispersed in most places to be useful. Average beach tides reflect a difference between high-tide and low-tide water levels of about one meter. A commercial tidal-power electricity-generating plant requires at least 5 meters difference between high and low tides for efficient generation of electricity, and a bay or inlet with a narrow opening that could be dammed to regulate the water flow in and out (figure 14.24). The proper conditions exist in very few places in the world. Tidal power is being used in a small way at several locations—near St. Malo, France, in the Netherlands, and in the former Soviet Union. Worldwide, the total potential of tidal power is estimated at only 2 percent of the energy potential of conventional hydropower.

Wind Energy

Because the winds are ultimately powered by the sun (chapter 9), wind energy can be regarded as a variant of solar energy. It is clean and, like sunshine, renewable indefinitely (at least for five billion years or so). Wind power has been utilized to some extent for more than two thousand years; the windmills of the Netherlands are probably the best-known historic example. Today, there is considerable interest in making more extensive use of wind power. Windmills currently are most widely used for pumping ground water and for generating electricity for individual homes and farms.

The cost of commercial wind power on a large scale is not now competitive with conventionally generated electricity, but that may simply be because windmills are not now mass-produced and are, therefore, relatively expensive. Technological design improvements to increase efficiency are also possible.

Wind energy shares certain limitations with solar energy. It is dispersed, not only in two dimensions but in three: It is spread out through the atmosphere. Wind is also erratic, highly variable in speed both regionally and locally. The regional variations in potential power supply are even more significant than they may appear from average wind velocities because windmill power generation increases as the cube of wind speed. So if wind velocity doubles, power output increases by a factor of 8 ($2 \times 2 \times 2$).

Figure 14.25 shows that most of the windiest places in the United States are rather far removed physically from most of the heavily populated and heavily industrialized areas. As with other physically localized power sources, the technological difficulty of long-distance transmission of electricity will limit wind power's near-term contribution in areas of high electricity consumption.

Even where average wind velocities are great, strong winds do not always blow. This presents the same storage problem as does solar electricity, and it has likewise not yet been solved satisfactorily. In the near future, wind-generated electricity might most effectively be used to supplement conventionally generated power when wind conditions are favorable, with conventional plants bearing most or all of the demand load at other times. Still, design improvements have increased the reliability of windmills; the newest wind turbines are available for power generation at least 95 percent of the time.

Winds blow both more strongly and more consistently at high altitudes. Tall windmills, perhaps a mile high, may provide more energy with less of a storage problem. However, such tall windmills would be technically more difficult to build and would require substantial quantities of material.

Figure 14.26 Wind-turbine array near Palm Springs, California. Different elements of the array can take advantage of various velocities of wind. Some of the smaller turbines are turning even in a light wind, while the larger turbines require stronger winds to drive them.   © The McGraw-Hill Companies, Inc./Doug Sherman, photographer.

The ultimate potential of wind energy is unclear. Certainly, blowing wind represents far more energy than we can use, but most of it cannot be harnessed. Most schemes for commercial wind-power generation of electric power involve "wind farms," concentrations of many windmills in a few especially favorable windy sites (figure 14.26). The Great Plains region is the most promising area for initial sizable efforts, as suggested by figure 14.25. The limits to wind-power use would then include the area that could be committed to windmill arrays, as well as the distance the electricity could be transmitted without excessive loss in the power grid. About one thousand 1-megawatt windmills would be required to generate as much power as a sizable conventional electric power plant. The windmills have to be spread out, or they block each other's wind flow. Spacing them at four windmills per square kilometer requires about 250 square kilometers (100 square miles) of land to produce energy equivalent to a 1,000-megawatt power plant. However, the land need not be devoted exclusively to the wind-powered generators. Farming and the grazing of livestock, common

activities in the Great Plains area, could continue on the same land concurrently with power generation. The storage problem, however, would still remain. Other concerns relating to wind farms include the aesthetic impact (the windmill arrays must naturally be very exposed); interference with and death to migrating birds; and disruption of communications—TV, radio, cellular telephones, perhaps even aircraft communications.

Industrial and consulting firms have projected that the United States might produce 25 to 50 percent of its electricity from wind power by the year 2000, but only with a major national program for wind-energy development. This percentage would represent a significant fraction of anticipated total energy needs, and it also would contribute to conserving nonrenewable fuels. However, no governmental program of any such scale is presently active or contemplated. In fact, in 1995, wind-powered electric generating facilities in the United States accounted for only about 3 percent of U.S. generating capacity, and only one percent of actual power generation. At that, the United States is a world leader in generation of electricity from wind power, although plans for wind-power development in Europe are more ambitious than those of the United States over the next decade.

Biomass

The term **biomass** technically refers to the total mass of all the organisms living on earth. In an energy context, the term has become a catchall for various ways of deriving energy from organisms or from their remains. Indirectly, biomass-derived energy is ultimately solar energy, since most biomass energy sources are plant materials, and plants need sunlight to grow. Biomass fuels could also be thought of as "unfossilized fuels" because they represent fuels derived from living or recent organisms rather than ancient ones, and hence have not been modified extensively by geologic processes acting over long periods of time.

The possibilities for biomass-derived energy are many. The use of wood as a fuel is increasing in popularity. Home use of wood stoves still is an insignificant part of total U.S. energy use, but it is increasing—20 to 25 percent of U.S. households burn at least some wood, and it is the primary source of heat in about 4 percent of households. Industrial use of wood as fuel is about triple household use.

Sometimes, using biomass energy sources means burning waste plant materials after a crop is harvested (see box 14.4). The combustible portion of urban refuse can be burned to provide heat for electric generating plants. Total energy derived in the United States from biomass fuels has been growing; it is now the second-largest source of renewable energy (recall figure 14.23). More extensive use of incineration as a means of solid-waste disposal, both in the United States and abroad, would further increase the use of biomass for energy worldwide.

Because certain plants manufacture flammable, hydrocarbon-rich fluids, the possibility of "raising" liquid fuel through farming is being evaluated. A variation on this theme involves microorganisms that produce oil-like substances: The organisms could be cultivated on ponds and their oil skimmed off. Such possibilities represent potential options for the future.

Biomass fuels are burned to release their energy, so they share the carbon-dioxide-pollution problems of fossil fuels. Some biomass fuels, like wood, also contribute particulate air pollutants. However, unlike the fossil fuels, biomass fuels are renewable. We can, in principle, produce and replenish them on the same time scale as that on which we consume them.

Alcohol as Fuel

One biomass fuel that has received special attention is alcohol. Initially, it was extensively developed mainly for incorporation into **gasohol.** Gasohol as originally created was a blend of 90% gasoline and 10% alcohol. The alcohol has historically been produced mainly from grains, such as corn. The use of potential food crops to make motor fuel has, to some, appeared frivolous in a hungry world. Improved technology, however, has made it possible to use the nonfood parts of plants to make the alcohol, thereby reducing that concern. The 90:10 mix of gasoline and alcohol does not represent a whole new fuel, but its use would extend by 10% the supply of gasoline, which would buy more time for alternative fuel development.

The higher the proportion of alcohol in the mix, of course, the further the gasoline can be stretched. The 90:10 mix, however, allows conventional gasoline engines to use it without modifications. There is, in principle, no reason why engines cannot be designed to run on 20%, 50%, or 100% alcohol, and, in fact, some vehicles have now been made to run solely on alcohol. The main problem is that the modified vehicles must have an assured supply of their special fuels: They cannot burn gasoline in a pinch. Until alcohol-rich fuels are widely available, most cars will not be designed to run on such special mixtures.

There seem to be fads in fuels as in fashions. In the late 1970s and early 1980s, gasohol use rose briskly. Besides having the appeal of being a partial alternative to OPEC oil, gasohol seemed to improve the performance of many car engines. It generally cost a bit more than straight unleaded gasoline, however. In the face of stabilizing fuel prices and worries about diverting food to fuel, its popularity waned. Several major oil companies then appeared to phase out gasohol at their service stations. However, it is back, or still there, in the guise of "super unleaded" or "super unleaded with ethanol." Under such names, and with ethanol production costs now partially subsidized in some farm-belt states, gasohol consumption continues, and consumption is likely to grow as expanded alcohol-producing operations bring down its price. In 1992, 3.2 billion liters (about 800 million gallons) of ethanol were produced in the United States—from corn, from other grains, from molasses and food wastes. Gasohol can help significantly in extending petroleum fuel supplies while transportation alternatives are developed. However, by 1995, alcohol fuels still accounted for only about 0.12 percent of energy consumed in the United States.

Government policy may encourage expanded use of alcohol and alcohol/gasoline blends. The Clean Air Act Amendments of 1990 include provisions for reduced vehicle emissions, and gasohol is somewhat cleaner-burning. The Energy Policy Act of 1992 is requiring gradual replacement of some government, utility-company, and other "fleet" vehicles with "alternative-fuel" vehicles, which would include vehicles powered by straight alcohol. (This is more immediately feasible for fleet vehicles because they can be refueled at central locations stocking their special fuels.) The same act extends an excise-tax break on ethanol-blend fuels. Large gains in ethanol consumption remain to be realized, however.

Biogas

Another biomass fuel growing in use could be called "gas from garbage." When broken down in the absence of oxygen, organic wastes yield a variety of gaseous products.

Box 14.4 New Energy Diet for a Small User: Hawaii

It may be some time before adequate alternatives to fossil fuels are developed for areas where energy use is intensive, such as northern U.S. cities dominated by heavy industry. Elsewhere, however, a judicious mix of renewable energy sources can already serve as the major source of energy. The Hawaiian Islands have made great strides in that direction.

The efforts have been motivated, in part, by the realities of geology: The islands have no native fossil-fuel resources, so conventional energy costs are high and the imported supplies are subject to disruption. Geology, however, has also contributed to the development of alternatives. The "hot spot" beneath the island of Hawaii and the active volcanism it causes make the area promising for the generation of geothermal power. A pilot geothermal-electricity plant (figure 1) now taps some of the hottest geothermal steam in the world (360° C, or 675° F). The total potential of the site is estimated at more than five times the island's current electricity needs. A drawback, of course, is the ever-present threat of damage from earthquakes or new eruptions, but the plant has been engineered to minimize such damage.

On Oahu, where most of Hawaii's people live, the potential for power generated from the very strong and consistent trade winds is substantial. Wind power may supply 10 percent of the state's energy within a decade. And about one-third of the state's electricity is already being provided by biomass—specifically, by burning plant wastes from sugar-cane processing, wastes that were previously dumped in the ocean or in landfills.

Expanded use of alternative energy sources could make Hawaii energy-self-sufficient by the year 2000 or so. However, it is not clear how rapidly the islands' potential will be developed. The very active East Rift zone of Kilauea is an especially attractive site for extensive geothermal development. Concerns relate to possible diversions of power by some of the more populous islands (Oahu, Maui) and to violation of sacred ground (figure 2). Recent court decisions have decreed that environmental impact statements (see chapter 18) must be prepared before substantial geothermal development can occur on the Big Island of Hawaii. The situation reinforces the realization that solving technological problems, like a need for energy, may involve many considerations beyond the strictly scientific and technical ones.

Figure 1 A geothermal power plant below Kilauea's East Rift zone taps steam at 350° C (about 660° F) from a depth of 2,000 meters (1.2 miles).   Photo by Hawaii Volcanoes Observatory, U.S. Geological Survey.

Figure 2 On the Hawaiian islands, native peoples recognize many sacred places, such as the site of this temple, or *heiau.* This one may be threatened only by the lava flows that have so far left it untouched; but some of the promising sites for geothermal-power development are also considered sacred.

Some of these are useless, or smelly, or toxic, but among them is methane (CH_4), the same compound that predominates in natural gas. Sanitary-landfill operations (see chapter 15) are suitable sites for methane production as organic wastes in the refuse decay. Landfill-derived gas can then be blended with purer, pipelined natural gas to extend the gas supply. (Straight landfill gas is too full of impurities to use alone.) Methane can also be produced from decaying manures. Where large quantities of animal waste are available, as for example on feedlots, the wastes may constitute a locally significant energy source. Certainly, it is better to extract energy from these wastes than simply to let them add to the area's pollution problems!

Overview

Summary

As we anticipate the end of the abundant supplies of petroleum and conventional natural gas, we find an almost bewildering variety of alternatives available. None is as versatile as liquid and gaseous petroleum fuels, and the potential of many is limited or unknown. Serious environmental pollution and other risks are associated with the alternatives most extensively developed so far—coal and nuclear fission power. Technological advances will be needed before renewable, environmentally more benign sources like geothermal, wind, and solar energy can make major overall contributions to meeting our energy needs, though each is already locally significant in some places, and each represents a more-than-adequate potential supply of energy. New options—fusion power, various biomass sources, and others—continue to be developed. The energy-use picture in the future probably will not be dominated by a single source as has been the case in the fossil-fuel era. Rather, a blend of sources with different strengths, weaknesses, and specialized applications is likely. Different nations, too, may make different choices, for reasons of geology or geography, economics, or differing environmental priorities (figure 14.27).

In the near term, given present economic, technological, and other practical limitations, it seems most likely that conventional oil and natural gas, their lifetimes perhaps extended by unconventional gas sources, will be replaced principally by solar energy for heating and by coal or coal derivatives for other purposes. Over the longer term, it clearly makes sense to develop renewable resources more fully and to make wider use of such options as solar and wind energy or the various biomass energy sources. Ultimately, fusion may be a major energy source, but it is technologically furthest from commercial reality. Time is required to improve the technologies involved with many of these sources, as well as to manufacture the necessary equipment in sufficient quantity to utilize them on a major scale. Certainly, we do not lack options, and, collectively, they represent a fully adequate solution to future energy needs.

In the meantime, as the alternatives are explored and developed, vigorous efforts to conserve energy would yield much-needed time to make a smooth transition from present to future energy sources. However, considerable pressures toward increasing energy consumption exist in countries at all levels of technological development, and there is currently little evidence of effective energy-conservation activities on a large scale.

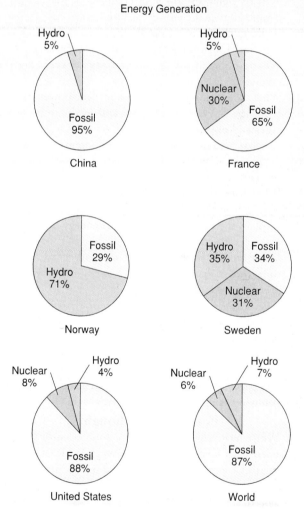

Energy Generation

China — Hydro 5%, Fossil 95%

France — Hydro 5%, Nuclear 30%, Fossil 65%

Norway — Fossil 29%, Hydro 71%

Sweden — Hydro 35%, Fossil 34%, Nuclear 31%

United States — Nuclear 8%, Hydro 4%, Fossil 88%

World — Nuclear 6%, Hydro 7%, Fossil 87%

Figure 14.27 Shares of fossil and non-fossil fuels in energy consumption for selected countries. Source: *IAEA Bulletin* 1993, No. 4, p. 23. Reprinted by permission.

Terms to Remember

active-solar heating 337
biomass 347
breeder reactor 329
chain reaction 327
core meltdown 329
decommissioning 334
fission 327

fusion 327
gasohol 347
geothermal energy 340
hot-dry-rock 343
passive-solar heating 337
photovoltaic cells 337

Exercises

Questions for Review

1. Briefly describe the nature of the fission chain reaction used to generate power in commercial nuclear power plants. How is the energy released utilized?
2. If the nuclear power option is pursued, breeder reactors and fuel reprocessing will be necessary. Why? What additional safety and security concerns are then involved?
3. What is "decommissioning" in a nuclear-power context?
4. Describe the fusion process, and evaluate its advantages and present limitations.
5. In what areas might solar energy potentially make the greatest contributions toward our energy needs? Explain.
6. What technological limitations do solar and wind energy presently share?
7. Explain the nature of geothermal energy and how it is extracted.
8. What factors restrict the use of geothermal energy in time and in space? How do hot-dry-rock geothermal areas expand its potential?
9. Assess the potential of (a) conventional hydropower and (b) tidal power to help solve impending energy shortages.
10. What are biomass fuels? Describe two examples.
11. Choose any two energy sources from this chapter and compare/contrast them in terms of the negative environmental impacts associated with each.

For Further Thought

1. While most alternative energy sources must be developed on a large scale, solar energy (at least for space heating) can be installed building by building. For your region, explore the feasibility and costs of conversion to solar heating.
2. Investigate the history of a commercial nuclear-power plant project, completed or pending. How long a history does the project have? What are/were the cost projections of the plant and the energy consumption projections on which those costs were justified? Have the projections held up? Has the plant suffered regulatory or other construction delays?
3. Consider having a home energy audit (often available free or at nominal cost through a local utility company) to identify ways to conserve energy.
4. Choose any nonfossil fuel, and investigate its development over recent decades, in the U.S., in another nation, or globally. How significant a factor is it in worldwide energy production currently? How do its economics compare with fossil fuels? What are its future prospects? (The Internet may provide particularly current data.)

Suggested Readings/References

Anspaugh, L. R., Catlin, R. J., and Goldman, M. 1988. The global impact of the Chernobyl reactor accident. *Science* 242 (16 December): 1513–19.

Booth, W. 1987. Postmortem on Three Mile Island. *Science* 238: 1342–45.

Crutis, F. A., ed. 1984. *Energy developments: New forms, renewables, conservation.* New York: Pergamon Press.

Edmonds, J., and Reilly, J. M. 1985. *Global energy: Assessing the future.* New York: Oxford University Press.

Edwards, M. 1994. Chernobyl. *National Geographic* (August): 101–15.

Fischett, M. A. 1986. The puzzle of Chernobyl. *IEEE Spectrum* 23 (July): 34–41.

Ford, D. F. 1982. *Three Mile Island.* New York: Viking Press.

Fowler, J. M. 1984. *Energy and the environment.* 2d ed. New York: McGraw-Hill.

Golay, M. W. 1990. Longer life for nuclear plants. *Technology Review* (May/June): 25–30.

Hall, C. A. S., Cleveland, C. J., and Kaufman, D. 1986. *Energy and resource quality: The ecology of the economic process.* New York: John Wiley & Sons.

Hoyle, F., and Hoyle, G. 1980. *Commonsense in nuclear energy.* San Francisco: W. H. Freeman.

Hunt, V. D. 1982. *Handbook of energy technology.* New York: Van Nostrand Reinhold.

International Atomic Energy Agency, Vienna, Austria. (This agency publishes a quarterly bulletin with articles about many aspects of nuclear power and related applications of radiochemistry. Some of the data for this chapter were taken from the March 1983, Supplement 1982, Autumn and Winter 1986, 1989 no. 2, 1990 no. 3, 1991 no. 2, and 1993 no. 1 and no. 4 issues.)

International Energy Agency. 1987. *Renewable sources of energy.* Paris: OECD.

Kendall, H. W., and Nadis, S. J., eds. 1980. *Energy strategies: Toward a solar future.* Cambridge, Mass.: Ballinger.

Kerr, R. A. 1991. Geothermal tragedy of the commons. *Science* 253 (12 July): 134–35.

McGowan, J. G. 1993. Tilting toward windmills. *Technology Review* (July): 39–46.

National Academy of Sciences. 1979. *Energy in transition, 1985–2010.* San Francisco: W. H. Freeman.

National Geographic Society. 1981. *Energy* (Special Report). Washington, D.C.: National Geographic Society.

Perlman, E. 1982. Kilowatts in paradise. *Science 82* (American Association for the Advancement of Science) (January/February): 78–81.

Pool, R. 1990. Cold fusion: Only the grin remains. *Science 250* (9 November): 754–55.

Ruedisili, L. C., and Firebaugh, M. W., eds. 1982. *Perspectives on energy.* 3d ed. New York: Oxford University Press.

Smith, C. 1995. Revisiting solar power's past. *Technology Review* (July): 38–47.

Stine, W. B., and Harrigan, R. W. 1985. *Solar energy fundamentals and design.* New York: John Wiley & Sons.

Tenenbaum, D. 1995. Tapping the fire down below. *Technology Review* (January): 38–47. (Pertains to geothermal power generation, at Los Alamos and in general.)

U.S. Department of Energy, Energy Information Administration. 1983. *Commercial nuclear power: Prospects for the U.S. and the world.* Washington, D.C.: U.S. Government Printing Office.

———. 1983. *Nuclear plant cancellations: Causes, costs, and consequences.* Washington, D.C.: U.S. Government Printing Office.

———. 1986. *Domestic uranium mining and milling industry.* Washington, D.C.: U.S. Government Printing Office.

———. 1992. *International energy annual 1991.* Washington, D.C.: U.S. Government Printing Office.

———. 1992. *Annual energy review 1991.* Washington, D.C.: U.S. Government Printing Office.

———. 1996. *Annual energy review 1995.* Washington, D.C.: U.S. Government Printing Office.

Whitmore, E. C., Jr., and Williams, M. E., eds. 1982. *Resources for the twenty-first century.* U.S. Geological Survey Professional Paper 1193.

World Resources Institute. 1992. *World Resources 1992–93.* New York: Oxford University Press.

World Resources Institute, United Nations Environment Programme, United Nations Development Programme, and The World Bank. 1996. *World Resources 1996–97.* (Chapter 12, "Energy and Materials.") New York: Oxford University Press.

Yergin, D. 1993. Energy policy: The ecumenical approach 1993–1997. *Geotimes* (March): 12–15. (See also related following short articles.)

A major U.S. Government source of energy statistics is the Department of Energy's Energy Information Administration, with home page at

http://www.eia.doe.gov/

Two particularly useful documents newly online at the time of publication are *Annual Energy Review 1995,* an extensive document of U.S. and some global data, at http://www.eia.doe.gov/emeu/aer95/pubpdf/aer95pdf.html and *Annual Energy Outlook 1996,* which makes some projections for energy consumption over the next two decades at

http://www.eia.doe.gov/oiaf/aeo96/demand96.html

The U.S Geological Survey's Energy Resource Surveys Program is accessible via the home page

http://www.energy.usgs.gov/

For data and graphics concerning worldwide use of nuclear fission power, try the International Atomic Energy Agency (IAEA) site at

http://www.iaea.or.at/worldatom/

The IAEA—which is sometimes described as the "Atoms for Peace Organization of the UN"—cosponsored the "One Decade after Chernobyl" conference in April 1996, highlights of which can be found at

http://www.iaea.or.at/worldatom/thisweek/preview/chernobyl/

Information on commercial U.S. reactors specifically—numbers, types, the licensing process, etc.—is available from the Nuclear Regulatory Commission at

http://www.nrc.gov/reactors.html#COMMSITES

A history of U.S. nuclear fission power (and other data) can be found at

http://www.ucm.com:80/ucm/info/cenuchis.htm#history

Natural Resources Canada (NR Canada) makes available contact e-mail addresses for many agency personnel handling a variety of energy and related policy issues, at

http://www.NRCan.gc.ca/htbin-post/x500services

Section 5

The two subjects of this section—waste disposal and pollution—are intimately related. Both concern the presence and handling of certain chemicals in our physical environment. There is, of course, nothing ominous about chemicals *per se*, despite the fact that the word prompts sinister associations in many people's minds. Water is a chemical; rocks are collections of chemicals; indeed, the human body is a very complex mass of chemicals. A *pollutant* sometimes is defined simply as a "chemical out of place," a substance found in sufficient concentration in some setting that it creates a nuisance or a hazard. Recently, the concept of pollution has been broadened to

include thermal pollution, the input of excess heat into a system (usually water). What might be called *biological pollution*, such as excessive growth of harmful microorganisms or algae, is often a further result of chemical or thermal pollution.

Some chemical deterioration of the environment is natural and occurs with or without human activities; natural oil seeps are just one example. Much of it, however, is just the result of careless waste disposal or of the lack of deliberate waste-disposal efforts. A basic principle to keep in mind in dealing with these subjects, credited to biologist Dr. Barry Commoner, may be expressed as, "Nothing ever goes

away." Sulfur gases and ash from a coal-fired power plant can be trapped by smokestack scrubbers rather than released into the air, but they must still be *put* somewhere afterward. A factory may not release toxic chemicals into a lake, but the chemicals have to go *someplace*, or else be thoroughly destroyed. Out of sight may be out of mind, but not out of environmental circulation. In a real sense, as many pollution problems are reduced, waste-disposal problems are increased. In chapters 15–17, we survey various waste-disposal problems and strategies and then consider some of the consequences of pollution, identifying several specific problems in the process.

Chapter 15

Waste Disposal

High-consumption technological societies also tend to generate copious quantities of wastes. In the United States alone, each person generates, on the average, more than 1.8 kilograms (about 4 pounds) of what is broadly called "garbage" every day. That represents an increase of more than 50 percent over 1960 per-capita waste production. Every five years, each average American generates a mass of waste equal to the mass of the Statue of Liberty! That, plus industrial, agricultural, and mineral wastes together amount to an estimated total of 4 billion tons of solid waste produced in the United States each year. Most of the 34 billion gallons of water withdrawn daily by public water departments end up as sewage-tainted wastewater, and more-concentrated liquid wastes are generated by industry. Each day, the question of where to put the growing accumulations of radioactive waste materials becomes more pressing. Proper, secure disposal of all these varied wastes is critical to minimizing environmental pollution. According to data from the Recycling Council of Ontario, North America has 8 percent of the world's population, consumes one-third of the world's resources—and produces almost half the world's non-organic garbage. In this chapter, we will survey various waste-disposal strategies and examine their pros and cons.

Exposed garbage attracts scavengers, as here at a waste-disposal site in England. © David Woodfall/Tony Stone Images

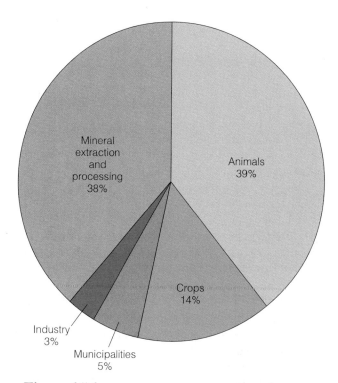

Figure 15.1 Principal sources of solid wastes. Note that industry and municipalities together account for less than 10% of the total. Data from: J. E. Fergusson, 1982. *Inorganic Chemistry and the Earth.* New York: Pergamon Press.

Solid Wastes—General

The principal sources of solid waste in the United States are shown in figure 15.1. More than 50 percent of the wastes are linked to agricultural activities, with the dominant component being waste from livestock. Most of this waste is not highly toxic except when contaminated with agricultural chemicals, nor is it collected for systematic disposal. The volume of the problem, therefore, is seldom realized.

The other major waste source is the mineral industry, which generates immense quantities of spoils, tailings, slag, and other rock and mineral wastes. Materials such as tailings and spoils are generally handled on-site, as, for example, when surface mines are reclaimed. The amount of waste involved makes long-distance transportation or sophisticated treatment of the wastes uneconomical. The weathering of mining wastes can be a significant water-pollution hazard, depending on the nature of the rocks, with metals and sulfuric acid among the principal pollutants. Shielding the pulverized rocks from rapid weathering with a soil cover is a common control/disposal strategy. In addition, certain chemicals used to extract metals during processing are toxic and require special handling in disposal like other industrial wastes.

Much of the attention in solid-waste disposal is devoted to the comparatively small amount of municipal waste. It is concentrated in cities, is highly visible, and must be collected, transported, and disposed of at some cost. The following section explores historical and present strategies for dealing with municipal wastes.

Nonmining industrial wastes likewise command a relatively large share of attention because many industrial wastes are highly toxic, and most of the highly publicized unsafe hazardous-waste disposal sites involve improper disposal of industrial chemical wastes (see boxes 15.1 and 15.2). Some of the principal industrial solid-waste sources are shown in figure 15.2.

While industrial wastes may supply the largest volumes of toxic materials, municipal waste is far from harmless. Aside from the organic materials like food waste and paper, a wide variety of poisons is used in every household: corrosive cleaning agents, disinfectants, solvents such as paint thinner and dry-cleaning fluids, insecticides and insect repellents, and so on. These toxic chemicals together represent a substantial, if more dilute, potential source of pollution if carelessly handled.

Municipal Waste Disposal

A great variety of materials collectively make up the solid-waste disposal problem that costs municipalities several billion dollars each year (figure 15.3). The complexity of the waste-disposal problem is thus compounded by the mix of different materials to be dealt with. The best disposal method for one kind of waste may not be appropriate for another.

Open Dumps

A long-established method for solid-waste disposal that demands a minimum of effort and expense has been the open dump site. Drawbacks to such facilities are fairly obvious, especially to those having the misfortune to live nearby. Open dumps are unsightly, unsanitary, and generally smelly; they attract rats, insects, and other pests; they are fire hazards. Surface water percolating through the trash can dissolve out, or leach, harmful chemicals that are then carried away from the dump site in surface runoff or through percolation into ground water. Trash may be scattered by wind or water. Some of the gases rising from the dump may be toxic.

All in all, open dumps are an unsatisfactory means of solid-waste disposal. Yet only two decades ago, three-fourths of solid wastes were still disposed of in open dumps. Even now, after concerted efforts to close open dump sites, probably close to half of solid wastes are so handled. The Environmental Protection Agency has estimated that there may still be several hundred thousand open dumps in the United States (figure 15.4).

Sanitary Landfills

Where open dumps have been phased out, they have most often been supplanted by **sanitary landfills** (figure 15.5). The basic method has been in use since the early twentieth century. In a sanitary-landfill operation, a layer of compacted trash is covered with a layer of earth at least once a day. The earth cover keeps out vermin and helps to confine

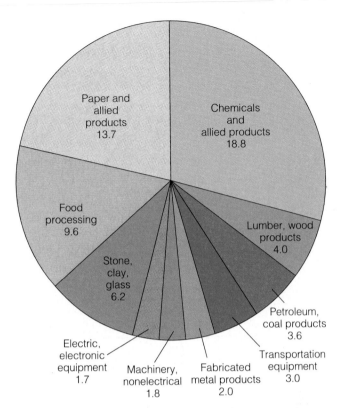

Figure 15.2 Principal industrial solid-waste sources; this is the breakdown of the 3% of figure 15.1   Source: Data from *Environmental Quality, 1986*, Council on Environmental Quality.

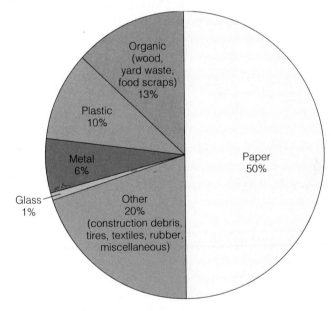

Figure 15.3 Typical composition of municipal solid waste (the 5% of figure 15.1); individual municipalities' waste composition may vary considerably.   Source: Data from W. L. Rathje, "Once and Future Landfills," *National Geographic*, May 1991, pp. 117–134.

the refuse. Landfills have generally been sited in low places—natural valleys, old abandoned gravel pits, or surface mines. When the site is full, a thicker layer of earth is placed on top, and the land can be used for other purposes, provided that the nature of the wastes and the design of the

Earth layers

Current day's trash

Previous day's trash

A

B

Figure 15.5 Sanitary landfills are planned and monitored. (*A*) Basic design of a sanitary landfill (partial cutaway view). The "sanitary" part is the covering of the wastes with soil each day. (*B*) Lowry landfill, Arapahoe County, Colorado. Photograph courtesy of USGS Photo Library, Denver, CO.

Figure 15.4 Unofficial, unsupervised, roadside open dump site. Photograph courtesy of USDA Soil Conservation Service.

landfill are such that leakage of noxious gases or toxins is minimal. The most suitable uses include parks, pastureland, parking lots, and other facilities not requiring much excavation. The city of Evanston, Illinois, built a landfill up into a hill, and the now-complete "Mount Trashmore" is a ski area. Golf courses built over old landfill sites are increasingly common. Attempts to construct buildings on old landfill sites may be complicated by the limited excavation possible because of refuse near the land surface, by the settling of trash as it later decomposes (which can put stress on the structures), and by the possibility of pollutants escaping from the landfill site.

Pollutants can escape from improperly designed landfills in a variety of ways. Gases are produced in decomposing refuse in a landfill just as in an open dump site, though the particular gases differ somewhat as a result of the exclusion of air from landfill trash. Initially, decomposition in a landfill, as in an open dump, proceeds aerobically, producing such products as carbon dioxide (CO_2) and sulfur dioxide (SO_2). When oxygen in the covered landfill is used up, anaerobic decomposition yields such gases as methane (CH_4) and hydrogen sulfide (H_2S). If the surface soil is permeable, these gases may escape through it. Sealing the landfill to the free escape of gases can serve a twofold purpose: reduction of pollution, and retention of useful methane as described in chapter 14. The potential buildup of excess gas pressure requires some venting of gas, whether deliberate or

otherwise. Where the quantity of usable methane is insufficient to be recoverable economically, the vented gases may be burned directly at the vent over the landfill site to break down noxious compounds.

If soil above or below a landfill is permeable, **leachate** (infiltrating water containing dissolved chemicals from the refuse) can escape to contaminate surface or ground waters (figure 15.6 and table 15.1). This is a particular problem with landfills so poorly sited that the regional water table reaches the base of the landfill during part or all of the year. It is also a problem with older landfills that were constructed without impermeable liners beneath. Increasing awareness particularly of the danger of groundwater pollution is leading to improvements in location and design of sanitary landfills. They are now ideally placed over rock or soil of limited permeability (commonly clay-rich soils, sediments, or sedimentary rocks, or unfractured bedrock) well above the water table. Where the soils are too permeable, liners of plastic or other waterproof material may be used to contain percolating leachate, or thick layers of low-permeability clay (several meters or more in thickness) may be placed beneath the site before infilling begins.

Figure 15.6 Leachate can escape from a poorly designed or poorly sited landfill to contaminate surface or ground waters. (*A*) Direct contamination of ground water; water table intersects landfill. (*B*) Leachate runs off over sloping land to pollute lake or stream below, despite impermeable material directly under landfill site. (*C*) Leachate infiltrates to water table through permeable materials under landfill site.

Unfortunately, there is another potential problem with landfills sealed below with low-permeability materials, if the layers above are fairly permeable. Infiltrating water from the surface will accumulate in the landfill as in a giant bathtub, and the leachate may eventually spill out and pollute the surroundings (figure 15.7). Use of low-permeability materials above as well as below the refuse minimizes this possibility. Alternatively, any accumulating leachate can be pumped out, but it then constitutes a further waste-disposal problem. Leachate problems are lessened in arid climates where precipitation is light, but unfortunately, most large U.S. cities are not located in such places.

A subtle pathway for the escape of toxic chemicals may be provided by plants growing on a finished landfill site. As plant roots take up water, they also take up chemicals dissolved in the water, some of which, depending on the nature of the refuse, may be toxic. This possibility indicates the need for caution in using an old landfill site for cropland or pastureland.

Table 15.1 Contents of Some Dissolved Constituents in Municipal Landfill Leachate

Substance	Concentration range (ppm)
copper	0–9
iron	0.2–5,500
lead	0–5
manganese	0.06–1,400
nitrogen (as nitrate)	0–1,300
phosphate	0–154
zinc	0–1,000

Source: Data from J. E. Fergusson, *Inorganic Chemistry and the Earth*.
Copyright © 1982 Pergamon Press, Oxford, England.
Note: The table does not include the many synthetic chemicals that may be found in various sites. Leachate composition can vary considerably from landfill to landfill, as the trash mix varies.

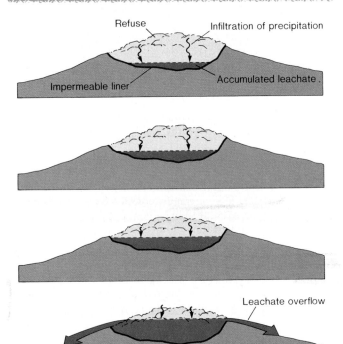

Figure 15.7 The "bathtub effect" caused by the accumulation of infiltrated leachate above impermeable liner. Overflow may not occur for months or years.

Both open dumps and landfills require commitment of land. A rule of thumb for a sanitary landfill for municipal wastes is that one acre is needed for each ten thousand people each year, if the landfill is filled to a depth of about 3 meters. For a large city, that represents considerable real estate. True, land used for a sanitary landfill can later be used for something else when the site is full, but then another site must be found, and another, and another. All the time, the population grows and spreads and also competes for land.

Table 15.2   Representative Products of Municipal Incineration

Material	Percentage of total
solids (ash, residue)	22
water vapor	64
other gases	14
Breakdown of other gases (except water vapor)	
nitrogen (N_2)	79.6%
oxygen (O_2)*	14.3%
carbon dioxide (CO_2)	6%
carbon monoxide (CO)	0.06%
hydrogen chloride (HCl)	5–500 ppm
nitrogen oxides (NO, NO_2, N_2O)	93 ppm
sulfur dioxide (SO_2)	22 ppm

Source: Data from J. E. Fergusson, *Inorganic Chemistry and the Earth.*
Copyright © 1982 Pergamon Press, Oxford, England.
*Note: The proportion of oxygen is substantially lower than in average atmosphere, as it is consumed during combustion.

More than half the cities in the United States are pressed for landfill facilities, with municipal solid waste piling up at over 150 million tons per year.

Local residents typically resist when new landfill sites are proposed, a phenomenon sometimes described by the acronym "NIMBY"—"Not In My Back Yard." (Other variants that have appeared on wall posters at the Environmental Protection Agency include NIMFYE (Not In My Front Yard Either), NIMEY (Not In My Election Year), and NOPE (Not On Planet Earth)—but landfills always have to be put *somewhere*.) Between the political/social constraints, and the increased geological constraints on landfill siting that have evolved as understanding of pollutant escape from disposal sites has improved, the number of potential sites for new landfill operations is limited. In states with high population density and limited landfill sites, the ratio of people to landfills may be 80,000 to 1 or higher, and obviously, the greater this ratio, the more rapidly the landfills fill. One state in this situation, New Jersey, has to ship half its 11 million tons/year of municipal waste to other states' landfills. In 1990, in a move to conserve dwindling landfill volume, the state of Illinois decreed that residents could no longer include landscape waste (grass clippings, leaves, brush, and so on) with trash collected for landfill disposal; the landscape waste is collected separately for composting.

Incineration

Incineration as a means of waste disposal provides a partial solution to the space requirements of landfills. However, it is an imperfect solution, since burning wastes contributes to air pollution, adding considerable carbon dioxide (CO_2) if nothing else (table 15.2). At moderate temperatures, incineration may also produce a variety of toxic gases, depending on what is burned. For instance, plastics when burned can release chlorine gas and hydrochloric acid, both of which are toxic and corrosive, or deadly hydrogen cyanide; combustion of sulfur-bearing organic matter releases sulfur dioxide (SO_2); and so on.

The technology of incineration has been improved in recent years. Modern very-high-temperature (up to 1,700° C, or 3,000° F) incinerators break down hazardous compounds, such as the complex organic compounds, into much less dangerous ones, especially carbon dioxide (CO_2) and water (H_2O). Still, individual chemical *elements* are not destroyed. Volatile toxic elements, mercury and lead among others, may escape with the waste gases. Much of the volume of urban waste is noncombustible, even at high temperatures. Refuse that will not burn must be separated before incineration or removed afterward and disposed of in some other way—often in a landfill site. Harmful nonvolatile, noncombustible substances in the wastes will be concentrated in the residue. If the potential toxicity of the residue is high, it may require handling comparable to that for toxic industrial wastes, raising the net cost of waste disposal considerably.

The most sophisticated high-temperature incinerators can cost up to $2,000 per ton of waste to operate and are best reserved for the destruction of limited quantities of highly toxic industrial wastes. Simpler, general-purpose municipal incinerators can be operated as cheaply as $75 per ton of waste and are most useful in reducing the total volume of solid wastes by burning the paper, wood, and other combustibles before ultimate disposal.

A further benefit of incineration can be realized if the heat generated thereby is recovered. For years, European cities have generated electricity using waste-disposal incinerators as the source of heat. The combined benefits of land conservation and energy production have led to extensive adoption of incineration in other nations (figure 15.8): In 1990, over 67% of municipal waste generated in Japan was burned; in France, 42%; in Sweden, 60%; in Switzerland, 80%. The United States has been slower to adopt this practice, probably because of more abundant supplies of petroleum (and perhaps land also); only 6% of U.S. municipal refuse, and only 4% in Canada, was burned in 1990.

Several U.S. cities, however, have put the considerable quantity of heat energy released by an incinerator to good use. The city of North Little Rock, Arkansas, turned to an incinerator when landfill space grew scarce. The steam-generating incinerator is saving the city an estimated $50,000/year, while reducing required additional landfill volume by 95 percent. In Kansas City, Kansas, the American Walnut Company is burning waste sawdust and wood chips in a steam-generating boiler, saving itself natural gas previously used to help dispose of the wastes and even generating surplus steam, which it sells to other local users. Such fringe benefits of incineration contribute to increasing acceptance of incineration as a waste-disposal strategy. There is still considerable public resistance to new incinerators, however, which is part of the reason that it may take three to five years

Figure 15.8 Proportions of municipal waste handled by various disposal methods in selected countries. Data from *Geology and the Environment in Western Europe*, ed. by G. I. Lumsden. Copyright © 1992 Clarendon Press, Oxford, England.

Ocean Dumping

A variant of land-based incineration, developed over the past decade, is shipboard incineration in the open ocean. Following combustion, unburned materials are simply dumped at sea. This method has been applied to stockpiles of particularly hazardous chemical wastes. A 1981 report of the Environmental Protection Agency described the technique as "promising" for a variety of reasons, making the statement that "it has a minimal impact on the environment by removing the destruction site far from populated areas so that emissions are absorbed by the oceans," and noting that offshore incinerators "not handicapped by emission control requirements that apply to land-based units" could be very cost-effective. The desirability of this method plainly depends on one's point of view. It does not much matter if carbon dioxide is added to the air over land or over water; it still contributes to the rising carbon dioxide levels in the atmosphere. True, dumping the solid residues at sea puts them out of the sight of people, but if toxic materials are present and left unburned, they contribute to the pollution of the oceans to which the world turns increasingly for food. As noted in chapter 16, scientists really do not yet know the

ultimate fate of many of the chemicals involved in water pollution. Perhaps this waste-disposal scheme should be embraced with some caution.

Ocean dumping *without* prior incineration has also been used for chemical wastes, municipal garbage, and other refuse. The potential for water pollution is apparent. In some cases, too, shifting currents have brought the waste back to shore rather than dispersing it in the oceans as intended. Increasing recognition of the dangers of dumping untreated wastes in the sea led to drastic curtailment of the practice by the Environmental Protection Agency in most areas over the last decade. In the late 1980s, public outcry over incidents of wastes (including needles and other medical wastes) washing up onto beaches and high bacterial counts and deaths of marine mammals combined to add pressure on Congress to do something. The Ocean Dumping Ban Act of 1988 decreed that ocean dumping of U.S. sewage sludge and industrial waste would cease after 1991. Growing numbers of nations are recognizing ocean dumping as a pollution problem and moving to end it. Britain, the other nation dumping large quantities of sewage sludge, will phase out the practice by 1998; it moved to end ocean dumping of its industrial waste by 1993. However, the oceans remain a dumping site for one very high-volume waste product: *dredge spoils*.

Figure 15.9 Concentration of pollutants in stream delta sediments. Pollutants that are solid or that attach themselves to solid particles accumulate along with the sediments and may become highly concentrated in dredge spoils. The ultimate sink for many such pollutants could be any one of several other depositional environments—along river systems, in estuaries, or on the continental shelf, for example.

Figure 15.10 Relative amounts of solid waste generated per capita in selected cities.   Data from *World Resources 1995–96,* World Resources Institute.

Dredge spoils are sediments dredged from reservoirs and waterways to enlarge capacity or improve navigation. They are dumped in the oceans at the rate of about 200 million tons a year. At first, this may seem harmless enough; it is "just dirt." However, dumping fine sediments may harm or destroy marine organisms that cannot survive in sediment-clouded waters. Also, many toxic chemicals dumped into water do not go into or stay in solution, but stick to the surfaces of sediment particles. Dumping sediments dredged from the mouth of a polluted river means dumping a load of such pollutants concentrated from the whole drainage basin (figure 15.9). When sediments dredged from fresher waters are dumped into saline waters, some adsorbed chemicals may be redissolved as a consequence of the change in water chemistry, posing a pollution threat even if the sediments themselves settle harmlessly to the bottom.

Reducing Waste Volume

The sheer volume of the solid-waste-disposal problem has led to a variety of attempts to reduce it. In industrialized countries, the problem can be especially acute (figure 15.10). Incineration disposes of some waste, leaving less to be dealt with by other means. Another volume-reduction strategy is compaction, either in individual homes with trash compactors or at large municipal compaction facilities. Less vol-

ume means less landfill space used or the slower filling of available sites. However, it also means no reuse of any potentially recoverable material and the slower decay of organic material (a concern if methane production is desired).

In that connection, it is perhaps worth noting that studies of landfill sites over the last two decades have revealed that decomposition in landfills takes place much more slowly than once believed, especially if the landfill is relatively dry. The University of Arizona's Garbage Project has found well-preserved food wastes two decades old, and readable newspapers and telephone books three or four decades old. Decomposition rates seem to be faster in wet waste—such as in the deeper layers of the world's largest landfill, New York City's Fresh Kills, which are below the water table—but modern landfills are most often designed to minimize water inflow. In the future, careful manipulation of water content and, perhaps, judicious addition of suitable microorganisms can maximize biodegradation (and production of useful methane) while minimizing the risks from toxic leachate.

Handling (Nontoxic) Organic Matter

Swine were once extensively used for garbage disposal, since they will consume a variety of food waste. Earlier in this century, it was economically feasible to collect food waste from larger municipalities and use it for hog feed. In 1941, more than 2 million tons of municipal garbage were so used. The feeding of raw garbage to swine led to outbreaks of disease in humans who subsequently ate the pork; consequently, beginning in the 1950s, the waste had to be heat-treated to destroy disease-producing organisms before being fed to the animals. This procedure was effective, but more costly. The feeding of garbage to swine declined in the ensuing decades and is practiced little today.

On-site disposal, for example with a home in-sink garbage disposal unit, is not really disposal at all. The practice merely diverts some organic matter to become part of the water-pollution problem. The organic-matter content of the water is increased (see chapter 16 for the consequences of this), and more of a load is placed on municipal sewage-treatment plants.

Organic matter can be turned to good use through composting, a practice long familiar to gardeners and farmers. Many kinds of plant wastes and animal manures can be handled this way. Partial decomposition of the organic matter by microorganisms produces a crumbly, brown material rich in plant nutrients. Finished compost is a very useful soil additive, improving soil structure and water-holding capacity, as well as adding nutrients. In the United States, composting is generally practiced only by individuals. In parts of Europe and Asia, demand for finished compost makes it economically practical to establish municipal composting facilities. The city of Auckland, New Zealand, began such a facility in 1960 after it became evident that remaining landfill sites were severely limited. The process is not without complications, however. Organic matter must be separated from glass, metal, and other noncompostables. Occasionally, chemical contamination of the refuse with weed-killers or other toxic substances makes some batches of compost unusable for agriculture. Still, sale of the finished compost helps to pay the cost of the composting operation in Auckland, and the volume of waste going to the landfill is reduced by a factor of about fifteen in the process.

Recycling

In chapter 12, we considered recycling metals in the context of mineral resources and noted that an energy saving can often be realized in the process. Recycling and reuse are also waste-reduction strategies. Glass is not made from scarce commodities, but just as quartz is a weathering-resistant mineral, silica-rich glass is virtually indestructible in dumps and landfills—and along roadsides. It can be broken up, but it does not readily dissolve or break down. Reuse of returnable glass bottles requires only about one-third the energy needed to make new bottles. Recycling all glass beverage containers would reduce by over 5 million tons per year the amount of glass contributing to the solid-waste disposal problem. (Glasses vary in composition, so, like alloys, scrap glasses cannot be mixed indiscriminately and reprocessed— but once a glass bottle is manufactured, it can be reused many times, if carefully handled.)

Imposing deposits on beverage containers also provides a financial incentive not to litter. In Oregon, passage of a mandatory-deposit law for beverage containers reduced roadside litter by up to 84%. Since Oregon pioneered the idea in 1971, several other states have followed suit. (However, voter resistance to the perceived inconvenience of having to return cans and bottles to recover deposits, and lobbying efforts by bottlers, have led to the defeat of "bottle bills" in some states, too.) Currently, the United States recycles about 10% of its glass; in countries such as Switzerland and the Netherlands, the proportion recycled approaches 50%.

Paper might also be recycled more extensively. In the United States, about 25% of the paper used is recycled; in Japan, the percentage is 50%. Paper recycling is easiest and most effective when a single type of paper—newspaper, for example, or computer printout—is collected in quantity.

That limits the variety of inks and other chemicals that must be handled during reprocessing. Mixtures of printed, waxed, and plasticized papers are somewhat harder to handle economically but can nevertheless be recycled. In 1975, the Environmental Protection Agency instituted paper-recycling efforts by government agencies in the Denver, Colorado, area, and the program has subsequently been expanded nationwide. Government recycling of 223,000 tons of high-grade paper fiber saves an estimated 4 million trees and $7.4 million in waste-disposal costs each year. The Recycling Council of Ontario estimates that every ton of newspapers recycled means about 18 trees, and 3 cubic meters of landfill space, saved. Moreover, making paper from the recycled fibers requires 60% less energy than does manufacturing paper from newly cut trees.

Plastics continue to be something of a problem, though much less of one than they were believed to be a decade or more ago. The same durability that makes them useful also makes them difficult to break down when no longer needed, except by high-temperature combustion. Some degradable plastics have been developed to break down in the environment after a period of exposure to sunlight, weather, and microbial activity, but these plastics are only suitable for applications where they need only hold together for a short time—for example, fast-food containers. (And if a long useful life for such a product is not actually required, perhaps it did not need to be made of plastic anyway.) Another difficulty in recycling plastics is similar to the problem with different steels: A mix of plastics, when reprocessed, is unlikely to have quite the right properties for any of the applications from which the various scrap plastics were derived. Still, the blend may be suitable for other uses—plastic piping, for example, or a shredded-plastic stuffing for upholstery.

One new approach to facilitating plastic recycling is to mark those plastics that can be more easily recycled, with the triangular symbol of three arrows head-to-tail that is widely used to represent recycling, and identify the basic type of plastic by a number (1, 2, 3, etc.) within the triangle. (Look on the bottom of a two-liter soda bottle, gallon plastic milk jug, or other plastic food container; see figure 15.11 for examples and key.) A remaining obstacle is that there must exist an identifiable market or demand for a *particular* plastic in order for its recycling to be economically feasible; so a given municipal waste hauler might be able to collect soda bottles and milk jugs for recycling, but not foam packing materials. Markets will vary by region. It should also be realized that plastic is typically not recycled into the same form—old soda bottle to new soda bottle, for instance. In this respect, plastic recycling differs from that of other materials. A recycled aluminum can is likely to come back as a new aluminum can, a recycled bottle as a refilled or remanufactured bottle. A plastic's properties tend to change during recycling, however, and recycled plastic may be less strong than new plastic. So a soda bottle may be recycled, but it will be transformed into something else—such as fiber for carpeting, plastic trash bags, "plastic lumber" for park benches.

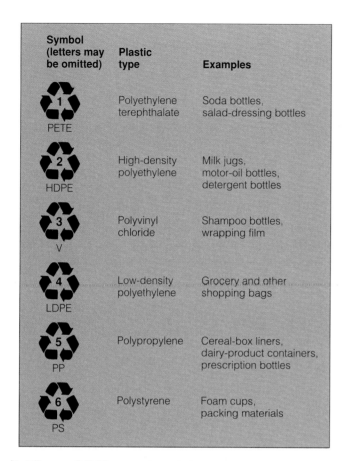

Symbol (letters may be omitted)	Plastic type	Examples
1 PETE	Polyethylene terephthalate	Soda bottles, salad-dressing bottles
2 HDPE	High-density polyethylene	Milk jugs, motor-oil bottles, detergent bottles
3 V	Polyvinyl chloride	Shampoo bottles, wrapping film
4 LDPE	Low-density polyethylene	Grocery and other shopping bags
5 PP	Polypropylene	Cereal-box liners, dairy-product containers, prescription bottles
6 PS	Polystyrene	Foam cups, packing materials

Figure 15.11 Use of standard symbols facilitates separation of distinct types of plastics for recycling. Sometimes the same symbol with a numeral "7" is used for "other." Certain products (e.g., milk jugs and soda bottles) are consistently made of one type of plastic; others (e.g., containers for detergents or dairy products) may be made of different types depending on the desired properties—clear container vs. opaque, flexible vs. rigid, and so on.

Recent research has explored converting plastics into other compounds, such as waxes, for reuse, or developing novel applications for chopped or shredded recycled plastic—for example, using it as fill for pillows, as insulation in clothing, or as an additive to fiberglass. There is particular interest at the moment in finding uses for the plastic-rich residue from the shredders used to chew up junked cars. Once the metals are extracted from the shredded product, most of the remaining 150 to 200 kilograms (300 to 400 pounds) of material per car is plastic. It might soon be possible to recover more than a million pounds of reusable plastic per year from scrap-auto shredders. The problem at present is finding the uses for the material that will make the whole process more economically practical.

There are additional general obstacles to recycling, some of which were mentioned earlier. Where recovery of materials from municipal refuse is desired, **source separation** is generally necessary. This means that individual homeowners, businesses, and other trash generators must sort that trash into categories—paper, plastic, glass, metal,

and so on—prior to collection. While this has been successful in some communities, it is not now widely practiced, although where legally mandated, it can work well once residents adjust to the new realities. Increasingly, communities are focusing on ways to recycle household trash that do *not* require extensive source separation.

Recycling may conflict in some measure with other waste-disposal objectives. For example, recycling combustible materials reduces the energy output of municipal incinerators used to generate power. Paper recyclers are already encountering this problem, and as uses are found for recycled plastics, the same difficulty may arise with those materials.

In the United States, waste recovery for recycling or composting is increasing—from less than 10 percent of municipal wastes in 1985 to over 17 percent in 1990. Still, there is clearly room for expansion of these activities.

Other Options

New and creative approaches to solid-waste problems are continually being developed. When Chicago's Edens Expressway was rebuilt in 1979–80, 335,000 tons of old pavement were crushed and reused in the new roadbed, thereby saving that much asphalt (petroleum) and crushed stone, while reducing Chicago's solid-waste disposal problem by 335,000 tons. The United States and other countries are developing "waste exchanges," whereby one industry's discarded waste becomes another's raw materials. Ideally, this both reduces waste disposal and saves the waste generator money: For example, an aluminum smelting firm in Maryland that had been paying $70 a ton in landfill fees to dispose of its leftover powdered fluorite found a taker in Pittsburgh who would pay $20 a ton; one company's waste isopropyl alcohol became another's cleaning solvent. The Canadian Waste Materials Exchange has dealt in such diverse materials as glove-leather scrap, oat hulls, and fish-processing waste.

Still, some materials are too difficult to reuse efficiently, for reasons already outlined. Others are toxic byproducts of industrial processes and are not themselves useful or are too toxic for safe handling during extensive reprocessing. These will continue to require disposal. Also, many of the highly toxic industrial wastes are liquids rather than solids, and these wastes may require somewhat different handling from solid wastes.

Liquid-Waste Disposal

Liquid wastes are primarily of two types: sewage, which is discussed in a subsequent section, and the more concentrated, highly toxic, liquid waste by-products of industrial processes—acids, bases, organic solvents, and so on. This section focuses on disposal strategies for the latter. Additional liquid-waste problems, however, do exist: A notable example is the problem of used oil. Presently, over 1 billion gallons of used lubricants derived from petroleum are generated in the United States each year; 40 percent of this waste

Box 15.1 Toxic-Waste Time Bombs? The Chaos at Love Canal

Toxic waste is a nonspecific term usually used to describe industrial wastes that are poisonous, carcinogenic, or otherwise harmful even in small doses. Many of these materials are so damaging that one can be seriously affected without even being aware of exposure to the toxic substance. Logically, great care should be taken in disposing of such materials. However, that has not always been the case, especially in the past when less was known about ground water, subsurface waste migration, and related subjects, and when legislative controls on waste disposal were far fewer.

Love Canal—a ditch 20 meters wide, 3 meters deep, 1,000 meters long—was built near Niagara Falls, New York, in the 1890s as part of a scheme to develop hydropower. Economic and political factors led to the project's abandonment. Proximity to developing hydroelectric generating facilities drew industry to the area, however. In 1905, the Hooker Electrochemical Company was started in Niagara Falls. It grew rapidly. In 1942, Hooker bought the abandoned canal and some adjoining land for a

waste-disposal site. Over the next ten years, an estimated 21,000 tons of various toxic chemicals were dumped there. Sometimes, the wastes caught fire; sometimes, they spilled on the workers; sometimes, the air filled with acrid fumes. As the canal filled up, it became overgrown with grass and weeds and came to look like part of the fields.

In 1953, for a nominal fee of $1, the Niagara Falls school board bought 16 acres of the land, including the old canal site. Interestingly, the deed contained a disclaimer absolving Hooker from all future damage claims in connection with the chemical wastes.

A school and playground were built on the site. Storm sewers drained excess water from the site into the Niagara River. Homes were built near the school. Most homeowners were unaware of the presence of the old dump site.

The buried chemicals increasingly intruded into life at Love Canal. Pits yawned in the ground surface as barrels below disintegrated. The feet of children who played barefoot in the fields sometimes became irritated and inflamed.

Oily chemicals fouled basement sump pumps. Ironically, it was the 1976 identification of insecticide-contaminated fish in Lake Ontario that finally triggered the first chemical surveys in the area. A 1977 study showed 21 of the 188 homes adjacent to the canal plagued with odors and with toxic chemical residues in basements. Vapor tests in 1978 revealed the presence of ten toxic compounds in the air, seven of them carcinogenic in animals, one in humans. Eventually, more than forty toxic organic compounds were found in soil or air samples at the site. Health surveys of the residents indicated, among other effects, above-average incidences of birth defects and spontaneous abortions among the residents. As frightened residents increasingly demanded information and action, more and broader tests were made, and state and federal governmental agencies became increasingly involved in the situation.

Late in 1978, the New York State health commissioner suggested that pregnant women, and children under age two, be moved out of the immediate area of the canal. It was decided that the dump site posed a sufficient

is poured into the ground or into storm drains, and the fate of another 20 percent is unknown. Much of this "waste" could be reclaimed and recycled, but at this time, little is.

Handling of liquid wastes has tended to follow one of two divergent paths. The *dilute-and-disperse* approach, based on the assumption that, if toxic substances are sufficiently diluted, they will be rendered harmless, has been the rationale behind much dumping into oceans and large lakes and rivers. With the increasing recognition of substances that are toxic even at levels below 1 ppb in water, including many of the complex organic solvents, agricultural chemicals, and others, the basic premise has been brought into question. Also, certain pollutants can accumulate in organisms and become more concentrated up a food chain (see chapter 16), which means that the chemicals can return to hazardous concentrations.

The opposite approach is the *concentrate-and-contain* alternative. Thoughtless disposal of concentrated wastes followed by inadequate containment has led to disasters like Love Canal (box 15.1; see also figure 15.12). In the past, some concentrated liquid industrial wastes have been dumped in trenches or pits directly and buried, while other wastes have been placed in metal or plastic containers and consigned to dumps or landfills. The disposal sites frequently were not evaluated with respect to their suitability as toxic-waste disposal sites and, over the longer term, the wastes were *not* contained. Metal drums rusted, plastic cracked and leaked, and wastes seeped out to contaminate ground water and soil.

Secure Landfills

For some years, waste-disposal specialists have believed that, in principle, it is possible to design a **secure landfill** site hazard to justify an immediate emergency cleanup effort, with more than one hundred families to be relocated while the work was in progress. When the cleanup was begun, in October 1978, the areal extent of leachate contamination of ground water and soil was unknown. The plan was to place a clay cap over the site to prevent escape of gases and to drain and confine leachate underground. Whether the scheme would really make the area safe was also unknown. The inference was that confining further leakage of polluted leachate would eliminate future problems. The planned construction was completed early in 1980, after some families were displaced for months and the rest were exposed to sickening fumes released during the work. The clay cap now covers 22 acres.

In May 1980, the Environmental Protection Agency released a study showing a high rate of chromosome damage among Love Canal area residents. Within a week, President Carter declared the situation there an emergency. A growing body of scientific data, including soil and water surveys, suggested that the contamination had

progressed so far and had become so pervasive that many homes were too acutely contaminated to be regarded as safe. In the summer of 1980, the federal government agreed to contribute $7.5 million as a grant and another $7.5 million in loans toward the state's $20-million-plus costs to address the situation. Most of that money was spent to buy properties of residents leaving the area: Over five hundred families had moved out by the summer of 1981, with hundreds more hoping to do so but still waiting for the government buy-out necessary to make it possible. The exact extent of the hazard for those remaining was indeterminate. The legal wrangling, embroiling chemical companies as well as governments, continued. A decade after the hazards of the area were realized, nearly a thousand lawsuits—many against Occidental Chemical Corporation, the successor to Hooker—remained unsettled.

The Environmental Protection Agency has estimated that thirty thousand hazardous-waste sites, many now forgotten, may exist in the United States. It has already compiled an

inventory of more than seventeen thousand such sites, thirteen thousand of them still active. The same agency reported in 1980 that only 10 percent of hazardous wastes were being disposed of in appropriately safe and secure ways. A two-year review completed in 1984 suggested that the ongoing rate of toxic-waste generation is as much as six times greater than the agency had previously estimated: on the order of 300 million tons per year.

Many past waste-disposal abuses were due to simple ignorance. Better understanding of the chemistry of water, air, and soil and of ground-water movement, together with improvements in hazardous-waste handling, can stop the creation of any more Love Canals. But as the environment is monitored more carefully, more hazards created in the past are uncovered: Times Beach, Missouri; Woburn, Massachusetts; Waukegan Harbor, Illinois; the list goes on and on. Identifying these sites and cleaning them up to the extent possible will cost far more than proper initial disposal would have cost. (See also box 15.2.)

for toxic liquid wastes. An example of a recommended design is shown in figure 15.13A. The wastes are put in sealed drums before disposal. Beneath the drums are layers of plastic and/or compacted clay to contain any unexpected leaks. Wells and piping are installed so that the ground water below and around the site can be checked periodically for any sign of leakage of the waste chemicals. Excess accumulating leachate can be pumped out before it leaks out. Such a system provides multiple safeguards against accidental environmental contamination, and the monitoring wells allow prompt detection of any leaks.

Unfortunately, a growing body of evidence indicates that no site is truly secure, even if conscientiously designed. Carefully compacted clay may be very low in permeability but is probably never completely impermeable, especially over long time intervals. Chemical and biologi-

cal reactions in the wastes and leachate can rupture or decompose plastic, and the stress caused by the weight of wastes and cover can fracture a clay liner. Even relatively innocuous municipal waste can prove hard to contain. When built, the "Mount Trashmore" landfill in Evanston, Illinois, was hailed for its state-of-the-art design. But monitoring wells have now revealed detectable leakage of at least a dozen volatile organic compounds deemed high-priority toxic pollutants by the Environmental Protection Agency, including benzene, toluene, vinyl chloride, and chloroform. Leakage from "secure" toxic-waste dumps, in which hundreds or thousands of barrels of concentrated toxic liquid chemicals are stored, has far more potential for harm. Historical response to detection of leakage from one toxic-waste dump has been to dig up as much of the hazardous material and contaminated soil as possible and

A

B

Figure 15.12   Careless toxic-waste disposal leads to pollution. (*A*) Studies of contaminant transport in ground water help evaluate risks posed by waste-disposal site close to a residential area. (*B*) Unburied waste drums corrode, leak; cleanup crews must wear protective clothing.   Photographs by S. C. Delaney, U.S. Environmental Protection Agency; courtesy of USGS Photo Library, Denver, CO.

A

B

Figure 15.13   (*A*) A secure landfill design for toxic-waste disposal, including provisions for leachate containment and for monitoring the chemistry of subsurface water nearby. (*B*) Drilling of new monitoring well at Lowry landfill.   Photograph courtesy of USGS Photo Library, Denver, CO.

transfer it to a more secure landfill. There is now some question as to whether a wholly different disposal method might be preferable.

Deep-Well Disposal

Another alternative for disposal of liquid industrial waste is injection into deep wells (figure 15.14). This method has been practiced since World War II. The rock unit selected to receive the wastes must be relatively porous and permeable (commonly, sandstone or fractured limestone), and it must be isolated by low-permeability layers (for example, shale) above and below. The subsurface geology must be known in sufficient detail that there is reasonable confidence that the disposal stratum remains isolated for some distance from the

well site in all directions. Information about that geology may be derived from many sources: direct drilling to obtain core samples that provide a vertical section of the rock units present; geophysical studies that provide data on depths to and thicknesses of different rock layers, and on distribution of ground water; geologic mapping on the basis of cores, surface outcrops, and geophysical data to interpolate between points sampled directly.

These disposal wells are hundreds to thousands of meters deep, far removed from the surface, and below the regional water table. The pore water in the disposal stratum should be brackish or saline water not suitable for a water supply. Where the well intersects any shallower aquifers that are or might be used for water supply, it must be snugly

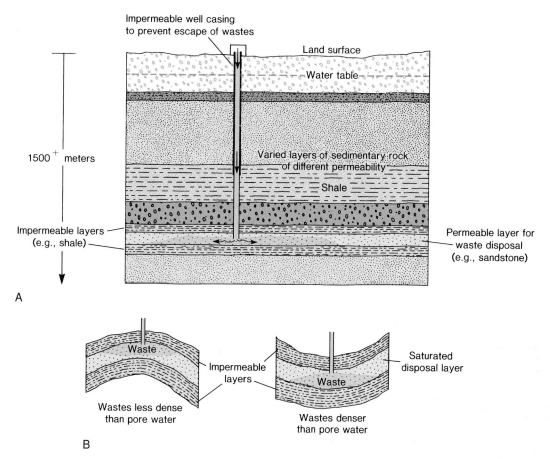

Figure 15.14 Deep-well disposal for liquid wastes. (*A*) Basic design: Wastes are placed in a deep permeable layer that is geographically isolated by low-permeability strata. (*B*) Containment of wastes assisted by geologic structures.

lined (cased) to prevent leakage of the wastes into those aquifers. Local well water is monitored to detect any accidental leaks promptly.

Movement of deep ground water is generally slow, and the assumption is that, by the time the toxic chemicals have migrated far enough laterally to reach a usable aquifer or another body of water, they will have become sufficiently diluted not to pose a threat. This presumes knowledge of the toxicity of the chemicals in low concentrations. When the wastes are more or less dense than the ground water they displace and not miscible with it, folds or other geologic structures may help to contain them and slow their spread, as oil traps contain petroleum (figure 15.14B). The behavior of chemicals that dissolve in the pore water is much less well understood. They can diffuse through the water more rapidly than the water itself moves, so that even if deep groundwater transport is slow, contaminant migration may not be.

Costs for deep-well disposal are comparable to or somewhat less than those associated with "secure" landfill sites. Wells of the order of 2,000 to 3,000 meters deep may cost $1 million to drill, which is similar to the initial construction cost of a secure landfill. Thereafter, well-disposal costs range from $15 to $100 per ton, while landfill costs range from $20 to $400 per ton. The rate of waste disposal in a deep well is limited by the permeability of the rocks of the disposal

stratum, while landfills have no equivalent limitation. The region's geology must be such that suitable strata exist for disposal by injection, while landfills can be constructed in a much greater variety of settings. Like landfills, deep injection wells may leak. Finally, as noted in chapter 4, deep-well waste injection may trigger earthquakes in faulted rocks. The various geologic constraints on possible deep-well disposal sites are thus more restrictive than those for landfill sites.

Other Strategies

There are now furnaces designed to incinerate fine streams of liquid. This permits the destruction of toxic liquid organic chemicals, often with no worse by-product than carbon dioxide (CO_2), and may ultimately prove the best way to deal with them. Some liquid wastes can be neutralized or broken down by chemical treatment, which may avoid the necessity for ultra-secure disposal altogether. As with solid wastes, it may even be possible to use certain liquid "wastes" via waste exchanges. The nitric acid used in quantity by the electronics industry to etch silicon wafers can be neutralized to produce calcium nitrate and then incorporated in high-grade fertilizers. Spent acid used in the steel industry is rich in dissolved iron and can be used at geothermal power plants to control hydrogen sulfide gas emissions, which react with the iron in solution to precipitate iron sulfides.

Box 15.2 Whose Garbage Was This? — Superfund to the Rescue!

Many of the nearly 55,000 hazardous-waste producers in the United States do not dispose of their own wastes directly. Instead, they contract the job to waste-disposal firms (about 14,000 of them), with the presumption that those firms will carry out the disposal appropriately and legally. Frequently, that has not happened, especially in the past. In other instances, companies have dumped or buried their own wastes on-site. Decades later, when the firm responsible is long gone, pollutants ooze from the land surface or are suddenly detected in ground water. Sometimes, the waste disposer can be found but denies that the problem is serious and/or declines to spend the money to clean up the site. Sometimes, the responsible company is no longer in business.

In response to these and other similar situations, in late 1980, Congress passed the Comprehensive Environmental Response, Compensation, and Liability Act, which, among other provisions, created what is popularly known as "Superfund." The original $1.6 billion trust fund was generated primarily through taxes on oil and on producers of forty-two specific hazardous chemicals. Superfund monies are intended to pay for immediate emergency cleanup of abandoned toxic-waste sites and of sites whose owners refuse to clean them up. It represents a major step toward prompt rectification of past

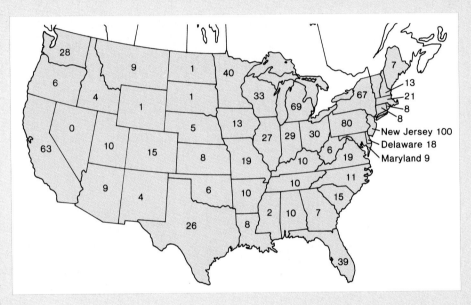

Figure 1   Distribution of the first 951 toxic-waste dump sites placed, or proposed for placement, on the EPA priority list for Superfund cleanup as of July 1987. Since then, numbers have continued to increase faster than cleanups have been completed. Source: U.S. Environmental Protection Agency.

waste-disposal abuses following recognition of the problem. Superfund was reauthorized and expanded in 1986, increasing authorization to $8.5 billion, with added taxes on petroleum (imported, 11.7 cents per barrel; domestic crude, 8.2 cents per barrel) and other chemicals. Another reauthorization in 1990 extended Superfund's taxing authority to 1994 and increased the fund to $11.5 billion.

Meanwhile, there was increasing realization that the size of the problem is far greater than originally believed. At the time Superfund was created, Congress envisioned cleanup of about 400 toxic-waste sites. By 1995, the Environmental Protection Agency was working with states to clean up over 40,000 sites. Over a thousand of these, most in the industrial northeast and upper Midwest, are identified as "priority" sites that are in particularly urgent need of cleanup (figure 1). The number of priority sites, too, continues to grow. Overall, EPA estimates

These and other means of handling toxic liquid industrial wastes will continue to be developed. What these methods generally have in common is that they are designed to deal either with specialized types of waste or with limited quantities. Volumetrically, at least, a far larger liquid-waste disposal problem is posed by that very commonplace material, sewage.

Sewage Treatment

Problems arising from organic matter in water include oxygen depletion and "algal bloom," described in the next chapter. These problems, and concern about the spread of disease through biological contamination of drinking-water supplies by disease-causing organisms, provide the motiva-

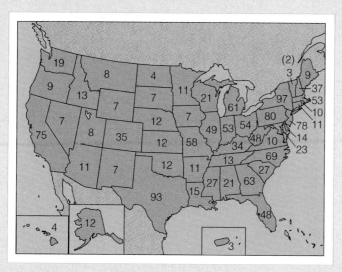

Figure 2 Completed removals during the first decade of "Superfund," 1980–1990. Source: Data from *Superfund Emergency Response Action*, Sixth Annual Report—Fiscal Year 1991, U.S. Environmental Protection Agency.

that 73 million people live within 4 miles of one or more Superfund sites.

Remedial action may be in progress at hundreds of sites each year; these include a number of sites with problems not acute enough to warrant placement on the National Priorities List. By 1995, over 3,300 removal actions—nearly 1,000 at NPL sites—had been conducted. These "removals" are short-term, emergency actions were "imminent and substantial" health risks are found to exist. Many such sites then require more comprehensive long-term cleanup action. As more and more problem areas are identified, and the extent of problems at these sites is more fully realized, it seems that even the expanded Superfund will be far from adequate to handle the staggering cleanup costs. The *average* cost of cleaning up a Superfund site is now over $30 million, not counting any associated litigation, and some projections now put eventual total cleanup costs of Superfund sites around *$1 trillion*. It has been suggested that additional funding might be provided by a tax on disposers of toxic waste, not just the generators of it. There is concern, however, that a tax added at the disposal end might tend to encourage more illicit dumping of toxic waste, ultimately compounding an already large problem.

Over time, concerns have been raised about the high cleanup costs, delays in some cleanup activities, costs assessed to parties other than those generating or disposing of wastes, and other issues. These led to an attempt in 1994 to "reform" Superfund, but this administration-led attempt failed in Congress, and Superfund was not reauthorized in that year, nor in the two years following. Some continuing funding was approved, but with a much-reduced budget for hazardous-waste-site cleanup, reflecting a shifting political climate and changes in the priorities given to such environmental issues as hazard remediation. Full reauthorization has yet to occur as this text is being written.

In the meantime, every bit of hazardous waste cleaned up and removed from one spot still has to be disposed of in another. . . . And Superfund deals only with abandoned or inoperative sites in need of cleanup. In the first decade of its operations, Superfund addressed cleanup operations at hundreds of sites on and off the "priority" list (figure 2). The EPA has also identified an estimated 2,500 *active* disposal sites that may require corrective action too. Waste minimization becomes increasingly appealing as these problematic sites multiply.

tion for effective sewage disposal. Much of the 36 billion gallons of water withdrawn for public water supplies in the United States each day winds up mixed with sewage. So does urban surface runoff water collected in storm drains, along with a portion of the water used in rural areas. Appropriate treatment strategy varies with population density and local geology.

Septic Systems

On an individual-user level, modern sewage treatment typically involves a septic system of some kind (figure 15.15). Wastes are first transferred to a settling tank in which solids settle out, to be broken down slowly through bacterial action. The remaining liquid with its dissolved organic-matter load is allowed to seep out through porous pipes into the soil

Figure 15.15 Basics of septic tank system: A settling tank for solids and the slow release of liquids into the soil of the leaching field for natural decomposition.

Table 15.3	Relationship Between Soil Character and Required Extent of Leaching Field

Soil type*	Absorption area needed (square meters per bedroom)
gravel or coarse sand	6.5
fine sand	8.3
sandy loam	10.6
clay loam	13.9
sandy clay	16.2
clay with minor sand or gravel	23.1
heavy clay	unsuitable

Source: Data from *Water in Environmental Planning*, by T. Dunne and L. B. Leopold, p. 187. Copyright © 1978 W. H. Freeman and Company.

*Arranged broadly in order of decreasing permeability from top to bottom of table

of the **leaching field** or **absorption field**. There, soil microorganisms and reaction with oxygen in the soil pore spaces can complete the breakdown of organic matter and destroy some disease-causing organisms. Passage through the soil, especially if it is fine-grained, also filters the liquid, removing remaining fine suspended solids and even the larger pathogenic organisms. Ideally, by the time any of the liquid reaches either the surface or groundwater supplies, it has been purified of biological and many chemical contaminants. Some compounds do remain in solution, however; nitrate is ordinarily the most significant potential pollutant among them. (Should the settling tank require pumping out, disposal of the septage can be a further problem, for few sites will accept it, and municipal sewage-treatment facilities are reluctant to take on the added load.)

There are several geologic requirements for a properly functioning septic system. The soil must be sufficiently permeable that the fluids will flow through it rather than merely backing up in the septic tank, but not so permeable that the flow into water supplies or out at the surface occurs before the wastes have been sufficiently purified. Ordinarily, the water table should be well below the level of the septic system: first, to avoid immediate groundwater contamination by raw sewage; second, because oxygen levels in saturated soil are often too low to permit rapid aerobic breakdown of the organic matter. There must be sufficient soil depth so that the wastewater will be adequately filtered by the time it reaches either surface or bedrock; this filtration usually requires at least 60 centimeters of soil above the pipes and 150 centimeters below. The leaching field should not extend to within about 15 meters of any body of surface water for similar reasons. (The exact dimensions required are controlled by soil characteristics.)

If a well is used to obtain drinking water on the same site, it must be far enough removed from the septic system that partially decomposed sewage does not reach the well, or else it should tap an aquifer that is isolated from the septic system by low-permeability material. Where many houses in a single area rely on septic systems, they must be spaced suf-

ficiently far apart so that they do not collectively saturate the soil with raw sewage and overwhelm the natural capacity of the soil and the microorganisms within it to handle the waste. The necessary spacing depends, in part, on the sizes of leaching fields to be accommodated. The required size of each leaching field is controlled, in turn, by soil permeability (table 15.3) and the number of persons to be served. Considerations such as these and assessments of the potential impact of the nitrate released from these systems lead local authorities to stipulate minimum lot sizes where septic tanks are to be used. Typical lot sizes might be one-half acre to one acre per dwelling, but again appropriate limits are controlled largely by local geology.

As is evident from the foregoing discussion, the sewage treatment associated with a septic system is entirely natural, and its thoroughness thus highly variable from site to site. One can enhance the suitability of the site—for example, by embedding the plumbing of the leaching field in "imported" sand if the soil is naturally too clay-rich to be suitable—but the natural soil chemistry, biology, and physical properties largely control the effectiveness of the system. Also, if toxic household wastes are dumped in the system, many will be untouched by the chemical processes and microbial activity that attack and decompose the organic wastes.

Municipal Sewage Treatment

In urbanized areas, population density is far too high to permit effective sewage treatment by septic systems. More than 75 percent of the U.S. population is now served instead by sewer systems. In fewer than 5 percent of the cases is completely untreated sewage released from such systems, but the completeness of municipal sewage treatment does vary. The basic steps involved are summarized in figure 15.16.

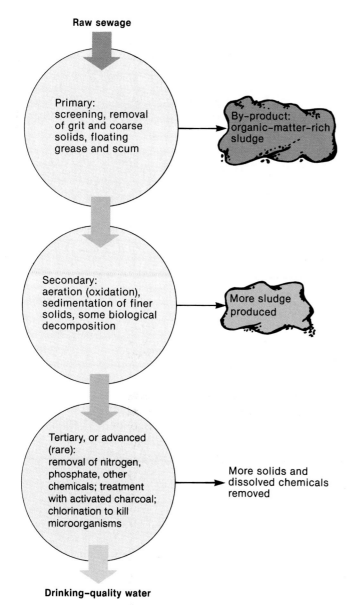

Raw sewage

Primary:
screening, removal
of grit and coarse
solids, floating
grease and scum

By-product:
organic-matter-rich
sludge

Secondary:
aeration (oxidation),
sedimentation of finer
solids, some biological
decomposition

More sludge
produced

Tertiary, or advanced
(rare):
removal of nitrogen,
phosphate, other
chemicals; treatment
with activated charcoal;
chlorination to kill
microorganisms

More solids and
dissolved chemicals
removed

Drinking-quality water

Figure 15.16 Schematic diagram of primary, secondary, and tertiary stages of municipal sewage treatment.

Primary treatment is usually physical processing only, involving removal of solids, including inorganic trash, fine sediment, and so on, that has been picked up by the wastewater. Oil or grease may have to be skimmed off the surface. Where industrial wastes are also present, some chemical treatment, such as neutralizing excess acidity or alkalinity, may be necessary.

Secondary treatment of the remaining liquid is mainly biological. Bacteria and fungi act on the dissolved and suspended organic matter to break it down. Sometimes, the water is also chlorinated to disinfect it when biological activity is completed. Primary plus secondary treatment together can reduce suspended solids and oxygen demand (see chapter 16) by about 90%, nitrogen by one-half, and phosphorus by one-third. According to the EPA, 20% of municipal sewage was given secondary treatment, and 54%, more than secondary treatment, in 1986.

A weakness in many municipal treatment systems, even those that practice secondary treatment, is that large amounts of rain or meltwater runoff in a short time can overload the capacity of treatment plants. At such times, some raw sewage may be released untreated. Also, even secondary treatment has little impact on dissolved minerals and potentially toxic chemicals, which remain in the water. Recently, the safety of chlorination has been questioned because, in some communities where hydrocarbon impurities were already present in the water supply, chlorination was found to lead to the formation of chlorinated hydrocarbons, such as chloroform. Some of these compounds may be carcinogenic. However, the chlorination procedure often can be adjusted to minimize the production of chlorinated hydrocarbons. The value of chlorination in the destruction of pathogenic organisms probably far outweighs the danger from this largely controllable problem.

After secondary treatment, wastewater can also be subjected to various kinds of tertiary, or advanced, treatment and purified sufficiently to be recycled as drinking water. Fine filtration, passage through activated charcoal, distillation, further chlorination, and various chemical treatments to remove dissolved minerals are some possible forms of tertiary treatment. Such thorough tertiary wastewater treatment may cost up to five times what primary and secondary treatment together do, which is why tertiary treatment is uncommon. Only 2 percent of the U.S. population is served by systems undertaking tertiary sewage treatment. It is now practiced mainly in a few communities where water shortages compel water recycling. On the other hand, tertiary treatment of all U.S. municipal wastewater would cost only a few tens of dollars per person per year. As a means of extending dwindling drinking-water supplies, tertiary treatment is likely to find increasing use in the future.

Wastewater treatment can yield other benefits besides drinking water. The city of Arcata, California uses sewage that has passed through primary treatment to supply water to marshes and ponds, where natural microbial action and plant growth purify the water further, and remove nutrients from it. Ultimately, the water purity is such that the output water can be used at a fish hatchery. The Tapia Water Reclamation Facility in northwestern Los Angeles County takes 10 million gallons of sewage per day and transforms it into water of high enough quality to use for irrigation.

A by-product of sewage treatment is quantities of sludge, which can present a solid-waste disposal problem. Chicago's municipal sewage treatment facilities process 1.4 billion gallons of sewage and generate approximately 600 tons of sludge (on a dry-weight basis) *every day*. What can be done with such wastes? One option, because sludge is rich in organic matter and nutrients, is to use it as fertilizer. In the early 1970s, for example, Chicago developed a plan for using some of its sludge to fertilize 11,000 acres of formerly strip-mined land during reclamation—thereby neatly solving several environmental problems at once. Austin, Texas composts and transforms sludge into "Dillo Dirt" with which parks and athletic fields are fertilized; it is also sold to gardeners. (The name comes from the

Box 15.3 Radiation— What Does It Do?

Alpha and beta particles and gamma rays all are types of ionizing radiation. That is, they can strip off electrons from atoms or split molecules into pieces. Depending on which particular atoms or molecules in an organism are affected and the intensity of the radiation dose, the results could include genetic mutations, cancer, tissue burns—or nothing significant at all.

How much damage is done by a given dose of radiation? How small a dose is harmless? Scientists do not know precisely. Most controlled radiation studies have been done on animals, not humans. The most definitive data on radiation effects on people come from accidental exposures of scientists or technicians to sizable doses of radiation or from observing the aftereffects of the explosion of atomic weapons in Japan during World War II. The reactor accident at Chernobyl is further expanding the data base.

The effects of low doses of radiation on humans are particularly hard to quantify for several reasons. One is that some of the consequences, such as slowly developing cancers, do not appear for many years after exposure. By that time, it is much harder to link cause and effect clearly than it is, for example, in the case of severe radiation burns resulting immediately from a massive dose. A related problem is that many of the results of radiation exposure—like cancer or mutations—can have other causes, too. There are hosts of known or suspected *chemical* carcinogens and mutagens; how does one uniquely identify the cause of each particular cancer or mutation?

Another uncertainty arises from the question of linearity in cause and effect. If a population is exposed to a certain dose of a certain type of radiation, and one hundred cases of cancer result, does that mean that one-tenth the exposure would induce ten cases of cancer, and so on? Or is there a minimum harmful level of radiation exposure, below which we are not at risk?

We are surrounded by radiation. Cosmic rays bombard us from space; naturally occurring radioactive elements decay in the soil beneath our feet and in the wood and concrete in our homes. Perhaps, in consequence, humans have developed an "immunity" to radiation doses that are not significantly above natural background levels. On the other hand, it may be that any dose of radiation is harmful, that natural radiation is regularly responsible for some of the cancers and other possibly radiation-caused ills not obviously attributable to other agents. Nobody can avoid *some* natural exposure to radiation (table 1). We increase our exposure with every medical X ray, every jet airplane ride, every journey to a high mountaintop, and every time we wear a watch with a luminous dial. Indeed, each of us is radioactive, a consequence of naturally occurring carbon-14. Does a small exposure to low-level radioactive waste represent a significantly increased risk? How small is small? Frustratingly, perhaps, there are no simple, direct answers to questions such as these. That uncertainty contributes to the controversy concerning radioactive-waste disposal.

Table 1 *Average Radiation Exposure of U.S. Population by Source*

Source	Percentage of exposure
cosmic	24%
terrestrial (rocks, soil)	32
medical (mainly X rays)	40
global nuclear fallout	2
miscellaneous (nuclear plant operation, wastes, etc.)	2

Source: Council on Environmental Quality.

city's mascot, the armadillo.) However, much urban sludge contains unacceptably high concentrations of heavy metals or other toxic industrial chemicals, which have to be removed with the organic solids, and which are too dispersed through the sludge for efficient extraction. This may limit the use of such sludge fertilizer to land not to be used for food crops or pastureland (parks, golf courses, and so on) and many even necessitate its disposal in landfill sites like other solid wastes.

In speaking of sewage treatment, it is worth keeping in mind the disparities between the industrialized and other nations. According to the World Resources Institute, it has been estimated that over 90 percent of the sewage in developing countries is discharged into surface waters with no treatment whatever; of India's 3,100-plus cities and towns, only 209 have even partial sewage treatment, and only 8 have complete wastewater collection and treatment.

Radioactive Wastes

Radioactive wastes differ somewhat in character from chemical wastes. Thus, proposals for disposal of solid and liquid radioactive wastes usually differ somewhat from the methods used for other wastes, although some radioactive materials are chemical toxins as well.

Some isotopes, some of the possible combinations of protons and neutrons in atomic nuclei, are basically unstable. Sooner or later, unstable nuclei decay. In doing so, they release radiation: alpha particles (nuclei of helium-4, with two protons, two neutrons, and a +2 charge), beta particles (electrons, with a –1 charge), or gamma rays (electromagnetic radiation, similar to X rays or microwaves but shorter in wavelength and more penetrating). A given decay may emit more than one type of radiation. Some aspects of the effects of radiation are discussed in box 15.3.

Radioactive decay is a statistical phenomenon. It is impossible to predict the instant at which an atom of a radioactive element will decay. However, over a period of time, a fixed percentage of a group of atoms of a given radioactive isotope (radioisotope) will decay. The idea is somewhat like flipping a coin: One cannot know beforehand whether a given flip will come up heads or tails, but over a great many flips, one can expect that half will come up heads, half tails.

Each different radioisotope has its own characteristic rate of decay, often described in terms of a parameter called **half-life.** The half-life of a radioisotope is the length of time required for half the atoms of that isotope initially present in a system to decay. The concept is illustrated in figure 15.17. Suppose, for example, that the half-life of a particular radioactive isotope is ten years and that, initially, there are 2,000 atoms of it present. After ten years, there will be about 1,000 atoms left; after ten more years, 500 atoms; after ten more years, 250; and so on. After five half-lives, only a few percent of the original amount is left; after ten half-lives or more, the fraction left is vanishingly small.

Half-lives of different radioactive elements vary from fractions of a second to billions of years, but for any specific radioactive isotope, the half-life is constant. Uranium-238 has a half-life of about 4.5 billion years; carbon-14, 5,730 years; radon-222, 3.82 days. The half-life of a given isotope is constant regardless of the physical or chemical state in which it exists, whether it is in a mineral or dissolved in water, in the atmosphere or deep in the crust at high temperature and pressure. Spontaneous radioactive decay cannot be accelerated or slowed down by chemical or physical means. An important consequence of this is that radioactive wastes will continue to be radioactive and each isotope will decay at its own rate regardless of how the wastes are handled. The wastes cannot be made to decay faster to get rid of the radioactive material more quickly, and they cannot be treated in any way to make them nonradioactive. This is a key difference between radioactive wastes and many chemical toxins; many of the latter can, through proper treatment, be broken down or neutralized, which increases the disposal options available for dealing with them.

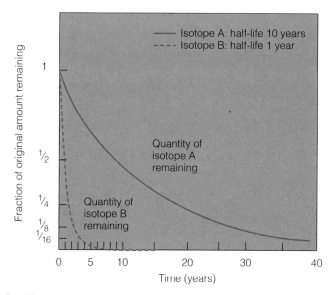

Figure 15.17   The phenomenon of radioactive decay: Each radioisotope decays at its own characteristic rate, defined by its half-life.

Nature of Radioactive Wastes

As noted in chapter 14, nuclear fission reactor wastes include many radioisotopes. The isotopes of most concern from the standpoint of radiation hazards are not those of very short half-life (if an isotope has a half-life of 10 seconds, it is virtually all gone in a few minutes) or very long half-life (if an isotope's half-life is 10 billion years, very little of it will be decaying over a human lifetime). It is the elements with half-lives of the order of years to hundreds of years that pose the greatest problems: They are radioactive enough to be significant hazards, yet will persist in the environment for some time.

Some of the waste radioactive isotopes are also toxic chemical poisons, so they are dangerous independently of their radioactivity, even if their half-lives are long. Plutonium-239, with a half-life of 24,000 years, is one example of such an isotope. Other isotopes pose special hazards because the corresponding elements are biologically concentrated, often in one particular organ, leading to the possibility of a concentrated radiation source in the body. Included in this category are such products as iodine-131 (half-life, 8 days), which would be concentrated, like any iodine, in the human thyroid gland; iron-59 (45 days), which would be concentrated in the iron-rich hemoglobin in blood; cesium-137 (30 years), which is concentrated in the nervous system; and strontium-90 (29 years) which, like all strontium, tends to be concentrated with calcium—in bones, teeth, and milk. (Before atmospheric nuclear weapons testing was banned, fallout over pastures resulted in elevated levels of strontium-90 in cows' milk.)

Radioactive wastes are often classified collectively as **low-level** or **high-level.** The division is an imprecise one. The low-level wastes are relatively low-radioactivity wastes, not requiring extraordinary disposal precautions. Volumetrically,

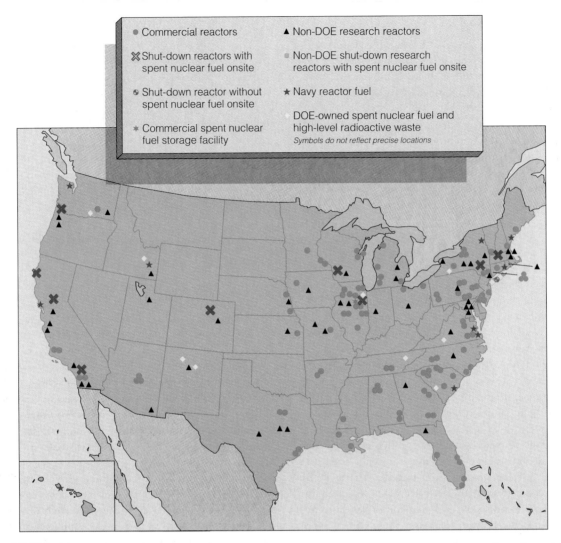

Figure 15.18 Locations (approximate) of spent nuclear fuel elements and high-level radioactive wastes in the U.S. awaiting disposal. Source: From U.S. Department of Energy, Office of Civilian Radioactive Waste Management.

they account for over 90 percent of radioactive wastes generated. Some low-level wastes are believed to be harmless even if released directly into the environment. In this category, for example, fall the routine small releases of radioactive gas from operating fission reactors. Liquid low-level wastes—from laundering of protective clothing, from decontamination processes, from floor drains—are often released also, diluted as necessary so that the concentration of radioactivity is suitably low. Solid low-level wastes, such as filters, protective clothing, and laboratory materials from medical and research laboratories, are commonly disposed of in landfills or held in temporary storage until the radioactivity has decreased enough that they can be consigned to a landfill. Currently, there are twenty temporary repositories and six commercial disposal sites for low-level wastes. For environmental and capacity reasons, including problems of waste leakage, some of the commercial disposal sites have been closed. Recent federal legislation has required states to take the responsibility for their own low-level wastes in the early 1990s (though

they may enter into multi-state compacts, and several of these exist). States that benefit most from nuclear power and research and other facilities using nuclear materials thus bear more of the burden of the disposal of the corresponding low-level wastes; and the transportation of those wastes is reduced. There is often concern over local disposal of low-level wastes (the NIMBY phenomenon again), and many states have not yet developed the mandated disposal sites; but these issues are minimal compared to concerns over the much more intensely radioactive high-level wastes.

Spent reactor fuel rods and by-products from their fabrication and reprocessing are examples of high-level wastes, for which various more-or-less elaborate disposal schemes have been proposed. Currently, they sit in temporary storage—often on-site at nuclear reactors and other nuclear facilities—awaiting permanent disposal (figure 15.18). The basic disposal problem with the chemically varied high-level wastes is how to isolate them from the biosphere with some confidence that they will stay isolated for thousands of years or longer. In 1985, the Environmental Protection Agency specified that high-level wastes be isolated in such a way as to cause fewer than one thousand deaths in ten thousand years (the expected fatality rate associated with unmined uranium ore). The sections that follow survey some of the many strategies proposed for dealing with these high-level wastes.

Historical Suggestions: Space, Ice, and Plate Tectonics

Some of the more exotic schemes proposed in the past for disposal of high-level radioactive wastes involved putting wastes in space, perhaps rocketing them into the sun. Certainly, that would get rid of the wastes permanently—if the rocket launch were successful. However, many launches are not successful or are aborted, and the payloads return to earth, sometimes in an uncontrolled fashion. Furthermore, since many spacecraft would be required, the costs would be extremely high, even if successful launches could somehow be guaranteed.

Others have proposed putting encapsulated wastes under thick ice sheets, with Antarctica a suggested site. High-level radioactive wastes generate heat. They could melt their way down through the ice sheet, to be well isolated eventually by the overlying blanket of ice, which would refreeze above. But glacier ice flows, and it flows at different rates at the top and bottom of a glacier. How would one keep track of waste movement through time? What if future climatic changes were to melt the ice sheet? For various reasons, disposal of radioactive wastes in Antarctica is presently prohibited by international treaty, and no other thinner ice sheets are seriously under consideration.

Subduction zones also have been considered as possible repositories for high-level wastes. The seafloor trenches above subduction zones are often sites of fairly rapid accumulation of sediments eroded from a continent. In theory, these sediments would cover the wastes initially, and the wastes would, in time, be carried down with the subducting plate into the mantle, well out of the way of surface-dwelling life.

Several objections suggest themselves. Subduction is a slow process; the wastes would sit on the sea floor for some time before disappearing into the mantle. Even at sedimentation rates of several tens of centimeters per year, large waste containers would take years to cover. Seawater, especially seawater warmed by decaying wastes, is highly corrosive. Might the wastes not begin to leak out into the ocean before burial or subduction was complete? How would the wastes be attached to the down-going plate to ensure that they would actually be subducted? Currently, the behavior of sediments in subduction zones is not fully understood. Geologic evidence suggests that some sediments are indeed subducted with down-going oceanic lithosphere, but some, on the other hand, are "scraped off" by the overriding plate and not subducted. The uncertainties appear to be too great at present for this scheme to be satisfactory.

Seabed Disposal

The deep-sea floor is a relatively isolated place, one with which humans are unlikely to make accidental contact. Much of this area is covered by layers of clay-rich sediment that may reach hundreds of meters in thickness, and studies have shown that the clays have remained stable for millions of years. Interest is growing in radioactive-waste disposal schemes that involve sinking waste canisters into these deep, stable clays (in areas well removed from both active plate boundaries and any seafloor resources, notably manganese nodules). The low permeability of clays means restricted water circulation in the sediments. Many of the elements in the wastes, even if they did escape from the containers, would be readily adsorbed by the fine clay particles. Any dissolved elements escaping into the ocean would be entering the dense, cold (1 to 4° C, or 34 to 39° F), saline, deep waters that mix only very slowly with near-surface water. By the time any leaked material reached inhabited waters, it would be greatly diluted, and many of the shorter-half-life isotopes would have decayed. The long-term geologic stability of the deep-sea floor is appealing, especially to nations having radioactive wastes to dispose of but not located in particularly stable areas of the continents, or having such high population densities that they lack remote areas for disposal sites. An international Seabed Working Group currently has this option under active study. It has already concluded that seabed disposal is technically feasible, but that more research on the long-term safety of the method is still needed.

Bedrock Caverns for Liquid Waste

Many of the high-level radioactive wastes are presently held in liquid form rather than solid, in part to keep the heat-producing radioactive elements dilute enough that they do not begin to melt their storage containers. These liquid wastes are currently stored in cooled underground tanks, holding up to one million gallons apiece, which from time to time have leaked spectacularly: At the federal radioactive-waste repository in Hanford, Washington, more than 300,000 gallons of high-level liquid wastes have leaked from their huge storage tanks over the last two decades. (Currently, remediation efforts are in process, including bioremediation through which microorganisms break down organic compounds and convert toxic metals to less-mobile compounds; but secure initial disposal is clearly preferable to later remediation.)

Caverns hollowed out of low-permeability, unfractured rocks—igneous rocks such as basalt or granite, for example—have been suggested as more secure, permanent disposal sites for high-level liquid wastes. The rocks' physical properties would have to be thoroughly investigated beforehand to ensure that the caverns could indeed contain the liquid effectively. The wells leading down to the caverns, through which wastes would be pumped, would have to be tightly cased to prevent shallower leaks. The long-term concern, however, is the geologic stability of the bedrock unit chosen. Faulting or fracturing of the rocks would create conduits through which the highly radioactive liquid could escape rapidly, perhaps to the surface or into nearby aquifers.

While it is possible to identify with a high degree of probability the areas that are likely to remain stable for thousands or millions of years, stability cannot be absolutely guaranteed.

Serious consideration has been given to solidifying liquid high-level wastes prior to disposal because of the much lower mobility of solids. Certain minerals and glasses have proven to be relatively resistant to leaching over centuries or even millennia in the natural environment. Vitrification—incorporation of the waste in a glassy matrix—would be a first step in many high-level waste disposal schemes. It would be followed by sealing the waste in canisters and then placing the canisters in some kind of bedrock cavern, old mine, or similar space. Solidification would not eliminate the uncertainty regarding bedrock stability, but it would render the wastes less susceptible to rapid escape if fracturing breached the storage volume.

Bedrock Disposal of Solid High-Level Wastes

A variety of bedrock types are being investigated as host rocks for solid high-level waste disposal. The general design of a bedrock disposal site or repository for such wastes is shown in figure 15.19. It involves the **multiple barrier** (or *multibarrier*) **concept:** surrounding solid waste with several different types of materials to create multiple obstructions to waste leakage or invasion by ground water. A major variable is the nature of the host rock. Each of the rock types highlighted in the following paragraphs has some particular characteristics that make it promising as a potential host rock, as well as some potential drawbacks.

Granite is a plentiful constituent of the continental crust. Granites are strong and very stable structurally following excavation. They have low porosity; the common minerals of granite (quartz, feldspars) are quite insoluble (in a temperate climate). A concern with respect to granite is the possible presence of fractures (natural or created during excavation) that would allow solutions to permeate and wastes, perhaps, to escape. One might expect that, below a depth of about 1 kilometer, the pressures would close any fractures. However, field measurements suggest that this is not necessarily the case. Tests on granite are underway at the Nevada Test Site; the United States and Sweden also have a joint testing program in progress at Stripa, Sweden.

Thick *basalts,* such as the Columbia River plateau basalts described in chapter 5, are another option. Some of these basalts are found in the area of the Hanford reservation in Washington state, where the Department of Energy already owns land and where radioactive materials have already been handled and stored for many years. Fresh, unfractured basalt is a strong rock. Basalt consists of high-temperature minerals and glass, able to withstand the elevated temperatures of high-level wastes, and with fairly high thermal conductivity, to allow that heat to be dispersed. However, basalts commonly contain porous zones of gas bubbles; they may also weather easily. Both phenomena contribute to zones of weakness. The Hanford basalts, in particular, are also extensively fractured in places, and those fractures might allow excessive

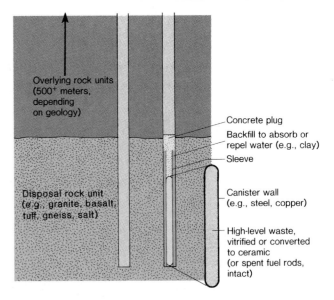

Figure 15.19 Schematic of a high-level-waste disposal site in bedrock: the multibarrier concept.

water flow through the system. Even the planes between successive flows may provide water conduits. Groundwater flow patterns in the Hanford basalts, which are interlayered with sedimentary rocks, are poorly understood.

Massive deposits of *tuff,* pyroclastic materials deposited during past large-scale explosive volcanic eruptions, are another possible host rock for solid high-level wastes. *Welded tuffs* are those that were so hot at the time of deposition that they fused; these deposits are not unlike basalts in many of their physical properties. But they are very brittle and easily fractured, which raises questions about their integrity with respect to water flow. Other tuffs have been extensively altered, producing abundant *zeolites.* These hydrous silicate minerals—the same group used in water softeners—are potentially valuable in a waste-disposal site for their ion-exchange capacity. They should tend to capture and hold wastes that might begin to migrate from a repository. However, zeolitic tuffs are weaker, more porous, and more permeable than welded tuffs; furthermore, zeolites begin to break down and dehydrate at relatively low temperatures (100 to 200° C, or about 200 to 400° F), which reduces their ion-exchange potential and also releases water that might facilitate waste migration.

Shale and other clay-rich sedimentary rocks are another option. As noted earlier in the seabed-disposal discussion, clays may adsorb migrating elements, providing a measure of containment for the waste. Shale is also low in permeability, and somewhat plastic under stress, though less so than salt (see the next paragraph). On the other hand, shaly rocks are weak, fracture readily, and may be interlayered with much more permeable sedimentary units. At elevated temperatures, clays, like zeolites, may decompose and dehydrate. Some clays are stable to higher temperatures than are most zeolites, but those that would most actively adsorb waste components tend to lose their adsorptive

The labels in the figure read:

Overlying rock units (500⁺ meters, depending on geology)

Disposal rock unit (e.g., granite, basalt, tuff, gneiss, salt)

Concrete plug

Backfill to absorb or repel water (e.g., clay)

Sleeve

Canister wall (e.g., steel, copper)

High-level waste, vitrified or converted to ceramic (or spent fuel rods, intact)

A

B

Figure 15.20 Waste Isolation Pilot Plant (WIPP) in New Mexico. (*A*) Plan of storage facility. (*B*) Waste canisters are emplaced, by remotely-controlled machinery, into holes bored into salt walls.   Sources: (A) After Office of Environmental Restoration and Waste Management, *Environmental Restoration and Waste Management: An Introduction*, U.S. Department of Energy. (B) Photograph courtesy of U.S. Department of Energy.

efficiency at those higher temperatures. So, overall, clays may not have clear advantages over zeolites in disposal sites.

In the United States, much consideration has been given to high-level waste disposal in thickly bedded *salt* deposits or salt domes. Salt has several special properties that may make it a particularly suitable repository. It has a high melting point, higher in fact than those of most rocks. Consequently, salt can withstand considerable heating by the radioactive wastes without melting. Although soluble under wet conditions, salt typically has low porosity and permeability. Under dry conditions, then, it can provide very "tight" storage with minimal leakage potential. Salt's low porosity and permeability result, in part, from its ability to flow plastically under pressure while remaining solid. It is, therefore, somewhat self-sealing. If it were fractured by seismic activity or by accumulated stress in the rocks, flow in the salt could seal the cracks. Most other rock types are more brittle under crustal conditions and do not so readily self-seal. Finally, evaporites are not particularly scarce resources. Using an old salt mine for radioactive-waste disposal would not be hardship in that context.

Some years ago, the federal government went so far as to start intensive work (and publicity) on an old salt mine at Lyons, Kansas, with nuclear-waste disposal in mind, and then it abandoned "Project Salt Vault" when the site was deemed unsuitable due to local human activity and the presence of numerous old wells in overlying strata. Other salt deposits are still under study. The Waste Isolation Pilot Plant site, in southwestern New Mexico, is a region of bedded salt deposits under investigation as a disposal site for high-level wastes generated by the military (figure 15.20). To date, however, there is by no means unanimous agreement that salt disposal is the way to go, let alone which site would be best. Only one of the three proposed western disposal sites for wastes of the nuclear utilities is a salt deposit (box 15.4).

Box 15.4 Wanted: One Disposal Site for High-Level Waste

Much of the delay in selection of a site for disposal of high-level radioactive waste has resulted from the need for thorough investigation of the geology of any proposed site. However, in the United States at least, there is also a political component to the delay. Simply put, nobody wants to host a high-level waste disposal site.

The Nuclear Waste Policy Act of 1982 provided for the establishment of two high-level waste disposal sites, one in the western United States, one in the East, both to be identified by the Department of Energy (DOE). The same Act created the Office of Civilian Radioactive Waste Management and established the Nuclear Waste Fund (NWF), supported by spent-fuel fee charges on waste generators, and a one-mill-per-kilowatt-hour fee on electricity. This sounds small, but to date, utilities have contributed billions to this fund; NWF income for 1991 was over $900 million. The NWF is to finance disposal-site selection and investigation. That investigation must be thorough, for the waste is to be isolated for at least ten thousand years. EPA guidelines specify that radiation containment be effective enough that no more than one thousand deaths result from exposure in that ten thousand years (which, as mentioned earlier, is the estimated risk from unmined uranium ore).

In 1986, DOE announced three candidates for the western site: Hanford, Washington (where disposal would be in basalt); Yucca Mountain, Nevada (ash-flow tuff); and Deaf Smith County, Texas (salt). All three states protested. So did Tennessee, where—at Oak Ridge—the waste-packaging site was to be built to prepare the wastes for disposal. Original

Figure 1 Proposed waste-disposal site at Yucca Mountain, Nevada. Photograph courtesy of U.S. Department of Energy.

plans would next have called for a billion-dollar testing program to be undertaken at each of the three proposed disposal sites, to be followed by selection of one to be *the* western site. Congress subsequently called a year-long halt to further action.

Then, in late 1987, Congress suddenly moved decisively to end the uncertainties—at least those of choice of site. It was decided that the Yucca Mountain site (figure 1) would be developed and investigation of the other sites halted. Resultant cost savings will be nearly $4 billion. The state of Nevada will receive $20 million per year and "special consideration" of certain research grants by way of compensation. If the intensive evaluation of the Yucca Mountain site yields

satisfactory results, construction of the repository is set to begin in 1998, with an opening date in the year 2003.

Properties of the site that contributed to its initial attractiveness include: (1) the tuff host rock, useful characteristics of which are described in the text; (2) generally arid climate, with no perennial streams and limited subsurface water percolation in the area; (3) low population density; (4) low regional water table, so the repository would be 200 to 400 meters above the water table; and (5) apparent geologic stability. The site is also located on lands already owned by the federal government and overlaps the Nevada Test Site previously used for nuclear-weapons testing. Site-investigation activities will include

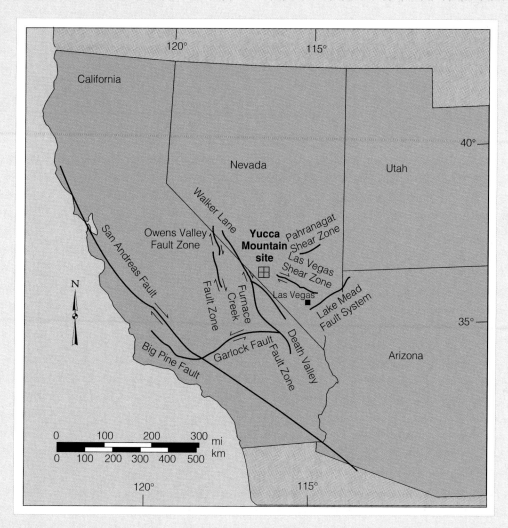

Figure 2 Major faults in the vicinity of Yucca Mountain site. Source: After *Site Characterization Plan Overview*, fig. 2–4, p. 20, U.S. Department of Energy, 1988.

much more detailed structural, hydrologic, and geochemical studies, and more detailed analysis of geologic-stability issues. Seismicity is an obvious concern: there are a number of faults in the area, some related to collapse of ancient volcanic calderas, some related to formation of the mountains, some related to plate motions (figure 2). However, current seismicity is low (figure 3), and evidence suggests limited local fault displacement for the last 10 million years; still, repository design will allow for earthquakes up to magnitude 6.8 or so, the largest plausible on the known faults. The most recent volcanism in the region is a cinder cone that may be as young as 20,000 years old, and basaltic volcanism is possible within the next 10,000 years or so. There are

—Continued

Box 15.4—Continued

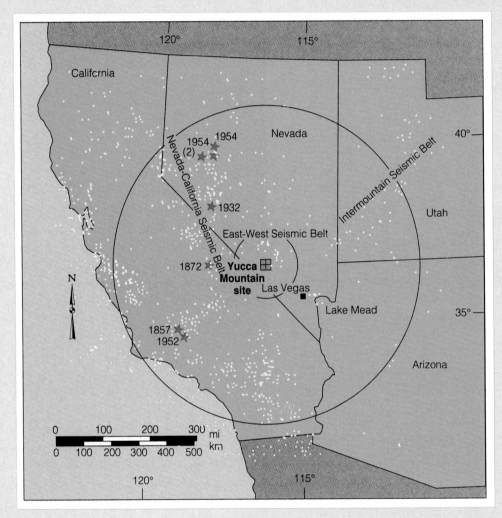

Figure 3 Seismicity in the southwestern United States, 1969–1978, showing earthquakes of magnitude 4 or higher. Stars denote a magnitude of 6.5 or higher.   Source: After *Site Characterization Plan Overview*, U.S. Department of Energy, 1988.

also questions about how likely the region is to stay as dry as it currently is.

The evaluation will be comprehensive. The site characterization *plan* is a nine-volume, 6000+-page docu-ment; the environmental impact statement is expected to be equally massive, or larger. As with Superfund, shifting political priorities have affected the Yucca Mountain project as well; the budget for site evaluation was reduced in 1996. The outcome of the studies is yet to come, and meanwhile, the high-level wastes remain on hold, in temporary storage.

No High-Level Radioactive Waste Disposal Yet

At least two dozen nations are making plans to dispose of high-level nuclear wastes (figure 15.21); but so far, *no* high-level radioactive wastes have been consigned to permanent disposal sites anywhere in the world by any nation. All this waste—some solid, some liquid—is in temporary storage awaiting selection and construction of disposal sites. Geologists can identify sites that have a very high probability of remaining geologically quiet and secure for many generations to come, but absolute guarantees are not possible. Uncertainties about the long-term integrity and geologic stability of any given site, coupled with increasing public resistance in regions with sites under consideration, have held up the disposal efforts. Worldwide, the accumulated wastes amount to millions of gallons of liquids, tens of thousands of tons of solids.

Radioactive-waste disposal remains an issue even if the nuclear-fission power option is not pursued. A growing number of nations in the industrialized world have accumulated high-level wastes, from past power generation, nuclear weapons production, and radiochemical research including medical applications. All of that waste must ultimately be put somewhere in permanent disposal. (The only alternative is to keep the wastes at the surface, closely monitored and guarded to protect against accident or sabotage, assuming that long-term political stability is more likely than geologic stability—a debatable assumption, surely.) But it remains an open question where the wastes will be put, and how soon.

Spent fuel management approaches in selected countries

	Deferred decision	Direct disposal	Repro-cessing
Argentina	•		•
Belgium			•
Brazil			•
Bulgaria	•		•
Canada		•	
China			•
Czech Republic	•		•
Finland	•	•	
France			•
Germany		•	•
Hungary	•		
India	•		•
Italy	•		•
Japan			•
Korea, Rep. of	•		
Lithuania	•		
Mexico	•		
Netherlands			•
Pakistan	•		
Russia			•
Slovak Republic	•		•
Slovenia	•		
South Africa	•		
Spain	•		•
Sweden		•	
Switzerland			•
United Kingdom	•		•
Ukraine	•		•
United States		•	

Note: Some countries have different spent fuel management approaches for different fuel types. In some countries one spent fuel management approach is presently being followed but future options applying different approaches are being evaluated.
Source: Data reported to the IAEA.

Figure 15.21 Approaches to management of spent nuclear fuel in different countries. From F. Takáts, et al, "Management of spent fuel from power and research reactors: International status and trends," in the *IAEA Bulletin 1993*, No. 3, pp. 18-22. Reprinted by permission.

Overview

Summary

Methods of solid-waste disposal include open dumping (being phased out for health reasons), use of landfills, incineration on land or at sea, and disposal in the oceans. Incineration contributes to air pollution but can also provide energy. Landfills require a commitment of space, although the land may be used for other purposes when filled. "Secure" landfills for toxic solid or liquid industrial wastes should isolate the hazardous wastes with low-permeability materials and include provisions for monitoring ground water chemistry to detect any waste leakage. Increasing evidence suggests that construction of wholly secure landfill sites is not possible. Deep-well disposal and incineration are alternative methods for handling toxic liquid wastes.

The principal liquid-waste problem in terms of volume is sewage. Low-population areas can rely on the natural sewage treatment associated with septic tanks, provided that soil and site characteristics are suitable. Municipal sewage treatment plants in densely populated areas can be constructed to purify water so well that it can even be recycled as drinking water, although most municipalities do not go to such expensive lengths.

A complicating feature of radioactive wastes is that they cannot be made nonradioactive; each radioisotope decays at its own characteristic rate regardless of how the waste is handled. Given the long half-lives of many isotopes in high-level wastes and the fact that some are chemical poisons as well as radiation hazards, long-term waste isolation is required. Disposal in bedrock, tuff, salt deposits, or on the deep-sea floor are among the methods under particularly serious consideration, but so far, no high-level radioactive wastes have been consigned to permanent disposal sites.

Terms to Remember

absorption field 370	low-level waste 373
half-life 373	multiple barrier concept 376
high-level waste 373	sanitary landfills 356
leachate 357	secure landfill 364
leaching field 370	source separation 363

Exercises

Questions for Review

1. What two kinds of activities generate the most solid wastes?
2. What advantages does a sanitary landfill have over an open dump? Describe three pathways through which pollutants may escape from a landfill site.
3. Landfills and incinerators have in common the fact that both can serve as energy sources. Explain.
4. Use of in-sink garbage disposal units in the home is sometimes described as "on-site disposal." Is this phrase accurate? Why or why not?
5. Compare the relative ease of recycling of (a) glass bottles, (b) paper, (c) plastics, (d) copper, and (e) steel, noting what factors make the practice more or less feasible in each case.
6. Compare the dilute-and-disperse and concentrate-and-contain philosophies of liquid-waste disposal.
7. Outline the relative merits and drawbacks of deep-well disposal and of incineration as disposal strategies for toxic liquid wastes.
8. What kinds of limitations restrict the use of septic tanks?
9. Municipal sewage treatment produces large volumes of sludge as a by-product. Suggest possible uses for this material, and note any factors that might restrict its use.
10. Evaluate the advantages, disadvantages, and possible concerns relating to disposal of high-level radioactive wastes in (a) subduction zones, (b) sediments on the deep-sea floor, (c) tuff, and (d) bedded salt.

For Further Thought

1. Where does your garbage go? How is it disposed of? Is any portion of it utilized in some useful way? If your community has a recycling program, find out what is recycled, in what quantities, and what changes have occurred over the last five or ten years in materials recycled and levels of participation.
2. For a period of one or several weeks, weigh all the trash you discard. If every one of the close to 265 million people in the United States or the 30 million in Canada discarded the same amount of trash, how much would be generated in a year? Suppose that each of the more than 5 billion people on earth did the same. How much garbage would that make? Is this likely to be a realistic estimate of actual world trash generation?
3. Investigate your community's sewage treatment. What level is it? What is the quality of the outflow? How does the storm-sewer system connect with the sewage-treatment system?
4. Choose a single Superfund NPL site, or several sites within a state or region and check the history and current status of cleanup activities there. A Web site including all the NPL (National Priorities List) sites by region, which would be a place to start, is at http://www.epa.gov/superfund/oerr/impm/products/nplsites/usmap.htm

Suggested Readings/References

Ackland, L. 1993. Radiation risks revisited. *Technology Review* (February/March): 56–61.

Boraiko, A. A. 1985. Hazardous waste: Storing up trouble. *National Geographic* 167 (March): 318–51.

Brookins, D. G. 1984. *Geochemical aspects of radioactive waste disposal.* New York: Springer-Verlag.

Council on Environmental Quality. 1984. *Environmental quality 1983.* Washington, D.C.: Council on Environmental Quality.

———. 1988. *Environmental quality 1986.* Washington, D.C.: Council on Environmental Quality, 17th Annual Report.

Dunne, T., and Leopold, L. B. 1978. *Water in environmental planning.* San Francisco: W. H. Freeman.

Environmental Protection Agency. 1981. Managing the wastes problem. In *Environment Midwest, 1981.* Washington, D.C.: Environmental Protection Agency.

Fergusson, J. E. 1982. *Inorganic chemistry of the earth.* New York: Pergamon Press.

Freeman, H. M. 1990. *Hazardous waste minimization.* New York: McGraw-Hill.

Gertler, N., and Ehrenfeld, J. R. 1996. A down-to-earth approach to clean production. *Technology Review* (February/March): 48–54.

Gonzales, S. 1982. Host rocks for radioactive waste disposal. *American Scientist* 70 (February): 191–200.

Hammer, M. J. 1986. *Water and wastewater technology.* 2d ed. New York: Wiley.

IAEA Bulletin. 1989, 1992, 1993. Vol. 31 no. 4, Vol. 34 no. 3, and Vol. 35 nos. 3 and 4 are devoted to various aspects of the nuclear-waste-disposal problem.

International Atomic Energy Agency. 1981. *The management of radioactive wastes.* Vienna: International Atomic Energy Agency.

Kathren, R. L. 1984. *Radioactivity in the environment.* New York: Harwood Academic Publishers.

Ketchum, B. H., Kester, D. R., and Park, P. K., eds. 1978. *Ocean dumping of industrial wastes.* New York: Plenum Press.

Lester, J. N. 1982. Sewage and sewage sludge treatment. Ch. 5 in *Pollution: Causes, effects, and control,* edited by R. M. Harrison. London: The Royal Society of Chemistry.

Levine, A. G. 1982. *Love Canal: Science, politics, and people.* Lexington, Mass.: D. C. Heath.

Lindgren, G. F. 1983. *Guide to managing industrial hazardous waste.* Ann Arbor, Mich.: Ann Arbor Science Publishers.

Morell, D., and Magorian, C. 1982. *Siting hazardous waste facilities.* Cambridge, Mass.: Ballinger.

Office of Civilian Radioactive Waste Management. 1992. *Annual Report to Congress.* U.S. Department of Energy.

O'Leary, P. R., Walsh, P. W., and Ham, R. K. 1988. Managing solid waste. *Scientific American* 259 (December): 36–42.

Organization for Economic Cooperation and Development. 1984. *Geological disposal of radioactive waste.* Paris: Organization for Economic Cooperation and Development.

Peterson, I. 1984. New life for old plastics. *Science News* 126:140–41.

Rathje, W. L. 1991. Once and future landfills. *National Geographic* (May): 117–34.

Robinson, A. L. 1987. Thirty ways to temporize on waste. *Science* 237 (7 August): 591–92.

Squires, D. F. 1983. *The ocean dumping quandary.* Albany, N.Y.: State University of New York Press.

Stone, R. F., Sagar, A. D., and Ashford, N. A. 1992. Recycling the plastic package. *Technology Review* (July): 49–56.

U.S. Department of Energy. 1988. *Site characterization plan overview, Yucca Mountain site, Nevada research and development area, Nevada.* Washington, D.C.: U.S. Government Printing Office.

Wilson, D. G. 1981. *Waste management—planning, evaluation, technologies.* Oxford, England: Clarendon Press.

Woods Hole Oceanographic Institution. 1990. Ocean disposal reconsidered. *Oceanus* 33 (Summer). This whole issue deals with aspects of waste disposal in the oceans.

World Resources Institute, United National Environment Programme, United Nations Development Programme, and The World Bank. 1996. *World Resources 1996–97.* (Chapter 3, "Urban Impacts on Natural Resources.") New York: Oxford University Press.

The U.S. Environmental Protection Agency Citizen's Information Page includes links to information on a range of topics, including many relating to pollution and waste disposal, at

> http://www.epa.gov/epahome/Citizen.html

EPA's "Consumer Handbook for Recycling Solid Waste" is available online at http://www.epa.gov/OSWRCRA/non-hw/reduce/catbook

Superfund has its own site at
> http://www.epa.gov/superfund/

A selection of Superfund fact sheets, collected under the heading "Setting the Record Straight," are available via
> gopher://gopher.epa.gov:70/00/Offices/Waste/OSWERSuper/facts

The Recycling Council of Ontario is a nonprofit organization founded in 1978 whose site offers information on waste generation and recycling in Ontario, in Canada, and in North America, at
> http://www.web.apc.org/rco/

An update on the status of U.S. radioactive waste disposal is available from the Nuclear Regulatory Commission at
> http://www.nrc.gov/NRC/radwaste.html

A variety of information on the Yucca Mountain Project and broader issues relating to radiation and radioactive waste disposal can be reached via
> http://www.ymp.gov/

The U.S. Navy Recycling Program home page is at
> http://www.infi.net/~calvers/

A technical report on computer modeling of a landfill, conducted at Oxford University and funded jointly by the U.S. Army Corps of Engineers and the U.K. Department of Environment, is available through
> http://info.ox.ac.uk/~ayoung/landfill.html

The Los Alamos National Laboratory Waste Management and Technologies Analytical Database System was developed to collect, organize, and update information on hazardous and radioactive wastes; see
> http://mwir.lnl.gov/

Chapter 16

Water Pollution

As noted in chapter 10, the more-or-less fresh ground water and surface water on the continents constitutes less than 1 percent of the water in the hydrosphere. Pollution has tainted much of the most accessible of this reserve: lakes, streams, water in shallow aquifers near major population centers. The number and variety of potentially harmful water pollutants are staggering. In this chapter, we review some important categories of pollutants and the nature of the problems each presents. We also look at some possible solutions to existing water-pollution problems. While it is true that substantial quantities of surface water and ground water are unpolluted by human activities, much of this water is far removed from population centers that require it, and the quality of much of the most accessible water has indeed been degraded.

Obviously polluted runoff water from mine dump, Leadville, Colorado. © David Hiser/Tony Stone Images

General Principles

Any natural water—rainwater, surface water, or ground water—contains dissolved chemicals (recall table 10.3). Some of the substances that find their way naturally into water are unhealthy to us or to other life forms, as, unfortunately, are some of the materials produced by modern industry, agriculture, and just people themselves.

Geochemical Cycles

All of the chemicals in the environment participate in geochemical cycles of some kind, similar to the rock cycle. Not all environmental cycles are equally well understood.

For example, a simplified natural cycle for calcium is shown in figure 16.1. Calcium is a major constituent of most common rock types. When weathered out of rocks, usually by solution of calcium carbonate minerals or by the breakdown of calcium-bearing silicates, calcium goes into solution in surface water or ground water. From there it may be carried into the oceans. Some is taken up in the calcite of marine organisms' shells, in which form it may later be deposited on the sea floor after the organisms die. Some is precipitated out of solution into limestone deposits, especially in warm, shallow water; other calcium-bearing sediments may also be deposited. In time, the sediments are lithified, then metamorphosed, or subducted and mixed with magma from the mantle, or otherwise reincorporated into the continental crust as the cycle continues. Subcycles also exist. For example, some mantle-derived calcium is leached directly into the oceans by the warmed seawater or other hydrothermal fluids acting on the fresh seafloor basalts at spreading ridges; after precipitation, it may be reintroduced into the mantle. Some dissolved weathering products are carried into continental waters—such as lakes—from which they may be directly redeposited in sediments.

Residence Time

Within the calcium cycle outlined, there are several important calcium reservoirs, including the mantle, continental crustal rocks, submarine limestone deposits, and the ocean, in which large quantities of calcium are dissolved. A measure of how rapidly the calcium cycles through each of these reservoirs is a parameter known as **residence time.** The technical definition of residence time is given by the relationship:

$$\text{Residence time (of a substance in a reservoir)} = \frac{\text{Capacity (of the reservoir to hold that substance)}}{\text{Rate of influx (of the substance into the reservoir)}}$$

Capacity, for a naturally occurring dissolved substance, reflects the concentration that can be achieved before the reservoir is saturated and the substance precipitated. In practical terms, residence time can also be described as the average length of time a substance remains in a system (or reservoir). This may be a more useful working definition for substances whose capacity is not limited in a given reservoir

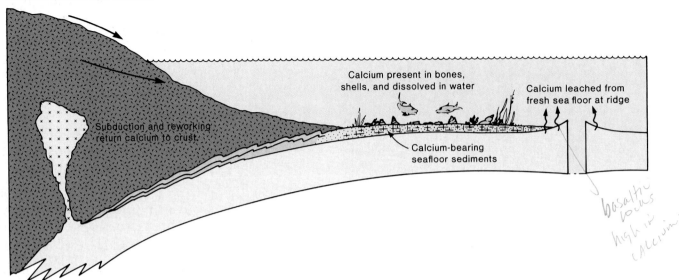

Calcium (derived from weathering of rocks) dissolved in runoff from continent

Calcium present in bones, shells, and dissolved in water

Calcium leached from fresh sea floor at ridge

Subduction and reworking return calcium to crust.

Calcium-bearing seafloor sediments

Figure 16.1 Simplified calcium cycle: Calcium weathered out of rocks is washed into bodies of water. Some remains in solution, some is taken up by organisms, and some is deposited in sediments. The deposited calcium is, in time, reworked through the rock cycle into new rock.

(the continental crust, for instance) or for substances that do not achieve saturation but are removed from the reservoir by decomposition after a time, as many synthetic chemicals are.

A nongeologic example may clarify the concept of residence time. Consider a hotel with one hundred single rooms. Some lodgers stay only one night, others stay for weeks. On the average, though, ten people check out each day, so (assuming this is a popular hotel that can always fill its rooms) ten new people check in each day. The residence time of lodgers in the hotel, then, is:

$$\text{Residence time} = \frac{100 \text{ persons}}{10 \text{ persons/day}} = 10 \text{ days}$$

In other words, people stay an average of ten days each at this hotel. A competing hotel in the next block has similar turnover of clientele, but a much lower occupancy rate—only 50 of its 100 rooms are typically occupied. Then if 10 people check in and out each day, the residence time of patrons in the second hotel is only (50 persons)/(10 persons/day) = 5 days; each patron stays only half as long, on average, in the second hotel.

Of various reservoirs in figure 16.1, the one in which the residence time of calcium can be calculated most directly is the ocean. The capacity of the ocean to hold calcium is basically controlled by the solubility of calcite. (If too much calcite has gone into solution, some is removed by precipitation.) The rate of influx can be estimated by measuring the amount of dissolved calcium in all major rivers draining into the ocean. The resulting calculated residence time of calcium in the oceans works out to about 1 million years. In other words, the average calcium atom normally floats around in the ocean for 1 million years from the time it is flushed or leached into the ocean until it is removed by precipitation in sediments or in shells.

Oceanic residence times for different elements vary widely (see table 16.1). Sodium, for instance, is put into the oceans less rapidly (in smaller quantity) than calcium, and the capacity of the ocean for sodium is much larger. (Sodium chloride, the main dissolved salt in the ocean, is very soluble; what sodium is removed is taken up by clay-rich sediments.) The residence time of sodium in the oceans, then, is longer than that of calcium: 100 million years. By contrast, iron compounds, as noted earlier, are not very soluble in the modern ocean. The residence time of iron in the ocean is only two hundred years; iron precipitates out rather quickly once it is introduced into the sea. Residence times, of course, differ for any given element in different reservoirs with different capacities and rates of influx.

Human activities may alter the natural figures somewhat, chiefly by altering the rate of influx. In the case of calcium, for instance, extensive mining of limestone for use in agriculture (to neutralize acid soil) and construction (for building stone and the manufacture of cement) increases the amount of calcium-rich material exposed to weathering; calcium salts are also used for salting roads, and these salts gradually dissolve away into runoff water. These additions are probably insufficient to alter appreciably the residence times of materials in very large reservoirs like the ocean. Locally, however, as in the case of a single lake or stream, they may change the rate of cycling and/or increase a given material's concentration throughout a cycle.

Residence Time and Pollution

In principle, one can speak of residence times of more complex chemicals, too, including synthetic ones, in natural systems. Here, there is the added complication that not enough is really known about the behavior of many synthetic chemicals

Table 16.1 Residence Times of Selected Major and Minor Elements in Seawater with Respect to Modern Influx Through Surface Runoff

Element	Concentration (ppm)	Residence time (years)
chlorine	18,980 (1.9%)	68,000,000
sodium	10,540 (1.0%)	100,000,000
magnesium	1,270	12,000,000
calcium	400	1,000,000
potassium	380	7,000,000
bromine	60	100,000,000
silicon	3.0	18,000
phosphorus	0.07	180,000
aluminum	0.01	100
iron	0.01	200
cadmium	0.00011	500,000
mercury	0.00003	80,000
lead	0.00003	400

Source: Data from J. E. Fergusson, *Inorganic Chemistry and the Earth.*
Copyright © 1982 Pergamon Press, Oxford, England.

in the environment. There is no long natural history to use for reference when estimating the capacities of different reservoirs for recently created compounds. Some compounds' residence times are limited by rapid breakdown into other chemicals, but breakdown rates for many are slow or unknown. So, too, is the ultimate fate of many synthetic chemicals. Uncertainties of this sort make evaluation of the seriousness of industrial pollution very difficult.

Where residence times of pollutants are known in particular systems, they indicate the length of time over which those pollutants will remain a problem in those reservoirs, and thus the rapidity with which the pollutants will be removed from that system. Suppose, for example, that a factory has been discharging a particular toxic compound into a lake. Some of the compound is drained out of the lake, by an outflowing stream. The residence time of the water, and thus of the toxic chemical in the lake, has been determined to be 20 years. If that is the average length of time the water and synthetic chemical remain in the lake, then if input of the toxin ceases, but flushing of water through the lake continues, then after one residence time (20 years), half of the chemical should have been removed from the lake; after a second 20 years, another half; and so on. One can then project, for instance, how long it would take for a safely low concentration to be reached. The calculations are more complex if additional processes are involved (if, for instance, some of the toxin decomposes, and some is removed into lake-bottom sediment), but the principles remain the same. However, substances removed from one system, unless broken down, will only be moved into another, where they may continue to pose a threat.

Point and Nonpoint Pollution Sources

Sources of pollution may be subdivided into point sources and nonpoint sources (figure 16.2). **Point sources,** as their name suggests, are sources from which pollutants are released at one readily identifiable spot: a sewer outlet, a steel mill, a septic tank, and so forth. **Nonpoint sources** are more diffuse; examples would include fertilizer runoff from farmland, acid drainage from an abandoned strip mine, or runoff of sodium or calcium chloride from road salts. The point sources are often easier to identify as potential pollution problems (figure 16.3). They are also easier to monitor systematically. It is a comparatively straightforward task to evaluate the quality of water from a single sewer outfall. Sampling runoff from a field in a representative way is more difficult.

The reverse problem—identifying the sources of water pollutants once a problem is recognized—can likewise be difficult. Sometimes, the source can be identified by the overall chemical "fingerprint" of the pollution, for rarely is a

Figure 16.2 Point sources (such as industrial plants and storm sewer outflow) and nonpoint sources (for example, fields and dispersed litter) of water pollution in a stream. Downward infiltration of contaminant-laden water from the nonpoint sources eventually also pollutes the ground water. Source: After USDA Soil Conservation Service.

Figure 16.3 Acid mine drainage at the surface is an obvious source of pollution.   Photograph courtesy of USGS Photo Library, Denver, CO.

Table 16.2 Principal Trace Metals in Industrial Wastewaters	
Industry	**Metals**
mining and ore processing	arsenic, beryllium, cadmium, lead, mercury, manganese, uranium, zinc
metallurgy, alloys	arsenic, beryllium, bismuth, cadmium, chromium, copper, lead, mercury, nickel, vanadium, zinc
chemical industry	arsenic, barium, cadmium, chromium, copper, lead, mercury, tin, uranium, vanadium, zinc
glass	arsenic, barium, lead, nickel
pulp and paper mills	chromium, copper, lead, mercury, nickel
textiles	arsenic, barium, cadmium, copper, lead, mercury, nickel
fertilizers	arsenic, cadmium, chromium, copper, lead, mercury, manganese, nickel, zinc
petroleum refining	arsenic, cadmium, chromium, copper, lead, nickel, vanadium, zinc

Source: Data from J. E. Fergusson, *Inorganic Chemistry and the Earth.* Copyright © 1982 Pergamon Press, Oxford, England.

single substance involved. Different pollutant sources—municipal sewage, a factory's wastewater, a farmer's particular type of fertilizer—are characterized by a specific mix of compounds, which may, in some cases, be identified in similar proportions in polluted water.

Industrial Pollution

Hundreds of new chemicals are created by industrial scientists each year. The rate at which new chemicals are developed makes it impossible to demonstrate the safety of the new chemicals as fast as they are invented. To "prove" a chemical safe, it would be necessary to test a wide range of doses on every major category of organism—including people—at several stages in the life cycle. That simply is not possible, in terms of time, money, or laboratory space. Tests are therefore usually made on a few representative populations of laboratory animals to which the chemical is believed most likely to be harmful, or perhaps on tissue cultures. Whether other organisms will turn out to be unexpectedly sensitive to the chemical will not be known until after they have been exposed to it in the environment. The completely safe alternative would be to stop developing new chemicals, of course, but that would deprive society of many important medicinal drugs, as well as materials to make life safer, more comfortable, or more pleasant. The magnitude of the mystery about the safety of industrial chemicals is emphasized by a recent National Research Council report. A committee compiled a list of nearly 66,000 drugs, pesticides, and other industrial chemicals and discovered that *no* toxicity data at all were available for 70 percent of them; a complete health hazard evaluation was possible for *only 2 percent*. In 1990, the American Chemical Society recorded the *ten millionth* entry in its registry of new chemicals created or

identified since 1957. Complete toxicity assessment for that many substances would obviously be a formidable task.

Inorganic Pollutants—Metals

Many of the inorganic industrial pollutants of particular concern are potentially toxic metals. Manufacturing, mining, and mineral-processing activities can all increase the influxes of these naturally occurring substances into the environment and locally increase concentrations from harmless to toxic levels. Examples of metal pollutants commonly released in wastewaters from various industries are shown in table 16.2.

The element mercury is a case in point. In nature, the small amounts of mercury (1 to 2 ppb) found in rocks and soil are weathered out into waters (where mercury is typically found in concentrations of a few ppb) or into the air (less than 1 ppb), from which it is quickly removed with precipitation. Mercury in water is in time removed into sediments. Due to its low solubility, its residence time in the oceans is about 80,000 years.

Mercury is one of the **heavy metals**, a group that includes lead, cadmium, plutonium, and others. A feature that the heavy metals have in common is that they tend to accumulate in the bodies of organisms that ingest them.

A

B

 Figure 16.4 (*A*) Schematic food chain in a body of water: Chemicals ingested by microorganisms are consumed by small fish, the larger fish eat the smaller fish, and so on up to birds and mammals higher up the chain. (*B*) If the chemicals are not readily excreted, they become more concentrated the higher one goes in the food chain. Accumulative toxins may become very highly concentrated in creatures at the top of the chain.

Therefore, their concentrations increase up a food chain (figure 16.4). Some marine algae may contain heavy metals at concentrations of up to one hundred times that of the water in which they are living. Small fish eating the algae develop higher concentrations of heavy metals in their flesh, larger fish who eat the smaller fish concentrate the metals still further, and so on up to fish-eating birds or mammals.

In most natural settings, heavy-metal accumulations in organisms are not very serious because the natural concentrations of these metals are low in waters and soils to begin with. The problems develop when human activities locally upset the natural cycle. Mining and processing heavy metals in particular can increase the rate at which the heavy metals weather out of rock into the environment, and these industries have also discharged concentrated doses of heavy metals into some bodies of water.

The toxic effects of heavy-metal poisoning are now well documented. In 1953, there was a major industrial discharge of mercury near Minamata, Japan. The Japanese rely heavily on fish in their diets, and fish concentrate mercury from the water. The fish at the top of that food chain contained up to fifty parts per *million* (not billion) mercury. By 1960, 43 people had died and 116 had been permanently affected by mercury poisoning.

Since then, officials occasionally have banned consumption of food fish relatively high in the food chain because the fish have been found to contain unacceptably high levels of mercury. Mercury acts on the central nervous system and the brain. It can cause loss of sight, feeling, and hearing, as well as nervousness, shakiness, and death. By the time the symptoms become apparent, the damage is irreversible. The effects of mercury poisoning may, in fact, have been recognized long before their cause. In earlier times, one common use of mercury was in the making of felt. Some scholars have suggested that the idiosyncrasies of the Mad Hatter of *Alice in Wonderland* may actually have been typical of hat makers of the period, whose frequent handling of felt led to chronic mercury poisoning, commonly known as "hatter's shakes."

Mercury has also been used as a fungicide. In Iraq, in 1971–72, severe famine drove peasants to eat mercury-treated seed grain instead of saving it to plant. Five hundred died, and an estimated seven thousand people were affected. (Incidentally, not all forms of mercury are equally deadly. It has positive medical applications in some compounds and doses. Also, the "silver" fillings that dentists put in teeth are really an amalgam of mercury and silver. It is principally as the compound methyl mercury that mercury is strongly toxic.)

Mercury is not the only toxic heavy metal. Lead poisoning is considered in chapter 17. Cadmium poisoning was most dramatically demonstrated in Japan when cadmium-rich mine wastes were dumped into the upper Zintsu River. The people used the water for irrigation of the rice fields and for domestic use. Many developed *itai-itai* (literally, "ouch-ouch") disease, cadmium poisoning characterized by abdominal pain, vomiting, and other unpleasant symptoms. Fear of the toxic-metal effects of plutonium is one concern of those worried about radioactive-waste disposal.

For some applications, substitutes for toxic metals can be found. Mercury is no longer used in the manufacture of felt, for instance. Certainly, now that the toxicity of these metals is realized, greater care can be taken with their disposal. Regular monitoring of water quality or of levels of potentially dangerous chemicals in fish or other organisms can also reveal contamination episodes before they have spread too far. Still, problems persist, for various reasons noted later.

Other Inorganic Pollutants

Some nonmetallic elements commonly used in industry are also potentially toxic to aquatic life, if not to humans. For example, chlorine is widely used to kill bacteria in municipal water and sewage treatment plants and to destroy various microorganisms that might otherwise foul the plumbing in power stations. Released in wastewater, it can also kill algae and harm fish populations.

Acids from industrial operations used to be a considerably greater pollution problem before stringent controls on their release were imposed. Acid mine drainage, however, remains a serious source of surface and groundwater pollution, especially in coal- and sulfide-mining areas.

The toxic effects of asbestos minerals were not manifested or well defined until long after their initial release into the environment by human activities. Asbestos minerals have been prized for decades for their fire-retardant properties and have been used in ceiling tiles and other building materials for years. Wastes from asbestos mining and processing were dumped into many bodies of water, including the Great Lakes—they were, after all, believed to be just inert, harmless mineral materials. By the time it was generally accepted that asbestos can be carcinogenic, asbestos workers had been exposed and segments of the public had been drinking asbestos-bearing waters for twenty years or more. The full impact of this sustained exposure is still being assessed. Growing concerns about possible dangers from asbestos and a proposed Environmental Protection Agency ban on certain asbestos products have continued to depress asbestos

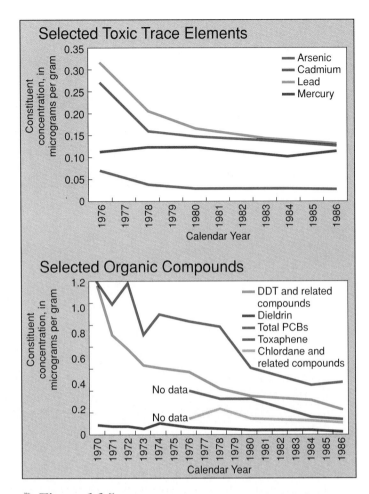

Figure 16.5 Decreases in concentrations of selected toxic elements (*A*) and organic compounds (*B*).   Source: After U.S. Geological Survey Water Supply Paper 2400.

production, but the residual effects of earlier production and disposal remain an ongoing problem, particularly for workers with high occupational exposure.

Unfortunately, the harmful effects of such inorganic pollutants are often slow to develop, which means that decades of testing might be required to reveal the dangers of such materials. The costs of such long-term testing are extremely high, generally prohibitively so.

Organic Compounds

The majority of new chemical compounds created each year are organic (carbon-containing) compounds. Many thousands of these compounds, naturally occurring and synthetic, are widely used as herbicides and pesticides, as well as in a variety of industrial processes. Their negative effects in organisms vary with the particular type of compound: Some are carcinogenic, some are directly toxic to humans or other organisms, and others make water unpalatable. Some also accumulate in organisms as the heavy metals do.

Oil spills, discussed in chapter 13, are one kind of organic-compound pollution. At least as much additional oil pollution occurs each year from the careless disposal of used

Table 16.3 Toxic Organic-Chemical Residues in Humans and Other Organisms*

	Year	DDT†	Dieldrin‡	PCBs§
Humans	1970	8.09	0.23	(not available)
	1973	6.09	0.22	79.8 ‖
	1976	4.68	0.15	98.1 ‖
	1979	3.14	0.11	98.2 ‖
Birds (Mississippi flyway)	1972	0.37	0.02	0.66
	1976	0.25	0.05	0.23
	1979	0.17	0.05	0.11
	1982	0.13	n.a.	0.14
Lake Michigan lake trout	1970	19.19	0.27	n.a.
	1973	9.96	0.27	18.93
	1976	5.65	0.30	18.68
	1980	4.74	0.34	9.93
	1984	2.20	0.38	4.48

Source: *Environmental Quality 1982*, pp. 287–289, Council on Environmental Quality, Washington, DC.

* Data are averages, expressed in ppm, based on animal weight, except in humans, in which the concentration is ppm in fatty tissues.

† Most uses banned 1972; see box 16.1 (data include DDT derivatives).

‡ A pesticide, banned 1974.

§ Production banned 1977.

‖ Percentage of tested population showing detectable levels.

crankcase oil, dumping of bilge from ships, and the runoff of oil from city streets during rainstorms. Underground tanks and pipelines may also leak, and drilling muds and waste brines (saline pore fluids) discarded in oil fields may be contaminated with petroleum. Oil spills into U.S. waters average over 10 million gallons per year.

Another type of organic-compound pollution is the result of the U.S. plastics industry's demand for production of nearly 7 billion pounds of vinyl chloride each year. Vinyl chloride vapors are carcinogenic, and it is not known how harmful traces of vinyl chloride in water may be.

Polychlorinated biphenyls (PCBs) were used for nearly twenty years as insulating fluid in electrical equipment and as plasticizers (compounds that help preserve flexibility in plastics). Laboratory tests revealed that PCBs in animals cause impaired reproduction, stomach and liver ailments, and other problems. PCB production in the United States was banned in 1977, but approximately 900 million pounds of PCBs had already been produced and some portion of that quantity released into the environment, where it remains.

Problems of Control

Even when the threat posed by a toxic agent is recognized and its production actually ceases, it may prove long-lived in the environment (figure 16.5 and table 16.3; see also box 16.1). Once released and dispersed, it may be impossible to clean up. The only course is to wait for its natural destruction.

Cost and efficiency are also factors in pollution control. Those substances not to be released with effluent water (or air) must be retained. This becomes progressively more difficult as complete cleaning of the effluent is approached. First, there is no wholly effective way to remove pollutants from wastewater before discharge. No chemical process is 100% efficient. Second, as cleaner and cleaner output is approached, costs skyrocket. If it costs $1 million to remove 90% of some pollutant from industrial wastewater, then removal of the next 9% (90% of the remaining 10% of the pollutant) costs an additional million dollars. To get the water 99.9% clean costs a total of $3 million, and so on. Some toxin always remains, and the costs escalate rapidly as "pure" water is approached. At what point is it no longer cost-effective to keep cleaning up? That depends on the toxicity of the pollutant and on the importance (financial or otherwise) of the process of which it is the product.

Furthermore, the toxins retained as a result of the cleaning process do not just disappear. They become part of the growing mass of industrial toxic waste requiring careful disposal. In 1984, U.S. industry generated more than 300 million tons of wastes classified as "hazardous" by the Environmental Protection Agency. In the same year, the U.S. Coast Guard reported 1,433 spills of hazardous wastes—3.8 million pounds of them—in and around U.S. waters.

Thermal Pollution

Thermal pollution, the release of excess or waste heat, is a by-product of the generation of power. The waste heat from automobile exhaust or heating systems contributes to thermal pollution of the atmosphere, but its magnitude is generally believed to be insignificant. Potentially more serious is the thermal pollution of water by electric generating plants and other industries using cooling water.

Box 16.1 DDT: Dream Cure to Disaster

The compound DDT (dichloro-diphenyl trichloroethane) was discovered to act as an insecticide during the 1930s. Its first wide use during World War II, killing lice, ticks, and malaria-bearing mosquitoes, unquestionably prevented a great deal of suffering and many deaths. In fact, the scientist who first recognized DDT's insecticidal effects, Paul Muller, was subsequently awarded the Nobel Prize in Medicine for his work. Farmers undertook wholesale sprayings with DDT to control pests in food crops. The chemical was regarded as a panacea for insect problems.

Then the complications began. One was that whole insect populations began to develop some resistance to DDT. Each time DDT was used, those individual insects with more natural resistance to its effects would survive in greater proportions than the population as a whole. The next generation would contain a higher proportion of insects with some inherited resistance, who would, in turn, survive the next spraying in greater numbers, and so on. This development of resistance resulted in the need for stronger and stronger applications of

DDT, which ultimately favored the development of ever-tougher insect populations that could withstand these higher doses. Insects' short breeding cycles made it possible for all of this to occur within a very few years.

Early on, DDT was also found to be quite toxic to tropical fish, which was not considered a big problem. Moreover, it was believed that DDT would break down fairly quickly in the environment. In fact, it did not and still does not. Out of 1 million tons of DDT that have been produced, an estimated two-thirds is still active. It is a persistent chemical in the natural environment. Also, like the heavy metals, DDT is an accumulative chemical. It is fat-soluble and builds up in the fatty tissues of humans and animals. Fish not killed by DDT nevertheless accumulated concentrated doses of it, which were passed on to fish-eating birds. Then another deadly effect of DDT was realized: It impairs calcium metabolism. In birds, this effect was manifested in the laying of eggs with very thin and fragile shells. Whole colonies of birds were wiped out, not because the adult birds died, but because not a single egg survived long

enough to hatch. Robins picked up the toxin from DDT-bearing worms. Whole species were put at risk. All of this prompted Rachel Carson to write *Silent Spring* in 1962, warning of the insecticide's long-term threat.

Finally, the volume of data demonstrating the toxicity of DDT to fish, birds, and valuable insects such as bees became so large that, in 1972, the U.S. Environmental Protection Agency banned its use—more or less. Use of DDT was still allowed on a few minor crops and in medical emergencies (to fight infestations of disease-carrying insects). Encouragingly, some wildlife populations have recovered since DDT use in the United States was sharply curtailed. Still, DDT continues to be applied extensively in other countries.

Should DDT not have been used? It has undoubtedly been a uniquely powerful tool for fighting insect pests transmitting serious diseases. Did the benefits outweigh the costs? Should the extent of DDT use have been more tightly restricted until its impacts were better understood? Questions such as these will continue to arise as long as scientists continue to invent new materials and new ways to use old ones.

Only part of the heat absorbed by cooling water can be extracted effectively and used constructively. The still-warm cooling water is returned to its source—most commonly a stream—and replaced by a fresh supply of cooler water. The resulting temperature increase usually exceeds normal seasonal fluctuations in the stream and results in a consistently higher temperature regime near the effluent source.

Fish and other cold-blooded organisms, including a variety of microorganisms, can survive only within certain temperature ranges. Excessively high temperatures may result in the wholesale destruction of organisms. More moderate increases can change the balance of organisms present. For example, green algae grow best at 30 to 35° C (86 to 95° F); blue-green algae, at 35 to 40° C (95 to 104° F). Higher temperatures thus favor the blue-green algae, which are a

poorer food source for fish and may be toxic to some. Many fish spawn in waters somewhat cooler than they can live in; a temperature rise of a few degrees might have no effect on existing adult fish populations but could seriously impair breeding and thus decrease future numbers. Furthermore, changes in water temperature change the rates of chemical reactions and critical chemical properties of the water, such as concentrations of dissolved gases.

Thermal pollution can be greatly reduced by holding water in cooling towers before release. (In practice, however, even where cooling towers are used, the released water is still at a somewhat elevated temperature.) The extent of thermal pollution is also restricted in both space and time. It adds no substances to the water that persist for long periods or that can be transported over long distances, and reduction in the

Figure 16.6 Concentration of cattle on a feedlot is a major source of organic waste. Photograph courtesy of USGS Photo Library, Denver, CO.

output of excess waste heat results in an immediate reduction in the magnitude of the continuing pollution problems.

Organic Matter

Nature and Impacts

Organic matter in general (as distinguished from the smaller subset of toxic organic compounds) includes a variety of materials, ranging from dead leaves settling in a stream to algae on a pond. Its most abundant and problematic form in the context of water pollution is human and animal wastes. Feedlots and other animal-husbandry activities create large concentrations of animal wastes (figure 16.6). Food-processing plants are other sources of large quantities of organic matter discharged in wastewater.

Sewage becomes mixed with a large quantity of wastewater not originally contaminated with it: Domestic wastewater is typically all channeled into one outlet, and in municipalities, domestic wastes, industrial wastes, and frequently even storm runoff collected in street drains all ultimately come together in the municipal sewer system. Moreover, sewage-treatment plants are commonly located beside a body of surface water, into which their treated water is discharged. To the extent that the wastewater is incompletely treated, the pollution of the surface water is increased.

Aside from its potential for spreading disease, organic matter creates another kind of water-pollution problem. In time, organic matter is broken down by microorganisms, especially bacteria. If there is ample oxygen in the water, the breakdown occurs **aerobically** (using oxygen), carried out by aerobic organisms. This depletes the dissolved oxygen supply in the water. Eventually, so much of the oxygen may be used up that further breakdown must proceed **anaerobically** (without oxy-

Figure 16.7 Effects of wastewater and organic matter on dissolved-oxygen content. (*A*) Typical oxygen sag curve. Below wastewater outflow, consumption of oxygen by decaying organic matter causes acute oxygen deficiency. Then, as organic matter is used up and simultaneous reoxygenation occurs, oxygen levels are gradually restored downstream. (*B*) Superimposed effects of two organic-matter sources: Each contributes a sag to the dissolved-oxygen curve.

gen). Anaerobic decay produces a variety of noxious gases, including hydrogen sulfide (H_2S), the toxic gas that smells like rotten eggs, and methane (CH_4), which is unhealthy and certainly of no practical use in a body of water. More importantly, anaerobic decay signals oxygen depletion. Animal life, including fish, requires oxygen to survive—no dissolved oxygen, no fish. When the dissolved oxygen is used up, the water is sometimes described as "dead," though anaerobic bacteria and some plant life may continue to live and even to thrive.

Biochemical Oxygen Demand

The organic-matter load in a body of water is described by a parameter known as **biochemical oxygen demand,** or BOD for short. The BOD of a system is a measure of the amount of oxygen required to break down the organic matter aerobically: the more organic matter, the higher the BOD. The BOD may far exceed the actual amount of dissolved oxygen in the water. As oxygen is depleted, more tends to dissolve in the water to restore chemical equilibrium, but, initially, this gradual reoxygenation lags behind oxygen consumption. Measurements of water chemistry along a stream below a sizable source of organic-waste matter characteristically define an **oxygen sag curve** (figure 16.7), a graph of

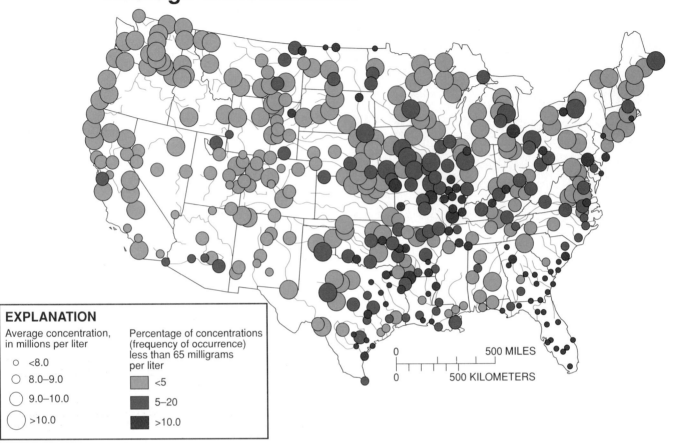

Dissolved Oxygen 1980–89 Average Concentration

EXPLANATION

Average concentration, in millions per liter

○ <8.0
○ 8.0–9.0
◯ 9.0–10.0
◯ >10.0

Percentage of concentrations (frequency of occurrence) less than 65 milligrams per liter

<5
5–20
>10.0

0 ——— 500 MILES
0 ——— 500 KILOMETERS

Figure 16.8 Dissolved-oxygen concentrations in U.S. surface waters in the 1980s. Low oxygen levels can be due to nutrient loading/eutrophication. Also, oxygen dissolves more readily in cold water than warm, accounting for depressed oxygen concentrations in the southeastern U.S.   Source: After U.S. Geological Survey Water Supply Paper 2400.

dissolved-oxygen content as a function of distance from the waste source that shows sharp depletion near the source, recovering downstream. Persistent oxygen depletion is more likely to occur in a body of standing water, such as a lake or reservoir, than in a fast-flowing stream. Flowing water is better mixed and circulated, and more of it is regularly brought into contact with the air, from which it can take up more oxygen. Climate also plays a role (figure 16.8).

Eutrophication

The breakdown of excess organic matter not only consumes oxygen, but it also releases a variety of compounds into the water, among them nitrates, phosphates, and sulfates. The nitrates and phosphates, in particular, are critical plant nutrients, and an abundance of them in the water strongly encourages the growth of plants, including algae. Excess fertilizer runoff from farmland can also put excessive nutrients into water. This development is known as **eutrophication** of the water; the water itself is then described as *eutrophic*. The exuberant algal growth that often accompanies or results

from eutrophication has also been called "algal bloom" and may appear as slimy green scum on the water (figure 16.9). Once such a condition develops, it in turn acts to continue to worsen the water quality (figure 16.10). Algal growth proceeds in the photic (well-lighted) zone near the water surface until the plants are killed by cold or crowded out by other plants. The dead plants sink to the bottom, where they, in turn, become part of the organic-matter load on the water, increasing the BOD and, as they decay, re-releasing nutrients into the water. Again, the consequences are most acute in very still water, where bottom waters do not readily circulate to the surface to take up oxygen. It should be noted that eutrophication can be a normal process in the development of a natural lake. Pollution, however, may accelerate or enhance it to the point that it becomes problematic.

Thorough waste treatment, as described in chapter 15, reduces the impact of organic matter, at least in terms of oxygen demand. However, as already noted, breakdown of the complex organic molecules releases into the water simpler compounds that may serve as nutrients. Human or

394

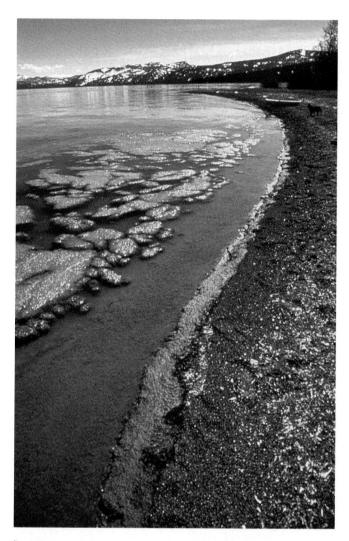

Figure 16.9 Algal bloom on Lake Tahoe, caused by an excess of nutrients in the water. In the past, Lake Tahoe was renowned for the exceptional clarity of its water.   Photograph by Belinda Rain, courtesy of EPA/National Archives.

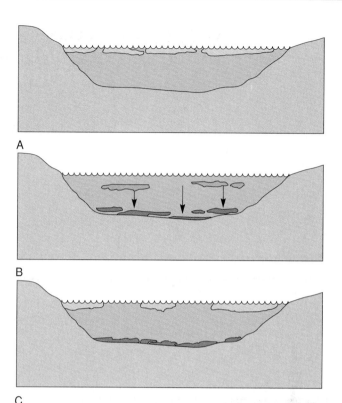

Figure 16.10 Schematic diagram showing effects of algal bloom on water quality. (*A*) Abundant growth of algae in sunlit shallow water when nutrients are abundant. (*B*) In colder weather, algae die and sink to the lake bottom. (*C*) The next growing season, more algae thrive at the surface while older material decays at the bottom, increasing BOD and releasing more nutrients to fuel new growth.

animal wastes are not the only agents in sewage that contribute to eutrophication, either. Phosphates from detergents are potential nutrients, too. The phosphates are added to detergents to enhance their cleaning ability, in part by softening the water. Realization of the harmful environmental effects of phosphates in sewage has led to restrictions on the amounts used in commercial detergents, but some phosphate component is still allowed in most places. Furthermore, even the best sewage treatment does not remove all of the dissolved nutrient (nitrate and phosphate) load.

Agricultural Pollution

Fertilizers

Some water-pollution problems resulting from agricultural activities are similar to ones already described. One is the problem of excess fertilizer runoff. The three principal con-

stituents of commercial fertilizers are nitrates, phosphates, and potash. When applied to or incorporated in the soil, they are not immediately taken up by plants. The compounds must be soluble for plants to use them, but that means that they can also dissolve in surface-water and groundwater bodies. These plant foods then contribute to eutrophication problems. The impact of fertilizer runoff is apparent from a recent study, in which, of eleven states reporting on the levels of eutrophication in their lakes, the states reporting the highest proportion of eutrophic lakes were farm-belt states. In the worst case, Iowa, 100 percent of the 107 lakes tested were eutrophic; in the next-worse case, Ohio, 84 percent of 119 lakes were eutrophic. There is a clear correlation with applications of fertilizer (figure 16.11).

Reduction in fertilizer applications to the minimum needed, perhaps coupled with the use of slow-release fertilizers, would minimize the harmful effects. Nitrates could alternatively be supplied by the periodic planting of legumes (a plant group that includes peas, beans, and clovers), on the roots of which grow bacteria that fix nitrogen in the soil. This practice reduces the need to apply soluble synthetic fertilizers.

Unfortunately, use of other natural fertilizers, like animal manures, fails to eliminate the water-quality problems associated with fertilizer runoff. Manures are, after all,

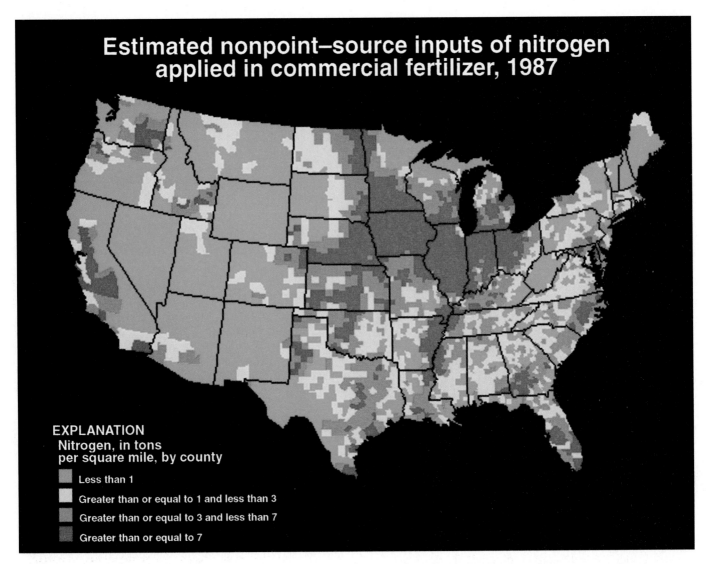

Estimated nonpoint–source inputs of nitrogen applied in commercial fertilizer, 1987

EXPLANATION
Nitrogen, in tons
per square mile, by county

- Less than 1
- Greater than or equal to 1 and less than 3
- Greater than or equal to 3 and less than 7
- Greater than or equal to 7

Figure 16.11   Nonpoint-source pollution from nitrogen in commercial fertilizer.   Source: After L. J. Puckett, U.S. Geological Survey National Water-Quality Assessment Program.

organic wastes. Thus, they contribute to the BOD of runoff waters in addition to adding nutrients that lead to eutrophication.

Likewise, runoff of wastes from domestic animals, especially from commercial feedlots, is a potential problem. One possible productive solution would be more extensive use of these wastes for the production of methane for fuel (see chapters 14 and 15). This has already been found to be economically practical where large numbers of animals are concentrated in one place.

Sediment Pollution

Some cases of high suspended-sediment load in water occur naturally (figure 16.12). However, these are commonly localized situations, and may be temporary as well. In many agricultural areas, sediment pollution of lakes and streams is the most serious water-quality problem, and is ongoing. Farmland and forest land together are believed to account

for about 75 percent of the 3 billion tons of sediment supplied annually to U.S. waterways. Of this total, the bulk of sediment is derived from farmland; sediment yields from forest land are typically low.

Sediment pollution not only causes water to be murky and unpleasant to look at, swim in, or drink, it reduces the light available to underwater plants, and blankets food supplies and the nests of fish, thus reducing fish and shellfish populations. Channels and reservoirs may be filled in, as noted in earlier chapters. The sediment also clogs water filters and damages power-generating equipment. Some sediment transport is a perfectly natural consequence of stream erosion, but agricultural development typically increases erosion rates by four to nine times unless agricultural practices are chosen carefully.

Chapter 11 examined some strategies for limiting soil loss from farmland. From a water-quality standpoint, an additional approach would be to use settling ponds below fields

A

B

Figure 16.12 (*A*) Sediment pollution in a stream. The stream at right has become clouded with sediment, probably because of bank collapse. Note the contrast with the clear tributary at left, and that the waters of the two streams do not immediately mix where they join. Yellowstone National Park. (*B*) Suspended glacial rock flour—another form of sediment pollution—contributes to the brilliant milky aqua of the water of Moraine Lake, Banff National Park, Alberta, Canada.

Surface runoff carries a suspended sediment load.

Sediment is dropped in still water of settling pond.

A

Outflow water is clear of suspended sediment.

B

Figure 16.13 Settling ponds enhance water quality and reduce soil loss. (*A*) Settling ponds can improve the quality of surface water by reducing sediment pollution and by trapping pollutants adsorbed onto the sediment particles. Removal of sediment from the water, however, may increase erosion below the settling pond. (*B*) Settling ponds below terraced farm fields trap eroded soil.   Photograph courtesy of USDA Soil Conservation Service.

subject to serious erosion by surface runoff (figure 16.13). These ponds do not stop the erosion, but by impounding the water, they cause the suspended soil to be dropped before the water is released to a lake or stream. Settling ponds can also be helpful below large construction projects, logging operations, or anywhere ground disturbance is accelerating erosion.

Herbicides and Pesticides

Many farmers are concerned that, if they adopt practices recommended to reduce soil erosion, such as shallower cultivation, they will merely have to increase their use of herbicides. Herbicides (chemical weed-killers) and pesticides (chemicals used to kill insect pests) are already a significant source of pollution in agriculture. Most such compounds now in use are complex organic compounds of the kinds described in the "Industrial Pollution" section earlier in the chapter and are in some measure toxic to humans or other life-forms. U.S. farmers use more than 500 million pounds of

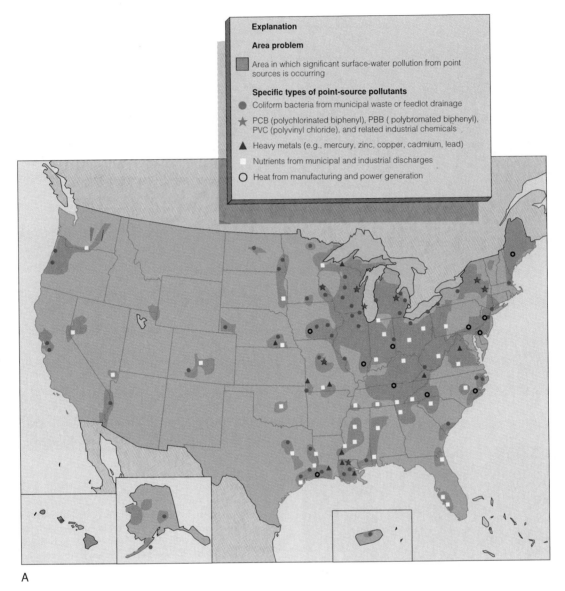

Explanation

Area problem

☐ Area in which significant surface-water pollution from point sources is occurring

Specific types of point-source pollutants

● Coliform bacteria from municipal waste or feedlot drainage

★ PCB (polychlorinated biphenyl), PBB (polybromated biphenyl), PVC (polyvinyl chloride), and related industrial chemicals

▲ Heavy metals (e.g., mercury, zinc, copper, cadmium, lead)

☐ Nutrients from municipal and industrial discharges

○ Heat from manufacturing and power generation

A

Figure 16.14 Surface-water pollution in the United States. (*A*) Point-source pollution. (*B*) Nonpoint-source pollution. (*C*) Eutrophication. In all cases, shaded areas represent sites of significant problems; symbols identify type(s).   Source: Modified from U.S. Water Resources Council, *The Nation's Water Resources 1975–2000*, vol. 1, pp. 61–63.

herbicides and pesticides each year, of which 90 percent are synthetic organic compounds. The use of these synthetic herbicides grew by over 40 percent just in the two-year period from 1979 to 1981.

The fiasco of DDT is just one instance of a much-heralded "miracle chemical" fallen out of favor (recall box 16.1). Over the years, a number of widely used agricultural chemicals have similarly been found to have toxic side effects on people, farm animals, and other organisms. Indeed, some are far *more* toxic than the now-banned DDT. Many are believed to break down rapidly and thus to have short residence times in the environment in general and food

in particular. Some have, however, proven quite persistent (recall, for example, table 16.3). As with many modern chemical creations, not enough is known about the long-term fate or effects of many of these compounds to be confident that herbicide and pesticide runoff is harmless, and dozens of such chemicals are in widespread use in agriculture (figure 16.14).

Agricultural chemicals themselves are not the sole problem. Many of the chemicals used in producing the end-use products are themselves toxic and persistent (dioxin, for instance); so are some of the early products of herbicide and pesticide decomposition, when and if they do begin to break down.

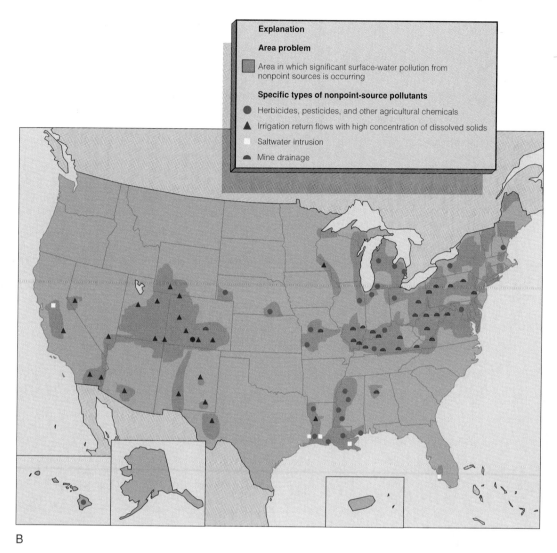

Explanation

Area problem

☐ Area in which significant surface-water pollution from nonpoint sources is occurring

Specific types of nonpoint-source pollutants

● Herbicides, pesticides, and other agricultural chemicals

▲ Irrigation return flows with high concentration of dissolved solids

☐ Saltwater intrusion

◓ Mine drainage

B

Figure 16.14—*Continued*

The potential risks could be reduced by eliminating routine applications of such chemicals as a sort of preventive medicine. Applying a remedy only when a problem is demonstrated to exist may initially reduce synthetic chemical use at the cost of crop yield. However, over the longer term, the results may be more economical, in part because the bugs or weeds will not as readily develop resistance to the substance being used.

A growing variety of nonchemical insect-reduction strategies are also finding increasing acceptance. Some are specific to a single pest. For instance, the larvae of a small wasp that is harmless to people and animals prey parasitically on tomato hornworms and kill them. A particular bacterium (*Bacillus thuringiensis*) attacks and destroys several types of borer insects, but again produces no known ill effects in humans, mammals, birds, or fish, and leaves no persistent chemical residues on food.

Other remedies may be more broadly applicable. An example is the practice of sterilizing a large population of a particular insect pest by irradiation and then releasing the sterilized insects into the fields. Assuming the usual number of matings, the next generation of pests should be smaller in number because some of the matings will have involved infertile insects. Other growers bait simple insect traps with scents carefully synthesized to simulate female insects, thereby luring the males to their doom. Procedures such as these offer promise for reducing the need for synthetic chemicals of uncertain or damaging environmental impact.

Reversing the Damage—Surface Water

All of the foregoing pollutants together have significantly affected surface waters over broad areas of the United States (figure 16.14). Reduction in pollutant input is one strategy

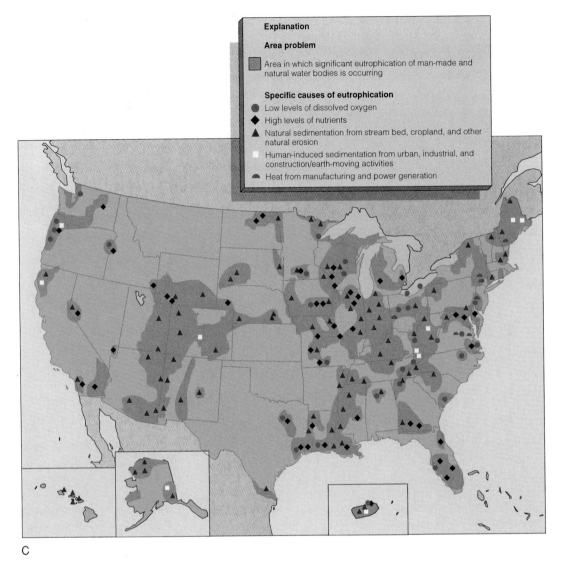

Explanation

Area problem

☐ Area in which significant eutrophication of man-made and natural water bodies is occurring

Specific causes of eutrophication

● Low levels of dissolved oxygen
◆ High levels of nutrients
▲ Natural sedimentation from stream bed, cropland, and other natural erosion
☐ Human-induced sedimentation from urban, industrial, and construction/earth-moving activities
⬬ Heat from manufacturing and power generation

C

Figure 16.14—*Continued*

for addressing such problems (as described in box 16.2), but it requires time to succeed, especially with relatively stagnant waters such as lakes. A variety of more aggressive treatment methods are available where indicated.

Dredging

Many water pollutants, including phosphates, toxic organics such as PCBs, and heavy metals, can become attached to the surfaces of fine sediment particles on a lake bottom. After a long period of pollutant accumulation, the bottom sediments may contain a large reserve of undesirable or toxic chemicals, which could, in principle, continue to be re-released into the water long after input of new pollutants ceased. Wholesale removal of the contaminated sediments by dredging takes the pollutants permanently out of the system. However, they still remain to be disposed of, commonly in landfills.

Dredging operations must be done carefully to minimize the amount of very fine material churned back into suspension in the water. Fine resuspended sediment, with its high surface-to-volume ratio, tends to contain the highest

concentrations of pollutants adsorbed onto grain surfaces. Also, if fine resuspended sediments remain suspended for some time, the increased water turbidity may be harmful to life in the lake. The process, moreover, can be expensive, with costs occasionally running over $10 per cubic meter of sediment extracted, but more commonly amounting to several dollars per cubic meter.

Dredging projects in the United States so far have tended to be small, usually involving less than 1 million cubic meters of sediment, and most are sufficiently recent that the long-term impacts are not known. However, the dredging of nutrient-laden sediments from Lake Trummen in Sweden during the early 1970s has been notably successful in reducing eutrophication by virtually eliminating nutrient recycling out of the sediments. In the near term, dredging may be used most in the United States for emergency cleanup of toxic wastes. When 250 gallons of PCBs spilled into the Dunwamish Waterway near Seattle in 1974, the EPA chose to dredge nearly four meters' thickness of contaminated sediment from the waterway

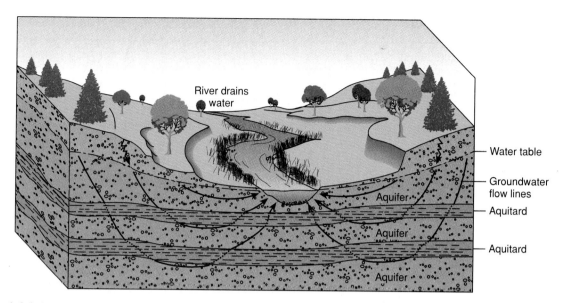

Figure 16.15 Groundwater flow rates may vary widely; in deep aquifers beneath multiple aquitards, centuries or more may elapse between infiltration in the recharge area, and extraction in a well.

and thereby succeeded in recovering 220 to 240 gallons of the spilled chemicals.

Physical Isolation or Chemical Treatment of Sediments

Another way to reduce the escape of contaminants from bottom sediments is to leave the sediments in place but to isolate them wholly or partially with a physical barrier. Over limited areas, like lagoons and small reservoirs, impermeable plastic liners have been placed on top of the sediments and held in place by an overlying layer of sand. The permanence of the treatment is questionable, and it has not been used on a large scale. Compacted clay layers of low permeability might, in principle, serve a similar purpose, but this treatment, to date, is a theoretical possibility only.

The addition of salts of aluminum, calcium, and iron to sediment changes the sediment chemistry, fixing phosphorous in the sediment and thereby reducing eutrophication. The technique has been used in small lakes in Wisconsin, Ohio, Minnesota, Washington, Oregon, and elsewhere. While not always successful, the method has often reduced phosphate levels in the water and corresponding algal growth. The cost compares favorably with the use of synthetic algicides to destroy the algae. However, careless overtreatment may prove toxic to fish, and, in any event, the treatment must be repeated every two to three years.

Decontamination; Aeration

Decontamination is most often used in response to toxic-waste spills. The specific treatment methods depend on the toxin involved. In 1974, a toxic organic-chemical herbicide washed into Clarksburg Pond in New Jersey from an adjacent parking lot. Many fish were killed, and local wildlife dependent on the pond for water were endangered. The contamination also threatened to spread to ground water by infiltration and to the Delaware River to which the pond water flows overland. The EPA set up an emergency water-cleaning operation. The herbicide was removed principally by activated charcoal through which the water was filtered. Subsequent tests showed the ground water to be uncontaminated, and within two years, fish were again plentiful in the pond.

Artificial aeration addresses oxygen depletion in a lake. There are several possible procedures, including bubbling air or oxygen up through the waters, thereby providing more oxygen to oxygen-starved deep waters, or simply circulating the water mechanically, cycling up deep waters to the surface where they can dissolve oxygen directly from the atmosphere. When successful, aeration can transform water from an anaerobic to an aerobic condition, to the great benefit of fish populations (a common reason for the attempt). However, when air rather than pure oxygen is bubbled through the water, an excess of dissolved nitrogen may develop in the water, which is toxic to some fish.

Groundwater Pollution

Groundwater pollution, whether from point or nonpoint sources, is especially insidious because it is not visible and often goes undetected for some time. Municipalities using well water must routinely test its quality. Homeowners relying on wells may be less anxious to go to the trouble or expense of testing, particularly if they are unaware of any potential danger. Also, in most instances, the passage of pollutants from their source into an aquifer used for drinking water is slow because that passage occurs by percolation of water through rock and soil, not by overland flow. There may therefore be a significant time lapse between the introduction of a pollutant into the system in one spot and its appearance in ground water elsewhere (figure 16.15). Conversely, groundwater pollution in

Box 16.2 "Who Killed Lake Erie?"

In 1969, NBC made a television documentary by the above title. Lake Erie was quite generally regarded as "dead": It was low in oxygen and high in pollution and algae. Its situation was a consequence partly of geology and partly of human use.

The lake is bordered by an especially large number of major cities (see figure 1). Each of these contributed both sewage and industrial wastes to the lake. Among the many industries in the region are the automotive manufacturers of Detroit, petrochemical and steel industries in Cleveland, glass and steel works in Toledo, paper mills in Erie, and chemical manufacturing in Buffalo. By the late 1960s, sewer systems serving about 9 million people on the U.S. side of the lake were discharging partially treated wastes into Lake Erie; another 2 million persons had septic-tank systems from which some wastes might have reached the lake.

Lake Erie's geometry and geography make it especially vulnerable to pollution. Although its area is large, the volume of water is low, especially in relation to the waste discharges. The lake is nearly 400 kilometers long and 80 kilometers wide, but its average depth is less than 20 meters. Some pollution was observed as early as the 1920s. Decreasing fish catches began to indicate that all was not well with the lake. Matters grew progressively worse until the early 1970s, when neighboring states were shocked into action by the lake's appalling condition.

Figure 1 Lake Erie and its heavily populated environs.

High-phosphate detergents were banned by several states, including Michigan, Indiana, and New York. Industrial waste dumping was brought under stricter control, and municipal sewage was more extensively treated before discharge to reduce phosphorous loading by nearly 50 million pounds.

It now appears that Lake Erie's obituary may have been premature. Airline pilots have observed a shrinking of the algal mats on the surface. Dissolved oxygen has been sufficiently restored that some game fish have been successfully restocked in the lake. Beaches closed for a decade or more have been reopened.

The cure is incomplete. Lake Erie drains through Niagara Falls; foaming water below the falls bespeaks residual pollution (figure 2A). Just downstream, sewer outfalls may contribute more pollution, sediment pollution from storm runoff at a minimum (figure 2B). The lake is still far from having the sparkling, pristine water it did when the retreating glaciers filled it twelve thousand years ago. That high quality of water is almost certainly irrecoverable. However, Lake Erie's condition is improving. Once a widely cited environmental disaster, Lake Erie is now acclaimed as a dramatic example of the kind of environmental recovery possible with concerted efforts.

A

B

Figure 2 Pollution below Niagara Falls, at the outflow of Lake Erie. (*A*) Foaming water below the falls. (*B*) Nearby sewer outfall just downstream.

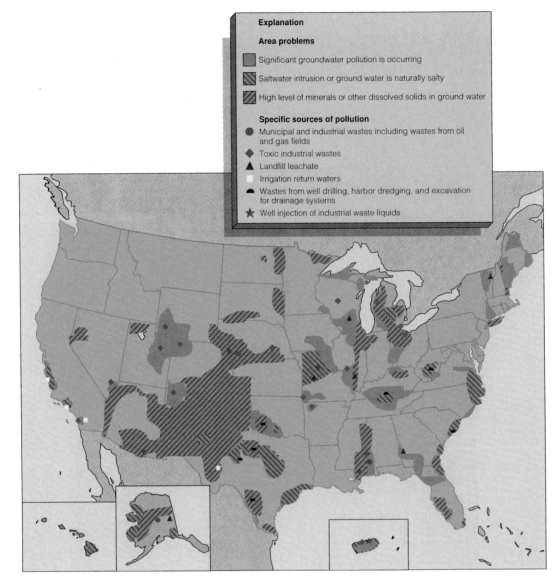

Figure 16.16 Groundwater pollution problems.   Source: Modified from U.S. Water Resources Council, *The Nation's Water Resources 1975–2000,* vol. 1, p. 65.

karst terranes, with their rapid drainage, may spread unexpectedly swiftly. The "out of sight, out of mind" aspect of ground water contributes to the problem; so does imperfect understanding of the interrelationships between surface and ground water.

Altogether, substantial groundwater-quality problems exist in the United States (figure 16.16)—some natural (for example, high dissolved mineral content), some due to human activities (pollution). The nature of the problems, and their causes, can vary greatly from place to place. Not surprisingly, pesticides (or their breakdown products) are common in ground water beneath agricultural areas, and more than 150 such compounds have been detected—though rarely in concentrations high enough to be felt

harmful. Nitrate concentrations, too, tend to be higher in ground water below agricultural areas than that under undeveloped areas (figure 16.17). On the other hand, pesticides and herbicides are not unique to agricultural regions; they may be used abundantly in residential areas, on golf courses and public gardens, and elsewhere. And where improvements in sewage treatment have been made to reduce ammonia concentrations, this has often been accomplished by converting the nitrogen in the ammonia to nitrate—thereby increasing nitrate concentrations in the water system. To some extent, susceptibility of ground water to nitrate pollution is a function of topography and soil character, not just total application of nitrate fertilizer (figure 16.18).

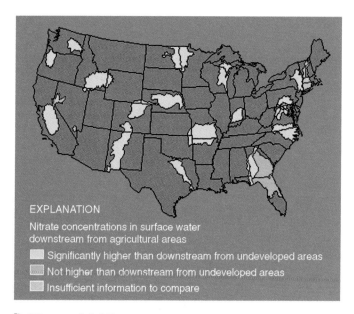

Figure 16.17 Nitrate concentrations in ground water are often higher under agricultural areas than under undeveloped ones.   Source: U.S. Geological Survey National Water Quality Assessment Program.

Groundwater pollution has sometimes appeared decades after the industry or activity responsible for it has ceased to operate and disappeared from sight and memory. For example, chemicals dumped or spilled into the soil long ago might not reach an aquifer for years. Even after the source has been realized, so large an area may have been contaminated that cleanup is impractical and/or prohibitively expensive. This is a problem with many old, abandoned toxic-waste dump sites. Groundwater pollution from nonpoint sources, like farmland, may also be so widespread that cleanup is not feasible. Some strategies for addressing groundwater contamination are explored later in this section. With ground water, however, it may be especially important to avoid or limit pollution initially, because it is typically more difficult to clean up afterward than is polluted surface water. So, for example, there is pressure to replace underground fuel storage tanks of metal (which may corrode, and leak) with tanks of more corrosion-resistant materials, such as plastic or fiberglass.

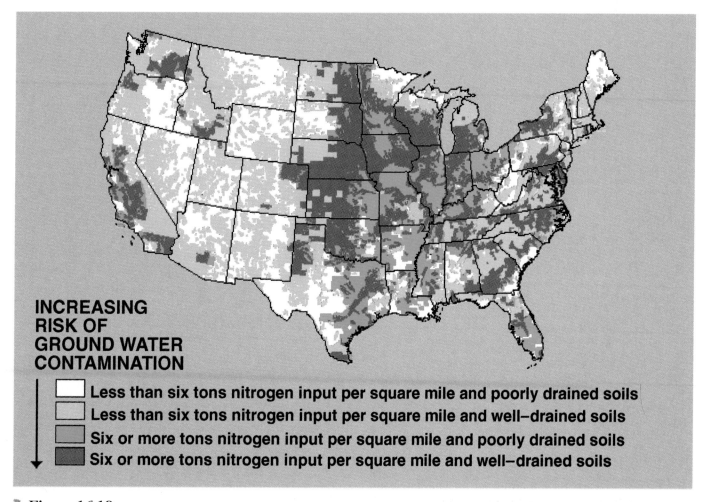

Figure 16.18 Susceptibility of ground water to nitrate pollution are related not only to nitrogen applied in fertilizer but to speed of drainage; "well-drained" soils are permeable soils through which water infiltrates readily.   Source: B. T. Nelson and B. C. Buddy, 1996, Fact Sheet FS-092-96 of the U.S. Geological Survey National Water Quality Assessment Program.

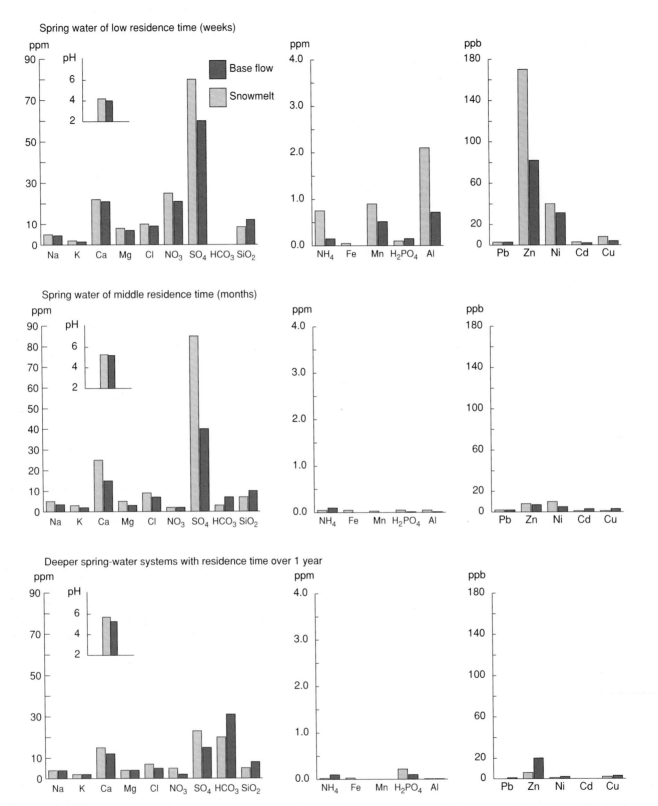

Figure 16.19 Concentrations of chemical constituents in springs of short, medium, and long water residence times. (Darker bars represent base flow; lighter bars, snowmelt.)   Redrawn from *Geology and the Environment in Western Europe*, ed. by G. I. Lumsden. Copyright © 1992 Oxford University Press, Oxford, England.

Pollution avoided is pollution that does not have to be treated. What is perhaps less obvious is the extent to which surface-water pollution may result in groundwater pollution too, given the interrelationships among water reservoirs in the hydrologic cycle.

The Surface–Ground Water Connection Explored

Given the nature of groundwater recharge, the process can readily introduce soluble pollutants along with recharge water. For example, agricultural pesticides, after all, are applied at the surface. But those pesticides neither used up nor broken down may remain dissolved in infiltrating water, resulting in pesticide contamination of ground water, as previously described. Acid mine drainage, storm-sewer runoff, and other surface-water pollutants are likewise potential groundwater pollutants.

Acid rain can be a particular problem, because many toxic metals and other substances are more soluble in more acidic water. How severely groundwater chemistry is affected is related, in part, to the residence time of the ground water in its aquifer before it is extracted through a well or spring. That is, if water residence time is short, as with the shallow aquifers in figure 16.15, there is less opportunity for chemical reactions in the aquifer to buffer or moderate the impacts of acid rain. Longer residence times may allow more moderation of the chemistry of the acid waters. This is well illustrated by German studies of impacts of acid rain (figure 16.19), which is locally a significant problem due to heavy reliance on the abundant German coal resources. Especially after major snowmelt events, there is rapid influx of acidified precipitation into groundwater systems. Springs fed by ground waters with short water residence times show not only high levels of sulfate (SO_4) and nitrate (NO_3) from sulfuric and nitric acid, but also elevated concentrations of a number of metals, due to solution in the acidic waters. Over time, the acidic waters can react, especially with aquifers containing dissolved carbonate; the acidity is reduced, along with the concentrations of many other dissolved constituents, while dissolved carbonate (shown as bicarbonate ion, HCO_3) increases. The chemical impact on springs fed by ground waters with intermediate (months) or long (years) water residence times in the aquifers is correspondingly much less.

Bearing in mind the potential of groundwater pollution from surface-water infiltration, one can also assess the vulnerability of a groundwater system to such pollution in terms of how well protected it is (table 16.4). Such an assessment may in turn suggest how closely the groundwater system should be monitored. For the sources of surface-water pollution are many. In addition to those already cited, there are a variety of casual pollution sources related to improper disposal: crankcase oil dumped into a storm sewer

Table 16.4   Relative Vulnerability to Pollution, by Hydrologic Setting

Vulnerability	Types of setting (examples)
Extreme	1. Bedrock aquifers crop out (especially in karst areas); or bedrock aquifers overlain by less than 3 meters of soil.
	2. Unconfined sand and gravel aquifers with unsaturated zone less than 3 meters thick
High	1. Bedrock aquifers overlain by 3^+ meters sand and gravel, 3–10 meters sandy till, or 3–5 meters clay or clay-rich till
	2. Unconfined sand and gravel aquifers with unsaturated zone 3^+ meters thick
Moderate	Unconfined aquifers overlain by 10^+ meters sandy till, or 5–10 meters clay-rich till, clay, or peat
Low	Confined aquifers overlain by 10^+ meters clay-rich till or low-permeability rock such as shale

Source: Data from *Geology and the Environment of Western Europe*, ed. by George I. Lumsden. Copyright © 1992 Oxford University Press, Oxford, England.

by a home mechanic, solvents and cleaning fluids poured down the drain by homeowners or dry cleaners or painters or furniture refinishers, oil that was deposited on roads by traffic flushed off and running into storm sewers in a heavy rain, even chemicals from a college science laboratory thoughtlessly flushed down a laboratory drain—all these, many simply reflecting individual carelessness, may pollute ground water.

Reversing the Damage— Ground Water

Earlier sections of this chapter noted various means of reducing the flow of pollutants into water. The approach of reducing or stopping the input of further pollutants, then waiting for natural processes to remove or destroy the pollutants already in the system, is often the *only* approach technically or economically possible in the case of contaminated ground water, which is relatively inaccessible. Systematic monitoring of ground water can be difficult without an extensive

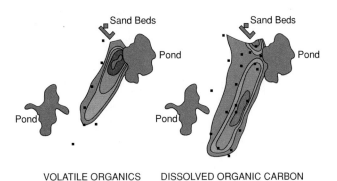

VOLATILE ORGANICS
Micrograms per Liter

DISSOLVED ORGANIC CARBON
Milligrams per Liter

Figure 16.20 A groundwater contaminant plume from a sewage treatment plant at Otis Air Force Base, Massachusetts, ca. 1984. Top diagram: Concentrations of nitrate and ammonia (nitrogen) and detergents in a longitudinal section of the plume; water flow is to the right. Bottom: Concentrations of volatile organics and dissolved organic carbon; water flow is toward lower left. Source: U.S. Geological Survey.

(and usually expensive) network of wells because, by the time the pollution problems are recognized, the contaminants have typically spread widely in the aquifer system (see, for example, figure 16.20 and box 16.3). The slow migration and limited mixing of most ground waters complicates *in situ* treatment of the problems; bear in mind the way ground water occurs, in cracks and pores dispersed throughout a large volume of rock. Most commonly, then, polluted ground water is only treated after it is extracted for use.

In Situ Decontamination

Treatment of contaminated ground water in place is possible only when the extent and nature of the pollution are well defined; the methods used are very site- and pollutant-specific.

For inorganic pollutants, such as heavy metals, immobilization is one strategy. Injection wells placed within the contaminated zone are used to add chemicals that cause the toxic substances to precipitate and thus to become stationary. The uncontaminated water can then be extracted for use.

Biological decomposition is effective on a broad range of organic compounds, many of which can be broken down by microorganisms. The process can be stimulated, in individual cases, by addition of oxygen or nutrients to accelerate growth of the microorganisms; additional microorganisms (indigenous or foreign) can be introduced to attack the particular compound(s). Not all organics can be eliminated in this way, and occasionally, objectionable residues affect water taste and odor. However, most of the twenty-four "priority organic pollutants" identified by the Environmental Protection Agency can be successfully attacked by biological means. With petroleum products that do not mix well with water, but tend to float on it (as with natural oil and gas deposits), wells that reach barely to water-table level may selectively pump out contaminant-rich fluid, leaving much purer water behind.

Decontamination after Extraction

Inorganic compounds can be removed from extracted ground water by adjusting its acidity (pH). Addition of alkalies may result in the precipitation of many heavy or toxic metals as hydroxides. Many of the same metals may be precipitated as sulfides, which are relatively insoluble, or as carbonates. All of these strategies yield a solid sludge (the precipitate) that contains the toxic metals and must still be disposed of.

Just as microorganisms can be used to treat contamination by organic toxins *in situ*, microbial activity can break down organisms in the water after it is extracted. The organisms can be mixed in bulk with the water; after treatment, the biomass must then be settled or filtered out. Alternatively, the organisms can be fixed on a solid substrate, and the contaminated water passed over them.

Air stripping encompasses a set of methods by which volatile organic pollutants are transferred from water into air and thus removed from the water. The mechanical details of the process vary, but each case involves some form of aeration of the water, followed by separation of the gas. The pollutants are still present in the gas phase and must be removed for alternate disposal.

Activated charcoal (activated carbon), used in a variety of filters, adsorbs organic compounds dissolved in ground water. Again, the compounds themselves remain intact and require disposal.

In short, treatment methods available *after* ground water is extracted are many, and span the same broad range as methods of treating surface-water pollution or purifying a municipal water supply. But as long as the polluted ground water remains in the ground, unless it is contained (which is not common, because of costs), the contaminants can go right on spreading, carried with the water flow or diffusing through the fluid.

Box 16.3 Groundwater Pollution and Control, Rocky Mountain Arsenal

One human activity can have multiple environmental impacts. We noted in chapter 4 that deep-well disposal of liquid waste at the Rocky Mountain Arsenal, near Denver, Colorado, was identified as the cause of a series of small earthquakes in the area. Careless surface disposal of toxic wastes at the same site also resulted in contamination of local ground water. Beginning in 1943, wastewater containing a variety of organic and inorganic chemicals was discharged into unlined surface ponds. The principal contaminants were the organics diisomethylphosphonate and dichloropentadiene.

The water table in the shallow aquifer in the area is locally only 2 to 5 meters below the ground surface. Water from this aquifer is widely used for irrigation and for watering livestock. Incidents of severe crop damage were reported in the early 1950s. To limit further contamination, an asphalt-lined disposal pond was constructed in 1956, but, in time, the liner leaked. Contaminants already in the groundwater system continued to spread. When complaints of renewed damage to crops and livestock were made in the early 1970s, the Colorado Department of Health investigated. They detected a variety of toxic organic compounds, some at concentrations of tens of parts per million, in water and soil on the arsenal property; some of these compounds were present, though in much lower concentrations, in water drawn from municipal supply wells off the site. The Department of Health ordered a cessation of the waste discharges and cleanup of the existing pollution, with provision to prevent further discharge of pollutants from the arsenal property.

Because groundwater contamination was already so widespread, the

Figure 1 Sketch of boundary control system for contaminated ground water at Rocky Mountain Arsenal.

only feasible approach to containing the further spread of pollutants was to halt them at the arsenal boundaries. This effort was combined with cleanup activity (figure 1). A physical barrier of clay-rich material was placed along the arsenal side of the boundary at several locations on the north and west sides, where the natural groundwater flow was outward from the arsenal. On the arsenal side of the barriers, wells extract the contaminated ground water. It is treated by a variety of processes, including filtration, chemical oxidation, air stripping, and passage through activated charcoal. The cleaned water is then reintroduced into the aquifer on the outward side of the barrier. The net groundwater flow

regime outside of the arsenal has been only minimally disrupted, and pollutants in nearby water supplies have been greatly reduced.

The efforts have not been inexpensive. By 1984, over $25 million had been spent on the control and decontamination activities, and the boundary barrier system will have to continue in operation indefinitely as the slowly migrating contaminants in the ground water continue to move toward the limits of the site. These activities illustrate both the decontamination and control success that can be achieved in selected cases, and the value (practical and economic) of appropriate planning to prevent contamination in the first place.

Overview

Summary

A variety of substances, some naturally occurring and some added by human activities, cycles through the hydrosphere. How long each substance spends in a particular reservoir is described by the residence time of that substance in the reservoir. The materials of greatest concern in the context of pollution are usually toxic ones with long residence times in the environment. Little is known about many synthetic chemicals—their movements, longevity in natural systems, and often even the extent of their toxicity.

Sediment pollution is a problem particularly in agricultural areas, where better erosion control could greatly reduce it. Organic matter, whether contributed by domestic sewage, agriculture, or industry, increases the biochemical oxygen demand (BOD) of the water, making it inhospitable to oxygen-breathing organisms. Where the water is rich in plant nutrients from organic wastes, fertilizer runoff, or other sources, such as phosphate from detergents, a eutrophic condition develops that encourages undesirably vigorous growth of algae and other plant life.

In addition to adding many new and sometimes-toxic chemicals to the environment, industrial activities can unbalance the natural cycles of other harmful substances, such as the heavy metals. These elements share with some other chemicals the tendency to accumulate in organisms and increase in concentration up the food chain, becoming a greater hazard higher up in the chain. Power plants may present a thermal, rather than a chemical, pollution threat.

Reduction in pollutant release into aqueous systems is the most common strategy for reducing pollution problems. As a general rule, pollutants from point sources are more easily controlled or confined than are pollutants from nonpoint sources. More aggressive physical and/or chemical treatments are sometimes used in water-quality restoration efforts, particularly in the case of small, still water bodies, or to combat toxic-chemical spills. However, even if harmful materials are contained rather than released into the environment, or are removed from water or underlying sediment, they still pose a waste-disposal problem. Groundwater pollution is a particular challenge, not only because it is often less readily detected, but because the nature of ground water makes it harder to treat *in situ.*

Terms to Remember

aerobic 393	heavy metals 388
anaerobic 393	nonpoint source 387
biochemical oxygen	oxygen sag curve 393
demand 393	point source 387
eutrophication 394	residence time 385

Exercises

Questions for Review

1. Explain the concept of residence time; illustrate it with respect to some dissolved constituent in seawater.
2. In what way do human activities most commonly alter the cycles of naturally occurring elements?
3. What characteristic of the so-called heavy metals causes them to be especially hazardous to humans and other animals high in food chains?
4. Why do potentially harmful health effects of organic compounds constitute a major area of concern?
5. What is BOD? How is it related to the oxygen sag curve often noted in streams below sources of organic-waste matter?
6. What is thermal pollution? From what activity does it principally originate? In what sense is it a less-worrisome kind of pollution than most types of chemical pollution?
7. Cite at least three possible sources of the nutrients that contribute to eutrophication of water; explain the concept. Why are eutrophic conditions generally considered undesirable?
8. What kinds of pollution can be reduced by the use of settling ponds? Explain.
9. Briefly describe two after-the-fact approaches to reducing water pollution, and note their limitations.

For Further Thought

1. As the largest freshwater lakes in the United States, the Great Lakes receive considerable attention and study. Find out the current status of pollution in Lake Erie, and compare it with any of the other Great Lakes.
2. Investigate what becomes of a local industry's wastewater. How is it treated before release? Where is it released? What pollutants remain and in what quantities?
3. The U.S. Environmental Protection Agency's Office of Water produces state fact sheets accessible online at http://www.epa.gov/OWOW/sec1/states.html
   Choose a state, investigate the status of its water quality, then use library or online resources to identify the causes of any pollution problems you identify.

Suggested Readings/References

Adler, T. 1996. The expiration of respiration. *Science News* 149 (10 February): 88–89. (Deals with oxygen depletion in water.)

Adriano, D. C. 1986. *Trace elements in the terrestrial environment.* New York: Springer-Verlag.

Brown, K. W. 1994. New horizons in soil remediation. *Geotimes* (September): 15–17.

Canter, L. W., and Knox, R. C. 1985. *Groundwater pollution control.* Chelsea, Mich.: Lewis.

Cleva, S. 1995. Let them eat cyanide. *Geotimes* (July): 5. (Reports on tests of bioremediation at sites of old mining/mineral-processing operations.)

Dunne, T., and Leopold, L. B. 1978. *Water in environmental planning.* San Francisco: W. H. Freeman.

Environmental Protection Agency. 1990. Saving the nation's great water bodies. *EPA Journal* 16 (November/December). (This whole issue is devoted to pollution and other water-related problems.)

Fergusson, J. E. 1982. *Inorganic chemistry and the earth.* New York: Pergamon Press.

Garrels, R. M., Mackenzie, F. T., and Hunt, C. A. 1975. *Chemical cycles and the global environment.* Los Altos, Calif.: William Kaufmann.

Gore, J. A. 1985. *The restoration of rivers and streams.* Ann Arbor, Mich.: Ann Arbor Science Publishers.

Griffin, R., Jr. 1991. Introducing NPS water pollution. *EPA Journal* 17 (November/December): 6–9.

Groundwater contamination. 1984. Washington, D.C.: National Academy Press.

Hodges, L. 1977. *Environmental pollution.* 2d ed. New York: Holt, Rinehart and Winston.

Lippmann, M., and Schlesinger, R. B. 1979. *Chemical contamination in the human environment.* New York: Oxford University Press.

Lumsden, G. I., ed. 1992. *Geology and the environment in Western Europe.* Oxford: Clarendon Press. (Section 5.5 is devoted to principles and case studies of groundwater-pollution problems.)

Molnia, B. F. ed. 1993. Ground-water cleanup vs. ground-water protection—where should the $$$ go? (Forum). *GSA Today* (July): 183–94.

Rinella, J. F., Hamilton, P. A., and McKenzie, S. W. 1993. Persistence of the DDT pesticide in the Yakima River Basin, Washington. U.S. Geological Survey Circular 1090.

Roush, W. 1995. Building a wall against toxic waste. *Science* 269 (28 July): 473.

Smith, R. A., Alexander, R. B., and Wolman, M. G. 1987. Water-quality trends in the nation's rivers. *Science* 235: 1607–15.

U.S. Council on Environmental Quality. 1984. *Environmental quality 1983.* Washington, D.C.: U.S. Council on Environmental Quality.

———. 1988. *Environmental quality 1986.* Washington, D.C.: U.S. Council on Environmental Quality.

U.S. Environmental Protection Agency. 1979. *Lake restoration.* Washington, D.C.: U.S. Environmental Protection Agency.

———. 1980. *National accomplishments in pollution control 1970–1980.* Washington, D.C.: U.S. Environmental Protection Agency.

———. 1983. *National water quality inventory 1982.* Washington, D.C.: U.S. Environmental Protection Agency.

A comprehensive (and continually expanding) source of water-quality data for the U.S.—through which a number of water-quality reports by individual investigators can be accessed—is collectively identified as the National Water Quality Assessment Program of the U.S. Geological Survey.
http://wwwrvares.er.usgs.gov/nawqa/nawqa_home.html
One of several available papers especially relevant to this chapter is "Stream-water quality in the conterminous U.S.—Status and trends of selected indicators during the 1980s," USGS Water Supply Paper 2400 by R. A. Smith et al., found at
http://h2o.usgs.gov/public/nwsum/sal/index.html
Another is a collection of data on pesticides in ground water at
http://water.wr.usgs.gov/pnsp/index.html

The "Issues and Topics" page of Environment Canada offers links to information on such topics as water quality and pollution, including selected environmental indicators; at
http://www.doe.ca/isseng.html
Specific information on pesticides and alternatives to them is found at
http://www.doe.ca/ENGLISH/POLLUTION/PESTICIDES/

The Environmental Protection Agency is responsible for the National Water Quality Inventory and the Report to Congress of the results. The latest as of this writing, "The Quality of Our Nation's Water 1994" can be accessed via
http://earth1.epa.gov/305b/execsum.html
(There are many links to more specialized reports from the summary.)

The Great Lakes Environmental Research Laboratories of the National Oceanic and Atmospheric Administration are at
http://www.glerl.noaa.gov/

A variety of water-related information is available from the U.S. Geological Survey's Water Resources Division; see
http://h2o.usgs.gov/

Chapter 17

Air Pollution

Donora, Pennsylvania, 1948: "The fog closed over Donora on the morning of Tuesday, October 26th. . . . By Thursday, it had stiffened . . . into a motionless clot of smoke. That afternoon, it was just possible to see across the street . . . the air began to have a sickening smell, almost a taste . . ." (Roueché, 1953, p. 175). By Friday, residents with asthma and other lung disorders began to find it difficult to breathe. Then more of the residents took sick, becoming nauseous, coughing and choking, suffering from headaches and abdominal pains. The fog persisted for five days. Altogether, nearly six thousand people were stricken by the polluted fog; twenty of them died.

Donora was a mill town with a steel plant, a wire plant, and a zinc and sulfuric acid plant among its industries. The litany of toxic chemicals in its air identified during later investigations included fluoride, chloride, hydrogen sulfide, sulfur dioxide, and cadmium oxide, along with soot and ash. These materials were all routinely released into the air, but not generally with such devastating effects.

The events of late October 1948 in Donora might be called an acute air-pollution episode—sudden, obvious, and dramatic. Other still more serious single episodes are known. For four days in early December 1952, weather conditions trapped high concentrations of smoke and sulfur gases from coal burning in homes and factories over London, England; an estimated thirty-five hundred to four thousand people died in consequence, with many more made ill. Acute pollution events of this kind are, fortunately, rare. The health impact of moderately elevated levels of pollutants found widely over the earth, especially near urban areas, is more difficult to assess.

This brilliant sunset over Great Salt Lake results from the natural air pollution of the smoke from nearby forest fires. © Doug Sherman/Geofile

Atmospheric Chemistry— Cycles and Residence Times

The atmosphere consists of three principal elements: Nitrogen comprises over 76% of the total; oxygen, about 23%; the inert gas argon, close to 1%. Everything else in the atmosphere together makes up much less than 1% of it.

Materials cycle through the atmosphere as they do through other natural reservoirs. One can speak of the residence times of gases or particles in the atmosphere just as one can discuss residence times of chemicals in the ocean (table 17.1). Oxygen, for example, is added to the atmosphere during photosynthesis by plants; it is removed by oxygen-breathing organisms, by solution in the oceans, by reaction with rocks during weathering, and by combustion. Its residence time in the atmosphere is estimated at 7 million years.

Carbon dioxide (CO_2) has an even more complex cycle (figure 17.1). It is added to the atmosphere by volcanic eruptions and as a product of respiration and combustion, and it is removed during photosynthesis and by solution in the oceans. It is further removed from the oceans by precipitation in carbonate sediments. The concentration of carbon dioxide in the atmosphere is about 350 ppm, far less than that of oxygen, and carbon dioxide has a correspondingly much shorter residence time, estimated at as little as four years. Some of the consequences of increased CO_2 content in the atmosphere were explored in chapter 9. One of the factors complicating projections of the greenhouse effect relates to imprecise knowledge of the carbon fluxes between pairs of reservoirs shown in figure 17.1. For example, consider the difficulty of estimating precisely the amount of CO_2 consumed by plants worldwide during photosynthesis. What might seem a far simpler problem, estimating net CO_2 flux between atmosphere and oceans, is not so simple after all, for the extent to which CO_2 dissolves in the oceans varies with temperature, and air and water temperatures in turn vary both regionally and seasonally. Still, the broad outline of the global carbon cycle, in terms of reservoirs and processes, is reasonably well understood, as shown.

The chemically inert gases, including argon and trace amounts of helium and neon, have virtually infinite residence times: they do not, by definition, react chemically with other substances, so they are not readily removed by natural processes. The geochemical cycle of nitrogen is so complex that its overall residence time in the atmosphere is not known, although residence times of individual compounds have been estimated (see table 17.1). As with many synthetic water pollutants, the fate and residence times of anthropogenic air pollutants, those added by human activity, often are also poorly known. The chlorofluorocarbon compounds, discussed later in this chapter, are just one case in point.

Table 17.1 Abundances and Residence Times of Some Elements and Gases in the Atmosphere

Substance	Average concentration (by weight)	Estimated residence time
nitrogen		
molecular (N_2)	76.6%	44 million years
ammonia (NH_3)	6 ppb	3–4 months
nitrous oxide (N_2O)	250 ppb	12–13 years
nitrogen dioxide (NO_2)	2 ppb	1–2 months
nitric acid (HNO_3)	unknown; low	2–3 weeks
oxygen (as O_2)	23.1%	7 million years
carbon		
methane (CH_4)	1.6 ppm	3.6 years
carbon dioxide (CO_2)	350 ppm	4 years
carbon monoxide (CO)	0.1 ppm	1–2 months
sulfur		
sulfur dioxide (SO_2)	0.2 ppb	hours or days
hydrogen sulfide (H_2S)	0.2 ppb	hours
sulfuric acid (H_2SO_4)	1 ppb	several days
mercury	0.001 ppb	60 days
lead	0.003 ppb	2 weeks

Sources: Data from R. M. Garrels, F. T. Mackenzie, and C. Hunt, *Chemical Cycles and the Global Environment,* copyright © 1975 William Kaufmann Inc., Los Altos, CA; U.S. Environmental Protection Agency data reported in *Environmental Quality 1985,* p. 325, 1984, Council on Environmental Quality, Washington, DC.

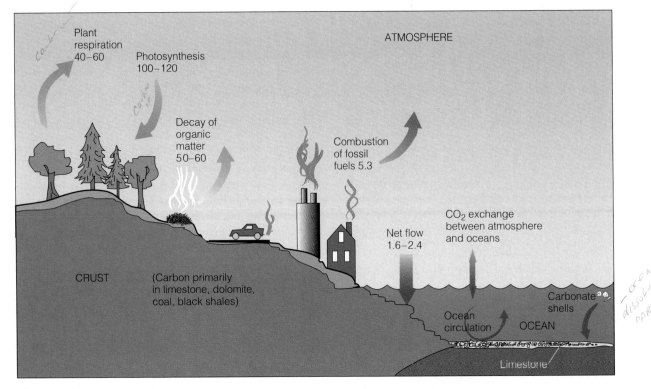

Figure 17.1 Reservoirs and fluxes in the global carbon cycle. There are also some flux changes resulting from changes in land-use practices (e.g., clearing of forests). Note relatively large uncertainties in some of the major fluxes. Numbers are in billions of tons.
Source: Data from Post, et al., 1990.

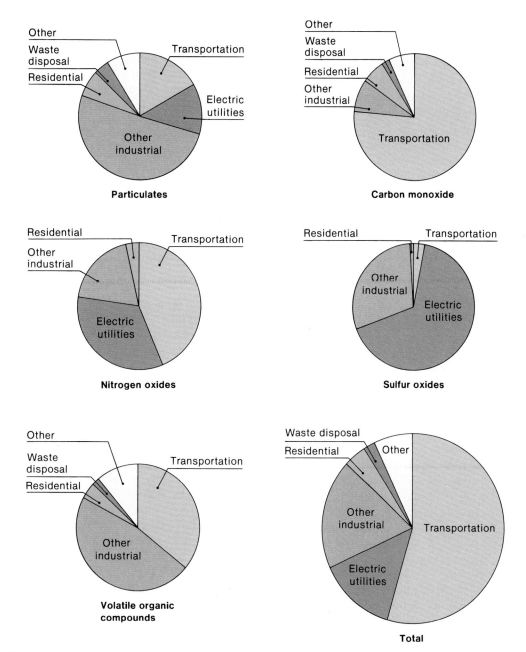

Figure 17.2 Principal sources of U.S. air pollutants.   Source: U.S. Environmental Protection Agency.

Costs of Air Pollution

Air pollution is costly, and not only in terms of health. The Council on Environmental Quality has estimated *direct* costs at $16 billion per year in the United States alone, including over $1 billion just for cleaning soiled items and $500 million in damage to crops and livestock. Worldwide, estimated costs of forest-product harvest reduction due to air pollution exceed $40 billion per year.

In addition, there are sizable costs in illness, medical expenses, absenteeism, and loss of production, which are somewhat harder to quantify. One estimate suggests that reduction of air-pollution levels by 50 percent in major urban areas would save more than $2 billion per year in health costs. Implicit in the decreased health costs is an increase in longevity

and a decrease in illness among those now adversely affected by air pollution. The financial and human considerations together are powerful incentives for trying to limit air pollution. As we will see later in the chapter, correspondingly large expenditures in this area have already been made.

Types and Sources of Air Pollution

Most air pollutants are either gases or particulates (fine, solid particles). The principal gaseous pollutants are oxides of carbon, nitrogen, and sulfur. They share some common sources but create distinctly different kinds of problems. The principal sources for each of the major pollutants are indicated in figure 17.2.

Particulates

The particulates include soot, smoke, and ash from fuel (mainly coal) combustion, dust released during industrial processes, and other solids from accidental and deliberate burning of vegetation. In other words, particulate air pollutants generated by human activity are largely derived from point sources. Estimates of the magnitude of global anthropogenic contributions vary widely, from about 35 million tons/year (two-thirds from combustion) to 180 million tons/year (mostly industrial). Additional particulates are added by many natural processes, including volcanic eruptions, natural forest fires, erosion of dust by wind, and blowing salt spray off the sea surface.

Overall, anthropogenic particulates appear to amount to less than 10 percent of the natural particulate air pollution. Furthermore, neither natural nor anthropogenic particulates have long residence times in the atmosphere. Generally, they are quickly removed by precipitation, usually within days or weeks. In rare cases, such as fine volcanic ash shot high into the atmosphere by violent eruptions, the finest, lightest material may stay in the air for as long as several years. Most frequently, however, particulate pollution is a local problem, most severe close to its source and for a short time only.

The nature of the problem(s) posed by particulate pollution depends somewhat on the nature of the particulates. Certainly, dense smoke is unsightly, whatever its makeup. In areas showered with ash from a power station or industrial plant, cleaning expenses increase. Fine rock and mineral dust of many kinds has been shown to be carcinogenic when inhaled. Many particulates are also chemically toxic. Coal ash may contain both heavy metals and uranium stuck to the particles, for example. Control of particulate pollution is thus a matter of both health and aesthetics.

Carbon Gases

The principal anthropogenic carbon gases are carbon monoxide and carbon dioxide. Carbon dioxide is not generally regarded as a pollutant *per se*. Some is naturally present in the atmosphere. It is essential to the life cycles of plants, and it is not unhealthful to humans, especially in the moderate concentrations in which it occurs (currently about 350 ppm). It is a natural end product of the complete combustion of carbon-bearing fuels:

$$\text{C} \quad + \quad \text{O}_2 \quad = \quad \text{CO}_2$$
$$\text{carbon} \qquad \text{oxygen} \qquad \text{carbon dioxide}$$

As such, carbon dioxide is continually added to the atmosphere through fossil-fuel burning, as well as by natural processes, including respiration by all oxygen-breathing organisms.

At one time, it was believed that natural geochemical processes would keep the atmospheric carbon dioxide level constant, that the oceans would serve as a "sink" in which any temporary excess would dissolve to make, ultimately, more carbonate sediments. It is now known that this is not true. As noted in chapter 9, atmospheric carbon dioxide levels have increased more than 15 percent in the last century because of heavy fossil-fuel use. An estimated 30 billion tons of anthropogenic carbon dioxide are added to the atmosphere each year. The principal concern, as explained in chapter 9, is that the increased carbon dioxide concentration causes increased greenhouse-effect heating of the atmosphere. Consequent potentially harmful impacts on global climate and civilization include accelerated melting of global ice sheets (with attendant sea-level rise) and additional pressure on world agriculture due to heat and drought, as already discussed in chapter 9.

Carbon monoxide (CO), while volumetrically far less important than carbon dioxide, is more immediately deadly. It is produced during the incomplete combustion of carbon-bearing materials, when less oxygen is available:

$$2\text{C} \quad + \quad \text{O}_2 \quad = \quad 2\text{CO}$$
$$\text{carbon} \qquad \text{oxygen} \qquad \text{carbon monoxide}$$

Anthropogenic carbon monoxide, nearly all of it from fossil-fuel burning, does not persist in the atmosphere for very long. Within a few months, carbon monoxide reacts with oxygen in the air to create more excess carbon dioxide. Actually, human activities add far less carbon monoxide to the atmosphere than do natural sources on a global basis. Nevertheless, it may constitute a serious *local* health hazard.

Carbon monoxide is an invisible, odorless, colorless, tasteless gas. Its toxicity to animals arises from the fact that it replaces oxygen in the hemoglobin in blood. The vital function of hemoglobin is to transport oxygen through the bloodstream. Carbon monoxide molecules can attach themselves to the hemoglobin molecule in the site that oxygen would normally occupy; moreover, carbon monoxide bonds there more strongly. The oxygen-carrying capacity of the blood is thereby reduced. As carbon monoxide builds up in the bloodstream, cells (especially brain cells) begin to die from the lack of oxygen, and, eventually, enough cells may fail and cause the death of the whole organism. The difficulty of detecting carbon monoxide, together with the sleepiness that is an early consequence of a reduced supply of oxygen to the brain, may explain the number of accidental deaths from carbon monoxide poisoning in inadequately ventilated spaces.

Carbon monoxide does not remain in the blood indefinitely. If someone in the early stages of carbon monoxide poisoning is removed to fresh air or, if possible, given concentrated medical oxygen, the carbon monoxide is gradually released (though some irrevocable brain damage may have occurred). The problem is to recognize in time what is taking place. Carbon monoxide can build up to dangerous levels wherever combustion is concentrated and air circulation is poor.

The single largest source of carbon monoxide in the atmosphere, by far, is the automobile (recall figure 17.2). Nonfatal cases of carbon monoxide poisoning have been widely reported in congested urban areas with high traffic density. (Traffic police officers in Tokyo have had to take periodic "oxygen breaks" to counteract the accumulation of carbon monoxide in their blood.) This is one argument in favor of cleaning up auto emissions and making engines burn fuel as efficiently as possible. Another is simply that, if an engine is producing carbon monoxide, it is wasting energy. Burning carbon completely to make carbon dioxide releases three times the energy that is produced from incomplete combustion to carbon monoxide. When fuel is in short supply, we should try to extract as much energy as possible from our fuels. Even if anthropogenic production of carbon monoxide could be reduced or eliminated, however, the problems associated with carbon dioxide remain.

Sulfur Gases

The principal sulfur gas produced through human activities is sulfur dioxide, SO_2. More than 50 million tons are emitted worldwide each year. About two-thirds of this amount is released from coal combustion in factories, power-generating plants, and, in some places, home heating units. Most of the rest is released during the refining and burning of petroleum. In laboratory experiments, root weights achieved by radishes dosed with sulfur dioxide gas for one to two days were lower by up to 90 percent compared to untreated plants, depending on the age of the treated plants at the time of exposure. Within a few days of its release into the atmosphere, sulfur dioxide reacts with water vapor and oxygen in the atmosphere to form sulfuric acid (H_2SO_4), which is a strong and highly corrosive acid. Much of this is scavenged out of the atmosphere in the form of acid rain (discussed later in the chapter) to contribute to acid runoff (figure 17.3). As long as it remains in the air, sulfuric acid is severely irritating to lungs and eyes.

Nitrogen Gases

The geochemistry of nitrogen oxides in the atmosphere is complex. Since nitrogen and oxygen are by far the most abundant elements in air, it is not surprising that, at the high temperatures found in engines and furnaces, they react to form nitrogen oxide compounds (principally NO and NO_2). Nitrogen monoxide (NO) can act somewhat like carbon monoxide in the bloodstream, though it rarely reaches toxic levels. In time, it reacts with oxygen to make nitrogen dioxide (NO_2). Nitrogen dioxide reacts with water vapor in air to make nitric acid (HNO_3), which is both an irritant and corrosive.

Combustion of various kinds adds approximately 50 million tons of nitrogen dioxide to the air each year, which is less than 10 percent of the nitrogen dioxide estimated to be produced by natural biological action. However, most anthropogenic nitrogen dioxide production is strongly concentrated in urban and industrialized areas and may create serious problems in those areas.

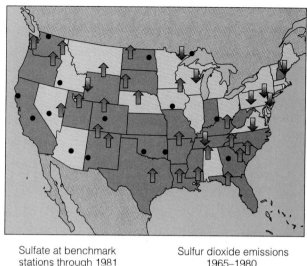

Sulfate at benchmark stations through 1981	Sulfur dioxide emissions 1965–1980
● Unchanged	▓ Increasing levels
⬆ Uptrend	░ Decreasing levels
⬇ Downtrend	

A

B

Figure 17.3 (*A*) Increased sulfate (SO_4) in surface waters may be attributable to sulfur dioxide (SO_2) emissions upwind. (*B*) Reduced sulfur emissions from 1980 to 1991 have been reflected in decline in acid precipitation.   Source: After U.S. Geological Survey.

The most damaging effect of nitrogen dioxide is its role in the production of photochemical smog, sometimes also called Los Angeles smog, after one city where it is common. Key factors in the formation of photochemical smog are high concentrations of nitrogen oxides and strong sunlight. Dozens of chemical reactions may be involved, but the critical one involving sunlight is the breakup of nitrogen dioxide (NO_2) to produce nitrogen monoxide (NO) and a free oxygen atom, which reacts with the common

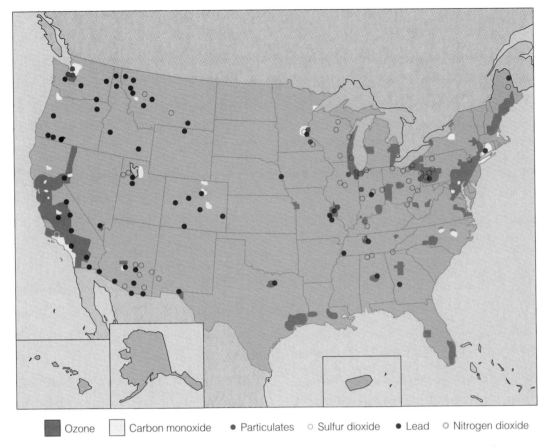

| ▪ Ozone | ▫ Carbon monoxide | • Particulates | ○ Sulfur dioxide | ● Lead | ◌ Nitrogen dioxide |

Figure 17.4 Areas identified by the Environmental Protection Agency as having "substandard" air quality in 1990. There is an evident correlation with urbanization.

oxygen molecule (O₂) to make *ozone* (O₃), a somewhat unusual molecule made up of three oxygen atoms bonded together. Ozone is a strong irritant to lungs, especially dangerous to those with lung ailments or those who are exercising and breathing hard in the polluted air. Significant adverse medical effects can result from ozone concentrations below 1 ppm. Ozone also inhibits photosynthesis in plants. The dual requirement of nitrogen dioxide plus sunlight to produce ozone at ground level explains why "ozone alerts" are more often broadcast in cities with heavy traffic, and during the summertime, when sunshine is abundant and strong. Areas described as "substandard" in air quality with respect to ozone are, for the most part, urbanized areas (figure 17.4).

Ozone and Chlorofluorocarbons (CFCs)

If ozone is so harmful, why has so much concern been expressed over the possible destruction of the "ozone layer"? The answer is that ozone at ground level, where it can interact with animals and plants, is a good example of the chemical-out-of-place definition of a pollutant. In the upper atmosphere, more than 15 kilometers above the surface, ultraviolet rays from the sun interact with ordinary oxygen to produce ozone. That ozone can absorb further

ultraviolet radiation, shielding the earth's surface from it. Ultraviolet (UV) radiation can cause skin cancer (it is what makes excessive tanning and sun exposure unhealthy). The presence of the ozone layer decreases that risk. Ultraviolet radiation spans a range of wavelengths shorter than visible light. UVA, the longest-wavelength UV, is not particularly damaging. UVC, the shortest, would be highly damaging, but it is largely absorbed by oxygen in the atmosphere. It is the intermediate-wavelength UVB, also potentially damaging, that is particularly absorbed by ozone. Scientists of the U.N. Environment Programme estimate that each 1 percent reduction in stratospheric ozone could result in a 3 percent increase in nonmelanoma skin cancers in light-skinned people, in addition to increased occurrence of melanoma, blindness related to development of cataracts, gene mutations, and immune-system damage. Because of the time-lag between exposure and development of medical problems, however, solid evidence of the link between destruction of stratospheric ozone and health effects is so far limited.

The ozone in the upper atmosphere can be destroyed by components of the exhaust of high-altitude aircraft like the SST and Concorde, and particularly by chlorofluorocarbon (CFC) compounds used as refrigerants and as

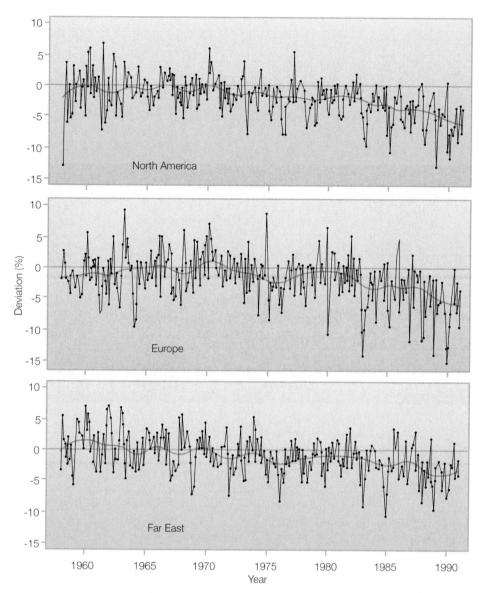

Figure 17.5 Stratospheric ozone has been declining perceptibly in recent years. These regional data for the period 1960 to 1990 were collected by ground-based monitoring stations. Satellite data are also helpful in monitoring global patterns and trends; see box 17.1. Redrawn with permission from R. Stolarski, et al., "Measured Trends in Stratospheric Ozone," *Science*, April 17, 1992, Vol. 256, p. 346. Copyright 1992 by the AAAS.

propellants in aerosol spray cans. Many manufacturers have reduced or eliminated chlorofluorocarbons in their propellants or substituted a manual spray pump instead. However, substitutes have not so readily been found for chlorofluorocarbon refrigerants. More than 1.4 billion pounds of CFCs were still produced in the United States in 1981 (down from nearly 1.8 billion pounds in 1974). Since 1931, close to 40 billion pounds have been produced worldwide, and of that total, an estimated 90 percent has been released into the atmosphere. At present, high-altitude aircraft are few, but the unknown magnitude of their impact on the ozone concentration may also be cause for concern.

The ozone-layer issue further illustrates both the need for more hard data and the difficulty of modeling complex geologic systems. Estimates made in the mid-1970s by the National Research Council, based principally on levels of chlorofluorocarbon release, projected a reduction of 15 to 18 percent in the ozone layer by the late 1970s. By 1983, improved atmospheric models and revised calculations reduced the projections to a decrease of 2 to 4 percent. Other models, taking into account the effects of other air pollutants like carbon dioxide, nitrous oxide, and methane, suggested that a small net increase in ozone was possible. Ultimately, however, a clear downward trend was demonstrated, both regionally and globally (figure 17.5).

Box 17.1 The Mysterious "Ozone Hole"

Figure 1 Map of ozone measurements over the Southern Hemisphere made by satellite each October since 1979 reveal pronounced and deepening ozone hole over Antarctica. Units are Dobson units, where 1 DU=0.001 cm thickness of ozone, distributed over the whole vertical column of atmosphere.   Data from NASA, Goddard Space Flight Center.

The distribution of ozone in the ozone layer (10 to 50 kilometers above the earth's surface) varies seasonally and with latitude. Ozone production rates are most rapid near the equator, as a consequence of the strong sunlight there, but generally, the total ozone in a vertical column of air increases with latitude (that is, increases toward the poles) as a result of the balance be-tween natural ozone production and destruction rates, and atmospheric cir-culation patterns.

In 1985, scientists monitoring at-mospheric chemistry recognized a thinning in the earth's ozone layer over Antarctica. From 1970 to 1994, the average quantity of ozone present, as measured in the month of October each year, dropped by more than one-third. The phenomenon was quickly dubbed an "ozone hole"; and it deep-ened year by year (figure 1).

Measurements showed unexpect-edly high concentrations of reactive chlorine compounds in the air of that region, compounds that could have been derived from anthropogenic chlorofluorocarbons (CFCs). The Antarctic "ozone hole" appeared to be

The chemistry of ozone destruction is complex. It ap-pears that CFCs remain in the atmosphere for about a cen-tury, slowly migrating upward and breaking down over time. When chlorine atoms are freed by this CFC decompo-sition, which seems to occur especially readily in polar clouds during the winter, the chlorine reacts with ozone to form chlorine monoxide (ClO). A strong negative correla-tion between ClO and ozone concentrations is an important piece of evidence linking CFCs to ozone destruction.

Within the last decade, the unexpected discovery of an "ozone hole" over Antarctica prompted intensive study of the ozone layer and factors that may be contributing to its demise (box 17.1). More recently, reduction in ozone has been detected over the northern polar regions during their winter too. The release of the reactive chlorine is catalyzed by fine ice particles in the stratosphere; thus the pronounced effect in polar regions. Other atmospheric components—in-cluding sulfate aerosols, as from Mount Pinatubo's erup-tion—can have similar effects, but they are of shorter dura-tion and not of as great significance.

Recognizing the obvious threat of CFCs, the global community has banded together to support the Montreal

Figure 2 While the link with CFCs is more firmly established for the Antarctic "ozone hole," there is demonstrable ozone depletion over the Northern Hemisphere as well.   Data from NASA, Goddard Space Flight Center.

the first clear evidence of significant depletion of atmospheric ozone as a result of human activities. Subsequent measurements documented decreases in stratospheric ozone at lower latitudes as well, and various studies strengthened the evidence that CFCs were involved in the depletion.

While the severe ozone depletion seemed to be localized over the very low-population Antarctic region, global concern was moderate. Then winter ozone depletions became marked in the Northern Hemisphere too (figure 2), which added some urgency to the research. Between CFC accumulation and the apparent contribution of Mount Pinatubo's sulfate-rich emissions, ozone levels between latitudes 30° and 60° N in late 1992 were about 10 percent below normal. Pinatubo aside, stratospheric ozone in the Northern Hemisphere has continued to decline. As this translates directly into increased transmission of potentially cancer-causing ultraviolet rays in these high-population areas of North America and Europe, pressure has been increasing on those nations that have not yet signed the Montreal Protocol.

Protocol on Substances that Deplete the Ozone Layer. First adopted in 1987, strengthened in 1990, and signed by more than 70 nations through mid-1991, the Montreal Protocol stipulates phaseout of production of CFCs and other ozone-depleting compounds by the year 2000. One remaining concern is that some populous developing countries have been reluctant to sign, so the overall effectiveness of the Montreal Protocol remains to be seen. Another concern is that even in developed nations, there is some delay in the CFC phaseout; and the compounds being substituted for CFCs themselves can contribute to ozone depletion, though not as strongly.

Lead

One air pollutant that has been greatly reduced is lead. Most of the lead once released into the atmosphere was emitted by automobile exhaust systems. While lead is not naturally a significant component of petroleum, one particular lead compound, tetraethyl lead, had been a gasoline additive since the 1940s as an antiknock agent to improve engine performance.

Lead is one of the heavy metals that accumulate in the body (see chapter 16). It has a variety of harmful effects, including brain damage in high concentrations. Mild lead poisoning in the nervous system can cause depression,

Lead Use in Gasoline and Average Blood Lead Levels

Lead in gas

Blood lead level

Total lead used per 6-month period (1000 tons)

Average blood lead levels (micrograms per deciliter)

1976 1977 1978 1979 1980

Figure 17.6 Declines in lead used in gasoline are directly reflected in lower lead levels in blood. Source: After L. Whiteman, "Trends to Remember: The Lead Phase Down," *EPA Journal*, May/June 1992, p. 38.

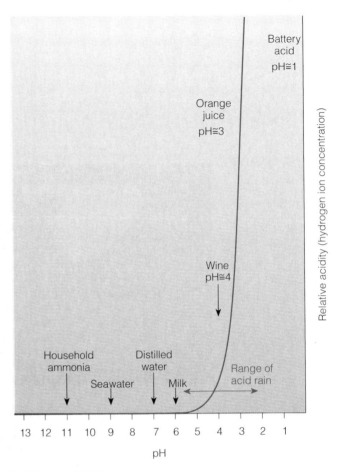

Relative acidity (hydrogen ion concentration)

Battery acid
pH≅1

Orange juice
pH≅3

Wine
pH≅4

Household ammonia

Seawater

Distilled water

Milk

Range of acid rain

13 12 11 10 9 8 7 6 5 4 3 2 1

pH

Figure 17.7 The pH scale is a logarithmic one; small differences in pH value translate into large differences in acidity. Range of acid precipitation is compared with typical pH values of common household substances.

nervousness, apathy, and other psychological disorders, as well as learning difficulties. Acute lead poisoning from exhaust fumes alone has not been documented. However, children in urban areas who breathe lead-laden air and also consume chips of old lead-based paint may indeed develop high and harmful lead levels in their blood. It has been estimated that 5 to 10 percent of inner-city children may suffer from lead poisoning.

Lead levels allowed in most paints have been greatly reduced, and lead can be left out of gasoline. Equally good engine performance can be obtained from higher-octane gasolines at the cost of a few more cents a gallon. (In the long run, in fact, leaded gasoline could be the more costly because some lead compounds produced when leaded gasoline is burned coat engine parts and decrease their useful life.) Also, illegal use of leaded gasoline in cars equipped with catalytic converters destroys the converters' effectiveness and results in the emission of more unburned hydrocarbons, carbon monoxide, and other pollutants. Beginning in the early 1970s, the Environmental Protection Agency began to mandate reductions in the lead content of gasoline, with lead phased out almost completely from automotive fuel by 1987. Lead consumed in gasoline dropped 70 percent from 1975 to 1982. Concentrations of atmospheric lead decreased

correspondingly, by about 65 percent over the same period, and this in turn was reflected in greatly reduced lead levels in blood (figure 17.6). The burning of gasoline will necessarily continue to generate several kinds of pollutants, as noted earlier, but lead-free gasoline should virtually eliminate lead as an air pollutant. Currently, a limited quantity of leaded fuel (containing less than 0.1 gram lead per gallon) is still available, primarily for the benefit of old engines in farm equipment that for technical reasons need some lead to protect their valves. But this is far below the average of 2.0 grams lead per gallon of the leaded gasolines of 1975; by 1996, a complete ban on leaded gasoline took effect.

Other Pollutants

Volatile, easily vaporized organic compounds are a major component of air pollution. Their principal sources are (1) transportation, especially automobiles emitting unburned gasoline, and (2) industrial processes, which release a variety of organics including unburned hydrocarbons and volatile organic solvents. Unburned hydrocarbons are not themselves highly toxic, but they play a role in the formation of photochemical smog and may react to form other compounds that irritate eyes and lungs.

A

B

Figure 17.8 (A) Once sharply defined sculptural details on this grotesque are being eaten away by acidic rainfall. (B) Cleopatra's Needle, a granite obelisk, dramatically demonstrates the effects of acid rain. It was carved around 1600 B.C., and its features remained sharply defined through 3500 years in the dry climate of Egypt (left photo). Just 75 years after it was moved to Central Park in New York City, it showed serious deterioration (right photo) due to the much more rapid weathering caused by wetter conditions compounded by acidity of rainfall.   Photographs: (A) Metropolitan Museum of Art; (B) Courtesy of New York City Parks Archives.

Other air pollutants are of more localized significance. Heavy metals other than lead, for example, can be a severe problem close to mineral-smelting operations. Such elements as mercury, lead, cadmium, zinc, and arsenic may be emitted either as vapors or attached to particulate emissions from smelters. While they are quickly removed from the atmosphere by precipitation, the metals may then accumulate in local soils, from which they are concentrated in organic matter, including growing plants. They can constitute a serious health hazard if accumulated in food or forage crops.

Finally, there are air pollutants of particular concern only indoors; see box 17.2.

Acid Rain

The Nature of Acid Rain

Acidity is reported on the pH scale. The pH of a solution is inversely proportional to the hydrogen-ion (H^+) concentration in the solution; the pH scale, like the Richter magnitude scale, is a logarithmic one (figure 17.7). Neutral liquids, such as pure water, have a pH of 7. Acid substances have pH values less than 7; the lower the number, the more acidic the solution. Typical pH values of common household acids like vinegar and lemon juice are in the range of pH 2 to 3. Alkaline solutions, like solutions of ammonia, have pH values greater than 7. All natural precipitation is actually somewhat acidic as a result of solution of gases (for instance, CO_2) that make acids (such as carbonic acid, H_2CO_3). **Acid rain,** then, is rain that is more acidic than normal. While many gases in the air contribute to the acidity of rain, discussions of acid rain focus on the sulfur gases that react to form atmospheric sulfuric acid.

As noted in chapter 13, acid rain can contaminate water supplies, stunt or kill plants and animals, and damage structures through its corrosive effects (figure 17.8). Impacts of acid rain on ground water were also explored in the last chapter. Acidic water more readily leaches potentially toxic metals from soils, transferring them to surface or groundwater supplies, and reduces the capacity of soils to neutralize further additions of acid. Acid water leaching nutrients from

Box 17.2 Indoor Air Pollution

Most people associate air pollution principally with city streets in heavily industrialized towns, or factories belching smoke. Scientists are increasingly coming to realize the hazards of air pollution in homes and offices. Some hazards are mineralogic; asbestos is a prime example. It has been widely used in ceiling materials because of its fire-retardant properties. Intact panels are harmless, but from damaged ones may fall fine asbestos fibers that may be carcinogenic when ingested in quantity.

Many other hazards have been created or intensified by our search for energy efficiency in response to dwindling fuel supplies and rising costs. For example, it was learned after the fact that blown-in foam insulation releases dangerous quantities of the volatile organic compound formaldehyde as it cures. (Formaldehyde is the pungent liquid in which creatures are commonly preserved for dissection by biology classes.) Many homes in which this insulation had been installed had to be torn apart later to remove it. The very act of insulating tightly has meant that quantities of toxic gases that might once have escaped harmlessly through cracks and small air leaks are now being sealed in, and their concentrations are building up. Among them are smoke and carbon monoxide (from furnaces, heaters, gas appliances, and cigarettes), and radon (figure 1).

Radon is a colorless, odorless, tasteless gas that also happens to be radioactive. It is produced in small quantities in nature by the decay of the natural trace elements uranium and thorium. Not only is radon itself radioactive, but it decays into radioactive isotopes of lead, bismuth, and other metals that stick to dust particles, which, in turn, may be inhaled, lodge in the lungs, and sometimes cause lung cancer. When radon is diluted in the open atmosphere, its hazards are low. When it is concentrated inside a building, the hazards are increased. The *average* indoor radon exposure in the United States represents a radiation hazard five times as great as all other natural radiation sources combined. Tightly sealed homes may have radon levels two hundred times higher, far above the levels found in uranium mines. The Environmental Protection Agency has estimated that radon levels in 1 home out of every 10 are high enough that corrective action should be undertaken, and tight sealing of all homes could lead to tens of thousands of new cases of lung cancer. For the *average* radon level in a home in the United States, the lifetime risk of death from radon-related lung cancer is estimated at 0.4 percent, or one chance in 250. Where high radon levels are present, the risk may be comparable to that from cigarette smoking.

How does all this radon get in? Uranium and thorium occur in most rocks and soils and in building materials made from them, like concrete or brick. Being a gas, radon can diffuse into the house through unsealed foundation walls or directly into unfinished crawl spaces from the soil. It can emanate from masonry walls. Radon seeps into ground water from aquifer rocks and thus can enter the house via the plumbing. Local geology and the details of a home's construction affect the severity of the radon hazard; see also chapter 20.

Given all the air pollutants that can accumulate inside the home, most experts now recommend that homes being built or remodeled with emphasis on insulation for energy efficiency also include an air exchanger designed to provide adequate ventilation without loss of appreciable heat. Anyone living in a very well insulated house might be wise to have local authorities test the air quality inside.

Various strategies are available to reduce indoor air pollution. For example, sealing masonry of foundations and plugging tightly around pipes entering the building will reduce incoming radon; filters can trap smoke and other particulates; improved ventilation can reduce buildup of radon and volatile organic compounds. At this point, there is still need for more complete assessment of the risks, establishment of appropriate indoor-air-quality standards, and consideration of how to promote/enforce those standards in individual homes.

Paint

Moth crystals

Kerosene heater

Fresh dry cleaning

Carpeting

Curtains

Plywood

Radon in water

Gas range

Fireplace

Wood stove

Foam padding

Radon from soil

Asbestos insulation

Radon from walls

Disinfectants
Pesticides
Cleaners
Solvents
Aerosols
Glues

Heating

Carbon monoxide

Figure 1 Sources of indoor air pollution are many. Sources of radon in the home include the soil, walls, and water.   Source: After U.S. Environmental Protection Agency, *EPA Journal*, Sept./Oct. 1990.

lake-bottom sediments may contribute to algal bloom. Though its source is air pollution, its consequences thus include subsequent water pollution. Considerable controversy currently exists both about the precise extent of the acid-rain problem and about its causes and, therefore, its cures. The controversy arises, in part, from incomplete scientific data. However, studies have found a correlation between rainfall acidity and potentially toxic levels of mercury in lakes in northern North America. Other metals that may be linked to acid rain include lead, zinc, selenium, copper, and aluminum. The accumulative properties of the heavy metals in turn raise concerns about toxic levels in fish and consequent risks to humans and other fish-eaters.

Regional Variations in Rainfall Acidity and Impacts

Rainfall is demonstrably more acidic in regions downwind from industrial areas releasing significant amounts of sulfur among their emissions. These observations form the basis for allegations of the need to control sulfur pollution. What is less clear is the precise role that anthropogenic sulfur plays in the sulfur cycle.

For one thing, there is little information about natural background levels of sulfur dioxide (SO_2) and related sulfur species. Historically, pure rainwater has been assumed to have a pH of about 5.6 in areas away from industrial sulfur sources. This is the pH to be expected as a consequence of the formation of carbonic acid due to natural carbon dioxide in the air. Recent research suggests that natural background pH may be somewhat lower as a result of other chemical reactions in the atmosphere and may vary from place to place as a consequence of localized conditions. Natural rainfall acidity may correspond to a pH as low as 5. However, air over all the globe is now polluted to some extent by human activities. Therefore, the chemistry of precipitation in wholly unpolluted air cannot be determined precisely enough to assess anthropogenic impacts fully. It does appear that, overall, more sulfate is added to the air by sea spray containing dissolved sulfate minerals than by human activities. Perhaps anthropogenic sulfate only locally aggravates an existing situation. As previously noted, recent studies have shown that samples of ancient air as much as several hundred thousand years old may be retrieved for analysis from gas bubbles trapped in the ice of long-lived continental glaciers. Such studies should contribute greatly to the understanding of long-term human impacts on atmospheric chemistry, in terms of acidity as well as "greenhouse gases."

It may, in fact, be the case that increases in rainfall acidity due to formation of nitric acid from nitrogen oxides formed in internal-combustion engines is a much more significant problem in and near urban areas. The focus on sulfuric acid in rain partially reflects the fact that sulfur gas pollution from stationary sources is more readily controllable,

and partially recognizes the more long-range and global impact of anthropogenic sulfuric acid, including the political aspects of the issue (see below).

The nature of the local geology on which acid rain falls can strongly influence the severity of the acid's impact. Chemical reactions between rain and rock are complex. Farmers and gardeners have known for years that certain kinds of rock and soil react to form acidic pore waters, while others yield alkaline water. For instance, because limestone reacts to make water more alkaline, it serves, to some extent, to neutralize acid waters. The effects of acid rain falling on carbonate-rich rocks and soils are therefore moderated. Conversely, granitic rocks and the soils derived from them are commonly more acidic in character. They cannot buffer or moderate the effects of acid rain; the acid rain just makes the surface and subsurface waters more acidic in such regions. Some operators of coal-fired plants in the Midwest have argued that the predominantly granitic soils of the Northeast are principally at fault for the existence of very acid lakes and streams in the latter region. Certainly, the geology does not *reduce* the problem. However, the rainfall in the Northeast, with an average pH of 4.2, is clearly acidified.

Nor is the northeastern United States the only area concerned about sulfuric acid generated in this country. The Canadian government has expressed great concern over the effects of acid rain caused by pollutants generated in the United States—especially the industrialized upper Midwest and Northeast—that subsequently drift northeast over eastern Canada. Much of that area, too, is underlain by rocks and soils highly sensitive to acid precipitation (figure 17.9) and is already subject to acid precipitation associated with Canadian sulfate sources (figure 17.10). In Europe, acid rain produced by industrial activity in central Europe and the United Kingdom falls on Scandinavia. Acid rain, then, has become an international political concern.

More information is needed about natural water chemistry in various (unpolluted) geologic settings before the role of acid rain can be evaluated precisely. The science of pinpointing sulfur sources is also becoming more sophisticated as means are developed to "fingerprint" those sources by the assemblage of gases present and their composition. This will allow cause-and-effect relationships to be determined more accurately. It can at least be said, however, that the effects of adding more sulfur dioxide to the air are likely to be negative. How negative is still a matter of debate. The report of the National Acid Precipitation Assessment Program, a ten-year effort by representatives of many federal agencies, took a relatively unconcerned position, to the dismay of many researchers studying effects of acid precipitation.

Negative impacts of acid rain are certainly well documented. Increasingly, damage from related phenomena—acid snow and even acid fog—are being recognized. In snowy regions with acid precipitation, rapid spring runoff of

A

High sensitivity—Mostly granite and granitic rocks

Intermediate-to-high sensitivity—Including volcanic rocks, ultramafic rocks, gabbro; shales and similar sedimentary rocks

Intermediate-to-low sensitivity—Calcareous and carbonate rocks interbedded or interspersed with more sensitive sedimentary, igneous, and metamorphic rocks

Low sensitivity—Limestone, dolomite, and metamorphic equivalents

B

High sensitivity—Areas where surface deposits are likely to be noncalcareous/ noncarbonate

Indeterminate or intermediate sensitivity— Areas with calcareous glacial deposits transported from major carbonate outcrops, or likely to have been sites of carbonate deposition from glacial lakes

Low sensitivity—Areas underlain by carbonate bedrock or by carbonate bedrock interbedded with noncarbonate bedrock; all surface deposits likely to be calcareous

→   Generalized direction of glacial transport

?   Areas where limits of glacial dispersal of carbonate bedrock are not well known

Figure 17.9 (*A*) Sensitivity of bedrock, and soils derived from bedrock, to acid precipitation in southeastern Canada. (*B*) Sensitivity of glacial and postglacial surface sediments to acid precipitation.   Source: After W. W. Shilts, *Sensitivity of Bedrock to Acid Precipitation: Modification by Glacial Processes*, Geological Survey of Canada, Paper 81–14, 1981.

Figure 17.10 Sulfur dioxide emissions in Canada, 1990.   Source: Data from Environment Canada.

Air Pollution and Weather

Thermal Inversion

Air-pollution problems are naturally more severe when air is stagnant and pollutants are confined. Particular atmospheric conditions can contribute to acute air-pollution episodes of the kinds mentioned at the beginning of this chapter. A frequent culprit is the condition known as a **thermal inversion** (figure 17.11). Within the lower atmosphere, air temperature normally decreases as altitude increases. This is why mountaintops are cold places, even though direct sunshine is stronger there. In a thermal inversion, there is a zone of relatively warmer air at some distance above the ground. That

is, going upward from the earth's surface, temperatures decrease for a time, then increase in the warmer layer (inversion of the normal pattern), then ultimately continue to decrease at still higher altitudes. Inversions may become established in a variety of ways. Warmer air moving in over an area at high altitude may move over colder air close to the ground, thus creating an inversion. Rapid cooling of near-surface air on a clear, calm night may also lead to a thermal inversion.

Most air pollutants, as they are released, are warmer than the surrounding air—exhaust fumes from automobiles, industrial smokestack gases, and so on. Warm air is less dense than colder air, so ordinarily, warm pollutant gases rise and keep rising and being dispersed through progressively cooler air above. When a thermal inversion exists, a warm-air layer becomes settled over a cooler layer. The warm pollutant gases rise only until they reach the warm-air layer. At that point, the pollutants are no longer less dense than the air above, so they stop rising. They are effectively trapped close to the ground and simply build up in the near-surface air. Sometimes, the cold/warm air boundary is so sharp that it is visible as a planar surface above the polluted

428

Warm pollutant-bearing gases rise through cooler, denser air.

Decreasing temperature

A

Increasing temperature

Trapped gases caught below warmer air layer

Decreasing temperature

B

Figure 17.11 Effect of thermal inversion in trapping air pollution. (A) Normal conditions permit warm pollutants to rise and disperse through cooler air above. (B) Inversion traps warm pollutant gases under a stable warm-air layer.

Figure 17.12 Thermal inversion traps smog over an oil field near San Francisco Bay at sunset. Note drill rigs at left center.   Photograph by Belinda Rain, courtesy of EPA/National Archives.

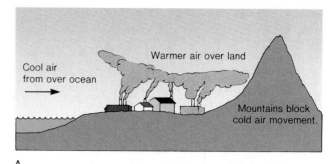

Cool air from over ocean

Warmer air over land

Mountains block cold air movement.

A

Warm air moving over land

Trapped cold air pocket in valley

B

Figure 17.13 Influence of topography on establishing and maintaining thermal inversions. (A) Cool sea breezes blow over Los Angeles and are trapped by the mountains to the east. (B) Cold air, once settled into a valley, may be hard to dislodge.

zone (figure 17.12). The Donora, Pennsylvania, episode mentioned at the beginning of the chapter was triggered by an inversion. Its effect on a freight train's smoke was described by physician R. W. Koehler: "They were firing up for the grade and the smoke was belching out, but it didn't rise. . . . It just spilled out over the lip of the stack like a black liquid, like ink or oil, and rolled down to the ground and lay there" (Roueché, 1953, 175).

Every one of the half-dozen major acute air-pollution episodes this century has been associated with a thermal inversion, as have many milder ones. Certainly, air pollution can be a health hazard in the absence of a thermal inversion. The inversion, however, concentrates the pollutants more strongly. Moreover, inversions can persist for a week or longer, once established, because the denser, cooler air near the ground does not tend to rise through the lighter, warmer air above.

Certain geographic or topographic settings are particularly prone to formation of thermal inversions (figure 17.13). Los Angeles has the misfortune to be located in such a spot. Cool air blowing across the Pacific Ocean moves over the land and is trapped by the mountains to the east beneath a layer of warmed air over the continent. Donora, Pennsylvania lies in a valley. When a warm front moves across the hilly terrain, some pockets of cold air may remain in the valleys, establishing a thermal inversion. Germany's coal-rich Ruhr Valley suffered so from pollution trapped by a prolonged inversion during the winter of 1984–85 that schools

Figure 17.14 Air-pollution plumes from the Chicago, Illinois, and Gary/Hammond, Indiana, area bring clouds and snow to the area to the northeast.   Photograph ©NASA.

and many factories had to be closed and the use of private automobiles banned until the air cleared. As with most weather conditions, nothing can be done about a thermal inversion except wait it out.

Impact on Precipitation

While weather conditions, such as thermal inversions, can affect the degree of air pollution, air pollution, in turn, can affect the weather in ways other than reducing visibility, modifying air temperature, and making rain more acidic. This is particularly true when particulates are part of the pollution. Water vapor in the air condenses most readily when it has something to condense on. This is the principle behind cloud seeding: Fine, solid crystals are spread through wet air, and water droplets form on and around these seed crystals. Particulate pollutants can perform a similar function.

This is dramatically illustrated in figure 17.14. Southwest of Lake Michigan lies the industrialized area of Chicago, Illinois, and Gary and Hammond, Indiana. The Gary/Hammond area, especially, is a major steel milling region. At the time this satellite photograph was taken, waste gases from the mills were being blown northeast

across the lake (note faint smoke plumes at bottom end of lake, just left of center). At first, they remained almost invisible, but as water vapor condensed, clouds appeared and, in this scene, snow crystals eventually formed. A large area of snow in Michigan is on a direct line with the pollution plumes from the southwest. The particulate pollutants in the exhaust gases facilitate the condensation and precipitation.

Toward Air-Pollution Control

Air-Quality Standards

Growing dissatisfaction with air quality in the United States led, in 1970, to the Clean Air Act Amendments, which empowered the Environmental Protection Agency to establish and enforce air-quality standards (see also chapter 18). The standards developed are designed to protect human health, clear the visible pollution from the air, and prevent crop and structural damage and other adverse effects. (See also box 17.3.)

Control Methods

Air-pollutant emissions can be reduced either by trapping the pollutants at the source or by converting dangerous compounds to less harmful ones prior to effluent release. A variety of technologies for cleaning air exist.

Where particulates are the major concern, filters can be used to clear the air, with the fineness of the filters adjusted to the size range of particles being produced. Filters may be up to 99.9 percent efficient, but they are commonly made of paper, fiber, or other combustible material. These cannot be used with very-high-temperature gases.

An alternate means of removing particulates is electrostatic precipitators or "scrubbers." In these, high voltages charge the particulates, which are then attracted to and caught on charged plates. Efficiencies are usually between 98 and 99.5 percent. These systems are sufficiently costly that they are practical only for larger operations. They are widely used at present in coal-fired electricity-generating plants.

Wet scrubbers are another possibility. In these scrubbers, the gas is passed through a stream or mist of clean water, which removes particulates and also dissolves out some gases. Wet chemical scrubbing, in which the gas is passed through a water-chemical slurry rather than pure water, is used particularly to clean sulfur gases out of coal-fired plant emissions. Currently, the most widely used designs use a slurry of either lime (calcium oxide, CaO) or limestone (CaCO$_3$), which reacts with and thus removes the sulfur gases. Wet scrubbing produces contaminated fluids, which present cleaning and/or waste-disposal problems.

Both carbon monoxide and the nitrogen oxides are best controlled by modifying the combustion process. Lower

Box 17.3 Air Pollution and Health: The Standard Air-Pollution Index

Prior to the mid-1970s, at least four-teen different indices for describing air quality were in use in the United States and Canada. None seemed especially helpful to the individual wanting to know just how unhealthy the air was. In 1975, a U.S. federal task force set about developing an index that could be related directly to human health impacts. The outcome was the Pollutant Standards Index (PSI).

The index reports air-pollution levels on a scale of 0 to 500. Separate scales have been established for each of five pollutants—carbon monoxide (CO), sulfur dioxide (SO_2), ozone (O_3), nitrogen dioxide (NO_2), and particulates (total suspended particulates, or TSP). For each, a value of 100 on the PSI scale equals the ambient air quality standard for that substance. Pollutant levels above that standard are regarded as unhealthful

in varying degrees. The actual concentrations of each pollutant that correspond to a given index value were adjusted so that a PSI rating of 100 to 200 could be described as "unhealthful"; 200 to 300, "very unhealthful"; over 300, "hazardous"; and so on. A summary of the PSI values and corresponding concentrations is given in table 1. Overall air quality should be reported in terms of the pollutant with the highest PSI rating at a given time. For example, if ozone is in the "very unhealthful" range while all other pollutants fall in the "moderate" range, the air should nevertheless be described as "very unhealthful." Where air quality is monitored, then, the PSI provides a measure of air pollution that can be understood by the public as a direct indicator of potential health hazard.

The Canadian government introduced its Clean Air Act in 1969, es-

tablishing the Canadian National Air Pollution Surveillance Network to track air pollution in Canada. National Ambient Air Quality Objectives (NAAQOs) were established for the same set of common pollutants as shown in table 1, and there is a Canadian Index of the Quality of Air (IQUA), with possible values of 0 to over 125, to provide a relative measure of overall air quality, with descriptors "good" (IQUA 0–25), "fair" (25–50), "poor" (50–100), and "very poor" (over 100). A study of trends in air quality in 17 major Canadian cities over the period 1979–1982 showed gratifying improvements: particulates, down 54%; SO_2, down 61%; CO, down 70%; and NO_2, down 38%; see NetNotes for the pertinent reference.

—Continued

combustion temperatures, for example, minimize the production of nitrogen oxides and moderate the production of carbon monoxide. Use of afterburners to complete combustion is another way to reduce carbon monoxide emission by converting it to carbon dioxide.

Combustion is an effective way to destroy organic compounds, including hydrocarbons, producing carbon dioxide and water. More complex scrubbing or collection procedures for organics also exist but are not discussed in detail here. Such measures are necessary for low-temperature municipal incinerators burning materials that might include toxic organic chemicals that can only be thoroughly decomposed in the high-temperature incinerators designed for toxic-waste disposal.

Automobile Emissions

As part of the efforts to reduce air pollution, the Environmental Protection Agency has imposed limits on permissible

levels of carbon monoxide and hydrocarbon emissions from automobiles. Many car manufacturers use catalytic converters to meet the standard. In a catalytic converter, a catalyst (commonly platinum or another platinum-group metal) enhances the oxidation, or combustion, of the hydrocarbons and carbon monoxide to water and carbon dioxide. In this process, catalytic converters are quite effective. Unfortunately, they also promote the production of sulfuric acid from the small quantities of sulfur in gasoline, as well as the formation of nitrogen oxides. Emissions of these compounds may be *increased* by the use of catalytic converters. While the increased sulfuric acid emission by vehicles using catalytic converters is minor compared to the output from coal-fired power plants, it nevertheless makes automobile exhaust more irritating to breathe where exhaust fumes are confined. The environmental impacts of the nitrogen oxides may be more negative than previously assumed, which may dictate a need to control these more carefully as well.

Box 17.3—Continued

Table 1 Pollutant Standards Index (PSI) Values, Health Effects, and Cautionary Statements

Index value	Air-quality level	Pollutant level					Health effect descriptor
		TSP*	SO₂*	CO†	O₃‡	NO₂‡	
500	significant harm	1,000	2,620	57.5	1,200	3,750	
400	emergency	875	2,100	46.0	1,000	3,000	hazardous
300	warning	625	1,600	34.0	800	2,260	
							very unhealthful
200	alert	375	800	17.0	400	1,130	
							unhealthful
100	NAAQS§	260	365	10.0	160	— ‖	
							moderate
50		75	80	5.0	80	— ‖	
							good
0		0	0	0	0	— ‖	

Hazardous, PSI values 400–500

Premature death of ill and elderly. Healthy people will experience adverse symptoms that affect their normal activity.

All persons should remain indoors, keeping windows and doors closed. All persons should minimize physical exertion and avoid traffic.

Hazardous, PSI values 300–400

Premature onset of certain diseases in addition to significant aggravation of symptoms and decreased exercise tolerance in healthy persons.

Elderly and persons with existing diseases should stay indoors and avoid physical exertion. General population should avoid outdoor activity.

Very unhealthful, PSI values 200–300

Significant aggravation of symptoms and decreased exercise tolerance in persons with heart or lung disease, with widespread symptoms in the healthy population.

Elderly and persons with existing heart or lung disease should stay indoors and reduce physical activity.

Unhealthful, PSI values 100–200

Mild aggravation of symptoms in susceptible persons, with irritation symptoms in the healthy population.

Persons with existing heart or respiratory ailments should reduce physical exertion and outdoor activity.

Source: U.S. Environmental Protection Agency, "Guideline for Public Reporting of Daily Air Quality," as cited in *A Recommended Air Pollution Index*, 1976, p. 46, Council on Environmental Quality, Washington, DC.

*Twenty-four-hour, micrograms/cu. m
†Eight-hour, milligrams/cu. m
‡One-hour, micrograms/cu. m
§National Ambient Air Quality Standard
‖ No index values reported at levels below those specified as "alert" level

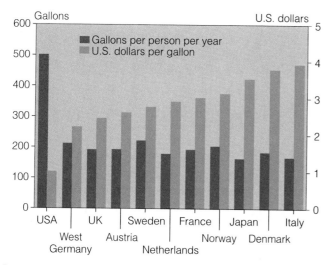

Figure 17.15 Worldwide, there is an apparent inverse correlation between gasoline price and gasoline consumption. Source: After World Resources Institute, *1992 Information Please Environmental Almanac*, p. 68.

Table 17.2 Pollution Emissions Estimates by Type in the United States

Pollutant	Emissions (millions of tons)			
	1940	1970	1980	1990
particulates	24.6	19.8	8.5	7.4
carbon monoxide	87.8	110.2	76.2	67.7
nitrogen oxides	7.4	19.9	20.4	19.4
sulfur oxides	19.9	31.2	23.2	21.0
volatile organics	18.8	27.8	22.7	17.6
Totals	158.5	208.9	151.0	133.1

Sources: U.S. Environmental Protection Agency data reported in *Environmental Quality 1983* and *Environmental Quality 1986*, C72–C74 for 1980–1984 data, Council on Environmental Quality, Washington, DC; *World Resources 1992–1993*, and *World Resources 1994–95*, World Resources Institute.

Improvements in fuel economy reduce all emissions simultaneously, of course, as less fuel is burned for a given distance traveled. The EPA-mandated fuel economy standards have been increasing year by year since 1978, to 27.5 miles per gallon (mpg) in 1985. New-car fuel economy has actually averaged somewhat higher than the standards on occasion; in 1981, for example, new cars averaged 25 mpg against a standard of 22 mpg. However, the average mileage of *all* cars on the road in the same year was estimated at only 15.5 mpg. Furthermore, the depressed fuel prices of the mid-1980s contributed to a rebound in popularity of larger, less-fuel-efficient cars, and some auto manufacturers began to have corresponding difficulty in meeting average new-car fuel economy standards; worldwide, it is clear that fuel cost and fuel consumption are inversely correlated (figure 17.15). Still, the stricter mileage requirements are improving overall average mileage, which had climbed to an estimated 20.9 mpg by 1990.

Cost and Effect

Expenditures for pollution abatement and control in the United States have risen steadily, from $6.5 billion in 1972 to an estimated $115 billion in 1990, according to EPA. This was about 40 percent of the size of the defense budget in the same year. It is clear from table 17.2 that considerable further progress could be made.

Air quality has improved somewhat in consequence (figure 17.16). The most pronounced reduction has been in lead, due to the widespread switch to unleaded gasoline. Substantial reductions in carbon monoxide, sulfur dioxide, and ozone are also apparent. The reduction in particulates is moderate, perhaps in part because there are many natural sources for these in addition to the anthropogenic ones. Nitrogen dioxide is also unabated; relatively little attention has been paid specifically to reducing nitrogen oxide emissions.

With reauthorization of the Clean Air Act Amendment in 1990 and resultant rulemaking (see chapter 18), further improvements have been seen. Phase I of the 1990 reauthorization targeted a number of sources of sulfur pollution, with a view to reducing acid rain. A 1995 study suggests that the reductions in sulfur emissions were having the desired effects. Sulfate and acidity (hydrogen ion concentration) both showed substantial reductions downwind of Phase I targets (figure 17.17A and B). Furthermore, it can be shown that this result is not an artifact of precipitation variations. It is true that over time, low precipitation tends to be reflected in higher concentrations of the dissolved pollutants, and this effect is locally evident in the data; see precipitation data in figure 17.17C. But other dissolved substances showed quite different concentration trends from sulfate and acidity; see, for example, nitrate (figure 17.17D).

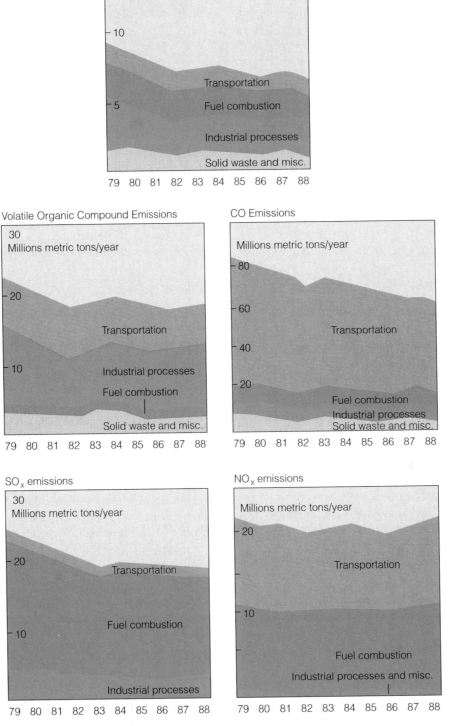

Figure 17.16 Pollution-control laws have brought about significant decreases in certain pollutants over time.   Source: U.S. Environmental Protection Agency, *EPA Journal*, Sept./Oct. 1990, pp. 16–17.

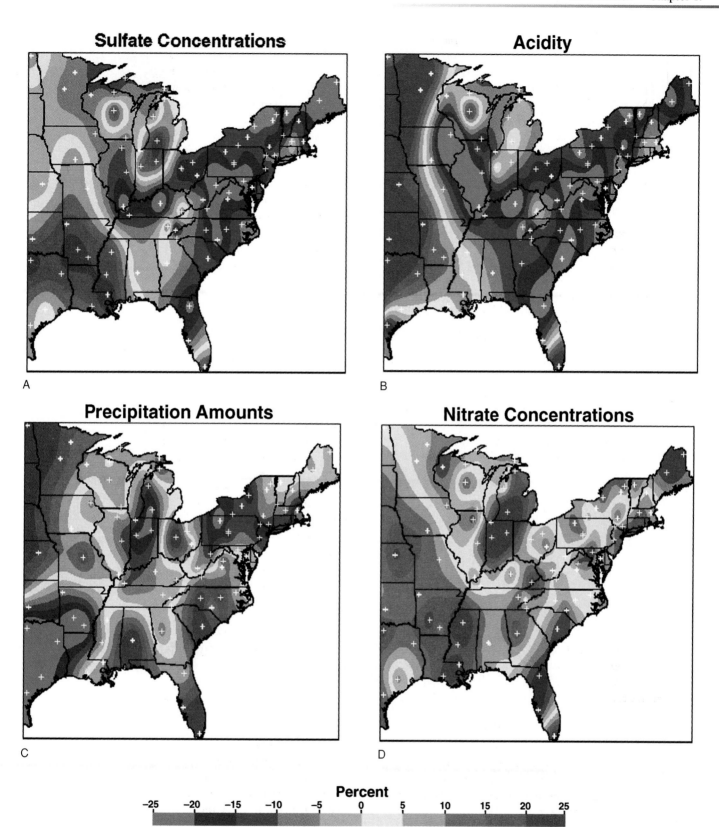

Sulfate Concentrations

A

Acidity

B

Precipitation Amounts

C

Nitrate Concentrations

D

Percent

−25   −20   −15   −10   −5   0   5   10   15   20   25

Figure 17.17 Departures in 1995 precipitation data for the eastern U.S. from 1983–94 trends. (All data in percentages.) (*A*) Sulfate concentrations. (*B*) Acidity (hydrogen-ion concentrations). (*C*) Precipitation amounts. (*D*) Nitrate concentrations.   Source: All maps from U.S. Geological Survey Open-File Report 96–0346.

Overview

Summary

The principal air pollutants produced by human activities are particulates, gaseous oxides of carbon, sulfur, or nitrogen, and organic compounds. The largest source of this pollution is combustion, and especially transportation. For some pollutants, anthropogenic contributions are far less than natural ones overall, but human inputs tend to be spatially concentrated. Control of air pollutants is complicated because so many of them are gaseous substances generated in immense volumes that are difficult to contain; and once released, they may disperse widely and rapidly in three dimensions in the atmosphere, thwarting efforts to recover or remove them. Climatic conditions can influence the severity of air pollution: Strong sunlight in urban areas produces photochemical smog, and temperature inversions in the atmosphere trap and concentrate pollutants. Air pollutants, in turn, may influence weather conditions, by contributing to acid rain and by providing particles on which water or ice can condense. Imposition of air-quality and emissions standards by the Environmental Protection Agency has led to measurable improvements in air quality over the last two decades, particularly with respect to pollutants for which anthropogenic sources are relatively important.

Terms to Remember

acid rain 423   thermal inversion 428

Exercises

Questions for Review

1. Describe the principal sources and sinks for atmospheric carbon dioxide.
2. Carbon monoxide is a pollutant of local, rather than global, concern. Explain.
3. What is the origin of the various nitrogen oxides that contribute to photochemical smog?
4. What is photochemical smog, and under what circumstances is this problem most severe?
5. Ozone is an excellent example of the chemical-out-of-place definition of a pollutant. Explain, contrasting the effects of ozone in the ozone layer with ozone at ground level.
6. What radiation hazard is associated with indoor air pollution, and why has it only recently become a subject of concern?
7. What pollutant species is believed to be primarily responsible for acid rain? What is the principal source of this pollutant?

8. Explain briefly why the seriousness of the problems posed by acid rain may vary with local geology.
9. Describe the phenomenon of a thermal inversion. Give an example of a geographic setting that might be especially conducive to the development of an inversion. Outline the role of thermal inversion in intensifying air-pollution episodes.
10. Federal emissions-control regulations for automobiles have led to improvements in air quality with respect to some but not all pollutants in auto exhaust systems. Explain briefly.

For Further Thought

1. Investigate in some detail the technology and economics associated with the control of pollutants from burning coal, and compare the apparent cost effectiveness of different technologies and the energy requirements of each.
2. Find out who, if anyone, monitors air quality or issues air-pollution hazard warnings in your area or, perhaps, in a nearby urban area. Keep track, over a period of months, of the weather conditions on days when warnings are issued, and see what patterns emerge.
3. Check the latest information on stratospheric ozone depletion. (The National Aeronautics and Space Administration or National Oceanic and Atmospheric Administration may be good sources.) Investigate the validity of claims made by those opposing CFC restrictions, that natural chlorine sources (seaspray, volcanic gases) can account for ozone depletion. (See NetNotes for information sources.)

Suggested Readings/References

Cicerone, R. J. 1987. Changes in stratospheric ozone. *Science* 237 (3 July): 35–42.

Council on Environmental Quality. 1976. *A recommended air pollution index*. Washington, D.C.: Council on Environmental Quality.

———. 1984. *Environmental quality 1983*. Washington, D.C.: Council on Environmental Quality.

Davies, B. E. 1983. Heavy metal contamination from base metal mining and smelting: Implications for man and his environment. In *Applied environmental geochemistry*, edited by I. Thornton, pp. 425–61. New York: Academic Press.

Fergusson, J. E. 1982. Air pollution and air pollutants. Chapters 8 and 9 in *Inorganic chemistry and the earth*. New York: Pergamon Press.

Garrels, R. M., Mackenzie, F. T., and Hunt, C. 1975. *Chemical cycles and the global environment*. Los Altos, Calif.: William Kaufmann.

Graedel, T. E., Hawkins, D. T., and Claxton, L. D. 1986. *Atmospheric chemical compounds*. New York: Academic Press.

Graves, C. K. 1980. Rain of troubles. *Science 80* (May): 74–79.

Indoor Air. 1993. *EPA Journal* 19 (October-December). This whole issue is devoted to indoor air pollution issues.

Krupnick, A. J., and Portney, P. R. 1991. Controlling urban air pollution: A benefit-cost assessment. *Science* 252 (26 April): 522–28.

Levi, B. G. 1992. Arctic measurements indicate the chilly prospect of ozone depletion. *Physics Today* (July): 17–19.

Lippmann, M., and Schlesinger, R. B. 1979. *Chemical contamination in the human environment.* New York: Oxford University Press.

Lynch, J. A., Bowersox, V. C., and Grimm, J. W. 1996. Trends in precipitation chemistry in the United States, 1983–94: An analysis of the effects in 1995 of Phase I of the Clean Air Act Amendments of 1990, Title IV. USGS Open-File Report 96–0346.

Makhijani, A., Bickel, A., and Makhijani, A. 1990. Still working on the ozone hole. *Technology Review* (May/June): 53–59.

Mohnen, V. A. 1988. The challenge of acid rain. *Scientific American* 259 (August): 30–38.

Nero, A. V., Jr. 1988. Controlling indoor air pollution. *Scientific American* 258 (May): 42–48.

———. 1989. Earth, air, radon, and home. *Physics Today* (April): 32–39.

Newell, R. E. 1971. The global circulation of atmospheric pollutants. *Scientific American* 224 (January): 32–42.

Otton, J. K. *The geology of radon.* U.S. Geological Survey.

Post, W. M., Peng, T., Emanuel, W. R., King, A. W., Dale, V. H., and DeAngelis, D. L. 1990. The global carbon cycle. *American Scientist* 310 (July-August): 310–26.

Raloff, J. 1991. Mercurial risks from acid's reign. *Science News* 139 (March 9): 152–56.

Roueché, B. 1953. The fog. In *Eleven blue men*, pp. 173–96. New York: Berkley Medallion Books. (A description of the Donora pollution episode.)

Rowland, F. S. 1989. Chlorofluorocarbons and the depletion of stratospheric ozone. *American Scientist* 77 (January-February): 36–45.

Schindler, D. W. 1988. Effects of acid rain on freshwater ecosystems. *Science* 239 (8 January): 149–56.

Seinfeld, J. H. 1989. Urban air pollution: State of the science. *Science* 243 (10 February): 745–52.

Turk, J. T. 1983. *An evaluation of trends in the acidity of precipitation of surface water in North America.* U.S. Geological Survey Water Supply Paper 2249.

U.S. Environmental Protection Agency. 1980. *National accomplishments in pollution control 1970–1980.* Washington, D.C.: U.S. Environmental Protection Agency.

Woestman, J. T., and Logothetis, E. M. 1995. Controlling automotive emissions. *The Industrial Physicist:* 20–24.

World Resources Institute. 1994. *World Resources 1994–95.* New York: Oxford University Press.

Comprehensive data on ozone and the underlying scientific issues is available through the Environmental Protection Agency at

http://www.epa.gov/docs/ozone/science/

"Air Quality Trends in Canadian Cities 1979–1992" is available online through Environment Canada; see

http://www.doe.ca/pdb/naqt/aqfact_e.html

USGS Open-File Report 96–0346 (see Lynch *et al.* in the reference list) analyzes effects of the Clean Air Act Amendments of 1970, and is found online at

http://h2o.usgs.gov/public/pubs/acidrain/

A review of many aspects of ozone chemistry, with illustrations, is available from the National Aeronautics and Space Administration; see

http://spso.gsfc.nasa.gov/NASA_FACTS/ozone/ozone.html

The 1995 Nobel Prize in Chemistry was awarded to scientists Paul Crutzen, Mario Molnia, and F. Sherwood Rowland for their researches in atmospheric chemistry particularly as related to ozone. The citation can be viewed online at

http://www.nobelse/announcement95-chemistry.html

The Environmental Protection Agency has a fact sheet on "Your car air conditioner and the CFC-12 Production Ban" at

http://www.epa.gov/ozone/title6/609/cons1.html

Environment Canada has prepared a State of the Environment Fact Sheet, "Environmental implications of the automobile," which synthesizes impacts of many types — including implications for mineral and energy resources, as well as for air pollution. This analysis is found at

http://199.212.18.12/folio.pgi/osoeeng/query=/doc/{2,0,0,0}?

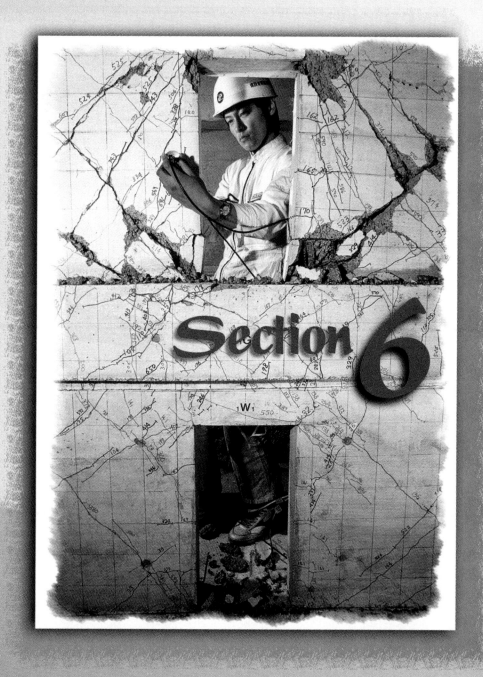

Technician assessing damage to a wall in a test of earthquake resistance, Tsukuba, Japan. © Paul Chesley/Tony Stone Images.

In this section, we move further afield from the more traditional areas of environmental geology, and also address some topics that integrate material from many of the more specific chapters in earlier sections of the book.

Volumes could be, and have been, written about environmental law. In chapter 18, we will very briefly examine several types of laws closely related to geologic subjects dealt with earlier: resources, pollution, and geologic hazards. Chapter 19 is concerned with the related areas of land-use planning and engineering geology. Sound land-use planning takes geology into account in trying to make the best use of each parcel of land. Sensible engineering of buildings, dams, bridges, and other structures likewise generally requires consideration of the constraints imposed by a variety of geologic phenomena and processes. Chapter 20 examines geomedicine, the study of relationships between geology and health. This is a comparatively new subdiscipline, but one likely to have increasing importance in the future.

Chapter

18

Environmental Law

Environmental laws of one sort or another have a long history. Many centuries ago, English kings restricted the burning of coal in London because the air pollution had become choking, and the penalty for violators could be execution! Penalties for breaking environmental laws in the United States today are more often financial than physical; sometimes, they are nonexistent. The number of such laws and the areas they regulate continue to increase. Through time, the philosophy of appropriate approaches to resource development, pollution control, and land use has changed, and the focus of environmental laws has shifted in response. A comprehensive treatment of environmental legislation is well beyond the scope of this book, but in this chapter, we will look at common themes of and problems with some environmental laws related to geologic matters.

The massive scale of the Trans-Alaska Pipeline, and the correspondingly large concerns about its potential environmental impacts, required extensive analysis before construction. © Tom and Pat Leeson/Photo Researchers, Inc.

Resource Law: Water

Given the importance of water as a resource and the increasing scarcity of high-quality water, it is not surprising that laws specifying water rights are necessary. What is perhaps unexpected is that the basic principles governing water rights vary not only from nation to nation, but also from place to place within a single country. Also, quite different principles may be applied to surface-water rights and groundwater rights, even though surface and subsurface waters are inevitably linked through the hydrologic cycle.

Surface-Water Law

Navigable streams are treated as community property, accessible to all, things that cannot be "appropriated" or assigned to individuals. To the extent that they can be viewed as "owned," ownership lies with the state, on behalf of the public. However, while one cannot own the stream, one may have rights to use its water.

The two principal approaches to surface-water rights in the United States are the *Riparian Doctrine* and the *Doctrine of Prior Appropriation*. The term *riparian* is derived from the Latin word for "bank," as in riverbank, and sums up the essence of the **Riparian Doctrine.** Whoever owns land adjacent to a body of surface water (lake, stream) has a right to use that water, and all those bordering on a given body of water have an equal right to that water. Provision is also generally made to the effect that the water must be used for "natural purposes" or "beneficial uses" and returned to the body of water from which it came in essentially the same amount and quality as when it was removed. The Riparian Doctrine, long the basis for English surface-water law, likewise became the basis for assigning surface-water rights in the eastern United States, where moisture is generally ample. Often implicit or explicit in surface-water laws based on the Riparian Doctrine is the concept that no one user's water use should substantially interfere with others' use—but where water is relatively plentiful, this is typically not a problem. On the other hand, strictly speaking, *only* landowners bordering the water body have any rights to its water.

In the western United States, where surface water is in shorter supply, the prevailing doctrine is that of **Prior Appropriation.** Under this scheme, whoever is first, historically, to use water from a given surface-water source has top-priority rights to that water. Users who begin to draw on the same water source later have subordinate rights, in the order in which they begin to use the water. One need not own any land at all to have water rights under this scheme. One's water rights under this doctrine are established and maintained by continuing to divert the water for "beneficial use." Future use is to be proportional to the quantity used when the rights were established. An advantage of this system is that it makes clearer who is entitled to what in times of shortage. Ideally, it ought also to discourage excessive settlement or development in arid regions, because latecomers have no guarantee of surface-water rights if and when demand exceeds supply. In practice, however,

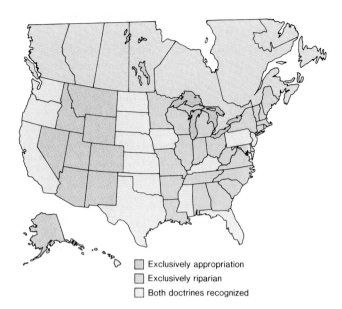

Exclusively appropriation
Exclusively riparian
Both doctrines recognized

Figure 18.1 Distribution of underlying principles of surface-water law applied in the United States. Source: Data from New Mexico Bureau of Mines Circular 95, 1968.

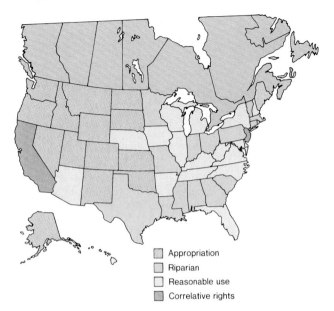

Appropriation
Riparian
Reasonable use
Correlative rights

Figure 18.2 Distribution of principles of groundwater law applied in the United States. Source: Data from New Mexico Bureau of Mines Circular 95, 1968.

it does not appear to have served as such a deterrent. Moreover, because failure to use the water for a period of time causes one to lose water rights under the appropriation doctrine, it may unintentionally lead to waste of water, to water use simply for the sake of preserving those rights rather than out of genuine need—not, perhaps, a wise arrangement in regions where water supplies are inadequate.

What constitutes "beneficial use" of water also varies regionally. Domestic water use is commonly accepted as beneficial. So is hydropower generation, which, in any event, does not appreciably consume water. Irrigation that is critical to agriculture may be regarded as a legitimate "beneficial use," but watering one's lawn might not be. Industrial water use is often a low-priority use. In some areas, a hierarchy of beneficial uses has been established. A typical sequence, from highest to lowest priority, might be: domestic water use, municipal supply, irrigation, watering livestock, diversion of water for power generation, use for mining or drilling for oil or gas, navigation, and recreation.

One measure of the degree of benefit that could be used in balancing competing industrial and agricultural interests would be the dollar value of the return expected on the proposed water investment. The resulting priority rankings would then depend on the specific industrial activity, crops, and so forth being compared.

Such prioritizing of usage complicates water-rights questions. An additional complication is that several states, usually those with both water-rich and water-poor areas, have tried to combine both the riparian and prior-appropriation principles (see figure 18.1). The result is generally described as the "California Doctrine," although each state's particular scheme is different. Some states—for example, California—have extended the principle of appropriation to give the *state* the right to appropriate surface

water and to transfer it from one region to another or otherwise redistribute it to maximize its beneficial use.

Groundwater Law

Groundwater law can be an even more confusing problem, in part because groundwater law was, to a great extent, developed before there was general appreciation of the hydrologic cycle, the movement of ground water, or its relationship to surface water. As with surface-water rights, too, different principles may govern groundwater rights, producing a similar patchwork distribution of water-rights laws across the country (figure 18.2). Two distinctly different principles have been applied. The so-called English Rule, or "Rule of Capture," gives property owners the right to all the ground water they can extract from under their own land. This ignores the reality of lateral movement of ground water. Ground water does not recognize property lines, of course, so heavy use by one landowner may deprive adjacent property owners of water, as ground water flows toward well sites. The American Rule, or "Rule of Reasonable Use," limits a property owner's groundwater use to beneficial use in connection with the land above, and it includes the idea that one person's water use should not be so great that it deprives neighboring property owners of water. However, as with surface-water laws, there is no consistent definition of what constitutes "beneficial use." These rules have then been variously combined with appropriation or riparian philosophies of water rights.

Some states, notably California, have specifically addressed the problem of several landowners whose properties overlie the same body of ground water (aquifer system), specifying that each is entitled to a share of the water that is proportional to his or her share of the overlying land (*correlative rights*). Many other states have introduced a prior-appropriation principle into the determination of groundwater rights.

One factor that complicates allocation of ground water is its comparative invisibility. The water in a lake or stream can be seen and measured. The extent, distribution, movement, and quantity of water in a given aquifer system may be so imperfectly known that it is difficult to recognize at what level one individual's water use may deprive others who draw on the same water source.

As an aside, note that similar problems arise with other reservoirs of fluids underground—specifically, oil and gas. Rights to petroleum from an oil field underlying the Iraq-Kuwait border were a point of contention between those nations prior to the 1990 invasion of Kuwait by Iraq.

Resource Law: Minerals and Fuels

Mineral Rights

For more than a century, U.S. federal laws have included provisions for the development of mineral and fuel resources. The fundamental underlying legislation, even today, is still the 1872 Mining Law. Its intent was to encourage exploitation of mineral resources by granting mineral rights and often land title to anyone who located a mineral deposit on federal land and invested some time and money in developing that deposit. Federal land could be converted to private ownership for $2.50 to $5 an acre—and as those prices were set in the legislation, they would remain in effect for any minerals still subject to the 1872 law, even decades later when inflation had raised land values. The individual (or company) did not necessarily even have to notify the government of the exact location of the deposit, nor was the developer required to pay royalties on the materials mined. The style of mining was not regulated, and there was no provision for reclamation or restoration of the land when mining operations were completed. What this amounted to was a colossal giveaway of minerals on public lands (and land too) in the interest of resource development.

Subsequent laws, particularly the Mineral Leasing Act of 1920 and the Mineral Leasing Act for Acquired Lands of 1947, began to restrict this giveaway. Certain materials, such as oil, gas, coal, potash, and phosphate deposits, were singled out for different treatment. Identification of such deposits on federally controlled land would not carry with it unlimited rights of exploitation. Instead, the extraction rights would be leased from the government for a finite period, and royalties had to be paid, compensating the public in some measure for the mineral or fuel wealth extracted from the public lands. Still, no requirements concerning the mining or any land reclamation were imposed, and the so-called hard-rock minerals (including iron, copper, gold, silver, and others) were exempted. The Outer Continental Shelf Lands Act of 1953 extended similar leasing provisions to offshore regions under U.S. jurisdiction.

Sparked by a 1989 General Accounting Office report recommending reform of some provisions of the 1872 Mining Law, intense debate has been taking place in Congress about possible reform of that law. Under reform proposals in a 1995 bill passed by the House of Representatives, rights to ore deposits still governed by the 1872 law would not be so freely granted, but would be treated more like the rights to oil, coal, and other materials already explicitly removed from the jurisdiction of the 1872 law. Fairer (nearer market) prices for land would have to be paid, as would royalties; and strict mine-reclamation standards would have to be met. Again, an obvious motive is to assure the public, in whose benefit federal lands are presumably held, fair value for mineral rights granted, a return on minerals extracted, and environmental protection. Industry advocated lower royalty rates, deduction of certain mineral-extraction expenses from the amount on which royalties would be paid, and other differences from the House-passed bill, backing a less radical reform passed by the Senate in 1996. As of this writing, both bills are still in committee, and prospects for reconciliation are viewed as poor. Until a revision is passed, however, the 1872 Mining Law, for all its perceived flaws, remains in effect.

Mine Reclamation

Growing dissatisfaction with the condition of many mined lands eventually led to the imposition of some restrictions on mining and post-mining activities in the form of the Surface Mining Control and Reclamation Act of 1977. This act applies only to coal, whether surface-mined or deep-mined, and attempts to minimize the long-term impact of coal mining on the land surface. Where farmland is disturbed by mining, for example, the land's productivity should be restored afterward. In principle, the approximate surface topography should also be restored. However, in practice, where a thick coal bed has been removed over a large area, there may be no nearby source of sufficient fill material to make this possible, so this provision may be disputed in specific areas. Similarly, the act stipulates that groundwater recharge capacity be restored, but particularly if an aquifer has been removed during mining, perfect restoration may not be possible. Still, even when a complete return to pre-mining conditions is not possible, this legislation and other laws patterned after it go a long way toward preserving the long-term quality of public lands and preventing them from becoming barren wastes (see chapter 12). The National Environmental Policy Act (discussed later in the chapter) has provided a vehicle for controlling mining practices for other kinds of mineral and fuel deposits.

Many states impose laws similar to or stricter than the federal resource and/or environmental laws on lands under state control. However, the impact of the federal laws alone should not be underestimated. Nearly one-third of the land in the United States, most of it in the west, is under federal control (figure 18.3). Indeed, part of the pressure for reforming the 1872 Mining Law comes from state land-management agents in the western states, where the great disparities between mineral-rights and mineral-development laws on federal and on state land create some tensions. For example, it may be economically advantageous under current laws for a company to seek mineral rights on, and mine, federal rather than adjacent state-controlled lands; but the states (like the federal government) receive no returns from mineral extraction on federal lands within their borders, while they do have to deal with problems such as air and water pollution associated with the mining and mineral processing.

International Resource Disputes

As noted in chapter 12, additional legal problems develop when it is unclear in whose jurisdiction valuable resources fall. International disputes over marine mineral rights have

intensified as land-based reserves have been depleted and as improved technology has made it possible to contemplate developing underwater mineral deposits.

Traditionally, nations bordering the sea claimed territorial limits of 3 miles (5 kilometers) outward from their coastlines. Realization that valuable minerals might be found on the continental shelves led some individual nations to assert their rights to the whole extent of the continental shelf. That turns out to be a highly variable extent, even for individual countries. Off the east coast of the United States, for example, the continental shelf may extend beyond 150 kilometers (about 100 miles) offshore. The continental shelf off the west coast is less than one-tenth as wide. Nations bordered by narrow shelves quite naturally found this approach unfair. Some consideration was then given to equalizing claims by extending territorial limits out to 200 miles offshore. That would include all of the continental shelves and a considerable area of sea floor as well. The biggest beneficiaries of such a scheme would be countries that were well separated from other nations and had long expanses of coastline. Closely packed island nations, at the other extreme, might not realize any effective increase in jurisdiction over the old 3-mile limit. Landlocked nations, of course, would continue to have no seabed territorial claims at all.

Law of the Sea and Exclusive Economic Zones

After eight years of intermittent negotiations collectively described as the "Third U.N. Conference on the Law of the Sea," an elaborate Law of the Sea Treaty emerged that attempted to bring some order out of the chaos. Resources are only one area addressed by the treaty. Territorial waters are extended to a 12-mile limit, and various navigational, fishing, and other rights are defined. **Exclusive Economic Zones** (EEZs) extending up to

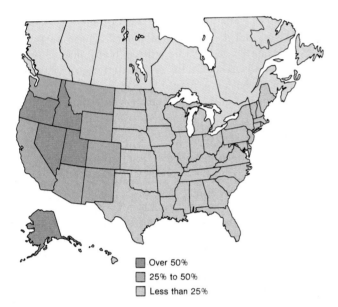

Figure 18.3   Federal land ownership in the United States.
Source: Data from New Mexico Bureau of Mines Circular 95, 1968.

Over 50%
25% to 50%
Less than 25%

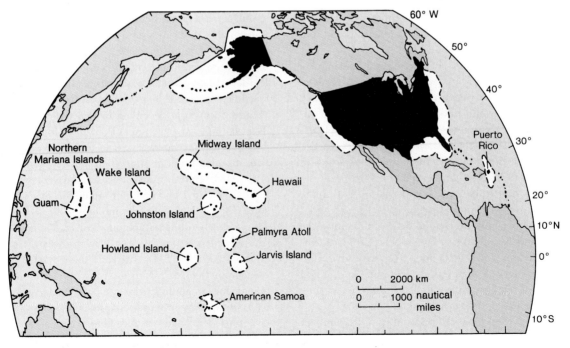

Figure 18.4   Exclusive Economic Zone of the United States.   Source: U.S. Geological Survey Annual Report 1983.

200 miles from the shorelines of coastal nations are established, within which those nations have exclusive rights to mineral-resource exploitation. These areas are not restricted to continental shelves only. Some deep-sea manganese nodules fall within Mexico's Exclusive Economic Zone; Saudi Arabia and the Sudan have the rights to the metal-rich muds on the floor of the Red Sea. Included in the EEZ off the west coast of the United States is a part of the East Pacific Rise spreading-ridge system, the Juan de Fuca rise, along which sulfide mineral deposits are actively forming. Areas of the ocean basins outside EEZs are under the jurisdiction of an International Seabed Authority. Exploitation of mineral resources in areas under the Authority's control requires payment of royalties to it, and the treaty includes some provision for financial and technological assistance to developing countries wishing to share in these resources.

The United States was one of only four nations (out of 141) to vote against the treaty, objecting principally to some of the provisions concerning deep-seabed mining. However, the United States has so far tended to abide by most of the treaty's provisions. In March 1983, President Reagan unilaterally proclaimed a U.S. Exclusive Economic Zone that extends to the 200-mile limit offshore. This move expands the undersea territory under U.S. jurisdiction to some 3.9 billion acres, more than $1\frac{1}{2}$ times the land area of the United States and its territories (figure 18.4).

So far, exploration of this new domain has been minimal. However, certain kinds of mineral and fuel potential are already recognized. For example, off the Atlantic coast, there is potential for finding oil and gas deposits in the thick sediment deposited as the Atlantic Ocean formed—though there is also considerable concern about environmental impacts of oil and gas exploration on the continental shelf. "Black smoker" hydrothermal vents and related hydrothermal activity on the Pacific continental margin have produced sulfide and other metal deposits. Sulfides may also be associated with the Pacific island margins (consider the islands' volcanic origins), and on the flanks of those islands, in many places, mineral crusts rich in cobalt and manganese have been formed (figure 18.5). The western and northern margins of Alaska are, like the Atlantic coast, thickly blanketed with sediment hosting known and potential petroleum deposits. Altogether, quite a variety of resources may be recoverable from the EEZ of the United States (figure 18.6).

Figure 18.5 Cobalt/manganese-rich crust formed on rocks of volcanic islands in the Pacific.   Photograph courtesy of Frank T. Manheim, U.S. Geological Survey, Office of Energy and Marine Biology, Woods Hole, MA 02543.

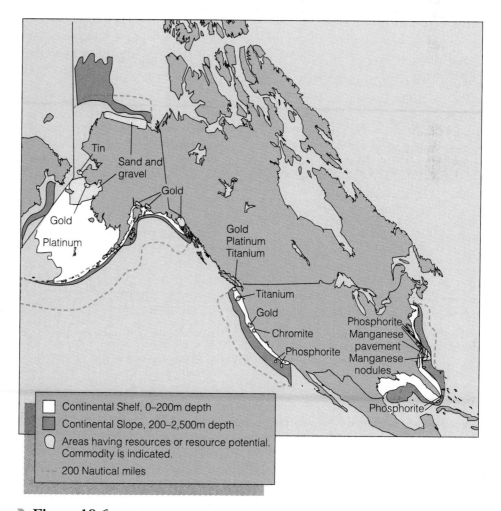

Figure 18.6 Possible mineral and rock resources in the North American EEZ of the United States.   Source: Adapted from B. A. McGregor and T. W. Ofield, *The Exclusive Economic Zone: An Exciting New Frontier*, U.S. Geological Survey, p. 7.

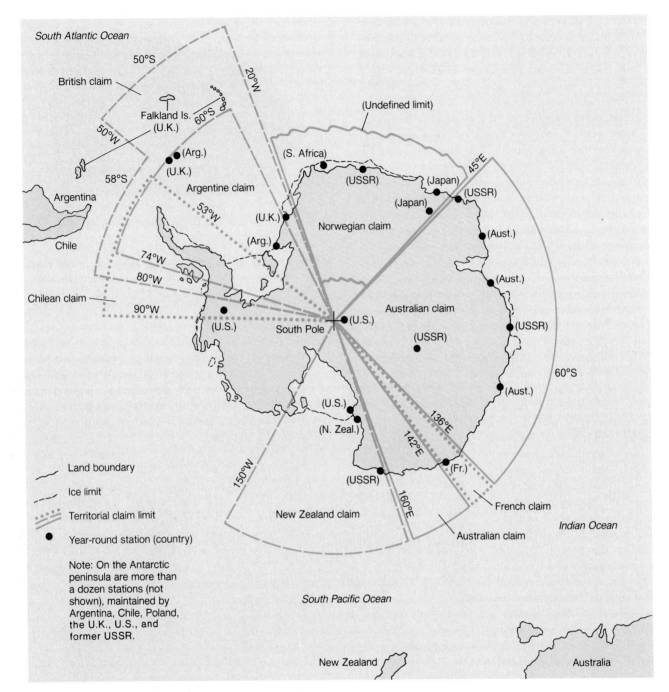

Figure 18.7 Conflicting land claims in Antarctica prior to the 1961 treaty. Note overlapping claims. Sources: Data from C. Craddock, et al., Geological Maps of Antarctica, *Antarctic Map Folio Series,* Folio 12, copyright 1970 by the American Geographical Society; and J. H. Zumberge, "Mineral Resources and Geopolitics in Antarctica," *American Scientist* 67, p. 68–76, 1979.

Antarctica

Antarctica has presented a jurisdictional problem not unlike that of the deep-sea floor. Prior to 1960, seven countries—Argentina, Australia, Chile, France, New Zealand, Norway, and the United Kingdom—had laid claim to portions of Antarctica, either on the grounds of their exploration efforts there or on the basis of geographic proximity (see figure 18.7). (Note the British claim based in part on the location of the Falkland Islands, site of a brief "war" between the

U.K. and Argentina.) Several of these claims overlapped. None was recognized by most other nations, including five (Belgium, Japan, South Africa, the United States, and the former Soviet Union) that had themselves conducted considerable scientific research in Antarctica.

After a history of disagreement punctuated by occasional violence, these twelve nations signed a treaty in 1961 that set aside all territorial claims to Antarctica. Various additional points were agreed upon: (1) the whole continent should remain open; (2) military activities, weapons testing, and nuclear-waste disposal should be banned there; and (3) every effort should be made to preserve the distinctive Antarctic flora and fauna. Other nations subsequently signed also; there are now twenty nations that are "consultative parties" to the Antarctic treaty.

The treaty did not address the question of mineral resources, however, in part because the extent to which minerals and petroleum might occur there was not realized, and in part because their exploitation in Antarctica was then economically quite impractical anyway. Not until 1972 was the question of minerals even raised. The issue continued to be a source of contention for some years. Finally, in 1988, after six years of negotiations, the consultative parties (including the United States) produced a convention to regulate mineral-resource development in Antarctica: the Convention for the Regulation of Antarctic Mineral Resources Activities (CRAMRA). Under this convention, no such activity—prospecting, exploration, or development—could occur without prior environmental-impact assessment; significant adverse impacts will not be permitted. Actual mineral-resource development would require unanimous consent of a supervisory commission composed mainly of representatives of the consultative parties.

In the years after CRAMRA's formulation, debate has continued. Sixteen of the twenty consultative parties must approve the treaty in order for it to take effect. Part of the reason for resistance to ratification was a concern that CRAMRA, by establishing guidelines for mineral-resource development, was effectively opening up the continent to such development (and possible resultant negative environmental impacts, safeguards notwithstanding). In the United States, this negative reaction to CRAMRA led to passage of the Antarctic Protection Act of 1990. Briefly, the Act prohibits Antarctic mineral prospecting and development by U.S. citizens and companies, and urges other nations to join in an indefinite ban on such activities, and consider banning them permanently. It remains to be seen to what extent the consultative parties will follow this lead.

Pollution and Its Control

A Clean Environment—By What Right?

At the outset of a discussion of laws regulating pollution, one might legitimately ask what legal basis or justification exists in the United States for antipollution legislation. Have we, in fact, any legal right to a clean environment? The primary Constitutional justification depends on an interpretation of the Ninth Amendment: "The enumeration in the Constitution of certain rights shall not be construed to deny or disparage others retained by the people." The argument made is that the right to a clean, healthful environment is one of those individual rights so basic as not to have required explicit protection under the Constitution.

This interpretation generally has not been upheld by the courts. However, those harmed or threatened by pollution have often sought relief through lawsuits alleging nuisance or negligence on the part of the polluter. Nuisance and negligence are varieties of *torts* (literally, "wrongs") that are violations of individual personal or property rights, punishable under civil law, and distinct from violations of basic public rights. The government has also exercised its authority to promote the general welfare by enacting legislation to restrict the spread of potentially toxic or harmful substances (pollutants).

Water Pollution

Water pollution ought not to have been a significant problem in the United States since the turn of the century when the Refuse Act of 1899 prohibited dumping or discharging refuse into any body of navigable water. Because most streams drain into larger streams, which ultimately become navigable, or into large lakes, this act presumably banned the pollution of most lakes and streams. Unfortunately, the enforcement of the law left much to be desired.

In succeeding decades, stricter and more specific anti-water-pollution laws have been enacted. The Federal Water Pollution Control Act of 1956 focused particularly on municipal sewage-treatment facilities. The much broader Water Quality Improvement Act of 1970 and the subsequent amendments in the Clean Water Act of 1977 also address oil spills and various chemical pollutants, from both point and nonpoint sources. One of the objectives was to eliminate the discharge of pollutants into navigable waters of the United States by 1985—something that was, in theory, mandated eighty-six years earlier. The Clean Water Act and amendments nominally required all municipal sewage-treatment facilities to undertake secondary treatment by 1983, although practice has lagged somewhat behind mandate; only about two-thirds of municipalities actually met that deadline. Another provision was that "best available technologies" to control pollutant releases by twenty-nine categories of industries were to be in place by July 1984. (This concept of "best available" treatment is discussed further later in the chapter.) All of this pollution control is expensive: In 1981, businesses spent $21 billion on water-pollution abatement in the United States to remove over 27 million tons of water pollutants.

The fundamental underlying objective is "to restore and maintain the chemical, physical, and biological integrity of the Nation's waters." To this end, the Environmental Protection Agency (EPA) is directed to establish water-quality criteria for various types of water uses and to monitor compliance with water-quality standards (table 18.1).

Locally, significant improvements in surface-water quality resulting from improved pollution control have certainly occurred, but as is evident from chapter 16, more remains to be done. A major limitation of water-pollution legislation generally is that it treats surface waters only, although there is growing realization of the urgency of groundwater protection as well as the link between surface and ground

Table 18.1 EPA's Primary Safe Drinking Water Standards for Chemical Components

Contaminants	Health effects	Max. allowed (ppm)	Sources
INORGANIC CHEMICALS			
Arsenic	Dermal and nervous system toxicity effects	0.05	Geological, pesticide residues, industrial waste, smelter operations
Barium	Circulatory system effects	1	
Cadmium	Kidney effects	0.01	Geological, mining and smelting
Chromium	Liver/kidney effects	0.05	
Lead	Central and peripheral nervous system damage, kidney effects, highly toxic to infants and pregnant women	0.05	Leaches from lead pipes and lead-based solder pipe joints
Mercury	Central nervous system disorders, kidney effects	0.002	Used in manufacture of paint, paper, vinyl chloride; used in fungicides; geological
Nitrate	Methemoglobinemia ("blue-baby syndrome")	10	Fertilizer, sewage, feedlots, geological
Selenium	Gastrointestinal effects	0.01	Geological, mining
Silver	Skin discoloration (argyria)	0.05	Geological, mining
Fluoride	Skeletal damage	4	Geological; additive to drinking water, toothpaste, foods processed with fluorinated water
ORGANIC CHEMICALS			
Endrin	Nervous system/kidney effects	0.0002	Insecticide used on cotton, small grains, orchards (canceled)
Lindane	Nervous system/kidney effects	0.004	Insecticide used on seed and soil treatments, foliage application, wood protection
Methoxychlor	Nervous system/kidney effects	0.1	Insecticide used on fruit trees, vegetables
2, 4-D	Liver/kidney effects	0.1	Herbicide used to control broadleaf weeds in agriculture; used on forests, range, pastures, and aquatic environments
2, 4, 5-TP Silvex	Liver/kidney effects	0.01	Herbicide (canceled in 1964)
Toxaphene	Cancer risk	0.005	Insecticide used on cotton, corn, grain
Benzene	Cancer	0.005	Fuel (leaking tanks); solvent commonly used in manufacture of industrial chemicals, pharmaceuticals, pesticides, paints, and plastics
Carbon tetrachloride	Possible cancer	0.005	Common cleaning agents, industrial wastes from manufacture of coolants
p-Dichlorobenzene	Possible cancer	0.075	Used in insecticides, mothballs, air deodorizers
1, 2-Dichloroethane	Possible cancer	0.005	Used in manufacture of insecticides, gasoline
1, 1-Dichloroethylene	Liver/kidney effects	0.007	Used in manufacture of plastics, dyes, perfumes, paints, synthetic chemicals
1, 1, 1-Trichloroethane	Nervous system problems	0.2	Used in manufacture of food wrappings, synthetic fibers
Trichloroethylene (TCE)	Possible cancer	0.005	Waste from disposal of dry cleaning materials and manufacture of pesticides, paints, waxes and varnishes, paint stripper, metal degreaser
Vinyl chloride	Cancer risk	0.002	Polyvinylchloride pipes and solvents used to join them, waste from manufacturing plastics and synthetic rubber
Total trihalomethanes (TTHM) (chloroform, bromoform, bromodichloromethane, dibromochloromethane)	Cancer risk	0.1	Primarily formed when surface water containing organic matter is treated with chlorine

Source: U.S. Environmental Protection Agency.

water. Recently, in fact, a polluter was successfully prosecuted for groundwater pollution by toxic wastes after the ground water seeped into and contaminated associated surface waters.

The fate of water-quality legislation in the near future has become increasingly cloudy. Since 1991, authorization and funding for many programs of the Clean Water Act and its later reauthorizations has lapsed. Repeated efforts since 1992 to enact further reauthorizations have failed. Recent versions of such bills, introduced in 1995 and 1996, have included controversial wetlands-protection provisions that are opposed by some private-property owners. Whether reauthorization will occur in the late 1990s remains to be seen.

Air Pollution

In the United States, air pollution was not addressed at all on the federal level until 1955, when Congress allocated $5 million to study the problem. The Clean Air Act of 1963 first empowered agencies to undertake air-pollution-control efforts. It was amended in 1965 to allow national regulation of motor-vehicle emissions. The Air Quality Act of 1965 required the establishment of air-quality standards to be based on the known harmful effects of various air pollutants. An overriding objective of this act and a series of 1970 amendments was to "protect and enhance the quality of the Nation's air resources so as to promote the public health and welfare and the productive capacity of its population" (Shaw 1976, 340), though most of the details of how this was to be accomplished were unspecified. Nevertheless, substantial progress has been made with respect to many pollutants, as noted in chapter 17.

The Clean Air Act was reauthorized in 1990. Various provisions of these Clean Air Act Amendments (CAAA) deal with many areas of air quality. The CAAA goes beyond the Montreal Protocol (chapter 17) in reducing production of chlorofluorocarbons and other compounds believed harmful to the ozone layer; provides for reductions in industrial emissions of nearly 200 airborne pollutants; requires further reduction of vehicle emissions; sets standards for improved air quality with respect to ozone, carbon monoxide, and particulates in urban areas; and provides for reduction of SO_2 emissions from power plants. The impact of the Phase I regulations on sulfate and acid rain was noted in the last chapter. Significant further air-quality improvements are expected to result; see, for example, figure 18.8. (Comprehensive assessments of reductions in other pollutants have not yet been completed.)

Waste Disposal

The Solid Waste Disposal Act of 1965 was intended mainly to help state and local governments to dispose of municipal solid wastes. Gradually, the emphasis shifted as it was recognized that the most serious problem was hazardous or toxic wastes. The Resource Conservation and Recovery Act of 1976 includes provisions for assisting state and local governments in developing solid-waste management plans, but it

A

B

Figure 18.8 Projected air-quality improvements anticipated from the Clean Air Act Amendments of 1990. (*A*) Urban air-quality improvements. (*B*) Projected reductions in cancer risks from airborne toxins. (Note dependence on "maximum achievable technology.")   Sources: (*A*) After J. Seitz, "Urban Air Quality: The Strategy," *EPA Journal,* Jan./Feb. 1991. (*B*) After L. Wegman, "Air Toxics: The Strategy," *EPA Journal,* Jan./Feb. 1991, U.S. Environmental Protection Agency.

also regulates waste-disposal facilities and the handling of hazardous waste and establishes cooperative efforts among governments at all levels to recover valuable materials and energy from solid waste. Hazardous wastes are defined and guidelines issued for their transportation and disposal.

The Environmental Protection Agency monitors compliance with the act's provisions. Development of waste-handling standards, issuance of permits to approved disposal facilities, and inspections and prosecutions for noncompliance with regulations are ongoing activities. In 1983, there were an estimated 54,000 producers and 13,000 transporters of hazardous waste in the United States. In part because of this, the Environmental Protection Agency has delegated many of its responsibilities under the Resource Conservation and Recovery Act to the states.

Amendments to the Resource Conservation and Recovery Act in 1984 considerably expanded the scope of the original legislation. Approximately one hundred thousand firms generating only small amounts of hazardous waste (less than 1,000 kilograms per month) are now subject to regulation under the act; previously, these small generators were exempt. Before 1984, a waste producer could request "delisting" as a generator of hazardous waste by demonstrating that its wastes no longer contained whatever hazardous chemicals (as identified by the EPA) had originally caused the operation to be listed; now, delisting requires that the waste producer demonstrate essentially that its wastes contain no identified hazardous chemicals at all. A whole new set of regulations has been imposed on landfills and underground storage tanks containing hazardous substances, which in many cases require expensive retrofitting of the disposal site to meet higher standards; landfill disposal of toxic liquid waste is banned. These and many other provisions clearly reflect growing recognition of the seriousness of the toxic-waste problem. Recognition of past carelessness in waste disposal, in turn, was the motivation behind the creation of "Superfund," discussed in chapter 15. A 1992 RCRA reauthorization attempt failed, but the legislation will be reintroduced, and broader pollution-control regulations, including provisions related to mining, could result. The details, however, are likely to depend on the political climate in which the reauthorization is passed.

The U.S. Environmental Protection Agency

In 1970, the U.S. Environmental Protection Agency was established in conjunction with the reorganization of many federal agencies. Among its primary responsibilities were the establishment and enforcement of air- and water-quality standards. Under the Toxic Substances Control Act of 1976, the EPA was given the authority to require toxicity testing of all chemical substances entering the environment and to regulate them as necessary. The EPA also has a responsibility to encourage research into environmental quality and to develop a body of personnel to carry out environmental monitoring and improvement. As the largest and best-funded

of the pollution-control agencies, it tends to have the most clout in the areas of regulation and enforcement. However, the scope of its responsibilities is broad, and expanding. The number and variety of situations with which the Environmental Protection Agency is expected to deal, restrictions imposed by limited budgets, and the varying emphases placed on different aspects of environmental quality by different EPA administrators are all factors that have contributed to limiting the agency's effectiveness. Legislation in the late 1980s and early 1990s tended to strengthen EPA's enforcement ability by establishing clear penalties for noncompliance with regulations. The CAAA, for example, allow EPA to impose penalties of up to $200,000 for violations, and even to pursue criminal sanctions against serious, knowing violations of CAAA. It is also true that efforts of the U.S. EPA are often supplemented by activities of state EPAs. But enforcement also requires staff and budget for investigations, legal actions, and so on. Sharp recent budget cuts for EPA, reflecting reduced priority in Congress for environmental-protection activities in general, will certainly reduce the EPA's ability to carry out its various mandates and monitoring and enforcement of activities.

Defining Limits of Pollution

Enforcement of antipollution regulations remains a major stumbling block, in part because of the terms of antipollution laws themselves. Consider, for example, an objective of "zero pollutant discharge" by a certain year. At first inspection, "zero" might seem a well-defined quantity. However, from the standpoint of practical science, it is not. Water quality must be measured by instruments, and each instrument has its own detection limits for each chemical. No commonly used chemical-analysis techniques for air or water can detect the occasional atom or molecule of a pollutant. A given instrument might, for example, be able to detect manganese down to a concentration of 1 ppm, lead down to 500 ppb (0.5 ppm), mercury to 150 ppb. Discharges of these elements below the instrument's detection limit will register as "zero" when the actual concentrations in the wastewater might be hundreds of parts per billion. How close one must actually come to zero discharge, then, depends strongly on the sensitivity of the particular analytical techniques and instruments used.

An alternative approach would be to specify a (detectable) maximum permissible concentration for pollutants discharged in wastewater or exhaust gases. But pollutants vary so widely in toxicity that one may be harmless at a concentration that for another is fatal. That implies a need to specify *different* permissible maxima for individual chemicals. Given the number and variety of chemicals involved, a staggering amount of research would be needed to determine scientifically the appropriate safe upper limit for each. Without all the necessary data, the Environmental Protection Agency and other agencies may set standards that later may be found to be too liberal or unnecessarily conservative. For some naturally occurring toxic elements, they have occasionally set standards stricter than the natural ambient air or water quality in an area. To meet the discharge standards, then, a company might be required to release wastewater cleaner than the water it took in! For instance, suppose the EPA guidelines specify a mercury concentration of 10 ppb or below in wastewater released. In an area with abundant mercury ore deposits, streams might naturally contain 100 ppb of mercury dissolved or leached from the ores. A company using that water, whether it added any mercury or not, could be required to remove some of the natural mercury to comply with the standards. A great deal of additional information, both about natural air and water quality, and about specific harmful effects of individual chemicals, is needed before safe and realistic limits can be set on all pollutant discharges.

No treatment process is perfectly efficient, and emissions-control standards must necessarily recognize this fact. For instance, 1982 incinerator standards established under the Resource Conservation and Recovery Act specify 99.99 percent destruction and/or removal efficiency for certain specified toxic organic compounds. While this is a high level of destruction, if 1 million gallons of some such material is incinerated in a given facility in a year, the permitted 0.01 percent would amount to a total of 100 gallons of the toxic waste released.

Imprecision in terminology remains a concern. The CAAA, for example, includes such expressions as "lowest achievable level" (of emissions; in the context of CFC emissions) and "nonessential" applications of CFCs and other controlled ozone-threatening substances (use in such applications banned beginning in 1992). In some instances, introduction of economic considerations further muddies the regulatory waters, both in the framing of laws and regulations, and in their implementation; see "Cost-Benefit Analysis" later in the chapter.

The Canadian Environmental Protection Act

During a period when U.S commitment to environmental protection—at least at the federal-government level—seems to be waning, it has been growing in neighboring Canada. The Canadian Environmental Protection Act (CEPA) was enacted in 1988, as a consolidation of previous provisions to control toxic substances, protecting health and the environment. Enforcement provisions were expanded and made more consistent, higher environmental-performance standards were set for the Canadian government's handling of its own lands, provisions were made for public input and for intergovernmental agreements. CEPA also mandated establishment of environmental-quality objectives, and regular assessments of and reports on the state of the Canadian environment.

CEPA included a requirement of a five-year review, recently completed. A number of accomplishments were identified. Some health benefits may be hard to isolate, but mandated phaseout of leaded gasoline has resulted in measurable drops in blood lead levels among Canadians, and

regulations have improved the handling of PCBs. Coordination between scientific-research agencies on one hand, and policymakers on the other, has been improved. Greater public awareness of environmental issues has increased public involvement in regulatory decision making, while greater consultation with industry at early stages has enhanced the satisfaction of industry with resultant regulations.

Looking ahead, those evaluating CEPA note that in the future, environmental protection will become an increasingly international venture, directed in part by intergovernmental agreements. They also noted "more burdensome fiscal pressures . . . in the 1990s" as a factor that will limit the funds available for environmental protection and require efficiency in implementation. Fiscal concerns in the U.S. have led to growing emphasis on cost-benefit analysis in environmental protection.

International Initiatives

Within the last decade, some notable steps have been taken toward international cooperation and collective action to address problems of global impact and concern. Some of these initiatives relate to problems that affect localized areas, widely distributed around the world—for example, the U.N. Convention to Combat Desertification, which at this writing was still in the negotiating stages. Others, however, recognize that all nations share some resources—the atmosphere especially—and no nation individually causes or can solve problems associated with pollution of such a common global resource (though to be fair, it is certainly true that some nations have historically contributed a disproportionately large share to that pollution).

One important outgrowth of the Earth Summit in Rio de Janeiro in 1992 was the United Nations Framework Convention on Climate Change. This Convention recognizes the role of atmospheric CO_2 and other greenhouse gases in global warming, and other possible climatic consequences. It calls for nations to report on their releases of greenhouse gases and efforts to reduce these, and to strive for sound management and conservation of greenhouse-gas "sinks" (plants, oceans), and calls upon developed nations to assist developing nations in such efforts. This Convention came into force following its ratification by over 50 nations in March 1994. Its very existence is significant, as is the promptness of ratification. The Montreal Protocol on Substances that Deplete the Ozone Layer, discussed in the previous chapter, is another example of such international commitments to pollution control for the common good.

These accords reflect a new concept in international law and diplomacy, the **precautionary principle.** Historically, nations' activities were not restricted or prohibited unless a direct causal link to some specific damage had been established. However, there are cases in which, by the time scientific certainty has been achieved, serious and perhaps irreversible damage may have been done. Under the precautionary principle, if there is reasonable likelihood that certain actions may result in serious harm, they may be restrained before great damage is done. So, given the known heat-trapping behavior of the greenhouse gases and documented increases in their atmospheric concentrations, nations agree to limit releases of greenhouse gases *before* substantial global warming or climate change has occurred; given the links between ozone depletion and increased UVB transmission, and between UV radiation and certain health risks such as skin cancer, they agree to stop releasing ozone-depleting compounds *before* significant, negative health impacts have occurred.

It is worth noting, however, that commitments in principle are one thing, actions another, and agreements such as these typically lack strong enforcement provisions. Where financial pressure or lack of technology make compliance difficult, nations' actions may lag behind their goals and promises.

Cost-Benefit Analysis

Problems of Quantification

Antipollution laws that specify using the "best available (treatment) technology economically achievable," or contain some similar clause that introduces the idea of economic feasibility, result in additional conflicts and enforcement problems. As noted in earlier chapters, costs for exhaust or wastewater treatment rise exponentially as cleaner and cleaner air and water are achieved. These costs can be defined and calculated with reasonable precision for a given mine, manufacturing plant, automobile, or whatever. Balanced against these cleanup costs are the adverse health effects or general environmental degradation caused by the pollutants. The latter are far harder to quantify and, sometimes, even to prove.

For example, the harmful effects of acid rain are recognized, and increased acidity of precipitation results, in part, from increased sulfur gas emissions. It is not possible, however, to say that, for each ton of sulfur dioxide released, X billion dollars worth of damage will be done to limestone and marble structures from acid rain, Y trees will have their growth stunted, or Z fish will die in an acidified lake. Even if such precise analysis were possible, what is the dollar value of a tree, a fish, or a statue? Conversely, scientists cannot say exactly how much environmental benefit would be derived from closing down a coal-fired power plant releasing large quantities of sulfur, and many people could suffer financial and practical hardship from the loss of that facility.

Similar debates arise in many areas. Allowing development on a barrier island may increase tourism and expand the local tax base, helping schools and allowing municipal improvements—but at what risk to the lives and property of those who would live on the vulnerable land? The dangers of increased atmospheric CO_2 are widely accepted, and fossil-fuel extraction may cause environmental disturbances of various sorts—but our energy consumption continues to

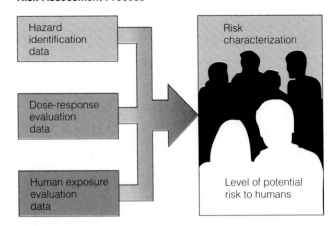

Risk-Assessment Process

Figure 18.9 Components of EPA's risk-assessment process. (For an explanation of dose-response curves, see chapter 20.)   Source: After D. E. Pattor, "The ABC's of Risk Assessment," *EPA Journal*, 19, Vol. 1, 1993, U.S. Environmental Protection Agency.

grow; clearly we tend to decide, implicitly at least, that the benefits outweigh the risks, perhaps, again, because the benefits are readily perceived, while the risks are very hard to quantify.

In the realm of pollution control, companies frequently contest strict regulation of their pollutant discharges on the grounds that meeting the standards is economically impossible and/or unreasonable. Each individual, each judge, and each jury has somewhat different ideas—of what is "reasonable"; of the value of clean air and water; of the importance of preserving wildlife, vegetation, natural topography, or geology; and of what constitutes an acceptably low health risk (assuming that that health risk can be quantified fairly precisely in the first place; see figure 18.9). Enforcement thus is very uneven and inconsistent. Virtually everyone agrees that, in principle, it is desirable to maintain a high-quality environment. But just what does that mean? What should it cost? And by whom should those costs be borne?

Cost-Benefit Analysis and the Federal Government

A month after assuming office in 1981, President Reagan issued Executive Order 12291, which decrees, in part, that a (pollution-control) regulation may be put forth only if the potential benefits to society outweigh the potential costs. A Regulatory Impact Analysis must be performed on any proposed new "major" regulation, which is, in turn, defined as one that is likely to result in either (1) an impact on the economy of $100 million or more per year; (2) a major cost increase to consumers, individual industries, governmental agencies, or geographic regions; or (3) significant adverse effects on competition, employment, investment, productivity, or the ability of U.S.-based corporations to compete with foreign-based concerns.

Box 18.1 Future Directions in Environmental Law: Solar-Access Law

As increasing numbers of people incorporate solar-energy devices and designs into buildings, concern about solar access correspondingly increases. For example, how can laws best protect the right of one property owner to use solar energy without unduly restricting the right of another to build a new structure or grow tall shade trees, or the right of a neighborhood to guard against construction that is unsightly or inharmonious with its surroundings?

There are many options. Solar-access impact could be an added consideration in environmental-impact analysis. Legislation to restrict public and/or private vegetation that affects solar access could be enacted. Zoning ordinances relating to building height, orientation, and positioning could be reexamined from the standpoint of solar access to see if the restrictions make incorporation of solar-design elements unreasonably difficult.

The extent to which solar access is a concern naturally varies with locality. So far, only in areas of strong insolation (for example, New Mexico and California) have states been moved to act. California, for instance, has gone so far as to limit the extent to which vegetation may shade solar collectors on adjacent properties and has provided for fines for violation of the law. In most places, however, the legal battle is being fought on a case-by-case basis as lawsuits are filed. If, as seems likely, development and use of solar-energy technology continue to expand, the need for workable, consistent solar-access laws will be emphasized.

This represented a major departure from previous policy. The Clean Water Act orders the Environmental Protection Agency to be sensitive to costs in establishing standards, but it does not require a cost-benefit analysis; the Clean Air Act prohibits considering costs in setting health standards. Cost-benefit analysis is still not the decisive factor in regulation, but it is at least taken into account.

The first area to which cost-benefit analysis was applied under the new order was the EPA's proposed effluent guidelines for iron and steel manufacturers. The agency evaluated the costs for various levels of enhanced pollutant removal and ultimately selected a set of treatment technologies and standards that, it was estimated, would increase total steel plant operating costs by less than 1%. Balanced against the costs were the benefits. In this case, the principal benefits anticipated were sufficient water-quality improvements to allow fishing and other recreational uses of the affected streams below the steel plants, analogous to existing uses above the plants. The analysis was necessarily limited in scale: Projections of costs and benefits were extrapolated from three specific rivers to the dozens affected over the industry; assumptions had to be made about future costs of pollution-control measures. Various combinations of parameters led to varying estimates of the likelihood that the benefits would outweigh the costs. The estimates ranged from a minimum of 44% to over 95%, although, for the great majority of cases, the probability was well over 50%. Therefore, it appeared probable that the measures were cost-effective. Cost-benefit analyses have since been undertaken with respect to other industries, including the organic-chemical, plastic, and synthetic-fiber industries, and more such analyses are to be anticipated in the future.

Laws Relating to Geologic Hazards

Laws or zoning ordinances restricting construction or establishing standards for construction in areas of known geologic hazards are a more modern development than antipollution legislation. Their typical objective is to protect persons from injury or loss through geologic processes, even when the individuals themselves may be unaware of the existence or magnitude of the risk. While such an objective sounds benign and practical enough, such laws may be vigorously opposed (see especially chapters 4 and 6). Opposition does not come only from real-estate investors who wish to maximize profits from land development in areas of questionable geologic safety. Individual homeowners also often oppose these laws designed to protect them, perhaps fearing the costs imposed by compliance with the laws. For example, engineering surveys may be required before construction can begin, or special building codes may have to be followed in places vulnerable to earthquakes or floods. People may also feel that it is their right to live where they like, accepting any natural risks. And some, of course, simply do not believe that whatever-it-is could possibly happen to *them*.

A

B

Figure 18.10 Hazard-mitigation legislation rarely requires privately owned buildings to be modified for greater resistance; the results can be unfortunate. (*A*) The first story of this two-story house completely collapsed in the Loma Prieta earthquake. (*B*) This apartment complex was severely damaged in the 1994 Northridge quake. (*A*) Photograph by D. Schwartz, (*B*) photograph by E.V. Leyendecker, both courtesy of U.S. Geological Survey.

Occasionally, lawsuits have been filed in opposition to zoning or land-use restrictions, on the grounds that such restrictions amount to a "taking"—depriving the property owner of the right to use that property freely—without compensation. Such action is prohibited by the Fifth Amendment to the Constitution. However, the courts have generally held that such restrictions are, instead, simply the government's exercise of its police power on behalf of the public good and that compensation is thus unwarranted.

Restrictive building codes can be strikingly successful in reducing further damage or loss of life. After a major earthquake near Long Beach, California, in 1933, the California legislature passed a regulation known as the Field Act, which created improved construction standards for school buildings. Many of those standards are still being followed. Their effectiveness was demonstrated in subsequent earthquakes, notably the major one in San Fernando in 1971. It was noted afterward that damage to school buildings from the earthquake was almost wholly limited either to schools built prior to 1933 or to those later buildings whose engineers had been permitted to deviate from strict compliance with the Field Act. The problem of pre-existing structures remains a major problem rarely addressed even by "good" building codes aimed at hazard reduction, as was illustrated yet again in the 1989 Loma Prieta earthquake, and the 1994 Northridge quake. There had been talk of "retrofitting" some older buildings, freeways, and bridges to improve safety in accordance with newer design specifications, but due to cost and inertia, many modifications went unmade, and structural failures resulted. Retrofitting of privately owned buildings may be even rarer (figure 18.10).

Controls on Construction in Areas of Landslide Hazard

Perhaps because the state is subject to so many geologic hazards, California and its municipalities have been among the leaders in passage of legislation designed to minimize the damage from geologic hazards. One successful effort in this regard has been the imposition of much more stringent regulations governing construction on potentially unstable, landslide-prone hillsides near Los Angeles. Over a ten-year period of rapid development (1952–62), increasingly strict requirements for slope and soil stability studies, site grading, and construction engineering were developed. The resulting reduction in the rate of landslide damage has been dramatic (table 18.2). Although the costs of structural damage to those homes actually damaged by landslides or soil failure has continued to be high (probably, in part, as the result of escalating property values), the number and percentage of individual units damaged have both dropped sharply, especially since 1963. Plainly, at least where the geologic hazards are moderate, the risk of damage from the hazards can be minimized by taking them into account and "designing around" them. This is a large part of engineering geology (see chapter 19).

Response to Earthquake Hazards

Not all hazard-mitigation legislation is equally effective, as evidenced by a few examples prompted by the 1971 San Fernando earthquake. Each of these bills is well-intentioned and represents a positive step toward reducing earthquake damage. Collectively, however, they also illustrate a number of weaknesses that limit their effectiveness.

Table 18.2   Comparison of Landslide Damage With and Without Restrictive Construction Guidelines in Los Angeles, California

	Prior to 1952	1952–62	1962–67
controls pertaining to landslide hazards	none	limited grading codes in place	much stricter requirements for engineering and soil studies; modern grading procedures required
sites constructed (approx.)	10,000	27,000	11,000
sites damaged (%)	1,040 (10.4%)	350 (1.3%)	17 (0.2%)
total damage to residences (approx.)	$3,300,000	$2,800,000	$80,000
average damage per site built	$330	$100	$7
average cost per site damaged	$3,200	$7,900	$4,700

Source: Data from J. F. Slosson, *The Role of Engineering Geology in Urban Planning*, Colorado Geological Survey Special Publication 1, 1969.

The Seismic Safety Element Bill (1971) requires that communities take seismic and related hazards (ground shaking, tsunamis, and so forth) into account in planning and regulating development. This is plainly sensible in a high-seismic-risk area. But the bill does not prohibit or regulate construction explicitly; public officials are left to decide what constitutes acceptable risk in siting and building. The qualifications of the person(s) evaluating the seismic risks are not stipulated either, and there is no penalty if the community actually ignores the bill altogether.

The Dam Safety Bill (1972), despite its name, has nothing directly to do with the engineering adequacy of dams. It requires dam owners to prepare maps showing the areas that would be flooded in the event of dam failure, and local authorities must prepare evacuation procedures for those areas threatened. Building in the areas at risk is not prohibited, and those living there may not necessarily be made aware of the dangers.

The Alquist-Priolo Geologic Hazards Act (1972; amended 1975) is a more specific and detailed bill. Under its provisions, the state geologist identifies "Special Studies Zones" along active faults. Proposed construction in these zones requires review by local authorities, who might, for instance, require that buildings not be placed within a certain distance (25, 50, or 100 meters, perhaps) of the fault trace proper. Of course, in the case of a major fault such as the San Andreas, which can cause substantial damage tens or hundreds of kilometers from the epicenter of a large earthquake, it would be of little use to worry about whether one's house was 10 or 100 meters from the fault line. Local officials are expected to consider the size of earthquakes and the extent of damage anticipated along each particular fault in deciding where or whether to allow building. Engineering and site-evaluation studies can be costly, and this prompted the 1975 amendment that *exempted* individual single-family homes. (Housing developments, apartments, and commercial buildings are still covered.) Once again, economic considerations vie with scientific ones in selecting a course of action.

Flood Hazards, Flood Insurance

When geologic catastrophes occur, federal disaster-relief funds are often part of the rescue/recovery/rebuilding operations. In other words, all taxpayers pay for the damage suffered by those who, through ignorance or by choice, have been living in areas of high geologic risks. This has struck some as inequitable and is part of the motivation behind the Flood Insurance Act (1968) and its successor, the Federal Flood Disaster Protection Act (1973). These measures provide for federally subsidized flood insurance for property owners in identified flood-hazard areas, whether in stream floodplains or in flood-prone coastal regions. Those most at risk, then, pay for insurance against their possible flood losses. The idea is that, in time, the flood insurance will replace after-the-fact disaster assistance in flood-prone areas. To "encourage" property owners to purchase the insurance, the legislation provides that it be required of those obtaining federally insured mortgages or mortgages through federally affiliated banks and lending institutions. For a community to remain eligible for the insurance program, it must, in turn, enact strict floodplain-zoning regulations. Initially, all proposed new construction in the floodplain must be carefully evaluated and structures designed so as to minimize potential flood damage. Later, after the Department of Housing and Urban Development has provided detailed floodplain maps, still stricter provisions are enacted that require floodproofing or elevation above the one-hundred-year flood level of new structures in the floodplain. The community may wish to include additional provisions, such as one requiring that any water-storage volume within the one-hundred-year floodplain eliminated or filled in by emplacement of new structures be compensated for by creation of at least an equal volume of additional floodwater storage elsewhere.

Figure 18.12 Landslide in Potrero Canyon, Pacific Palisades area, California after 7 inches of rain in January 1956. Later building codes made newer structures safer but left old ones in areas of known risk.   Photograph by J. T. McGill, U.S. Geological Survey.

Problems with Geologic-Hazard Mitigation Laws

A pervasive weakness in all kinds of geologic-hazard mitigation laws, including all of the examples discussed in this section, is the fact that they typically apply only to *new* structures. Only new schools in California need conform to Field Act building codes; only new hillside homes need be built using modern slope-stabilization methods; only new federally insured mortgages on flood-prone properties mandate that the property owner carry flood insurance; and so on. Where considerable development has preceded the legislation, those affected by the new law in the near term may be a very small percentage of those at risk: In the 1993 Mississippi River basin floods, many thousands of victims were uninsured, aware of neither the need for nor the availability of flood insurance. Where new construction is banned in very-high-risk areas, older structures nevertheless persist. A small move toward eliminating past poor practice would be a regulation requiring that, if an older structure in a geologically high-risk area were destroyed—whether by geologic processes or another means, such as fire—it could be rebuilt only if the new structure conformed to the newer laws for floodproofing, fault-trace setback, slope grading, and so forth. Even modest proposals of this kind, however, commonly meet with vocal citizen opposition.

Thus, laws designed to reduce the risk of damage from geologic hazards have several problems in common. The basic scientific information on which to base sensible regulations may be lacking. Laws may be poorly written, failing to specify fully what must be done and by whom, or they may be weakened by exceptions or omissions that exempt many from their provisions. A major omission, and a very common one, is that existing structures in areas at risk may continue unaffected (figure 18.12).

Figure 18.11 Development on Hatteras Island, North Carolina, is extremely vulnerable to flooding, particularly during storms.   Photograph © Frank M. Hanna/Visuals Unlimited.

The requisite flood-hazard maps can be slow in coming. As mentioned in chapter 6, detailed records collected over long periods, and careful analysis, are required to construct accurate maps indicating precisely the areas to be affected by fifty-year, one-hundred-year, or two-hundred-year floods. In many communities, detailed maps will not be available for years, and, meanwhile, development goes on. The laws do not completely ban further building in the flood-prone area, and as already noted, if more buildings—"flood-proof" or not—are placed in a floodplain, their very presence may in some measure increase flood stages.

An unintended and unfortunate side effect of the availability of subsidized flood insurance has been to encourage people to rebuild, sometimes several times, in severely flood-prone areas, and in a sense has encouraged continued development in such areas. The costs to the government can then far exceed the premiums collected. In some unstable coastal zones, this has happened to such an extent that Congress has voted to remove the most severely threatened areas from the program (figure 18.11; see also chapter 7).

The National Environmental Policy Act (1969)

The National Environmental Policy Act (NEPA) established environmental protection as an important national priority and provided for the creation of the Council on Environmental Quality in the Executive Office of the President. The **environmental impact statement** (EIS) is the most visible outgrowth of the NEPA. In this section, we will consider briefly what goes into an EIS, what the point is, and how well the legislation's objectives are being met.

The NEPA actually pertains only to federal agencies and their actions. Whenever a federal agency proposes legislation or "other major Federal action" (which can include anything from policy changes to construction projects that are wholly or partially supported by federal funds) that can significantly affect the quality of the human environment, an EIS must be prepared. Many states have adopted similar legislation related to projects involving funding or approval by state agencies, so the scope of the environmental impact statement process has been considerably broadened. The preparation of environmental impact statements should ensure that, before an agency acts, it considers as fully as possible the potential environmental ramifications of its action, in order to make the best possible decisions. The discussion that follows is restricted to federally mandated EISs.

The NEPA specifies that an environmental impact statement should include:

1. A description of the proposed action, its purpose, and why it is needed
2. A discussion of various alternatives (including the proposed action)
3. An indication of the environment to be affected and the environmental consequences anticipated
4. Lists of preparers of the statement and those agencies, organizations, or persons to whom copies of the statement are being sent

Additional supplementary material may also be included. The statement is expected to "rigorously explore and objectively evaluate all reasonable alternatives" (an analysis that presumably will ultimately supply appropriate justification for the particular course of action preferred by the agency). Possible environmental consequences might include not only those that are in some sense geologic (for example, altering natural processes like stream flow or runoff, causing air or water pollution, inevitable consumption of energy or other resources), but also biological ones (loss of habitat, destruction of organisms) and social ones (affecting urban quality, employment, historic or cultural resources). Cost-benefit analyses of various alternatives may be included if the agency is taking them into account.

After a draft environmental impact statement is prepared, a variety of others, ranging from other federal agencies to the general public, are invited to comment. In fact, comments should be actively solicited from all persons or organizations that are particularly likely to be interested or affected. The

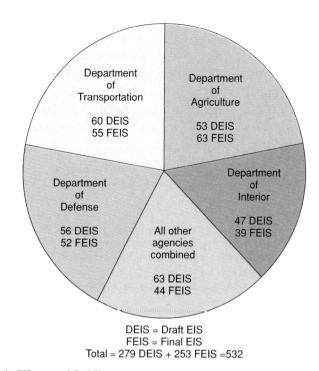

DEIS = Draft EIS
FEIS = Final EIS
Total = 279 DEIS + 253 FEIS =532

Figure 18.13 Distribution of EIS filings with the U.S. EPA in 1994, by agency.   Data from Council on Environmental Quality, 1995.

agency preparing the EIS is then expected to respond to those comments, which might mean modifying the proposed action or the analysis in the EIS or considering additional alternatives.

Over the first decade following passage of the NEPA, an average of more than one thousand environmental impact statements each year were prepared by federal agencies. Lawsuits were filed contesting about 10 percent of this number of projects. In many such cases, it was alleged that the EIS was inadequate, incomplete, or inaccurate. In other cases, the charge was that an EIS ought to have been prepared but was not (the responsible agency felt that no significant environmental impact was involved). Close to half the lawsuits involved citizen or environmental groups as plaintiffs. The number of environmental impact statements filed annually began to decline in the early 1980s; but the number of lawsuits rose, peaking at 157 in 1982; by the mid-1980s, the number of cases filed began to drop off sharply in turn.

As might be expected, the number of cases filed against a particular agency is broadly proportional to the number of EISs that it files. A small number of agencies account for most of the EISs. In 1994, for example, 80 percent of EISs were filed by just four agencies (figure 18.13): the Department of Defense, (especially in connection with U.S. Army Corps of Engineers and other projects related to navigation, dredging/filling, and watershed protection/flood control); Department of Agriculture (especially in connection with forestry and range management, watershed protection/flood control, and use of pesticides and herbicides); Department of the Interior (especially relative to recreation areas, resource management, and mining and oil/gas drilling activities); and

Box 18.2 EIS Examples: To Build and Not to Build

The Chevron Phosphate Project alluded to in the text, which involved construction of a plant to produce phosphate fertilizer (with phosphate from existing mines), and considerable supporting and transport facilities, raised a number of concerns. Chief among these were air and water quality, resource consumption (water), resource displacement (vegetation), visual impact, possible impacts on both wildlife and grazing, socioeconomic impacts, and possible local slope-stability problems. In response to the draft EIS, 24 "substantive" comment letters (and others expressing opinions) were submitted; the comments came from a great variety of sources, including the U.S. Department of Transportation in Washington, D.C.; the Department of the Army; the Sierra Club; the U.S. Geological Survey; the Office of the Governor and the Division of State History of the State of Utah (through which portions of slurry pipeline would pass); Pacific Power and Light Company; the U.S. Fish and Wildlife Service; and several private citizens. Point-by-point responses to concerns were made. Examples: Water use by the plant would

decrease the flow of the Green River by about half a percent, and increase salinity of water behind Imperial Dam by about 1 ppm. While the slurry pipeline might cost jobs for 40 truck drivers and about 20 other workers, a net gain of 30 jobs was projected overall. The loss of several thousand acres of deer, elk, and antelope habitat was judged negligible given the total wildlife habitat in the region. There was an unresolved issue about possible adverse impact of wastewater on two endangered fish species, the Colorado squawfish and the humpback chub. Construction would cause unquantifiable losses of fossils preserved in the rocks, but an on-site paleontologist would be available to judge whether fossils of particular importance were in jeopardy. On balance, the responsible agency—the Bureau of Land Management—supported the project, judging negative impacts to be moderate or manageable.

By contrast, one might expect a proposal to designate a portion (about 11 percent, or 300,000 acres) of an Alaskan national preserve (Bering Land Bridge) as a wilderness area to be relatively noncontroversial. Not so.

Alternatives included no wilderness designation, and wilderness designation for most or all of the 2,690,179 acres under consideration. Briefly, wilderness designation would preserve the land's natural condition, banning permanent roads, commercial enterprises, permanent structures, and motor vehicles, among other intrusions (see figure 1). Thus, wilderness areas cannot be extensively developed for tourism, mining, or many other activities, and that factor entered into some of the concerns expressed, being an undeniable economic consequence. Other issues related to village-to-village travel needs of native peoples, recreational uses of the land, impacts on wildlife, subsistence land and resource use by native peoples, and impacts on historic and prehistoric cultural resources. Some 63 comment letters arrived, and again the commenters were a diverse group, including: U.S. Bureau of Mines; Alaska Department of Labor; the National Parks and Conservation Association; the Shishmaref Native Corporation; the Federal Highway Administration; the Reindeer Herders Association; the Alaska Miners Association; Pacific

Department of Transportation (mostly relating to road construction). The latter two agencies were among the most frequent defendants in EIS lawsuits every year from 1975 to 1985. Agriculture and the Corps of Engineers appear on the frequent-defendant list less often, joined at various times by the Departments of Defense and of Housing and Urban Development, the Federal Energy Regulatory Commission, Nuclear Regulatory Commission, and the EPA.

The usefulness of environmental impact statements has been limited by a number of factors. We have already noted in the context of antipollution laws the difficulty of applying cost-benefit analysis to situations where not all factors have well-defined costs associated with them. Another potential problem is the impartiality (or lack of it) of EIS preparers or

proposers. An agency preparing an EIS for a project of its own has already selected a proposed course of action, and it might, consciously or otherwise, be inclined to present that action in an unduly favorable light. Where the proposed action is to be carried out by an individual or corporation—such as a company needing a federal permit to mine on federal lands—the EIS may, in part, draw on information supplied by the proposer, which could again be biased. There are also only general statutory requirements that the analysis in an EIS be thorough, which is no guarantee that all the appropriate kinds of specialists will be involved in the preparation of a given EIS. Audits of the accuracy with which EIS preparers predicted environmental impacts have often shown, in retrospect, rather low accuracy.

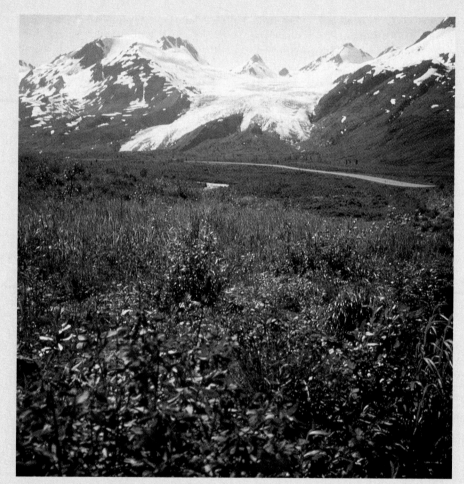

Legal Foundation; and others. The proposed action was believed to be a viable compromise between preservation and development (the commentors having taken the full range of positions from all-wilderness to no-wilderness). With the bulk of the land *not* designated as wilderness, further development of tourism, mining, etc., would be possible. Current use of the proposed 11 percent was low and known mineral or other exploitable resources minimal, so removing that land from prospective development should not cause great harm—yet it did still represent a sizeable reserve of wilderness, being, as one letter-writer pointed out, about half the size of Rhode Island!

These are just two of many examples of the complexity of the EIS process.

Figure 1 Preservation of undeveloped land, like this in Denali National Park, Alaska, may protect habitat and scenery, but those benefits may have to be balanced against prospects for discovery of needed resources, economic development, possible pollution, and other potential impacts.

A major purpose of the National Environmental Policy Act is to allow the public to be informed about and involved in decision making related to the environment. To prevent environmental impact statements from becoming prohibitively long and verbose, the NEPA stipulates that the length should normally be 150 pages or less, or up to 300 pages for especially complex projects. However, "tiering" is allowed, whereby one EIS may be written for each level of decisionmaking of a large or multistage project. One might then have to read through several statements of 150–300 pages each to comment intelligently on a particular project. Many people do not have the time to do that much reading, so they may not become involved at all. (For that matter, not all the relevant employees of the federal agencies involved may do their EIS homework adequately either, for the same reason.)

Nor can one always project the length of the EIS from the size of geographic area affected or the apparent complexity of the project. For example, a proposal to designate a portion of the Bering Land Bridge National Preserve as a wilderness area resulted in a 300-page final EIS. This was longer than the final EIS for a project in Wyoming that involved building a phosphate fertilizer plant, phosphate slurry pipeline, railroad spur, microwave communication system, and electrical power transmission system, and relocating a road! See box 18.2 for some highlights of these analyses, and box 18.3 for an example of one of the larger-scale projects for which environmental impact assessment has been done.

Box 18.3 An Exercise in Impact Analysis on a Grand Scale: The Trans-Alaska Pipeline

The discovery of economically important reserves of oil along Alaska's North Slope raised the immediate question of how to transport that oil to refineries in the lower forty-eight states. It was proposed that a pipeline be constructed to carry the oil south to the port of Valdez, from which tankers could move the oil further south. The planned route (figure 1A) crossed over 1,000 kilometers of federal lands, which necessitated a permit from the government; hence the requirement of an environmental impact statement.

Various possible alternatives to the proposed route included alternate pipeline routes across Alaska, transportation by some combination of pipeline, rail, and highway across both Alaska and Canada, and tanker transport directly from the North Slope (figure 1B). Transport exclusively by tanker was ultimately deemed impractical, for the climate of northern Alaska is harsh, and the harbors and surrounding seas are blocked by ice much of the year. Schemes involving pipelines across Alaska and subsequent tanker transport posed threats to both terrestrial and marine environments; routes involving only pipelines through Alaska and Canada would leave marine life intact but would disturb far more land. A pipeline through Canada would also necessarily be under the jurisdiction of a foreign government, although generally good U.S.-Canada relations did not suggest that this would be a likely source of problems.

The possible environmental impacts were many and varied. Construction and operation of overland pipelines could disturb the ground, water, vegetation, fish, and wildlife. Land must be committed to the project, which might cost wildlife habitats or inhibit migration of land animals. Warm oil in an underground pipeline could melt frozen ground, with possible subsidence of and damage to the pipeline in the resulting mud. There would be some thermal pollution from the heat released by the warm oil and a possibility of oil spills onto land or into freshwater lakes. The pipeline would cross major fault zones; southern Alaska, in particular, is subject to severe earthquakes that might rupture a pipeline.

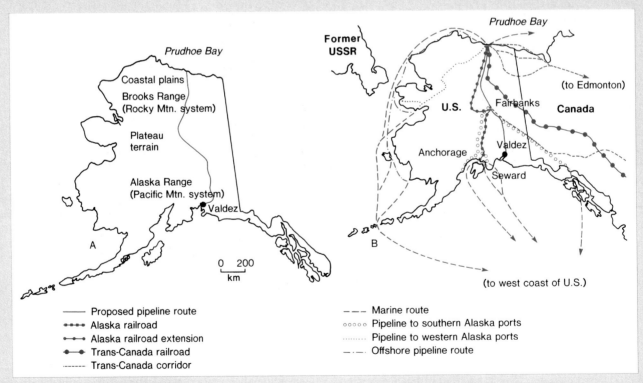

Figure 1 (*A*) Route of the Trans-Alaska Pipeline. (*B*) Proposed alternative means for transport of the oil. Source: After D. A. Brew, *Environmental Impact Analysis: The Example of the Trans-Alaska Pipeline,* U.S. Geological Survey Circular 695, 1974.

The use of tankers would mean possible marine oil spills also, and heavy tanker traffic could affect commercial fishing operations. On the plus side, pipeline construction would create jobs, besides providing access to major oil reserves. The influx of money and work crews and the impacts of a pipeline on the physical environment, however, would also significantly affect the lives of Alaskan natives, perhaps negatively.

On balance, the proposed route was deemed to have the least overall adverse impact (figure 2). Principal unavoidable impacts were: disturbances of terrain, fish, wildlife, and some human environments; some pollution at the port of Valdez from oil discharge from treatment of tanker ballast water; and secondary effects from influx of more people into the region. (We now know that tanker accidents can also be a significant problem.)

The analysis was a large and complex task, reflected in the size of the final environmental impact statement, which was six volumes long! However, such detailed analysis allowed the anticipation and minimization of many potential problems and the reduction or elimination of a variety of negative environmental effects. (Some examples are cited in chapter 19.)

Source: D. A. Brew, *Environmental Impact Analysis: The Example of the Trans-Alaska Pipeline*, U.S. Geological Survey Circular 695, 1974.

Figure 2   The size of the pipeline project, and character of much of the land it crossed, made extensive impact analysis necessary.
© Tom and Pat Leeson/Photo Researchers, Inc.

A major loophole of the NEPA, in the eyes of many environmental groups, is that statements must be prepared only when the anticipated environmental impact of a proposed project is "significant," where significance is evaluated by the federal agency involved in proposing or permitting the action in question. For each project for which an EIS is filed, many more have been deemed not to warrant one. In short, as with other kinds of environmental laws, the NEPA has some very desirable objectives, but practice has often fallen short of the ideal.

The Canadian Environmental Assessment Act (CEAA) became effective in January 1995. Like NEPA, CEAA strives to assure "an open and balanced process to assess the environmental effects of projects requiring federal action or decisions," with public input at an early stage. Unlike NEPA, CEAA identifies four levels of scrutiny, the degree of assessment required for each project dependent upon the size and complexity of the project and its probable impacts, and also provides for mediation as a means of resolving conflicts among parties concerned with a particular project. A new agency, the Canadian Environmental Assessment Agency, has been created to oversee assessments conducted under CEAA. It will be interesting to see whether this Canadian analogue to the U.S. NEPA will prove more effective at avoiding some of NEPA's problems.

Overview

Summary

Environmental laws relating to geologic matters include laws relating to the right to exploit certain resources, laws designed to limit pollution and other kinds of environmental deterioration, and laws intended to force individuals and agencies to take geologic hazards into account in development and construction. They may fall short of their goals for a variety of reasons. Terms may be inadequately defined or scientifically meaningless, making consistent enforcement difficult. Substantial economic pressure may force pursuit of a course that may be less desirable in purely scientific terms, or a rigorous cost-benefit analysis may be impossible. There may be individual or institutional resistance or bias to combat, or a particular law may recognize so many exceptions that the majority of cases ultimately end up unregulated. These and other difficulties, many without straightforward solutions, will continue to complicate efforts to develop environmental legislation. The future may see increasing development of intergovernmental agreements on critical global environmental issues.

Terms to Remember

environmental impact statement 457	Prior Appropriation 441
Exclusive Economic Zone 444	Riparian Doctrine 441
precautionary principle 451	

Exercises

Questions for Review

1. Compare and contrast the basic concepts of the riparian and appropriation doctrines underlying much surface-water law.
2. Why are groundwater rights inherently somewhat more difficult to define than surface-water rights?
3. What was the principal objective behind early (nineteenth-century) federal mineral-resource laws? How has the emphasis shifted over the last century?
4. What are exclusive economic zones? Give two examples of types of mineral resources they might encompass.
5. Discuss some of the difficulties of defining and achieving "zero pollutant discharge."
6. Recent changes in federal policy have changed the degree of emphasis put on cost-benefit analysis in setting pollution-control standards. Explain briefly.
7. Cite two common problems with construction or zoning restrictions that limit their effectiveness, particularly in densely populated areas.
8. Summarize the kinds of information included in an environmental impact statement.
9. Cite and explain at least two features of the EIS process that may limit its effectiveness.

For Further Thought

1. Seek out the environmental impact statement for a local project (offices of the Army Corps of Engineers, Department of the Interior, state highway department, or other governmental agency involved in development/construction projects may be able to suggest sources of information). Note what kinds of impacts are projected and how their importance is evaluated. Is the EIS comprehensible to you as an interested individual? Would you suggest any areas in which more thorough analysis seems to be needed?
2. Consider what restrictions, if any, you as a legislator might place on development in a floodplain, on landslide-prone hills, or in active fault zones. If your community is subject to such geologic hazards, look up existing ordinances to see how extensive their restrictions are, and inquire about how routinely variances and exceptions are granted.

3. Pick an item of environmental legislation, and investigate its current status. You might choose the reauthorization of the Clean Water Act, Clean Air Act Amendment, or CERCLA, or revisions of the 1872 Mining Law; the review of the Canadian Environmental Protection Act; or the ratification status of one of the U.N. climate conventions. See NetNotes for some possible sources of current data with which to start.

Suggested Readings/References

Arbuckle, J. G., et al. 1985. *Environmental law handbook.* 8th ed. Rockville, Md.: Government Institutes.

Brew, D. A. 1974. *Environmental impact analysis: The example of the proposed Trans-Alaska Pipeline.* U.S. Geological Survey Circular 695.

Bureau of Land Management. 1983. *Final environmental impact statement on the Chevron Phosphate Project.* U.S. Department of the Interior.

Council on Environmental Quality. 1980. *Environmental quality 1979.* Washington, D.C.: U.S. Government Printing Office.

———. 1984. *Environmental quality 1983.* Washington, D.C.: U.S. Government Printing Office.

———. 1988. *Environmental quality 1986.* Washington, D.C.: U.S. Government Printing Office.

Evans, J. E. 1994. A course in geology and public policy. *Journal of Geological Education* 42: 10–16. (Includes references to many information sources.)

Fawcett, J. E. S., and Parry, A. 1981. *Law and international resource conflicts.* Oxford, England: Clarendon Press.

Firestone, D. B., and Reed, F. C. 1983. *Environmental law for non-lawyers.* Ann Arbor, Mich.: Ann Arbor Science.

Freedman, W. 1987. *Federal statutes on environmental protection.* New York: Quorum Books.

Hayes, G. B. 1979. *Solar access law.* Cambridge, Mass.: Ballinger.

Heer, J. E., Jr., and Haggerty, D. J. 1977. *Environmental assessments and statements.* New York: Van Nostrand Reinhold.

Hoban, T. M., and Brooks, R. O. 1987. *Green justice: The environment and the courts.* Boulder, Colo.: Westview Press.

Jain, R. K., Urban, L. V., and Stacey, G. S. 1981. *Environmental impact analysis.* New York: Van Nostrand Reinhold.

Keating, M. 1993. *The Earth Summit's agenda for change.* Geneva, Switzerland: Centre for Our Common Future.

McGregor, B. A., and Offield, T. W. 1987. *The Exclusive Economic Zone: An exciting new frontier.* U.S. Geological Survey.

McKelvey, V. E. 1982. Law of the sea: Implications for the United States. *Geotimes* (October): 21–22.

National Park Service. 1988. *Final environmental impact statement: Wilderness recommendation, Bering Land Bridge National Preserve, Alaska.* U.S. Department of the Interior.

Orloff, N. 1978. *The environmental impact statement process.* Washington, D.C.: Information Resources Press.

Rosenbaum, W. A. 1985. *Environmental politics and policy.* Washington, D.C.: Congressional Quarterly Press.

Shaw, B. 1976. *Environmental law.* St. Paul, Minn.: West.

Sloan, I. J. 1979. *Environment and the law.* 2d ed. Dobbs Ferry, N.Y.: Oceana Publications.

Tank, R. W. 1983. *Legal aspects of geology.* New York: Plenum Press.

Teclaff, L. A., and Utton, A. E., eds. 1974. *International environmental law.* New York: Praeger.

U.S. Environmental Protection Agency. 1991. The new Clean Air Act: What it means to you. *EPA Journal* 17 (January/February).

———. 1991. Setting environmental priorities: The debate about risk. *EPA Journal* 17 (March/April).

———. 1993. Profiles in risk assessment. *EPA Journal* 19 (January/February/March).

World Resources Institute. 1992. *The 1992 Information Please Environmental Almanac.* Boston: Houghton Mifflin.

World Resources Institute, United Nations Environment Programme, United Nations Development Programme, and The World Bank. 1996. *World Resources 1996–97.* (Chapter 14, "Atmosphere and Climate.") New York: Oxford University Press.

Zumberge, J. H. 1979. Mineral resources and geopolitics in Antarctica. *American Scientist* 67: 68–76.

The American Geological Institute's governmental-affairs arm tracks U.S. environmental legislation and provides links to other information sources. See AGI's home page at
http://jei.umd.edu/agi/agi.html

NEPANET provides up-to-date information on EISs filed with the U.S. Environmental Protection Agency; see
http://ceq.eh.doe.gov/nepa/nepanet.htm

Materials from the Canadian Environmental Assessment Agency can be found via
http://www.ceaa.gc.ca:80/
This site offers links to comprehensive discussions of the Canadian Environmental Assessment Act.

For more complete information on the review of the Canadian Environmental Protection Act, see
http://www.doe.ca/cepa

History of and current information on the status of the Convention to Combat Desertification can be found via
http://www.iisd.ca/linkages/desert.html

The USGS Water Resources Division maintains "Environmental Impact Analysis Data Links" on a variety of topics; see at
http://h2o.usgs.gov/public/eap/env_data.html
Its Environmental Affairs Program (EAP) includes information on NEPA generally, and EIS's referred to USGS for comment specifically, accessible via the EAP home page at
http://h2o.usgs.gov/public/eap/eappage.html

A wealth of information on the Montreal Protocol specifically and ozone generally is available via the Ozone Secretariat of the U.N. Environment Programme at
http://unep.org/unep/secretar/ozone/home.html
Comprehensive information on the UN Framework Convention on Climate Change and implementation activities can be found at
http://www.unep.ch/iucc.html

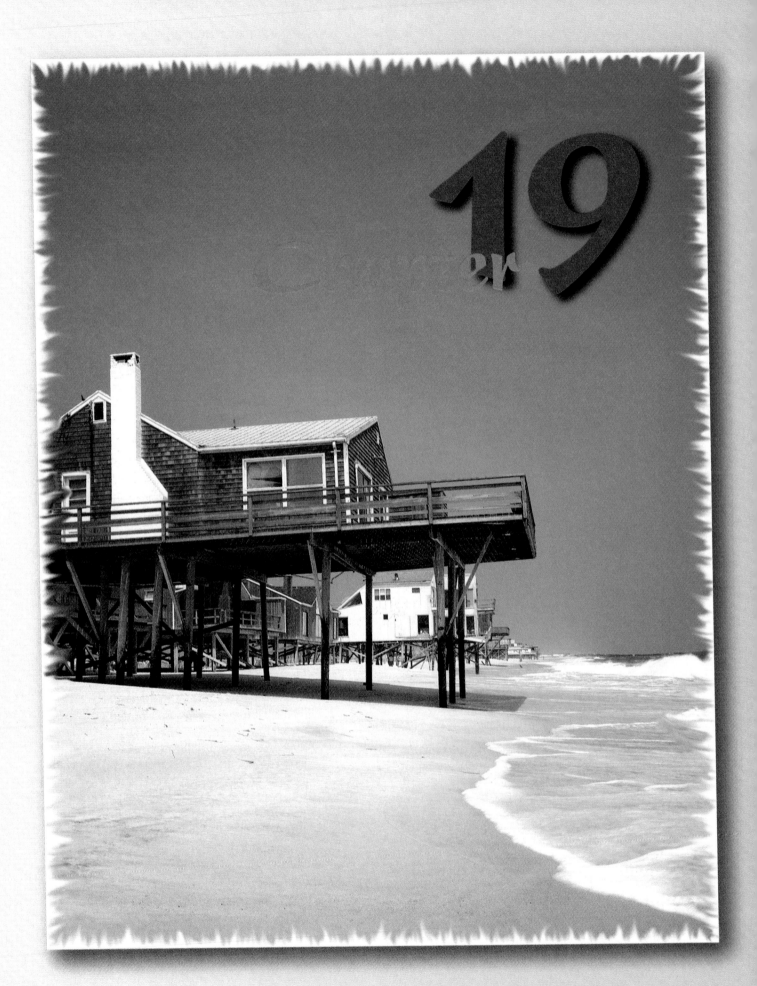

Chapter 19

Land-Use Planning and Engineering Geology

The purpose of land-use planning is to make the best, most sensible, practical, safe, efficient use of each parcel of land. Because such decisions are based, in part, on geologic considerations, land-use planning and engineering geology overlap. In general, however, land-use planning involves a much larger body of geologic and nongeologic facts and concerns. These may, for instance, include potential economic or practical benefits from a given use of the land and possible negative environmental or aesthetic impacts. Often, land-use planning takes the form of assessing the suitability of a particular parcel of land for a particular purpose and proceeds somewhat like an environmental-impact-statement assessment.

A frequent problem in land-use planning, as in other areas such as cost-benefit analysis in pollution control or evaluation of energy options, is that individual judgments about the relative importance or value of different considerations are involved. A strip mine may be beautiful to an unemployed miner, an uninterrupted river may be more precious to a camper than a dam and hydroelectric plant, and so on. This chapter does not attempt to make such judgments but instead reviews briefly some principles and practices of land-use planning.

Civil engineering, in some sense, goes back to the oldest human communities, to the irrigation canals of Egypt dug in 2400 B.C., to the pyramids, to the Great Wall of China, and to the aqueducts and roads of ancient Rome. The building of many ancient structures required some understanding of earth materials. Mistakes resulted from gaps in that understanding. The Leaning Tower of Pisa leans, not by the design of its builders and not because of any structural flaw within the edifice proper, but because it was built on unstable soils, some of which flowed out from under it; the tilt has increased over the centuries.

Questionable land-use planning may result in a need for extra care in design and engineering to minimize damage from hazards such as beach erosion; Westhampton Beach, Long Island, New York.
© Rafael Macia/Photo Researchers, Inc.

Over the last two hundred years, it has become increasingly possible to incorporate geologic principles and considerations into construction plans. The paramount concern of the modern engineering geologist is to take a site's geology fully into account in designing a structure so that the structure will be safe and stable. Increasingly, engineering geologists also consider the impact that the building or structure, in turn, will have on the geologic environment.

Land-Use Planning — Why?

One motive behind land-use planning is safety: Some land is unstable and unsuitable for certain kinds of structures, so the geology has restricted the possible uses to which that land can safely be put. However, much of the motivation for land-use planning arises out of the current reality that a large and growing population occupies a fixed expanse of real estate. When the population was much less, it mattered less, in a sense, whether a particular parcel of land was misused. If a farm's soil eroded away or became infertile or baked hard, if a stream dried up or became polluted, one could move on to a new, pristine spot and begin again.

Conversion of Rural Land

Centuries ago, there was ample space for all. That is becoming less true all the time. According to the U.S. Department of Agriculture, each year in the United States some 2 million acres of rural land are converted to other specific uses. The breakdown is as follows:

urban development	420,000 acres
airports, highways	160,000
reservoirs and flood control	420,000
wilderness areas, parks, and so on	1,000,000
Total	2,000,000 acres

Admittedly, 2 million acres may not seem like a lot compared to the 2,300 million acres of land in the United States. Year after year, however, it adds up. And the whole 2,300 million acres in the United States are not ultimately available for conversion, either, for it is on much of that rural land that food crops and livestock (and food for export) are produced.

Some Considerations in Planning

It is becoming increasingly important to consider not only what can safely be done with a given piece of land, but the optimum uses to which the land can be put. Many of the considerations involved are geologic ones similar to those outlined at the beginning of this chapter. Additional geologic constraints might include the resource potential of a particular piece of land or its susceptibility to pollution, erosion, or other disruption if certain kinds of land uses are

465

permitted. Again, biological/ecological, economic, and political factors may enter into the decisions ultimately made. So may aesthetic factors. For example, as noted in chapter 14, Yellowstone National Park represents a tremendous geothermal resource. However, it is also a great scenic resource and important wildlife habitat, and these considerations are deemed important enough to override the potential gain from exploiting the geothermal energy. In early oil-boom days, on the other hand, oil was so sought-after that considerations of whether or not one might want a derrick in one's front yard were often overlooked (figure 19.1). The Bingham Canyon copper mine in Utah, one of the world's largest open-pit mines, has irrevocably altered some 7,500 acres of undeveloped western land, but it has yielded, on average, over $9 million per acre in metallic minerals.

Land-Use Options

Particularly in densely populated areas, there is growing interest not only in making the best use of each piece of land but, if possible, in using the same parcel of land for several different purposes. The strategies involved are called *multiple use* and *sequential use*.

Multiple Use

Multiple use, as the name implies, means using the same land for two or more purposes at the same time. An example is shown in figure 19.2. At first glance, the area appears just to be a ball field. However, note that it is built into a basin. When it rains—when the field would not be used as a ball field anyway—the field acts as a recharge basin, catching fresh rainwater and allowing it to infiltrate slowly down to the seriously depleted aquifers under Long Island. So the same land serves two purposes: one recreational, the other directed toward conserving/increasing groundwater resources. Other examples of multiple land use would be the generation of wind power from windmill arrays in fields used simultaneously for farmland or grazing land, or underground mining deep below the surface while urban development proceeds above.

Sequential Use

The alternative approach of **sequential use** involves using land for two or more different purposes, one after another. Because the different land uses need not be compatible, a greater variety of combinations is possible. Abandoned underground mines, if dry and adequately ventilated, can be used for warehouse space, manufacturing, or even office space. This is being done, for example, in the Kansas City area, in abandoned underground limestone quarries. The old quarries now include nearly 1 million cubic meters of frozen-food storage space (about one-tenth of the

U.S. total), and the rock's insulating properties make the practice very energy-efficient. The great strength of rock floors makes possible the use of heavy manufacturing equipment; the controlled humidity underground is an advantage to print shops and a sailboat factory located there. Rock

Figure 19.1 Drilling for oil, Long Beach, California, 1901. Photograph by C.W. Hayes, USGS Photo Library, Denver, CO.

Figure 19.2 Multiple land use: This ball field doubles as a recharge basin for Long Island aquifers. Photograph courtesy of USGS Photo Library, Denver, CO.

walls are, of course, fireproof, and also contain noise well. On a smaller scale, abandoned mine space at Wampum, Pennsylvania, has been converted to office and laboratory space for the mining company.

Not all abandoned underground mines are equally suitable for subsequent occupation. The rock structure must be sufficiently strong for safety and not prone to deterioration through slow weathering with time. Even if above the water table, the space must be protected by overlying impermeable rock from infiltration of subsurface water from above. Flat-lying rock strata facilitate conversion to occupied space. There must be no likelihood of quantities of dangerous gases present; old limestone mines would be safe in this respect, while old coal mines, with possible associated methane present, probably would not be.

Alternatively, abandoned mines could be used for waste disposal (recall the proposal in chapter 15 to place high-level radioactive wastes in abandoned salt mines). Strip mines have been suggested as possible landfill sites when mining is completed; this would follow resource extraction with waste disposal, after which the land could be covered, regraded, and put to another use. Denver, Colorado, has carried out a scheme of this sort: Old gravel pits were used for sanitary landfill, and then, after they had been filled, the Denver Coliseum and its parking lots were constructed over the top. (Using such permeable materials for landfill may or may not be advisable in a given situation.)

An alternative post-landfill use might be the extraction of methane gas for energy. An abandoned quarry can be flooded to make a recreational lake. In several places, completed landfills have had golf courses developed atop them. Many such sequences of activities can be imagined; another example is shown in figure 19.3. The advantage they share is land conservation. If one piece of land can be made to do double or triple duty, that much less land must be used for or disturbed by human activities overall. Further, it becomes easier to avoid feeling forced to use the least stable or most vulnerable areas at all.

Maps as a Planning Tool

Many kinds of information, geologic and otherwise, go into comprehensive land-use planning. Much of this information can most quickly and clearly be examined in map form.

Any geological property or process that varies from place to place, including topography or steepness of slopes, bedrock geology, surficial materials or soil types, depth to water table, rates of cliff erosion, and so on, can be represented. Maps can also show locations of hazards past or present, such as fault zones, floodplains, and landslides. Figures 19.4 and 19.5 illustrate some examples. Nongeologic factors—vegetation, present population density, or present land use,

A

B

Figure 19.3 Aerial photographs of an area of St. Louis Park, Hennepin County, Minnesota (A) pre-1973, during operation of a coal-tar-distillation and wood-preserving facility, and (B) during subsequent redevelopment. Photographs courtesy of USGS Photo Library, Denver, CO.

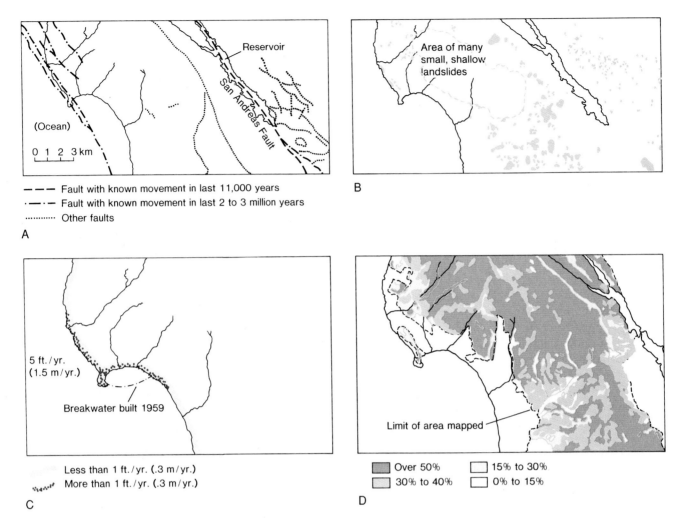

Figure 19.4 Map representation of several kinds of geologic considerations. (*A*) Fault zones. (*B*) Landslides (shaded). (*C*) Historic coastal cliff erosion rates. (*D*) Slope steepness.   Source: Modified from B. Atwater, "Land-Use Controls Arising from Erosion of Seacliffs, Landsliding, and Fault Movement," in U.S. Geological Survey Professional Paper 950, 1979.

for instance—may also lend themselves to representation in map form (figure 19.6). See also appendix B, which explores various kinds of maps and imagery in greater detail.

A land-use planner can then seek sites for particular land uses based on whatever set of criteria seems most appropriate. Maps make it possible to see quickly where several different conditions are satisfied. In siting a major interstate highway, for example, a planner might seek gently sloping terrain not underlain by expansive clay soils or active fault zones, where the present population density is low. A survey of an appropriate set of maps permits the planner to find potentially suitable sites swiftly. Conversely, maps can aid in long-term land-use planning even when conversion of rural land is not immediately contemplated. If the location of a stream's one-hundred-year floodplain or a major fault zone is known, restrictive zoning ordinances can prevent unwise development before it occurs, or special construction requirements for those areas can be imposed.

Of course, the suitability of a particular area for a specific land use cannot always be determined in a clear-cut, yes-or-no fashion. A land-use planner may instead need to consider a number of alternative sites for a project, each of which is less than fully desirable in some different way. Or there may be degrees of suitability for some purpose, such as housing developments, considering a variable like slope stability. In the latter case, the planner might stipulate different levels of intensity of the same land use in different areas. A higher density of homes might be permitted on gently sloping terrain than on steeper hillsides, for instance. See also box 19.1.

In recent years, computers have played an increasingly important role in the land-use planning process because of their capacity to manipulate large volumes of quantitative information. A map can be broken down into a grid of numbers, each data point representing a particular property (slope, soil type, and so forth) over some area

Figure 19.5 Map showing various functions (land uses) to be preserved in the vicinity of Reutlingen/Tübingen, Germany. Shading indicates number of functions of concern in a given region; numbers indicate type(s) of function(s). Redrawn from E. Stenestad and G. Sustrac, "The Role of Geoscience in Planning and Development," in *Geology and the Environment in Western Europe*, ed. by G. I. Lumsden, copyright © 1992, by permission of Oxford University Press, Oxford, England.

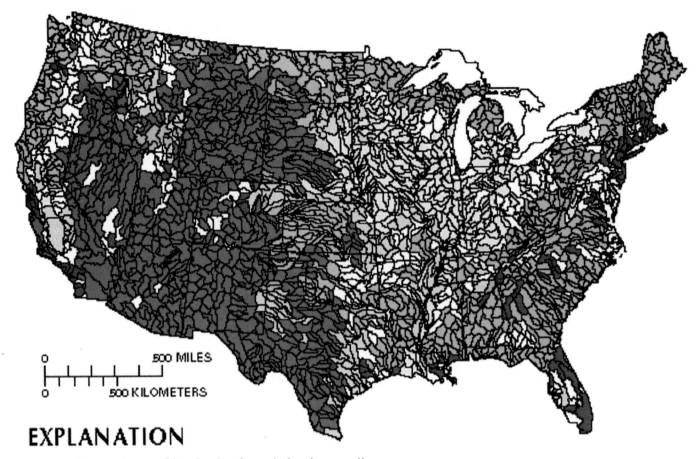

EXPLANATION

—— Boundary of hydrologic cataloging unit

LAND USE SHOWN ON MAP	CRITERIA FOR DETERMINING LAND USE (Land cover, in percent; crop type, in percent; population density, in persons per square mile; total water withdrawals for domestic use, in million gallons per day)	
AGRICULTURE Wheat	Land cover: >40 crop and pasture <40 forest <10 urban	Crop type: >50 wheat <20 corn and soybeans
Corn and soybeans	Land cover: >40 crop and pasture <40 forest <10 urban	Crop type: >50 corn and soybeans <20 wheat
Mixed	Land cover: >40 crop and pasture <40 forest <10 urban	Crop type: <50 wheat and corn and soybeans
URBAN	Land cover: <30 crop and pasture Population: >100	Total water withdrawals for domestic use: >6
FOREST	Land cover: >50 forest <40 agriculture <10 urban	
RANGE	Land cover: >50 range and barren land <40 agriculture <40 forest <10 urban	
OTHER	All land cover not meeting any of the above criteria	

Figure 19.6 U.S. land-use classifications combine elements of nature of cover, urbanization, and water use.   From U.S. Geological Survey Water Supply Paper 2400.

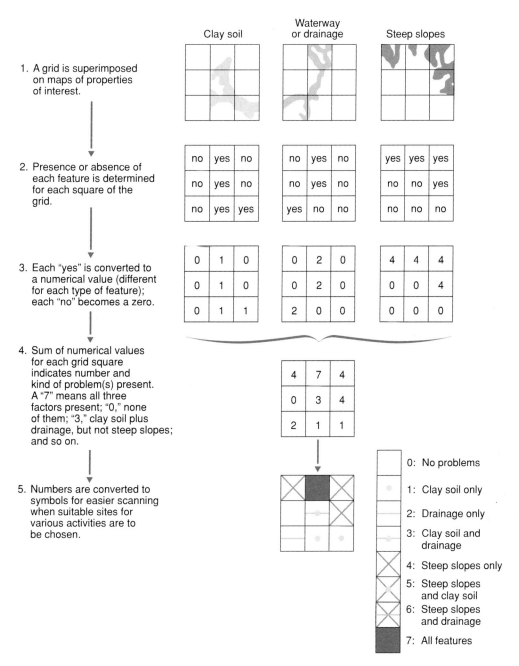

1. A grid is superimposed on maps of properties of interest.

2. Presence or absence of each feature is determined for each square of the grid.

3. Each "yes" is converted to a numerical value (different for each type of feature); each "no" becomes a zero.

4. Sum of numerical values for each grid square indicates number and kind of problem(s) present. A "7" means all three factors present; "0," none of them; "3," clay soil plus drainage, but not steep slopes; and so on.

5. Numbers are converted to symbols for easier scanning when suitable sites for various activities are to be chosen.

Figure 19.7 Digitized maps can represent data in a form that both human planners and computers can use. In this example, for each parameter (clay soil, water drainage, steep slopes), each point (area) has one of two values: Either the area is underlain by clay or it is not; either the slopes are steeper than a certain grade or they are not; and so on. In a more complex scheme, there might be several possible values for each parameter.   Source: A. J. Froelich, et al., "Planning a New Community in an Urban Setting: Lehigh," U.S. Geological Survey Professional Paper 950, pp. 69–89, 1979.

(1 square kilometer, 10 square kilometers, or whatever) (figure 19.7). The computer can then combine as many kinds of information as desired, each represented on a separate array, and produce a composite measure for each point of the grid to indicate the overall suitability of that area for the land use under consideration, which is most often urbanization/housing development (figure 19.8). The same basic geologic or nongeologic data can be combined in different ways for different purposes, by weighing various factors differently. In mapping suitability of land for wildlife habitat or refuge, for example, abundant surface and near-surface water might be a positive factor and the presence of expansive clay soils largely irrelevant. A land-use planner looking at the same region with high-density housing in mind might well regard both factors as negative. The data-processing programs can be adjusted accordingly to produce summary data tailored to particular objectives.

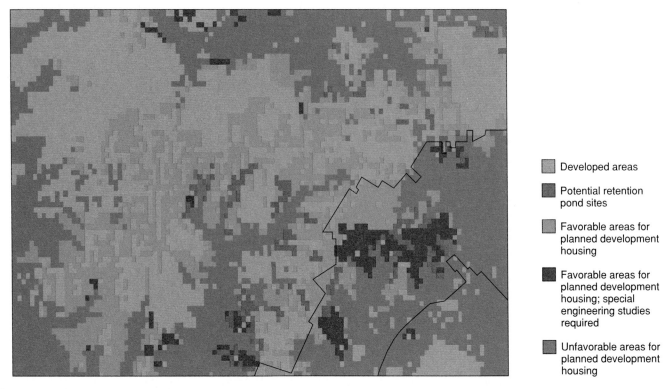

Developed areas

Potential retention pond sites

Favorable areas for planned development housing

Favorable areas for planned development housing; special engineering studies required

Unfavorable areas for planned development housing

Figure 19.8 The computer can rapidly combine digitized data into a composite map for land-use planning, showing more- and less-suitable areas for development or any other intended purpose.   Source: A. J. Froelich, et al., "Planning a New Community in an Urban Setting: Lehigh," U.S. Geological Survey Professional Paper 950, pp. 69–89, 1979.

Maps as tools have limitations, too. One is the matter of scale: The information presented must be given in sufficient detail that features of interest will, in fact, show up in the data. On a map on which 1 centimeter represents 1 kilometer of distance on the earth's surface (1:100,000 scale), all kinds of relatively small features—cliffs, sinkholes, small streams, and so on—might not appear at all; yet these features could be of great concern to a prospective home owner. (The astronauts had the same problem with early lunar landings. Images of the moon made from earth, covering very large expanses of the moon's surface, were used to direct them to "fairly flat" areas. But when those proposed landing areas could be seen close up on a finer scale, many were found to be strewn with boulders several meters across, which definitely presented a hazard to smooth and safe landings!)

Sometimes, the difficulty is that the information itself is only available on a gross scale. Maybe the only maps of vegetation or surficial geology were made from satellite images, for instance. Sometimes, the data are simply unavailable. Producing many kinds of topographic, geologic, or other earth-science maps means compiling many observations or measurements, which takes time, personnel, and funds. Even in the United States, which is rather thoroughly mapped by world standards, not all types of information have been obtained for all areas. In these cases,

land-use decisions may have to be based on incomplete data, perhaps supplemented by broad, reconnaissance-type surveys of the area under study. Sometimes, too, the problem is not so much lack of data as conflict among competing options; see box 19.2.

Engineering Geology— Some Considerations

Many of the geologic factors that go into site evaluation have been surveyed in earlier chapters. Obviously, every geologic factor or process is not equally important to every project, but a major construction project—for example, a housing development—might require consideration of a very broad range of geologic matters. Some are outlined below.

What rock types are present? Are they strong enough to support the proposed structure(s)? Do they fracture easily? Are there structural features within the rocks—folds, faults, bedding planes—that make the rocks' properties nonuniform and that should be taken into account? How porous and permeable are the rocks? Over the longer term, are they unusually prone to erosion or weathering?

Many of the same questions about porosity, permeability, strength, and stability might be asked about the soils. Are any slopes likely to give way to landslides? Does the

Hazard	——— Lives lost - - - - Property damage, dollars	
Floods	more than 85 $1,200 million	
Earthquakes	more than 8 $100 million	
Landslides	unknown $1,000 million	
Coastal erosion	unknown $300 million	
Expansive soil	unknown $2,200 million	
Others*	unknown $100 million	

*Includes volcanic eruptions, tsunamis, subsidence, creep, and other phenomena.

A

B

Figure 19.9 Natural hazards take lives and destroy property. (*A*) Relative magnitude of loss of life and property damage in the United States from various geologic hazards. Note that expansive soils cause nearly as much property damage as all other listed hazards combined. (*B*) Failure of masonry structure on unstable soil.   (*A*) Source: U.S. Geological Survey Professional Paper 950, 1979. (*B*) Photograph courtesy of USDA Soil Conservation Service.

construction itself have the potential to trigger slides? How cohesive is the soil? How compressible and prone to settling? How elastic? Do the soils tend to expand and contract as moisture content varies? Are they subject to hydrocompaction? This problem of expansive clay soils as a construction hazard is not a trivial one (figure 19.9). Their failures take few, if any, lives, by contrast with more obvious hazards like floods and earthquakes, but the total structural damage to homes, commercial buildings, pavements, and utilities in the United States each year amounts to over

$2.2 billion, approximately equalling the total costs of all other geologic hazards combined.

What about water? What quality and quantity of surface water or ground water is available? What are the surface runoff patterns? Does part of the site now serve as a recharge area for an aquifer system? If septic tanks are planned, are soil properties and topography appropriate for them?

Then there are the possible catastrophic hazards. Earthquakes and volcanoes are significant hazards in relatively few places, but where they are, they need to be taken

Box 19.1 Land-Use Planning and the Siting of Nuclear Power Plants

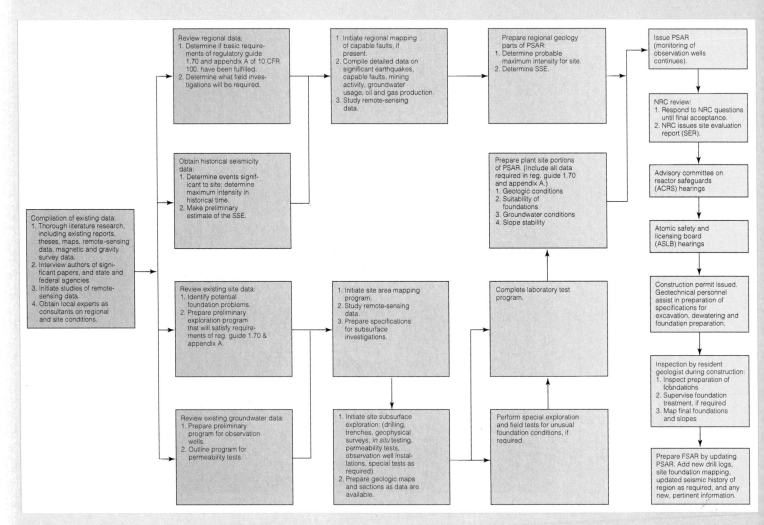

Figure 1   The typical sequence of activities involved in the evaluation of a nuclear power plant site.   After M. J. Adair, "Geologic Evaluation of a Site for a Nuclear Power Plant," *Reviews in Engineering Geology*, Vol. IV, 1979, p. 28, Geological Society of America.

The siting of nuclear fission power plants is a sensitive matter, for obvious reasons. The planning process, which involves both state and federal agencies, is correspondingly complex (figure 1).

It is made still more complex by the fact that—nongeologic considerations aside—the geologic constraints may be poorly understood. This is, to some extent, the case in New England, at least with respect to seismic hazards (figure 2). While the region now is generally quiet seismically, significant earthquakes are reported to have occurred in the eighteenth century. The region itself is a passive continental margin, not an active plate boundary, and the reasons for the so-called Cape Ann earthquakes are not obvious. This introduces an element of uncertainty about the extent to which seismic hazards can or should be taken into account in siting power plants in the region.

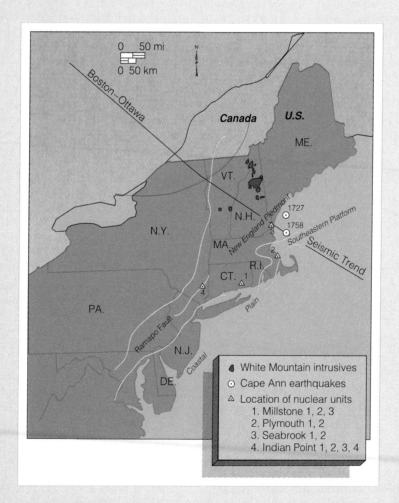

Figure 2 Relation of selected New England power plant sites to historic seismic activity (Cape Ann earthquakes), much older igneous activity ("White Mountain intrusives"), and structural boundaries in the region.   Source: After J. F. Davis, "The State-Federal Partnership in the Siting of Nuclear Power Plants," *Reviews in Engineering Geology*, Vol. IV, p. 52, Geological Society of America.

Box 19.2 The Federal Government and Land-Use Planning

The federal government controls a third of the land in the United States. The proportion, however, varies widely among states, from less than 1% in the states of the northeast and midcontinent to 73% of Utah and Arizona, 85% of Nevada, and, at one time, over 95% of Alaska. The impact of federal land-use policies is therefore felt in very different degrees in different states.

Federal land-use policies have changed through time. As with mineral-resource laws (chapter 18), federal emphasis was initially on resource development in preference to preservation. Early preservation efforts were limited. The first of the national parks, Yellowstone, was established in 1872; the first national forests were created two decades later. Still, resource development continued to be a high priority until about the middle of the twentieth century.

Federal lands can be broadly divided into two types: those intended primarily for preservation (including national parks and wilderness areas) and those on which multiple and, one hopes, compatible land uses can be allowed (for example, national forests). On the latter lands, additional uses beyond recreation or habitat preservation might include livestock grazing, mining, logging, and exploration and drilling for petroleum. Problems arise when the multiple uses allowed turn out not to be compatible after all. Livestock may outcompete wildlife for limited food; overgrazing, careless timbering, or even just too many tourists passing through may accelerate soil erosion and loss. Because the preservation function has sometimes been less successful when multiple land uses are allowed, many groups have tried to pressure the federal government to put more of its lands into the highly protected categories.

In 1978, President Carter did that with 107 million acres of Alaska. Under the Alaska Lands Act, that land—almost one-third of the state—was designated as national monuments, wilderness areas, or similarly protected land. Partially as a result of that action, mining and petroleum development now are prohibited or severely restricted on 40 percent of Alaska's total land. Some hailed the decision as a forward-looking move to preserve dwindling wilderness. Oil and mining interests were very unhappy at the reduction in land available for exploration and warned of possible future resource shortages. State residents who had hoped to benefit from more jobs or from taxes on those companies' profits from resource exploitation were likewise not pleased. The state, in turn, wanted more control over what was to be done with its lands.

Box 18.2 illustrated similar conflicts in two quite specific cases. Often, in fact, controversies arising in connection with the EIS process reflect significantly different opinions about appropriate or optimal land uses for the area in question. So, what's a government to do? What *does* constitute managing the public lands in everyone's best interests?

---

into account in building siting and/or design (figure 19.10). Landslides and floods potentially affect many more areas. Do sinkholes occur in the area? Has there ever been underground mining in the area?

Long as this list of geologic concerns is, it does not exhaust the possibilities. Consider the case of the Trans-Alaska Pipeline mentioned in chapter 18. At 1,300 kilometers long, it spans a variety of geologic settings and potential problems. Much of the terrain it crosses is rugged, including two major mountain ranges (the Brooks Range and the Alaska Range). It also crosses about eight hundred streams, the largest of which is the Yukon River. All of those streams freeze in winter, and some thaw partially in summer, complicating the engineering. A number of faults crisscross the area, and the southern end of the pipeline, the city of Valdez, is close to the epicenter of the 1964 Alaskan earthquake. The pipeline, therefore, was designed to take the stress not only of thermal expansion and contraction but also of earthquakes and ground displacement (figure 19.11); near the Valdez end, it was built to withstand earthquakes up to Richter magnitude 7.5.

A further construction headache peculiar to very cold climates is permafrost. In temperate climates, the upper part of the soil may freeze in the bitterest cold of winter, but it thaws during the warmer part of the year. In arctic climates, at high latitudes or very high altitudes, the winter freeze is so deep and pervasive that summer's thaw does not penetrate the whole frozen zone, and it may barely thaw the surface. There is then a layer of more or less permanently frozen ground called **permafrost.** The top of the frozen zone is sometimes referred to as a **permafrost table,** and it behaves similarly to a water table, rising and falling with seasonal climatic variations (figure 19.12).

As long as the permafrost stays frozen, it makes a fairly solid base for structures. If it is disturbed and warmed, as by construction, some of the ice melts. The deeper soil is still frozen, however, so depending on the topography, the water may drain slowly or not at all. The result is a mucky,

A

B

Figure 19.10 Studies of performance of structures in earthquake-prone regions yield information useful in improving structural designs. (*A*) This light-steel-framed industrial building in Ganjeh, 30 km from the epicenter of the 1990 Iranian earthquake, survived without structural damage; however, only 2 percent of buildings in this area were built with steel framing or reinforced concrete. (*B*) Total collapse of a mid-rise building in Rasht, 60 km from the epicenter of the same earthquake, shows its concrete to have been poorly, inadequately reinforced.   Photographs by M. Mehrain, Dames and Moore, courtesy of NOAA.

A

B

Figure 19.11 Trans-Alaska pipeline design takes into account the need to allow for both thermal expansion and contraction as temperatures rise and fall, and possible displacement resulting from earthquakes. (*A*) "Kinks" in the pipeline, and loose coupling to supports (see also figure 19.14), help accomplish this. (*B*) The result is that in places, the pipeline zigzags across the land; here, it also dives under the river in the foreground.

Figure 19.12 In cold climates, a permanently frozen soil zone, permafrost, may be found beneath the land surface.

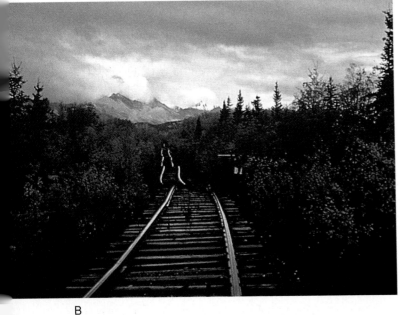

Figure 19.13 Permafrost affects construction work and transportation systems. (*A*) Melted permafrost produces a sodden mass of waterlogged soil like that surrounding this tractor. (*B*) Differential subsidence of railroad tracks due to partial thawing of permafrost, Copper River region, Alaska. This track had to be abandoned in 1938.   Photograph (*A*) by T. L. Péwé, courtesy of O. J. Ferrians, USGS Photo Library, Denver, CO. Photograph (*B*) by L. A. Yehle, courtesy of O. J. Ferrians, USGS Photo Library, Denver, CO.

sodden mass of saturated soil that is difficult to work in or with and that is structurally weak (figure 19.13). The work crews on the Trans-Alaska Pipeline were faced with trying to install, under very unstable conditions, structural members that would be quite adequately, solidly supported *after* construction was done and the melted permafrost had refrozen. Even the oil in the pipeline posed a problem. The oil is warm (about 60° C, or 140° F) as it comes up from deep in the earth, and its temperature stays well above freezing as it flows through the pipeline. The warm oil could thaw a zone within the permafrost immediately around the pipeline, and the sagging of the pipeline in the resultant mud could lead to pipeline rupture. Ultimately, buried sections of the Trans-Alaska Pipeline were actually refrigerated in places to keep the permafrost frozen! Elsewhere, cooling systems were installed to cool the supports of aboveground sections of pipeline, to keep them firmly in place (figure 19.14).

Case Histories, Old and New

This section briefly outlines several case histories, in partial illustration of the range of problems encountered in engineering geology and some approaches to their solution.

The Leaning Tower of Pisa

The Leaning Tower of Pisa was built in several phases between A.D. 1173 and 1370. It began to tilt even before it was completed. The tilt has been self-reinforcing—that is, as the structure began to lean, more and more pressure was concentrated on the lower side, causing more flow of the unstable clay layers and more tilt. The 55-meter-high tower now leans more than 6 meters out of plumb. Over the first half of this century, the tilt increased about 0.15 percent, corresponding to an average rate of 165 millimeters (6.5 inches) per century. More disturbing were the early 1970s measurements that suggested that the tilt rate had increased to nearly twice that amount.

A nearby cathedral had also settled in the soft clays and had suffered enough structural damage that it had to be torn down and rebuilt. Because the Leaning Tower itself is still intact, there is some hope that it can be saved if ways can be found to halt (or better, reduce) the leaning. Suggestions have included physical methods, such as boring and selectively removing some material from the north (high) side of the foundation to reduce the tilt, and chemical methods, such as treating the clays to stiffen them and prevent further sinking of the settled south side.

Recent efforts are particularly encouraging. In 1993, 900 tons of lead were placed on the high side of the Tower's base to anchor and counterweight it. Four years later, measurements indicated that no further tilting had occurred.

The Panama Canal

The idea of a canal across the Isthmus of Panama was suggested as early as 1528 by one of Cortez's men, but it was not until 1880 that a French company began excavation. The initial design, patterned after the Suez Canal, was for a sea-level cut from Atlantic to Pacific, with no locks—this

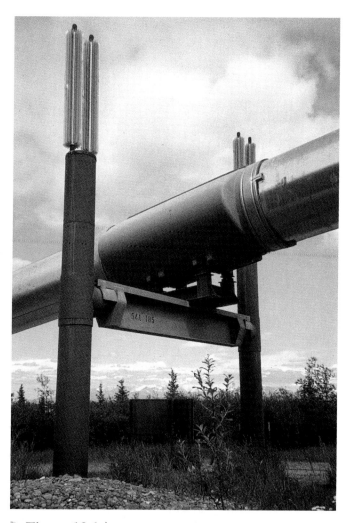

Figure 19.14 Close-up of pipeline support system shows how pipeline can shift from side to side between "bumpers" on support posts. Fins on tops of posts are part of a system designed to cool the underground sections of the posts. Coolant fluid circulates vertically, and the fins expose maximum surface area to the cool air—working much as fins in a radiator do—to counteract the warming effects of oil or of sunlight heating the ground.

Figure 19.15 Geologic factors complicated construction of the Panama Canal. (*A*) Dipping beds sliding into a canal reduce the canal's capacity, requiring more excavation. (*B*) Removing the weight of the rocks in the canal may allow buckling of rocks below, which are under pressure from the rocks on the canal's sides.

being the preference not of the engineers, but of those financing and directing the project. The Suez Canal, however, was excavated mainly through sand and soil, not rock, and in a climate with much less rainfall to promote sliding. Also, because of the higher elevation of the Isthmus of Panama, a much larger volume of material would have to be excavated for a sea-level canal across Panama.

Little or no investigation of the geology of the Panama Canal site was made before excavation began. Such investigation would have revealed layers of young volcanic rocks, lava flows, and pyroclastic deposits, interbedded with some shale and sandstone. Because the rocks dip in many places toward the canal, excavation removed the support from some of these rock layers, which then tended to slide (figure 19.15A). Sliding was facilitated by very high rainfall, which averages 215 centimeters (85 inches) a year. Elsewhere, as the canal was dug, the weight of the rocks on the side caused flow and buckling of the rocks at the bottom of the excava-

tion, which might rise 10 meters or more, requiring redigging to open the canal and removal of material from the sides to relieve some of the pressure (figure 19.15B).

Added to the geologic obstacles were heat, humidity, and rampant diseases, such as the mosquito-spread yellow fever, that caused many deaths among the workers. After seven frustrating years, the French company was bankrupt. A five-year lull in excavation was followed by renewed efforts by a second French firm, but it, too, later abandoned the attempt.

The United States acquired the rights to take over the project in 1903. At that point, the whole project design was reconsidered, and instead of a sea-level canal, a more complex design was chosen. A river was dammed to create a lake in mid-isthmus, and locks were to be built on either side to carry ships, in several stages, from sea level up to lake level and back down to sea level on the other side. One technical advantage of the lock design was that it would reduce the total volume of excavation required. An ecological benefit was that the lock design would better separate the Atlantic and Pacific oceans; a concern with a sea-level canal and associated free flow of water and organisms between oceans was the possible impact on various ecosystems.

Even so, finishing the canal wasn't easy (figure 19.16). A geologist was not invited to examine the excavation and associated

Figure 19.16 Excavation of the Panama Canal, 1907. Courtesy Panama Canal Commission.

landslides until 1910, by which time many large slides were already beyond control. Sliding and excavation, in fact, continued long past the time of the canal's nominal completion and substantially increased its costs. Early consideration of the geology of the canal zone might not have eliminated the sliding problems, especially considering the lesser understanding of rock mechanics at the time of construction. However, some of the instability problems could have been anticipated and reduced, and the costs could certainly have been projected more accurately. As it was, the canal was not completed until 1914.

At that, it has continued to evolve. Parts were widened in the mid-1900s, and a new phase of widening, begun in 1992, is scheduled for completion by 2004. The geology remains a challenge; better understanding of it, coupled with better equipment, makes modern excavation more manageable.

The Rotterdam Subway, Holland

In 1964, a tunnel system for rapid-transit lines was to be constructed in Rotterdam, Holland. Like much of the reclaimed land in the Netherlands, Rotterdam actually lies below sea level. The city is underlain by several meters of marine clay deposits and peat below the fill; these, in turn, are underlain by 15 to 25 meters of coarse sand and gravel. The sand/gravel layer is the more stable, and most buildings are supported by pilings driven down to that layer, 17 meters (about 55 feet) below the surface. The water table, as might be expected, is extremely high, nowhere more than 2 meters below the ground surface.

The presence of such shallow ground water was the main complicating factor in subway construction. It made the unconsolidated clays and fill too unstable to tunnel through. On the other hand, doing enough pumping to dig out a dry trench, build the tunnel, and then cover it over would have involved displacing large volumes of ground water and altering the water table elsewhere.

An ingenious solution was found. The trenches for the tunnel were allowed to stay flooded during excavation. Concrete piles were driven below the completed trenches to the sand and gravel layer to support the weight of the eventual tunnel. The tunnel itself was built on dry land as huge sections of concrete tube 10 to 15 meters wide, 6 to 10 meters deep, and 45 to 90 meters long. The sections of tube were sealed at the ends, ballasted with water, and moved into position along the flooded trenches. There they were sunk to rest on the support piles, assembled, and sealed together. Excavated sediment was replaced around and above them, and when the tunnel was complete, it was pumped dry. A little geological forethought allowed construction in saturated sediments with minimal disturbance of the ground water.

Tower Latino Americano, Mexico City

Modern engineering practice often involves basing foundations of large buildings on bedrock for maximum stability. In many places, however, the bedrock is far too deep to reach

A

B

Figure 19.17 Some 21 million acre-feet of water are impounded in Lake Mead behind Hoover Dam. (*A*) Close-up of the dam. (*B*) Regional overview gives a better appreciation of the volume of water impounded in the reservoir.   Photograph (*A*) by E. E. Herzog, USGS Photo Library, Denver, CO.

economically. Sometimes, it is even hard to reach a strong layer of sedimentary rock. Special care must be taken in designing building foundations in such settings.

Mexico City is underlain by expandable clays formed from volcanic ash deposits, interlayered with sands. The weakness of these sedimentary layers was a factor in the seriousness of damage from the 1985 earthquakes. The original modern water table was very high, so the sediments are water-rich. In recent years, however, heavy pumping of ground water has lowered the water table and aggravated structural damage by contributing to surface subsidence.

Before the Tower Latino Americano was built in Mexico City, test holes were drilled 70 meters deep so that the underlying materials could be thoroughly examined. A reasonably strong sand layer was found 33.5 meters below the surface, so the concrete slab on which the structure was built was supported on piles driven down to that sand layer. By doing this, the engineers avoided having the weight of the building resting on the uppermost two layers of volcanic clays, which have been the cause of much differential settling and foundation damage to other buildings.

So far, no settling of the tower has occurred. Its stability is in marked contrast to the situation of the nearby Palace of Fine Arts. Subsidence due to groundwater withdrawal and differential settling into the expansive volcanic clays have caused parts of this second building to sink more than 3 meters (nearly 10 feet), and the structural damage has been considerable.

Dams and Their Failures

When a building cracks and crumbles due to settling or subsidence, the repairs may be costly, but the toll rarely includes lives. Even when a bridge or tunnel collapses, only those few persons on or in it at the time are affected. A catastrophic dam failure, on the other hand, can destroy whole towns and take thousands of lives in a matter of minutes; one dam can impound a tremendous volume of water (figure 19.17). The motivation for intelligent and careful design and siting of dams is thus particularly strong. Unfortunately, past practice has frequently fallen short of the ideal. In chapter 8, we examined the Vaiont Reservoir disaster. In this section, we will look briefly at another well-documented dam disaster. Neither need have occurred; careful geological investigation beforehand would have shown that neither site was suitable.

The St. Francis Dam

The St. Francis Dam was built in California, about 70 kilometers north of Los Angeles. The reservoir it impounded was principally intended for a water supply. The dam, 60 meters high and over 150 meters wide, was completed in 1926. The valley walls on one side of the dam were made of coarse sandstones and other sedimentary rocks. On the other side, the rocks were schists, mica-rich metamorphic rocks that tended to break along parallel planes that sloped toward the dam and reservoir. The contact between the two rock types, over which the dam was built, was a fault. The presence of that fault was known—and even mapped—before the dam was built.

On 12 March 1928, the dam abruptly failed. Several hundred people were drowned; about $10 million in damage was also done. The reasons for the failure became clear from subsequent laboratory tests, and the fault zone did not actually appear to have been a significant factor. The rocks of the valley walls had seemed to be sufficiently strong when dry. But the sedimentary rocks on one side of the dam were rich in gypsum, which is very soluble. When a fist-sized sample of one of the sedimentary rocks was placed in water, however, it bubbled out air, soaked up water, and disintegrated, in less than an hour, into a heap of sand and clay. This was hardly an appropriate rock type with which to contain a large reservoir of water! Apparently, these weak rocks had collapsed first, taking out one side of the dam. This reduced the support for the rest of the system. As the water began to pour out, it further eroded the base of the schists. They then slid, and the other side of the dam collapsed. Somewhat astonishingly, the central section of the dam remained standing (figure 19.18).

As with the Vaiont Dam case, a little more careful investigation in advance could have averted disaster. In the case of the St. Francis Dam, just testing the reaction of the wall rocks to immersion in water should have told the engineers everything they needed to know about site suitability!

Other Examples

These two dam disasters are not, alas, the only ones ever to have occurred. Between 1864 and 1876, some one hundred dams in the United States failed because reservoir waters undermined their foundations. In 1900, the Austin, Texas, dam failed when water seeping through cracks in the foundation rocks lubricated clays and shales, and a 150-meter length of dam broke loose and slid 20 meters downstream. A Spanish dam failed in 1959 because of differential settling of supporting buttresses. In the same year, a French dam collapsed when the foundations slipped due to fractures in the gneissic rocks below. The list could go on and on.

Dams, more dramatically than most other structures, thus underscore the need for careful application of the principles of engineering geology before and during construction. They may also underscore the limits of our understanding. The Baldwin Hills reservoir was built near Los Angeles, California, between 1947 and 1951. The geological setting was thoroughly studied and (so it was believed) adequately taken into account. The reservoir was underlain by an active fault zone; the design incorporated a seismic safety factor double that was used by engineers in other earthquake-prone areas. Drainage was provided to prevent saturation of the foundations. The design was conservative, and provision was made for monitoring after construction. Nevertheless, on 14 December 1963, a differential slip of more than 10 centimeters occurred along a fault, water began to scour its way out, and two hours later, the dam

Figure 19.18   Failure of the St. Francis Dam, California: The St. Francis Dam after its collapse. Note the water line on the surrounding hills, marking the water level in the reservoir before failure.   Photograph by H. T. Stearns, USGS Photo Library, Denver, CO.

was abruptly breached. Clearly, the designers did not understand the geology as well as they thought.

Geology and geography may, in fact, lead to the building of dams on faults. The zone of weakness created by a major fault zone may become a topographic low along which streams flow and lakes accumulate. This is a natural site, then, for a dam and reservoir, but damming such a stream necessarily involves placing a dam on or very close to the fault. The San Andreas is only one of many faults marked, in part, by the presence of natural sag ponds and artificial reservoirs (figure 19.19).

The majority of the approximately 50,000 dams built in the United States over the last hundred years were not built on faults, but they may nevertheless suffer from inadequate design. All dams are designed to survive any imaginable flood load, typically through provision of a *spillway* that allows water to flow over or around the dam unimpeded when the water level in the reservoir becomes too high. The worst-case scenario for which a spillway is constructed is commonly a one-thousand-year flood. Still, as noted in chapter 6, it may be difficult to project just what volume of water that (or any other) flood represents, and changing land-use patterns may be altering the flood-frequency curves also. The 1993 flooding produced some unpleasant surprises in this regard in the Mississippi River basin. Dam failures have occurred, too, from dam washout as the sides or face of the spillway were scoured away by floodwaters. In 1981, the U.S. Army Corps of Engineers surveyed dams in this country. They identified 8,639 of these dams as ones whose failure would cause serious property damage or loss of life. Of that number, one-third (2,884) were identified as unsafe, and in over 80 percent of these cases, the problem was an inadequate or poorly designed spillway.

Thus, dam construction becomes increasingly complex and sophisticated (figure 19.20), aided by computers, laboratory tests, field studies, and scale-model experiments (box 19.3).

Figure 19.19 Sag ponds and reservoirs abound along the San Andreas Fault; here, San Andreas and Crystal Springs reservoir. Photograph by R. E. Wallace, USGS Photo Library, Denver, CO.

A

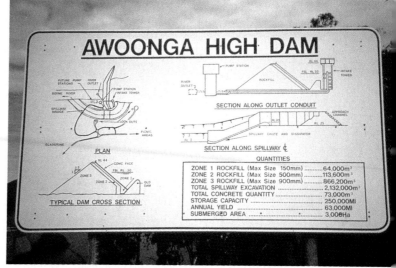

B

Figure 19.20 Awoonga High Dam in Australia reveals modern construction methods. (*A*) View of dam and reservoir; spillway in foreground. (*B*) Nearby billboard illustrates dam structure and presents some pertinent data.

Box 19.3 Testing, Testing—Modeling Nature and Testing Materials for Greater Safety

Many scale-model experiments take place in the laboratory. Nowadays, in fact, much "testing" is done using sophisticated computer models rather than by building physical models at all. However, it is often still useful to conduct physical experiments and observe the results. The photographs in figure 1 show a 1:500-scale model of the Swift Dam and Reservoir on the south flank of Mount St. Helens. The model (figure 1A) was built by Western Canada Hydraulic Laboratories for the dam's owners, Pacific Power and Light Company of Portland, Oregon.

High-density muds like those used in drilling oil wells simulated volcanic mudflows entering the reservoir from the creeks that feed it. Figures 1B and 1C show the progress of one such flow, from the model Swift Creek, into the reservoir. With each test, maximum wave height at the model dam and overall reservoir response to mudflow input were closely observed. At the request of the State of Washington, maximum mudflow volumes corresponding to about half the reservoir volume were tested. Under these conditions, waves from single flows, and the final reservoir height, nearly reached the dam crest.

After the 1994 Northridge earthquake, engineers were rudely surprised to discover the response—and failure—of some structures, believed adequately designed. The Federal Emergency Management Agency and California Office of Emergency Services committed $2.3 million to studies to address some of the problems (figure 2). After examining a number of failures (figure 2A), the engineers considered various modifications (figure 2B), investigated computer simulations of behavior of different designs (2C), and conducted stress tests on full-size models of steel beams and joints (2D). Following these initial studies, the engineers proposed interim guidelines for design of new buildings and repair/strengthening of existing structures. Several more years of work are planned to refine the data and recommendations.

Experiments such as these can play a key role in engineering safer structures in high-risk areas.

Figure 1 Completed model reservoir being filled with water to normal depth prior to test runs (*A*). Mudflow scaled to correspond to 44,000 acre-feet is released from model Swift Creek into reservoir (*B*); as flow progresses into reservoir (*C*), water height increases and waves lap at the dam.   Photographs by J. E. Peterson, USGS Photo Library, Denver, CO.

A

SAC-Connection Assembly Tests
Repaired & Modified Connections – Moderate Size

Repair Only

Repair & B. Haunch
Girder bottom flange
not welded

Repair, & B. Haunch
Girder bottom flange
welded

SAC Steel Program

FEMA/OES

B

C

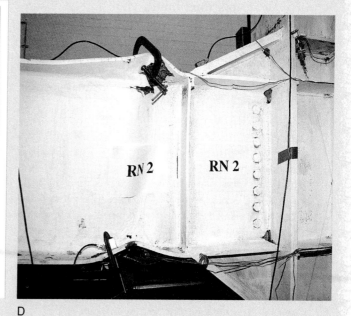

D

Figure 2 After examination revealed damage from the Northridge earthquake (*A*), engineers considered design alternatives (*B*), computed theoretical responses (*C*), and tested the modified designs (*D*). Photographs (A & D) courtesy Federal Emergency Management Agency and the SAC Joint Venture (B & C) from the Federal Emergency Management Agency and the SAC Joint Venture. Reprinted by permission.

Overview

Summary

Engineering geology is concerned with making structures as safe and stable as possible, given various kinds of geologic hazards and potential problems. Land-use planning encompasses the same concerns as engineering geology, combined with additional geologic and nongeologic considerations. The aim of the land-use planner is to make the best possible use of limited land, taking all these factors into account. Using the same land for several purposes, either simultaneously or sequentially, is one approach to conserving the land resource. The success of both geological-engineering and land-use-planning efforts is heavily dependent on the accuracy and completeness of the data available to the individuals carrying out these tasks.

Terms to Remember

multiple use 466   permafrost table 476
permafrost 476   sequential use 466

Exercises

Questions for Review

1. Describe the concepts of *multiple use* and *sequential use*, and give an example of each.
2. Maps of geologic or other factors are useful tools in land-use planning, but their usefulness can be limited by practical problems. Describe two such possible limitations.
3. How can the computer's ability to process large volumes of data assist in the planning process? How can programs be adjusted for different planning objectives?
4. Cite at least ten geologic considerations that might be important in siting an apartment building. How many of the same considerations would be relevant to siting a parking lot?
5. What is permafrost, and why is construction made more difficult by its presence?
6. How can engineers often minimize problems posed by unstable clays in near-surface rock and soil layers?
7. Dams are not infrequently built over faults. Why?

For Further Thought

1. Most city or county planning offices have long-range plans for development in undeveloped areas. Seek out such plans for your area, if available, and investigate what kinds of considerations (geologic or otherwise) went into those plans.

2. Make a walking tour of your neighborhood or of another area to look for signs of careless or thoughtless engineering practice—buildings showing severe cracking or other structural damage from ground failure beneath, serious erosion of steep slopes, large puddles accumulated after rain because roads or buildings dam surface runoff, and so on. Is any particular problem especially common?

3. Explore the Internet for an example of geographic representation of geologic, cultural, or other data relevant to land-use decisions, and examine data at different scales, if available. See NetNotes for some places you might start; see the SAST site or Federal Geographic Data Committee site, or sites relating to geographic information systems (GIS).

Suggested Readings/References

Allen, F. 1996. Inside the Panama Canal. *Invention & Technology* (Fall): 9–24.

Anderson, J. G. C., and Trigg, C. F. 1976. *Case histories in engineering geology.* London: Elek Science.

Arnold, C., and Reitherman, R. 1982. *Building configuration and seismic design.* New York: John Wiley & Sons.

Atwater, B. 1979. Land-use controls arising from erosion of seacliffs, landsliding, and fault movement. In U.S. Geological Survey Professional Paper 950, 11–19.

Betz, F., Jr., ed. 1975. *Environmental geology.* Benchmark Papers in Geology No. 25. Stroudsburg, Pa.: Dowden, Hutchinson & Ross.

Bingler, E. C. 1987. Siting the SSC: A geologic view. *Geotimes* (November): 8–9.

Froelich, A. J., Garnaas, A. D., and Van Driel, J. N. 1979. Planning a new community in an urban setting: Lehigh. In U.S. Geological Survey Professional Paper 950, 69–89.

Hatheway, A. W., and McClure, C. R., Jr., eds. 1979. *Geology in the siting of nuclear power plants.* Boulder, Colo.: Geological Society of America.

Hunt, R. E. 1986. *Geotechnical engineering analysis and evaluation.* New York: McGraw-Hill.

Jackson, R. H. 1981. *Land use in America.* New York: John Wiley & Sons.

Kiersch, G. A., Cleaves, A. B., Adams, W. M., Pincus, H. J., and Ferguson, H. F., eds. 1974. *Engineering geology case histories 6–10.* Boulder, Colo.: Geological Society of America.

Legget, R. F. 1973. *Cities and geology.* New York: McGraw-Hill.

Lumsden, G. I., ed. 1992. *Geology and the environment in western Europe.* Oxford: Clarendon Press.

Nisbet, I. C. T. 1979. Alaskan land: Owners can't be users. *Technology Review* (March/April): 16–17.

Petroski, H. 1993. The Panama Canal. *American Scientist* 81 (January/February): 13–18.

———. 1996. The Petronas Twin Towers. *American Scientist* 84
    (July-August): 322–26. (Describes construction of the
    world's tallest building, in Malaysia.)
Rahn, P. H. 1986. *Engineering geology: An environmental approach.* New
    York: Elsevier.
Robinson, G. D., and Spieker, A. M., eds. 1978. *Nature to be
    commanded.* U.S. Geological Survey Professional Paper 950.
Sloan, I. J. 1979. *Environment and the law.* Dobbs Ferry, N.Y.:
    Oceana Publications.
Stauffer, T. P., Sr. 1975. Occupance and use of underground space
    in the greater Kansas City area. In *Environmental Geology,*
    edited by F. Betz, Jr., pp. 193–203. Benchmark Papers in
    Geology No. 25. Stroudsburg, Pa.: Dowden, Hutchinson
    & Ross.
Walter, R. C. S. 1971. *Dam geology.* 2d ed. London: Butterworth.

The U.S. Geological Survey offers an introduction to
geographic information systems at
    http://info.er.usgs.gov/research/gis/title.home

Geomatics Canada, a division of Natural Resources Canada
responsible for geographic information systems and
mapping, is accessible via
    http://abbott.ccm.emr.ca/

The Federal Geographic Data Committee site assembles
references to a wide variety of data in map form; see
    http://fgdc.er.usgs.gov/gdc/html/fgdc.html

At the Scientific Assessment and Strategy Team (SAST) site
(see chapter 6) are results of detailed studies of the Upper
Mississippi River basin and various characteristics related to
flood hazards; see
    http://edcwww2.cr.usgs.gov/sast-home.html

The Earthquake Engineering Research Institute's home
page, with links to much information on earthquakes and
earthquake response, is found at
    http://www.eeri.org

The Canadian Association of Earthquake Engineering site
gives an index of projects and links to other earthquake-
related sites; see
    http://www.civeng.carleton.ca/CAEE/

The U.S. Army Corps of Engineers engages in a variety of
projects. The Corps home page is at
    http://www.usace.army.mil/
    Its environmental-restoration projects are described at
    http://www.mrd.usace.army.mil/mrded-h/access/
    statement.html
    Superfund-related restorations specifically can be
    found at
    http://www.mrd.usace.army.mil/mrded-h/access/
    epaproj.html

Through the Coastal Engineering Page, one can access
information about various coastal-engineering projects
(including some pictures). See at
    http://www.coastal.udel.edu/coastal.html

Chapter 20

Geomedicine

I n earlier chapters (particularly chapters 16 and 17, which dealt with pollution), we considered some of the adverse health effects of anthropogenic substances and the human activities that have caused a redistribution of some naturally occurring substances. Geomedicine, however, encompasses not only pollution, but also investigation of the broader relationships between the natural geologic environment and the health of or occurrence of disease in the people, animals, and plants living in that environment.

The composition of the soil as well as of the water that runs through the irrigation ditches will influence the chemistry of these lettuces, grown near Yuma, Arizona; regional differences in trace elements in food and water may in turn be reflected in regional variations in health. © Doug Sokell/Tom Stack & Associates

Table 20.1 Principal Chemical Constituents of the Human Body*

Element	Percentage of body weight
oxygen	61
carbon	23
hydrogen	10
nitrogen	2.6
calcium	1.4
phosphorus	1.1
Total	99.1

Source: R. M. Parr, "Trace Elements in Human Milk," *International Atomic Energy Agency Bulletin* 25, 2 (1983): 8.

*All other elements—led by sulfur, potassium, sodium, and chlorine—make up the other 0.9 percent of the body.

Basic Principles of Geomedicine

Trace Elements

Just as most of the earth's crust is composed of only a few major elements, so, too, are most organisms, though the elements involved are, for the most part, different. More than 99 percent of the human body is made of the six elements oxygen, carbon, hydrogen, nitrogen, calcium, and phosphorus (table 20.1). In fact, we are about 60 percent water. Most other elements do occur in the body, as they do in rocks, but only at very low concentrations, as **trace elements.** Trace elements may be defined loosely as those found in concentrations of about 10 to 100 ppm or less. The major constituents of the human body are obviously critical to life and health. However, to a great extent, geomedicine is the study of the effects of the presence or absence of the trace elements, or compounds present in equally minute amounts.

Geomedicine has roots that go back centuries; papers appeared in the nineteenth century noting geographical patterns in distribution of disease. However, many areas of the subject are relatively new—for several related reasons. One is that it was necessary to develop instruments sufficiently sophisticated and sensitive to measure extremely low concentrations of elements before their distribution could be determined or studied. Another reason is that it has historically been easier to recognize the effects of the few major constituents than of any one of the many minor ones in people and in other organisms. To take an agricultural example, farmers realized that nitrogen, phosphorus, and potassium are key nutrients required for plant growth long before it became apparent that the trace element boron is required for the proper growth of beets. The

Table 20.2 Essential Trace Elements in Humans

Element	Average body concentration (ppm)	Year recognized as necessary	Function	Some effects of deficiency in humans
iron	60	17th century	oxygen transport	anemia
iodine	0.2	1850	component of thyroid hormones	goiter
copper	1	1928	interacts with iron; in some enzymes	anemia, bone changes; possible elevated cholesterol
manganese	0.2	1931	involved in metabolism	not known
zinc	33	1934	involved in metabolism	depressed growth, slow healing
cobalt	0.02	1935	in vitamin B_{12}	seen as B_{12} deficiency
molybdenum	0.1	1953	in some enzymes	not known
selenium		1957	in enzymes; interacts with heavy metals	selenium deficiency heart disease
chromium	0.03	1959	involved with insulin action	insulin resistance; lowered glucose tolerance
tin*	0.2	1970	not known	not known
vanadium*		1971	not known	not known
fluorine	37	1971	important to proper tooth and bone growth	increased tooth decay; osteoporosis
silicon	260	1972	calcification; may function in connective tissue	not known
nickel	0.1	1976	interacts with iron absorption	not known
arsenic*	18	1977	not known	not known
cadmium*	0.7	1977	not known	not known

Source: R. M. Parr, "Trace Elements in Human Milk," *International Atomic Energy Agency Bulletin* 25, 2 (1983): 8.

*Need in humans not directly established; inferred from animal experiments

importance of iron in the human diet has been realized since the seventeenth century, but for many of the essential trace elements listed in table 20.2, recognition has come only within the last few decades.

The number of trace elements and their varying effects from organism to organism can make the isolation and identification of the function of any one a slow and complex task. A further complication is that the same element can have quite different effects, depending on the concentration in which it occurs or is consumed. This is a well-recognized phenomenon with many drugs. For instance, small doses of the drug digitalin, found in foxglove plants, can be helpful in treating heart problems, but large doses can be fatal, as any murder-mystery fan who has encountered a foxglove-leaves-in-the-salad plot is aware.

Dose-Response Curves

Such variability in effects of trace elements can be described by a **dose-response curve,** on which the positive or negative effects of a trace element are plotted as a function of dosage. Several hypothetical examples are shown in figure 20.1. In case A, the element in question would have no known beneficial or harmful effects in low concentrations but would be toxic in high concentrations and fatal in very high doses. Lead and mercury might be examples of this kind of trace element. In case B, the element is needed to avoid deficiency, but benefits only increase up to a point, after which no particular benefit or harm is derived from consuming higher levels. An element like calcium might be an example: It is necessary to the proper development of bones and teeth, and it would be hard to "overdose" on this major element in the

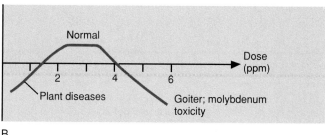

Figure 20.2 Generalized dose-response curves for copper (*A*) and molybdenum (*B*), combining responses of both plants and animals.

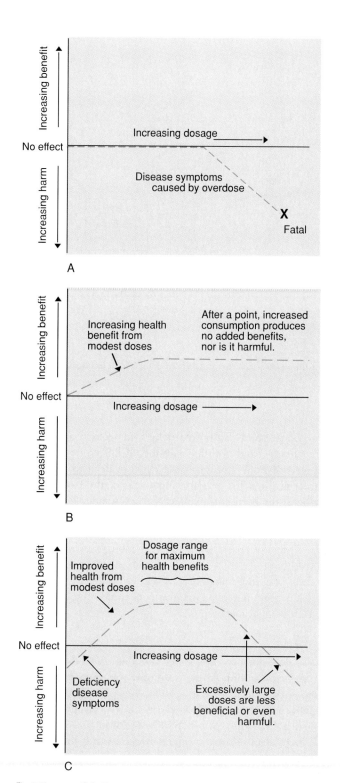

Figure 20.1 Possible dose-response curves. (*A*) No known need; element toxic at higher concentrations. (*B*) Element essential in moderate amounts; no known harm from high doses. (*C*) Trace amounts of element required; toxic in high doses.

body. Case C, perhaps the most common among the essential trace elements of table 20.2, is one in which a small amount of the element is necessary for optimum health, but too much can be toxic or even fatal. Copper and molybdenum are examples (figure 20.2). We will look at additional examples in more detail shortly.

As a practical matter, there are limits on the extent to which detailed dose-response curves can be constructed for humans. The limits are related to the difficulties in isolating the effects of individual trace elements (see later in this chapter) and in having representative human populations exposed to the full range of doses of interest. Controlled experiments on plant or animal populations are possible, but there is then the difficulty of trying to extrapolate the results to humans with any confidence. It is also true that different individuals, and persons of different ages, may respond somewhat differently to any chemical, trace elements included, so any dose-response curve must either be specific to a particular population, or it must be only an approximation.

Geology, Trace Elements, and Health

Controls on Elemental Intake

The ultimate source of the body's trace elements is the earth—generally, rocks. The various pathways through which the trace elements find their way into the body are summarized schematically in figure 20.3. The

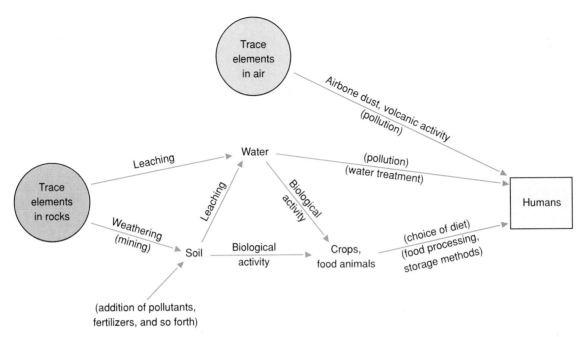

Figure 20.3   Pathways through which trace elements enter the body, including processes through which trace-element concentrations may be modified. Human impacts are in parentheses.

concentrations of trace elements in rocks vary by rock type and location and are a fundamental control on the availability of trace elements to humans (table 20.3).

Trace-element concentrations are modified by a variety of natural processes and deliberate and accidental human activities. Rocks weather into soil, gaining, or more often losing, some of their chemical constituents. Soils may lose some elements by leaching; agricultural chemicals and pollutants may be added. Crops selectively remove from the soil the elements they require for growth; animals intensify this selectivity by consuming only certain parts of plants for food. Methods of processing and storing food further change its composition, and we then exercise preferences in diet, consuming only some of the available range of foods and nutrients. The water we drink contains trace elements leached from rock and soil and may also have been polluted or chemically treated. The air we breathe also is a source of a few trace elements, outgassed from the earth, released during volcanic eruptions, or added through pollution.

Superimposed on geologic processes are what might be called "cultural factors," such as dietary choices and pollution, that affect our elemental intake. Examples exist not only in the present but in the past. Ancient Roman aristocrats used cosmetics containing lead and piped their water through lead pipes. It has been hypothesized that the fall of the Roman Empire can be attributed at least in part to widespread lead poisoning and correspondingly decreased fertility among the upper classes.

Within the last few centuries, the idea that geology might be related to health was initially suggested by geo-graphic correlation. Both Kentucky and Ireland are famed for their horses, and both have abundant limestone bedrock; high levels of bone-strengthening calcium derived from the limestone in the soils, grasses, and water of the two regions might be a key factor in explaining the animals' vigor. The frequency of occurrence of certain diseases has seemed to vary by location, to be more common in some places than others. In a few cases, from the coincidence of distinctive chemical characteristics of rock or soil and the distribution of disease, inferences have been drawn that the two might be related. Further controlled experiments could then be undertaken to test the hypotheses.

Iodine

Iodine is necessary to the proper functioning of the thyroid gland. A lack of iodine leads to thyroid enlargement and the deficiency disease known as goiter. Until the last century, the occurrence of goiter was particularly common in the northern half of the United States, an area known as the "goiter belt." Subsequent analysis revealed that the soils in that area were relatively low in iodine; consequently, the crops grown and the animals grazing on those lands constituted an iodine-deficient diet for people living there. When their diets were supplemented with iodine, the occurrence of goiter decreased.

Iodine is not difficult to incorporate into the diet, once the need is identified. Saltwater fish are naturally high in iodine because they live in the salty oceans in which iodine salts are dissolved. Moreover, for half a century, it has been common practice to "iodize" table salt, adding a little sodium

Table 20.3 Variations in Composition Among Common Rock Types*

Element	Average concentration				
	Granite	Basalt	Sandstone	Shale	Carbonate
silicon (%)	32.3	24.0	36.5	27.1	2.4
aluminum (%)	7.7	8.8	3.0	9.7	0.5
iron (%)	2.7	8.6	1.0	4.7	0.4
magnesium (%)	0.6	4.5	0.7	1.5	4.8
calcium (%)	1.6	6.7	3.9	2.2	30.0
sodium (%)	2.8	1.9	0.3	1.0	0.04
potassium (%)	3.3	0.8	1.1	2.7	0.3
titanium (ppm)	2,300	9,000	4,600	1,500	400
phosphorus (ppm)	700	1,400	750	170	400
manganese (ppm)	600	2,000	850	10–100	1,100
chromium (ppm)	25	200	100	35	11
copper (ppm)	20	100	50	1–10	4
lead (ppm)	20	8	20	7	9
nickel (ppm)	8	160	80	2	20
cobalt (ppm)	5	45	20	0.3	0.1

Sources: A. Brownlow, *Geochemistry*, copyright © 1979 Prentice-Hall, Inc., Englewood Cliffs, NJ; K. B. Krauskopf, *Introduction to Geochemistry*, copyright © 1979 McGraw-Hill, Inc., New York, NY; and B. Mason and C. B. Moore, *Principles of Geochemistry*, copyright © 1982 John Wiley & Sons, New York, NY.

*Data are presented for major elements and selected minor elements.

iodide or potassium iodide salt to the sodium chloride. Routine use of iodized salt has virtually eliminated goiter without the need for other special dietary precautions.

There may be a geologic reason for the dearth of iodine in northern U.S. soils, aside from any deficiencies in the bedrocks. The northern part of the country was the portion most affected by continental glaciation during the last Ice Age. When the ice melted, huge volumes of water must have drained through the soils on their way to the sea. Iodine salts are quite soluble; perhaps, then, these northern soils had much of their iodine leached out by the glacial meltwaters. (In the midcontinent region, the lack of iodine may be compounded by distance from the ocean; sea spray puts some iodine into the air, from which it falls on nearby land.)

The regional link between soil composition and goiter was particularly clear because, at the time the connection was observed, most people were eating locally grown food. Nowadays, in this and many other developed countries, food production is concentrated in a few areas, and the bulk of most people's food is not locally grown but is brought in from many places. This will make it more difficult to see links between soil chemistry and health in the future, except in countries or areas where most food is still locally produced.

Because extremely high doses of iodine also cause abnormal thyroid function, iodine is an example of the "type C" trace element in figure 20.1. (Excess iodine intake through normal dietary sources, however, is extremely improbable.)

Fluorine

Water chemistry, like soil chemistry, can influence health. Perhaps the best-documented example of a positive relationship between a trace element in water and improved health is the effect of fluorine in reducing the incidence and severity of dental caries (tooth decay). Teeth are composed of a calcium phosphate mineral—apatite. Incorporation of some fluoride into its crystal structure makes the apatite harder and, therefore, more resistant to decay. This, too, was recognized initially on the basis of regional differences in the occurrence of tooth decay.

Fluorine is present in average continental crust at a concentration of several hundred ppm; its concentration in natural waters is much lower, typically only a few ppm. Even such seemingly low concentrations make a demonstrable difference in tooth-decay statistics. Persons whose drinking water contains fluoride concentrations even as low as 1 ppm show a reduction in tooth-decay incidence. This corresponds to a fluoride intake of 1 milligram per liter of water consumed; a milligram is one thousandth of a gram, and a

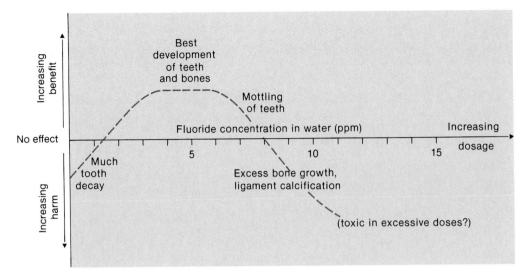

Figure 20.4   Generalized dose-response curve for fluoride.

gram is about the mass of one raisin. Fluoride may also be important in reducing the incidence of osteoporosis, a degenerative loss of bone mass commonly associated with old age. Normal total daily fluoride intake ranges from 0.3 to 5 milligrams. (Up to 3 milligrams may be contributed by food, but this varies widely with diet and with the soil chemistry where crops are grown.)

Fluoride is not an unmixed blessing. Where natural water supplies are unusually high in fluoride, so that two to eight times the normal dose of fluoride is consumed, teeth may become mottled with dark spots. The teeth are still quite decay resistant, however; the spots are only a cosmetic problem. Of more concern is the effect of very high fluoride consumption (twenty to forty times the normal dose), which may trigger abnormal, excess bone development (bone sclerosis) and calcification of ligaments. Fluoride, then, might be characterized by the dose-response curve shown in figure 20.4.

The prospect of reduced tooth decay is the principal motive behind fluoridation of public water supplies where natural fluoride concentrations are low. Because extremely high doses of fluoride might be toxic to humans (fluorine is an ingredient of some rat poisons), there has been some localized opposition to fluoridation programs. However, the concern is a misplaced one, for the difference between a beneficial and toxic dose is enormous. The usual concentration achieved through purposeful fluoridation is 1 ppm (1 milligram per liter). A lethal dose of fluoride would be about 4 grams (4,000 milligrams). To take in that much fluoride through fluoridated drinking water, one would not only have to consume 4,000 liters (about 1,000 gallons) of water, but would have to do so within a day or so, because fluoride is not accumulative in the body like some other elements, such as the heavy metals. Even to consume as much as 20 milligrams per day of fluoride, which might cause bone sclerosis, would require the consumption of 20 liters (5 gallons) of water, every day, for a period of time. That, too, is most improbable.

Zinc

Zinc is one of the heavy-metal elements which, nevertheless, appears to be a critical trace-element nutrient, recognized as such since the 1930s. Zinc-deficiency symptoms can take many forms, including dwarfism, dermatitis (skin disease), loss of taste sensitivity, and delay in the rate at which wounds heal. Optimal zinc consumption should fall in the range of 5 to 40 milligrams per day. At doses above 150 milligrams per day, zinc excess may cause anemia; the very much higher dose of about 6,000 milligrams per day is lethal in humans.

Zinc is present at concentrations of about 70 ppm in average crustal rock, but the concentration varies widely with geology. So far, no regional correlation between geology (such as zinc content of water or soil) and the occurrence of zinc deficiency has been found in the United States. However, there is some concern that deficiencies may become more widespread in the future as certain incidental sources of zinc, particularly galvanized items and water pipes, are phased out of our lives. In the absence of dietary supplements, we are becoming increasingly dependent on plants as sources of necessary zinc, and these do show clear regional variations in zinc content. In general, low-zinc sandy soils and soils rich in calcium carbonate produce more zinc-deficient plants. A single plant species might show a range in zinc content from only a few ppm on a zinc-poor soil to several thousand ppm elsewhere. While regional zinc deficiencies closely correlated with local geology might be avoided by interregional food transport, widespread zinc deficiencies could result if large proportions of some important food plants were grown in low-zinc regions.

Selenium

Selenium is quite a rare metal, for which excess is more often a concern than is deficiency. Its concentration in most rocks and soils is less than one-tenth part per million, and normal selenium intake in humans is 0.006 to 0.2 milligrams per day. Lack of selenium has been shown to cause abnor-

Figure 20.5 Kesterson Wildlife Refuge, with selenium-contaminated ponds. These would be an attractive habitat for waterfowl, were it not for the toxicity of the waters.   Photograph courtesy of USGS Photo Library, Denver, CO.

malities in many plants and animals; selenium deficiency symptoms in humans are not well documented except in the Keshan region of southeastern China, where the most selenium-deficient soils known are found. On the other hand, selenium toxicity (when more than about 5 milligrams per day is consumed) is well recognized, reflected in development of cancers, malformation of nails and hair, depression, nervousness, and other symptoms in humans.

While soil chemistry is a major control on selenium uptake in plants, different species have been shown to vary widely in the degree to which they absorb selenium from the soil. The "locoweed" known in the western United States, actually a group of plant species of the genus *Astragalus*, concentrates selenium strongly. The informal name derives from the selenium poisoning that occurs in livestock that graze on it: The animals develop a disease known as "blind staggers." Other selenium-concentrating plants are used in prospecting for ores, including uranium ores. Among human food crops, plants that concentrate sulfur—which include cabbage, mustard, and onions—also concentrate the geochemically similar selenium.

Where levels of soluble selenium in the soil are high, most varieties of vegetation grown on that soil will also be high in selenium, whether those varieties particularly concentrate selenium or not. In the early 1930s, one farm in South Dakota, known locally as the Reed farm, had a history of frequent changes of ownership. In time, each new owner usually filed a lawsuit charging misrepresentation against the former owner. Chicken eggs did not hatch or the chicks were malformed; cattle and horses became sick. Eventually, the trouble was traced to selenium-rich crops, especially wheat, grown on the farm. After the problem was identified, the federal government bought up the most seriously affected farms in the region.

Fortunately, the number of places in the world having soils that routinely produce deadly selenium-rich crops is small. Selenium is somewhat concentrated in volcanic rocks and in the soils derived from them (as in Hawaii, for instance) as well as in certain shales rich in sulfide. The desert plume plant of the southwestern United States, in fact, *requires* selenium-rich soil, so its presence is a clue to the possible hazards of the soil chemistry. Selenium is also more readily leached from the pedocal soils common to the western United States. In early 1985, it was discovered that runoff drained from farmland in California's San Joaquin Valley and dumped in ponds in the Kesterson Wildlife Refuge (figure 20.5) was causing deaths and deformities among the local duck and fish populations. The culprit appears to be selenium in the water, leached from the farm soils.

Interestingly, the problems became acute only as a result of human activities. The area's shale-derived soils were naturally high in selenium (figure 20.6), but in this normally dry area, the selenium was ordinarily relatively immobile (figure 20.7). Only with the extensive flooding of the land for the cultivation of rice were high levels of selenium leached, washed into the ponds of Kesterson, and there concentrated by evaporation, increasing the flux to those surface-water reservoirs, disrupting the established geochemical cycle, and creating a hazard to wildlife. Selenium, like the heavy metals, can be accumulative in the food chain, so selenium toxicity symptoms were noted in birds (at the top of an algae → insects → fish → birds chain) before toxic selenium concentrations were achieved in the water of Kesterson.

Radon

The problem of radon as a pollutant of indoor air was discussed in chapter 17. A fundamental control on the extent of

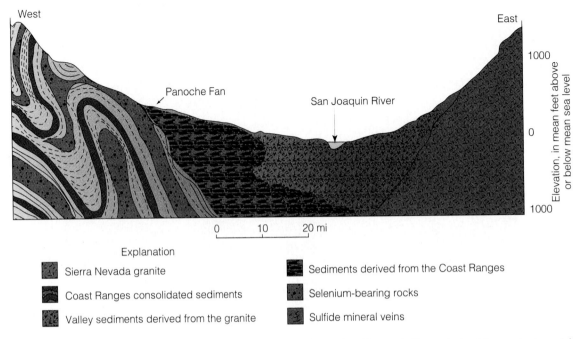

West   East

Panoche Fan   San Joaquin River

Elevation, in mean feet above or below mean sea level

1000

0

1000

0   10   20 mi

Explanation

Sierra Nevada granite   Sediments derived from the Coast Ranges

Coast Ranges consolidated sediments   Selenium-bearing rocks

Valley sediments derived from the granite   Sulfide mineral veins

Figure 20.6 Simplified cross section of San Joaquin Valley. Selenium-rich sediments/soils are derived from selenium-rich sedimentary rocks of the Coast Ranges.   Source: After U.S. Geological Survey 1985 Annual Report.

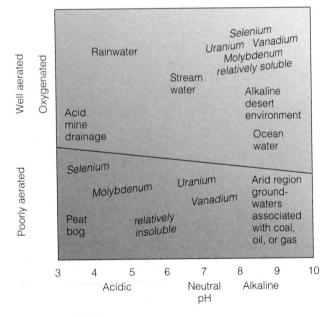

Figure 20.7 Solubility of selenium and associated elements (uranium, molybdenum, vanadium) is related to pH and to oxygen availability in the environment. These elements are very soluble in waters in the normally dry, alkaline soils of the arid areas of the southwestern United States.   Source: After U.S. Geological Survey 1985 Annual Report.

the hazard is the underlying regional geology, and especially the concentration of uranium in the rocks (figure 20.8). Rocks that may have relatively high uranium contents include granites, silicic volcanic rocks, phosphate-rich sedimentary rocks, sulfide-rich shales, and metamorphic rocks formed from any of these.

The quantity of a radioactive material may be expressed in units of *curies*, where one curie is the activity (number of radioactive decay events per unit of time) of one gram of pure radium. (The unit is named for Marie Curie, who discovered radium.) One curie = 37,000,000,000 disintegrations (decays) per second. This would be an extremely intense and hazardous level of radiation; environmental radiation levels are more often measured in picocuries (pCi), where a picocurie is one trillionth of a curie, or .037 disintegrations per second, or about 133 radioactive decay events per hour. Figure 20.9 illustrates the range of radon activities in ground water, air in soil, and indoor and outdoor air.

The potentially carcinogenic effects of radon exposure were recognized earlier in populations exposed to high radiation levels (e.g., uranium miners). As with any radiation exposure, it is difficult to assess precisely the risk represented by low doses. Given the concentrating effects of well-sealed houses relative to outdoor air, it is at least clear that the risk increases in such structures, and clearly the higher the natural radiation levels resulting from the particular geologic setting, the higher the risk associated with concentrating that radon.

Cases in Which Connections Are Less Clear

In the previous examples, a clear relationship could be established between geochemistry and health because a specific chemical substance could be identified with particular deficiency or disease symptoms. On the frontiers of medical geology, considerable efforts are being made to establish the causative links more firmly where they are presently unknown.

1. Great Salt Lake: Water absorbs gamma rays so it shows as no data area on the map.

2. Nebraska Sand hills: Wind has separated the lighter quartz sand from the clay and heavier minerals that usually contain uranium.

3. The Black Hills: A core of granites and metamorphic rocks high in radioactivity is surrounded by less radioactive sedimentary rocks and gives a distinctive pattern.

4. Pleistocene glacial deposits: The area has low surface radioactivity, but uranium occurs just below the surface. Thus it has a high radon potential.

5. Deposits of glacial Lake Agassiz: Clay and silt from a prehistoric glacial lake have higher radioactivity than glacial drift surrounding it.

6. Ohio State: Uranium-bearing black shale with a narrow outcrop zone was scooped up and spread over a large area in west-central Ohio by glaciers.

7. Reading Prong: Uranium-rich metamorphic rocks and numerous fault zones produce high radon in indoor air and in groundwater.

8. Appalachian Mountains: Granites contain elevated uranium, particularly in fault zones. Black shales and soils above limestone also contain moderate to high levels of uranium.

9. Chattanooga and New Albany Shales: Uranium-bearing black shales in Ohio, Kentucky, and Indiana have a distinctive outcrop pattern clearly defined by radioactivity.

10. Outer Atlantic and Gulf Coastal Plan: This area of unconsolidated sands, silts, and clays has one of the lowest radon potentials in the United States.

11. Phosphatic rocks, Florida: These rocks are high in phosphate and associated uranium.

12. Inner Gulf Coastal Plain: This area of the Inner Coastal Plain has sands containing glauconite, a mineral high in uranium.

13. Rocky Mountains: Granites and metamorphic rocks in these ranges contain more uranium than sedimentary rocks to the east, resulting in high radon in indoor air and in ground water.

14. Basin and Range: Granitic and volcanic rocks in the ranges, alternating with basins filled with alluvium shed from the ranges, give this area a generally high radioactivity.

15. Sierra Nevada: Granites containing high uranium, particularly in east-central California, show as red areas.

16. Northwest Pacific Coastal Mountains and Columbia Plateau: This area of volcanic basalts is low in uranium.

Figure 20.8 Radioactivity map of the United States reveals uranium-rich rocks and soils that may represent potential radon-hazard areas.   Source: After J. K. Otton, *The Geology of Radon*, U.S. Geological Survey, 1992.

Radioactivity and Tobacco

Cigarette smoking has been associated with the increased occurrence of lung cancer. While certain chemical components in tobacco have certainly been shown to be damaging, it has also been suggested that agents of geologic origin may indirectly be partly responsible also. The phosphate fertilizers spread on cropland tend to be somewhat en-riched in uranium, as a consequence of the natural geo-chemistry of uranium and phosphate. As noted in chapter 17, one product of uranium decay is radon gas. It is hy-pothesized that this naturally occurring radon gas rises from the soil to be adsorbed onto tobacco leaves, where it decays, in turn, to solid radioactive products, including iso-topes of lead and bismuth (figure 20.10). These radioactive particles would be inhaled with the smoke and could lodge in the lungs, where the radiation from their decay could cause cancer. There is a good deal of speculation in this scenario, but it is plausible and worth testing further.

Regional Variations in Heart Disease

In a number of instances, geographic variations in occurrences of particular symptoms are known, but there is either no known parallel pattern in the geology or no clear connection recognized between the known geologic variables and the particular health problem. For example, it has been known for two decades that the hardness of public water supplies shows regional variations, rooted in geology, that are broadly similar to the variations in death rate from certain types of heart disease (figure 20.11). In the central United States, where limestone-rich aquifer systems contribute to relatively hard water, coronary heart disease death rates are generally lower. Along the Atlantic coast, the water is much softer, and the heart-disease death rates are relatively high. Why?

So far, no single answer has emerged. Is there something beneficial in hard water that protects against heart disease in some way not yet recognized? Soft water is usually more acidic than hard water, and acids more readily leach metals from pipes—does that result in consumption of one or more metals that somehow cause heart disease? Many people living in hard-water areas use water softeners, which add salt to the water, and sodium has been implicated in cardiovascular disease; but not everyone in hard-water areas drinks softened water. Does the water chemistry, in fact, have any causative link with heart disease at all? Not necessarily; the apparent connection might be pure geographic coincidence. Perhaps the stress of living in East Coast cities rather than in more rural areas of the Midwest is a causative factor. Perhaps no one has even proposed the right answer yet. Still, the similar geographic patterns in heart disease and water hardness are there, and they are intriguing. Also, the same kind of geographic correlation between these two particular parameters has been observed in twenty-seven studies in eight countries, which strengthens the impression that some genuine cause-and-effect relationship between water chemistry and the disease exists.

Figure 20.9   Ranges of radioactivity due to radon in natural settings—ground water, air in soil pores, outdoor air—and in the home. Note that the radioactivity scale is exponential, so radiation levels in high-radon homes may be hundreds of times what is typical in outdoor air.

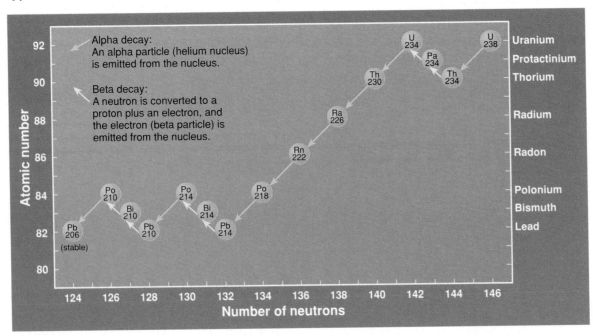

Figure 20.10   Uranium decays through a series of radioactive isotopes, eventually to stable lead isotopes. This is the main decay path for uranium-238. (See chapter 2 for review of isotope notation.) Radon, a chemically inert gas, is nevertheless radioactive; among its decay products are radioactive isotopes of bismuth and lead, heavy metals that accumulate in the body.

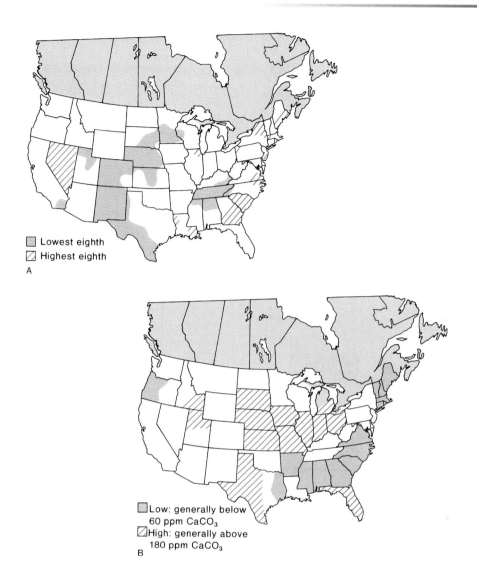

Figure 20.11 The relationship between deaths caused by heart disease and water-supply hardness is suggested by these maps showing regional variations in each parameter. (*A*) Death rates from cardiovascular heart disease, white males ages forty-five to sixty-four, 1950–61. (*B*) Public water-supply hardness in the contiguous United States.   Source: (A) Data from H. I. Sauer and F. R. Brand, M.D. "Geographic Patterns in the Risk of Dying," Geological Society of America Memoir 123, 1971. (B) Data from B. A. Anderson, "Calcium-Carbonate Hardness of Public Water Supplies in the Conterminous United States," Geological Society of America Memoir 123, 1971.

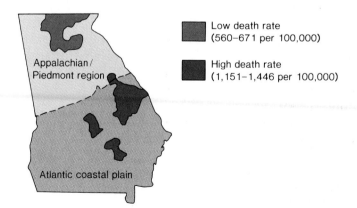

Figure 20.12 Regional variations in heart-disease death rates in Georgia appear to be correlated with local geology for reasons not yet determined.   Source: After H. I. Shacklette, et al., "Geochemical Environments and Cardiovascular Mortality Rates in Georgia," U.S. Geological Survey Professional Paper 547C, 1970.

Cardiovascular Disease in Georgia

Another puzzling geographic pattern is the regional variation in cardiovascular-disease death rates in Georgia, which corresponds to variation in the underlying geology (figure 20.12). The counties with the lowest death rates are all in the Appalachian highlands in the northern part of the state. The counties with the highest death rates lie wholly or partially within the Atlantic coastal plain to the south. This observation inspired exhaustive geochemical investigations to try to find a geologic cause for the pattern.

Soils, plants, trees, and food crops were analyzed for thirty different elements and the analyses compared between regions. In general, the soils of the coastal plain are more weathered and more extensively leached. The highland soils are significantly higher in aluminum, barium, calcium, chromium, copper, iron, potassium, magnesium, manganese, niobium, phosphorus, titanium, and vanadium, and somewhat

Figure 20.13 Variations in soil type in coastal Iran show a distribution suggesting a possible link with incidence of esophageal cancer. Source: Data from J. Kmet and E. Mahboui, "Esophageal Cancer in the Caspian Littoral of Iran: Initial Studies," *Science*, vol. 175, Feb. 25, 1972, pp. 846–853. Copyright © 1972 by the AAAS.

lower in zirconium. Several of these elements (chromium, copper, manganese, and vanadium) are believed to have some beneficial effect on the cardiovascular system. However, no simple relationship was found between soil chemistry and the chemistry of vegetables grown on that soil, and it is, after all, the vegetables that are consumed, not the soil. The sources of drinking water in the region were so numerous and varied, including both public and individual supplies, surface and subsurface sources, that systematic sampling and analysis could not be undertaken. Yet the water might be a significant factor. More sampling and further studies to confirm the influence of some of these trace elements on cardiovascular function are needed before any straightforward explanation of the death-rate pattern in terms of geology can be shown.

This study illustrates one of the special problems faced by medical geologists as opposed to laboratory scientists. In the lab, it is possible to control experiments, to vary one parameter, such as the concentration of one chemical element, at a time. In nature, dozens of chemical characteristics are likely to be varying at once from place to place. It becomes correspondingly far more difficult to pick out the influence of any one. The situation is further complicated by interactions between trace elements, as will be shown shortly.

Other Intriguing Patterns

Other diseases show strong geographic variations that might be tied to geology. In Iran, the rate of occurrence of cancer of the esophagus shows a regional pattern that strikingly mimics the distribution of soil types (figure 20.13). Whether the soil chemistry has, in fact, any direct influence on the incidence of the cancer is not known. Elsewhere, possible geochemical causes have been proposed to explain regional variations in the frequency of occurrence of other forms of cancer and of gout. The incidence of stomach cancer in Wales, for example,

Table 20.4 Apparent Relationship Between Age of Rocks and Heart Disease

Age of surface and underlying rocks	Countries	Mean ischemic heart disease rate, 1967 (per 100,000 population)
Precambrian (over 600 million years)	Sweden, Finland, Denmark, Scotland	314
Early Paleozoic (600–300 million years)	Norway, England, Northern Ireland	291
Late Paleozoic (300–180 million years)	Ireland, Austria, Hungary, France, Germany, Netherlands, Belgium, Czechoslovakia, Poland, Spain, Portugal, Romania	175
Mesozoic or younger (under 180 million years)	Italy, Yugoslavia, Bulgaria, Greece, Switzerland	159

Source: R. Masironi, "World Health Organization Studies in Geochemistry and Health," *Geochemistry and the Environment*, vol. 2, 132–138, National Research Council, 1977.

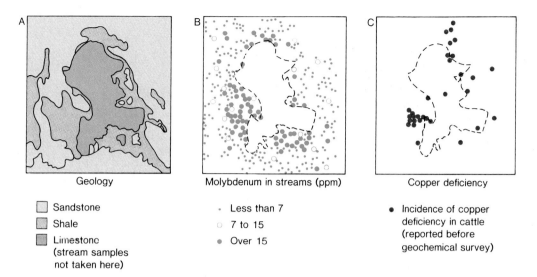

Geology

Sandstone
Shale
Limestone
(stream samples
not taken here)

Molybdenum in streams (ppm)

- Less than 7
○ 7 to 15
● Over 15

Copper deficiency

● Incidence of copper
deficiency in cattle
(reported before
geochemical survey)

Figure 20.14 Interaction between trace elements is illustrated by a link between molybdenum content of streams (controlled by geology) and incidence of copper deficiency in cattle in central England. (*A*) Geology of central England. (*B*) Molybdenum content of streams in central England. (*C*) Incidence of copper deficiency in cattle in central England, reported before geochemical survey. Source: After J. S. Webb, "Regional Geochemical Reconnaissance in Medical Geography," Geological Society of America Memoir 123, 1971.

seems to be correlated with the amount of organic matter in the soils. Whether this is a consequence of the tendency of organic matter to concentrate some potentially toxic metals or of some other factor is not presently known.

There also are correlations with a less obvious geochemical basis. A recent study of death rates from one type of heart disease in Europe suggested a correlation with the age of the rocks in the region (table 20.4). A given age group may comprise a wide variety of rock types, and the same rock type may be found in regions of several ages. Thus, geology cannot be related in any simple way to the pattern of death rates. Certainly, why the *age* of a rock, by itself, should make any particular difference to the health of those living above it is not at all apparent. In this case, it may be easy, intuitively, to dismiss geology—or at least rock age—as a probable causative factor. Where spatial correlations with chemical factors are observed, however, it is more difficult to establish or refute a connection.

There are also situations in which expected regional correlations are absent. There are clear regional variations in radon levels in rocks and soils, and consequently in homes. Yet so far, corresponding consistent regional variations in occurrence of lung cancer have not been clearly documented, perhaps because these variations are masked by other factors causing lung cancer, such as cigarette smoking.

Further Complicating Factors

Cause-and-Effect or Coincidence?

Medical geologists must distinguish those correlations that arise because of cause-and-effect from apparent correlations that do not. A tobacco company pointed out several years ago during an advertising campaign that more people die in

bed than anywhere else, but that hardly means that it is dangerous to go to bed. One might observe a much higher percentage of gray-haired listeners at a symphony concert featuring classical music than at a rock concert, but it would surely be scientifically irresponsible to conclude that classical music causes gray hair. Many obviously silly examples could be constructed to prove the point that coincidence in time or space of two things does not at all indicate a cause-and-effect relationship between them. Similarly, the fact that the geology of a particular area may seem to be correlated with a higher or lower incidence of a particular disease does not by itself prove any causal connection between the two.

Trace-Element Interactions

As noted earlier, genuine cause-and-effect relationships may be especially hard to recognize in natural geologic systems, in which numerous mineralogic and geochemical parameters vary simultaneously and in different ways. The picture is further clouded because the effect of a particular element often is modified in the presence of other elements. An example is shown in figure 20.14. The incidence of copper deficiency in cattle in central England appeared strongly correlated with underlying geology, with most cases in or near areas underlain by shales. Yet the shales and the soils derived from them were not markedly poor in copper. They *were* rich in molybdenum, as were many streams in that area, and this was the source of the problem. Apparently, increased intake of molybdenum increased the need for copper in the animals' diets. Animals drinking from less-molybdenum-rich streams might show no signs of copper deficiency even when the levels of copper in their blood were relatively low.

Other known and suspected interactions abound. Plants growing in high-phosphorus soils (including those enriched by fertilizers) are frequently deficient in zinc. Uptake of high levels of zinc by plants may limit the extent to which

501

Figure 20.15 A simplified periodic table of the elements. Elements with atomic numbers higher than that of uranium (92) are not naturally occurring.

the plants can absorb cadmium; thus, application of zinc to cadmium-rich soils may be helpful in controlling a potentially harmful accumulation of high cadmium concentrations in food crops. Sulfur may "compete with" selenium during uptake in plants, but experimental results are inconsistent: Applying sulfur-bearing fertilizers to low-selenium soils has depressed the levels of selenium in plants even further, but on the other hand, efforts to prevent accumulation of high selenium levels in plants growing on selenium-rich soils by applying sulfur have not worked well.

In animals, metabolic processes involving magnesium, potassium, and calcium have been shown to be interrelated, and evidence indicates that the same is true in humans. Thus a deficiency in one element may upset the balance of another. Copper is essential to the proper metabolism of iron; molybdenum may also affect iron metabolism. Plant fiber may inhibit the absorption of zinc in the human digestive tract. In plants, phosphorus may inhibit uptake of zinc, causing zinc deficiency in plants (and higher organisms, including humans, consuming the plants); this can be a particular problem because phosphorus is a major ingredient in fertil-

izer. Zinc and cadmium may compete in humans as well as in plants; hence, increased zinc consumption may afford some protection against cadmium. Selenium may protect against the harmful effects of some mercury compounds (but, of course, selenium in turn is potentially toxic and thus should not be consumed indiscriminately).

The list of established and suspected interactions is extended year by year. Some of the interactions—especially competition among elements—can be anticipated from the elements' positions in the periodic table (figure 20.15). Elements in the same column (sulfur (S) and selenium (Se), column VIa; zinc (Zn) and cadmium (Cd), column IIIB; calcium (Ca) and magnesium (Mg), column IIa) have similar electronic structures and may therefore show similar bonding characteristics. Other apparent interactions between less similar elements (copper and molybdenum, for example) obviously involve more complex chemistry and biochemistry and are not so readily identified or understood. The growing recognition of trace-element interactions is one reason why doctors warn against consuming large doses of mineral supplements containing a few specific elements. Aside from any

Figure 20.16 Chrysotile asbestos, or "white asbestos," accounts for about 95% of asbestos mined and used in the U.S. © The McGraw-Hill Companies, Inc./Bob Coyle, photographer

potential toxic effects of the particular elements consumed, there is concern that chemical imbalances of other elements could be created, with serious health consequences.

Distinguishing Risk from Risk Perception

Costs of asbestos abatement over the last decade or so are estimated to have totaled tens of billions of dollars for commercial buildings alone. Yet much of this expenditure may have addressed a nonexistent or insignificant risk. It is quite true that studies have documented increased incidence of certain cancers and other health problems among miners and other workers with high occupational exposure to some kinds of asbestos. However, it does not follow that all asbestos everywhere is dangerous.

First, there are six different silicate minerals that are called "asbestos," for the term is used to describe various minerals with a fibrous habit. By far the most common—accounting for an estimated 95 percent of asbestos mined and used in the United States—is *chrysotile*, a hydrous chain silicate (figure 20.16). Yet there is little or no evidence that chrysotile poses a health risk to the general population; indeed, one case study of a school in Ambler, Pennsylvania, where for over a century, schoolchildren have been exposed to a 150,000-ton pile of chrysotile-containing material beside the school, showed not a single documented case of asbestos-caused illness. Even where occupationally exposed workers had somewhat elevated incidence of lung disease, the possible effect of asbestos could not be clearly separated from effects of the workers' smoking. Some of the rarer asbestos minerals show clearer

risk; but asbestos-abatement legislation does not make mineralogical distinctions among asbestos varieties.

Part of the push for asbestos abatement, too, comes from the unverified "one-fiber" theory—the assumption that a single asbestos fiber may cause, for instance, lung cancer, or other asbestos-related disease. But fibrous minerals, like dust, are everywhere: it is estimated that, breathing average outdoor air, a normal human would inhale nearly four thousand asbestos fibers a day! This is not to say that occupational exposures for those who work with asbestos materials should not be regulated in the interest of workers' health. The risks to the general public from asbestos, however, would seem to be far lower than commonly perceived, and much of the investment in "abatement" of chrysotile-containing material, in particular, may be far out of proportion to the risks that it actually presents.

Impacts of Human Activities

Human activities are modifying trace-element concentrations in many ways, some of which were already considered in this and the previous two chapters. Soil pollution with metals (lead, zinc, copper, cadmium, arsenic, uranium, and others) is common near smelters and other ore-processing facilities, and the pollution may lead not only to water pollution but also to contamination of plants growing nearby. Soils and plants growing near busy highways frequently showed elevated lead levels due to pollution from car exhaust. Consideration of the possibility of substituting manganese compounds for now-banned lead in gasoline has, in turn, raised concerns about possible manganese poisoning: Nerve and brain damage from excess exposure are known to occur in some miners and other workers exposed to high levels of manganese. Agricultural practices, including modification of the hydrologic cycle (as by irrigation with attendant redistribution of leachable elements) and application of natural or synthetic fertilizers that may add key trace elements, modify geochemical cycles by changing chemical fluxes and residence times. These and other modifications of natural elemental cycles mask the basic geologic factors influencing health and disease, making the medical geologist's task more difficult.

And new environmental health problems continue to emerge. One of the newest is the discovery that certain pesticides can mimic hormones in the bodies of humans and other animals, causing anomalous development of various kinds. Some appear to suppress the immune system, lowering an organism's resistance to disease. Others are believed to cause reproductive problems, or anomalies in the development of embryos. While over 60 percent of pesticide sales are currently in the U.S., Western Europe, and Japan, pesticide use is increasing far more rapidly in developing nations, where mortality rates are already high. Many agencies are now calling for more extensive and detailed studies of the toxicity of these pesticides.

Overview

Summary

In a growing number of instances, the geology and geochemistry of the environment have been shown to bear on the chemistry and health of plants, animals, and people. Many of these effects are independent of human activities such as those causing pollution. Often, they are recognized initially through geographic variations in occurrence of a particular disease. The identification of significant natural geochemical or mineralogic factors is complicated because cultural factors are superimposed on natural geologic and biologic systems, by complex interactions between trace elements in which the presence or abundance of one modifies the effect of another, and by the difficulty of isolating the effects of a single chemical substance from the dozens of geochemical parameters that vary simultaneously with local geology. Better understanding of the medical importance of naturally occurring trace elements could lead to the elimination of many instances of regionally chronic excess or deficiency diseases having a geologic basis. The health hazards of synthetic chemicals in the environment are not always well determined. There are currently numerous examples of geographic variations in disease occurrence that *may* arise from geologic factors but for which the specific cause is speculative or wholly unknown. Conversely, there are cases of overreaction to a perceived geologic health hazard, as with asbestos abatement.

Terms to Remember

dose-response curve 490   trace elements 489

Exercises

Questions for Review

1. What are trace elements? Are the trace elements in humans the same as those in rocks? Explain.
2. Draw a hypothetical dose-response curve and explain it.
3. Outline the pathways through which trace elements enter the body, noting both natural processes and human activities that alter the elements' concentrations along the way.
4. Links between regional soil chemistry and health may nowadays be easier to recognize in less-developed countries. Why?
5. An apparent connection between water hardness and the incidence of heart disease is well documented in many countries. Suggest several possible explanations.
6. Does a correlation in time or space between two factors or events indicate a cause-and-effect relationship between them? Explain.
7. The concentration of one trace element may alter the effect of another. Describe an example. How does this complicate geomedicine?

For Further Thought

1. Examine the label of a package of multivitamin-plus-mineral tablets. How many of the essential trace elements of table 20.2 are included? Are any additional elements included for which a need has not yet been clearly established?
2. Investigate the possibility of unusual trace-element concentrations in soils in your area. (The county Extension Service or an agricultural college may have relevant data.) Are any of the anomalies believed to be particularly beneficial or potentially harmful? If so, what (if anything) is being done about them?

Suggested Readings/References

Adriano, D. C. 1986. *Trace elements in the terrestrial environment.* New York: Springer-Verlag.

Anderson, B. A. 1971. Calcium-carbonate hardness of public water supplies in the conterminous United States. In *Environmental geochemistry in health and disease,* edited by H. L. Cannon and H. C. Hopps, pp. 151–53. Geological Society of America Memoir 123.

Beeson, K. C., and Matrone, G. 1976. *The soil factor in nutrition.* New York: Marcel Dekker.

Brookins, D. G. 1981. *Earth resources, energy, and the environment.* Columbus, Ohio: Charles E. Merrill.

Calabrese, E. J. 1981. *Nutrition and environmental health.* New York: Wiley-Interscience.

Cannon, H. L., and Hopps, H. C., eds. 1970. *Geochemical environment in relation to health and disease.* Geological Society of America Special Paper 140.

——. 1971. *Environmental geochemistry in health and disease.* Geological Society of America Memoir 123.

The citizen's guide to geologic hazards. 1993. American Institute of Professional Geologists. (See "Asbestos," pp. 23–28, and "Radon," pp. 29–37.)

Draggan, S., Cohrssen, J. J., and Morrison, R. E., eds. 1987. *Environmental impacts on human health.* New York: Praeger.

——. 1987. *Geochemical and hydrologic processes and their protection.* New York: Praeger.

Feder, G. L. 1985. Environmental influence of selenium in waters of the western United States. *U.S. Geological Survey 1985 Annual Report,* 5–8.

Fergusson, J. E. 1982. Inorganic materials in the biosphere. Chapter 13 in *Inorganic chemistry and the earth,* pp. 334–59. New York: Pergamon Press.

Gunter, M. E. 1994. Asbestos as a metaphor for teaching risk perception. *Journal of Geological Education* 42: 17–24.

Hurley, P. M. 1982. *Living with nuclear radiation.* Ann Arbor, Mich.: University of Michigan Press.

Lippmann, M., and Schlesinger, R. B. 1979. *Chemical contamination in the human environment.* New York: Oxford University Press.

Masironi, R. 1977. World Health Organization studies in geochemistry and health. In *Geochemistry and the environment,* vol. 2. Washington, D.C.: National Research Council.

McLachlan, J. A., and Arnold, S. F. 1996. Environmental estrogens. *American Scientist* 84 (September-October): 452–61.

National Research Council. 1977. The relation of other selected trace elements to health and disease. In *Geochemistry and the environment,* vol. 2. Washington, D.C.: National Research Council.

Oehme, F. W., ed. 1979. *Toxicity of heavy metals in the environment.* New York: Marcel Dekker.

Otton, J. K. 1992. *The geology of radon.* Washington, D.C.: U.S. Geological Survey.

Parr, R. 1983. Trace elements in human milk. *International Atomic Energy Agency Bulletin* 25 (2): 7–15.

Pier, S. M., and Bang, K. M. 1980. The role of heavy metals in human health. Chapter 11 in *Environment and health,* edited by N. M. Trieff, pp. 366–409. Ann Arbor, Mich.: Ann Arbor Science Publishers.

Raloff, J. 1994. The gender benders. *Science News* 145 (8 January): 24–28.

Sauer, H. I., and Brand, F. R. 1971. Geographic patterns in the risk of dying. In *Environmental geochemistry in health and disease,* edited by H. L. Cannon and H. C. Hopps, pp. 131–50. Geological Society of America Memoir 123.

Shacklette, H. I., Sauer, H. I., and Miesch, A. T. 1970. *Geochemical environments and cardiovascular mortality rates in Georgia.* U.S. Geological Survey Professional Paper 547C.

Skinner, H. C. W., and Ross, R. 1994. Minerals and cancer. *Geotimes* (January): 13–15.

Speidel, D. H., and Agnew, A. F. 1982. *The natural geochemistry of our environment.* Boulder, Colo.: Westview Press.

Villanueva, W. 1982. The cartography of death. *Discover* (September): 82–85.

Webb, J. S. 1971. Regional geochemical reconnaissance in medical geography. In *Environmental geochemistry in health and disease,* edited by H. L. Cannon and H. C. Hopps, pp. 31–42. Geological Society of America Memoir 123.

The U.S. Agency for Toxic Substances and Disease Registry is a good source of fact sheets that respond to frequently asked questions (FAQs) about various health hazards, including geologic ones. Their home page, through which one can also access a new newsletter, *Hazardous Substances and Public Health,* is at

http://atsdr1.atsdr.cdc.gov:8080/atsdrhome.html
The FAQ fact sheets are accessible more directly via
http://atsdr1.atsdr.cdc.gov:8080/toxfaq.html

The Citizen Information Page of the U.S. Environmental Protection Agency includes information sheets on potential health hazards such as lead and radon, at

http://www.epa.gov/epahome/Citizen.html
EPA's Citizen Guide to Radon is found at
http://www.epa.gov/docs/RadonPubs/citguide.txt.html

The U.S Geological Survey offers a variety of information on radon at

http://sedwww.cr.usgs.gov:8080/radon/radonhome.html

A variety of environmental-medicine case studies are accessible via

http://uhs.bsd.uchicago.edu/case.studies/index.html

Canada's Environmental Health directorate, a division of Health Canada, can be reached at

http://www.hwc.ca/dataehd/English/opening.htm

Canada is a major supplier of chrysotile asbestos. The Canadian-based Asbestos Institute is at

http://www.asbestos-institute.ca/

Highlights of a World Resources Institute report, "Pesticides and the Immune System," can be found online at

http://www.wri.org/wri/health/pis-high.html

Appendix A
Geologic Time, Geologic Process Rates

Introduction

Much of the understanding of geologic processes, including the rates at which they occur and, therefore, the kinds of impacts they may have on human activities, is made possible through the development of various means of "telling time" in geologic systems. In this appendix, we will explore several of these methods. A final section examines some applications of geologic age determinations to the study of process rates.

Relative Dating

Arranging Events in Order

Before any ways of establishing numerical ages of rocks or geologic events were known, it was sometimes at least possible to place a sequence of events in the proper order. Among the earliest recognized efforts in this direction were those of Nicolaus Steno. In 1669, he set forth two very basic principles that could be applied to sedimentary rocks. The first, the **Principle of Superposition** (figure A.1), pointed out that, in an undisturbed pile of sediments or sedimentary rocks, those on the bottom were deposited first, followed in succession by the layers above them, ending with the youngest on top. (Today, this idea may seem so obvious as not to be worth stating, but at the time it represented a real step forward in thinking logically about rocks.) The second, the **Principle of Original Horizontality,** was based on the observation that sediments are deposited in approximately horizontal, flat-lying layers. Therefore, if one comes upon sedimentary rocks in which the layers are folded or dipping steeply, they must have been displaced or deformed after deposition and solidification into rock (figure A.2).

In later centuries, these ideas were extended to work igneous rocks into such sequences (figure A.3). If an igneous rock cuts across layers of sedimentary rocks, the sedimentary rocks must have been there first, the igneous rock introduced later. This is sometimes called the *Principle* (or *Law*) of *Crosscutting Relationships.* Often, there is a further clue to the correct sequence: The hotter igneous rock may have "baked," or metamorphosed, the sedimentary rocks immediately adjacent to it, so again the igneous rock must have come second. If an igneous rock contains pieces of other rock types (figure A.4), those pieces must have been picked up as solid chunks by the invading magma, so the rocks making up the inclusions must predate the host rock (*Principle of Inclusion*).

Such geologic common sense can be applied in many ways. If a strongly metamorphosed sedimentary rock is over-

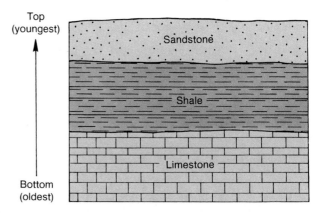

Figure A.1 The Principle of Superposition: The rocks on the bottom of a sequence of undisturbed sedimentary rocks were deposited first, and the depositional ages become younger higher in the sequence.

Figure A.2 The Principle of Original Horizontality indicates that the sequence of events shown here in the Grand Canyon was: Deposition of the lower layers; tilting and erosion; deposition of the still-flat-lying upper layers.   © The McGraw-Hill Companies, Inc./Doug Sherman, photographer

lain by a completely unmetamorphosed one, for instance, the metamorphism must have occurred after the first sedimentary rock formed but before the second. Quite complex sequences of geologic events can be unraveled by taking into account such principles (see figure A.5 for an example).

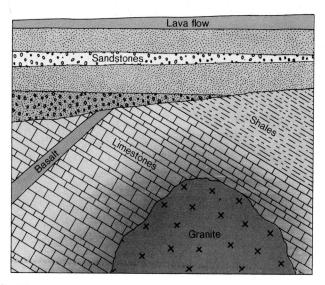

Figure A.3 An igneous rock crosscutting a sedimentary sequence or other rock must postdate the rocks it cuts across.

Rocks adjacent to intruding magma may also be metamorphosed by its heat.

Figure A.5 Sorting out relative ages of rock in an outcrop: The limestones must be oldest (Principle of Superposition), followed by the shales. The granite intrusion and basalt must both be younger than the limestone they crosscut. It is not possible to tell whether the igneous rocks predate or postdate the shales or to determine whether the sedimentary rocks were tilted before or after the igneous rocks were emplaced. After the limestones and shales were tilted, they were eroded, and then the sandstones were deposited on top. Finally, the lava flow covered the entire sequence.

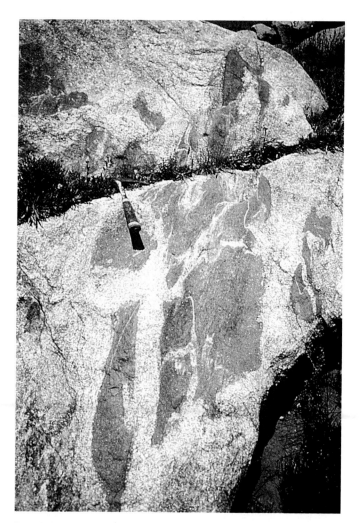

Figure A.4 The darker rock is a schist that was invaded by the magma that formed the (lighter-colored) granite, which picked up pieces of schist. The schist is older than the granite.

Correlation

Fossils play a role in the determination of relative ages, too. The concept that fossils could be the remains of older life-forms dates back at least to the ancient Greeks, but for some time it fell out of favor. The idea was seriously revived in the 1700s, and around the year 1800, William Smith put forth the **Law of Faunal Succession.** The basic principle involved was that, through time, life-forms change, old ones disappear from and new ones appear in the fossil record, but the same form is never exactly duplicated at two different times in history. This principle, in turn, implies that, when one finds exactly the same type of fossil preserved in two rocks, even if the rocks are quite different compositionally and geographically widely separated, they should be the same age. Smith's law thus allowed age *correlation* between rock units exposed in different places (figure A.6). A limitation on its usefulness is that it can be applied only to rocks in which fossils are preserved, which are almost exclusively sedimentary rocks.

The foregoing ideas were all useful in clarifying age relationships among rock units. They did not, however, help answer questions like: How old is this granite? How long did it take to deposit this limestone? How recently, and over how long a period, did this apparently extinct volcano erupt? Has this fault been active in modern times?

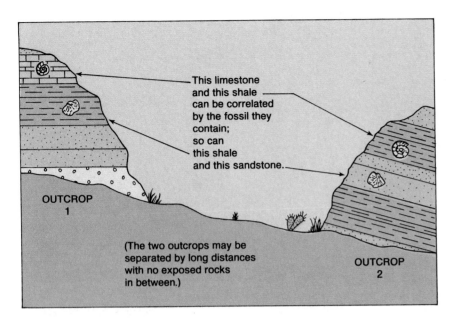

This limestone and this shale can be correlated by the fossil they contain; so can this shale and this sandstone.

OUTCROP 1

(The two outcrops may be separated by long distances with no exposed rocks in between.)

OUTCROP 2

Figure A.6 Similarity of fossils suggests similarity of ages even in different and quite widely separated rocks.

Uniformitarianism

Geologic processes were the focus of another early worker — the physician, farmer, and part-time geologist James Hutton. Hutton is widely credited with developing and popularizing the concept of **uniformitarianism.** Uniformitarianism is sometimes described briefly by the phrase "the present is the key to the past." What Hutton meant by this is that because the same physical laws have operated throughout the earth's history, by studying present, active geologic processes and their products, geologists can infer much about how rocks were formed and changed in the past. Today, a volcanic eruption and the hardening of flowing magma into basaltic rock suggest the source of an ancient basalt flow even if geologists cannot now find the volcano from which it erupted. A broad, white sandy beach or a windswept desert might offer some useful conclusions about the origins of sandstones. Retreating ice sheets in Greenland and Antarctica leave moraines behind, which help geologists to understand the significance of the moraines found in the Midwest and Canada.

Uniformitarianism has sometimes been misunderstood to mean that the *rates* of geologic processes have been the same through time as well, that by observing processes like erosion, sedimentation, and seafloor spreading as they now occur, one can know the rate at which they occurred in the past. Unfortunately, this is not true, and was not really part of uniformitarianism as originally conceived. Volcanism may have involved the same types of processes throughout geologic history, but it might have been much more active early in the earth's history, before the earth cooled appreciably, than it is now; rates of chemical sedimentation, as well as the

proportions of different minerals in the sediments, could have been very different in the past, when the temperature and chemistry of the oceans were also different; and so on. In short, even uniformitarianism does not allow numerical answers to questions about geologic process rates.

How Old Is the Earth?

Geologists and nongeologists alike have been fascinated for centuries with the very basic question of the earth's age. Many have attempted to answer it, but with little success until the twentieth century.

Early Efforts

One of the earliest widely publicized determinations of the age of the earth was the seventeenth-century work of Archbishop Ussher of Ireland. He painstakingly and literally counted up the generations in the Bible and arrived at a date of 4004 B.C. for the formation of the earth. This very young age was hard for many geologists to accept; the complex geology of the modern earth seemed to require far longer to develop.

Even less satisfactory from that point of view was the estimate of the philosopher Immanuel Kant. He tried to find a maximum possible age for the earth, assuming that the sun had always shone down on earth and that the sun's tremendous energy output was due to the burning of some sort of conventional fuel. But a mass of fuel the size of the sun would burn up in only one thousand years, given the rate at which the energy is released, plainly an impossible result in view of several millennia of recorded human history. Kant, of course, knew nothing of nuclear fusion.

The Nineteenth-Century View

About 1800, Georges L. L. de Buffon attacked the problem from another angle. He assumed that the earth was initially molten, modeled it as a ball of iron, and calculated how long it would take this quantity of iron to cool to present surface temperatures: 75,000 years. To most geologists, this still was not nearly long enough.

Around 1850, physicist Hermann L. F. von Helmholtz took the approach of supposing that the sun's luminosity was due to infall of particles into its center, converting gravitational potential energy to heat and light. This gave a maximum age for the sun (and presumably earth) of 20 to 40 million years. Again, his assumptions were wrong, so his answer was also.

In the late 1800s, Lord William T. Kelvin reworked Buffon's calculations, modeling the earth more realistically in terms of rock properties rather than those of metallic iron.

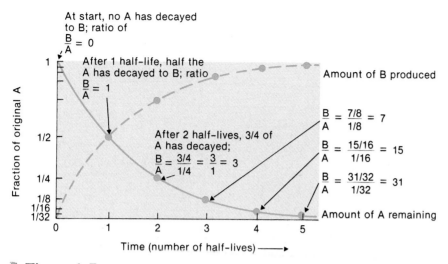

Figure A.7 Decay of radioactive parent A proceeds with complementary accumulation of daughter B.

The Discovery of Radioactivity

Henri Becquerel did not set out to solve any geologic problems nor even to discover radioactivity. He was interested in a curious property of some uranium salts: If exposed to light, the salts continued to emit light for a while afterward even in a dark room (phosphoresce). He had put a vial of uranium salts away in a drawer on top of some photographic plates well wrapped in black paper. When he next examined the photographic plates, they were fogged with a faint image of the vial, as if they had somehow been exposed to light right through the paper. The uranium was emitting something that could pass through light-opaque materials. We now know that uranium is one of several substances that are naturally radioactive, that undergo spontaneous decay. Once the phenomenon of radioactive decay was reasonably well understood, physicists and geoscientists began to realize that it could be a very useful tool for investigating earth's history.

Radioactive Decay and Dating

As noted in chapter 15, one key property of any particular radioisotope is that it decays at a constant, characteristic rate, with a distinct half-life, which can be determined in the laboratory. One can then, in principle, use the relative amounts of a decaying isotope (parent) and the product isotope into which it decays (daughter) to find the age of the sample. Figure A.7 illustrates the fundamental idea. Suppose that parent isotope A decays to daughter B with a half-life of 1 million years. In a rock that contained some A and no B when it formed, A will gradually decay and B will accumulate. After 1 million years (one half-life of A), half of the A initially present will have decayed to yield an equal number of atoms of B. After another half-life, half of the remaining half will have decayed, so three-fourths of the initial amount of A will have been converted to B and one-fourth will remain; the ratio of B to A will be 3:1. After another million years, there will be seven atoms of B for every atom of A, and so on. If a sample contains 24,000 atoms of B and 8,000 atoms of A, then, assuming no B was present when the rock formed and noting the 3:1 ratio of B to A, we would conclude that the sample was two half-lives of A (in this case, 2 million years) old.

The foregoing is, of course, an oversimplification of the dating of natural geologic samples. Many samples contain some of the daughter isotope as well as the parent at the time of formation; and correction for this must be made. Many

Interestingly, he, too, arrived at an age of 20 to 40 million years for the earth. What he did not take into account, because natural radioactivity was then unknown, was that some heat is continually being *added* to the earth's interior through radioactive decay, so it has taken longer to cool down to its present temperature regime.

The calculations went on, and something was wrong with each. In 1893, U.S. Geological Survey geologist Charles D. Walcott tried to compute the total thickness of the sedimentary rock record throughout geologic history and, dividing by typical modern sedimentation rates, to estimate how long that pile of sediments would have taken to accumulate. His answer was 75 million years. Walcott was hampered in his efforts by several factors, including the gaps in many sedimentary rock sequences (periods during which no sedimentation occurred or some sediments were eroded away). He also had no good idea of the total thickness of sediments deposited in the time before organisms capable of preservation as fossils became widespread, which turns out to be most of earth's history.

In 1899, physicist John Joly published calculations based on the salinity of the oceans. Taking the total dissolved load of salts delivered to the seas by rivers, and assuming that the ocean was initially pure water, he determined that it would take about 100 million years for the present concentrations of salts to be reached. Aside from Joly's assumption that rates of weathering and salt input into the oceans throughout earth's history were constant, he did not consider that the buildup of salts is slowed by the removal of some of the dissolved material, for example as chemical sediments.

So the debate continued, frequently heated, until the discovery of radioactivity provided a much more powerful and accurate tool with which to undertake a solution.

work from bottom up

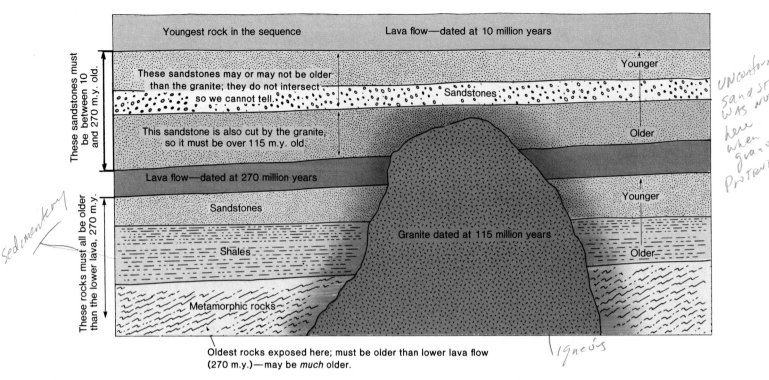

sedimentary

igneous

unconformity sandstone was not here when granite protruded.

Figure A.8 "Absolute" and relative dating together can put constraints on ages of units not dateable directly.

other samples have not remained chemically closed throughout their histories; they have gained or lost atoms of the parent or daughter isotope of interest, which makes the calculated date incorrect. In the case of samples with a complex history—an igneous rock that has been metamorphosed once or twice and also weathered, for example—it can be difficult both to obtain a date and to decide which event, if any, is being dated! For a somewhat more thorough discussion of the complexities of dating of geologic systems, refer to the text by D. L. Eicher (1976) included in the reference list at the end of this appendix.

Choice of an Isotopic System

Several conditions must be satisfied by an isotopic system to be used for dating geologic materials. The parent isotope chosen must be abundant enough in the sample for its quantity to be measurable. Either the daughter isotope must not normally be incorporated into the sample initially, or there must be a means of correcting for the amount initially present. The half-life of the parent must be appropriate to the age of the event being dated: long enough that some parent atoms are still present but short enough that some appreciable decay and accumulation of daughter atoms have occurred. A radioisotope with a half-life of ten days would be of no use in dating a million-year-old rock; the parent isotope would have decayed away completely long ago, and there would be no way to tell how long ago. Conversely, a radioisotope with a half-life of 10 billion years would be useless for dating a fresh lava flow because the atoms of the

daughter isotope that would have accumulated in the rock would be too few to be measurable. Only about half a dozen isotopes are widely used in dating geologic samples. They include both uranium-235 and uranium-238 (with half-lives of 700 million years and 4.5 billion years, respectively), potassium-40 (1.3 billion years), rubidium-87 (49 billion years), and carbon-14 (5,730 years, meaning that carbon-14 is useful only for quite young samples, such as remains from prehistoric civilizations).

Radiometric and Relative Ages Combined

Radiometric dates (dates determined using radioisotopic methods) are sometimes imprecisely called "absolute" ages to distinguish them from the relative ages determined as described earlier. For various reasons, accurate radiometric dates cannot be determined for many rocks and fossils, so "absolute" and relative dating methods are often used in conjunction (figure A.8). Undateable sedimentary rocks crosscut by an igneous intrusion must at least be older than the igneous rock, which may be dateable; a fossil found in a rock sandwiched between two dateable lava flows has its age bracketed by the ages of the flows; and so on.

Many samples and events cannot be accurately dated, but radiometric dating has nonetheless proven an extremely powerful tool for studying the earth's history. Because the earth is geologically still very active, no rocks have been preserved unchanged since its formation. However, the dating of meteorites and moon rocks, coupled with the supposition that the whole solar system formed at the same time, has led

Table A.1 The Phanerozoic Time Scale

Era	Period	Epoch	Start of interval*	Distinctive life-forms
	Quaternary	Holocene	0.1	modern humans
		Pleistocene	2	Stone-Age humans
		Pliocene	5	
		Miocene	24	flowering plants common
		Oligocene	37	ancestral pigs, apes
		Eocene	58	ancestral horses, cattle
Cenozoic	Tertiary	Paleocene	66	
	Cretaceous		144	dinosaurs become extinct; flowering plants appear
	Jurassic		208	birds, mammals appear
Mesozoic	Triassic		245	dinosaurs, first modern corals appear
	Permian		286	rise of reptiles, amphibians
	Carboniferous		360	coal forests; first reptiles, winged insects
	Devonian		408	first amphibians, trees
	Silurian		438	first land plants, coral reefs
	Ordovician		505	first fishlike vertebrates
Paleozoic	Cambrian		570	first widespread fossils

*Dates, in millions of years before the present, from compilation by the Geological Society of America for the Decade of North American Geology.

to the deduction of an age of about 4.5 billion years for the earth. The oldest samples from the continents are close to 4 billion years old. Sea floor is much more readily destroyed/recycled by plate tectonics; the oldest seafloor samples recovered are only about 200 million years old.

The Geologic Time Scale

Particularly in the days before radiometric dating, it was necessary to establish some subdivisions of earth history by which particular intervals of time could be indicated. This initially was done using those rocks with abundant fossil remains. The resultant time scale was subsequently refined with the aid of radiometric dates.

The Law of Faunal Succession and the appearance and disappearance of particular fossils in the sedimentary rock record were first used to mark the boundaries of the time units. The principal divisions were the *eras* — **Paleozoic, Mesozoic,** and **Cenozoic,** meaning, respectively, "ancient life," "intermediate life," and "recent life." The eras were subdivided into *periods* and the periods into *epochs*. The names of these smaller time units were assigned in various ways. Often, they were named for the location of the *type section,* the place where rocks of that age were well exposed and that time unit was first defined. For example, the Jurassic period is named for the Jura Mountains of France. Other periods are named on the basis of some characteristic of the type rocks. The Cretaceous period derives its name from the Latin *creta* ("chalk"), for the chalky strata of that age in

southern England and northern Europe; rocks of Carboniferous age commonly include coal beds. All of the relatively unfossiliferous rocks that seemed to be older (lower in the sequence) than the Cambrian period were lumped together as pre-Cambrian, later formally named **Precambrian.**

With the advent of radiometric dating, numbers could be attached to the units and boundaries of the time scale. It became apparent that, indeed, geologic history spanned long periods of time and that the most detailed part of the scale, the **Phanerozoic** (Cambrian and later), was by far the shortest, comprising less than 15 percent of earth history. The approximate time framework of the Phanerozoic is shown in table A.1.

The Precambrian has continued to pose something of a problem. A unit that spans 4 billion years of time seems to demand subdivision — but, in the virtual absence of fossils, on what basis? The principal division that has been made splits the Precambrian into two nearly equal halves — the *Archean* ("ancient") and *Proterozoic* ("pre-life"). The Archean spans the time from the earth's formation to 2,500 million years ago, the Proterozoic from 2,500 million years ago to the start of the Cambrian (570 million years ago). Universal agreement has not yet been reached on how to subdivide these two-billion-year blocks of time further.

Geologic Process Rates

The rates at which geologic processes occur can be estimated in a variety of ways, many of which rely on radiometric dating techniques.

Examples of Rate Determination

The rate of seafloor spreading away from a particular spreading ridge can be found by dividing the age of a sample of sea floor into its present distance from the ridge. If a 10-million-year-old seafloor sample is now 400 kilometers from the ridge, then, on average, it has been moved away from the ridge at a rate of

$$\frac{400 \text{ km}}{10,000,000 \text{ yr}} \times \frac{1,000 \text{ m}}{1 \text{ km}} \times \frac{100 \text{ cm}}{1 \text{ m}} = \frac{4 \text{ cm}}{\text{yr}}$$

The minimum rate of uplift of rocks in a mountain range might be estimated from the age of marine sedimentary rocks in the mountains, once deposited under water, now high above sea level. Such uplift rates are typically 1 centimeter per year or less, often much less. Beaches formed on Scandinavian coastlines during the last Ice Age have been rising since the ice sheets melted and their mass was removed from the land. From the beach deposits' ages and present elevation, uplift rates can be approximated. Typical rates of this postglacial rebound are on the order of 1 centimeter per year. Radiometric dating has shown that a small volcano may stay active for over 100,000 years, a major volcanic center for 1 to 10 million years. To build a large mountain range may take 100 million years.

Rates of continental erosion due to weathering can be deduced from the loads of major rivers draining the continents, dividing the volume of rock those loads represent by the surface area of the corresponding drainage basin(s). The North American continent is being leveled by erosion at an average rate of 0.03 millimeters per year, or about a tenth of an inch per century. The larger the area over which such measurements are made, however, the greater the potential for large local deviations from the average due to special local soil or weather conditions.

The Danger of Extrapolation

One must be somewhat cautious about extrapolating present process rates too far into the past or future. This point is admirably illustrated by the following passage from Mark Twain's *Life on the Mississippi*, in which he speculates on the implications of the shortening of the Mississippi River by the cutoff of meanders:

"In the space of one hundred and seventy-six years the lower Mississippi has shortened itself two hundred and forty-two miles. This is an average of trifle over one mile and a third per year. Therefore, any calm person, who is not blind or idiotic, can see that in the Old Oolitic Silurian Period just a million years ago next November, the Lower Mississippi River was upwards of one million three hundred thousand miles long, and stuck out over the Gulf of Mexico like a fishing rod. And by the same token any person can see that seven hundred and forty-two years from now the Lower Mississippi will be only a mile and three-quarters long, and Cairo and New Orleans will have joined their streets together, and be plodding along comfortably under a single mayor and a mutual board of aldermen. There is something fascinating about science. One gets such wholesale returns of conjecture out of such a trifling investment of fact."

Overview

Summary

Before the discovery of natural radioactivity, only relative age determinations for rocks and geologic events were possible. Field relationships and fossil correlations were the principal methods used for this purpose. Radiometric dating has made possible quantitative age measurements, although not all geologic materials can be so dated. Radiometric ages, in turn, have been used to explore the rates at which different kinds of geologic processes occur and have considerably advanced understanding of earth's development. However, while observations of present geologic processes can be used to understand past geologic history, it cannot be assumed that all of those processes have proceeded at rates comparable to those presently observed.

Terms to Remember

Cenozoic 511
Law of Faunal
  Succession 507
Mesozoic 511
Paleozoic 511
Phanerozoic 511
Precambrian 511
Principle of Original
  Horizontality 506
Principle of Superposition 506
uniformitarianism 508

Suggested Readings/References

Cloud, P. 1988. *Oasis in space: Earth history from the beginning.* New York: W. W. Norton.
Dickin, A. P. 1995. *Radiogenic isotope geology.* New York: Cambridge University Press.
Eicher, D. L. 1976. *Geologic time.* 2d ed. Englewood Cliffs, N.J.: Prentice-Hall.
Faure, G. 1986. *Principles of isotope geology.* 2d. ed. New York: John Wiley & Sons.
Johns, R. B., ed. 1986. *Biological markers in the sedimentary record.* New York: Elsevier.
McLaren, D. J. 1978. *Dating and correlation, a review.* American Association of Petroleum Geologists, Studies in Geology #6.
Moorbath, S. 1977. The oldest rocks and the growth of continents. *Scientific American* 236 (March): 92–104.

Appendix B
Introduction to Topographic and Geologic Maps and Satellite Imagery

Maps and Scale

Many kinds of information can be presented in map form. Topographic and geologic maps are the two kinds most frequently used by geologists. This section presents a brief introduction to both types of maps.

A point to note at the outset is that, in the United States, topographic and geologic maps commonly carry English, rather than metric, units. Many of these maps were drawn before there was any move toward adoption of metric units in the United States. The task of redrawing maps to convert to metric units is formidable, particularly with topographic maps. (A metric map series is being prepared by the U.S. Geological Survey, but it will be many years before its completion.)

A basic feature of any map is the map **scale,** a measure of the size of the area represented by the map. Map scales are reported as ratios—1:250,000, 1:62,500, and so on; or equivalently in words, for example, "one to two hundred fifty thousand." On a 1:250,000 scale map, a distance of 1 inch on the map represents 250,000 inches (almost 4 miles) in reality; on a 1:62,500 scale map, 1 inch equals about 1 mile of actual distance. The larger the scale factor, the more real distance is represented by a given distance on the map. As a result, the fineness of detail that can be represented on a map is reduced as the scale factor is increased. The choice of scale factor often involves a compromise between minimizing the size of the map (for convenience of use) or the number of maps required to cover the area, and maximizing the amount of detail that can be shown.

Topographic Maps

Topographic maps primarily represent the form of the earth's surface. Selected other features, both natural and artificial, may be included for information. Once one becomes accustomed to reading them, topographic maps can make excellent navigational aids.

Contour Lines, Contour Intervals

The problem of representing three-dimensional features on a two-dimensional map is addressed through the use of **contour lines.** A contour line is a line connecting points of equal elevation, measured in feet or meters above or below sea level. The **contour interval** of a map is the difference in elevation between successive contour lines. For instance, if the contour interval is 10 feet, contour lines are drawn at elevations of 1,000 feet, 1,010 feet, 1,020 feet, 1,030 feet, and so on (in whatever range of elevations is appropriate to that map). If the contour interval is 20 feet, contours would be drawn at elevations of 1,000 feet, 1,020 feet, 1,040 feet, and so on. The contour interval chosen for a particular map depends on the overall amount of relief in the map area. If the terrain is very flat, as in a midwestern floodplain, a 10-foot or even 5-foot contour interval may be appropriate to depict what relief there is. In a rugged mountain terrain, like the Rockies, the Cascades, or the Sierras, with several thousand feet of vertical relief, using a 10-foot contour interval would make for a very cluttered map, thick with contour lines; a 50-foot or even 100-foot contour interval may be used in such a case.

The relationship between the spacing of contours and the steepness of a slope can be seen in figure B.1. A comparison of the actual relief, as shown in figure B.1A, and the resultant arrangement of contour lines on the corresponding topographic map, shown in figure B.1B, illustrates that, on a given map, the more closely spaced the contours, the steeper the corresponding terrain. In other words, if the map shows many closely spaced contours, this means that there is a great deal of vertical relief over a limited horizontal distance. Note also that contours run across the face of a slope; the upslope and downslope directions are perpendicular to the contours.

Contours do close, although they may not do so on any one given map. A series of concentric closed contours indicates a hill. If there is a local depression, so that the same contour is encountered twice, once going up in elevation and again within the depression, the repetition of the contour within the depression is marked by a hachured contour, as shown in figure B.2.

Where contour lines cross a stream valley, they point upstream, toward higher elevations (figure B.3). Contour lines corresponding to different elevations should not ordinarily cross each other (that would imply that the same point has multiple elevations). The only exception to this would occur in the case of an overhanging cliff, where a range in elevations exists at one spot on the map.

Other Features on Standard Topographic Maps

The most extensive topographic maps of the United States have been compiled by the U.S. Geological Survey (USGS). The USGS has adopted a uniform set of symbols for various

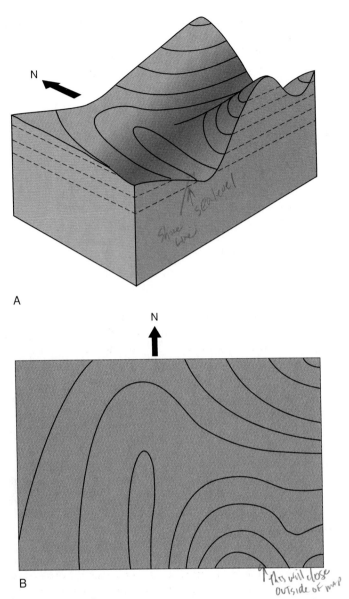

Figure B.1 Contours and relief. (*A*) Perspective view of a hilly area, with contours superimposed for reference. (*B*) The same contours as they would appear on a topographic map.

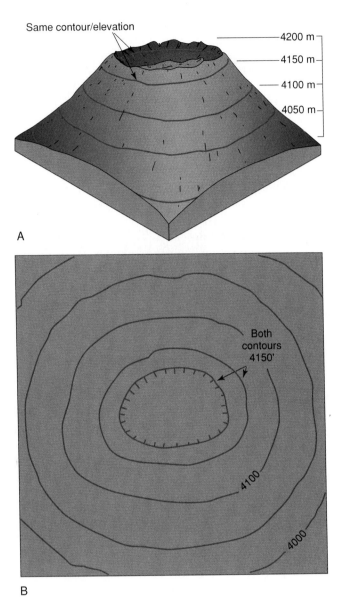

Figure B.2 Volcanic caldera with summit depression. (*A*) Actual relief. (*B*) As the volcano would appear on a topographic map.

kinds of features, which facilitates the reading of topographic maps. Some of these symbols are illustrated in figure B.4.

Contour lines are drawn in brown. Major contours (typically, those contours corresponding to multiples of 100 feet, except on maps with large contour intervals) are drawn more boldly and labeled with the corresponding elevation. This is especially convenient in rugged terrain, where it would otherwise be easy to lose count of numerous closely spaced contours. Roads are drawn in black and/or red. Town limits are shown in light red, and the names of towns, cities, and mountains, or any other labels, are printed in black. Watery features are drawn in blue: lakes and perennial streams in solid blue, intermittent streams in dashed blue lines, swampy areas represented by blue symbols resembling tufts of low grasses. The background color for wooded areas is green; for open fields, brushy areas, deserts,

bare rock, or any area lacking overhead vegetative cover, the background color is white. Revisions based on aerial photography ("photorevisions") are shown in purple.

Obtaining Topographic Maps

U.S. Geological Survey maps are sometimes identified as "15-minute" or "7½-minute" *quadrangle* maps. Such a designation means that the map area is a rectangle corresponding to so many minutes of latitude and longitude. Because it covers a smaller area, a 7½-minute quadrangle map shows more detail than a 15-minute quadrangle. Many parts of the United States have been mapped at both scales; for others, only 15-minute quadrangles are available. Individual maps are named for a town, mountain, or other prominent geographic feature within the map area: for example, "Cooke City 15-minute quadrangle," "Cutoff Mountain 7½-minute

A

B

Figure B.3 Contours crossing stream valleys. (*A*) Relief with superimposed contours. (*B*) On topographic map, the contours point upstream.

quadrangle." There are also special maps covering areas of particular interest, such as individual national parks.

Topographic maps may be ordered directly from the U.S. Geological Survey at nominal cost (variable with map size and scale). If you are uncertain of what map(s) you want, you can request free *index maps* of the corresponding state(s), along with price lists, before ordering. For example, if you were interested in obtaining quadrangle maps of selected parts of Yellowstone National Park, you would request a topographic index map of Wyoming and then choose the appropriate quadrangle(s) from that. The index map shows what part of the state is covered by which quadrangle map, the name by which each map is identified, and the date of the map or last revision thereof. A cautionary note: The same place name may be used both for a 15-minute quadrangle and for a $7\frac{1}{2}$-minute quadrangle

within it. Be alert to this when ordering, and specify quadrangle size if necessary.

Topographic maps can be obtained from two principal locations. For maps of areas east of the Mississippi River, contact:

Branch of Distribution
U.S. Geological Survey
1200 S. Eads Street
Arlington, Virginia 22202

For areas west of the Mississippi River, maps can be obtained from:

Branch of Distribution
U.S. Geological Survey
Box 25286 Federal Center
Denver, Colorado 80225

Geologic Maps

Topographic maps show the form of the land surface; geologic maps are one way of representing the underlying geology. The most common kind of geologic map is a map of **bedrock geology,** which shows the geology as it would appear with soil stripped away. (Other maps may show distribution of different soil types, glacial deposits, or other features of surface geology.) From a well-prepared geologic map, aspects of the subsurface structure can often be deduced.

Basic Concepts Related to Geologic Maps

Fundamental to making a geologic map is identifying a suitable set of **map units.** These may, for example, be individual sedimentary rock formations, distinguishable lava flows, or metamorphic rock units. The main requirement for a map unit is that it be identifiable by the presence or absence of some characteristic(s), and thus distinct from other map units chosen. The mapper then marks which of the map units are found at each place where the rocks are exposed.

Additional information, such as the orientation of beds or the location of contacts between map units, may also be recorded. Where obvious, contacts are drawn as solid lines; where only inferred, as dashed lines. (If one finds granite at point *A* and limestone at point *B* a short distance away, one can infer a contact between the granite and limestone somewhere between *A* and *B*, even if the exact spot is covered by soil.)

How easy it is to produce an accurate and complete geologic map depends on several factors (aside from the competence of the mapper!). Bedrock exposure is one. Thick soil, water or swamps, and vegetation can all obscure what lies below. In some areas, the geology is well exposed only along stream valleys; in others, the rocks are completely exposed, and mapping is greatly simplified. In glaciated areas, too, one need not only deal with the cover of glacial till (assuming that is not the material of interest) but also be cautious about mistakenly identifying a large, partially

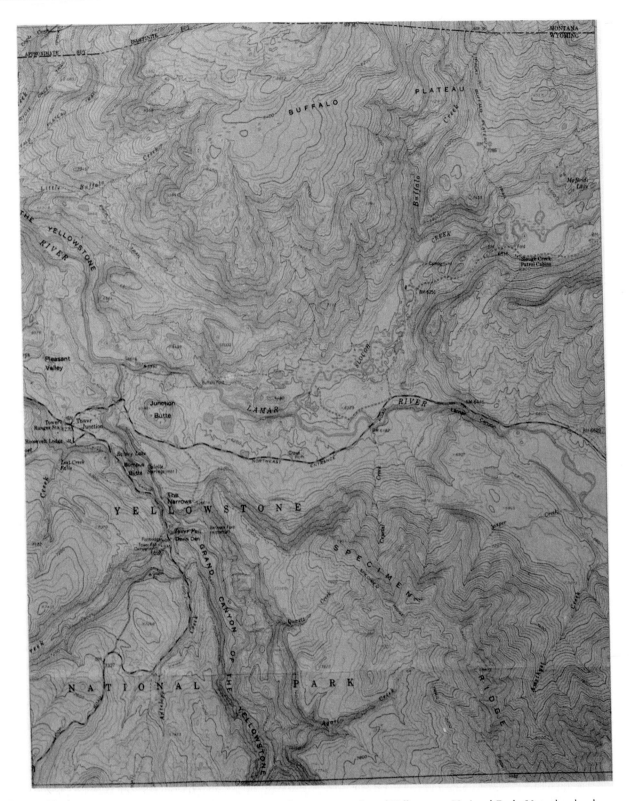

Figure B.4 Representative section of standard topographic map: a portion of Yellowstone National Park. Note the closely spaced contours along the Grand Canyon of the Yellowstone and edges of plateaus, the shapes of contours that cross streams, the closed contours around hills (for example, Junction Butte), and the swampy flat area (along Slough Creek).   Source: United States Geological Survey.

Figure B.5   Sample of geologic map with key: Simplified bedrock geology of Wisconsin.   Source: After M. E. Ostrom, Wisconsin Geological and Natural History Survey, April 1981.

Figure B.6   Selected standard map symbols for various rock types.

buried bit of glacial debris as a small exposure of bedrock. And some areas are simply much more complicated, geologically, than others.

On a geologic map, different units are represented in different colors for clarity (figure B.5). Units of similar age may be shown in different shades of the same color. The map is accompanied by a key showing all the map units,

arranged in chronological order (insofar as their ages are known), with the youngest at the top. Ordinarily, a brief description of each map unit is given; alternatively, standard patterns may be used to indicate the general rock type (figure B.6). Each unit is also assigned a symbol. The first part of the symbol consists of one or two letters corresponding to the unit's age (generally, the geologic era or period;

Figure B.7   The presence of the Grand Canyon allows much better knowledge of the area's subsurface geology. Geologic map shows different rock units in different colors. Note monotony outside canyon.   Used with permission of Grand Canyon Natural History Association.

see appendix A). This is usually followed by one to three lowercase letters corresponding to the rock type or unit's name (if any). For example, "p€qm" might be used for an unnamed Precambrian quartz monzonite, "Dl" for the Devonian Littleton Formation, "Qa" for Quaternary alluvium along a stream channel. These symbols are useful for distinguishing units mapped in similar colors, as well as for providing a general indication of the age of each unit directly on the map.

Interpretation from Geologic Maps—Examples

Interpretation of geologic maps likewise varies in difficulty, depending on the fundamental complexity of the geology and the extent of exposure (completeness of map). Sometimes, too, how much of the geology can be seen depends not only on cover or lack of it, but on topography. That is, in rugged terrain with considerable vertical relief, one has more of a three-dimensional look at the geology than in flat terrain. For example, in the vicinity of the Grand Canyon, the Paleozoic sedimentary sequence is quite flat-lying. So, for the most part, is the land surface. Outside the canyon, only the Kaibab limestone is exposed, forming a nearly level plateau and giving no evidence of what lies below. The cutting of the canyon has exposed many more rock units, down to the Precambrian. This is apparent on a geologic map (figure B.7).

The Grand Canyon has a regular, layer-cake geology that has a straightforward interpretation. Where the geology is more complex, patterns of repetition of units, and orienta-

tion of beds, must be used to interpret the structure. For example, the map pattern of figure B.8 shows circular exposures of rock units. All units dip away from the center of the circle, and the key shows that the oldest rocks are at the center. This, then, is a dome.

When sets of rocks are repeated, this may indicate the presence of a fault. In the map pattern shown in figure B.9, the same sequence of units is repeated; moreover, all are dipping in the same direction. It is difficult to create such a pattern by folding. A fault, however, can account for the result (figure B.9). Offset of features that are otherwise continuous is another sign of possible faulting (figure B.10).

Cross Sections

Interpreting structure from geologic map patterns and rock orientations requires some practice and the ability to visualize in three dimensions. Often, the maker of a geologic map assists the map reader by supplying one or more geologic **cross sections.** A cross section is a three-dimensional interpretation of the geology seen at the surface. The line along which the cross section is drawn is indicated on the map. The cross section uses the same map units and symbols as the map proper and attempts to show the geometric relationships inferred to exist among those units—folds, faults, intrusive relationships, and so on.

A cross section is drawn by starting with a topographic profile along the chosen line and marking on it the geology as seen from the surface (figure B.11A). A structural interpretation that is consistent with all the known data is then

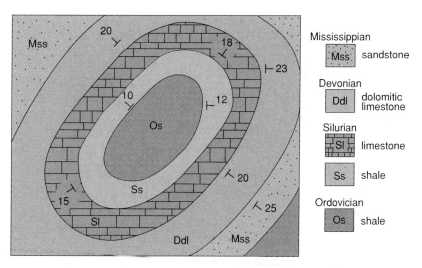

Figure B.8 A dome as it would appear on a geologic map.

Faulting of tilted sediments

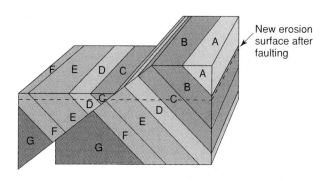

New erosion surface after faulting

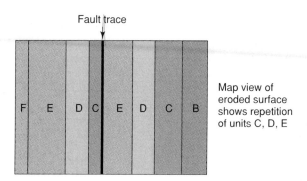

Fault trace

Map view of eroded surface shows repetition of units C, D, E

Figure B.9 Faulting can produce repetition of map units. Top surface of top block is the original surface exposure of units; bottom diagram is map result after faulting and erosion to the dashed surface of center diagram.

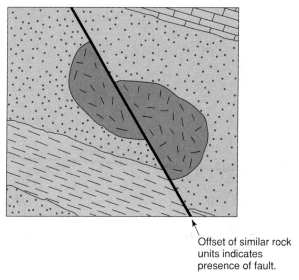

Offset of similar rock units indicates presence of fault.

Figure B.10 Igneous body and surrounding rocks offset by faulting.

devised (figure B.11B). Depending on the complexity of the geology and the completeness of exposure on which the original geologic map has been based, it may or may not be possible to develop a unique structural interpretation for the observed map pattern. If it is not, several plausible alternatives might be presented. Cross sections can be very valuable in evaluating site suitability for construction or other purposes.

Obtaining Geologic Maps

A variety of geologic maps is available from the U.S. Geological Survey, through the distribution offices listed earlier, in the section on topographic maps. As with the latter, index maps are available to assist in selecting the geologic map(s) of interest.

Geologic mapping is more time-consuming than topographic mapping, and the U.S. Geological Survey has not

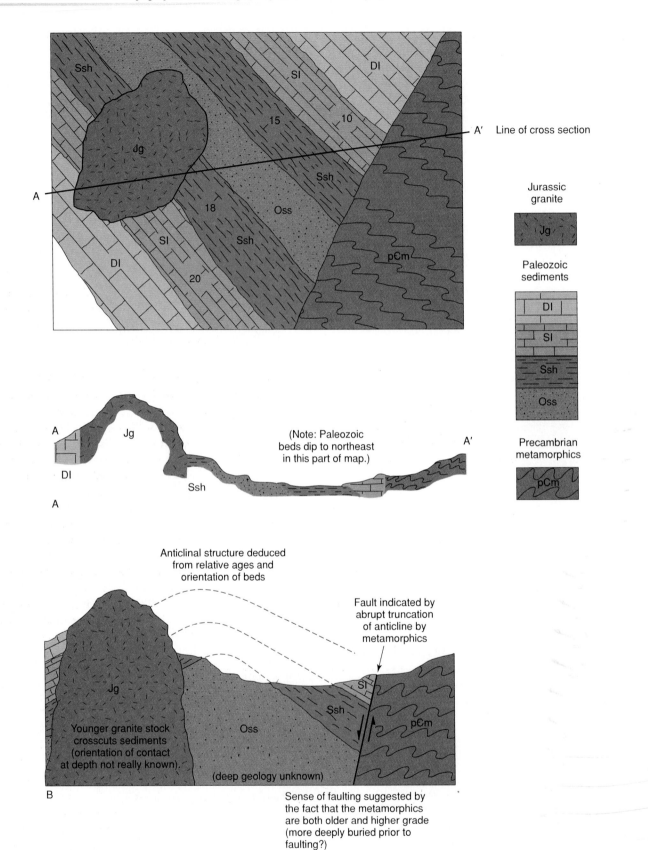

A' Line of cross section

Jurassic
granite

Jg

Paleozoic
sediments

Dl

Sl

Ssh

Oss

Precambrian
metamorphics

pЄm

(Note: Paleozoic
beds dip to northeast
in this part of map.)

Anticlinal structure deduced
from relative ages and
orientation of beds

Fault indicated by
abrupt truncation
of anticline by
metamorphics

Younger granite stock
crosscuts sediments
(orientation of contact
at depth not really known).

(deep geology unknown)

Sense of faulting suggested by
the fact that the metamorphics
are both older and higher grade
(more deeply buried prior to
faulting?)

Figure B.11   Constructing a cross section. (*A*) Geology as seen at surface is sketched onto a topographic profile. (*B*) The pattern is interpreted, and a set of structures consistent with the pattern seen at the surface is sketched in.

had the personnel or funding to produce geologic maps of every part of the United States. However, many states have very active geological surveys that have undertaken extensive mapping programs within their home states. The information office of the state of interest should be able to supply information about the state geological survey.

Once one has developed an interest in geology, one may begin asking, "What's that?" when observing rocks while traveling, especially in unfamiliar areas. For a quick answer, it may be helpful to consult a *geological highway map*. The American Association of Petroleum Geologists has produced a set of such maps that collectively cover all of the contiguous United States. Given that each map covers several states, the level of detail possible on each map is necessarily limited, but the maps do provide an overview of a region's geology. Information, prices, and maps can be obtained from:

American Association of Petroleum Geologists
P.O. Box 979
Tulsa, Oklahoma 74101

Remote Sensing and Satellite Imagery

Remote sensing methods encompass all of those means of examining planetary features that do not involve direct contact. Instead, these methods rely on detection, recording, and analysis of wave-transmitted energy—visible light, infrared radiation, and others. Aerial photography, in which standard photographs of relatively large regions of the earth are taken from aircraft, is one example. Radar mapping of surface topography, using airplanes or spacecraft, is another. Still another involves analyzing the light reflected from the surface of a body. In the case of many planets, remotely sensed data may be the only kind readily available. In the case of the earth, remote sensing, especially using satellites, is a quick and efficient way to scan broad areas, to examine regions having such rugged topography or hostile climate that they cannot easily be explored on foot or with surface-based vehicles, and to view areas to which ground access is limited for political reasons. Probably the best-known and most comprehensive earth satellite imaging system is the one initiated in 1972, known as Landsat while under the control of the federal government, renamed Eosat when later "privatized."

The Landsat satellites orbit the earth in such a way that images can be made of each part of the earth. Each orbit is slightly offset from the previous one, with the areas viewed on one orbit overlapping the scenes of the previous orbit. Each satellite makes 14 orbits each day; complete coverage of the earth takes 18 days. Therefore, images of any given area should be available every 18 days, although in practice, shifting distributions of clouds obscure the surface some part of the time at any point. Five Landsat satellites have been launched; a sixth is planned.

The sensors in the Landsat satellites do not detect all wavelengths of energy reflected from the surface. They do not take photographs in the conventional sense. They are particularly sensitive to selected green and red wavelengths in the visible light spectrum and to a portion of the infrared (invisible heat radiation, with wavelengths somewhat longer than those of red light). These wavelengths were chosen particularly because plants reflect light most strongly in the green and the infrared. Later Landsat satellites (4 and 5) included sensors sensitive to more wavelengths in the visible part of the electromagnetic spectrum. Different plants, rocks, and soils reflect different proportions of radiation of different wavelengths. Even the same feature may produce a somewhat different image under different conditions. Wet soil differs from dry; sediment-laden water looks different from clear; a given variety of plant may reflect a different spectrum of radiation depending on what trace elements it has concentrated from the underlying soil or how vigorously it is growing. Landsat images can be powerful mapping tools.

Landsat Images and Applications

A common format for Landsat imagery is photographic prints at 1:1,000,000 scale. At that scale, a 23-centimeter (9-inch) print covers 34,225 square kilometers (13,225 square miles). The smallest features that can be distinguished in the image are about 30 meters (100 feet) in size, which gives some idea of the quality of the resolution. Multiple images can be joined into mosaics covering whole countries or continents.

Images are typically presented either in black and white or as *false-color composites*. The latter are produced by projecting the data for individual spectral regions through colored filters and superimposing the results. The false-color images are so named because the resulting pictures, though superficially resembling color photographs, do not present all features in the colors they would appear to the human eye. The most striking difference is in vegetation, which appears in shades of red, not green. Rock and soil usually show as white, blue, yellow, or brown, depending on composition. Water is blue to bluish-black; snow and ice are white. Examples of false-color Landsat images are used throughout this text. Landsat image data can also be further processed by computer to produce images in more "realistic" (expected) colors or to enhance particular features by emphasizing certain wavelengths of radiation.

Dozens of applications of Landsat imagery exist in the natural sciences—for example, basic geologic mapping, identification of geologic structures, and resource exploration. Because Landsat scans the same area repeatedly over time, seasonal changes (figure B.12) and the progress and extent of occasional events such as flooding (as shown in box 6.2) can be monitored. Landsat imagery also can be used to monitor development of and changes in surface features such as stream channels and currents (figure B.13). It is helpful in

521

A

B

C

Figure B.12 Seasonal variations in Utah Lake. (*A*) In early August, some incoming sediment clouds the lake. (*B*) By mid-September, dry conditions have greatly reduced the extent of sediment input; water appears dark and clear. (*C*) The high runoff of spring, and the corresponding increased sediment input, is reflected in the very turbid water in this image, taken in late May.   © NASA.

Figure B.13 Swirling currents in James Bay, Canada, are marked by patterns of suspended sediment.   © NASA.

Figure B.14 Spaceborne radar image of a part of Guangdong province, China. This is a mineral-rich area, and radar imagery may assist in prospecting for additional deposits. © NASA

identifying patterns of land use and in monitoring the progress of crops and the extent of damage to vegetation from fires, insects, or disease.

In all cases, Landsat imagery is especially useful when some *ground truth* can be obtained. Ground truth is information gathered by direct surface examination (best done at the time of imaging, if vegetation is involved), which can provide critical confirmation of interpretations based on remotely sensed data.

An overview of global diversity as seen through Landsat imagery and a survey of many applications of remotely sensed data are presented in *Mission to Earth: Landsat Views the World* by N. M. Short, P. D. Lowman, Jr., S. C. Freden, and W. A. Firch, Jr., published by NASA in 1976, and available through the U.S. Government Printing Office. Canadian features are analyzed in more detail in *Landsat Im-*

ages of Canada—A Geological Appraisal by V. R. Slaney, Geological Survey of Canada Paper 80–15, published in 1981, and available from the Canadian Government Printing Office. Individual Landsat images can be ordered from EROS Data Center, Sioux Falls, South Dakota, 57198.

Radar Imaging

A still newer technique, or set of techniques, is the use of imaging radar. It has been likened to taking a photograph with a flash camera. Just as such a camera generates light that illuminates the pictures it takes, so imaging radar sends pulses of microwaves that bounce off the surface or scene being imaged. The techniques of processing the returned signal have

Figure B.15 Three-dimensional radar image of Long Valley Caldera, California, with Crowley Lake in the foreground.   © NASA

become increasingly refined since NASA and the Jet Propulsion Laboratory initiated the SEASAT imaging radar system in 1978 (figure B.14). Some radar-imaging devices are flown on jets, and some have been carried on the space shuttle.

A basic radar image of the Long Valley Caldera is shown in figure B.15. Some of the possible enhancements of the imagery are shown in figure B.16. One of the advantages of radar imaging (by comparison with visible-light images, for example) is that microwave radiation is penetrating enough that it is unaffected by cloud cover, heavy rain, or other such conditions that can obscure a visual image. Therefore, it can be used to monitor such phenomena as shifting lahars on Mount Pinatubo during monsoon season (figure B.17). For information on availability of radar images, contact:

Radar Data Center
Mail Stop 300-233
Jet Propulsion Laboratory
4800 Oak Grove Drive
Pasadena, CA 91109

Figure B.16 Examples of enhanced radar imagery of several kinds. Image (*A*) shows contrasting terrain styles—dark grey, smooth valley floors; brighter areas, hilly regions of rock outcrops and pine forest. (Dark patch at lower left is Lake Crowley.) Image (*B*) illustrates radar interferometry; color bands highlight elevation changes. From this is derived topography as shown in (*C*), with contour lines (heavier contours are at intervals of 250 meters, about 820 feet), and coloring reflecting increments of elevation. (Image is rotated relative to (*A*) and (*B*).) The image in (*D*), similar in style to B.15, is of the northeast rim of the caldera, looking northwest. © NASA

Figure B.17 Years after the 1991 explosive eruptions, lahars can still be triggered by monsoon rains on Mount Pinatubo. Ash from those eruptions is shown in red, lahars in black. Comparison of April (left) and October (right) 1994 images shows evolution of lahars and their deposits, and helps in hazard evaluation.   © NASA

Terms to Remember

bedrock geology 515	map unit 515
contour interval 513	remote sensing 521
contour lines 513	scale (map) 513
cross section 518	

Information about the U.S. Geological Survey Photographic Library and the availability of its products may be found at
> http://www.usgs.gov/fact-sheets/photographic-library/ photographic-library.html

To contact USGS map dealers, see
> http://www-nmd.usgs.gov/esic/usimage/dealers.html

A large collection of Landsat imagery is available through the EROS data center, at
> http://edcwww.cr.usgs.gov
> The Landsat satellites and their imagery are described at
> http://edcwww.cr.usgs.gov/glis/hyper/guide/landsat
> Or see information on the Landsat thematic mapper at
> http://edcwww.cr.usgs.gov/glis/hyper/guide/landsat_tm

The home page for NASA/JPL imaging radar can be found at
> http://southport.jpl.nasa.gov/

Additional imagery information and related links can be reached at the United Nations Environment Programme site for UNEP/GRID; see
> http://grid2.cr.usgs.gov/links/data.html

Appendix C
Mineral and Rock Identification

Mineral Identification

Table C.2 lists many of the more common minerals. Representative chemical formulas are provided for reference. Some appear complex because of opportunities for solid solution; some have been simplified by limiting the range of compositions represented, although additional elemental substitutions are possible.

A few general identification guidelines and comments:

Hardness is an approximate measure of how readily a mineral scratches, or is scratched by, other minerals. Values of hardness range from 1 (softest) to 10 (hardest), and are measured against the ten reference minerals of the Mohs hardness scale (table C.1). For example, a mineral that scratches quartz and is scratched by topaz would have a hardness of $7\frac{1}{2}$.

Luster describes the surface sheen of a mineral sample, and the terms (such as metallic, pearly, earthy) are self-explanatory. *Cleavage* is a mineral's tendency to break preferentially along certain planes of weakness in the crystal structure.

Minerals showing metallic luster are usually sulfides (or native metals, but these are much rarer). Native metals have been omitted from the table. Those few that are likely to be encountered, such as native copper or silver, may be identified by their resemblance to household examples of the same metals.

Of the nonmetals, the silicates are generally systematically harder than the nonsilicates. Hardnesses of silicates are typically over 5, with exceptions principally among the sheet silicates; many of the nonsilicates, such as sulfates and carbonates, are much softer.

Distinctive luster, cleavage, or other identifying properties are listed under the column "Other Characteristics." In a few cases, this column notes restrictions on the occurrence of certain minerals as a possible clue to identification; for example, "found only in metamorphic rocks," or "often found in pegmatites."

Rock Identification

One approach to rock identification is to decide whether the sample is igneous, sedimentary, or metamorphic and then look at the detailed descriptions in the corresponding chapter. How does one identify the basic rock type? Here are some general guidelines:

1. Glassy or vesicular rocks are volcanic.
2. Coarse-grained rocks with tightly interlocking crystals are likely to be plutonic, especially if they lack foliation.
3. Coarse-grained sedimentary rocks differ from plutonic rocks in that the grains in the sedimentary rocks tend to be more rounded and to interlock less closely. A breccia does have angular fragments, but the fragments in a breccia are typically rock fragments, not individual mineral crystals.
4. Rocks that are not very cohesive, that crumble apart easily into individual grains, are generally clastic sedimentary rocks. One other possibility would be a poorly consolidated volcanic ash, but this should be recognizable by the nature of the grains, many of which will be glassy shards. (Note, however, that extensive weathering can make even a granite crumble.)
5. More cohesive, fine-grained sedimentary rocks may be distinguished from fine-grained volcanics because sedimentary rocks are generally softer and more likely to show a tendency to break along bedding planes. Phenocrysts, of course, indicate a (porphyritic) volcanic rock.
6. Foliated metamorphic rocks are distinguished by their foliation (schistosity, compositional banding). Also, rocks containing abundant mica, garnet, or amphibole are commonly metamorphic rocks.
7. Nonfoliated metamorphic rocks, like quartzite and marble, resemble their sedimentary parents but are harder, denser, and more compact. They may also have a shiny or glittery appearance on broken surfaces, due to recrystallization during metamorphism.

Once a preliminary determination of category (igneous/sedimentary/metamorphic) has been made, table C.3 can be used in conjunction with the appropriate text chapter to identify the rock type. (Keep in mind, however, that the key and text focus on relatively common rock types.)

Table C.1 The Mohs Hardness Scale

Mineral	Assigned hardness	
talc	1	
gypsum	2	
		← (fingernail)
calcite	3	
		← (copper penny)
fluorite	4	
apatite	5	
		← (glass, pocketknife blade)
orthoclase	6	
quartz	7	
topaz	8	
corundum	9	
diamond	10	

Hardnesses of some common materials are indicated for reference.

Table C.2 A Brief Mineral Identification Key

Mineral	Formula	Color	Hardness	Other characteristics
amphibole (e.g., hornblende)	$(Na,Ca)_2(Mg,Fe,Al)_5Si_8O_{22}(OH)_2$	green, blue, brown, black	5 to 6	often forms needlelike crystals; two good cleavages forming 120-degree angle
apatite	$Ca_5(PO_4)_3(F,Cl,OH)$	usually yellowish	5	crystals hexagonal in cross section
azurite	$Cu_3(CO_3)_2(OH)_2$	vivid blue	$3\frac{1}{2}$ to 4	often associated with malachite
barite	$BaSO_4$	colorless	3 to $3\frac{1}{2}$	high specific gravity, 4.5 (denser than most silicates)
beryl	$Be_3Al_2Si_6O_{18}$	aqua to green	$7\frac{1}{2}$ to 8	usually found in pegmatites
biotite (a mica)	$K(Mg,Fe)_3AlSi_3O_{10}(OH)_2$	black	$5\frac{1}{2}$	excellent cleavage into thin sheets
bornite	Cu_5FeS_4	iridescent blue, purple	3	metallic luster
calcite	$CaCO_3$	variable; colorless if pure	3	effervesces in weak acid
chalcopyrite	$CuFeS_2$	brassy yellow	$3\frac{1}{2}$ to 4	
chlorite	$(Mg,Fe)_3(Si,Al)_4O_{10}(OH)_2$	light green	2 to $2\frac{1}{2}$	cleaves into small flakes
cinnabar	HgS	red	$2\frac{1}{2}$	earthy luster; may show silver flecks
corundum	Al_2O_3	variable; colorless in pure form	9	most readily identified by its hardness
covellite	CuS	blue	$1\frac{1}{2}$ to 2	metallic luster
dolomite	$CaMg(CO_3)_2$	white or pink	$3\frac{1}{2}$ to 4	powdered mineral effervesces in acid
epidote	$Ca_2FeAl_2Si_3O_{12}(OH)$	green	6 to 7	glassy luster
fluorite	CaF_2	variable; often green or purple	4	cleaves into octahedral fragments; may fluoresce in ultraviolet light
galena	PbS	silver-gray	$2\frac{1}{2}$	metallic luster; cleaves into cubes
garnet	$(Ca,Mg,Fe)_3(Fe,Al)_2Si_3O_{12}$	variable; often dark red	7	glassy luster
graphite	C	dark gray	1 to 2	streaks like pencil lead
gypsum	$CaSO_4 \cdot 2H_2O$	colorless	2	
halite	$NaCl$	colorless	$2\frac{1}{2}$	salty taste; cleaves into cubes
hematite	Fe_2O_3	red or dark gray	$5\frac{1}{2}$ to $6\frac{1}{2}$	red-brown streak regardless of color
kaolinite	$Al_2Si_2O_5(OH)_4$	white	2	earthy luster

Mineral	Formula	Color	Hardness	Other characteristics
kyanite	Al_2SiO_5	blue	5 to 7	found in high-pressure metamorphic rocks; often forms bladelike crystals
limonite	$Fe_4O_3(OH)_6$	yellow-brown	2 to 3	earthy luster; yellow-brown streak
magnetite	Fe_3O_4	black	6	strongly magnetic
malachite	$Cu_2CO_3(OH)_2$	green	$3\frac{1}{2}$ to 4	often forms in concentric rings of light and dark green
molybdenite	MoS_2	dark gray	1 to $1\frac{1}{2}$	cleaves into flakes; more metallic luster than graphite
muscovite (a mica)	$KAl_3Si_3O_{10}(OH)_2$	colorless	2 to $2\frac{1}{2}$	excellent cleavage into thin sheets
olivine	$(Fe,Mg)_2SiO_4$	yellow-green	$6\frac{1}{2}$ to 7	glassy luster
phlogopite	$KMg_3AlSi_3O_{10}(OH)_2$	brown	$2\frac{1}{2}$ to 3	mica closely resembling biotite
plagioclase feldspar	$(Na,Ca)(Al,Si)_2Si_2O_8$	white to gray	6	may show fine striations on cleavage surfaces
potassium feldspar	$KAlSi_3O_8$	white; often stained pink or aqua	6	two good cleavages forming a 90-degree angle; no striations
pyrite	FeS_2	yellow	6 to $6\frac{1}{2}$	metallic luster; black streak
pyroxene (e.g., augite)	$(Na,Ca,Mg,Fe,Al)_2Si_2O_6$	usually green or black	5 to 7	two good cleavages forming a 90-degree angle
quartz	SiO_2	variable; commonly colorless or white	7	glassy luster; conchoidal fracture
serpentine	$Mg_3Si_2O_5(OH)_4$	green to yellow	3 to 5	waxy or silky luster; may be fibrous (asbestos)
sillimanite	Al_2SiO_5	white	6 to 7	occurs only in metamorphic rocks; often forms needlelike crystals
sphalerite	ZnS	yellow-brown	$3\frac{1}{2}$ to 4	glassy luster
staurolite	$Fe_2Al_9Si_4O_{20}(OH)_2$	brown	7 to $7\frac{1}{2}$	found in metamorphic rocks; elongated crystals may have crosslike form
sulfur	S	yellow	$1\frac{1}{2}$ to $2\frac{1}{2}$	sulfurous odor
sylvite	KCl	colorless	2	cleaves into cubes; salty taste, but more bitter than halite
talc	$Mg_3Si_4O_{10}(OH)_2$	white to green	1	greasy or slippery to the touch
tourmaline	$(Na,Ca),(Li,Mg,Al)(Al,Fe,Mn)_6(BO_3)_3Si_6O_{18}(OH)_4$	black, red, green	7 to $7\frac{1}{2}$	elongated crystals, triangular in cross section; conchoidal fracture

Table C.3   A Key to Aid in Rock Identification

Igneous Rocks

I. Extremely coarse-grained: Rock is a pegmatite. (Most pegmatites are granitic, with or without exotic minerals.)

II. Phaneritic (coarse enough that all grains are visible to the naked eye).
   A. Significant quartz visible; only minor mafic minerals: *granite*.
   B. No obvious quartz; feldspar (light-colored) and mafic minerals (dark) in similar amounts: *diorite*.
   C. No quartz; rock consists mostly of mafic minerals: *gabbro*.
   D. No visible quartz *or* feldspar: Rock is ultramafic.

III. Porphyritic with fine-grained groundmass: Go to part IV to describe groundmass (using phenocryst compositions to assist); adjective "porphyritic" will preface rock name.

IV. Aphanitic (grains too fine to distinguish easily with the naked eye).
   A. Quartz is visible or rock is light in color (white, cream, pink): probably *rhyolite*.
   B. No visible quartz; medium tone (commonly gray or green); if phenocrysts are present, commonly plagioclase, pyroxene, or amphibole: probably *andesite*.
   C. Rock is dark, commonly black; any phenocrysts are olivine or pyroxene: *basalt*
   D. Rock is glassy and massive: *obsidian* (regardless of composition).
   E. Rock consists of gritty mineral grains, ash and glass shards: *ignimbrite* (welded tuff).

Sedimentary Rocks

I. Rock consists of visible shell fragments or of oolites: *limestone*.

II. Rock consists of interlocking grains with texture somewhat like that of igneous rock and is light in color: probable chemical sedimentary rock.
   A. Tastes like table salt: *halite*.
   B. No marked taste; hardness of 2 (if grains are large enough to scratch); does not effervesce: *gypsum*.
   C. Effervesces in weak HCl: *limestone* (calcite).
   D. Effervesces weakly in HCl, only if scratched; *dolomite*.

III. Rock consists of grains apparently cemented or compacted together: probable clastic sedimentary rock.
   A. Coarse grains (several millimeters or more in diameter), perhaps with a finer matrix: *conglomerate* if the grains are rounded, *breccia* if they are angular.
   B. Sand-sized grains; gritty feel: sandstone. If predominantly quartz grains, *quartz sandstone;* if roughly equal proportions of quartz and feldspar, *arkose;* if many rock fragments, and perhaps a fine-grained matrix, *greywacke.*
   C. Grains too fine to see readily with the naked eye: *mudstone*. If rock shows lamination, and a tendency to part along parallel planes, *shale.*

IV. Relatively dense; compact, dark, no visible grains; massive texture, conchoidal fracture: *chert* (silica).

Metamorphic Rocks

I. Nonfoliated; compact texture with interlocking grains: identified by predominant mineral(s).
   A. If quartz-rich, perhaps with a sugary appearance: *quartzite*.
   B. If calcite or dolomite (identified by effervescence, hardness): *marble*.
   C. Rock consists predominantly of amphiboles: *amphibolite*.

II. Foliated: classified mainly by texture.
   A. Very fine-grained; pronounced rock cleavage along parallel planes, to resemble flagstones: *slate*.
   B. Fine-grained; slatelike, but with glossy cleavage surfaces: *phyllite*.
   C. Coarser grains; obvious foliation, commonly defined by prominent mica flakes, sometimes by elongated crystals like amphiboles: *schist*.
   D. Compositional or textural banding, especially with alternating light (quartz, feldspar) and dark (ferromagnesian) bands: *gneiss*.

Appendix D
Units of Measurement — Conversions

Length

1 cm = 0.394 in
1 m = 39.37 in = 1.09 yd
1 km = 0.621 mi
1 in = 2.54 cm
1 yd = 0.914 m
1 mi = 1,760 yd = 1.61 km

Area

1 sq cm = 0.155 sq in
1 sq m = 1.20 sq yd = 1,550 sq in
1 sq km = 0.386 sq mi
1 sq in = 6.45 sq cm
1 sq yd = 1,296 sq in = 0.836 sq m
1 sq mi = 2.59 sq km
1 acre = 4,840 sq yd = 4,047 sq m

Volume

1 cu cm = 0.061 cu in
1 cu m = 1.31 cu yd
1 cu km = 0.240 cu mi
1 cu in = 16.4 cu cm
1 cu yd = 0.765 cu m
1 cu mi = 4.17 cu km

Liquid Volume

1 ml = 0.0338 fl oz
1 liter = 1.06 qt
1 fl oz = 29.6 ml
1 qt = 0.946 liter
1 gal = 4 qt = 3.78 liter
1 acre-foot = 326,000 gal = 1,220 cu m

Weight/Mass

1 g = 0.0353 oz
1 kg = 2.20 lb
1 metric ton = 1,000 kg = 2,200 lb
1 oz (avoirdupois) = 28.4 g
1 lb = 454 g = 0.454 kg
1 ton = 2,000 lb = 909 kg
1 troy oz = 1.10 oz avoirdupois = 31.2 g

Energy

1 cal (calorie) = amount of heat required to raise
    temperature of 1 ml of water by 1° C
1 Btu (British thermal unit) = amount of heat required
    to raise temperature of 1 lb of water by 1° F
1 Btu = 252 cal
1 quad = 1 quadrillion Btu =
    1,000,000,000,000,000 Btu

Average Energy Contents of Various Fossil Fuels

Fuel	Calories	Btu
1 barrel crude oil	1,460,000,000	5,800,000
1 ton coal	5,650,000,000	22,400,000
1 cu ft natural gas	257,000	1,020

Explanation of Prefixes and Their Values

deci-: one-tenth	1 deciliter (dl) = 0.1 liter
centi-: one-hundredth	1 centimeter (cm) = 0.01 meter
milli-: one-thousandth	1 milliliter (ml) = 0.001 liter
kilo-: one thousand	1 kilometer (km) = 1,000 meters

Glossary

A

ablation The loss of glacier ice by melting or evaporation.

abrasion Erosion by wind-transported sediment or by the scraping of rock fragments frozen in glacial ice.

absorption field *See* leaching field.

acid rain Rain that is more acidic (has lower pH) than normal precipitation.

active margin A continental margin at which there is significant volcanic and earthquake activity; commonly, a convergent plate margin.

active-solar heating system A solar-heating system that uses mechanical devices, such as pumps.

active volcano A volcano with a record of eruption within recent history.

aerobic Using or consuming oxygen.

aftershocks Earthquakes that follow the main shock when a fault has slipped; of magnitude equal to or lower than the main shock.

A horizon Usually top zone in soil; also called the zone of leaching.

algal bloom An overly exuberant growth of algae in eutrophic water.

alluvial fan A wedge-shaped sediment deposit left where a tributary flows into a more slowly flowing stream, or where a mountain stream flows into a desert.

alpine (valley) glacier A glacier occupying a valley in mountainous terrain.

anaerobic Describes that which occurs without using, or in the absence of, oxygen.

andesite Volcanic rock intermediate in composition between basalt and rhyolite.

angle of repose The maximum slope angle at which a given unconsolidated material is stable.

anion An ion with a net negative charge.

anthracite The hardest of naturally occurring coals.

anthropogenic Produced by human activity.

aquifer Rock that is sufficiently porous and permeable to be useful as a source of water.

aquitard Rock of low permeability, through which water flows very slowly.

artesian system A confined aquifer system in which ground water can rise above its aquifer under its own pressure.

asthenosphere Partially molten, "weak" zone within the upper mantle immediately below the lithosphere.

atom The smallest particle into which a chemical element can be subdivided.

atomic mass number The sum of the number of protons and the number of neutrons in an atomic nucleus.

atomic number The number of protons in an atomic nucleus; characteristic of a particular element.

B

banded iron formation A sedimentary rock consisting of alternating iron-rich and iron-poor bands, found in Precambrian rocks, that may serve as an ore of iron.

barrier islands Long, low, narrow islands parallel to a coastline that protect the coastline somewhat from wave action.

basalt A volcanic rock rich in ferromagnesian minerals; relatively low in silica.

base level The lowest elevation to which a stream can cut down; for most streams, this is the level of the body of water into which they flow, such as another stream, lake, or ocean.

beach Gently sloping shoreline area washed over by waves.

bed load Material moved along a streambed by flowing water.

bedrock geology The geology as it would appear with the overlying soil and vegetative cover stripped away.

B horizon Soil layer found directly below the soil's A horizon; also known as the zone of accumulation or the zone of deposition.

biochemical oxygen demand (BOD) Quantity of oxygen required for aerobic decomposition of organic matter in a system.

biogas Methane derived from decaying organic matter.

biomass (energy) Energy derived from living organisms or from organic matter.

biosphere The sum of all living things on earth.

bituminous A form of coal that is softer than anthracite but harder than lignite.

BOD *See* biochemical oxygen demand.

body waves Seismic waves that pass through the earth's interior; includes P waves and S waves.

braided stream A stream with multiple channels that divide and rejoin.

breeder reactor A reactor in which new fissionable material is produced in quantity at the same time as energy is generated.

brittle Describes materials that tend to rupture before appreciable plastic deformation has occurred.

C

caldera A large, bowl-shaped summit depression in a volcano; may be formed by explosion or collapse.

calving The formation of icebergs by the breakup of a glacier flowing out over water.

capacity (stream) The load that a stream can carry.

carbonate Nonsilicate mineral containing carbonate groups (CO_3), carbon and oxygen in the proportions of one atom of carbon to three atoms of oxygen.

carrying capacity The ability of a system (or the whole earth) to sustain its population in reasonably healthy and comfortable conditions.

catalyst A substance that promotes chemical reactions.

cation An ion with a net positive charge.

Cenozoic Geologic era spanning the time from 66 million years ago to the present.

chain reaction (nuclear) The process during which fission of one nucleus triggers fission of others, which, in turn, induces fission in others, and so on.

channelization The modification of a stream channel, such as deepening or straightening of the channel, usually with the objective of reducing flood hazards.

chemical sediment Sediment formed at low temperature by direct precipitation from solution.

chemical weathering The breakdown of minerals by chemical reaction with water, with other chemicals dissolved in water, or with gases in the air.

C horizon Soil layer found directly below the soil's B horizon; consists of coarsely broken bedrock.

cinder cone A volcano built of cinders and other pyroclastics piled up around the volcanic vent.

clastic Broken or fragmented; describes sediments or sedimentary rocks that are formed from fragments of pre-existing rocks or minerals.

coal A solid, carbon-rich fuel formed from the remains of land plants through the effects of heat and pressure in the earth's crust.

composite volcano *See* stratovolcano.

composting A method of handling (nontoxic) organic waste matter by which it is converted into a beneficial soil additive through controlled decay.

compound A chemical combination of two or more elements, in specific proportions, having a distinctive set of resultant physical properties.

compressive stress Stress tending to compress an object.

concentration factor (ore) The concentration of a metal in a given ore deposit divided by its average concentration in the continental crust.

conditional resources *See* subeconomic resources.

cone of depression A broadly conical depression of the water table or potentiometric surface caused by pumped groundwater withdrawal.

confined aquifer An aquifer overlain by an aquitard or aquiclude.

contact metamorphism Local metamorphism adjacent to a cooling magma body.

continental drift The concept that the continents have moved about over the earth's surface.

continental glacier A large glacier covering extensive land area; also known as an *ice cap* or *ice sheet;* may be several kilometers thick.

contour interval The difference in elevation between successive contour lines on a map.

contour line A line joining points of equal elevation on a map.

contour plowing Plowing so that rows run horizontally around the curve of a slope, rather than up- and downslope.

convection cells Circulating masses of material driven by temperature differences (hot material rises, then moves laterally, cools, sinks, and is reheated to rise again); convection cells in the asthenosphere may be the driving force behind plate tectonics.

convergent plate boundary Plate boundary at which lithospheric plates are moving toward each other; for example, a subduction zone or continental collision zone.

core The innermost zone of the earth; composed largely of iron.

core meltdown A possible nuclear reactor accident resulting from loss of core coolant and subsequent overheating.

covalent bonding Bonding involving sharing of electrons between atoms.

creep (fault) The slow, gradual slip along a fault zone without major, damaging earthquakes.

creep (rock or soil) Slow, gradual downslope movement of unstable surficial materials (as contrasted with more abrupt landslides).

crest The maximum stage reached during a flood event.

cross section An interpretation of geology and structure in the third (vertical) dimension commonly based on rock exposures and attitudes at the surface; drawn in a particular vertical plane.

crust The outermost compositional zone of the earth; composed predominantly of relatively low-density silicate minerals.

crystalline Describes materials possessing a regular, repeating internal arrangement of atoms.

cumulative reserves The total reserves, including materials already consumed or exploited.

Curie temperature The temperature above which a magnetic material loses its magnetism; different for each such material.

cut bank Steep stream bank being eroded by lateral migration of meanders.

D

debris avalanche A mixed flow of rock, soil, vegetation, or other materials.

decommissioning (nuclear plant) The shutdown of a nuclear reactor at the end of its safe, useful life; includes the disposal of radioactive parts.

deep-well disposal A method for the disposal of toxic liquid waste by injection into deep, porous, permeable strata confined by impermeable rocks.

deflation The wholesale removal of loose sediment by wind erosion.

delta A fan-shaped deposit of sediment formed at a stream's mouth.

desert A barren region incapable of supporting appreciable life.

desertification The process by which marginally habitable arid lands are converted to desert; typically accelerated by human activities.

desert pavement A desert surface produced by the combined effects of wind erosion and overland surface-water runoff, in which the larger rocks are exposed by the selective removal of fine sediment; these rocks, in turn, protect finer material below from erosion.

differentiation (earth) Process by which zones of different composition developed from an initially homogeneous earth.

discharge (stream) The amount of water flowing past a given point per unit time.

dissolved load Sum of dissolved material transported by a stream.

divergent plate boundary A boundary along which lithospheric plates are moving apart; for example, seafloor spreading ridges and continental rift zones.

dome (volcanic) A compact, steep-sided structure built of very viscous, silicic lava emitted from a central pipe or vent.

dormant volcano A volcano with no recent eruptive history but that still looks relatively fresh and unweathered; may become active again in future.

dose-response curve A curve showing the beneficial or harmful effects of a trace element as a function of its dosage.

doubling time The length of time required for a population to double in size.

downstream flood A flood affecting a large area of drainage basin or a large stream system; typically caused by prolonged rain or rapid regional snowmelt.

drainage basin The region from which surface water drains into a particular stream.

dredge spoils Sediment dredged from waterways to improve navigation or to increase water capacity.

drift Sediment transported and deposited by a glacier; *see also* till, outwash.

drowned valley Along a coastline, a stream valley that is partially flooded by seawater as a consequence of land sinking and/or sea level rising.

ductile Describes material that readily undergoes plastic deformation.

dune A low mound or ridge of sediment (usually sand) deposited by wind.

dust bowl Any region severely affected by drought and wind erosion; when capitalized (Dust Bowl), refers specifically to a large region of the central United States so affected in the early 1930s.

E

earthquake Ground displacement and energy release associated with the sudden motion of rocks along a fault.

earthquake cycle The concept that there is a periodic quality about occurrence of major earthquakes on a given fault zone, with repeated cycles of stress buildup, rupture, and relaxation of stress through smaller aftershocks.

EEZ *See* Exclusive Economic Zone.

EIS *See* environmental impact statement.

elastic deformation Deformation proportional to applied stress, from which the affected material will return to its original size and shape when the stress is removed.

elastic limit The stress above which a material will cease to deform elastically.

elastic rebound Phenomenon whereby stressed rocks snap back elastically after an earthquake to their pre-stress condition.

electron A subatomic particle with an electrical charge of –1; generally found orbiting an atomic nucleus.

end moraine A ridge of till accumulated at the end of a glacier.

enhanced recovery Any method used to increase the amount of oil or gas recovered from a petroleum reservoir.

environmental impact statement (EIS) An analysis of the environmental impacts to be anticipated from a proposed action and its alternatives; mandated by the National Environmental Policy Act and legislation patterned after it.

eolian Deposited or shaped by wind action.

epicenter The point on the earth's surface directly above the focus of an earthquake.

equilibrium line Line on the surface of a glacier at which accumulation just equals ablation.

estuary A body of water along a coastline that contains a mix of fresh and salt water.

eutrophication The development of high nutrient levels (especially, high concentrations of nitrates and phosphates) in water; may lead to algal bloom.

evaporite A sedimentary mineral deposit formed when shallow or inland seas dry up; also, the minerals commonly deposited in such an environment.

Exclusive Economic Zone (EEZ) A zone extending to 200 miles offshore from a nation's coast, within which the 1982 Law of the Sea Treaty recognizes that nation's exclusive right to resource exploitation.

exponential growth Growth characterized by a constant percentage increase per unit time.

extinct volcano A volcano that has no recent eruptive history and appears very weathered in appearance; not expected to erupt again.

F

fall (rock) Mass wasting by free-fall of material not always in contact with the ground underneath.

fault Planar break in rock along which one side has moved relative to the other.

ferromagnesian A term describing silicates containing significant amounts of iron and/or magnesium; these minerals are usually dark-colored.

fission (atomic) The process by which atomic nuclei are split into smaller fragments.

fissure eruption The eruption of lava from a crack in the lithosphere, rather than from a central vent.

flood Condition in which stream stage is above channel bank height.

flood-frequency curve A graph of stream stage or discharge as a function of recurrence interval (or annual probability of occurrence).

floodplain A flat region or valley floor surrounding a stream channel, formed by meandering and sediment deposition, into which the stream overflows during flooding.

flow Mass wasting in which materials move in chaotic fashion.

focus The point of first break on a fault during an earthquake.

foliation Parallel alignment of linear or platy minerals in a rock.

fossil fuels Hydrocarbon fuels formed from organic matter.

fusion (nuclear) The process by which atomic nuclei combine to produce larger nuclei.

G

gasification Any process by which coal is converted to a gaseous hydrocarbon fuel.

gasohol A blend of gasoline with alcohol, usually in proportions close to 9:1.

geopressurized natural gas Gas dissolved in deep pore waters.

geopressurized zones Deep aquifers under unusually high pressure, exceeding normal hydrostatic (fluid) pressure.

geothermal energy Energy derived from the internal heat of the earth; its use usually requires a near-surface heat source, such as young igneous rock, and nearby circulating subsurface water.

glacier A mass of ice that moves or flows over land under its own weight.

glass Solid (especially silicate) lacking a regular internal crystal structure.

gradient The slope (steepness) of a stream channel along its length.

granite Coarse-grained plutonic igneous rock, typically rich in quartz and feldspars.

greenhouse effect The warming of the atmosphere, especially as due to the increased concentration of carbon dioxide derived from the burning of fossil fuels.

ground water Water in the zone of saturation, below the water table.

H

half-life The length of time required for half of an initial quantity of a radioactive isotope to decay.

halide Compound of one or more metals plus a halogen element (fluorine, chlorine, iodine, or bromine).

hardness (mineral) The ability of a mineral to resist scratching.

hard water Water containing substantial quantities of dissolved calcium, magnesium, and/or iron.

heavy metals A group of dense metals, including mercury, lead, cadmium, plutonium, and others, that share the characteristic of being accumulative in organisms and tending to become increasingly concentrated in organisms higher up a food chain.

high-level (radioactive) waste Waste sufficiently radioactive to require special handling in disposal.

hot-dry-rock Geothermal resource area in which geothermal gradients are high but indigenous ground water is lacking.

hot spot An isolated center of volcanic activity; often not associated with a plate boundary.

hydrocarbon Any compound consisting of carbon and hydrogen.

hydrocompaction The process by which a sediment settles, cracks, and becomes compacted when wetted.

hydrograph A graph of stream stage or discharge against time.

hydrologic cycle The cycle through which water in the hydrosphere moves; includes such processes as evaporation, precipitation, and surface and groundwater runoff.

hydrosphere All water at and near the earth's surface that is not chemically bound in rocks.

hydrothermal ores The ores deposited by circulating warm fluids in the earth's crust.

hypothesis A conceptual model or explanation for a set of data, measurements, or observations.

hypothetical resources That quantity of a resource material expected to be found in areas in which like deposits are known to exist.

I

ice age Any period of extensive continental glaciation; when capitalized (Ice Age), the phrase refers to the most recent such episode, from 2 million to 10,000 years ago.

igneous rock A rock formed or crystallized from a magma.

infiltration The process by which surface water sinks into the ground.

intensity (earthquake) A measure of the damaging effects of an earthquake at a particular spot; commonly reported on the Modified Mercalli Scale.

ion Atom that has gained or lost electrons, so that it has a net electrical charge.

ionic bonding Bonding due to attraction between oppositely charged ions.

island arc Chain of volcanic islands formed parallel to a subduction zone, on the overriding plate.

isotopes Atoms of a given chemical element having the same atomic number but different atomic mass numbers.

K

karst topography Topography characterized by abundant formation of underground solution cavities and sinkholes; commonly underlain by limestone.

kerogen A waxy solid hydrocarbon in oil shale.

kimberlites Igneous rocks that occur as pipelike intrusive bodies that probably originated in the mantle.

L

lahar A volcanic mudflow deposit formed from hot ash and water, the latter often derived from melting snow on a snow-capped or glaciated volcano.

landslide A general term applied to a rapid mass-wasting event.

laterite An extreme variety of pedalfer soil that is highly leached; common in tropical climates.

lava Magma that flows out at the earth's surface.

Law of Faunal Succession The concept that life-forms change through time and that therefore each given fossil organism corresponds to only one interval of time.

leachate Water containing dissolved chemicals; applied particularly to fluids escaping from waste-disposal sites.

leaching The removal of elements or compounds by dissolution.

leaching field A network of porous pipes and surrounding soil from which septic-tank effluent is slowly released.

levees Raised banks along a stream channel that tend to contain the water during high-discharge events.

lignite The softest of the coals.

liquefaction (coal) Any process by which coal is converted into a liquid hydrocarbon fuel.

liquefaction (soil) A quicksand condition arising in wet soil shaken by seismic waves; soil loses its strength as particles lose contact with each other.

lithification Conversion of unconsolidated sediment into cohesive rock.

lithosphere The solid, outermost zone of the earth, including the crust and a portion of the upper mantle; approximately 50 kilometers thick under the oceans and commonly more than 100 kilometers thick beneath the continents.

littoral drift Sand movement along the length of a beach.

load (stream) The total quantity of material transported by a stream: sum of bed load, suspended load, and dissolved load.

locked fault A section of fault along which friction prevents creep in response to stress.

loess Wind-deposited sediment composed of fine particles (typically 0.01 to 0.06 millimeters in diameter).

longitudinal profile Diagram of elevation of a streambed along its length.

longshore current Net movement of water parallel to a coastline, arising when waves and currents approach the shore at an oblique angle.

low-level (radioactive) waste Wastes that are sufficiently low in radioactivity that they can be released safely into the environment or disposed of with minimal precautions.

M

magma A naturally occurring silicate melt, which may also contain mineral crystals, dissolved water, or gases.

magmatic deposits (ore) Those deposits associated with magma emplacement or crystallization.

magnitude (earthquake) Measure of earthquake size; usually reported using the Richter magnitude scale.

manganese nodules Lumps of manganese and iron oxides and hydroxides, with other metals, found on the sea floor.

mantle The zone of the earth's interior between crust and core; rich in ferromagnesian silicates.

map unit A distinct, identifiable rock unit used in preparing a geologic map.

mass movement *See* mass wasting.

mass wasting The downslope movement of material due to gravity; also known as mass movement.

meanders The curves or bends in a stream channel.

mechanical weathering The physical breakup of rock or mineral grains by surface processes.

Mesozoic Geologic era from 245 to 66 million years ago.

metamorphic rock Rock that is changed in form (deformed and/or recrystallized) through the effects of heat and/or pressure.

methane hydrate A crystalline solid of natural gas and water molecules, found in arctic regions and marine sediments.

milling Erosion by the grinding action of sand-laden waves on a coast.

mineral A naturally occurring, inorganic, solid element or compound with a definite composition or range in compositions, usually having a regular internal crystal structure.

moraine Landform made of till.

mouth (stream) The point where a stream ends; where it reaches its base level.

multiple barrier concept Waste-disposal approach that involves several mechanisms or materials for isolating the waste from the environment; often used in the context of high-level radioactive wastes.

multiple use The land-use practice in which a given piece of land is used for two or more purposes simultaneously.

N

native element Nonsilicate mineral consisting of a single chemical element.

natural gas Gaseous hydrocarbons, especially methane (CH_4).

neutron An electrically neutral subatomic particle with a mass approximately equal to one atomic mass unit; generally found within an atomic nucleus.

nonpoint source A diffuse source of pollutants, such as runoff from farmland or drainage from a strip mine.

nonrenewable Not being replenished or formed at any significant rate on a human timescale.

nucleus The center of an atom, containing protons and neutrons.

O

oil Any of various liquid hydrocarbon compounds.

oil shale A sedimentary rock containing the waxy solid hydrocarbon kerogen.

ore A rock in which a valuable or useful metal occurs at a concentration sufficiently high to make it economically practical to mine.

outwash Glacial sediment moved and redeposited by meltwater.

oxbows Old meanders now cut off or abandoned by a stream.

oxide Nonsilicate mineral containing oxygen combined with one or more metals.

oxygen sag curve A graph depicting oxygen depletion followed by reoxygenation in a stream system below a source of organic waste matter; caused by aerobic decay of the organic matter.

P

paleomagnetism The "fossil magnetism" preserved in rocks formed in the past.

Paleozoic Geologic era from 570 to 245 million years ago.

Pangaea The name given to the ancient "supercontinent" that existed some 200 million years ago.

particulates (pollution) Solid particles suspended in air; includes soot, ash, and dust.

passive margin A geologically quiet continental margin, lacking significant volcanic or seismic activity.

passive-solar heating system A heating system using solar energy and thermal reservoirs but involving no power-consuming machinery or moving parts.

pedalfer A moderately leached soil rich in residual iron and aluminum oxide minerals.

pedocal A soil in which calcium carbonate and other readily soluble minerals are retained; characteristic of drier climates.

pegmatite A very coarsely crystalline igneous rock.

periodic table The regular arrangement of chemical elements in a chart that reflects patterns of chemical behavior related to the electronic structure of atoms.

permafrost A condition found in cold climates, wherein ground remains frozen year-round at some depth below the surface.

permafrost table The top of the permanently frozen (permafrost) zone.

permeability A measure of how readily fluid can flow through a rock, sediment, or soil.

Phanerozoic The time from 570 million years ago to the present.

photovoltaic cells (solar cells) Devices that convert solar radiation directly to electricity.

phreatic eruption A violent, explosive volcanic eruption like a steam-boiler explosion, occurring when subsurface water is heated and converted to steam by hot magma underground.

phreatic zone *See* zone of saturation.

placers The ores concentrated by stream or wave action on the basis of mineral densities and/or resistance to weathering.

plastic deformation Permanent strain in material stressed beyond the elastic limit; the material will not return to its original dimensions when the stress is removed.

plate tectonics The theory that holds that the rigid lithosphere is broken up into a series of movable plates.

plutonic Describes an igneous rock crystallized well below the earth's surface; typically coarse-grained.

point bar A sedimentary feature built in a stream channel, on the inside of a meander or anywhere the water slows.

point source A single, concentrated, identifiable source of pollutants, such as a sewer outfall or factory smokestack.

polar-wander curve A plot of apparent magnetic pole positions at various times in the past relative to a continent, assuming the continent's position to have been fixed on the earth.

porosity Proportion of void space in rock, sediment, or soil.

potentiometric surface A feature analogous to a water table, but applied to confined aquifers; indicates the height to which the water's pressure would raise the water if the water were unconfined.

Precambrian The time from the formation of the earth to 570 million years ago (the start of the Cambrian period).

precautionary principle A concept in international law/diplomacy under which nations' activities may be restricted if there is reasonable likelihood that those activities may be harmful; significant damage (as to the environment) need not already have occurred.

precursor phenomena Phenomena that precede an earthquake, volcanic eruption or other natural event, which may be used to predict the upcoming event.

Principle of Original Horizontality The concept that sediments are deposited in approximately horizontal, flat-lying layers.

Principle of Superposition The concept that, in an undisturbed pile of sedimentary rocks, those on the bottom were deposited first and those on top, last.

Prior Appropriation The principle of surface-water law by which users of water from a given source have priority rights to it on the basis of relative time of first use.

proton Subatomic particle with a charge of +1 and a mass of approximately one atomic mass unit; generally found within an atomic nucleus.

P waves Compressional seismic body waves.

pyroclastic flow Denser-than-air flow of hot gas and ash from a volcano.

pyroclastics The hot fragments of rock and magma emitted during an explosive volcanic eruption.

Q

quick clay The sediment formed from glacial rock flour deposited in a marine setting, weakened by subsequent flushing with fresh pore water.

R

rain shadow A dry zone landward of a coastal mountain range, which is caused by loss of moisture from air passing over the mountains.

recharge The processes of infiltration and migration by which ground water is replenished.

recrystallization Atomic rearrangement to form new crystals in a rock while it remains in the solid state; often, the grain sizes in the rock after recrystallization are coarser than in the rock originally.

recurrence interval The average length of time between floods of a given size along a particular stream.

recycling The recovery and reprocessing or reuse of materials already used previously in construction, manufacturing, and so on; a method of conserving resources by minimizing the need for the development of new resources.

regional metamorphism Metamorphism on a large scale, involving increased heat and pressure; often associated with mountain building.

regolith Surficial sediment deposit formed in place, whether or not capable of supporting life. (*See also* soil.)

remote sensing Investigation without direct contact, as by using aerial or satellite photography, radar, etc.

renewable Capable of replenishment or regeneration on a human timescale.

reprocessing (nuclear fuel) The treatment of used fuel elements to extract plutonium or other fissionable material to make new fuel elements.

reserves That quantity of a (resource) material that has been found and is recoverable economically with existing technology.

residence time The average length of time a substance persists in a system; defined as (capacity)/(rate of influx).

resources (general) Those things important or necessary to human life and civilization; also, the total quantity of a given (resource) commodity on earth, discovered and undiscovered, that might ultimately become part of the reserves.

retention pond A large basin designed to catch surface runoff to prevent its flow directly into a stream.

rhyolite Silica-rich volcanic rock; the volcanic compositional equivalent of granite.

Richter magnitude scale A logarithmic scale used for reporting the size of earthquakes; Richter magnitude is proportional to amount of ground shaking (displacement) at the epicenter.

Riparian Doctrine The principle of surface-water law by which all landowners bordering a body of water have equal rights to it.

rock A solid, cohesive aggregate of one or more minerals.

rock cycle The concept that rocks are continually subject to change and that any rock may be transformed into another type of rock through an appropriate geologic process.

rockfall *See* fall (rock).

rock flour A fine sediment of pulverized rock produced by glacial erosion.

rupture Breakage or failure of material under stress.

S

saltation The process by which particles are moved in short jumps over the ground surface or streambed by wind or water.

saltwater intrusion A process by which salt water replaces fresh ground water when the fresh water is being used more rapidly than it is being recharged; especially common in coastal areas.

sanitary landfill A disposal site for solid or contained liquid waste; in simplest form, a dump site at which wastes are covered with layers of earth daily or more often.

scale (map) The ratio of a unit of length on the map to the corresponding actual horizontal distance represented.

scientific method Means of discovering scientific principles by formulating hypotheses, making predictions from them, and testing the predictions.

scrubbers (electrostatic precipitators) Devices that remove pollutant gases and particulates from exhaust gases of power or manufacturing plants.

seafloor spreading The process by which new lithosphere is created at spreading ridges as plates of oceanic lithosphere move apart.

secure landfill A sanitary landfill designed to contain toxic chemical wastes; typically includes one or more impermeable liners and often is monitored by nearby wells.

sediment Surface accumulation of loose, unconsolidated mineral or rock particles.

sedimentary rock A rock formed from sediments at low temperature.

seismic gap A section of an active fault along which few earthquakes are occurring, in contrast to adjacent fault segments; presumably, a locked section of fault.

seismic shadow zone An effect of the earth's liquid outer core, which blocks S waves and thus casts a "seismic shadow" on the opposite side of the earth from the site of a major earthquake.

seismic waves The form in which most energy is released during earthquakes; divided into body waves and surface waves.

seismograph An instrument used to measure ground motion caused by seismic waves (and other disturbances).

sensitive clay A material similar in behavior to *quick clay*, but derived from different materials, such as volcanic ash.

septic tank A sewage-disposal device involving the slow dispersal of wastes into soil for natural aerobic breakdown.

sequential use The land-use principle by which the same land is used for two or more different purposes, one after another.

settling pond A pond in which sediment-laden surface runoff is impounded to allow sediment to settle out before the water is released.

shearing stress Stress that tends to cause different parts of an object to slide past each other across a plane; with respect to mass movements, stress tending to pull material downslope.

shear strength The ability of a material to resist shearing stress.

shield volcano A low, flat, gently sloping volcano built from many flows of fluid, low-viscosity basaltic lava.

silicate Mineral containing silicon and oxygen and usually one or more additional elements.

sinkhole A circular depression in the ground surface commonly caused by collapse into an underground cavern formed by solution.

slide A form of mass wasting in which a relatively coherent mass of material moves downslope along a well-defined surface.

slip face The downwind side of a dune, on which material tends to assume the slope of the angle of repose of the sediment of the dune.

slump A slide moved only a short distance, often with a rotational component to the movement.

soil The accumulation of unconsolidated rock and mineral fragments and organic matter formed in place at the earth's surface; capable of supporting life.

soil moisture The water in the soil in the zone of aeration, or vadose zone.

sorting A measure of the uniformity of a sediment with respect to particle size and/or density.

source (stream) The place where the stream originates.

source separation The sorting of (waste) material by type prior to collection, usually to

facilitate the recycling of individual materials or to ready material for a particular disposal strategy, such as incineration.

speculative resources That quantity of a resource material that may be found in areas where similar deposits are not already known to exist.

spoil banks Piles of waste rock and soil left behind by surface mining, especially strip-mining.

stage (stream) The height (elevation) of a stream surface at a given point along the stream's length; usually expressed as elevation above sea level.

strain Deformation resulting from the application of stress.

stratovolcano A composite volcanic cone built of interlayered lava flows and pyroclastic materials.

stream A body of flowing water confined within a channel.

stress A force applied to an object.

striations Parallel grooves in a rock surface cut when a glacier containing rock debris flows over that rock.

strip-mining The surface-mining technique in which the cover rock or soil is removed from a near-surface mineral or fuel deposit, and the deposit is then extracted by open excavation, in a series of parallel strips.

subduction zone A convergent plate boundary at which oceanic lithosphere is being pushed beneath another plate (continental or oceanic) into the asthenosphere.

subeconomic resources Those resources already found that cannot be exploited economically with existing technology; also known as conditional resources.

subsurface water The water in the hydrosphere below the ground surface.

sulfate Nonsilicate mineral containing sulfate (SO_4) groups, each made up of one atom of sulfur and four atoms of oxygen.

sulfide Nonsilicate mineral containing sulfur but no oxygen.

surface processes The geologic processes acting principally at the earth's surface; they involve the effects of wind, water, ice, and gravity.

surface water Liquid water above the ground surface.

surface waves The seismic waves that travel along the earth's surface.

surge (storm) Localized increase in water level of an ocean or large lake; caused by extreme low pressure associated with major storms.

suspended load (stream) The material that is light or fine enough to be moved along suspended in the stream, supported by the flowing water.

S waves Shear seismic body waves; do not propagate through liquids.

synfuels Synthetic fuels; also fuels derived from alternative fossil-fuel sources like oil shale, tar sand, and liquefied and gasified coal.

T

tailings The piles of crushed waste rock created as a by-product of mineral processing.

talus Accumulated debris from rockfalls and rockslides.

tar sand Sedimentary rock, usually sand, containing thick, tarlike, heavy oils.

tectonics The study of large-scale movement and deformation of the earth's crust.

tensile stress Stress tending to pull an object apart.

terminal moraine The end moraine marking the farthest advance of a glacier.

theory A generally accepted explanation for a set of data or observations; its validity has usually been tested by the scientific method.

thermal inversion The condition in which air temperature increases, rather than decreases, with increasing altitude; the overlying layer of warmer air may then trap warm, rising, pollutant-laden gases.

till Poorly sorted sediment deposited by melting glacial ice.

topsoil The topmost portion of soil, richest in organic matter.

trace elements Those elements present in a system at very low concentrations, typically 10 to 100 ppm or less.

transform fault A fault between offset segments of a spreading ridge, along which two plates move horizontally in opposite directions.

tsunami A seismic sea wave, generated by a major earthquake in or near an ocean basin; sometimes incorrectly called a "tidal wave."

unconfined aquifer An aquifer not overlain directly by an aquitard or aquiclude.

uniformitarianism The concept that the same basic physical laws have operated throughout the earth's history, and that therefore, by studying present geologic processes and their products, we can infer much about geologic processes operative in the past.

upstream flood Flood affecting only localized sections of a stream system; caused by such events as an intense, local cloudburst or a dam failure; typically brief in duration.

vadose zone *See* zone of aeration.

ventifact Literally, "wind-made" rock, shaped by wind erosion.

volcanic Describes an igneous rock formed at or near the earth's surface.

volcanic dome *See* dome (volcanic).

water table The top of the zone of saturation.

wave base Depth at which the movement of water molecules in waves becomes negligible.

wave-cut platform Steplike surface cut in rock by wave action at sea or lake level.

wave refraction The deflection of waves as they approach shore.

weathering The physical and/or chemical breakdown of rocks and minerals.

well sorted Describes sediments displaying uniform particle size and/or density.

Z

zone of accumulation *See* B horizon.

zone of aeration A partly saturated zone of rock or soil above the water table; also known as the vadose zone.

zone of deposition *See* B horizon.

zone of leaching *See* A horizon.

zone of saturation The region of rock or soil in which pore spaces are completely filled with liquid; also known as the phreatic zone.

Index